Function Notation (3-3, 3-7, 3-8)

$f(x)$ — Value of f at x
$(f \circ g)(x) = f[g(x)]$ — Composite function
$f^{-1}(x)$ — Value of inverse of f at x

Linear Equations and Functions (2-1, 3-2, 3-4)

$y = mx + b$ — Slope–intercept form
$(y - y_1) = m(x - x_1)$ — Point–slope form
$f(x) = mx + b$ — Linear function
$y = b$ — Horizontal line
$x = a$ — Vertical line

Polynomial and Rational Forms (1-2, 1-3, 1-4, 2-8, 3-4, 3-6, 4-1–4-7)

$f(x) = ax^2 + bx + c$ — Quadratic function

$f(x) = a_n x^n + a_{n-1} x^{n-1} + \cdots + a_1 x + a_0,$ $a_n \neq 0,$ n a nonnegative integer — Polynomial function

$f(x) = \dfrac{p(x)}{q(x)},$ p and q polynomial functions, $q(x) \neq 0$ — Rational function

Exponential and Logarithmic Functions (5-1–5-5)

$f(x) = b^x,$ $b > 0, b \neq 1$ — Exponential function
$f(x) = \log_b x,$ $b > 0, b \neq 1$ — Logarithmic function
$y = \log_b x$ if and only if $x = b^y,$ $b > 0, b \neq 1$

Variation (App. B)

$y = kx,$ $y \neq 0$ — Direct
$y = k/x,$ $k \neq 0$ — Inverse
$w = kxy,$ $k \neq 0$ — Joint

Matrices and Determinants (6-1, 7-1–7-7)

$\begin{bmatrix} a & b & c \\ d & e & f \end{bmatrix}$ Matrix

$\begin{vmatrix} a & b & c \\ d & e & f \\ g & h & i \end{vmatrix}$ Determinant

Arithmetic Sequence (8-3)

$a_1, a_2, \ldots, a_n, \ldots$
$a_n - a_{n-1} = d$ — Common difference
$a_n = a_1 + (n-1)d$ — nth-term formula
$S_n = a_1 + \cdots + a_n = \dfrac{n}{2}[2a_1 + (n-1)d]$ — Sum of n terms
$S_n = \dfrac{n}{2}(a_1 + a_n)$

Geometric Sequence (8-4)

$a_1, a_2, \ldots, a_n, \ldots$
$\dfrac{a_n}{a_{n-1}} = r$ — Common ratio
$a_n = a_1 r^{n-1}$ — nth-term formula
$S_n = a_1 + \cdots + a_n = \dfrac{a_1 - a_1 r^n}{1 - r},$ $r \neq 1$ — Sum of n terms
$S_n = \dfrac{a_1 - r a_n}{1 - r},$ $r \neq 1$
$S_\infty = a_1 + a_2 + \cdots = \dfrac{a_1}{1 - r},$ $|r| < 1$ — Sum of infinitely many terms

Factorial and Binomial Formulas (8-5)

$n! = n(n-1) \cdots 2 \cdot 1,$ $n \in N$ — n factorial
$0! = 1$
$\dbinom{n}{r} = \dfrac{n!}{r!(n-r)!},$ $0 \leq r \leq n$
$(a + b)^n = \displaystyle\sum_{k=0}^{n} \binom{n}{k} a^{n-k} b^k,$ $n \geq 1$ — Binomial formula

Permutations and Combinations (10-1)

For $0 \leq r \leq n$:

$P_{n,r} = \dfrac{n!}{(n-r)!}$ — Permutation

$C_{n,r} = \dbinom{n}{r} = \dfrac{n!}{r!(n-r)!}$ — Combination

(Continued on back endpaper)

COLLEGE ALGEBRA

Barnett & Ziegler's Precalculus Series

College Algebra:

This book is the same as COLLEGE ALGEBRA WITH TRIGONOMETRY without the three chapters on trigonometry.

College Algebra with Trigonometry:

This book is the same as COLLEGE ALGEBRA with three chapters of trigonometry added. Comparing COLLEGE ALGEBRA WITH TRIGONOMETRY with PRECALCULUS, COLLEGE ALGEBRA WITH TRIGONOMETRY has more intermediate algebra review, starts trigonometry with angles and right triangles, and does not include as much material on analytic geometry.

Precalculus:

This book differs from COLLEGE ALGEBRA WITH TRIGONOMETRY in that PRECALCULUS starts at a higher level, placing intermediate algebra review in the appendix; starts trigonometry with the unit circle and circular functions; and has more analytic geometry topics.

Books by Barnett—Kearns—Ziegler

Barnett—Kearns: ELEMENTARY ALGEBRA: STRUCTURE AND USE, 5th edition

Barnett—Kearns: ALGEBRA: AN ELEMENTARY COURSE, 2d edition

Barnett—Kearns: INTERMEDIATE ALGEBRA: STRUCTURE AND USE, 4th edition

Barnett—Kearns: ALGEBRA: AN INTERMEDIATE COURSE, 2d edition

Barnett—Ziegler: COLLEGE ALGEBRA, 5th edition

Barnett—Ziegler: COLLEGE ALGEBRA WITH TRIGONOMETRY, 5th edition

Barnett—Ziegler: PRECALCULUS: FUNCTIONS AND GRAPHS, 3d edition

Also Available from McGraw-Hill

Schaum's Outline Series in Mathematics & Statistics

Most outlines include basic theory, definitions, and hundreds of solved problems and supplementary problems with answers.

Titles on the Current List Include:
Advanced Calculus
Advanced Mathematics
Analytic Geometry
Beginning Calculus
Boolean Algebra
Calculus, 3d edition
Calculus for Business, Economics, & the Social Sciences
Calculus of Finite Differences & Difference Equations
College Algebra
College Mathematics, 2d edition
Complex Variables
Descriptive Geometry
Differential Equations
Differential Geometry
Discrete Math
Elementary Algebra, 2d edition
Essential Computer Math
Finite Mathematics
Fourier Analysis
General Topology
Geometry, 2d edition
Group Theory
Laplace Transforms
Linear Algebra, 2d edition
Mathematical Handbook of Formulas & Tables
Matrix Operations
Modern Abstract Algebra
Modern Elementary Algebra
Modern Introductory Differential Equations
Numerical Analysis, 2d edition
Partial Differential Equations
Probability
Probability & Statistics
Real Variables
Review of Elementary Mathematics
Set Theory & Related Topics
Statistics, 2d edition
Technical Mathematics
Tensor Calculus
Trigonometry, 2d edition
Vector Analysis

FIFTH EDITION
COLLEGE ALGEBRA

Raymond A. Barnett
Merritt College

Michael R. Ziegler
Marquette University

McGraw-Hill, Inc.
New York St. Louis San Francisco Auckland Bogotá Caracas Lisbon
London Madrid Mexico Milan Montreal New Delhi Paris San Juan
Singapore Sydney Tokyo Toronto

COLLEGE ALGEBRA

1 2 3 4 5 6 7 8 9 0   VNH   VNH   9 0 9 8 7 6 5 4 3 2

ISBN 0-07-004995-5

This book was set in Times Roman by Waldman Graphics, Inc.
The editors were Michael Johnson, Karen M. Minette, and David A.
Damstra; the design was done by Caliber/Phoenix Color Corp.;
the production supervisor was Leroy A. Young.
Von Hoffmann Press, Inc., was printer and binder.

Library of Congress Cataloging-in-Publication Data

Barnett, Raymond A.
    College algebra / Raymond A. Barnett, Michael R. Ziegler.—
  5th ed.
        p.   cm.
    Includes index.
    ISBN 0-07-004995-5
    1. Algebra.   I. Ziegler, Michael R.   II. Title.
QA154.2.B35   1993
512.9—dc20   92-11134

ABOUT THE AUTHORS

Raymond A. Barnett, a native of California and educated in California, received his B.A. in mathematical statistics from the University of California at Berkeley and his M.A. in mathematics from the University of Southern California. He has been a member of the Merritt College Mathematics Department, and was chairman of the department for four years.

Associated with four different publishers, Raymond Barnett has authored or coauthored eighteen textbooks in mathematics, most of which are still widely used. In addition to international English editions, a number of the books have been translated into Spanish. Co-authors include Michael Ziegler, Marquette University; Thomas Kearns, Northern Kentucky University; Charles Burke, City College of San Francisco; and John Fujii, Merritt College.

Michael R. Ziegler received his B.S. from Shippensburg State College and his M.S. and Ph.D. from the University of Delaware. After completing postdoctoral work at the University of Kentucky, he was appointed to the faculty of Marquette University where he currently holds the rank of Professor in the Department of Mathematics, Statistics, and Computer Science.

Dr. Ziegler has published over a dozen research articles in complex analysis and has coauthored ten undergraduate mathematics textbooks with Raymond A. Barnett.

CONTENTS

PREFACE

The fifth edition of *College Algebra* is **written for student comprehension**. The exposition is informal but mathematically correct. Each concept is illustrated with a worked-out example which is followed by a matched problem and answer. Thus, students are actively involved in the learning process.

College Algebra is one of three books in a precalculus series. The three books are briefly compared on page ii. Many chapters in the book are independent of one another and may be rearranged as desired to suit a particular preference or course outline. A chapter dependency chart can be found on page xv.

Principal Changes from the Fourth Edition

Improvements in this edition evolved out of the generous response from users of the fourth edition and from reviewers of the manuscript form of the fifth edition. Fundamental to a book's growth and effectiveness are classroom use and feedback, of which COLLEGE ALGEBRA, now in its fifth edition, has had a substantial amount. New to this edition are:

New Design. New accessible, *full color* design and expanded art program (250 new illustrations) make the book more visually appealing and reinforce concepts.

Type style for exponents and case fractions has been improved for increased clarity and all figures, examples and application problems now have *captions*.

Arrows have been added to graphs where appropriate to indicate that the graph continues beyond the portion shown.

Graphics Calculator. Problems requiring the graphics calculator have been added to many exercise sets. These problems are clearly identified with the icon 🖩 and may be omitted, if desired, without loss of continuity. For those instructors that want calculator material tied more closely to the text, **a graphics calculator supplement** is available.

Examples. Responding to user's requests, more illustrative examples of the more difficult types of problems and applications found in the exercise sets were added in the text where appropriate.

Exercises. Exercise sets were reviewed very carefully and the variety and quantity of A and B level exercises were increased. The number and variety of **applications** have also been increased and the existing applications have been brought up to date.

Review. Each **chapter review** now includes a more comprehensive section-by-section review and the answers to the **chapter review exercises** are keyed to the corresponding text section in which a related topic is discussed.

Cumulative review exercises have been added to the text. The answers are keyed to the corresponding text section in which the related topic is discussed.

Cautions. Common student errors are now more prominently displayed as CAUTIONS and there are more of them.

Exposition. The **readability** of the book has been one of its major strengths. In order to make it even better, the **exposition** was carefully reviewed and changes were made where appropriate to increase clarity without sacrificing mathematical integrity. A *nonthreatening*, informal style is used throughout.

New Topics.   The section on graphing rational functions has been returned from the appendix to Chapter 3. The chapter on polynomial functions has been moved to Chapter 4, and now contains the section on partial fractions that was formerly in the appendix, and a new section on approximating zeros. Many sections in the book have been substantially rewritten.

A detailed list of chapter by chapter changes is available from the publisher by request.

Important Features Retained from the Fourth Edition

1. An **informal style** is used for exposition, statements of some definitions, and proofs of theorems.

2. **Examples with matched problems.** Each concept is illustrated with one or more examples, and following each example is a parallel problem with an answer. This encourages an active involvement in the learning process and improves understanding of concepts.

3. The text includes **more than 4,000 carefully selected and graded problems**. The exercises are divided into A, B, and C groupings, with the A problems easy and routine, and the C problems a mixture of theoretical and difficult mechanics. In short, the text is designed so that an average or below average student will be able to experience success and a very capable student will be challenged.

4. The subject matter is related to the real world through many carefully selected **realistic applications** from the physical sciences, business and economics, life sciences, and social sciences. Thus, the most skeptical student should be convinced that mathematics is really useful.

5. **Answers** to all chapter review exercises and cumulative review exercises, keyed to appropriate sections, are in the back of the book. Answers to all other odd-numbered problems are also in the back of the book.

Student Aids

1. **CAUTION screens** appear throughout the text where student errors naturally occur (see Sections 1-2 and 1-3).

2. **Think boxes** (dashed boxes) are used to enclose steps that are usually performed mentally (see Sections 1-2 to 1-4).

3. **Screened boxes** are used to highlight important definitions, theorems, results, and step-by-step processes throughout the text.

4. **Annotation** of examples and developments is found throughout the text to help students through critical stages (see Sections 1-2 to 1-4).

5. **Functional use of four colors** guides students through critical steps (see Sections 1-2 and 1-4), increases the clarity of graphs (see Chapter 3), and improves the ease of text use through carefully selected color screens, heads, captions, and so on.

6. **Calculator steps** are often included (see Sections 1-5 and 1-6).

7. **Calculus-related problems** are often flagged with a special icon shown at the beginning of the preface (see Sections 1-2 to 1-4).

8. **Graphic calculator-related problems** are flagged with a special icon shown at the beginning of the preface (see Sections 3-1 and 3-4).

9. **Chapter reviews** include a concise section-by-section review of the chapter and a comprehensive set of review exercises with answers keyed to relevant sections in which a topic is discussed.

10. **Cumulative review** exercise sets appear in the book, each after an appropriate group of chapters. Answers are keyed to relevant sections in which a topic is discussed.

11. **Summaries of formulas and symbols** (keyed to the sections in which they are introduced) are found inside the front and back covers of the text for convenient reference.

Student Supplements

1. A **solutions manual** is available at a nominal cost through a bookstore. The manual includes detailed solutions to most review exercises and most other odd-numbered problems.

2. Course **videotapes**, keyed to the text, are available for student use through your institution's learning center or lab.

3. A text-specific **graphing calculator enhancement** for TI-81, Casio fx-7700G, and TI-85, by Carolyn L. Meitler, helps students use the power of a graphing calculator to learn mathematics. Examples illustrate how to use the calculator with topics covered in the textbook.

4. **Interactive tutorial software** package is available for use with IBM, IBM compatibles, and Macintosh computers.

Instructor Aids

1. A unique **computer-generated test system** is available to instructors without cost. The system allows the instructor to create tests using algorithmically generated test questions and those from a standard testbank. This testing system enables the instructor to choose questions either manually or randomly by section, question type, difficulty level, and other criteria. This system is available for IBM, IBM compatible, and Macintosh computers.

2. A **printed and bound test bank** is also available. This bank is a hard-copy listing of the questions found in the standard testbank.

3. An **instructor's resource manual** provides sample tests, transparency masters, and additional teaching suggestions and assistance.

4. An **instructor's solutions manual** contains the even solutions for all exercises as well as the answers to all problems.

5. A text-specific **graphing calculator enhancement** for TI-81, Casio fx-7700G, and TI-85, by Carolyn L. Meitler, helps teachers use the power of a graphing calculator to learn mathematics. Examples illustrate how to use the calculator with topics covered in the textbook.

For further information, please contact your local McGraw-Hill College Division sales representative.

Error Check

Because of the careful checking and proofing by a number of mathematics instructors (acting independently), the authors and publisher believe this book to be substantially error-free. For any errors remaining, the authors would be grateful if they were sent to Mathematics Editor, College Division, 27th Floor, McGraw-Hill, Inc., 1221 Avenue of the Americas, New York, NY 10020.

Acknowledgments

The preparation of a book requires the effort and skills of many people in addition to the authors. We would like to extend particular thanks to several very competent people: mathematics editor Michael Johnson for his continued support and encouragement; editing supervisor David Damstra for his care in guiding the book through editing and production; Karen M. Minette for coordinating the supplements package; Fred Safier of City College of San Francisco for his careful checking of the exercise sets and his skillful preparation of the solutions manuals that accompany the text; Carolyn Meitler of Concordia University, Wisconsin for producing a very useful graphic calculator supplement; and Stephen J. Merrill and Robert E. Mullins of Marquette University for the very careful checking of the whole book including the exercise sets.

We wish to thank all the reviewers for their many helpful suggestions and comments. (It is this process of classroom use, feedback, and adjustment that produces an increasingly effective book for both students and instructors.) In particular, we wish to thank Edwin J. Buman, Jr., Creighton University; Donald W. Flage, Prince George's Community College; Barbara Glass, Sussex County Community College; Jack L. Hadley, Jackson State Community College, Norma F. James, New Mexico State University; David Price, Tarrant County Junior College; Elise Price, Tarrant County Junior College; Richard Semmler, Northern Virginia Community College; and William V. Thayer, St. Louis Community College at Meramec.

Raymond A. Barnett
Michael R. Ziegler

CHAPTER DEPENDENCIES

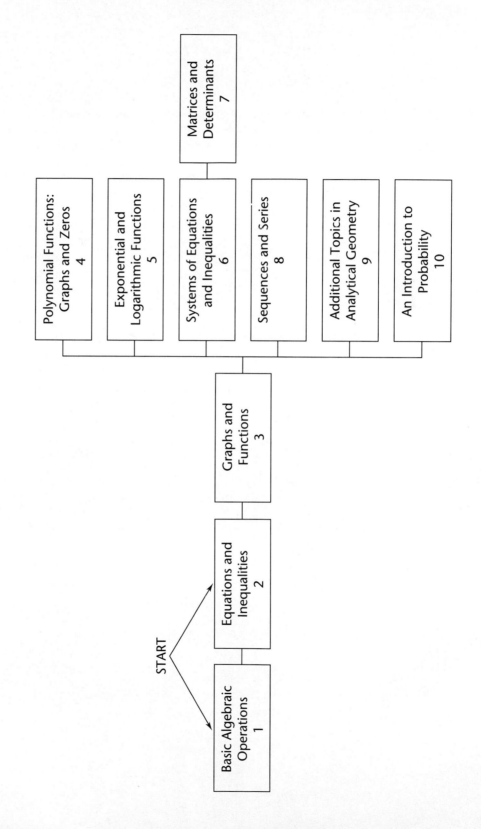

TO THE STUDENT

The following suggestions are made to help you get the most out of this book and your efforts.

As you study the text we suggest a five-step process. For each section:

1. Read a mathematical development.
2. Work through the illustrative example.
3. Work the matched problem.
4. Review the main ideas in the section.
5. Work the assigned exercise at the end of the section.

Repeat the 1-2-3 cycle until the section is finished.

All of this should be done with a scientific calculator and plenty of paper, pencils, and a wastebasket at hand. In fact, no mathematics text should be read without pencil and paper in hand; mathematics is not a spectator sport. Just as you cannot learn to swim by watching someone else swim, you cannot learn mathematics by simply reading worked examples—you must work problems, lots of them.

If you have difficulty with the course, then, in addition to doing the regular assignments, spend more time on the examples and matched problems and work more A exercises, even if they are not assigned. If you find the course too easy, then work more C exercises and applied problems, even if they are not assigned.

Raymond A. Barnett
Michael R. Ziegler

REMARKS ON CALCULATOR USE

In many places in the text calculator steps for new types of calculations are shown (similar to those steps shown here). These are only aids. Try the calculation without the aid; then use the aid if you get stuck.

Hand calculators are of two basic types relative to their internal logic (the way they compute): algebraic and reverse Polish notation (RPN). Throughout the book we will identify algebraic calculator steps with "A" and reverse Polish notation calculator steps with "P." Let's see how each type of calculator would compute

$$\frac{(5)(3)(2) \ - \ (7)(6)}{2(11)}$$

Press   Display

A: $\boxed{5}\ \boxed{\times}\ \boxed{3}\ \boxed{\times}\ \boxed{2}\ \boxed{-}\ \boxed{7}\ \boxed{\times}\ \boxed{6}\ \boxed{=}\ \boxed{\div}\ \boxed{2}\ \boxed{\div}\ \boxed{11}\ \boxed{=}$ $\boxed{-0.54545455}$

P: $\boxed{5}\ \boxed{\text{ENTER}}\ \boxed{3}\ \boxed{\times}\ \boxed{2}\ \boxed{\times}\ \boxed{7}\ \boxed{\text{ENTER}}\ \boxed{6}\ \boxed{\times}\ \boxed{-}\ \boxed{2}\ \boxed{\div}\ \boxed{11}\ \boxed{\div}$ $\boxed{-0.54545455}$

Some people prefer the algebraic logic and others prefer the Polish. Which is better is still being debated. The answer seems to rest with the type of problems encountered and with individual preferences.

In any case, irrespective of the type of calculator you own, it is essential that you read the user's manual for your own calculator. A large variety of calculators are on the market, and each is slightly different from the others. Therefore, it is important that you take time to read the manual. Do not try to read and understand everything the calculator can do; this will only tend to confuse you. Read only those sections that pertain to the operations you are or will be using; then return to the manual as necessary when you encounter new operations.

Problems requiring the use of a graphic calculator are clearly identified with the icon .

It is important to remember that *a calculator is not a substitute for thinking*. It can save you a great deal of time in certain types of problems, but you still must know how and when to use it.

Raymond A. Barnett
Michael R. Ziegler

COLLEGE ALGEBRA

Basic Algebraic Operations

1

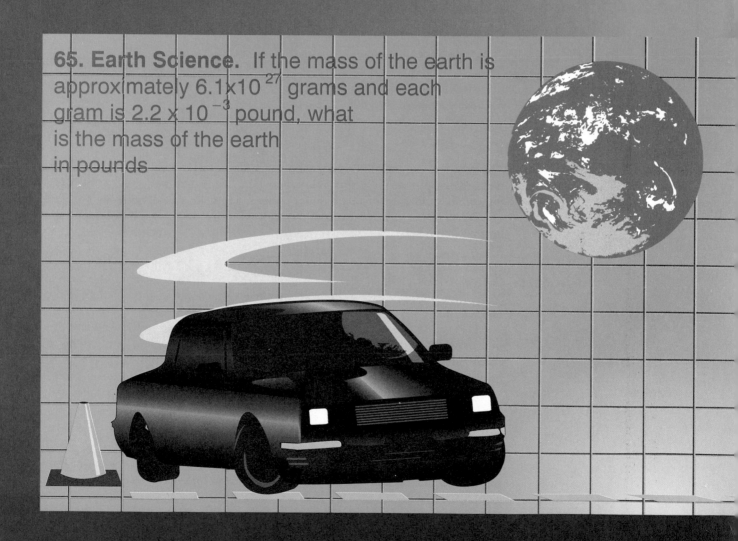

65. Earth Science. If the mass of the earth is approximately 6.1×10^{27} grams and each gram is 2.2×10^{-3} pound, what is the mass of the earth in pounds

Algebra is often referred to as "generalized arithmetic." In arithmetic we deal with the basic arithmetic operations of addition, subtraction, multiplication, and division performed on specific numbers. In algebra we continue to use all that we know in arithmetic, but, in addition, we reason and work with symbols that represent one or more numbers. In this chapter we review some important basic algebraic operations usually studied in earlier courses. The material may be studied systematically before commencing with the rest of the book or reviewed as needed.

SECTION 1-1   ## Algebra and Real Numbers

- Sets
- The Set of Real Numbers
- The Real Number Line
- Basic Real Number Properties
- Further Properties
- Fraction Properties

The rules for manipulating and reasoning with symbols in algebra depend, in large measure, on properties of the real numbers. In this section we look at some of the important properties of this number system. To make our discussions here and elsewhere in the text clearer and more precise, we first introduce a few useful notions about sets.

● **Sets**   Georg Cantor (1845–1918) developed a theory of sets as an outgrowth of his studies on infinity. His work has become a milestone in the development of mathematics.

Our use of the word "set" will not differ appreciably from the way it is used in everyday language. Words such as "set," "collection," "bunch," and "flock" all convey the same idea. Thus, we think of a **set** as a collection of objects with the important property that we can tell whether any given object is or is not in the set.

Each object in a set is called an **element**, or **member**, of the set. Symbolically,

$a \in A$ means   "a is an element of set A" $3 \in \{1, 3, 5\}$

$a \notin A$ means   "a is not an element of set A" $2 \notin \{1, 3, 5\}$

Capital letters are often used to represent sets and lowercase letters to represent elements of a set.

A set is **finite** if the number of elements in the set can be counted and **infinite** if there is no end in counting its elements. A set is **empty** if it contains no elements. The empty set is also called the **null** set and is denoted by $\varnothing$. It is important to observe that the empty set is *not* written as $\{\varnothing\}$.

A set is usually described in one of two ways—by **listing** the elements between braces, { }, or by enclosing within braces a **rule** that determines its elements. For example, if D is the set of all numbers x such that $x^2 = 4$, then using the listing method we write

$$D = \{-2, 2\} \quad \text{Listing method}$$

or, using the rule method we write

$$D = \{x \mid x^2 = 4\} \quad \text{Rule method}$$

Note that in the rule method, the vertical bar | represents "such that," and the entire symbolic form $\{x \mid x^2 = 4\}$ is read, "The set of all x such that $x^2 = 4$."

The letter x introduced in the rule method is a *variable*. In general, a **variable** is a symbol that is used as a placeholder for the elements of a set with two or more elements. This set is called the **replacement set** for the variable. A **constant**, on the other hand, is a symbol that names exactly one object. The symbol "8" is a constant, since it always names the number eight.

If each element of set A is also an element of set B, we say that A is a **subset** of set B, and we write

$$A \subset B \quad \{1, 5\} \subset \{1, 3, 5\}$$

Note that the definition of a subset allows a set to be a subset of itself.

Since the empty set $\varnothing$ has no elements, every element of $\varnothing$ is also an element of any given set. Thus, the empty set is a subset of every set. For example,

$$\varnothing \subset \{1, 3, 5\} \quad \text{and} \quad \varnothing \subset \{2, 4, 6\}$$

If two sets A and B have exactly the same elements, the sets are said to be **equal**, and we write

$$A = B \quad \{4, 2, 6\} = \{6, 4, 2\}$$

Notice that the order of listing elements in a set does not matter.

We can now begin our discussion of the real number system. Additional set concepts will be introduced as needed.

● **The Set of Real Numbers**

The real number system is the number system you have used most of your life. Informally, a **real number** is any number that has a decimal representation. Table 1 on the next page describes the set of real numbers and some of its important subsets. Figure 1 illustrates how these sets of numbers are related to each other.

FIGURE 1 Real numbers and important subsets.

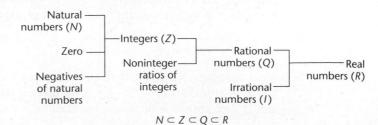

TABLE 1   The Set of Real Numbers

Symbol	Name	Description	Examples
N	Natural numbers	Counting numbers (also called positive integers)	$1, 2, 3, \ldots$
Z	Integers	Natural numbers, their negatives, and 0	$\ldots, -2, -1, 0, 1, 2, \ldots$
Q	Rational numbers	Numbers that can be represented as a/b, where a and b are integers and $b \neq 0$; decimal representations are repeating or terminating	$-4, 0, 1, 25, \frac{-3}{5}, \frac{2}{3}, 3.67, -0.33\overline{3},$* 5.272727
I	Irrational numbers	Numbers that can be represented as nonrepeating and nonterminating decimal numbers	$\sqrt{2}, \pi, \sqrt[3]{7}, 1.414213\ldots,$ $2.71828182\ldots$
R	Real numbers	Rational numbers and irrational numbers	

*The overbar indicates that the number (or block of numbers) repeats indefinitely.

FIGURE 2 A real number line.

• **The Real Number Line**   A one-to-one correspondence exists between the set of real numbers and the set of points on a line. That is, each real number corresponds to exactly one point, and each point to exactly one real number. A line with a real number associated with each point, and vice versa, as in Figure 2, is called a **real number line**, or simply a **real line**. Each number associated with a point is called the **coordinate** of the point. The point with coordinate 0 is called the **origin**. The arrow on the right end of the line indicates a positive direction. The coordinates of all points to the right of the origin are called **positive real numbers**, and those to the left of the origin are called **negative real numbers**. The real number 0 is neither positive nor negative.

• **Basic Real Number Properties**   We now take a look at some of the basic properties of real numbers. (See the box on the next page.)

You are already familiar with the **commutative properties** for addition and multiplication. They indicate that the order in which the addition or multiplication of two numbers is performed doesn't matter. For example,

$$4 + 5 = 5 + 4 \quad \text{and} \quad 4 \cdot 5 = 5 \cdot 4$$

Is there a commutative property relative to subtraction or division? That is, does $x - y = y - x$ or does $x \div y = y \div x$ for all real numbers x and y (division by 0 excluded)? The answer is no, since, for example,

$$7 - 5 \neq 5 - 7 \quad \text{and} \quad 6 \div 3 \neq 3 \div 6$$

Basic Properties of the Set of Real Numbers

Let R be the set of real numbers, and let x, y, and z be arbitrary elements of R.

Addition Properties

Closure: $x + y$ is a unique element in R.

Associative: $(x + y) + z = x + (y + z)$

Commutative: $x + y = y + x$

Identity: 0 is the additive identity; that is, $0 + x = x + 0 = x$ for all x in R, and 0 is the only element in R with this property.

Inverse: For each x in R, x is its unique additive inverse; that is, $x + (-x) = (-x) + x = 0$, and $-x$ is the only element in R relative to x with this property.

Multiplication Properties

Closure: xy is a unique element in R.

Associative: $(xy)z = x(yz)$

Commutative: $xy = yx$

Identity: 1 is the multiplicative identity; that is, for x in R, $(1)x = x(1) = x$, and 1 is the only element in R with this property.

Inverse: For each x in R, $x \neq 0$, $1/x$ is its unique multiplicative inverse; that is, $x(1/x) = (1/x)x = 1$, and $1/x$ is the only element in R relative to x with this property.

Combined Property

Distributive: $x(y + z) = xy + xz$ $\qquad$ $(x + y)z = xz + yz$

When computing

$$2 + 5 + 3 \quad \text{or} \quad 2 \cdot 5 \cdot 3$$

why don't we need parentheses to indicate which two numbers are to be added or multiplied first? The answer is to be found in the **associative properties**. These properties allow us to write

$$(2 + 5) + 3 = 2 + (5 + 3) \quad \text{and} \quad (2 \cdot 5) \cdot 3 = 2 \cdot (5 \cdot 3)$$

so it doesn't matter how we group numbers relative to either operation. Is there an associative property for subtraction or division? The answer is no, since, for example,

$$(8 - 4) - 2 \neq 8 - (4 - 2) \quad \text{and} \quad (8 \div 4) \div 2 \neq 8 \div (4 \div 2)$$

Evaluate both sides of these equations to see why.

> **Conclusion**
>
> Relative to addition, **commutativity** and **associativity** permit us to change the order of addition at will and insert or remove parentheses as we please. The same is true for multiplication, but not for subtraction and division.

What number added to a given number will give that number back again? What number times a given number will give that number back again? The answers are 0 and 1, respectively. Because of this, 0 and 1 are called the **identity elements** for the real numbers. Hence, for any real numbers x and y,

$$7 + 0 = 7 \qquad 0 + (x + y) = x + y \qquad \text{0 is the additive identity.}$$

$$1 \cdot 6 = 6 \qquad 1(x + y) = x + y \qquad \text{1 is the multiplicative identity.}$$

We now consider **inverses**. For each real number x, there is a unique real number $-x$ such that $x + (-x) = 0$. The number $-x$ is called the **additive inverse** of x, or the **negative** of x. For example, the additive inverse of 4 is -4, since $4 + (-4) = 0$. The additive inverse of -4 is $-(-4) = 4$, since $-4 + [-(-4)] = 0$. It is important to remember:

$-x$ is not necessarily a negative number; it is positive if x is negative and negative if x is positive.

For each number $x \neq 0$, there is a unique real number $1/x$ such that $x(1/x) = 1$. The number $1/x$ is called the **multiplicative inverse** of x, or the **reciprocal** of x. For example, the multiplicative inverse of 7 is $\frac{1}{7}$, since $7(\frac{1}{7}) = 1$. Also note that 7 is the multiplicative inverse of $\frac{1}{7}$.

We now turn to a real number property that involves both multiplication and addition. Consider the two computations:

$$3(4 + 2) = 3(6) = 18$$
$$3(4) + 3(2) = 12 + 6 = 18$$

Thus,

$$3(4 + 2) = 3(4) + 3(2)$$

and we say that multiplication by 3 *distributes* over the sum $(4 + 2)$. In general, multiplication **distributes** over addition in the real number system. Two more illustrations are given below:

$$2(x + y) = 2x + 2y \qquad (3 + 5)x = 3x + 5x$$

EXAMPLE 1   Using Real Number Properties

Which real number property justifies the indicated statement?

Statement	**Property Illustrated**
(A) $(7x)y = 7(xy)$	Associative ($\cdot$)
(B) $a(b + c) = (b + c)a$	Commutative ($\cdot$)
(C) $(2x + 3y) + 5y = 2x + (3y + 5y)$	Associative ($+$)
(D) $(x + y)(a + b) = (x + y)a + (x + y)b$	Distributive
(E) If $a + b = 0$, then $b = -a$.	Inverse ($+$)

Problem 1*   Which real number property justifies the indicated statement?

(A) $4 + (2 + x) = (4 + 2) + x$ (B) $(a + b) + c = c + (a + b)$
(C) $3x + 7x = (3 + 7)x$ (D) $(2x + 3y) + 0 = 2x + 3y$
(E) If $ab = 1$ and $a \neq 0$, then $b = 1/a$.

● **Further Properties**   Subtraction and division can be defined in terms of addition and multiplication, respectively:

DEFINITION 1   **Subtraction and Division**

For all real numbers a and b:

Subtraction: $a - b = a + (-b)$ $(-5) - (-3) = (-5) + (3) = -2$

Division: $b\overline{)a} = a \div b = \dfrac{a}{b} = a\left(\dfrac{1}{b}\right)$ $b \neq 0$ $3 \div 2 = 3\left(\dfrac{1}{2}\right)$

Thus, to subtract b from a, add the negative of b to a. To divide a by b, multiply a by the reciprocal of b. Note that division by 0 is not defined, since 0 does not have a reciprocal. It is important to remember:

Division by 0 is never allowed.

The following properties of negatives can be proved using the preceding properties and definitions.

*Answers to matched problems in a given section are found near the end of the section, before the exercise set.

Theorem 1

Properties of Negatives

For all real numbers a and b:

1. $-(-a) = a$

2. $(-a)b = -(ab) = a(-b) = -ab$

3. $(-a)(-b) = ab$

4. $(-1)a = -a$

5. $\dfrac{-a}{b} = -\dfrac{a}{b} = \dfrac{a}{-b}$ $b \neq 0$

6. $\dfrac{-a}{-b} = -\dfrac{-a}{b} = -\dfrac{a}{-b} = \dfrac{a}{b}$ $b \neq 0$

We now state an important theorem involving 0.

Theorem 2

Zero Properties

For all real numbers a and b:

1. $a \cdot 0 = 0$

2. $ab = 0$ if and only if $a = 0$ or $b = 0$ or both

EXAMPLE 2   Using Negative and Zero Properties

Which real number property or definition justifies each statement?

Statement	Property or Definition Illustrated
(A) $3 - (-2) = 3 + [-(-2)] - 5$	Subtraction (Definition 1 and Theorem 1, part 1)
(B) $-(-2) = 2$	Negatives (Theorem 1, part 1)
(C) $-\dfrac{-3}{2} = \dfrac{3}{2}$	Negatives (Theorem 1, part 6)
(D) $\dfrac{5}{-2} = -\dfrac{5}{2}$	Negatives (Theorem 1, part 5)
(E) If $(x - 3)(x + 5) = 0$, then either $x - 3 = 0$ or $x + 5 = 0$.	Zero (Theorem 2, part 2)

Problem 2   Which real number property or definition justifies each statement?

(A) $\dfrac{3}{5} = 3\left(\dfrac{1}{5}\right)$ (B) $(-5)(2) = -(5 \cdot 2)$ (C) $(-1)3 = -3$

(D) $\dfrac{-7}{9} = -\dfrac{7}{9}$ (E) If $x + 5 = 0$, then $(x - 3)(x + 5) = 0$.

● **Fraction Properties**   Recall that the quotient $a \div b$, $b \neq 0$, written in the form a/b is called a **fraction**. The quantity a is called the **numerator** and the quantity b is the **denominator**.

Theorem 3   **Fraction Properties**

For all real numbers $a, b, c, d,$ and k (division by 0 excluded):

1. $\dfrac{a}{b} = \dfrac{c}{d}$ if and only if $ad = bc$

$\dfrac{4}{6} = \dfrac{6}{9}$ since $4 \cdot 9 = 6 \cdot 6$

2. $\dfrac{ka}{kb} = \dfrac{a}{b}$

$\dfrac{7 \cdot 3}{7 \cdot 5} = \dfrac{3}{5}$

3. $\dfrac{a}{b} \cdot \dfrac{c}{d} = \dfrac{ac}{bd}$

$\dfrac{3}{5} \cdot \dfrac{7}{8} = \dfrac{3 \cdot 7}{5 \cdot 8}$

4. $\dfrac{a}{b} \div \dfrac{c}{d} = \dfrac{a}{b} \cdot \dfrac{d}{c}$

$\dfrac{2}{3} \div \dfrac{5}{7} = \dfrac{2}{3} \cdot \dfrac{7}{5}$

5. $\dfrac{a}{b} + \dfrac{c}{b} = \dfrac{a + c}{b}$

$\dfrac{3}{6} + \dfrac{5}{6} = \dfrac{3 + 5}{6}$

6. $\dfrac{a}{b} - \dfrac{c}{b} = \dfrac{a - c}{b}$

$\dfrac{7}{8} - \dfrac{3}{8} = \dfrac{7 - 3}{8}$

7. $\dfrac{a}{b} + \dfrac{c}{d} = \dfrac{ad + bc}{bd}$

$\dfrac{2}{3} + \dfrac{3}{5} = \dfrac{2 \cdot 5 + 3 \cdot 3}{3 \cdot 5}$

Answers to Matched Problems
1. (A) Associative ($+$)   (B) Commutative ($+$)   (C) Distributive   (D) Identity ($+$)
   (E) Inverse ($\cdot$)
2. (A) Division (Definition 1)   (B) Negatives (Theorem 1, part 2)
   (C) Negatives (Theorem 1, part 4)   (D) Negatives (Theorem 1, part 5)
   (E) Zero (Theorem 2, part 1)

EXERCISE 1-1   *All variables represent real numbers.*

A

In Problems 1–8, indicate true (T) or false (F).

1. $4 \in \{3, 4, 5\}$ **2.** $6 \in \{2, 4, 6\}$ **3.** $3 \notin \{3, 4, 5\}$ **4.** $7 \notin \{2, 4, 6\}$

5. $\{1, 2\} \subset \{1, 3, 5\}$ **6.** $\{2, 6\} \subset \{2, 4, 6\}$ **7.** $\{7, 3, 5\} \subset \{3, 5, 7\}$ **8.** $\{7, 3, 5\} = \{3, 5, 7\}$

In Problems 9–14, replace each question mark with an appropriate expression that will illustrate the use of the indicated real number property.

9. Commutative property $(+)$: $x + 7 = ?$

10. Commutative property $(\cdot)$: $uv = ?$

11. Associative property $(\cdot)$: $x(yz) = ?$

12. Associative property $(+)$: $3 + (7 + y) = ?$

13. Identity property $(+)$: $0 + 9m = ?$

14. Identity property $(\cdot)$: $1(u + v) = ?$

In Problems 15–26, each statement illustrates the use of one of the following properties or definitions. Indicate which one.

Commutative $(+, \cdot)$	*Identity $(+, \cdot)$*	*Division*
Associative $(+, \cdot)$	*Inverse $(+, \cdot)$*	*Negatives (Theorem 1)*
Distributive	*Subtraction*	*Zero (Theorem 2)*

15. $x + ym = x + my$

16. $7(3m) = (7 \cdot 3)m$

17. $7u + 9u = (7 + 9)u$

18. $-\dfrac{u}{-v} = \dfrac{u}{v}$

19. $(-2)(\frac{1}{-2}) = 1$

20. $8 - 12 = 8 + (-12)$

21. $w + (-w) = 0$

22. $5 \div (-6) = 5(\frac{1}{-6})$

23. $3(xy + z) + 0 = 3(xy + z)$

24. $ab(c + d) = abc + abd$

25. $\dfrac{-x}{-y} = \dfrac{x}{y}$

26. $(x + y) \cdot 0 = 0$

B

Write each set in Problems 27–32 using the listing method; that is, list the elements between braces. If the set is empty, write $\varnothing$.

27. $\{x \mid x$ is an even integer between -3 and $5\}$

28. $\{x \mid x$ is an odd integer between -4 and $6\}$

29. $\{x \mid x$ is a letter in "status"$\}$

30. $\{x \mid x$ is a letter in "consensus"$\}$

31. $\{x \mid x$ is a month starting with B$\}$

32. $\{x \mid x$ is a month with 32 days$\}$

In Problems 33–40, each statement illustrates the use of one of the following properties or definitions. Indicate which one.

Commutative $(+, \cdot)$	*Identity $(+, \cdot)$*	*Division*
Associative $(+, \cdot)$	*Inverse $(+, \cdot)$*	*Negatives (Theorem 1)*
Distributive	*Subtraction*	*Zero (Theorem 2)*

33. $(3x + 5) + 7 = 7 + (3x + 5)$

34. $(5x)(7y) = 5[x(7y)]$

35. $(3x + 2) + (x + 5) = 3x + [2 + (x + 5)]$

36. $(x + 3)(x + 5) = (x + 3)x + (x + 3)5$

37. $x(x - y) + y(x - y) = (x + y)(x - y)$

38. $\dfrac{-7}{-(m - n)} = \dfrac{7}{m - n}$

39. $(2x - 3)(x + 5) = 0$ if and only if $2x - 3 = 0$ or $x + 5 = 0$.

40. If $x(3x - 7) = 0$, then either $x = 0$ or $3x - 7 = 0$.

41. If $ab = 0$, does either a or b have to be 0?

42. If $ab = 1$, does either a or b have to be 1?

43. Indicate which of the following are true:
  (A) All natural numbers are integers.
  (B) All real numbers are irrational.
  (C) All rational numbers are real numbers.

44. Indicate which of the following are true:
  (A) All integers are natural numbers.
  (B) All rational numbers are real numbers.
  (C) All natural numbers are rational numbers.

45. Give an example of a rational number that is not an integer.

46. Give an example of a real number that is not a rational number.

47. Given the sets of numbers N (natural numbers), Z (integers), Q (rational numbers), and R (real numbers), indicate to which set(s) each of the following numbers belongs:
(A) -3 (B) 3.14 (C) π (D) $\frac{2}{3}$

48. Given the sets of numbers N, Z, Q, and R (see Problem 47), indicate to which set(s) each of the following numbers belongs:
(A) 8 (B) $\sqrt{2}$ (C) -1.414 (D) $\frac{-5}{2}$

In Problems 49 and 50, use a calculator to express each number as a decimal fraction to the capacity of your calculator (refer to the user's manual for your calculator). Observe the repeating decimal representation of the rational numbers and the apparent nonrepeating decimal representation of the irrational numbers.

49. (A) $\frac{8}{9}$ (B) $\frac{3}{11}$ (C) $\sqrt{5}$ (D) $\frac{11}{8}$

50. (A) $\frac{13}{6}$ (B) $\sqrt{21}$ (C) $\frac{7}{16}$ (D) $\frac{29}{111}$

51. Indicate true (T) or false (F), and for each false statement find real number replacements for a and b that will illustrate its falseness. For all real numbers a and b:
(A) $a + b = b + a$ (B) $a - b = b - a$
(C) $ab = ba$ (D) $a \div b = b \div a$

52. Indicate true (T) or false (F), and for each false statement find real number replacements for a, b, and c that will illustrate its falseness. For all real numbers a, b, and c:
(A) $(a + b) + c = a + (b + c)$
(B) $(a - b) - c = a - (b - c)$
(C) $a(bc) = (ab)c$
(D) $(a \div b) \div c = a \div (b \div c)$

C

53. If $A = \{1, 2, 3, 4\}$ and $B = \{2, 4, 6\}$, find:
(A) $\{x \mid x \in A \text{ or } x \in B\}$
(B) $\{x \mid x \in A \text{ and } x \in B\}$

54. If $F = \{-2, 0, 2\}$ and $G = \{-1, 0, 1, 2\}$, find:
(A) $\{x \mid x \in F \text{ or } x \in G\}$
(B) $\{x \mid x \in F \text{ and } x \in G\}$

55. If $c = 0.151515\ldots$, then $100c = 15.1515\ldots$ and

$$100c - c = 15.1515\ldots - 0.151515\ldots$$

$$99c = 15$$

$$c = \tfrac{15}{99} = \tfrac{5}{33}$$

Proceeding similarly, convert the repeating decimal $0.090909\ldots$ into a fraction. (All repeating decimals are rational numbers, and all rational numbers have repeating decimal representations.)

56. Repeat Problem 55 for $0.181818\ldots$

57. To see how the distributive property is behind the mechanics of long multiplication, compute each of the following and compare:

Long
Multiplication
$\quad$ 23
$\underline{\times 12}$

Use of the
Distributive Property
$23 \cdot 12$
$\quad = 23(2 + 10)$
$\quad = 23 \cdot 2 + 23 \cdot 10 =$

58. For a and b real numbers, justify each step using a property in this section.

Statement		Reason
1. $(a + b) + (-a) = (-a) + (a + b)$		1.
2.	$= [(-a) + a] + b$	2.
3.	$= 0 + b$	3.
4.	$= b$	4.

SECTION 1-2 Polynomials: Basic Operations

- Natural Number Exponents
- Polynomials
- Combining Like Terms
- Addition and Subtraction
- Multiplication
- Combined Operations
- Application

In this section we review the basic operations on *polynomials*, a mathematical form encountered frequently throughout mathematics. We start the discussion with a brief review of natural number exponents. Integer and rational exponents and their properties will be discussed in detail in subsequent sections.

● **Natural Number Exponents**

The definition of a **natural number exponent** is given below:

DEFINITION 1   **Natural Number Exponent**

For n a natural number and a any real number:

$$a^n = \underbrace{a \cdot a \cdot \cdots \cdot a}_{n \text{ factors of } a} \qquad 2^4 = \underset{4 \text{ factors of } 2}{2 \cdot 2 \cdot 2 \cdot 2}$$

Also, the **first property of exponents** is stated as follows:

Theorem 1   **First Property of Exponents**

For any natural numbers m and n, and any real number a:

$$a^m a^n = a^{m+n} \qquad (3x^5)(2x^7) \;\; \boxed{= 3 \cdot 2 x^{5+7}}^{\,*} = 6x^{12}$$

● **Polynomials**   **Algebraic expressions** are formed by using constants and variables and the algebraic operations of addition, subtraction, multiplication, division, raising to powers,

---

*Throughout the book, dashed boxes—called **think boxes**—are used to represent steps that are usually performed mentally.

and taking roots. Some examples are

$$\sqrt[3]{x^3 + 5} \qquad 5x^4 + 2x^2 - 7$$

$$x + y - 7 \qquad (2x - y)^2$$

$$\frac{x - 5}{x^2 + 2x - 5} \qquad 1 + \frac{1}{1 + \dfrac{1}{x}}$$

An algebraic expression involving only the operations of addition, subtraction, multiplication, and raising to natural number powers on variables and constants is called a **polynomial**. Some examples are

$$2x - 3 \qquad 4x^2 - 3x + 7$$

$$x - 2y \qquad 5x^3 - 2x^2 - 7x + 9$$

$$5 \qquad x^2 - 3xy + 4y^2$$

$$0 \qquad x^3 - 3x^2y + xy^2 + 2y^7$$

In a polynomial, a variable cannot appear in a denominator, as an exponent, or within a radical. Accordingly, a **polynomial in one variable** x is constructed by adding or subtracting constants and terms of the form ax^n, where a is a real number and n is a natural number. A **polynomial in two variables** x and y is constructed by adding and subtracting constants and terms of the form $ax^m y^n$, where a is a real number and m and n are natural numbers. Polynomials in three or more variables are defined in a similar manner.

Polynomial forms can be classified according to their *degree*. If a term in a polynomial has only one variable as a factor, then the **degree of that term** is the power of the variable. If two or more variables are present in a term as factors, then the **degree of the term** is the sum of the powers of the variables. The **degree of a polynomial** is the degree of the nonzero term with the highest degree in the polynomial. Any nonzero constant is defined to be a **polynomial of degree 0**. The number 0 is also a polynomial but is not assigned a degree.

EXAMPLE 1   **Polynomials and Nonpolynomials**

(A) Polynomials in one variable:

$$x^2 - 3x + 2 \qquad 6x^3 - \sqrt{2}x - \tfrac{1}{3}$$

(B) Polynomials in several variables:

$$3x^2 - 2xy + y^2 \qquad 4x^3y^2 - \sqrt{3}xy^2z^5$$

(C) Nonpolynomials:

$$\sqrt{2x} - \frac{3}{x} + 5 \qquad \frac{x^2 - 3x + 2}{x - 3} \qquad \sqrt{x^2 - 3x + 1}$$

(D) The degree of the first term in $6x^3 - \sqrt{2}x - \frac{1}{3}$ is 3, the degree of the second term is 1, the degree of the third term is 0, and the degree of the whole polynomial is 3.

(E) The degree of the first term in $4x^3y^2 - \sqrt{3}xy^2$ is 5, the degree of the second term is 3, and the degree of the whole polynomial is 5.

Problem 1   (A) Which of the following are polynomials?

$$3x^2 - 2x + 1 \qquad \sqrt{x - 3} \qquad x^2 - 2xy + y^2 \qquad \frac{x - 1}{x^2 + 2}$$

(B) Given the polynomial $3x^5 - 6x^3 + 5$, what is the degree of the first term? The second term? The whole polynomial?

(C) Given the polynomial $6x^4y^2 - 3xy^3$, what is the degree of the first term? The second term? The whole polynomial?

In addition to classifying polynomials by degree, we also call a single-term polynomial a **monomial**, a two-term polynomial a **binomial**, and a three-term polynomial a **trinomial**.

$$\frac{5}{2}x^2y^3 \qquad \text{Monomial}$$

$$x^3 + 4.7 \qquad \text{Binomial}$$

$$x^4 - \sqrt{2}x^2 + 9 \qquad \text{Trinomial}$$

● **Combining Like Terms**   We start with a word about *coefficients*. A constant in a term of a polynomial, including the sign that precedes it, is called the **numerical coefficient**, or simply, the **coefficient**, of the term. If a constant doesn't appear, or only a + sign appears, the coefficient is understood to be 1. If only a − sign appears, the coefficient is understood to be −1. Thus, given the polynomial

$$2x^4 - 4x^3 + x^2 - x + 5 \qquad 2x^4 + (-4)x^3 + 1x^2 + (-1)x + 5$$

the coefficient of the first term is 2, the coefficient of the second term is −4, the coefficient of the third term is 1, the coefficient of the fourth term is −1, and the coefficient of the last term is 5.

At this point, it is useful to state two additional distributive properties of real numbers that follow from the distributive properties stated in Section 1-1.

Additional Distributive Properties

1. $a(b - c) = (b - c)a = ab - ac$

2. $a(b + c + \cdots + f) = ab + ac + \cdots + af$

Two terms in a polynomial are called **like terms** if they have exactly the same variable factors to the same powers. The numerical coefficients may or may not be the same. Since constant terms involve no variables, all constant terms are like terms. If a polynomial contains two or more like terms, these terms can be combined into a single term by making use of distributive properties. Consider the following example:

$$5x^3y - 2xy - x^3y - 2x^3y \;\begin{aligned} &= 5x^3y - x^3y - 2x^3y - 2xy \\ &= (5x^3y - x^3y - 2x^3y) - 2xy \\ &= (5 - 1 - 2)x^3y - 2xy \end{aligned}$$

$$= 2x^3y - 2xy$$

It should be clear that free use has been made of the real number properties discussed earlier. The steps done in the dashed box are usually done mentally, and the process is quickly mechanized as follows:

Like terms in a polynomial are combined by adding their numerical coefficients.

EXAMPLE 2   Simplifying Polynomials

Remove parentheses and combine like terms:

(A) $2(3x^2 - 2x + 5) + (x^2 + 3x - 7)$ $\quad = 2(3x^2 - 2x + 5) + 1(x^2 + 3x - 7)$
Think

$$= 6x^2 - 4x + 10 + x^2 + 3x - 7$$
$$= 7x^2 - x + 3$$

(B) $(x^3 - 2x - 6) - (2x^3 - x^2 + 2x - 3)$

$$= 1(x^3 - 2x - 6) + (-1)(2x^3 - x^2 + 2x - 3)$$ Be careful with
Think   the sign here.

$$= x^3 - 2x - 6 - 2x^3 + x^2 - 2x + 3$$
$$= -x^3 + x^2 - 4x - 3$$

(C) $[3x^2 - (2x + 1)] - (x^2 - 1) = [3x^2 - 2x - 1] - (x^2 - 1)$ Remove inner
parentheses first.
$$= 3x^2 - 2x - 1 - x^2 + 1$$
$$= 2x^2 - 2x$$

Problem 2   Remove parentheses and combine like terms:

(A) $3(u^2 - 2v^2) + (u^2 + 5v^2)$
(B) $(m^3 - 3m^2 + m - 1) - (2m^3 - m + 3)$
(C) $(x^3 - 2) - [2x^3 - (3x + 4)]$

● **Addition and Subtraction**

Addition and subtraction of polynomials can be thought of in terms of removing parentheses and combining like terms, as illustrated in Example 2. Horizontal and vertical arrangements are illustrated in the next two examples. You should be able to work either way, letting the situation dictate the choice.

EXAMPLE 3   **Adding Polynomials**

Add: $x^4 - 3x^3 + x^2,$ $-x^3 - 2x^2 + 3x,$ and $3x^2 - 4x - 5$

Solution   Add horizontally:

$$(x^4 - 3x^3 + x^2) + (-x^3 - 2x^2 + 3x) + (3x^2 - 4x - 5)$$
$$= x^4 - 3x^3 + x^2 - x^3 - 2x^2 + 3x + 3x^2 - 4x - 5$$
$$= x^4 - 4x^3 + 2x^2 - x - 5$$

Or vertically, by lining up like terms and adding their coefficients:

$$
\begin{array}{r}
x^4 - 3x^3 + x^2 \\
- x^3 - 2x^2 + 3x \\
3x^2 - 4x - 5 \\
\hline
x^4 - 4x^3 + 2x^2 - x - 5
\end{array}
$$

Problem 3   Add horizontally and vertically:

$$3x^4 - 2x^3 - 4x^2,   x^3 - 2x^2 - 5x,   \text{and}   x^2 + 7x - 2$$

EXAMPLE 4   **Subtracting Polynomials**

Subtract: $4x^2 - 3x + 5$ from $x^2 - 8$

Solution $(x^2 - 8) - (4x^2 - 3x + 5)$ or $\begin{array}{r} x^2 \quad\quad - 8 \\ -4x^2 + 3x - 5 \\ \hline -3x^2 + 3x - 13 \end{array}$ ← Change signs and add.

$= x^2 - 8 - 4x^2 + 3x - 5$

$= -3x^2 + 3x - 13$

Problem 4   Subtract: $2x^2 - 5x + 4$ from $5x^2 - 6$

CAUTION   When you use a horizontal arrangement to subtract a polynomial with more than one term, you must enclose the polynomial in parentheses. Thus, to subtract $2x + 5$ from $4x - 11$, you must write

$$4x - 11 - (2x + 5)   \text{and not}   4x - 11 - 2x + 5$$

● **Multiplication**   Multiplication of algebraic expressions involves the extensive use of distributive properties for real numbers, as well as other real number properties.

EXAMPLE 5   Multiplying Polynomials

Multiply: $(2x - 3)(3x^2 - 2x + 3)$

Solution $(2x - 3)(3x^2 - 2x + 3) \quad = 2x(3x^2 - 2x + 3) - 3(3x^2 - 2x + 3)$

$$= 6x^3 - 4x^2 + 6x - 9x^2 + 6x - 9$$
$$= 6x^3 - 13x^2 + 12x - 9$$

Or, using a vertical arrangement,

$$
\begin{array}{r}
3x^2 - 2x + 3 \\
2x \quad - 3 \\
\hline
6x^3 - 4x^2 + 6x \phantom{{}- 9} \\
- 9x^2 + 6x - 9 \\
\hline
6x^3 - 13x^2 + 12x - 9
\end{array}
$$

Problem 5   Multiply: $(2x - 3)(2x^2 + 3x - 2)$

Thus, to multiply two polynomials, multiply each term of one by each term of the other, and combine like terms.

Products of certain binomial factors occur so frequently that it is useful to develop procedures that will enable us to write down their products by inspection. To find the product $(2x - 1)(3x + 2)$, we will use the popular **FOIL method**. We multiply each term of one factor by each term of the other factor as follows:

$$
\begin{array}{cccc}
\text{F} & \text{O} & \text{I} & \text{L} \\
\text{First} & \text{Outer} & \text{Inner} & \text{Last} \\
\text{product} & \text{product} & \text{product} & \text{product} \\
\downarrow & \downarrow & \downarrow & \downarrow
\end{array}
$$

$$(2x - 1)(3x + 2) = 6x^2 + 4x - 3x - 2$$

The inner and outer products are like terms and hence combine into one term. Thus,

$$(2x - 1)(3x + 2) = 6x^2 + x - 2$$

To speed up the process, we combine the inner and outer product mentally.

Products of certain binomial factors occur so frequently that it is useful to remember formulas for their products. The following formulas are easily verified by multiplying the factors on the left using the FOIL method:

Special Products

1. $(a - b)(a + b) = a^2 - b^2$

2. $(a + b)^2 = a^2 + 2ab + b^2$

3. $(a - b)^2 = a^2 - 2ab + b^2$

EXAMPLE 6   **Multiplying Binomials**

Multiply:

(A) $(2x - 3y)(5x + 2y) = 10x^2 - 11xy - 6y^2$
(B) $(3a - 2b)(3a + 2b) = 9a^2 - 4b^2$
(C) $(5x - 3)^2 = 25x^2 - 30x + 9$
(D) $(m + 2n)^2 = m^2 + 4mn + 4n^2$

Problem 6   Multiply:

(A) $(4u - 3v)(2u + v)$ (B) $(2xy + 3)(2xy - 3)$
(C) $(m + 4n)(m - 4n)$ (D) $(2u - 3v)^2$
(E) $(6x + y)^2$

CAUTION   Remember to include the sum of the inner and outer terms when using the
FOIL method to square a binomial. That is,

$$(x + 3)^2 \neq x^2 + 9 \quad (x + 3)^2 = x^2 + 6x + 9$$

● **Combined**   We now consider several examples that use all the operations just discussed. Before
Operations   considering these examples, it is useful to summarize order-of-operation con-
ventions pertaining to exponents, multiplication and division, and addition and
subtraction.

Order of Operations

1. Simplify inside the innermost grouping first, then the next innermost, and
so on.

$$2[3 - (x - 4)] = 2[3 - x + 4]$$
$$= 2(7 - x) = 14 - 2x$$

> **2.** Unless grouping symbols indicate otherwise, apply exponents before multiplication or division is performed.
>
> $$2(x - 2)^2 = 2(x^2 - 4x + 4) = 2x^2 - 8x + 8$$
>
> **3.** Unless grouping symbols indicate otherwise, perform multiplication and division before addition and subtraction. In either case, proceed from left to right.
>
> $$5 - 2(x - 3) = 5 - 2x + 6 = 11 - 2x$$

EXAMPLE 7   **Combined Operations**

Perform the indicated operations and simplify:

$$
\begin{aligned}
\text{(A)} \quad 3x - \{5 - 3[x - x(3 - x)]\} &= 3x - \{5 - 3[x - 3x + x^2]\} \\
&= 3x - \{5 - 3[-2x + x^2]\} \\
&= 3x - \{5 + 6x - 3x^2\} \\
&= 3x - 5 - 6x + 3x^2 \\
&= 3x^2 - 3x - 5
\end{aligned}
$$

$$
\begin{aligned}
\text{(B)} \quad (x - 2y)(2x + 3y) - (2x + y)^2 &= 2x^2 - xy - 6y^2 - (4x^2 + 4xy + y^2) \\
&= 2x^2 - xy - 6y^2 - 4x^2 - 4xy - y^2 \\
&= -2x^2 - 5xy - 7y^2
\end{aligned}
$$

$$
\begin{aligned}
\text{(C)} \quad (2m + 3n)^3 &= (2m + 3n)(2m + 3n)^2 \\
&= (2m + 3n)(4m^2 + 12mn + 9n^2) \\
&= 8m^3 + 24m^2n + 18mn^2 + 12m^2n + 36mn^2 + 27n^3 \\
&= 8m^3 + 36m^2n + 54mn^2 + 27n^3
\end{aligned}
$$

Problem 7   Perform the indicated operations and simplify:

(A) $2t - \{7 - 2[t - t(4 + t)]\}$ (B) $(u - 3v)^2 - (2u - v)(2u + v)$
(C) $(4x - y)^3$

• Application

EXAMPLE 8   **Volume of a Cylindrical Shell**

A plastic water pipe with a hollow center is 100 inches long, 1 inch thick, and has an inner radius of x inches (see the figure on the next page). Write an algebraic expression in terms of x that represents the volume of the plastic used to construct the pipe. Simplify the expression. [*Recall:* The volume V of a right circular cylinder of radius r and height h is given by $V = \pi r^2 h$.]

Solution   A right circular cylinder with a hollow center is called a **cylindrical shell**. The volume of the shell is equal to the volume of the cylinder minus the volume of the hole. Since the radius of the hole is x inches and the pipe is 1 inch thick, the radius of the cylinder is $x + 1$ inches. Thus, we have

$$\begin{pmatrix} \text{Volume of} \\ \text{shell} \end{pmatrix} = \begin{pmatrix} \text{Volume of} \\ \text{cylinder} \end{pmatrix} - \begin{pmatrix} \text{Volume of} \\ \text{hole} \end{pmatrix}$$

$$\begin{aligned} \text{Volume} &= \pi(x + 1)^2 100 - \pi x^2 100 \\ &= 100\pi(x^2 + 2x + 1) - 100\pi x^2 \\ &= 100\pi x^2 + 200\pi x + 100\pi - 100\pi x^2 \\ &= 200\pi x + 100\pi \end{aligned}$$

Problem 8   A plastic water pipe is 200 inches long, 2 inches thick, and has an outer radius of x inches. Write an algebraic expression in terms of x that represents the volume of the plastic used to construct the pipe. Simplify the expression.

Answers to Matched Problems
1. (A) $3x^2 - 2x + 1, x^2 - 2xy + y^2$ (B) $5, 3, 5$ (C) $6, 4, 6$
2. (A) $4u^2 - v^2$ (B) $-m^3 - 3m^2 + 2m - 4$ (C) $-x^3 + 3x + 2$
3. $3x^4 - x^3 - 5x^2 + 2x - 2$ 4. $3x^2 + 5x - 10$ 5. $4x^3 - 13x + 6$
6. (A) $8u^2 - 2uv - 3v^2$ (B) $4x^2y^2 - 9$ (C) $m^2 - 16n^2$ (D) $4u^2 - 12uv + 9v^2$
   (E) $36x^2 + 12xy + y^2$
7. (A) $-2t^2 - 4t - 7$ (B) $-3u^2 - 6uv + 10v^2$ (C) $64x^3 - 48x^2y + 12xy^2 - y^3$
8. Volume $= 200\pi x^2 - 200\pi(x - 2)^2 = 800\pi x - 800\pi$

EXERCISE 1-2

A

Problems 1–8 refer to the following polynomials:
(a) $2x^3 - 3x^2 + x + 5$ (b) $2x^2 + x - 1$ (c) $3x - 2$

1. What is the degree of (a)?

2. What is the degree of (b)?

3. Add (a) and (b).

4. Add (b) and (c).

5. Subtract (b) from (a).

6. Subtract (c) from (b).

7. Multiply (a) and (c).

8. Multiply (b) and (c).

In Problems 9–30, perform the indicated operations and simplify.

9. $2(x - 1) + 3(2x - 3) - (4x - 5)$

10. $2(u - 1) - (3u + 2) - 2(2u - 3)$

11. $2y - 3y[4 - 2(y - 1)]$

12. $4a - 2a[5 - 3(a + 2)]$

13. $(m - n)(m + n)$

14. $(a + b)(a - b)$

15. $(4t - 3)(t - 2)$

16. $(3x - 5)(2x + 1)$

17. $(3x + 2y)(x - 3y)$

18. $(2x - 3y)(x + 2y)$

19. $(2m - 7)(2m + 7)$

20. $(3y + 2)(3y - 2)$

21. $(6x - 4y)(5x + 3y)$

22. $(3m + 7n)(2m - 5n)$

23. $(3x - 2y)(3x + 2y)$

24. $(4m + 3n)(4m - 3n)$

25. $(4x - y)^2$

26. $(3u + 4v)^2$

27. $(a + b)(a^2 - ab + b^2)$

28. $(a - b)(a^2 + ab + b^2)$

29. $(3x + 2)^2$

30. $(4x + 3y)^2$

B _____

In Problems 31–46, perform the indicated operations and simplify.

31. $2x - 3\{x + 2[x - (x + 5)] + 1\}$

32. $m - \{m - [m - (m - 1)]\}$

33. $2\{3[a - 4(1 - a)] - (5 - a)\}$

34. $5b - 3\{-[2 - 4(2b - 1)] + 2(2 - 3b)\}$

35. $(2x^2 + x - 2)(x^2 - 3x + 5)$

36. $(x^2 - 2xy + y^2)(x^2 + 2xy + y^2)$

37. $(h^2 + hk + k^2)(h^2 - hk + k^2)$

38. $(n^2 + 2n + 1)(n^2 - 4n - 3)$

39. $(2x - 1)^2 - (3x + 2)(3x - 2)$

40. $(3a - b)(3a + b) - (2a - 3b)^2$

41. $(m - 3n)(m + 8n) + (m + 6n)(m + 4n)$

42. $(y - 2)(y + 1) + (y - 3)(y + 4)$

43. $(2m - n)^3$

44. $(x - 2y)^3$

45. $(2h + 3k)^3$

46. $(3a + 2b)^3$

∫ *Problems 47–52 are calculus-related. Perform the indicated operations and simplify.*

47. $5(x + h) - 4 - (5x - 4)$

48. $6(x + h) + 2 - (6x + 2)$

49. $3(x + h)^2 + 2(x + h) - (3x^2 + 2x)$

50. $4(x + h)^2 - 5(x + h) - (4x^2 - 5x)$

51. $-2(x + h)^2 - 3(x + h) + 7 - (-2x^2 - 3x + 7)$

52. $-(x + h)^2 + 4(x + h) - 9 - (-x^2 + 4x - 9)$

53. Subtract the sum of the first two polynomials from the sum of the last two: $3m^2 - 2m + 5$, $4m^2 - m$, $3m^2 - 3m - 2$, $m^3 + m^2 + 2$

54. Subtract the sum of the last two polynomials from the sum of the first two: $2x^2 - 4xy + y^2$, $3xy - y^2$, $x^2 - 2xy - y^2$, $-x^2 + 3xy - 2y^2$

C _____

In Problems 55–58, perform the indicated operations and simplify.

55. $2(x - 2)^3 - (x - 2)^2 - 3(x - 2) - 4$

56. $(2x - 1)^3 - 2(2x - 1)^2 + 3(2x - 1) + 7$

57. $-3x\{x[x - x(2 - x)] - (x + 2)(x^2 - 3)\}$

58. $2\{(x - 3)(x^2 - 2x + 1) - x[3 - x(x - 2)]\}$

59. If you are given two polynomials, one of degree m and the other of degree n, $m > n$, what is the degree of the sum?

60. What is the degree of the product of the two polynomials in Problem 59?

_____

APPLICATIONS

61. Geometry. The width of a rectangle is 5 centimeters less than its length. If x represents the length, write an algebraic expression in terms of x that represents the perimeter of the rectangle. Simplify the expression.

62. Geometry. The length of a rectangle is 8 meters more than its width. If x represents the width of the rectangle, write an algebraic expression in terms of x that represents its area. Change the expression to a form without parentheses.

★**63. Coin Problem.** A parking meter contains nickels, dimes, and quarters. There are 5 fewer dimes than nickels, and 2 more quarters than dimes. If x represents the number of nickels, write an algebraic expression in terms of x that represents the value of all the coins in the meter in cents. Simplify the expression.

★**64. Coin Problem.** A vending machine contains dimes and quarters only. There are 4 more dimes than quarters. If x represents the number of quarters, write an algebraic expression in terms of x that represents the value of all the coins in the vending machine in cents. Simplify the expression.

65. Packaging. A spherical plastic container for designer wristwatches has an inner radius of x centimeters (see the figure). If the plastic shell is 0.3 centimeter thick, write an algebraic expression in terms of x that represents the volume of the plastic used to construct the container. Simplify the expression. [*Recall:* The volume V of a sphere of radius r is given by $V = \frac{4}{3}\pi r^3$.]

66. Packaging. A cubical container for shipping computer components is formed by coating a metal mold with

polystyrene. If the metal mold is a cube with sides x centimeters long and the polystyrene coating is 2 centimeters thick, write an algebraic expression in terms of x that represents the volume of the polystyrene used to construct the container. Simplify the expression. [*Recall:* The volume V of a cube with sides of length t is given by $V = t^3$.]

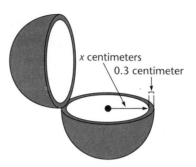

x centimeters

0.3 centimeter

SECTION 1-3   **Polynomials: Factoring**

- Factoring—What Does It Mean?
- Common Factors and Factoring by Grouping
- Factoring Second-Degree Polynomials
- More Factoring

● **Factoring—What Does It Mean?**

A **factor of a number** is one of two or more numbers whose product is the given number. Similarly, a **factor of an algebraic expression** is one of two or more algebraic expressions whose product is the given algebraic expression. For example,

$$30 = 2 \cdot 3 \cdot 5 \qquad \text{2, 3, and 5 are each factors of 30.}$$

$$x^2 - 4 = (x - 2)(x + 2) \qquad (x - 2) \text{ and } (x + 2) \text{ are each factors of } x^2 - 4.$$

The process of writing a number or algebraic expression as the product of other numbers or algebraic expressions is called **factoring**. We start our discussion of factoring with the positive integers.

An integer such as 30 can be represented in a factored form in many ways. The products

$$6 \cdot 5 \qquad (\tfrac{1}{2})(10)(6) \qquad 15 \cdot 2 \qquad 2 \cdot 3 \cdot 5$$

all yield 30. A particularly useful way of factoring positive integers greater than 1 is in terms of *prime* numbers.

DEFINITION 1   **Prime and Composite Numbers**

An integer greater than 1 is **prime** if its only positive integer factors are itself and 1. An integer greater than 1 that is not prime is called a **composite number**. The integer 1 is neither prime nor composite.

Prime numbers: 2, 3, 5, 7, 11, 13
Composite numbers: 4, 6, 8, 9, 10, 12

A composite number is said to be **factored completely** if it is represented as a product of prime factors. The only factoring of 30 given above that meets this condition is $30 = 2 \cdot 3 \cdot 5$.

EXAMPLE 1   **Factoring a Composite Number**

Write 60 in completely factored form.

Solution

$$60 = 6 \cdot 10 = 2 \cdot 3 \cdot 2 \cdot 5 = 2^2 \cdot 3 \cdot 5$$

or

$$60 = 5 \cdot 12 = 5 \cdot 4 \cdot 3 = 2^2 \cdot 3 \cdot 5$$

or

$$60 = 2 \cdot 30 = 2 \cdot 2 \cdot 15 = 2^2 \cdot 3 \cdot 5$$

Problem 1   Write 180 in completely factored form.

Notice in Example 1 that we end up with the same prime factors for 60 irrespective of how we progress through the factoring process. This illustrates an important property of integers:

Theorem 1   **The Fundamental Theorem of Arithmetic**

Each integer greater than 1 is either prime or can be expressed uniquely, except for the order of factors, as a product of prime factors.

We can also write polynomials in completely factored form. A polynomial such as $2x^2 - x - 6$ can be written in factored form in many ways. The products

$$(2x + 3)(x - 2) \qquad 2(x^2 - \tfrac{1}{2}x - 3) \qquad 2(x + \tfrac{3}{2})(x - 2)$$

all yield $2x^2 - x - 6$. A particularly useful way of factoring polynomials is in terms of prime polynomials.

DEFINITION 2	**Prime Polynomials**

A polynomial of degree greater than 0 is said to be **prime** relative to a given set of numbers if: (*1*) all of its coefficients are from that set of numbers; and (*2*) it cannot be written as a product of two polynomials, excluding 1 and itself, having coefficients from that set of numbers.

Relative to the set of integers:

$x^2 - 2$ is prime
$x^2 - 9$ is not prime, since $x^2 - 9 = (x - 3)(x + 3)$

[*Note:* The set of numbers most frequently used in factoring polynomials is the set of integers.]

A nonprime polynomial is said to be **factored completely relative to a given set of numbers** if it is written as a product of prime polynomials relative to that set of numbers.

Our objective in this section is to review some of the standard factoring techniques for polynomials with integer coefficients. In Chapter 4 we will treat in detail the topic of factoring polynomials of higher degree with arbitrary coefficients.

● **Common Factors and Factoring by Grouping**

The next example illustrates the use of the distributive properties in factoring.

EXAMPLE 2   Factoring Out Common Factors

Factor out, relative to the integers, all factors common to all terms:

(A) $2x^3y - 8x^2y^2 - 6xy^3$ (B) $2x(3x - 2) - 7(3x - 2)$

Solutions   (A) $2x^3y - 8x^2y^2 - 6xy^3$ $\;\boxed{= (2xy)x^2 - (2xy)4xy - (2xy)3y^2}$

$= 2xy(x^2 - 4xy - 3y^2)$

(B) $2x(3x - 2) - 7(3x - 2)$ $\;\boxed{= 2x(3x - 2) - 7(3x - 2)}$

$= (2x - 7)(3x - 2)$

Problem 2   Factor out, relative to the integers, all factors common to all terms:

(A) $3x^3y - 6x^2y^2 - 3xy^3$ (B) $3y(2y + 5) + 2(2y + 5)$

EXAMPLE 3   **Factoring Out Common Factors**

∫ Factor completely relative to the integers:

$$4(2x + 7)(x - 3)^2 + 2(2x + 7)^2(x - 3)$$

Solution
$$4(2x + 7)(x - 3)^2 + 2(2x + 7)^2(x - 3)$$
$$= 2(2x + 7)(x - 3)[2(x - 3) + (2x + 7)]$$
$$= 2(2x + 7)(x - 3)(2x - 6 + 2x + 7)$$
$$= 2(2x + 7)(x - 3)(4x + 1)$$

Problem 3   Factor completely relative to the integers:

$$4(2x + 5)(3x + 1)^2 + 6(2x + 5)^2(3x + 1)$$

Some polynomials can be factored by first grouping terms in such a way that we obtain an algebraic expression that looks something like Example 2B. We can then complete the factoring by the method used in that example.

EXAMPLE 4   **Factoring by Grouping**

Factor completely, relative to the integers, by grouping:

(A) $3x^2 - 6x + 4x - 8$ (B) $wy + wz - 2xy - 2xz$
(C) $3ac + bd - 3ad - bc$

Solutions   (A) $3x^2 - 6x + 4x - 8$
$\quad = (3x^2 - 6x) + (4x - 8)$ Group the first two and last two terms.
$\quad = 3x(x - 2) + 4(x - 2)$ Remove common factors from each group.
$\quad = (3x + 4)(x - 2)$ Factor out the common factor $(x - 2)$.

(B) $wy + wz - 2xy - 2xz$
$\quad = (wy + wz) - (2xy + 2xz)$ Group the first two and last two terms—be careful of signs.
$\quad = w(y + z) - 2x(y + z)$ Remove common factors from each group.
$\quad = (w - 2x)(y + z)$ Factor out the common factor $(y + z)$.

(C) $3ac + bd - 3ad - bc$

In parts (A) and (B) the polynomials are arranged in such a way that grouping the first two terms and the last two terms leads to common factors. In this problem neither the first two terms nor the last two terms have a common factor. Sometimes rearranging terms will lead to a factoring by grouping. In this case, we interchange the second and fourth terms to obtain a problem comparable to part (B), which can be factored as follows:

$$
\begin{aligned}
3ac - bc - 3ad + bd &= (3ac - bc) - (3ad - bd) \\
&= c(3a - b) - d(3a - b) \\
&= (c - d)(3a - b)
\end{aligned}
$$

Problem 4   Factor completely, relative to the integers, by grouping:

(A) $2x^2 + 6x + 5x + 15$ (B) $2pr + ps - 6qr - 3qs$
(C) $6wy - xz - 2xy + 3wz$

• Factoring Second-Degree Polynomials

We now turn our attention to factoring second-degree polynomials of the form

$$2x^2 - 5x - 3 \quad \text{and} \quad 2x^2 + 3xy - 2y^2$$

into the product of two first-degree polynomials with integer coefficients. The following example will illustrate an approach to the problem.

EXAMPLE 5   **Factoring Second-Degree Polynomials**

Factor each polynomial, if possible, using integer coefficients:

(A) $2x^2 + 3xy - 2y^2$ (B) $x^2 - 3x + 4$ (C) $6x^2 + 5xy - 4y^2$

Solutions   (A) $2x^2 + 3xy - 2y^2 = (2x + \ \ y)(x - \ \ y)$ Put in what we know. Signs must be opposite. (We can reverse this choice if we get $-3xy$ instead of $+3xy$ for the middle term.)

Now, what are the factors of 2 (the coefficient of y^2)?

$$
\begin{array}{c|l}
\dfrac{2}{1 \cdot 2} & (2x + y)(x - 2y) = 2x^2 - 3xy - 2y^2 \\[2mm]
2 \cdot 1 & (2x + 2y)(x - y) = 2x^2 \qquad\ \ - 2y^2
\end{array}
$$

The first choice gives us $-3xy$ for the middle term—close, but not there—so we reverse our choice of signs to obtain

$$2x^2 + 3xy - 2y^2 = (2x - y)(x + 2y)$$

(B) $x^2 - 3x + 4 = (x - \quad)(x - \quad)$ Signs must be the same because the third term is positive and must be negative because the middle term is negative.

$$\frac{4}{2 \cdot 2}$$

$2 \cdot 2$	$(x - 2)(x - 2) = x^2 - 4x + 4$
$1 \cdot 4$	$(x - 1)(x - 4) = x^2 - 5x + 4$
$4 \cdot 1$	$(x - 4)(x - 1) = x^2 - 5x + 4$

No choice produces the middle term; hence $x^2 - 3x + 4$ is not factorable using integer coefficients.

(C) $6x^2 + 5xy - 4y^2 = (\quad x + \quad y)(\quad x - \quad y)$

$\uparrow \quad \uparrow \quad \uparrow \quad \uparrow$
$?\quad ?\quad ?\quad ?$

The signs must be opposite in the factors, because the third term is negative. We can reverse our choice of signs later if necessary. We now write all factors of 6 and of 4:

$$\frac{6}{2 \cdot 3} \qquad \frac{4}{2 \cdot 2}$$

$$3 \cdot 2 \qquad 1 \cdot 4$$

$$1 \cdot 6 \qquad 4 \cdot 1$$

$$6 \cdot 1$$

and try each choice on the left with each on the right—a total of 12 combinations that give us the first and last terms in the polynomial $6x^2 + 5xy - 4y^2$. The question is: Does any combination also give us the middle term, $5xy$? After trial and error and, perhaps, some educated guessing among the choices, we find that $3 \cdot 2$ matched with $4 \cdot 1$ gives us the correct middle term. Thus,

$$6x^2 + 5xy - 4y^2 = (3x + 4y)(2x - y)$$

If none of the 24 combinations (including reversing our sign choice) had produced the middle term, then we would conclude that the polynomial is not factorable using integer coefficients.

Problem 5   Factor each polynomial, if possible, using integer coefficients:

(A) $x^2 - 8x + 12$ (B) $x^2 + 2x \mp 5$
(C) $2x^2 + 7xy - 4y^2$ (D) $4x^2 - 15xy - 4y^2$

● **More Factoring**   The factoring formulas listed below will enable us to factor certain polynomial forms that occur frequently.

Special Factoring Formulas

1. $u^2 + 2uv + v^2 = (u + v)^2$ **Perfect Square**
2. $u^2 - 2uv + v^2 = (u - v)^2$ **Perfect Square**
3. $u^2 - v^2 = (u - v)(u + v)$ **Difference of Squares**
4. $u^3 - v^3 = (u - v)(u^2 + uv + v^2)$ **Difference of Cubes**
5. $u^3 + v^3 = (u + v)(u^2 - uv + v^2)$ **Sum of Cubes**

The formulas in the box can be established by multiplying the factors on the right.

EXAMPLE 6   **Using Special Factoring Formulas**

Factor completely relative to the integers:

(A) $x^2 + 6xy + 9y^2$ (B) $9x^2 - 4y^2$ (C) $8m^3 - 1$ (D) $x^3 + y^3z^3$

Solutions   (A) $x^2 + 6xy + 9y^2 = x^2 + 2(x)(3y) + (3y)^2 = (x + 3y)^2$

(B) $9x^2 - 4y^2 = (3x)^2 - (2y)^2 = (3x - 2y)(3x + 2y)$

(C) $8m^3 - 1 = (2m)^3 - 1^3$
$= (2m - 1)[(2m)^2 + (2m)(1) + 1^2]$
$= (2m - 1)(4m^2 + 2m + 1)$

(D) $x^3 + y^3z^3 = x^3 + (yz)^3$
$= (x + yz)(x^2 - xyz + y^2z^2)$

Problem 6   Factor completely relative to the integers:

(A) $4m^2 - 12mn + 9n^2$ (B) $x^2 - 16y^2$ (C) $z^3 - 1$ (D) $m^3 + n^3$

CAUTION   Note that we did not list a special factoring formula for the sum of two squares. In general,

$$u^2 + v^2 \neq (au + bv)(cu + dv)$$

for any choice of real number coefficients a, b, c, and d. In the next chapter we will see that $u^2 + v^2$ can be factored using complex numbers.

We complete this section by considering factoring that involves combinations of the preceding techniques as well as a few additional ones. Generally speaking:

> When asked to factor a polynomial, we first take out all factors common to all terms, if they are present, and then proceed as above until all factors are prime.

EXAMPLE 7   Combining Factoring Techniques

Factor completely relative to the integers:

(A) $18x^3 - 8x$ (B) $x^2 - 6x + 9 - y^2$
(C) $4m^3n - 2m^2n^2 + 2mn^3$ (D) $2t^4 - 16t$
(E) $2y^4 - 5y^2 - 12$

Solutions   (A) $18x^3 - 8x = 2x(9x^2 - 4)$
$$= 2x(3x - 2)(3x + 2)$$

(B) $x^2 - 6x + 9 - y^2$
$$= (x^2 - 6x + 9) - y^2 \qquad \text{Group the first three terms.}$$
$$= (x - 3)^2 - y^2 \qquad \text{Factor } x^2 - 6x + 9.$$
$$= [(x - 3) - y][(x - 3) + y] \qquad \text{Difference of squares}$$
$$= (x - 3 - y)(x - 3 + y)$$

(C) $4m^3n - 2m^2n^2 + 2mn^3 = 2mn(2m^2 - mn + n^2)$

(D) $2t^4 - 16t = 2t(t^3 - 8)$
$$= 2t(t - 2)(t^2 + 2t + 4)$$

(E) $2y^4 - 5y^2 - 12 = (2y^2 + 3)(y^2 - 4)$
$$= (2y^2 + 3)(y - 2)(y + 2)$$

Problem 7   Factor completely relative to the integers:

(A) $3x^3 - 48x$ (B) $x^2 - y^2 - 4y - 4$
(C) $3u^4 - 3u^3v - 9u^2v^2$ (D) $3m^4 - 24mn^3$
(E) $3x^4 - 5x^2 + 2$

Answers to Matched Problems
1. $2^2 \cdot 3^2 \cdot 5$ 2. (A) $3xy(x^2 - 2xy - y^2)$ (B) $(3y + 2)(2y + 5)$
3. $2(2x + 5)(3x + 1)(12x + 17)$
4. (A) $(2x + 5)(x + 3)$ (B) $(p - 3q)(2r + s)$ (C) $(3w - x)(2y + z)$
5. (A) $(x - 2)(x - 6)$ (B) Not factorable using integers   (C) $(2x - y)(x + 4y)$
   (D) $(4x + y)(x - 4y)$
6. (A) $(2m - 3n)^2$ (B) $(x - 4y)(x + 4y)$ (C) $(z - 1)(z^2 + z + 1)$
   (D) $(m + n)(m^2 - mn + n^2)$
7. (A) $3x(x - 4)(x + 4)$ (B) $(x - y - 2)(x + y + 2)$ (C) $3u^2(u^2 - uv - 3v^2)$
   (D) $3m(m - 2n)(m^2 + 2mn + 4n^2)$ (E) $(3x^2 - 2)(x - 1)(x + 1)$

EXERCISE 1-3

A

In Problems 1–8, factor out, relative to the integers, all factors common to all terms.

1. $6x^4 - 8x^3 - 2x^2$

2. $6m^4 - 9m^3 - 3m^2$

3. $10x^3y + 20x^2y^2 - 15xy^3$

4. $8u^3v - 6u^2v^2 + 4uv^3$

5. $5x(x + 1) - 3(x + 1)$

6. $7m(2m - 3) + 5(2m - 3)$

7. $2w(y - 2z) - x(y - 2z)$

8. $a(3c + d) - 4b(3c + d)$

In Problems 9–16, factor completely relative to the integers.

9. $x^2 - 2x + 3x - 6$

10. $2y^2 - 6y + 5y - 15$

11. $6m^2 + 10m - 3m - 5$

12. $5x^2 - 40x - x + 8$

13. $2x^2 - 4xy - 3xy + 6y^2$

14. $3a^2 - 12ab - 2ab + 8b^2$

15. $8ac + 3bd - 6bc - 4ad$

16. $3pr - 2qs - qr + 6ps$

In Problems 17–28, factor completely relative to the integers. If a polynomial is prime relative to the integers, say so.

17. $2x^2 + 5x - 3$

18. $3y^2 - y - 2$

19. $x^2 - 4xy - 12y^2$

20. $u^2 - 2uv - 15v^2$

21. $x^2 + x - 4$

22. $m^2 - 6m - 3$

23. $25m^2 - 16n^2$

24. $w^2x^2 - y^2$

25. $x^2 + 10xy + 25y^2$

26. $9m^2 - 6mn + n^2$

27. $u^2 + 81$

28. $y^2 + 16$

B

In Problems 29–44, factor completely relative to the integers. If a polynomial is prime relative to the integers, say so.

29. $6x^2 + 48x + 72$

30. $4z^2 - 28z + 48$

31. $2y^3 - 22y^2 + 48y$

32. $2x^4 - 24x^3 + 40x^2$

33. $16x^2y - 8xy + y$

34. $4xy^2 - 12xy + 9x$

35. $6s^2 + 7st - 3t^2$

36. $6m^2 - mn - 12n^2$

37. $x^3y - 9xy^3$

38. $4u^3v - uv^3$

39. $3m^3 - 6m^2 + 15m$

40. $2x^3 - 2x^2 + 8x$

41. $m^3 + n^3$

42. $r^3 - t^3$

43. $c^3 - 1$

44. $a^3 + 1$

$\int$ *Problems 45–50 are calculus-related. Factor completely relative to the integers.*

45. $6(3x - 5)(2x - 3)^2 + 4(3x - 5)^2(2x - 3)$

46. $2(x - 3)(4x + 7)^2 + 8(x - 3)^2(4x + 7)$

47. $5x^4(9 - x)^4 - 4x^5(9 - x)^3$

48. $3x^4(x - 7)^2 + 4x^3(x - 7)^3$

49. $2(x + 1)(x^2 - 5)^2 + 4x(x + 1)^2(x^2 - 5)$

50. $4(x - 3)^3(x^2 + 2)^3 + 6x(x - 3)^4(x^2 + 2)^2$

In Problems 51–56, factor completely relative to the integers. In polynomials involving more than three terms, try grouping the terms in various combinations as a first step. If a polynomial is prime relative to the integers, say so.

51. $(a - b)^2 - 4(c - d)^2$

52. $(x + 2)^2 - 9y^2$

53. $2am - 3an + 2bm - 3bn$

54. $15ac - 20ad + 3bc - 4bd$

55. $3x^2 - 2xy - 4y^2$

56. $5u^2 + 4uv - 2v^2$

C

In Problems 57–72, factor completely relative to the integers. In polynomials involving more than three terms, try grouping the terms in various combinations as a first step. If a polynomial is prime relative to the integers, say so.

57. $x^3 - 3x^2 - 9x + 27$

58. $x^3 - x^2 - x + 1$

59. $a^3 - 2a^2 - a + 2$

60. $t^3 - 2t^2 + t - 2$

61. $4(A + B)^2 - 5(A + B) - 6$

62. $6(x - y)^2 + 23(x - y) - 4$

63. $m^4 - n^4$

64. $y^4 - 3y^2 - 4$

65. $s^4t^4 - 8st$

66. $27a^2 + a^5b^3$

67. $m^2 + 2mn + n^2 - m - n$

68. $y^2 - 2xy + x^2 - y + x$

69. $18a^3 - 8a(x^2 + 8x + 16)$

70. $25(4x^2 - 12xy + 9y^2) - 9a^2b^2$

71. $x^4 + 2x^2 + 1 - x^2$

72. $a^4 + 2a^2b^2 + b^4 - a^2b^2$

SECTION 1-4 Rational Expressions: Basic Operations

- Reducing to Lowest Terms
- Multiplication and Division
- Addition and Subtraction
- Compound Fractions

We now turn our attention to fractional forms. A quotient of two algebraic expressions, division by 0 excluded, is called a **fractional expression**. If both the numerator and denominator of a fractional expression are polynomials, the fractional expression is called a **rational expression**. Some examples of rational expressions are the following (recall, a nonzero constant is a polynomial of degree 0):

$$\frac{x - 2}{2x^2 - 3x + 5} \qquad \frac{1}{x^4 - 1} \qquad \frac{3}{x} \qquad \frac{x^2 + 3x - 5}{1}$$

In this section we discuss basic operations on rational expressions, including multiplication, division, addition, and subtraction.

Since variables represent real numbers in the rational expressions we are going to consider, the properties of real number fractions summarized in Section 1-1 play a central role in much of the work that we will do.

Even though not always explicitly stated, we always assume that variables are restricted so that division by 0 is excluded.

• Reducing to Lowest Terms

We start this discussion by restating the **fundamental property of fractions** (from Theorem 3 in Section 1-1):

> **Fundamental Property of Fractions**
>
> If a, b, and k are real numbers with $b, k \neq 0$, then
>
> $$\frac{ka}{kb} = \frac{a}{b} \qquad \frac{2 \cdot 3}{2 \cdot 4} = \frac{3}{4} \qquad \frac{(x-3)2}{(x-3)x} = \frac{2}{x}$$
>
> $$x \neq 0, \; x \neq 3$$

Using this property from left to right to eliminate all common factors from the numerator and the denominator of a given fraction is referred to as **reducing a fraction to lowest terms**. We are actually dividing the numerator and denominator by the same nonzero common factor.

Using the property from right to left—that is, multiplying the numerator and the denominator by the same nonzero factor—is referred to as **raising a fraction to higher terms**. We will use the property in both directions in the material that follows.

We say that a rational expression is **reduced to lowest terms** if the numerator and denominator do not have any factors in common. Unless stated to the contrary, factors will be relative to the integers.

EXAMPLE 1   Reducing Rational Expressions

Reduce each rational expression to lowest terms.

(A) $\dfrac{x^2 - 6x + 9}{x^2 - 9} = \dfrac{(x-3)^2}{(x-3)(x+3)}$ Factor numerator and denominator completely. Divide numerator and denominator by $(x - 3)$; this is a valid operation as long as $x \neq 3$ and $x \neq -3$.

$= \dfrac{x - 3}{x + 3}$

(B) $\dfrac{x^3 - 1}{x^2 - 1} = \dfrac{\overset{1}{\cancel{(x - 1)}}(x^2 + x + 1)}{\underset{1}{\cancel{(x - 1)}}(x + 1)}$ Dividing numerator and denominator by $(x - 1)$ can be indicated by drawing lines through both $(x - 1)$'s and writing the resulting quotients, 1's.

$= \dfrac{x^2 + x + 1}{x + 1}$

Problem 1   Reduce each rational expression to lowest terms.

(A) $\dfrac{6x^2 + x - 2}{2x^2 + x - 1}$ (B) $\dfrac{x^4 - 8x}{3x^3 - 2x^2 - 8x}$

EXAMPLE 2   Reducing a Rational Expression

Reduce the following rational expression to lowest terms.

$$\frac{6x^5(x^2 + 2)^2 - 4x^3(x^2 + 2)^3}{x^8} = \frac{2x^3(x^2 + 2)^2[3x^2 - 2(x^2 + 2)]}{x^8}$$

$$= \frac{2x^3(x^2 + 2)^2(x^2 - 4)}{x^8}$$

$$= \frac{2(x^2 + 2)^2(x - 2)(x + 2)}{x^5}$$

Problem 2   Reduce the following rational expression to lowest terms.

$$\frac{6x^4(x^2 + 1)^2 - 3x^2(x^2 + 1)^3}{x^6}$$

CAUTION   Remember to always factor the numerator and denominator first, then divide out any *common factors*. Do not indiscriminately eliminate *terms* that appear in both the numerator and the denominator. For example,

$$\frac{2x^3 + y^2}{y^2} \neq \frac{2x^3 + y^2}{y^2} = 2x^3 + 1$$

Since the term y^2 is not a factor of the numerator, it cannot be eliminated. In fact, $(2x^3 + y^2)/y^2$ is already reduced to lowest terms.

● **Multiplication**   Since we are restricting variable replacements to real numbers, multiplication and
  and Division   division of rational expressions follow the rules for multiplying and dividing real number fractions (Theorem 3 in Section 1-1).

Multiplication and Division

If a, b, c, and d are real numbers with b, $d \neq 0$, then:

1. $\dfrac{a}{b} \cdot \dfrac{c}{d} = \dfrac{ac}{bd}$ $\dfrac{2}{3} \cdot \dfrac{x}{x - 1} = \dfrac{2x}{3(x - 1)}$

2. $\dfrac{a}{b} \div \dfrac{c}{d} = \dfrac{a}{b} \cdot \dfrac{d}{c}$ $c \neq 0$ $\dfrac{2}{3} \div \dfrac{x}{x - 1} = \dfrac{2}{3} \cdot \dfrac{x - 1}{x}$

EXAMPLE 3   Multiplying and Dividing Rational Expressions

Perform the indicated operations and reduce to lowest terms.

(A) $\dfrac{10x^3y}{3xy + 9y} \cdot \dfrac{x^2 - 9}{4x^2 - 12x} = \dfrac{\overset{5x^2}{\cancel{10x^3y}}}{\underset{3 \cdot 1}{\cancel{3y}(x \cancel{+ 3})} } \cdot \dfrac{\overset{1 \cdot 1}{\cancel{(x - 3)}(x \cancel{+ 3})}}{\underset{2 \cdot 1}{\cancel{4x}(x \cancel{- 3})}}$ Factor numerators and denominators; then divide any numerator and any denominator with a like common factor.

$\qquad\qquad\qquad\qquad = \dfrac{5x^2}{6}$

(B) $\dfrac{4 - 2x}{4} \div (x - 2) = \dfrac{\overset{1}{\cancel{2}}(2 - x)}{\underset{2}{\cancel{4}}} \cdot \dfrac{1}{x - 2}$ $x - 2$ is the same as $\dfrac{x - 2}{1}$.

$\qquad\qquad\qquad\quad = \dfrac{2 - x}{2(x - 2)} = \dfrac{-\overset{-1}{\cancel{(x - 2)}}}{2\underset{1}{\cancel{(x - 2)}}}$ $b - a = -(a - b)$, a useful change in some problems.

$\qquad\qquad\qquad\quad = -\dfrac{1}{2}$

(C) $\dfrac{2x^3 - 2x^2y + 2xy^2}{x^3y - xy^3} \div \dfrac{x^3 + y^3}{x^2 + 2xy + y^2}$

$\qquad = \dfrac{\overset{2}{\cancel{2x}}(\overset{1}{\cancel{x^2 - xy + y^2}})}{\underset{y}{\cancel{xy}}\underset{1}{(x + y)}(x - y)} \cdot \dfrac{\overset{1}{\cancel{(x + y)^2}}}{\underset{1}{\cancel{(x + y)}}(\underset{1}{\cancel{x^2 - xy + y^2}})}$

$\qquad = \dfrac{2}{y(x - y)}$

Problem 3   Perform the indicated operations and reduce to lowest terms.

(A) $\dfrac{12x^2y^3}{2xy^2 + 6xy} \cdot \dfrac{y^2 + 6y + 9}{3y^3 + 9y^2}$

(B) $(4 - x) \div \dfrac{x^2 - 16}{5}$

(C) $\dfrac{m^3 + n^3}{2m^2 + mn - n^2} \div \dfrac{m^3n - m^2n^2 + mn^3}{2m^3n^2 - m^2n^3}$

● **Addition and Subtraction**   Again, because we are restricting variable replacements to real numbers, addition and subtraction of rational expressions follow the rules for adding and subtracting real number fractions (Theorem 3 in Section 1-1).

Addition and Subtraction

For a, b, and c real numbers with $b \neq 0$:

1. $\dfrac{a}{b} + \dfrac{c}{b} = \dfrac{a + c}{b}$ $\dfrac{x}{x - 3} + \dfrac{2}{x - 3} = \dfrac{x + 2}{x - 3}$

2. $\dfrac{a}{b} - \dfrac{c}{b} = \dfrac{a - c}{b}$ $\dfrac{x}{2xy^2} - \dfrac{x - 4}{2xy^2} = \dfrac{x - (x - 4)}{2xy^2}$

   Thus, we add rational expressions with the same denominators by adding or subtracting their numerators and placing the result over the common denominator. If the denominators are not the same, we raise the fractions to higher terms, using the fundamental property of fractions to obtain common denominators, and then proceed as described.

   Even though any common denominator will do, our work will be simplified if the least common denominator (LCD) is used. Often, the LCD is obvious, but if it is not, the steps in the box describe how to find it.

The Least Common Denominator (LCD)

The LCD of two or more rational expressions is found as follows:

1. Factor each denominator completely.

2. Identify each different prime factor from all the denominators.

3. Form a product using each different factor to the highest power that occurs in any one denominator. This product is the LCD.

EXAMPLE 4   **Adding and Subtracting Rational Expressions**

Combine into a single fraction and reduce to lowest terms.

(A) $\dfrac{3}{10} + \dfrac{5}{6} - \dfrac{11}{45}$ (B) $\dfrac{4}{9x} - \dfrac{5x}{6y^2} + 1$

(C) $\dfrac{x + 3}{x^2 - 6x + 9} - \dfrac{x + 2}{x^2 - 9} - \dfrac{5}{3 - x}$

Solutions   (A) To find the LCD, factor each denominator completely:

$$\left. \begin{array}{r} 10 = 2 \cdot 5 \\ 6 = 2 \cdot 3 \\ 45 = 3^2 \cdot 5 \end{array} \right\} \text{LCD} = 2 \cdot 3^2 \cdot 5 = 90$$

Now use the fundamental property of fractions to make each denominator 90:

$$\frac{3}{10} + \frac{5}{6} - \frac{11}{45} = \frac{9 \cdot 3}{9 \cdot 10} + \frac{15 \cdot 5}{15 \cdot 6} - \frac{2 \cdot 11}{2 \cdot 45}$$

$$= \frac{27}{90} + \frac{75}{90} - \frac{22}{90}$$

$$= \frac{27 + 75 - 22}{90} = \frac{80}{90} = \frac{8}{9}$$

(B) $\left.\begin{array}{l} 9x = 3^2x \\ 6y^2 = 2 \cdot 3y^2 \end{array}\right\}$ LCD $= 2 \cdot 3^2xy^2 = 18xy^2$

$$\frac{4}{9x} - \frac{5x}{6y^2} + 1 = \frac{2y^2 \cdot 4}{2y^2 \cdot 9x} - \frac{3x \cdot 5x}{3x \cdot 6y^2} + \frac{18xy^2}{18xy^2}$$

$$= \frac{8y^2 - 15x^2 + 18xy^2}{18xy^2}$$

(C) $\dfrac{x+3}{x^2 - 6x + 9} - \dfrac{x+2}{x^2 - 9} - \dfrac{5}{3-x} = \dfrac{x+3}{(x-3)^2} - \dfrac{x+2}{(x-3)(x+3)} + \dfrac{5}{x-3}$

Note: $-\dfrac{5}{3-x} = -\dfrac{5}{-(x-3)} = \dfrac{5}{x-3}$ We have again used the fact that $a - b = -(b - a)$.

The LCD $= (x - 3)^2(x + 3)$. Thus,

$$\frac{(x+3)^2}{(x-3)^2(x+3)} - \frac{(x-3)(x+2)}{(x-3)^2(x+3)} + \frac{5(x-3)(x+3)}{(x-3)^2(x+3)}$$

$$= \frac{(x^2 + 6x + 9) - (x^2 - x - 6) + 5(x^2 - 9)}{(x-3)^2(x+3)}$$ Be careful of sign errors here.

$$= \frac{x^2 + 6x + 9 - x^2 + x + 6 + 5x^2 - 45}{(x-3)^2(x+3)}$$

$$= \frac{5x^2 + 7x - 30}{(x-3)^2(x+3)}$$

Problem 4   Combine into a single fraction and reduce to lowest terms.

(A) $\dfrac{5}{28} - \dfrac{1}{10} + \dfrac{6}{35}$

(B) $\dfrac{1}{4x^2} - \dfrac{2x+1}{3x^3} + \dfrac{3}{12x}$

(C) $\dfrac{y-3}{y^2-4} - \dfrac{y+2}{y^2 - 4y + 4} - \dfrac{2}{2-y}$

• Compound Fractions   A fractional expression with fractions in its numerator, denominator, or both is called a **compound fraction**. It is often necessary to represent a compound fraction as a **simple fraction**—that is (in all cases we will consider), as the quotient of two polynomials. The process does not involve any new concepts. It is a matter of applying old concepts and processes in the right sequence. We will illustrate two approaches to the problem, each with its own merits, depending on the particular problem under consideration.

EXAMPLE 5   **Simplifying Compound Fractions**

Express as a simple fraction reduced to lowest terms:

$$\frac{\dfrac{2}{x} - 1}{\dfrac{4}{x^2} - 1}$$

Solution   *Method 1.* Multiply the numerator and denominator by the LCD of all fractions in the numerator and denominator—in this case, x^2. (We are multiplying by $1 = x^2/x^2$).

$$\frac{x^2\left(\dfrac{2}{x} - 1\right)}{x^2\left(\dfrac{4}{x^2} - 1\right)} = \frac{x^2\dfrac{2}{x} - x^2}{x^2\dfrac{4}{x^2} - x^2} = \frac{2x - x^2}{4 - x^2} = \frac{x(2 - x)^{\,1}}{(2 + x)(2 - x)_{\,1}}$$

$$= \frac{x}{2 + x}$$

Method 2. Write the numerator and denominator as single fractions. Then treat as a quotient.

$$\frac{\dfrac{2}{x} - 1}{\dfrac{4}{x^2} - 1} = \frac{\dfrac{2 - x}{x}}{\dfrac{4 - x^2}{x^2}} = \frac{2 - x}{x} \div \frac{4 - x^2}{x^2} = \frac{2 - x}{x}^{\,1} \cdot \frac{x^2}{(2 - x)(2 + x)}^{\,x}$$

$$= \frac{x}{2 + x}$$

Problem 5   Express as a simple fraction reduced to lowest terms. Use the two methods described in Example 5.

$$\frac{1 + \dfrac{1}{x}}{x - \dfrac{1}{x}}$$

EXAMPLE 6 Simplifying Compound Fractions

Express as a simple fraction reduced to lowest terms:

$$\frac{\dfrac{y}{x^2} - \dfrac{x}{y^2}}{\dfrac{y}{x} - \dfrac{x}{y}}$$

Solution   Using the first method described in Example 5, we have

$$\frac{x^2y^2\left(\dfrac{y}{x^2} - \dfrac{x}{y^2}\right)}{x^2y^2\left(\dfrac{y}{x} - \dfrac{x}{y}\right)} = \frac{x^2y^2\dfrac{y}{x^2} - x^2y^2\dfrac{x}{y^2}}{x^2y^2\dfrac{y}{x} - x^2y^2\dfrac{x}{y}} = \frac{y^3 - x^3}{xy^3 - x^3y} = \frac{\overset{1}{(\cancel{y-x})}(y^2 + xy + x^2)}{xy\underset{1}{(\cancel{y-x})}(y + x)}$$

$$= \frac{y^2 + xy + x^2}{xy(y + x)}$$

Problem 6   Express as a simple fraction reduced to lowest terms. Use the first method described in Example 5.

$$\frac{\dfrac{a}{b} - \dfrac{b}{a}}{\dfrac{a}{b} + 2 + \dfrac{b}{a}}$$

EXAMPLE 7 Simplifying Compound Fractions

Express as a simple fraction reduced to lowest terms:

$$2 - \frac{1}{2 - \dfrac{2}{2 + \dfrac{1}{x}}} = 2 - \frac{1}{2 - \dfrac{x \cdot 2}{x\left(2 + \dfrac{1}{x}\right)}}$$ First, write $\dfrac{2}{2 + \dfrac{1}{x}}$ as a simple fraction.

$$= 2 - \frac{1}{2 - \dfrac{2x}{2x + 1}}$$

$$= 2 - \frac{(2x + 1) \cdot 1}{(2x + 1)\left(2 - \dfrac{2x}{2x + 1}\right)}$$

$$= 2 - \frac{2x + 1}{4x + 2 - 2x} = 2 - \frac{2x + 1}{2x + 2}$$

$$= \frac{4x + 4 - 2x - 1}{2x + 2} = \frac{2x + 3}{2x + 2}$$

Problem 7 Express as a simple fraction reduced to lowest terms:

$$2 - \cfrac{1}{2 - \cfrac{1}{1 - \cfrac{1}{x}}}$$

Answers to Matched Problems

1. (A) $\dfrac{3x + 2}{x + 1}$ (B) $\dfrac{x^2 + 2x + 4}{3x + 4}$ 2. $\dfrac{3(x^2 + 1)^2(x + 1)(x - 1)}{x^4}$

3. (A) $2x$ (B) $\dfrac{-5}{x + 4}$ (C) mn

4. (A) $\dfrac{1}{4}$ (B) $\dfrac{3x^2 - 5x - 4}{12x^3}$ (C) $\dfrac{2y^2 - 9y - 6}{(y - 2)^2(y + 2)}$ 5. $\dfrac{1}{x - 1}$

6. $\dfrac{a - b}{a + b}$ 7. $\dfrac{x - 3}{x - 2}$

EXERCISE 1-4

A

In Problems 1–22, perform the indicated operations and reduce answers to lowest terms.
Represent any compound fractions as simple fractions reduced to lowest terms.

1. $\left(\dfrac{d^5}{3a} \div \dfrac{d^2}{6a^2}\right) \cdot \dfrac{a}{4d^3}$

2. $\dfrac{d^5}{3a} \div \left(\dfrac{d^2}{6a^2} \cdot \dfrac{a}{4d^3}\right)$

3. $\dfrac{2y}{18} - \dfrac{-1}{28} - \dfrac{y}{42}$

4. $\dfrac{x^2}{12} + \dfrac{x}{18} - \dfrac{1}{30}$

5. $\dfrac{3x + 8}{4x^2} - \dfrac{2x - 1}{x^3} - \dfrac{5}{8x}$

6. $\dfrac{4m - 3}{18m^3} + \dfrac{3}{4m} - \dfrac{2m - 1}{6m^2}$

7. $\dfrac{2x^2 + 7x + 3}{4x^2 - 1} \div (x + 3)$

8. $\dfrac{x^2 - 9}{x^2 - 3x} \div (x^2 - x - 12)$

9. $\dfrac{m + n}{m^2 - n^2} \div \dfrac{m^2 - mn}{m^2 - 2mn + n^2}$

10. $\dfrac{x^2 - 6x + 9}{x^2 - x - 6} \div \dfrac{x^2 + 2x - 15}{x^2 + 2x}$

11. $\dfrac{1}{a^2 - b^2} + \dfrac{1}{a^2 + 2ab + b^2}$

12. $\dfrac{3}{x^2 - 1} - \dfrac{2}{x^2 - 2x + 1}$

13. $m - 3 - \dfrac{m - 1}{m - 2}$

14. $\dfrac{x + 1}{x - 1} - 1$

15. $\dfrac{5}{x - 3} - \dfrac{2}{3 - x}$

16. $\dfrac{3}{a - 1} - \dfrac{2}{1 - a}$

17. $\dfrac{2}{y + 3} - \dfrac{1}{y - 3} + \dfrac{2y}{y^2 - 9}$

18. $\dfrac{2x}{x^2 - y^2} + \dfrac{1}{x + y} - \dfrac{1}{x - y}$

19. $\dfrac{1 - \dfrac{y^2}{x^2}}{1 - \dfrac{y}{x}}$

20. $\dfrac{1 + \dfrac{3}{x}}{x - \dfrac{9}{x}}$

21. $\dfrac{\dfrac{1}{m} + 1}{m + 1}$

22. $\dfrac{\dfrac{1}{m^2} - 1}{\dfrac{1}{m} + 1}$

B _____

$\int$ *Problems 23–28 are calculus-related. Reduce each fraction to lowest terms.*

23. $\dfrac{6x^3(x^2 + 2)^2 - 2x(x^2 + 2)^3}{x^4}$

24. $\dfrac{4x^4(x^2 + 3) - 3x^2(x^2 + 3)^2}{x^6}$

25. $\dfrac{2x(1 - 3x)^3 + 9x^2(1 - 3x)^2}{(1 - 3x)^6}$

26. $\dfrac{2x(2x + 3)^4 - 8x^2(2x + 3)^3}{(2x + 3)^8}$

27. $\dfrac{-2x(x + 4)^3 - 3(3 - x^2)(x + 4)^2}{(x + 4)^6}$

28. $\dfrac{3x^2(x + 1)^3 - 3(x^3 + 4)(x + 1)^2}{(x + 1)^6}$

In Problems 29–44, perform the indicated operations and reduce answers to lowest terms. Represent any compound fractions as simple fractions reduced to lowest terms.

29. $\dfrac{y}{y^2 - y - 2} - \dfrac{1}{y^2 + 5y - 14} - \dfrac{2}{y^2 + 8y + 7}$

30. $\dfrac{x^2}{x^2 + 2x + 1} + \dfrac{x - 1}{3x + 3} - \dfrac{1}{6}$

31. $\dfrac{9 - m^2}{m^2 + 5m + 6} \cdot \dfrac{m + 2}{m - 3}$

32. $\dfrac{2 - x}{2x + x^2} \cdot \dfrac{x^2 + 4x + 4}{x^2 - 4}$

33. $\dfrac{x + 7}{ax - bx} + \dfrac{y + 9}{by - ay}$

34. $\dfrac{c + 2}{5c - 5} - \dfrac{c - 2}{3c - 3} + \dfrac{c}{1 - c}$

35. $\dfrac{x^2 - 16}{2x^2 + 10x + 8} \div \dfrac{x^2 - 13x + 36}{x^3 + 1}$

36. $\left(\dfrac{x^3 - y^3}{y^3} \cdot \dfrac{y}{x - y}\right) \div \dfrac{x^2 + xy + y^2}{y^2}$

37. $\dfrac{x^2 - xy}{xy + y^2} \div \left(\dfrac{x^2 - y^2}{x^2 + 2xy + y^2} \div \dfrac{x^2 - 2xy + y^2}{x^2y + xy^2}\right)$

38. $\left(\dfrac{x^2 - xy}{xy + y^2} \div \dfrac{x^2 - y^2}{x^2 + 2xy + y^2}\right) \div \dfrac{x^2 - 2xy + y^2}{x^2y + xy^2}$

39. $\left(\dfrac{x}{x^2 - 16} - \dfrac{1}{x + 4}\right) \div \dfrac{4}{x + 4}$

40. $\left(\dfrac{3}{x - 2} - \dfrac{1}{x + 1}\right) \div \dfrac{x + 4}{x - 2}$

41. $\dfrac{\dfrac{1}{x} + \dfrac{1}{y}}{x + y}$

42. $\dfrac{c - d}{\dfrac{1}{c} - \dfrac{1}{d}}$

43. $\dfrac{1 + \dfrac{2}{x} - \dfrac{15}{x^2}}{1 + \dfrac{4}{x} - \dfrac{5}{x^2}}$

44. $\dfrac{\dfrac{x}{y} - 2 + \dfrac{y}{x}}{\dfrac{x}{y} - \dfrac{y}{x}}$

$\int$ *Problems 45–48 are calculus-related. Perform the indicated operations and reduce answers to lowest terms. Represent any compound fractions as simple fractions reduced to lowest terms.*

45. $\dfrac{\dfrac{1}{x + h} - \dfrac{1}{x}}{h}$

46. $\dfrac{\dfrac{1}{(x + h)^2} - \dfrac{1}{x^2}}{h}$

47. $\dfrac{\dfrac{(x + h)^2}{x + h + 2} - \dfrac{x^2}{x + 2}}{h}$

48. $\dfrac{\dfrac{2x + 2h + 3}{x + h} - \dfrac{2x + 3}{x}}{h}$

C _____

In Problems 49–54, perform the indicated operations and reduce answers to lowest terms. Represent any compound fractions as simple fractions reduced to lowest terms.

49. $\dfrac{y - \dfrac{y^2}{y - x}}{1 + \dfrac{x^2}{y^2 - x^2}}$

50. $\dfrac{\dfrac{s^2}{s - t} - s}{\dfrac{t^2}{s - t} + t}$

51. $2 - \dfrac{1}{1 - \dfrac{2}{a + 2}}$

52. $1 - \dfrac{1}{1 - \dfrac{1}{1 - \dfrac{1}{x}}}$

53. $1 - \dfrac{1}{1 - \dfrac{1}{1 - \dfrac{1}{1 - \dfrac{1}{x}}}}$

54. $1 + \dfrac{1}{1 + \dfrac{1}{1 + \dfrac{1}{1 + x}}}$

In Problems 55 and 56, a, b, c, and d represent real numbers.

55. (A) Prove that d/c is the multiplicative inverse of c/d $(c, d \neq 0)$.

   (B) Use part (A) to prove that

   $$\frac{a}{b} \div \frac{c}{d} = \frac{a}{b} \cdot \frac{d}{c} \qquad b, c, d \neq 0$$

56. Prove that

$$\frac{a}{b} + \frac{c}{b} = \frac{a + c}{b} \qquad b \neq 0$$

SECTION 1-5 Integer Exponents

- Integer Exponents
- Scientific Notation

The French philosopher/mathematician René Descartes (1596–1650) is generally credited with the introduction of the very useful exponent notation "x^n." This notation as well as other improvements in algebra may be found in his *Geometry*, published in 1637.

In Section 1-2 we introduced the natural number exponent as a short way of writing a product involving the same factors. In this section we will expand the meaning of exponent to include all integers, so that exponential forms of the following types will all have meaning:

$$7^5 \qquad 5^{-4} \qquad 3.14^0$$

• Integer Exponents Definition 1 generalizes exponent notation to include 0 and negative integer exponents.

DEFINITION 1 a^n, n an integer and a a real number

1. For n a positive integer:

$$a^n = \underbrace{a \cdot a \cdot \cdots \cdot a}_{n \text{ factors of } a} \qquad\qquad 3^5 = 3 \cdot 3 \cdot 3 \cdot 3 \cdot 3$$

2. For $n = 0$:

$$a^0 = 1 \qquad a \neq 0 \qquad\qquad 132^0 = 1$$
$$0^0 \text{ is not defined}$$

3. For n a negative integer:

$$a^n = \frac{1}{a^{-n}} \qquad a \neq 0 \qquad\qquad 7^{-3} \;\boxed{= \frac{1}{7^{-(-3)}}}\; = \frac{1}{7^3}$$

Note: In general, it can be shown that for *all* integers n

$$a^{-n} = \frac{1}{a^n} \qquad\qquad a^{-5} = \frac{1}{a^5} \qquad a^{-(-3)} = \frac{1}{a^{-3}}$$

EXAMPLE 1 Using the Definition of Integer Exponents

Write each part as a decimal fraction or using positive exponents.

(A) $(u^3v^2)^0 = 1 \quad u \neq 0, \quad v \neq 0$

(B) $10^{-3} = \dfrac{1}{10^3} = \dfrac{1}{1{,}000} = 0.001$

(C) $x^{-8} = \dfrac{1}{x^8}$

(D) $\dfrac{x^{-3}}{y^{-5}} = \dfrac{x^{-3}}{1} \cdot \dfrac{1}{y^{-5}} = \dfrac{1}{x^3} \cdot \dfrac{y^5}{1} = \dfrac{y^5}{x^3}$

Problem 1 Write parts (A)–(D) as decimal fractions and parts (E) and (F) with positive exponents.

(A) 636^0 (B) $(x^2)^0 \quad x \neq 0$ (C) 10^{-5}

(D) $\dfrac{1}{10^{-3}}$ (E) $\dfrac{1}{x^{-4}}$ (F) $\dfrac{u^{-7}}{v^{-3}}$

The basic properties of integer exponents are summarized in Theorem 1. The proof of this theorem involves *mathematical induction,* which is discussed in Chapter 11.

Theorem 1 **Properties of Integer Exponents**

For n and m integers and a and b real numbers:

1. $a^m a^n = a^{m+n}$ $a^5 a^{-7} = a^{5+(-7)} = a^{-2}$

2. $(a^n)^m = a^{mn}$ $(a^3)^{-2} = a^{(-2)3} = a^{-6}$

3. $(ab)^m = a^m b^m$ $(ab)^3 = a^3 b^3$

4. $\left(\dfrac{a}{b}\right)^m = \dfrac{a^m}{b^m} \quad b \neq 0$ $\left(\dfrac{a}{b}\right)^4 = \dfrac{a^4}{b^4}$

5. $\dfrac{a^m}{a^n} = \begin{cases} a^{m-n} \\ \dfrac{1}{a^{n-m}} \end{cases} \quad a \neq 0$

$\dfrac{a^3}{a^{-2}} = a^{3-(-2)} = a^5$

$\dfrac{a^3}{a^{-2}} = \dfrac{1}{a^{-2-3}} = \dfrac{1}{a^{-5}}$

EXAMPLE 2   Using Exponent Properties

Simplify using exponent properties, and express answers using positive exponents only.*

(A) $(3a^5)(2a^{-3}) \;\boxed{= (3 \cdot 2)(a^5 a^{-3})} \; = 6a^2$

(B) $\dfrac{6x^{-2}}{8x^{-5}} \;\boxed{= \dfrac{3x^{-2-(-5)}}{4}} \; = \dfrac{3x^3}{4}$

(C) $-4y^3 - (-4y)^3 = -4y^3 - (-4)^3 y^3 \;\boxed{= -4y^3 - (-64)y^3}$

$\qquad\qquad\qquad\qquad = -4y^3 + 64y^3 = 60y^3$

Problem 2   Simplify using exponent properties, and express answers using positive exponents only.

(A) $(5x^{-3})(3x^4)$ (B) $\dfrac{9y^{-7}}{6y^{-4}}$ (C) $2x^4 - (-2x)^4$

CAUTION

Be careful when using the relationship $a^{-n} = \dfrac{1}{a^n}$:

$$ab^{-1} \neq \frac{1}{ab} \qquad\qquad ab^{-1} = \frac{a}{b} \quad \text{and} \quad (ab)^{-1} = \frac{1}{ab}$$

$$\frac{1}{a+b} \neq a^{-1} + b^{-1} \qquad \frac{1}{a+b} = (a+b)^{-1} \quad \text{and} \quad \frac{1}{a} + \frac{1}{b} = a^{-1} + b^{-1}$$

Do not confuse properties 1 and 2 in Theorem 1:

$$a^3 a^4 \neq a^{3\cdot4} \qquad a^3 a^4 = a^{3+4} = a^7 \qquad \text{property 1, Theorem 1}$$

$$(a^3)^4 \neq a^{3+4} \qquad (a^3)^4 = a^{3\cdot4} = a^{12} \qquad \text{property 2, Theorem 1}$$

---

*By "simplify" we mean eliminate common factors from numerators and denominators and reduce to a minimum the number of times a given constant or variable appears in an expression. We ask that answers be expressed using positive exponents only in order to have a definite form for an answer. Later (in this section and elsewhere) we will encounter situations where we will want negative exponents in a final answer.

From the definition of negative exponents and the five properties of exponents, we can easily establish the following properties, which are used very frequently when dealing with exponent forms:

Theorem 2

Further Exponent Properties

For a and b any real numbers and m, n, and p any integers (division by 0 excluded):

1. $(a^m b^n)^p = a^{pm} b^{pn}$ **2.** $\left(\dfrac{a^m}{b^n}\right)^p = \dfrac{a^{pm}}{b^{pn}}$

3. $\dfrac{a^{-n}}{b^{-m}} = \dfrac{b^m}{a^n}$ **4.** $\left(\dfrac{a}{b}\right)^{-n} = \left(\dfrac{b}{a}\right)^n$

Proof We prove properties 1 and 4 in Theorem 2 and leave the proofs of 2 and 3 to you.

1. $(a^m b^n)^p = (a^m)^p (b^n)^p$ property 3, Theorem 1

$\qquad\quad = a^{pm} b^{pn}$ property 2, Theorem 1

4. $\left(\dfrac{a}{b}\right)^{-n} = \dfrac{a^{-n}}{b^{-n}}$ property 4, Theorem 1

$\qquad\quad = \dfrac{b^n}{a^n}$ property 3, Theorem 2

$\qquad\quad = \left(\dfrac{b}{a}\right)^n$ property 4, Theorem 1

EXAMPLE 3 **Using Exponent Properties**

Simplify using exponent properties, and express answers using positive exponents only.

(A) $(2a^{-3}b^2)^{-2} = 2^{-2}a^6 b^{-4} = \dfrac{a^6}{4b^4}$

(B) $\left(\dfrac{a^3}{b^5}\right)^{-2} = \dfrac{a^{-6}}{b^{-10}} = \dfrac{b^{10}}{a^6}$ or $\left(\dfrac{a^3}{b^5}\right)^{-2} = \left(\dfrac{b^5}{a^3}\right)^2 = \dfrac{b^{10}}{a^6}$

(C) $\dfrac{4x^{-3}y^{-5}}{6x^{-4}y^3} = \dfrac{2x^{-3-(-4)}}{3y^{3-(-5)}} = \dfrac{2x}{3y^8}$

(D) $\left(\dfrac{m^{-3}m^3}{n^{-2}}\right)^{-2} = \left(\dfrac{m^{-3+3}}{n^{-2}}\right)^{-2} = \left(\dfrac{m^0}{n^{-2}}\right)^{-2} = \left(\dfrac{1}{n^{-2}}\right)^{-2} = \dfrac{1}{n^4}$

(E) $(x + y)^{-3} = \dfrac{1}{(x + y)^3}$

Problem 3   Simplify using exponent properties, and express answers using positive exponents only.

(A) $(3x^4y^{-3})^{-2}$ (B) $\left(\dfrac{x^2}{y^4}\right)^{-3}$ (C) $\dfrac{6m^{-2}n^3}{15m^{-1}n^{-2}}$

(D) $\left(\dfrac{x^{-3}}{y^4y^{-4}}\right)^{-3}$ (E) $\dfrac{1}{(a-b)^{-2}}$

In simplifying exponent forms there is often more than one sequence of steps that will lead to the same result (see Example 3B). Use whichever sequence of steps makes sense to you.

EXAMPLE 4   **Simplifying a Compound Fraction**

Express as a simple fraction reduced to lowest terms:

$$\frac{x^{-2} - y^{-2}}{x^{-1} + y^{-1}} = \frac{\dfrac{1}{x^2} - \dfrac{1}{y^2}}{\dfrac{1}{x} + \dfrac{1}{y}} = \frac{x^2y^2\left(\dfrac{1}{x^2} - \dfrac{1}{y^2}\right)}{x^2y^2\left(\dfrac{1}{x} + \dfrac{1}{y}\right)}$$

$$= \frac{y^2 - x^2}{xy^2 + x^2y} = \frac{\overset{1}{(y-x)}\cancel{(y+x)}}{xy\underset{1}{\cancel{(y+x)}}}$$

$$= \frac{y-x}{xy}$$

Problem 4   Express as a simple fraction reduced to lowest terms:

$$\frac{x - x^{-1}}{1 - x^{-2}}$$

● **Scientific Notation**   Scientific work often involves the use of very large numbers or very small numbers. For example, the average cell contains about 200,000,000,000,000 molecules, and the diameter of an electron is about 0.000 000 000 0004 centimeter. It is generally troublesome to write and work with numbers of this type in standard decimal form. The two numbers written here cannot even be entered into most calculators as they are written. With exponents now defined for all integers, it is possible to express any decimal form as the product of a number between 1 and 10 and an integer power of 10; that is, in the form

$$a \times 10^n \qquad 1 \le a < 10, n \text{ an integer, } a \text{ in decimal form}$$

A number expressed in this form is said to be in **scientific notation**.

EXAMPLE 5   Scientific Notation

Each number is written in scientific notation:

$$7 = 7 \times 10^0 \qquad\qquad 0.5 = 5 \times 10^{-1}$$
$$720 = 7.2 \times 10^2 \qquad\qquad 0.08 = 8 \times 10^{-2}$$
$$6{,}430 = 6.43 \times 10^3 \qquad\qquad 0.000\ 32 = 3.2 \times 10^{-4}$$
$$5{,}350{,}000 = 5.35 \times 10^6 \qquad 0.000\ 000\ 0738 = 7.38 \times 10^{-8}$$

Can you discover a rule relating the number of decimal places the decimal point is moved to the power of 10 that is used?

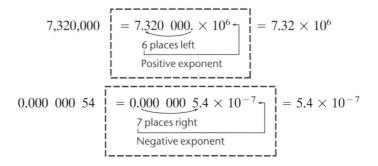

Problem 5   (A) Write each number in scientific notation: 430; 23,000; 345,000,000; 0.3; 0.0031; 0.000 000 683

(B) Write in standard decimal form: 4×10^3; 5.3×10^5; 2.53×10^{-2}; 7.42×10^{-6}

Most scientific and business calculators express very large and very small numbers in scientific notation. Read the instruction manual for your calculator to see how numbers in scientific notation are entered into your calculator. Numbers in scientific notation are displayed in most calculators as follows:

Calculator Display	Number Represented
$5.427493 \qquad -17$	$5.427\ 493 \times 10^{-17}$
$2.359779 \qquad 12$	$2.359\ 779 \times 10^{12}$

EXAMPLE 6   Using Scientific Notation on a Calculator

Write each number that cannot be entered directly into your calculator in decimal form in scientific notation; then carry out the computation using your calculator. (Refer to the user's manual accompanying your calculator for the procedure.) Express the answer to three significant digits* in scientific notation.

$$\frac{(25.32)(325,100,000,000)}{(0.0803)(0.000\ 000\ 000\ 000\ 0871)} = \frac{(25.32)(3.251 \times 10^{11})}{(0.0803)(8.71 \times 10^{-14})}$$

$$= \boxed{1.176920 \qquad 27} \quad \text{Calculator display}$$

$$= 1.18 \times 10^{27} \quad \text{To three significant digits}$$

Calculator Operations

A: $\boxed{25.32}\ \boxed{\times}\ \boxed{3.251}\ \boxed{\text{EE}}\ \boxed{11}\ \boxed{=}\ \boxed{\div}\ \boxed{0.0803}\ \boxed{\div}\ \boxed{8.71}\ \boxed{\text{EE}}\ \boxed{14}\ \boxed{1/-}\ \boxed{-}$

P: $\boxed{25.32}\ \boxed{\text{ENTER}}\ \boxed{3.251}\ \boxed{\text{EE}}\ \boxed{11}\ \boxed{\times}\ \boxed{0.0803}\ \boxed{\div}\ \boxed{8.71}\ \boxed{\text{EE}}\ \boxed{14}\ \boxed{+/-}\ \boxed{\div}$

Problem 6   Repeat Example 6 for: $\dfrac{(0.371)(0.000\ 000\ 006\ 932)}{(532)(62,600,000,000)}$

EXAMPLE 7   Measuring Time with an Atomic Clock

An atomic clock that counts the radioactive emissions of cesium is used to provide a precise definition of a second. One second is defined to be the time it takes cesium to emit 9,192,631,770 cycles of radiation. How many of these cycles will occur in 1 hour? Express the answer to five significant digits in scientific notation.

Solution   Since most calculators won't accept 10-digit numbers in either decimal or scientific notation, we round the number of cycles in a second to five significant digits before entering it into a calculator.

$$(9,192,631,770)(60^2) = (9.1926 \times 10^9)(3,600)$$
$$= 3.3093 \times 10^{13}$$

Problem 7   Refer to Example 7. How many of these cycles will occur in 1 year? Express the answer to five significant digits in scientific notation.

Answers to Matched Problems
1. (A) 1   (B) 1   (C) 0.000 01   (D) 1,000   (E) x^4 (F) v^3/u^7
2. (A) $15x$ (B) $3/(2y^3)$ (C) $-14x^4$

*For those not familiar with the meaning of *significant digits*, see Appendix A for a brief discussion of this concept.

3. (A) $y^6/(9x^8)$ (B) y^{12}/x^6 (C) $2n^5/(5m)$ (D) x^9 (E) $(a - b)^2$

4. x

5. (A) 4.3×10^2; 2.3×10^4; 3.45×10^8; 3×10^{-1}; 3.1×10^{-3}; 6.83×10^{-7}

   (B) $4{,}000$; $530{,}000$; 0.0253; $0.000\ 007\ 42$

6. 7.72×10^{-23} 7. 2.8990×10^{17}

EXERCISE 1-5   *All variables are restricted to prevent division by 0.*

A

Simplify Problems 1–16 and write the answers using positive exponents only.

1. $y^{-5}y^5$

2. x^3x^{-3}

3. $(2x^2)(3x^3)(x^4)$

4. $(2x^5)(3x^7)(4x^2)$

5. $(3x^3y^{-2})^2$

6. $(2cd^2)^{-3}$

7. $\left(\dfrac{ab^3}{c^2d}\right)^4$

8. $\left(\dfrac{x^2y}{2w^2}\right)^3$

9. $\dfrac{10^{23} \cdot 10^{-11}}{10^{-3} \cdot 10^{-2}}$

10. $\dfrac{10^{-13} \cdot 10^{-4}}{10^{-21} \cdot 10^3}$

11. $\dfrac{4x^{-2}y^{-3}}{2x^{-3}y^{-1}}$

12. $\dfrac{2a^6b^{-2}}{16a^{-3}b^2}$

13. $\left(\dfrac{n^{-3}}{n^{-2}}\right)^{-2}$

14. $\left(\dfrac{x^{-1}}{x^{-8}}\right)^{-1}$

15. $\dfrac{8 \times 10^3}{2 \times 10^{-5}}$

16. $\dfrac{18 \times 10^{12}}{6 \times 10^{-4}}$

Write the numbers in Problems 17–22 in scientific notation.

17. $32{,}250{,}000$

18. $4{,}930$

19. 0.085

20. 0.017

21. $0.000\ 000\ 0729$

22. $0.000\ 592$

In Problems 23–28, write each number in standard decimal form.

23. 5×10^{-3}

24. 4×10^{-4}

25. 2.69×10^7

26. 6.5×10^9

27. 5.9×10^{-10}

28. 6.3×10^{-6}

B

Simplify Problems 29–42, and write the answers using positive exponents only. Write compound fractions as simple fractions.

29. $\dfrac{27x^{-5}x^5}{18y^{-6}y^2}$

30. $\dfrac{32n^5n^{-8}}{24m^{-7}m^7}$

31. $\left(\dfrac{x^4y^{-1}}{x^{-2}y^3}\right)^2$

32. $\left(\dfrac{m^{-2}n^3}{m^4n^{-1}}\right)^2$

33. $\left(\dfrac{2x^{-3}y^2}{4xy^{-1}}\right)^{-2}$

34. $\left(\dfrac{6mn^{-2}}{3m^{-1}n^2}\right)^{-3}$

35. $\left[\left(\dfrac{u^3v^{-1}w^{-2}}{u^{-2}v^{-2}w}\right)^{-2}\right]^2$

36. $\left[\left(\dfrac{x^{-2}y^3t}{x^{-3}y^{-2}t^2}\right)^2\right]^{-1}$

37. $(x + y)^{-2}$

38. $(a^2 - b^2)^{-1}$

39. $\dfrac{1 + x^{-1}}{1 - x^{-2}}$

40. $\dfrac{1 - x}{x^{-1} - 1}$

41. $\dfrac{x^{-1} - y^{-1}}{x - y}$

42. $\dfrac{u + v}{u^{-1} + v^{-1}}$

Problems 43–48 are calculus-related. Write each problem in the form $ax^p + bx^q$ or $ax^p + bx^q + cx^r$, where a, b, and c are real numbers and p, q, and r are integers. For example,

$$\frac{2x^4 - 3x^2 + 1}{2x^3} \;\left[\; = \frac{2x^4}{2x^3} - \frac{3x^2}{2x^3} + \frac{1}{2x^3} \;\right] = x - \frac{3}{2}x^{-1} + \frac{1}{2}x^{-3}$$

43. $\dfrac{4x^2 - 12}{2x}$

44. $\dfrac{6x^3 + 9x}{3x^3}$

45. $\dfrac{5x^3 - 2}{3x^2}$

46. $\dfrac{7x^5 - x^2}{4x^5}$

47. $\dfrac{2x^3 - 3x^2 + x}{2x^2}$

48. $\dfrac{3x^4 - 4x^2 - 1}{4x^3}$

Evaluate Problems 49–52 to three significant digits using scientific notation where appropriate and a calculator.

49. $\dfrac{(32.7)(0.000\ 000\ 008\ 42)}{(0.0513)(80,700,000,000)}$

50. $\dfrac{(4,320)(0.000\ 000\ 000\ 704)}{(835)(635,000,000,000)}$

51. $\dfrac{(5,760,000,000)}{(527)(0.000\ 007\ 09)}$

52. $\dfrac{0.000\ 000\ 007\ 23}{(0.0933)(43,700,000,000)}$

In Problems 53–58, use y^x or a comparable key on your calculator to evaluate each of the following problems to five significant digits. (Read the instruction book accompanying your calculator.)

53. $(23.8)^8$

54. $(-302)^7$

55. $(-302)^{-7}$

56. $(23.8)^{-8}$

57. $(9,820,000,000)^3$

58. $(0.000\ 000\ 000\ 482)^{-4}$

C

Simplify Problems 59–64, and write the answers using positive exponents only. Write compound fractions as simple fractions.

59. $\dfrac{12(a + 2b)^{-3}}{6(a + 2b)^{-8}}$

60. $\dfrac{4(x - 3)^{-4}}{8(x - 3)^{-2}}$

61. $\dfrac{xy^{-2} - yx^{-2}}{y^{-1} - x^{-1}}$

62. $\dfrac{b^{-2} - c^{-2}}{b^{-3} - c^{-3}}$

63. $\left(\dfrac{x^{-1}}{x^{-1} - y^{-1}}\right)^{-1}$

64. $\left[\dfrac{u^{-2} - v^{-2}}{(u^{-1} - v^{-1})^2}\right]^{-1}$

APPLICATIONS

65. Earth Science. If the mass of the earth is approximately 6.1×10^{27} grams and each gram is 2.2×10^{-3} pound, what is the mass of the earth in pounds?

66. Biology. In 1929 Vernadsky, a biologist, estimated that all the free oxygen of the earth weighs 1.5×10^{21} grams and that it is produced by life alone. If 1 gram is approximately 2.2×10^{-3} pound, what is the weight of the free oxygen in pounds?

★**67. Computer Science.** Today's fastest computers can perform a single operation in 10^{-8} second, and the next generation of computers working with tiny supercon-

ducting devices are expected to be able to perform a single operation in 10^{-10} second. How many operations will each type of computer be able to perform in 1 second? In 1 minute?

★**68. Computer Science.** If electricity travels in a computer circuit at the speed of light (1.86×10^5 miles per second), how far will electricity travel in the superconducting computer (see Problem 67) in the time it takes it to perform one operation? (Size of circuits is a critical problem in computer design.) Give the answer in miles, feet, and inches (1 mile = 5,280 feet). Compute answers to two significant digits.

69. Economics. If in 1986 individuals in the United States paid about $349,000,000,000 in federal income taxes and the population was about 242,000,000, estimate to three significant digits the average amount of tax paid per person. Write your answer in scientific notation and in standard decimal form.

70. Economics. If the gross national product (GNP) was about $4,240,000,000,000 in the United States in 1986 and the population was about 242,000,000, estimate to three significant digits the GNP per person. Write your answer in scientific notation and in standard decimal form.

SECTION **1-6**   **Rational Exponents**

- Roots of Real Numbers
- Rational Exponents

We now know what symbols such as 3^5, 2^{-3}, and 7^0 mean; that is, we have defined a^n, where n is any integer and a is a real number. But what do symbols such as $4^{1/2}$ and $7^{2/3}$ mean? In this section we will extend the definition of exponent to the rational numbers. Before we can do this, however, we need a precise knowledge of what is meant by "a root of a number."

• Roots of Real Numbers

Perhaps you recall that a **square root** of a number b is a number c such that $c^2 = b$, and a **cube root** of a number b is a number d such that $d^3 = b$.

What are the square roots of 9?

$$3 \text{ is a square root of } 9, \text{ since } 3^2 = 9.$$

$$-3 \text{ is a square root of } 9, \text{ since } (-3)^2 = 9.$$

Thus, 9 has two real square roots, one the negative of the other.

What are the cube roots of 8?

$$2 \text{ is a cube root of } 8, \text{ since } 2^3 = 8.$$

And 2 is the only real number with this property. In general:

DEFINITION 1   **Definition of an *n*th Root**

For a natural number n and a and b real numbers:

$$a \text{ is an } n\text{th root of } b \text{ if } a^n = b$$

3 is a fourth root of 81, since $3^4 = 81$

How many real square roots of 4 exist? Of 5? Of -9? How many real fourth roots of 5 exist? Of -5? How many real cube roots of 27 are there? Of -27? The following important theorem (which we state without proof) answers these questions.

Theorem 1	**Number of Real *n*th Roots of a Real Number *b***		
		n even	**n odd**
	b positive	Two real *n*th roots −3 and 3 are both fourth roots of 81	One real *n*th root 2 is the only real cube root of 8
	b negative	No real *n*th root −9 has no real square roots	One real *n*th root −2 is the only real cube root of −8

Thus, 4 and 5 have two real square roots each, and −9 has none. There are two real fourth roots of 5 and none for −5. And 27 and −27 have one real cube root each. What symbols do we use to represent these roots? We turn to this question now.

● **Rational Exponents** If all exponent properties are to continue to hold even if some of the exponents are rational numbers, then

$$(5^{1/3})^3 = 5^{3/3} = 5 \quad \text{and} \quad (7^{1/2})^2 = 7^{2/2} = 7$$

Since Theorem 1 states that the number 5 has one real cube root, it seems reasonable to use the symbol $5^{1/3}$ to represent this root. On the other hand, Theorem 1 states that 7 has two real square roots. Which real square root of 7 does $7^{1/2}$ represent? We answer this question in the following definition.

DEFINITION 2 **$b^{1/n}$, Principal *n*th Root**

For *n* a natural number and *b* a real number,

$$b^{1/n} \text{ is the } \textbf{principal } \textit{n}\textbf{th root of } b$$

defined as follows:

1. If *n* is even and *b* is positive, then $b^{1/n}$ represents the positive *n*th root of *b*.

   $16^{1/2} = 4$ not −4 and 4.
   $-16^{1/2} = -4$ $-16^{1/2}$ and $(-16)^{1/2}$ are not the same.

2. If *n* is even and *b* is negative, then $b^{1/n}$ does not represent a real number. (More will be said about this case later.)

   $(-16)^{1/2}$ is not real.

3. If *n* is odd, then $b^{1/n}$ represents the real *n*th root of *b* (there is only one).

   $32^{1/5} = 2$ $(-32)^{1/5} = -2$

4. $0^{1/n} = 0$ $0^{1/9} = 0$ $0^{1/6} = 0$

*In this section we limit our discussion to real roots of real numbers. After the real numbers are extended to the complex numbers (see Section 2-5), additional roots may be considered. For example, it turns out that 1 has three cube roots: in addition to the real number 1, there are two other cube roots of 1 in the complex number system.

EXAMPLE 1   **Principal *n*th Roots**

(A) $9^{1/2} = 3$
(B) $-9^{1/2} = -3$ Compare parts (B) and (C).
(C) $(-9)^{1/2}$ is not a real number.   (D) $27^{1/3} = 3$
(E) $(-27)^{1/3} = -3$ (F) $0^{1/7} = 0$

Problem 1   Find each of the following:

(A) $4^{1/2}$ (B) $-4^{1/2}$ (C) $(-4)^{1/2}$
(D) $8^{1/3}$ (E) $(-8)^{1/3}$ (F) $0^{1/8}$

How should a symbol such as $7^{2/3}$ be defined? If the properties of exponents are to hold for rational exponents, then $7^{2/3} = (7^{1/3})^2$; that is, $7^{2/3}$ must represent the square of the cube root of 7. This leads to the following general definition:

DEFINITION 3 $b^{m/n}$ **and** $b^{-m/n}$**, Rational Number Exponent**

For m and n natural numbers and b any real number (except b cannot be negative when n is even):

$$b^{m/n} = (b^{1/n})^m \quad \text{and} \quad b^{-m/n} = \frac{1}{b^{m/n}}$$

$$4^{3/2} = (4^{1/2})^3 = 2^3 = 8 \quad 4^{-3/2} = \frac{1}{4^{3/2}} = \frac{1}{8} \quad (-4)^{3/2} \text{ is not real}$$

$$(-32)^{3/5} = [(-32)^{1/5}]^3 = (-2)^3 = -8$$

We have now discussed $b^{m/n}$ for all rational numbers m/n and real numbers b. It can be shown, though we will not do so, that all five properties of exponents listed in Theorem 1 in Section 1-5 continue to hold for rational exponents as long as we avoid even roots of negative numbers. With the latter restriction in effect, the following useful relationship is an immediate consequence of the exponent properties:

Theorem 2   **Rational Exponent Property**

For m and n natural numbers and b any real number (except b cannot be negative when n is even):

$$b^{m/n} = \begin{cases} (b^{1/n})^m \\ (b^m)^{1/n} \end{cases} \quad 8^{2/3} = \begin{cases} (8^{1/3})^2 \\ (8^2)^{1/3} \end{cases}$$

To see why b cannot be negative when n is even, consider what happens when we relax that restriction. Can you resolve the following contradiction?

$$-1 = (-1)^{2/2} = [(-1)^2]^{1/2} = 1^{1/2} = 1$$

The second member of the equality chain, $(-1)^{2/2}$, involves an even root of a negative number, which is not real. Thus, we see that the properties of exponents do not necessarily hold when we are dealing with nonreal quantities unless further restrictions are imposed. One such restriction is to require all rational exponents be reduced to lowest terms.

EXAMPLE 2   Using Rational Exponents

Simplify, and express answers using positive exponents only. All letters represent positive real numbers.

(A) $8^{2/3} = (8^{1/3})^2 = 2^2 = 4$ or $8^{2/3} = (8^2)^{1/3} = 64^{1/3} = 4$

(B) $(-8)^{5/3} = [(-8)^{1/3}]^5 = (-2)^5 = -32$

(C) $(3x^{1/3})(2x^{1/2}) = 6x^{1/3+1/2} = 6x^{5/6}$

(D) $\left(\dfrac{4x^{1/3}}{x^{1/2}}\right)^{1/2} = \dfrac{4^{1/2}x^{1/6}}{x^{1/4}} = \dfrac{2}{x^{1/4-1/6}} = \dfrac{2}{x^{1/12}}$

(E) $(u^{1/2} - 2v^{1/2})(3u^{1/2} + v^{1/2}) = 3u - 5u^{1/2}v^{1/2} - 2v$

Problem 2   Simplify, and express answers using positive exponents only. All letters represent positive real numbers.

(A) $9^{3/2}$ (B) $(-27)^{4/3}$ (C) $(5y^{3/4})(2y^{1/3})$ (D) $(2x^{-3/4}y^{1/4})^4$

(E) $\left(\dfrac{8x^{1/2}}{x^{2/3}}\right)^{1/3}$ (F) $(2x^{1/2} + y^{1/2})(x^{1/2} - 3y^{1/2})$

EXAMPLE 3   Evaluating Rational Exponential Forms with a Calculator

Evaluate to four significant digits using a calculator. (Refer to the instruction book for your particular calculator to see how exponential forms are evaluated.)

(A) $11^{3/4}$ (B) $3.1046^{-2/3}$ (C) $(0.000\ 000\ 008\ 437)^{3/11}$

Solutions   (A) First change $\frac{3}{4}$ to the standard decimal form 0.75; then evaluate $11^{0.75}$ using $\boxed{y^x}$ (or a comparable key) on your calculator.

$$11^{3/4} = 6.040$$

A: $\boxed{11}\ \boxed{y^x}\ \boxed{(}\ \boxed{3}\ \boxed{\div}\ \boxed{4}\ \boxed{)}\ \boxed{=}$

P: $\boxed{11}\ \boxed{\text{ENTER}}\ \boxed{3}\ \boxed{\text{ENTER}}\ \boxed{4}\ \boxed{\div}\ \boxed{y^x}$

(B) $3.1046^{-2/3} = 0.4699$

A: $\boxed{3.1046}\ \boxed{y^x}\ \boxed{(}\ \boxed{2}\ \boxed{\div}\ \boxed{3}\ \boxed{+/-}\ \boxed{)}\ \boxed{=}$

P: $\boxed{3.1046}\ \boxed{\text{ENTER}}\ \boxed{2}\ \boxed{\text{ENTER}}\ \boxed{3}\ \boxed{\div}\ \boxed{+/-}\ \boxed{y^x}$

(C) $(0.000\ 000\ 008\ 437)^{3/11} = (8.437 \times 10^{-9})^{3/11}$
$$= 0.006\ 281$$

Problem 3   Evaluate to four significant digits using a calculator.

(A) $2^{3/8}$ (B) $57.28^{-5/6}$ (C) $(83{,}240{,}000{,}000)^{5/3}$

EXAMPLE 4   Simplifying Fractions Involving Rational Exponents

Write the following expression as a simple fraction reduced to lowest terms and without negative exponents:

$$\frac{(1 + x^2)^{1/2}(2x) - x^2(\frac{1}{2})(1 + x^2)^{-1/2}(2x)}{1 + x^2}$$

Solution   The negative exponent indicates the presence of a fraction in the numerator. Multiply numerator and denominator by $(1 + x^2)^{1/2}$ to eliminate the negative exponent and simplify.

$$\frac{(1 + x^2)^{1/2}(2x) - x^2(\frac{1}{2})(1 + x^2)^{-1/2}(2x)}{1 + x^2} \cdot \frac{(1 + x^2)^{1/2}}{(1 + x^2)^{1/2}}$$

$$= \frac{2x(1 + x^2) - x^3}{(1 + x^2)^{3/2}} = \frac{2x + 2x^3 - x^3}{(1 + x^2)^{3/2}} = \frac{2x + x^3}{(1 + x^2)^{3/2}}$$

$$= \frac{x(2 + x^2)}{(1 + x^2)^{3/2}}$$

Problem 4   Write the following expression as a simple fraction reduced to lowest terms and without negative exponents:

$$\frac{x^2(\frac{1}{2})(1 + x^2)^{-1/2}(2x) - (1 + x^2)^{1/2}(2x)}{x^4}$$

EXERCISE $1\text{-}6$ *All variables represent positive real numbers unless otherwise stated.*

A

In Problems 1–12, evaluate each expression that results in a rational number.

1. $16^{1/2}$ 2. $64^{1/3}$ 3. $16^{3/2}$ 4. $16^{3/4}$

5. $-36^{1/2}$ 6. $32^{3/5}$ 7. $(-36)^{1/2}$ 8. $(-32)^{3/5}$

9. $(\frac{4}{25})^{3/2}$ 10. $(\frac{8}{27})^{2/3}$ 11. $9^{-3/2}$ 12. $8^{-2/3}$

Simplify Problems 13–20, and express answers using positive exponents only.

13. $y^{1/5}y^{2/5}$ 14. $x^{1/4}x^{3/4}$ 15. $d^{2/3}d^{-1/3}$ 16. $x^{1/4}x^{-3/4}$

17. $(y^{-8})^{1/16}$ 18. $(x^{-2/3})^{-6}$ 19. $(8x^3y^{-6})^{1/3}$ 20. $(4u^{-2}v^4)^{1/2}$

B

Simplify Problems 21–30, and express answers using positive exponents only.

21. $\left(\dfrac{a^{-3}}{b^4}\right)^{1/12}$ 22. $\left(\dfrac{m^{-2/3}}{n^{-1/2}}\right)^{-6}$ 23. $\left(\dfrac{4x^{-2}}{y^4}\right)^{-1/2}$ 24. $\left(\dfrac{w^4}{9x^{-2}}\right)^{-1/2}$

25. $\left(\dfrac{8a^{-4}b^3}{27a^2b^{-3}}\right)^{1/3}$ 26. $\left(\dfrac{25x^5y^{-1}}{16x^{-3}y^{-5}}\right)^{1/2}$ 27. $\dfrac{8x^{-1/3}}{12x^{1/4}}$ 28. $\dfrac{6a^{3/4}}{15a^{-1/3}}$

29. $\left(\dfrac{a^{2/3}b^{-1/2}}{a^{1/2}b^{1/2}}\right)^2$ 30. $\left(\dfrac{x^{-1/3}y^{1/2}}{x^{-1/4}y^{1/3}}\right)^6$

In Problems 31–38, multiply, and express answers using positive exponents only.

31. $2m^{1/3}(3m^{2/3} - m^6)$ 32. $3x^{3/4}(4x^{1/4} - 2x^8)$

33. $(a^{1/2} + 2b^{1/2})(a^{1/2} - 3b^{1/2})$ 34. $(3u^{1/2} - v^{1/2})(u^{1/2} - 4v^{1/2})$

35. $(2x^{1/2} - 3y^{1/2})(2x^{1/2} + 3y^{1/2})$ 36. $(5m^{1/2} + n^{1/2})(5m^{1/2} - n^{1/2})$

37. $(x^{1/2} + 2y^{1/2})^2$ 38. $(3x^{1/2} - y^{1/2})^2$

In Problems 39–46, evaluate to four significant digits using a calculator. (Refer to the instruction book for your calculator to see how exponential forms are evaluated.)

39. $15^{5/4}$ 40. $22^{3/2}$ 41. $103^{-3/4}$ 42. $827^{-3/8}$

43. $2.876^{8/5}$ 44. $37.09^{7/3}$ 45. $(0.000\ 000\ 077\ 35)^{-2/7}$ 46. $(491{,}300{,}000{,}000)^{7/4}$

∫ *Problems 47–52 are calculus-related. Write each problem in the form $ax^p + bx^q$, where a and b are real numbers and p and q are rational numbers. For example,*

$$\frac{2x^{1/3} + 4}{4x} \boxed{= \frac{2x^{1/3}}{4x} + \frac{4}{4x} = \frac{1}{2}x^{1/3-1} + x^{-1}} = \frac{1}{2}x^{-2/3} + x^{-1}$$

47. $\dfrac{12x^{1/2} - 3}{4x^{1/2}}$

48. $\dfrac{x^{2/3} + 2}{2x^{1/3}}$

49. $\dfrac{3x^{2/3} + x^{1/2}}{5x}$

50. $\dfrac{2x^{3/4} + 3x^{1/3}}{3x}$

51. $\dfrac{x^2 - 4x^{1/2}}{2x^{1/3}}$

52. $\dfrac{2x^{1/3} - x^{1/2}}{4x^{1/2}}$

C

In Problems 53–56, m and n represent positive integers. Simplify and express answers using positive exponents.

53. $(a^{3/n}b^{3/m})^{1/3}$

54. $(a^{n/2}b^{n/3})^{1/n}$

55. $(x^{m/4}y^{n/3})^{-12}$

56. $(a^{m/3}b^{n/2})^{-6}$

57. Find a real value of x such that:
   (A) $(x^2)^{1/2} \neq x$ (B) $(x^2)^{1/2} = x$

58. Find a real value of x such that:
   (A) $(x^2)^{1/2} \neq -x$ (B) $(x^2)^{1/2} = -x$

∫ *Problems 59–62 are calculus-related. Simplify by writing each expression as a simple fraction reduced to lowest terms and without negative exponents.*

59. $\dfrac{(2x - 1)^{1/2} - (x + 2)(\frac{1}{2})(2x - 1)^{-1/2}(2)}{2x - 1}$

60. $\dfrac{(x - 1)^{1/2} - x(\frac{1}{2})(x - 1)^{-1/2}}{x - 1}$

61. $\dfrac{2(3x - 1)^{1/3} - (2x + 1)(\frac{1}{3})(3x - 1)^{-2/3}(3)}{(3x - 1)^{2/3}}$

62. $\dfrac{(x + 2)^{2/3} - x(\frac{2}{3})(x + 2)^{-1/3}}{(x + 2)^{4/3}}$

APPLICATIONS

63. Economics. The number of units N of a finished product produced from the use of x units of labor and y units of capital for a particular Third World country is approximated by

$$N = 10x^{3/4}y^{1/4} \quad \text{Cobb–Douglas equation}$$

Estimate how many units of a finished product will be produced using 256 units of labor and 81 units of capital.

64. Economics. The number of units N of a finished product produced by a particular automobile company where x units of labor and y units of capital are used is approximated by

$$N = 50x^{1/2}y^{1/2} \quad \text{Cobb–Douglas equation}$$

Estimate how many units will be produced using 256 units of labor and 144 units of capital.

65. Braking Distance. R. A. Moyer of Iowa State College found, in comprehensive tests carried out on 41 wet pavements, that the braking distance d (in feet) for a particular automobile traveling at v miles per hour was given approximately by

$$d = 0.0212v^{7/3}$$

Approximate the braking distance to the nearest foot for the car traveling on wet pavement at 70 miles per hour.

66. Braking Distance. Approximately how many feet would it take the car in Problem 65 to stop on wet pavement if it were traveling at 50 miles per hour? (Compute answer to the nearest foot.)

SECTION 1-7 **Radicals**

- From Rational Exponents to Radicals, and Vice Versa
- Properties of Radicals
- Simplifying Radicals
- Sums and Differences
- Products
- Rationalizing Operations

What do the following algebraic expressions have in common?

$$2^{1/2} \qquad 2x^{2/3} \qquad \frac{1}{x^{1/2} + y^{1/2}}$$

$$\sqrt{2} \qquad 2\sqrt[3]{x^2} \qquad \frac{1}{\sqrt{x} + \sqrt{y}}$$

Each vertical pair represents the same quantity, one in rational exponent form and the other in *radical form*. There are occasions when it is more convenient to work with radicals than with rational exponents, or vice versa. In this section we see how the two forms are related and investigate some basic operations on radicals.

● **From Rational Exponents to Radicals, and Vice Versa**

We start this discussion by defining an *n*th-root radical:

DEFINITION 1

$\sqrt[n]{b}$, *n*th-Root Radical

For n a natural number greater than 1 and b a real number, we define $\sqrt[n]{b}$ to be the **principal *n*th root of *b*** (see Definition 2 in Section 1-6); that is,

$$\sqrt[n]{b} = b^{1/n}$$

If $n = 2$, we write $\sqrt{b}$ in place of $\sqrt[2]{b}$.

$\sqrt{25} \;\boxed{= 25^{1/2}} = 5 \qquad \sqrt[5]{32} \;\boxed{= 32^{1/5}} = 2$

$-\sqrt{25} \;\boxed{= -25^{1/2}} = -5 \qquad \sqrt[5]{-32} \;\boxed{= (-32)^{1/5}} = -2$

$\sqrt{-25}$ is not real $\qquad \sqrt[4]{0} = 0^{1/4} = 0$

The symbol $\sqrt{}$ is called a **radical**, n is called the **index**, and b is called the **radicand**.

As stated above, it is often an advantage to be able to shift back and forth between rational exponent forms and radical forms. The following relationships, which are direct consequences of Definition 1 and Theorem 2 in Section 1-6, are useful in this regard:

Rational Exponent/Radical Conversions

For m and n positive integers ($n > 1$), and b not negative when n is even,

$$b^{m/n} = \begin{cases} (b^m)^{1/n} = \sqrt[n]{b^m} \\ (b^{1/n})^m = (\sqrt[n]{b})^m \end{cases} \qquad 2^{2/3} = \begin{cases} \sqrt[3]{2^2} \\ (\sqrt[3]{2})^2 \end{cases}$$

Note: Unless stated to the contrary, all variables in the rest of the discussion are restricted so that all quantities involved are real numbers.

EXAMPLE 1   Rational Exponents/Radical Conversions

Change from rational exponent form to radical form.

(A) $x^{1/7} = \sqrt[7]{x}$

(B) $(3u^2v^3)^{3/5} = \sqrt[5]{(3u^2v^3)^3}$ or $(\sqrt[5]{3u^2v^3})^3$ The first is usually preferred.

(C) $y^{-2/3} = \dfrac{1}{y^{2/3}} = \dfrac{1}{\sqrt[3]{y^2}}$ or $\sqrt[3]{y^{-2}}$ or $\sqrt[3]{\dfrac{1}{y^2}}$

Change from radical form to rational exponent form.

(D) $\sqrt[5]{6} = 6^{1/5}$ (E) $-\sqrt[3]{x^2} = -x^{2/3}$ (F) $\sqrt{x^2 + y^2} = (x^2 + y^2)^{1/2}$

Problem 1   Change from rational exponent form to radical form.

(A) $u^{1/5}$ (B) $(6x^2y^5)^{2/9}$ (C) $(3xy)^{-3/5}$

Change from radical form to rational exponent form.

(D) $\sqrt[4]{9u}$ (E) $-\sqrt[7]{(2x)^4}$ (F) $\sqrt[3]{x^3 + y^3}$

• Properties of Radicals   The process of changing and simplifying radical expressions is aided by the introduction of several properties of radicals that follow directly from exponent properties considered earlier.

Theorem 1	**Properties of Radicals**

For n a natural number greater than 1, and x and y positive real numbers:

1. $\sqrt[n]{x^n} = x$ $\sqrt[3]{x^3} = x$

2. $\sqrt[n]{xy} = \sqrt[n]{x}\sqrt[n]{y}$ $\sqrt[5]{xy} = \sqrt[5]{x}\sqrt[5]{y}$

3. $\sqrt[n]{\dfrac{x}{y}} = \dfrac{\sqrt[n]{x}}{\sqrt[n]{y}}$ $\sqrt[4]{\dfrac{x}{y}} = \dfrac{\sqrt[4]{x}}{\sqrt[4]{y}}$

EXAMPLE 2   **Simplifying Radicals**

Simplify:

(A) $\sqrt[5]{(3x^2y)^5} = 3x^2y$

(B) $\sqrt{10}\sqrt{5} = \sqrt{50} = \sqrt{25\cdot 2} = \sqrt{25}\sqrt{2} = 5\sqrt{2}$

(C) $\sqrt[3]{\dfrac{x}{27}} = \dfrac{\sqrt[3]{x}}{\sqrt[3]{27}} = \dfrac{\sqrt[3]{x}}{3}$ or $\dfrac{1}{3}\sqrt[3]{x}$

Problem 2   Simplify:

(A) $\sqrt[7]{(u^2 + v^2)^7}$ (B) $\sqrt{6}\sqrt{2}$ (C) $\sqrt[3]{\dfrac{x^2}{8}}$

CAUTION

In general, properties of radicals can be used to simplify terms raised to powers, not sums of terms raised to powers. Thus, for x and y positive real numbers,

$$\sqrt{x^2 + y^2} \neq \sqrt{x^2} + \sqrt{y^2} = x + y$$

but

$$\sqrt{x^2 + 2xy + y^2} = \sqrt{(x + y)^2} = x + y$$

• Simplifying Radicals   The properties of radicals provide us with the means of changing algebraic expressions containing radicals to a variety of equivalent forms. One form that is often useful is a *simplified form*. An algebraic expression that contains radicals is said to be in **simplified form** if all four of the conditions listed in the following definition are satisfied.

DEFINITION 2 **Simplified (Radical) Form**

1. No radicand (the expression within the radical sign) contains a factor to a power greater than or equal to the index of the radical.
   For example, $\sqrt{x^5}$ violates this condition.

2. No power of the radicand and the index of the radical have a common factor other than 1.
   For example, $\sqrt[6]{x^4}$ violates this condition.

3. No radical appears in a denominator.
   For example, $y/\sqrt{x}$ violates this condition.

4. No fraction appears within a radical.
   For example, $\sqrt{\frac{3}{5}}$ violates this condition.

EXAMPLE 3 Finding Simplified Form

Express radicals in simplified form.

(A) $\sqrt{12x^3y^5z^2} = \sqrt{(4x^2y^4z^2)(3xy)}$ Condition 1 is not met.

$= \sqrt{(2xy^2z)^2(3xy)}$ $x^{pm}y^{pn} = (x^my^n)^p$

$= \sqrt{(2xy^2z)^2}\sqrt{3xy}$ $\sqrt[n]{xy} = \sqrt[n]{x}\sqrt[n]{y}$

$= 2xy^2z\sqrt{3xy}$ $\sqrt[n]{x^n} = x$

(B) $\sqrt[3]{6x^2y}\sqrt[3]{4x^5y^2}$ $= \sqrt[3]{(6x^2y)(4x^5y^2)}$ $\sqrt[n]{x}\sqrt[n]{y} = \sqrt[n]{xy}$

$= \sqrt[3]{24x^7y^3}$

$= \sqrt[3]{(8x^6y^3)(3x)}$ Condition 1 is not met.

$= \sqrt[3]{(2x^2y)^3(3x)}$ $x^{pm}y^{pn} = (x^my^n)^p$

$= \sqrt[3]{(2x^2y)^3}\sqrt[3]{3x}$ $\sqrt[n]{xy} = \sqrt[n]{x}\sqrt[n]{y}$

$= 2x^2y\sqrt[3]{3x}$ $\sqrt[n]{x^n} = x$

(C) $\sqrt[6]{16x^4y^2} = [(4x^2y)^2]^{1/6}$ Condition 2 is not met.

$= (4x^2y)^{2/6}$ Note the convenience of using rational exponents.

$= (4x^2y)^{1/3}$

$= \sqrt[3]{4x^2y}$

(D) $\sqrt[3]{\sqrt{27}} = [(3^3)^{1/2}]^{1/3}$

$$\boxed{= (3^3)^{1/6} = 3^{3/6}} = 3^{1/2} = \sqrt{3}$$

Problem 3 Express radicals in simplified form.

(A) $\sqrt{18x^5y^2z^3}$ (B) $\sqrt[4]{27a^3b^3}\sqrt[4]{3a^5b^3}$ (C) $\sqrt[9]{8x^6y^3}$ (D) $\sqrt{\sqrt[3]{4}}$

● **Sums and Differences**   Algebraic expressions involving radicals often can be simplified by adding and subtracting terms that contain exactly the same radical expressions. We proceed in essentially the same way as we do when we combine like terms in polynomials. The distributive property of real numbers plays a central role in this process.

EXAMPLE 4   **Combining Like Terms**

Combine as many terms as possible:

(A) $5\sqrt{3} + 4\sqrt{3} \;\boxed{= (5 + 4)\sqrt{3}} = 9\sqrt{3}$

(B) $2\sqrt[3]{xy^2} - 7\sqrt[3]{xy^2} \;\boxed{= (2 - 7)\sqrt[3]{xy^2}} = -5\sqrt[3]{xy^2}$

(C) $3\sqrt{xy} - 2\sqrt[3]{xy} + 4\sqrt{xy} - 7\sqrt[3]{xy} \;\boxed{= 3\sqrt{xy} + 4\sqrt{xy} - 2\sqrt[3]{xy} - 7\sqrt[3]{xy}}$

$$= 7\sqrt{xy} - 9\sqrt[3]{xy}$$

Problem 4   Combine as many terms as possible:

(A) $6\sqrt{2} + 2\sqrt{2}$ (B) $3\sqrt[5]{2x^2y^3} - 8\sqrt[5]{2x^2y^3}$
(C) $5\sqrt[3]{mn^2} - 3\sqrt{mn} - 2\sqrt[3]{mn^2} + 7\sqrt{mn}$

● **Products**   We will now consider several types of special products that involve radicals. The distributive property of real numbers plays a central role in our approach to these problems.

EXAMPLE 5   **Multiplication with Radical Forms**

Multiply and simplify:

(A) $\sqrt{2}(\sqrt{10} - 3) = \sqrt{2}\sqrt{10} - \sqrt{2}\cdot 3 = \sqrt{20} - 3\sqrt{2} = 2\sqrt{5} - 3\sqrt{2}$

(B) $(\sqrt{2} - 3)(\sqrt{2} + 5) = \sqrt{2}\sqrt{2} - 3\sqrt{2} + 5\sqrt{2} - 15$
$$= 2 + 2\sqrt{2} - 15$$
$$= 2\sqrt{2} - 13$$

(C) $(\sqrt{x} - 3)(\sqrt{x} + 5) = \sqrt{x}\sqrt{x} - 3\sqrt{x} + 5\sqrt{x} - 15$
$$= x + 2\sqrt{x} - 15$$

(D) $(\sqrt[3]{m} + \sqrt[3]{n^2})(\sqrt[3]{m^2} - \sqrt[3]{n}) = \sqrt[3]{m^3} + \sqrt[3]{m^2 n^2} - \sqrt[3]{mn} - \sqrt[3]{n^3}$
$$= m - \sqrt[3]{mn} + \sqrt[3]{m^2 n^2} - n$$

Problem 5   Multiply and simplify:

(A) $\sqrt{3}(\sqrt{6} - 4)$ (B) $(\sqrt{3} - 2)(\sqrt{3} + 4)$

(C) $(\sqrt{y} - 2)(\sqrt{y} + 4)$ (D) $(\sqrt[3]{x^2} - \sqrt[3]{y^2})(\sqrt[3]{x} + \sqrt[3]{y})$

● **Rationalizing Operations**   We now turn to algebraic fractions involving radicals in the denominator. Eliminating a radical from a denominator is referred to as **rationalizing the denominator**. To rationalize the denominator, we multiply the numerator and denominator by a suitable factor that will rationalize the denominator—that is, will leave the denominator free of radicals. This factor is called a **rationalizing factor**. The following special products are of use in finding some rationalizing factors (see Examples 6C, D):

$$(a - b)(a + b) = a^2 - b^2 \tag{1}$$

$$(a - b)(a^2 + ab + b^2) = a^3 - b^3 \tag{2}$$

$$(a + b)(a^2 - ab + b^2) = a^3 + b^3 \tag{3}$$

EXAMPLE 6   **Rationalizing Denominators**

Rationalize denominators.

(A) $\dfrac{3}{\sqrt{5}}$ (B) $\sqrt[3]{\dfrac{2a^2}{3b^2}}$ (C) $\dfrac{\sqrt{x} + \sqrt{y}}{3\sqrt{x} - 2\sqrt{y}}$ (D) $\dfrac{1}{\sqrt[3]{m} + 2}$

Solutions   (A) $\sqrt{5}$ is a rationalizing factor for $\sqrt{5}$, since $\sqrt{5}\sqrt{5} = \sqrt{5^2} = 5$. Thus, we multiply the numerator and denominator by $\sqrt{5}$ to rationalize the denominator:

$$\frac{3}{\sqrt{5}} = \frac{3\sqrt{5}}{\sqrt{5}\sqrt{5}} = \frac{3\sqrt{5}}{5}$$

(B) $\sqrt[3]{\dfrac{2a^2}{3b^2}} = \dfrac{\sqrt[3]{2a^2}}{\sqrt[3]{3b^2}} = \dfrac{\sqrt[3]{2a^2}\sqrt[3]{3^2b}}{\sqrt[3]{3b^2}\sqrt[3]{3^2b}} \quad \boxed{= \dfrac{\sqrt[3]{2\cdot 3^2a^2b}}{\sqrt[3]{3^3b^3}}} \quad = \dfrac{\sqrt[3]{18a^2b}}{3b}$

(C) Special product (1) suggests that if we multiply the denominator $3\sqrt{x} - 2\sqrt{y}$ by $3\sqrt{x} + 2\sqrt{y}$, we will obtain the difference of two squares and the denominator will be rationalized.

$$\frac{\sqrt{x} + \sqrt{y}}{3\sqrt{x} - 2\sqrt{y}} = \frac{(\sqrt{x} + \sqrt{y})(3\sqrt{x} + 2\sqrt{y})}{(3\sqrt{x} - 2\sqrt{y})(3\sqrt{x} + 2\sqrt{y})}$$

$$\boxed{= \frac{3\sqrt{x^2} + 2\sqrt{xy} + 3\sqrt{xy} + 2\sqrt{y^2}}{(3\sqrt{x})^2 - (2\sqrt{y})^2}}$$

$$= \frac{3x + 5\sqrt{xy} + 2y}{9x - 4y}$$

(D) Special product (3) above suggests that if we multiply the denominator $\sqrt[3]{m} + 2$ by $(\sqrt[3]{m})^2 - 2\sqrt[3]{m} + 2^2$, we will obtain the sum of two cubes and the denominator will be rationalized.

$$\frac{1}{\sqrt[3]{m} + 2} = \frac{1[(\sqrt[3]{m})^2 - 2\sqrt[3]{m} + 2^2]}{(\sqrt[3]{m} + 2)[(\sqrt[3]{m})^2 - 2\sqrt[3]{m} + 2^2]}$$

$$\boxed{= \frac{\sqrt[3]{m^2} - 2\sqrt[3]{m} + 4}{(\sqrt[3]{m})^3 + 2^3}}$$

$$= \frac{\sqrt[3]{m^2} - 2\sqrt[3]{m} + 4}{m + 8}$$

Problem 6 Rationalize denominators.

(A) $\dfrac{6}{\sqrt{2x}}$ (B) $\dfrac{10x^3}{\sqrt[3]{4x}}$ (C) $\dfrac{\sqrt{x} + 2}{2\sqrt{x} + 3}$ (D) $\dfrac{1}{1 - \sqrt[3]{y}}$

4. (A) $8\sqrt{2}$ (B) $-5\sqrt[5]{2x^2y^3}$ (C) $3\sqrt[3]{mn^2} + 4\sqrt{mn}$

5. (A) $3\sqrt{2} - 4\sqrt{3}$ (B) $2\sqrt{3} - 5$ (C) $y + 2\sqrt{y} - 8$ (D) $x + \sqrt[3]{x^2y} - \sqrt[3]{xy^2} - y$

6. (A) $\dfrac{3\sqrt{2x}}{x}$ (B) $5x^2\sqrt[3]{2x^2}$ (C) $\dfrac{2x + \sqrt{x} - 6}{4x - 9}$ (D) $\dfrac{1 + \sqrt[3]{y} + \sqrt[3]{y^2}}{1 - y}$

EXERCISE $1\text{-}7$ *Unless stated to the contrary, all variables are restricted so that all quantities involved are real numbers.*

A

In Problems 1–8, change to radical form. Do not simplify.

1. $m^{2/3}$

2. $n^{4/5}$

3. $6x^{3/5}$

4. $7y^{2/5}$

5. $(4xy^3)^{2/5}$

6. $(7x^2y)^{5/7}$

7. $(x + y)^{1/2}$

8. $x^{1/2} + y^{1/2}$

In Problems 9–16, change to rational exponent form. Do not simplify.

9. $\sqrt[5]{b}$

10. $\sqrt{c}$

11. $5\sqrt[4]{x^3}$

12. $7m\sqrt[5]{n^2}$

13. $\sqrt[5]{(2x^2y)^3}$

14. $\sqrt[9]{(3m^4n)^2}$

15. $\sqrt[3]{x} + \sqrt[3]{y}$

16. $\sqrt[3]{x + y}$

In Problems 17–32, write in simplified form.

17. $\sqrt[3]{-8}$

18. $\sqrt[3]{-27}$

19. $\sqrt{9x^8y^4}$

20. $\sqrt{16m^4y^8}$

21. $\sqrt[4]{16m^4n^8}$

22. $\sqrt[5]{32a^{15}b^{10}}$

23. $\sqrt{8a^3b^5}$

24. $\sqrt{27m^2n^7}$

25. $\sqrt[3]{24x^4y^7}$

26. $\sqrt[4]{24x^5y^8}$

27. $\sqrt[4]{m^2}$

28. $\sqrt[10]{n^6}$

29. $\sqrt[5]{\sqrt[3]{xy}}$

30. $\sqrt{\sqrt[4]{5x}}$

31. $\sqrt[3]{9x^2}\sqrt[3]{9x}$

32. $\sqrt{2x}\sqrt{8xy}$

In Problems 33–40, rationalize denominators, and write in simplified form.

33. $\dfrac{1}{\sqrt{5}}$

34. $\dfrac{1}{\sqrt{7}}$

35. $\dfrac{6x}{\sqrt{3x}}$

36. $\dfrac{12y^2}{\sqrt{6y}}$

37. $\dfrac{2}{\sqrt{2} - 1}$

38. $\dfrac{4}{\sqrt{6} - 2}$

39. $\dfrac{\sqrt{2}}{\sqrt{6} + 2}$

40. $\dfrac{\sqrt{2}}{\sqrt{10} - 2}$

B

In Problems 41–52, write in simplified form.

41. $x\sqrt[5]{36x^7y^{11}}$

42. $2a\sqrt[3]{8a^8b^{13}}$

43. $\dfrac{\sqrt[4]{32m^7n^9}}{2mn}$

44. $\dfrac{\sqrt[5]{32u^{12}v^8}}{uv}$

45. $\sqrt[6]{a^4(b - a)^2}$

46. $\sqrt[8]{3^6(u + v)^6}$

47. $\sqrt[3]{\sqrt[4]{a^9b^3}}$

48. $\sqrt{\sqrt[6]{x^8y^6}}$

49. $\sqrt[3]{2x^2y^4}\sqrt[3]{3x^5y}$

50. $\sqrt[4]{4m^5n}\sqrt[4]{6m^3n^4}$

51. $\sqrt[3]{a^3 + b^3}$

52. $\sqrt{x^2 + y^2}$

In Problems 53–64, rationalize denominators and write in simplified form.

53. $\dfrac{\sqrt{2m}\sqrt{5}}{\sqrt{20m}}$

54. $\dfrac{\sqrt{6}\sqrt{8c}}{\sqrt{18c}}$

55. $\dfrac{4a^3b^2}{\sqrt[3]{2ab^2}}$

56. $\dfrac{8x^3y^5}{\sqrt[3]{4x^2y}}$

57. $\sqrt[4]{\dfrac{3y^3}{4x^3}}$

58. $\sqrt[5]{\dfrac{4x^2}{16y^3}}$

59. $\dfrac{3\sqrt{y}}{2\sqrt{y}-3}$

60. $\dfrac{5\sqrt{x}}{3-2\sqrt{x}}$

61. $\dfrac{2\sqrt{5}+3\sqrt{2}}{5\sqrt{5}+2\sqrt{2}}$

62. $\dfrac{3\sqrt{2}-2\sqrt{3}}{3\sqrt{3}-2\sqrt{2}}$

63. $\dfrac{x^2}{\sqrt{x^2+9}-3}$

64. $\dfrac{-y^2}{2-\sqrt{y^2+4}}$

Problems 65–68 are calculus-related. Rationalize the numerators; that is, perform operations on the fractions that eliminate radicals from the numerators. (This is a particularly useful operation in some problems in calculus.)

65. $\dfrac{\sqrt{t}-\sqrt{x}}{t-x}$

66. $\dfrac{\sqrt{x}-\sqrt{y}}{\sqrt{x}+\sqrt{y}}$

67. $\dfrac{\sqrt{x+h}-\sqrt{x}}{h}$

68. $\dfrac{\sqrt{2+h}+\sqrt{2}}{h}$

In Problems 69–80, evaluate to four significant digits using a calculator. (Read the instruction booklet accompanying your calculator for the process required to evaluate $\sqrt[n]{x}$.)

69. $\sqrt{0.049\ 375}$

70. $\sqrt{306.721}$

71. $\sqrt[5]{27.0635}$

72. $\sqrt[8]{0.070\ 144}$

73. $\sqrt[7]{0.000\ 000\ 008\ 066}$

74. $\sqrt[12]{6,423,000,000,000}$

75. $\sqrt[3]{7}+\sqrt[3]{7}$

76. $\sqrt[5]{4}+\sqrt[3]{4}$

77. $\sqrt[3]{\sqrt[4]{2}}$ and $\sqrt[12]{2}$

78. $\sqrt[3]{\sqrt{5}}$ and $\sqrt[6]{5}$

79. $\dfrac{1}{\sqrt[3]{4}}$ and $\dfrac{\sqrt[3]{2}}{2}$

80. $\dfrac{1}{\sqrt[3]{5}}$ and $\dfrac{\sqrt[3]{25}}{5}$

C

For what real numbers are Problems 81–84 true?

81. $\sqrt{x^2}=-x$

82. $\sqrt{x^2}=x$

83. $\sqrt[3]{x^3}=x$

84. $\sqrt[3]{x^3}=-x$

In Problems 85–88, rationalize denominators.

85. $\dfrac{1}{\sqrt[3]{a}-\sqrt[3]{b}}$

86. $\dfrac{1}{\sqrt[3]{m}+\sqrt[3]{n}}$

87. $\dfrac{1}{\sqrt{x}-\sqrt{y}+\sqrt{z}}$

[*Hint:* Start by multiplying numerator and denominator by $(\sqrt{x}-\sqrt{y})-\sqrt{z}$.]

88. $\dfrac{1}{\sqrt{x}+\sqrt{y}-\sqrt{z}}$

Problems 89 and 90 are calculus-related. Rationalize numerators.

89. $\dfrac{\sqrt[3]{x+h}-\sqrt[3]{x}}{h}$

90. $\dfrac{\sqrt[3]{t}-\sqrt[3]{x}}{t-x}$

91. Show that $\sqrt[kn]{x^{km}}=\sqrt[n]{x^m}$ for k, m, and n natural numbers greater than 1.

92. Show that $\sqrt[m]{\sqrt[n]{x}}=\sqrt[mn]{x}$ for m and n natural numbers greater than 1.

APPLICATIONS

93. Physics—Relativistic Mass. The mass M of an object moving at a velocity v is given by

$$M=\dfrac{M_0}{\sqrt{1-\dfrac{v^2}{c^2}}}$$

where M_0 = mass at rest and c = velocity of light. The mass of an object increases with velocity and tends to infinity as the velocity approaches the speed of light. Show that M can be written in the form

$$M=\dfrac{M_0c\sqrt{c^2-v^2}}{c^2-v^2}$$

94. **Physics—Pendulum.** A simple pendulum is formed by hanging a bob of mass M on a string of length L from a fixed support (see the figure). The time it takes the bob to swing from right to left and back again is called the **period** T and is given by

$$T = 2\pi\sqrt{\frac{L}{g}}$$

where g is the gravitational constant. Show that T can be written in the form

$$T = \frac{2\pi\sqrt{gL}}{g}$$

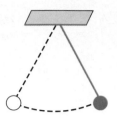

Chapter 1 Review

1-1   ALGEBRA AND REAL NUMBERS

A **set** is a collection of objects called **elements** or **members** of the set. Sets are usually described by **listing** the elements or by stating a **rule** that determines the elements. A set may be **finite** or **infinite**. A set with no elements is called the **empty set** or the **null set** and is denoted $\varnothing$. A **variable** is a symbol that represents unspecified elements from a **replacement set**. A **constant** is a symbol for a single object. If each element of set A is also in set B, we say A is a **subset** of B and write $A \subset B$.

Real numbers:

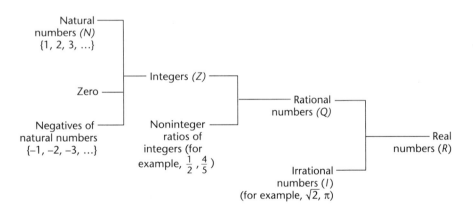

Real number line:

Basic real number properties include **associative properties**: $x + (y + z) = (x + y) + z$ and $x(yz) = (xy)z$; **commutative properties**: $x + y = y + x$ and $xy = yx$; **identities**: $0 + x = x + 0 = x$ and $(1)x = x(1) = x$; **inverses**: $-x$ is the additive inverse or **negative** of x and, if $x \neq 0$, $1/x$ is the multiplicative inverse or **reciprocal** of x; and **distributive property**: $x(y + z) = xy + xz$. **Subtraction** is defined by $a - b = a + (-b)$ and **division**

by $a/b = a(1/b)$. Division by 0 is never allowed. Additional properties include **properties of negatives**:

1. $-(-a) = a$

2. $(-a)b = -(ab) = a(-b) = -ab$

3. $(-a)(-b) = ab$

4. $(-1)a = -a$

5. $\dfrac{-a}{b} = -\dfrac{a}{b} = \dfrac{a}{-b} \qquad b \neq 0$

6. $\dfrac{-a}{-b} = -\dfrac{-a}{b} = -\dfrac{a}{-b} = \dfrac{a}{b} \qquad b \neq 0$

zero properties:

1. $a \cdot 0 = 0$.

2. $ab = 0 \qquad$ if and only if $\qquad a = 0 \qquad$ or $\qquad b = 0 \qquad$ or both.

and **fraction properties** (division by 0 excluded):

1. $\dfrac{a}{b} = \dfrac{c}{d} \qquad$ if and only if $ad = bc$ $\qquad$ **2.** $\dfrac{ka}{kb} = \dfrac{a}{b}$

3. $\dfrac{a}{b} \cdot \dfrac{c}{d} = \dfrac{ac}{bd}$ $\qquad$ **4.** $\dfrac{a}{b} \div \dfrac{c}{d} = \dfrac{a}{b} \cdot \dfrac{d}{c}$ $\qquad$ **5.** $\dfrac{a}{b} + \dfrac{c}{b} = \dfrac{a+c}{b}$

6. $\dfrac{a}{b} - \dfrac{c}{b} = \dfrac{a-c}{b}$ $\qquad$ **7.** $\dfrac{a}{b} + \dfrac{c}{d} = \dfrac{ad+bc}{bd}$

1-2 POLYNOMIALS: BASIC OPERATIONS

For n and m natural numbers and a any real number:

$$a^n = a \cdot a \cdot \cdots \cdot a \ (n \text{ factors of } a) \qquad \text{and} \qquad a^m a^n = a^{m+n}$$

An **algebraic expression** is formed by using constants and variables and the operations of addition, subtraction, multiplication, division, raising to powers, and taking roots. A **polynomial** is an algebraic expression formed by adding and subtracting constants and terms of the form ax^n (one variable), $ax^n y^m$ (two variables), and so on. The **degree of a term** is the sum of the powers of all variables in the term, and the **degree of a polynomial** is the degree of the nonzero term with highest degree in the polynomial. Polynomials with one, two, or three terms are called **monomials, binomials**, and **trinomials**, respectively. **Like terms** have exactly the same variable factors to the same powers and can be combined by adding their **coefficients**. Polynomials can be **added, subtracted**, and **multiplied** by repeatedly applying the distributive property and combining like terms. The **FOIL method** is used to multiply two binomials. **Special products** obtained using the FOIL method are:

1. $(a - b)(a + b) = a^2 - b^2$ $\qquad$ **2.** $(a - b)^2 = a^2 - 2ab + b^2$

3. $(a + b)^2 = a^2 + 2ab + b^2$

1-3 POLYNOMIALS: FACTORING

A number or algebraic expression is **factored** if it is expressed as a product of other numbers or algebraic expressions, which are called **factors**. An integer greater than 1 is a **prime**

number if its only positive integer factors are itself and 1, and a **composite number** otherwise. Each composite number can be **factored uniquely into a product of prime numbers**. A polynomial is **prime** relative to a given set of numbers (usually the set of integers) if (1) all its coefficients are from that set of numbers, and (2) it cannot be written as a product of two polynomials, excluding 1 and itself, having coefficients from that set of numbers. A nonprime polynomial is **factored completely relative to a given set of numbers** if it is written as a product of prime polynomials relative to that set of numbers. **Common factors** can be factored out by applying the distributive properties. **Grouping** can be used to identify common factors. Second-degree polynomials can be factored by trial and error. The following special factoring formulas are useful:

1. $u^2 + 2uv + v^2 = (u + v)^2$ Perfect Square

2. $u^2 - 2uv + v^2 = (u - v)^2$ Perfect Square

3. $u^2 - v^2 = (u - v)(u + v)$ Difference of Squares

4. $u^3 - v^3 = (u - v)(u^2 + uv + v^2)$ Difference of Cubes

5. $u^3 + v^3 = (u + v)(u^2 - uv + v^2)$ Sum of Cubes

There is no factoring formula relative to the real numbers for $u^2 + v^2$.

1-4 RATIONAL EXPRESSIONS: BASIC OPERATIONS

A **fractional expression** is the ratio of two algebraic expressions, and a **rational expression** is the ratio of two polynomials. The rules for adding, subtracting, multiplying, and dividing real number fractions (see Section 1-1 in this review) all extend to fractional expressions with the understanding that **variables are always restricted to exclude division by zero**. Fractions can be **reduced to lowest terms** or **raised to higher terms** by using the fundamental property of fractions:

$$\frac{ka}{kb} = \frac{a}{b} \qquad \text{with } b, \ k \neq 0$$

A rational expression is **reduced to lowest terms** if the numerator and denominator do not have any factors in common relative to the integers. The **least common denominator** (LCD) is useful for adding and subtracting fractions with different denominators and for reducing **compound fractions** to **simple fractions**.

1-5 INTEGER EXPONENTS

$a^n = a \cdot a \cdot \cdots \cdot a$ (n factors of a) for n a positive integer, $a^0 = 1$ ($a \neq 0$), and $a^n = 1/a^{-n}$ for n a negative integer ($a \neq 0$). 0^0 is not defined.

    Properties of integer exponents (division by 0 excluded):

1. $a^m a^n = a^{m+n}$ **2.** $(a^n)^m = a^{mn}$ **3.** $(ab)^m = a^m b^m$

4. $\left(\dfrac{a}{b}\right)^m = \dfrac{a^m}{b^m}$ **5.** $\dfrac{a^m}{a^n} = a^{m-n} = \dfrac{1}{a^{n-m}}$

    Further exponent properties (division by 0 excluded):

1. $(a^m b^n)^p = a^{pm} b^{pn}$ **2.** $\left(\dfrac{a^m}{b^n}\right)^p = \dfrac{a^{pm}}{b^{pn}}$ **3.** $\dfrac{a^{-n}}{b^{-m}} = \dfrac{b^m}{a^n}$

4. $\left(\dfrac{a}{b}\right)^{-n} = \left(\dfrac{b}{a}\right)^n$

Scientific notation:

$$a \times 10^n \qquad 1 \le a < 10$$

n an integer, a in decimal form.

1-6   RATIONAL EXPONENTS

For n a natural number and a and b real numbers:

$$a \text{ is an } n\text{th root of } b \text{ if } a^n = b$$

The **principal nth root** of b is denoted by $b^{1/n}$. If n is odd, b has one real nth root which is the principal nth root. If n is even and $b > 0$, b has two real nth roots and the positive nth root is the principal nth root. If n is even and $b < 0$, b has no real nth roots.

Rational number exponents (even roots of negative numbers excluded):

$$b^{m/n} = (b^{1/n})^m = (b^m)^{1/n} \qquad \text{and} \qquad b^{-m/n} = \frac{1}{b^{m/n}}$$

1-7   RADICALS

An **nth root radical** is defined by $\sqrt[n]{b} = b^{1/n}$, where $b^{1/n}$ is the principal nth root of b, $\sqrt{}$ is a **radical**, n is the **index**, and b is the **radicand**. Rational exponents and radicals are related by

$$b^{m/n} = (b^m)^{1/n} = \sqrt[n]{b^m} = (b^{1/n})^m = (\sqrt[n]{b})^m$$

Properties of radicals ($x > 0, y > 0$):

1. $\sqrt[n]{x^n} = x$ **2.** $\sqrt[n]{xy} = \sqrt[n]{x}\sqrt[n]{y}$ **3.** $\sqrt[n]{\dfrac{x}{y}} = \dfrac{\sqrt[n]{x}}{\sqrt[n]{y}}$

A radical is in **simplified form** if:

1. No radicand contains a factor to a power greater than or equal to the index of the radical.

2. No power of the radicand and the index of the radical have a common factor other than 1.

3. No radical appears in a denominator.

4. No fraction appears within a radical.

Algebraic fractions containing radicals are **rationalized** by multiplying numerator and denominator by a **rationalizing factor** often determined by using a special product formula.

Chapter 1 Review Exercise

Work through all the problems in this chapter review and check answers in the back of the book. Answers to all review problems are there, and following each answer is a number in italics indicating the section in which that type of problem is discussed. Where weaknesses show up, review appropriate sections in the text.

A

1. For $A = \{1, 2, 3, 4, 5\}$, $B = \{1, 2, 4\}$, and $C = \{4, 1, 2\}$, indicate true (T) or false (F):
   (A) $3 \in A$ (B) $5 \notin C$ (C) $B \in A$
   (D) $B \subset A$ (E) $B \neq C$ (F) $A \subset B$

2. Replace each question mark with an appropriate expression that will illustrate the use of the indicated real number property:
   (A) Commutative $(\cdot)$: $x(y + z) = ?$
   (B) Associative $(+)$: $2 + (x + y) = ?$
   (C) Distributive: $(2 + 3)x = ?$

Problems 3–7 refer to the following polynomials:
   (a) $3x - 4$ (b) $x + 2$ (c) $3x^2 + x - 8$ (d) $x^3 + 8$

3. Add all four.

4. Subtract the sum of (a) and (c) from the sum of (b) and (d).

5. Multiply (c) and (d).

6. What is the degree of (d)?

7. What is the coefficient of the second term in (c)?

In Problems 8–11, perform the indicated operations and simplify.

8. $5x^2 - 3x[4 - 3(x - 2)]$

9. $(3m - 5n)(3m + 5n)$

10. $(2x + y)(3x - 4y)$

11. $(2a - 3b)^2$

In Problems 12–14, write each polynomial in a completely factored form relative to the integers. If the polynomial is prime relative to the integers, say so.

12. $9x^2 - 12x + 4$ **13.** $t^2 - 4t - 6$ **14.** $6n^3 - 9n^2 - 15n$

In Problems 15–18, perform the indicated operations and reduce to lowest terms. Represent all compound fractions as simple fractions reduced to lowest terms.

15. $\dfrac{2}{5b} - \dfrac{4}{3a^3} - \dfrac{1}{6a^2b^2}$

16. $\dfrac{3x}{3x^2 - 12x} + \dfrac{1}{6x}$

17. $\dfrac{y - 2}{y^2 - 4y + 4} \div \dfrac{y^2 + 2y}{y^2 + 4y + 4}$

18. $\dfrac{u - \dfrac{1}{u}}{1 - \dfrac{1}{u^2}}$

Simplify Problems 19–24, and write answers using positive exponents only. All variables represent positive real numbers.

19. $6(xy^3)^5$

20. $\dfrac{9u^8v^6}{3u^4v^8}$

21. $(2 \times 10^5)(3 \times 10^{-3})$

22. $(x^{-3}y^2)^{-2}$

23. $u^{5/3}u^{2/3}$

24. $(9a^4b^{-2})^{1/2}$

25. Change to radical form: $3x^{2/5}$

26. Change to rational exponent form: $-3\sqrt[3]{(xy)^2}$

Simplify Problems 27–31, and express answers in simplified form. All variables represent positive real numbers.

27. $3x\sqrt[3]{x^5y^4}$

28. $\sqrt{2x^2y^5}\sqrt{18x^3y^2}$

29. $\dfrac{6ab}{\sqrt{3a}}$

30. $\dfrac{\sqrt{5}}{3-\sqrt{5}}$

31. $\sqrt[8]{y^6}$

B _____

32. Write using the listing method:

$\{x \mid x \text{ is an odd integer between } -4 \text{ and } 2\}$

In Problems 33–38, each statement illustrates the use of one of the following real number properties or definitions. Indicate which one.

Commutative $(+, \cdot)$	Distributive	Associative $(+, \cdot)$
Division	Negatives	Zero
Inverse $(+, \cdot)$	Identity $(+, \cdot)$	Subtraction

33. $(-3) - (-2) = (-3) + [-(-2)]$

34. $3y + (2x + 5) = (2x + 5) + 3y$

35. $(2x + 3)(3x + 5) = (2x + 3)3x + (2x + 3)5$

36. $3 \cdot (5x) = (3 \cdot 5)x$

37. $\dfrac{a}{-(b-c)} = -\dfrac{a}{b-c}$

38. $3xy + 0 = 3xy$

39. Indicate true (T) or false (F):
   (A) An integer is a rational number and a real number.
   (B) An irrational number has a repeating decimal representation.

40. Give an example of an integer that is not a natural number.

41. Given the algebraic expressions:
   (a) $2x^2 - 3x + 5$ (b) $x^2 - \sqrt{x-3}$ (c) $x^{-3} + x^{-2} - 3x^{-1}$ (d) $x^2 - 3xy - y^2$
   (A) Identify all second-degree polynomials.
   (B) Identify all third-degree polynomials.

In Problems 42–46, perform the indicated operations and simplify.

42. $(2x - y)(2x + y) - (2x - y)^2$

43. $(m^2 + 2mn - n^2)(m^2 - 2mn - n^2)$

44. $5(x + h)^2 - 7(x + h) - (5x^2 - 7x)$

45. $-2x\{(x^2 + 2)(x - 3) - x[x - x(3 - x)]\}$

46. $(x - 2y)^3$

In Problems 47–53, write in a completely factored form relative to the integers.

47. $(4x - y)^2 - 9x^2$

48. $2x^2 + 4xy - 5y^2$

49. $6x^3y + 12x^2y^2 - 15xy^3$

50. $(y - b)^2 - y + b$

51. $3x^3 + 24y^3$

$3x^2 - 5x - 2$

52. $y^3 + 2y^2 - 4y - 8$

53. $2x(x - 4)^3 + 3x^2(x - 4)^2$

In Problems 54–58, perform the indicated operations and reduce to lowest terms. Represent all compound fractions as simple fractions reduced to lowest terms.

54. $\dfrac{3x^2(x + 2)^2 - 2x(x + 2)^3}{x^4}$

55. $\dfrac{m - 1}{m^2 - 4m + 4} + \dfrac{m + 3}{m^2 - 4} + \dfrac{2}{2 - m}$

56. $\dfrac{y}{x^2} \div \left(\dfrac{x^2 + 3x}{2x^2 + 5x - 3} \div \dfrac{x^3y - x^2y}{2x^2 - 3x + 1} \right)$

57. $\dfrac{1 - \dfrac{1}{1 + \dfrac{x}{y}}}{1 - \dfrac{1}{1 - \dfrac{x}{y}}}$

58. $\dfrac{a^{-1} - b^{-1}}{ab^{-2} - ba^{-2}}$

In Problems 59–64, perform the indicated operations, simplify and write answers using positive exponents only. All variables represent positive real numbers.

59. $\left(\dfrac{8u^{-1}}{2^2u^2v^0} \right)^{-2} \left(\dfrac{u^{-5}}{u^{-3}} \right)^3$

60. $\dfrac{5^0}{3^2} + \dfrac{3^{-2}}{2^{-2}}$

61. $\left(\dfrac{27x^2y^{-3}}{8x^{-4}y^3} \right)^{1/3}$

62. $(a^{-1/3}b^{1/4})(9a^{1/3}b^{-1/2})^{3/2}$

63. $(x^{1/2} + y^{1/2})^2$

64. $(3x^{1/2} - y^{1/2})(2x^{1/2} + 3y^{1/2})$

65. Convert to scientific notation and simplify:

$$\dfrac{0.000\ 000\ 000\ 52}{(1,300)(0.000\ 002)}$$

Evaluate Problems 66–73 to four significant digits using a calculator.

66. $\dfrac{(20,410)(0.000\ 003\ 477)}{0.000\ 000\ 022\ 09}$

67. 0.1347^5

68. $(-60.39)^{-3}$

69. $82.45^{8/3}$

70. $(0.000\ 000\ 419\ 9)^{2/7}$

71. $\sqrt[5]{0.006\ 604}$

72. $\sqrt[3]{3 + \sqrt{2}}$

73. $\dfrac{2^{-1/2} - 3^{-1/2}}{2^{-1/3} + 3^{-1/3}}$

In Problems 74–82, perform the indicated operations and express answers in simplified form. All radicands represent positive real numbers.

74. $-2x\sqrt[5]{36x^7y^{11}}$

75. $\dfrac{2x^2}{\sqrt[3]{4x}}$

76. $\sqrt[5]{\dfrac{3y^2}{8x^2}}$

77. $\sqrt[9]{8x^6y^{12}}$

78. $\sqrt{\sqrt[3]{4x^4}}$

79. $(2\sqrt{x} - 5\sqrt{y})(\sqrt{x} + \sqrt{y})$

80. $\dfrac{3\sqrt{x}}{2\sqrt{x} - \sqrt{y}}$

81. $\dfrac{2\sqrt{u} - 3\sqrt{v}}{2\sqrt{u} + 3\sqrt{v}}$

82. $\dfrac{y^2}{\sqrt{y^2 + 4} - 2}$

83. Rationalize the numerator: $\dfrac{\sqrt{t} - \sqrt{5}}{t - 5}$

84. Write in the form $ax^p + bx^q$, where a and b are real numbers and p and q are rational numbers:

$$\dfrac{4\sqrt{x} - 3}{2\sqrt{x}}$$

C

85. Write the repeating decimal $0.545454\ldots$ in the form a/b reduced to lowest terms, where a and b are positive integers. Is the number rational or irrational?

86. If $M = \{-4, -3, 2\}$ and $N = \{-3, 0, 2\}$, find:
(A) $\{x \mid x \in M$ or $x \in N\}$
(B) $\{x \mid x \in M$ and $x \in N\}$

87. Evaluate $x^2 - 4x + 1$ for $x = 2 - \sqrt{3}$.

88. Simplify: $x(2x - 1)(x + 3) - (x - 1)^3$

89. Factor completely with respect to the integers:

$$4x(a^2 - 4a + 4) - 9x^3$$

90. Simplify:

$$\left(x - \dfrac{1}{1 - \dfrac{1}{x}} \right) \div \left(\dfrac{x}{x + 1} - \dfrac{x}{1 - x} \right)$$

In Problems 91–94, simplify and express answers using positive exponents only (m is an integer greater than 1).

91. $\dfrac{8(x - 2)^{-3}(x + 3)^2}{12(x - 2)^{-4}(x + 3)^{-2}}$

92. $\left(\dfrac{a^{-2}}{b^{-1}} + \dfrac{b^{-2}}{a^{-1}} \right)^{-1}$

93. $(x^{1/3} - y^{1/3})(x^{2/3} + x^{1/3}y^{1/3} + y^{2/3})$

94. $\left(\dfrac{x^{m^2}}{x^{2m-1}} \right)^{1/(m-1)} \qquad m > 1$

95. Rationalize the denominator: $\dfrac{1}{1 - \sqrt[3]{x}}$

96. Rationalize the numerator: $\dfrac{\sqrt[3]{t} - \sqrt[3]{5}}{t - 5}$

97. Write in simplified form: $\sqrt[n+1]{x^{n^2}x^{2n+1}} \qquad n > 0$

APPLICATIONS

98. Construction. A circular fountain in a park includes a concrete wall that is 3 feet high and 2 feet thick (see the figure). If the inner radius of the wall is x feet, write an algebraic expression in terms of x that represents the volume of the concrete used to construct the wall. Simplify the expression.

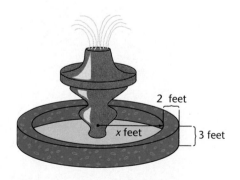

2 feet

x feet

} 3 feet

99. Energy Consumption. In 1984 the total amount of energy consumed in the United States was equivalent to 2,257,000,000,000 kilograms of coal (consumption of all forms of energy is expressed in terms of coal for purposes of comparison) and the population was about 235,000,000. Estimate to three significant digits the average energy consumption per person. Write your answer in scientific notation and in standard decimal notation.

Equations and Inequalities

2

One of the important uses of algebra is the solving of equations and inequalities. In this chapter we look at techniques for solving linear and nonlinear equations and inequalities. In addition, we consider a number of applications that can be solved using these techniques. Additional techniques for solving polynomial equations will be discussed in Chapter 4.

SECTION **2-1**   **Linear Equations**

- Equations
- Solving Linear Equations
- Some Final Observations on Linear Equations

• Equations   An **algebraic equation** is a mathematical statement that relates two algebraic expressions involving at least one variable. Some examples of equations with x as the variable are

$$3x - 2 = 7 \qquad \frac{1}{1 + x} = \frac{x}{x - 2}$$

$$2x^2 - 3x + 5 = 0 \qquad \sqrt{x + 4} = x - 1$$

The **replacement set**, or **domain**, for a variable is defined to be the set of numbers that are permitted to replace the variable.

Assumption   **On Domains of Variables**

Unless stated to the contrary, we assume that the domain for a variable is the set of those real numbers for which the algebraic expressions involving the variable are real numbers.

For example, the domain for the variable x in the expression

$$2x - 4$$

is R, the set of all real numbers, since $2x - 4$ represents a real number for all replacements of x by real numbers. The domain of x in the equation

$$\frac{1}{x} = \frac{2}{x - 3}$$

is the set of all real numbers except 0 and 3. These values are excluded because the left member is not defined for $x = 0$ and the right member is not defined for $x = 3$. The left and right members represent real numbers for all other replacements of x by real numbers.

The **solution set** for an equation is defined to be the set of elements in the domain of the variable that makes the equation true. Each element of the solution set is called a **solution**, or **root**, of the equation. To **solve an equation** is to find the solution set for the equation.

An equation is called an **identity** if the equation is true for all elements from the domain of the variable. An equation is called a **conditional equation** if it is true for certain domain values and false for others. For example,

$$2x - 4 = 2(x - 2) \quad \text{and} \quad \frac{5}{x^2 - 3x} = \frac{5}{x(x - 3)}$$

are identities, since both equations are true for all elements from the respective domains of their variables. On the other hand, the equations

$$3x - 2 - 5 \quad \text{and} \quad \frac{2}{x - 1} = \frac{1}{x}$$

are conditional equations, since, for example, neither equation is true for the domain value 2.

Knowing what we mean by the solution set of an equation is one thing; finding it is another. To this end we introduce the idea of equivalent equations. Two equations are said to be **equivalent** if they both have the same solution set for a given replacement set. A basic technique for solving equations is to perform operations on equations that produce simpler equivalent equations, and to continue the process until an equation is reached whose solution is obvious.

Application of any of the properties of equality given in Theorem 1 will produce equivalent equations.

Theorem 1

Properties of Equality

For a, b, and c any real numbers:

1. If $a = b$, then $a + c = b + c$.   **Addition Property**
2. If $a = b$, then $a - c = b - c$.   **Subtraction Property**
3. If $a = b$, then $ca = cb$, $c \neq 0$.   **Multiplication Property**
4. If $a = b$, then $\dfrac{a}{c} = \dfrac{b}{c}$, $c \neq 0$.   **Division Property**
5. If $a = b$, then either may replace the other   **Substitution Property**
in any statement without changing the
truth or falsity of the statement.

● **Solving Linear Equations**

We now turn our attention to methods of solving *first-degree*, or *linear*, *equations* in one variable.

DEFINITION 1   **Linear Equation in One Variable**

Any equation that can be written in the form

$$ax + b = 0 \qquad a \neq 0 \qquad \textbf{Standard Form}$$

where a and b are real constants and x is a variable, is called a **linear**, or **first-degree**, **equation** in one variable.

$5x - 1 = 2(x + 3)$ is a linear equation, since it can be written in the standard form $3x - 7 = 0$.

EXAMPLE 1   Solving a Linear Equation

Solve $5x - 9 = 3x + 7$ and check.

Solution   We use the properties of equality to transform the given equation into an equivalent equation whose solution is obvious.

$$5x - 9 = 3x + 7 \qquad \text{Original equation}$$
$$5x - 9 + 9 = 3x + 7 + 9 \qquad \text{Add 9 to both sides.}$$
$$5x = 3x + 16 \qquad \text{Combine like terms.}$$
$$5x - 3x = 3x + 16 - 3x \qquad \text{Subtract } 3x \text{ from both sides.}$$
$$2x = 16 \qquad \text{Combine like terms.}$$
$$\frac{2x}{2} = \frac{16}{2} \qquad \text{Divide both sides by 2.}$$
$$x = 8 \qquad \text{Simplify.}$$

The solution set for this last equation is obvious:

Solution set: {8}

And since the equation $x = 8$ is equivalent to all the preceding equations in our solution, {8} is also the solution set for all these equations, including the original equation. [*Note:* If an equation has only one element in its solution set, we generally use the last equation (in this case, $x = 8$) rather than set notation to represent the solution.]

Check
$$5x - 9 = 3x + 7 \qquad \text{Original equation}$$
$$5(8) - 9 \stackrel{?}{=} 3(8) + 7 \qquad \text{Substitute } x = 8.$$
$$40 - 9 \stackrel{?}{=} 24 + 7 \qquad \text{Simplify each side.}$$
$$31 \stackrel{\checkmark}{=} 31 \qquad \text{A true statement}$$

Problem 1   Solve and check: $7x - 10 = 4x + 5$

EXAMPLE 2 Solving a Linear Equation

Solve $3x - 2(2x - 5) = 2(x + 3) - 8$ and check.

Solution

$$3x - 2(2x - 5) = 2(x + 3) - 8 \quad \text{Original equation}$$

$$3x - 4x + 10 = 2x + 6 - 8 \quad \text{Clear parentheses.}$$

$$-x + 10 = 2x - 2 \quad \text{Combine like terms.}$$

$$-3x = -12 \quad \text{Subtract } 2x \text{ and 10 from both sides.}$$

$$x = 4 \quad \text{Divide both sides by } -3.$$

Check

$$3x - 2(2x - 5) = 2(x + 3) - 8$$

$$3(4) - 2[2(4) - 5] \overset{?}{=} 2[(4) + 3] - 8$$

$$6 \overset{\checkmark}{=} 6$$

Problem 2 Solve and check: $2(3 - x) - (3x + 1) = 8 - 2(x + 2)$

EXAMPLE 3 Using the LCD to Clear Fractions in an Equation

Solve: $\dfrac{x + 1}{3} - \dfrac{x}{4} = \dfrac{1}{2}$

Solution

If we can find a number that is divisible by each denominator, then we can use the multiplication property in Theorem 1 to clear the equation of fractions. The smallest positive integer that is divisible by each denominator is the least common denominator (LCD) of all fractions within the equation. (For a discussion of LCD's and how to find them, see Section 1-4.) In this example it is easy to see that the LCD, the smallest positive integer divisible by 3, 4, and 2, is 12. Thus,

$$\frac{x + 1}{3} - \frac{x}{4} = \frac{1}{2}$$

$$12\left(\frac{x + 1}{3} - \frac{x}{4}\right) = 12 \cdot \frac{1}{2} \quad \begin{array}{l}\text{Multiply both sides by 12, the LCD.}\\ \text{This and the next step usually can be done mentally.}\end{array}$$

$$12 \cdot \frac{x + 1}{3} - 12 \cdot \frac{x}{4} = 6$$

$$4(x + 1) - 3x = 6 \quad \text{The equation is now free of fractions.}$$

$$4x + 4 - 3x = 6$$

$$x = 2 \quad \text{The solution set is } \{2\}.$$

The check is left to the reader.

Problem 3   Solve: $\dfrac{x}{5} - \dfrac{x-2}{2} = \dfrac{3}{4}$

CAUTION

A very common error occurs about now—students tend to confuse *algebraic expressions* involving fractions with *algebraic equations* involving fractions.

Consider these two problems:

(A) Solve: $\dfrac{x}{2} + \dfrac{x}{3} = 10$ (B) Add: $\dfrac{x}{2} + \dfrac{x}{3} + 10$

The problems look very much alike but are actually very different. To solve the equation in (A) we multiply both sides by 6 (the LCD) to clear the fractions. This works so well for equations that students want to do the same thing for problems like (B). The only catch is that (B) is not an equation, and the multiplication property of equality does not apply. If we multiply (B) by 6, we simply obtain an expression 6 times as large as the original! Compare the following:

(A) $\qquad \dfrac{x}{2} + \dfrac{x}{3} = 10$ $\qquad$ (B) $\dfrac{x}{2} + \dfrac{x}{3} + 10$

$$6 \cdot \dfrac{x}{2} + 6 \cdot \dfrac{x}{3} = 6 \cdot 10$$

$$= \dfrac{3 \cdot x}{3 \cdot 2} + \dfrac{2 \cdot x}{2 \cdot 3} + \dfrac{6 \cdot 10}{6 \cdot 1}$$

$$3x + 2x = 60$$

$$= \dfrac{3x}{6} + \dfrac{2x}{6} + \dfrac{60}{6}$$

$$5x = 60$$

$$x = 12$$

$$= \dfrac{5x + 60}{6}$$

Some equations involving variables in a denominator can be transformed into linear equations. We may proceed in essentially the same way as in the preceding example; however, we must exclude any value of the variable that will make a denominator 0. With these values excluded, we may multiply through by the LCD even though it contains a variable, and, according to Theorem 1, the new equation will be equivalent to the old.

EXAMPLE 4   Solving an Equation with a Variable in the Denominator

Solve: $\dfrac{x}{2x - 4} - \dfrac{2}{3} = \dfrac{7 - 2x}{3x - 6}$

Solution

$$\dfrac{x}{2x - 4} - \dfrac{2}{3} = \dfrac{7 - 2x}{3x - 6}$$

$$\dfrac{x}{2(x - 2)} - \dfrac{2}{3} = \dfrac{7 - 2x}{3(x - 2)} \qquad \text{Factor denominators.}$$

Since division by 0 is never allowed, we must restrict the domain of x to exclude 2. With this restriction, we can multiply both sides by the LCD, $6(x - 2)$, to obtain

$$6(x - 2)\,\frac{x}{2(x - 2)} - 6(x - 2)\,\frac{2}{3} = 6(x - 2)\,\frac{7 - 2x}{3(x - 2)} \qquad x \neq 2$$

$$3x - 4x + 8 = 14 - 4x$$

$$3x = 6$$

$$x = 2$$

However, since 2 is not in the domain of x in the original equation, the original equation has no solution. (Note that when 2 replaces x in the left and right sides of the original equation, neither is defined.) Therefore, the solution set is $\varnothing$.

Problem 4 Solve: $\dfrac{x - 3}{2x - 2} = \dfrac{1}{6} - \dfrac{2 + x}{3x - 3}$

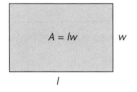

$A = lw$ w

l

FIGURE 1 Area of a rectangle.

We frequently encounter equations involving more than one variable. For example, if l and w are the length and width of a rectangle, respectively, the area of the rectangle is given by (see Figure 1)

$$A = lw$$

Depending on the situation, we may want to solve this equation for l or w. To solve for w, we simply consider A and l to be constants and divide both sides by l to obtain

$$w = \frac{A}{l} \qquad l \neq 0$$

EXAMPLE 5 Solving an Equation with More Than One Variable

Solve for P in terms of the other variables: $A = P + Prt$

Solution

$$A = P + Prt \qquad \text{Think of } A,\ r,\ \text{and } t \text{ as constants.}$$

$$A = P(1 + rt) \qquad \text{Factor to isolate } P.$$

$$\frac{A}{1 + rt} = P \qquad \text{Divide both sides by } 1 + rt.$$

$$P = \frac{A}{1 + rt} \qquad \text{Restriction: } 1 + rt \neq 0$$

Problem 5 Solve for F in terms of C: $C = \frac{5}{9}(F - 32)$

● **Some Final Observations on Linear Equations**

It can be shown that any equation that can be written in the form

$$ax + b = 0 \qquad a \neq 0 \qquad\qquad (1)$$

with no restrictions on x, has exactly one solution, and the solution can be found as follows:

$$ax + b = 0$$

$$ax = -b \qquad \text{Subtraction property of equality}$$

$$x = \frac{-b}{a} \qquad \text{Division property of equality}$$

Requiring $a \neq 0$ in equation (1) is an important restriction, because without it we are able to write equations with first-degree members that have no solutions or infinitely many solutions. For example,

$$2x - 3 = 2x + 5$$

has no solution, and

$$3x - 4 = 5 + 3(x - 3)$$

has infinitely many solutions. Try to solve each equation to see what happens.

Answers to Matched Problems
1. $x = 5$ **2.** $x = \frac{1}{3}$ **3.** $x = \frac{5}{6}$ **4.** No solution (solution set is $\varnothing$)   **5.** $F = \frac{9}{5}C + 32$

EXERCISE 2-1

A _____

In Problems 1–20, solve each equation.

1. $3(x + 2) = 5(x - 6)$

2. $5x + 10(x - 2) = 40$

3. $5 + 4(t - 2) = 2(t + 7) + 1$

4. $5x - (7x - 4) - 2 = 5 - (3x + 2)$

5. $x(x + 2) = x(x + 4) - 12$

6. $x(x + 4) - 2 = x^2 - 4(x + 3)$

7. $3 - \dfrac{2x - 3}{3} = \dfrac{5 - x}{2}$

8. $\dfrac{x - 2}{3} + 1 = \dfrac{x}{7}$

9. $5 - \dfrac{2x - 1}{4} = \dfrac{x + 2}{3}$

10. $\dfrac{x + 3}{4} - \dfrac{x - 4}{2} - \dfrac{3}{8}$

11. $0.1(x - 7) + 0.05x = 0.8$

12. $0.4(x + 5) - 0.3x = 17$

13. $0.3x - 0.04(x + 1) = 2.04$

14. $0.02x - 0.5(x - 2) = 5.32$

15. $\dfrac{3x}{2} - \dfrac{2 - x}{10} = \dfrac{5 + x}{4} + \dfrac{4}{5}$

16. $\dfrac{2x - 3}{9} - \dfrac{x + 5}{6} = \dfrac{3 - x}{2} + 1$

17. $\dfrac{1}{m} - \dfrac{1}{9} = \dfrac{4}{9} - \dfrac{2}{3m}$

18. $\dfrac{2}{3x} + \dfrac{1}{2} = \dfrac{4}{x} + \dfrac{4}{3}$

19. $\dfrac{5x}{x+5} = 2 - \dfrac{25}{x+5}$

20. $\dfrac{3}{2x-1} + 4 = \dfrac{6x}{2x-1}$

B

In Problems 21–30, solve each equation.

21. $\dfrac{2x}{10} - \dfrac{3-x}{14} = \dfrac{2+x}{5} - \dfrac{1}{2}$

22. $\dfrac{3x}{24} - \dfrac{2-x}{10} = \dfrac{5+x}{40} - \dfrac{1}{15}$

23. $\dfrac{1}{3} - \dfrac{s-2}{2s+4} = \dfrac{s+2}{3s+6}$

24. $\dfrac{n-5}{6n-6} = \dfrac{1}{9} - \dfrac{n-3}{4n-4}$

25. $\dfrac{3x}{2-x} + \dfrac{6}{x-2} = 3$

26. $5 - \dfrac{2x}{3-x} = \dfrac{6}{x-3}$

27. $\dfrac{5t-22}{t^2-6t+9} - \dfrac{11}{t^2-3t} - \dfrac{5}{t} = 0$

28. $\dfrac{5}{x-3} = \dfrac{33-x}{x^2-6x+9}$

29. $\dfrac{1}{x^2-x-2} - \dfrac{3}{x^2-2x-3} = \dfrac{1}{x^2-5x+6}$

30. $\dfrac{10}{x} - \dfrac{22}{3x-x^2} = \dfrac{10x-44}{x^2-6x+9}$

In Problems 31–34, use a calculator to solve each equation to 3 significant digits.

31. $3.142x - 0.4835(x-4) = 6.795$

32. $0.0512x + 0.125(x-2) = 0.725x$

33. $\dfrac{2.32x}{x-2} - \dfrac{3.76}{x} = 2.32$

34. $\dfrac{6.08}{x} + 4.49 = \dfrac{4.49x}{x+3}$

In Problems 35–42, solve for the indicated variable in terms of the other variables.

35. $a_n = a_1 + (n-1)d$ for d (arithmetic progressions)

36. $F = \frac{9}{5}C + 32$ for C (temperature scale)

37. $\dfrac{1}{f} = \dfrac{1}{d_1} + \dfrac{1}{d_2}$ for f (simple lens formula)

38. $\dfrac{1}{R} = \dfrac{1}{R_1} + \dfrac{1}{R_2}$ for R_1 (electric circuit)

39. $A = 2ab + 2ac + 2bc$ for a (surface area of a rectangular solid)

40. $A = 2ab + 2ac + 2bc$ for c

41. $y = \dfrac{2x-3}{3x+5}$ for x

42. $x = \dfrac{3y+2}{y-3}$ for y

C

In Problems 43–47, solve for x.

43. $\dfrac{x - \dfrac{1}{x}}{1 + \dfrac{1}{x}} = 3$

44. $\dfrac{x - \dfrac{1}{x}}{x + 1 - \dfrac{2}{x}} = 1$

45. $\dfrac{x + 1 - \dfrac{2}{x}}{1 - \dfrac{1}{x}} = x + 2$

46. $\dfrac{1}{x + 1 - \dfrac{x}{1 - \dfrac{1}{x}}} = 1 - x$

47. $\dfrac{x}{x + \dfrac{2}{3 + \dfrac{4}{x}}} = 1$

48. Solve for y in terms of x: $\dfrac{y}{1-y} = \left(\dfrac{x}{1-x}\right)^3$

49. Solve for x in terms of y: $y = \dfrac{a}{1 + \dfrac{b}{x+c}}$

50. Let m and n be real numbers with m larger than n. Then there exists a positive real number p such that $m = n + p$. Find the fallacy in the following argument:

$$m = n + p$$
$$(m - n)m = (m - n)(n + p)$$
$$m^2 - mn = mn + mp - n^2 - np$$
$$m^2 - mn - mp = mn - n^2 - np$$
$$m(m - n - p) = n(m - n - p)$$
$$m = n$$

SECTION 2-2 Applications Involving Linear Equations

- A Strategy for Solving Word Problems
- Number and Geometric Problems
- Rate–Time Problems
- Mixture Problems

• A Strategy for Solving Word Problems

A great many practical problems can be solved using algebraic techniques—so many, in fact, that there is no one method of attack that will work for all. However, we can formulate a strategy that will help you organize your approach.

> **Strategy for Solving Word Problems**
>
> 1. Read the problem carefully—several times if necessary; that is, until you understand the problem, know what is to be found, and know what is given.
>
> 2. Let one of the unknown quantities be represented by a variable, say x, and try to represent all other unknown quantities in terms of x. This is an important step and must be done carefully.
>
> 3. If appropriate, draw figures or diagrams and label known and unknown parts.
>
> 4. Look for formulas connecting the known quantities to the unknown quantities.
>
> 5. Form an equation relating the unknown quantities to the known quantities.
>
> 6. Solve the equation and write answers to *all* questions asked in the problem.
>
> 7. Check and interpret all solutions in terms of the original problem—not just the equation found in step 5—since a mistake may have been made in setting up the equation in step 5.

The examples in this section contain worked-out solutions to a variety of word problems. It is suggested that you cover up a solution, try solving the problem yourself, and uncover just enough of a solution to get you going again in case you get stuck. After successfully completing an example, try the matched problem. After completing the section in this way, you will be ready to attempt the fairly large variety of applications in Exercise 2-2.

● **Number and Geometric Problems**   The first examples introduce the process of setting up and solving word problems in a simple mathematical context. Following these, the examples are of a more substantive nature.

EXAMPLE 1   **Setting Up and Solving a Word Problem**

Find four consecutive even integers such that the sum of the first three exceeds the fourth by 8.

Solution   Let x = the first even integer, then

$$x \qquad x + 2 \qquad x + 4 \qquad \text{and} \qquad x + 6$$

represent four consecutive even integers starting with the even integer x. (Remember, even integers increase by 2.) The phrase, "the sum of the first three exceeds the fourth by 8" translates into an equation:

$$\text{Sum of the first three} = \text{Fourth} + \text{Excess}$$
$$x + (x + 2) + (x + 4) = (x + 6) + 8$$
$$3x + 6 = x + 14$$
$$2x = 8$$
$$x = 4$$

The four consecutive integers are 4, 6, 8, and 10.

Check

$$
\begin{array}{rl}
4 + 6 + 8 = \quad 18 & \text{Sum of first three} \\
\underline{-8} & \text{Excess} \\
10 & \text{Fourth}
\end{array}
$$

Problem 1   Find three consecutive odd integers such that 3 times their sum is 5 more than 8 times the middle one.

EXAMPLE 2   **Using a Diagram in the Solution of a Word Problem**

If one side of a triangle is one-third the perimeter, the second side is 7 meters, and the third side is one-fifth the perimeter, what is the perimeter of the triangle?

Solution   Let p = the perimeter. Draw a triangle and label the sides, as shown in the margin. Then

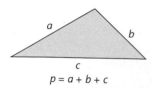

$p = a + b + c$

$$p = \frac{p}{3} + \frac{p}{5} + 7$$

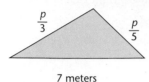

7 meters

$$15 \cdot p = 15 \cdot \frac{p}{3} + 15 \cdot \frac{p}{5} + 15 \cdot 7$$

$$15p = 5p + 3p + 105$$

$$7p = 105$$

$$p = 15$$

The perimeter is 15 meters.

Check

$$\frac{p}{3} = \frac{15}{3} = 5 \qquad \text{Side 1}$$

$$\frac{p}{5} = \frac{15}{5} = 3 \qquad \text{Side 2}$$

$$\underline{\hphantom{\frac{p}{5} = \frac{15}{5} = }\ 7} \qquad \text{Side 3}$$

$$15 \text{ meters} \qquad \text{Perimeter}$$

Problem 2   If one side of a triangle is one-fourth the perimeter, the second side is 7 centimeters, and the third side is two-fifths the perimeter, what is the perimeter?

● **Rate–Time Problems**   There are many types of rate–time problems and distance–rate–time problems. In general, if Q is the quantity of something produced (kilometers, words, parts, and so on) in T units of time (hours, years, minutes, seconds, and so on), then the formulas given in the box are relevant.

Quantity–Rate–Time Formulas

$$R = \frac{Q}{T} \qquad \text{Rate} = \frac{\text{Quantity}}{\text{Time}}$$

$$Q = RT \qquad \text{Quantity} = (\text{Rate})(\text{Time})$$

$$T = \frac{Q}{R} \qquad \text{Time} = \frac{\text{Quantity}}{\text{Rate}}$$

If Q is distance D, then

$$R = \frac{D}{T} \qquad D = RT \qquad T = \frac{D}{R}$$

[*Note:* R is an average or uniform rate.]

EXAMPLE 3   **A Distance–Rate–Time Problem**

The distance along a shipping route between San Francisco and Honolulu is 2,100 nautical miles. If one ship leaves San Francisco at the same time another leaves Honolulu, and if the former travels at 15 knots* and the latter at 20 knots, how long will it take the two ships to rendezvous? How far will they be from Honolulu and San Francisco at that time?

Solution   Let T = number of hours until both ships meet. Draw a diagram and label known and unknown parts. Both ships will have traveled the same amount of time when they meet.

20 knots ⟶   Meeting   ⟵ 15 knots

H ————————————————————×———————————————————— SF

$D_1 = 20T$ $D_2 = 15T$

$$\begin{pmatrix}\text{Distance ship 1}\\\text{from Honolulu}\\\text{travels to}\\\text{meeting point}\end{pmatrix} + \begin{pmatrix}\text{Distance ship 2}\\\text{from San Francisco}\\\text{travels to}\\\text{meeting point}\end{pmatrix} = \begin{pmatrix}\text{Total distance}\\\text{from Honolulu}\\\text{to San Francisco}\end{pmatrix}$$

$$D_1 \quad + \quad D_2 \quad = \quad 2{,}100$$
$$20T \quad + \quad 15T \quad = \quad 2{,}100$$
$$35T \quad = \quad 2{,}100$$
$$T \quad = \quad 60$$

Therefore, it takes 60 hours, or 2.5 days, for the ships to meet.

$$\text{Distance from Honolulu} = 20 \cdot 60 = 1{,}200 \text{ nautical miles}$$
$$\text{Distance from San Francisco} = 15 \cdot 60 = 900 \text{ nautical miles}$$

Check $$1{,}200 + 900 = 2{,}100 \text{ nautical miles}$$

Problem 3   An old piece of equipment can print, stuff, and label 38 mailing pieces per minute. A newer model can handle 82 per minute. How long will it take for both pieces of equipment to prepare a mailing of 6,000 pieces? [*Note:* The mathematical form is the same as in Example 3.]

EXAMPLE 4   **A Distance–Rate–Time Problem**

An excursion boat takes 1.5 times as long to go 360 miles up a river than to return. If the boat cruises at 15 miles per hour in still water, what is the rate of the current?

*15 knots means 15 nautical miles per hour. There are 6,076.1 feet in 1 nautical mile.

Solution   Let

$$x = \text{Rate of current (in miles per hour)}$$

$$15 - x = \text{Rate of boat upstream}$$

$$15 + x = \text{Rate of boat downstream}$$

$$\text{Time upstream} = (1.5)(\text{Time downstream})$$

$$\frac{\text{Distance upstream}}{\text{Rate upstream}} = (1.5)\frac{\text{Distance downstream}}{\text{Rate downstream}} \qquad \text{Recall: } T = \frac{D}{R}$$

$$\frac{360}{15 - x} = (1.5)\frac{360}{15 + x}$$

$$\frac{360}{15 - x} = \frac{540}{15 + x}$$

$$360(15 + x) = 540(15 - x) \qquad \begin{array}{l}\text{Multiply both sides by}\\ (15 - x)(15 + x).\end{array}$$

$$5{,}400 + 360x = 8{,}100 - 540x$$

$$900x = 2{,}700$$

$$x = 3$$

Therefore, the rate of the current is 3 miles per hour. The check is left to the reader.

Problem 4   A jetliner takes 1.2 times as long to fly from Paris to New York (3,600 miles) as to return. If the jet cruises at 550 miles per hour in still air, what is the average rate of the wind blowing in the direction of Paris from New York?

EXAMPLE 5   **A Quantity–Rate–Time Problem**

An advertising firm has an old computer that can prepare a whole mailing in 6 hours. With the help of a newer model the job is complete in 2 hours. How long would it take the newer model to do the job alone?

Solution   Let x = time (in hours) for the newer model to do the whole job alone.

$$\left(\begin{array}{l}\text{Part of job completed}\\ \text{in a given length of time}\end{array}\right) = (\text{Rate})(\text{Time})$$

$$\text{Rate of old model} = \frac{1}{6} \text{ job per hour}$$

$$\text{Rate of new model} = \frac{1}{x} \text{ job per hour}$$

$$\begin{pmatrix} \text{Part of job completed} \\ \text{by old model} \\ \text{in 2 hours} \end{pmatrix} + \begin{pmatrix} \text{Part of job completed} \\ \text{by new model} \\ \text{in 2 hours} \end{pmatrix} = 1 \text{ whole job}$$

$$\begin{pmatrix} \text{Rate of} \\ \text{old model} \end{pmatrix}\begin{pmatrix} \text{Time of} \\ \text{old model} \end{pmatrix} + \begin{pmatrix} \text{Rate of} \\ \text{new model} \end{pmatrix}\begin{pmatrix} \text{Time of} \\ \text{new model} \end{pmatrix} = 1 \qquad \text{Recall: } Q = RT$$

$$\frac{1}{6}(2) \qquad + \qquad \frac{1}{x}(2) \qquad\qquad = 1$$

$$\frac{1}{3} + \frac{2}{x} = 1$$

$$x + 6 = 3x$$

$$-2x = -6$$

$$x = 3$$

Therefore, the new computer could do the job alone in 3 hours.

Check   Part of job completed by old model in 2 hours $= 2(\frac{1}{6}) = \frac{1}{3}$
Part of job completed by new model in 2 hours $= 2(\frac{1}{3}) = \frac{2}{3}$
Part of job completed by both models in 2 hours $= \underline{ 1}$

Problem 5   Two pumps are used to fill a water storage tank at a resort. One pump can fill the tank by itself in 9 hours, and the other can fill it in 6 hours. How long will it take both pumps operating together to fill the tank?

● **Mixture Problems**   A variety of applications can be classified as mixture problems. Even though the problems come from different areas, their mathematical treatment is essentially the same.

EXAMPLE 6   **A Mixture Problem**

A specialty coffee shop wishes to blend a $6-per-pound coffee with an $11-per-pound coffee to produce a blend that will sell for $8 per pound. How much of each should be used to produce 300 pounds of the new blend?

Solution   Let

$$x = \text{Amount of \$6-per-pound coffee used}$$

$$300 - x = \text{Amount of \$11-per-pound coffee used}$$

Value before Blending   Value after Blending

$$\left(\begin{array}{c}\text{Value of}\\ \$6\text{-per-pound}\\ \text{coffee used}\end{array}\right) \quad + \quad \left(\begin{array}{c}\text{Value of}\\ \$11\text{-per-pound}\\ \text{coffee used}\end{array}\right) \quad = \quad \left(\begin{array}{c}\text{Total value}\\ \text{of 300 pounds}\\ \text{of the blend}\end{array}\right)$$

$$6x \quad + \quad 11(300 - x) \quad = \quad 8(300)$$
$$6x + 3{,}300 - 11x = 2{,}400$$
$$-5x = -900$$
$$x = 180$$
$$300 - x = 120$$

Use 180 pounds of the $6 coffee and 120 pounds of the $11 coffee. The check is left to the reader.

Problem 6   A concert brought in $82,160 on the sale of 5,000 tickets. If tickets sold for $12 and $20, how many of each were sold?

EXAMPLE 7   **A Mixture Problem**

How many liters of a mixture containing 80% alcohol should be added to 5 liters of a 20% solution to yield a 30% solution?

Solution   Let x = amount of 80% solution used.

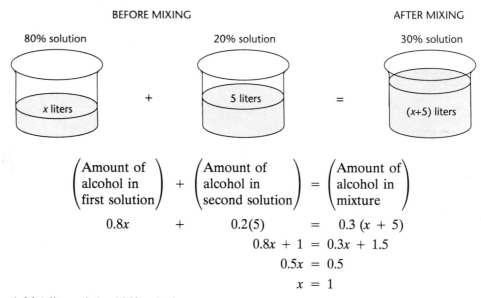

BEFORE MIXING

80% solution   20% solution   AFTER MIXING

30% solution

x liters   +   5 liters   = $(x+5)$ liters

$$\left(\begin{array}{c}\text{Amount of}\\ \text{alcohol in}\\ \text{first solution}\end{array}\right) \quad + \quad \left(\begin{array}{c}\text{Amount of}\\ \text{alcohol in}\\ \text{second solution}\end{array}\right) \quad = \quad \left(\begin{array}{c}\text{Amount of}\\ \text{alcohol in}\\ \text{mixture}\end{array}\right)$$

$$0.8x \quad + \quad 0.2(5) \quad = \quad 0.3\,(x + 5)$$
$$0.8x + 1 = 0.3x + 1.5$$
$$0.5x = 0.5$$
$$x = 1$$

Add 1 liter of the 80% solution.

Check

	Liters of solution	Liters of alcohol	Percent alcohol
First solution	1	$0.8(1) = 0.8$	80
Second solution	5	$0.2(5) = 1$	20
Mixture	6	1.8	$1.8/6 = 0.3$, or 30%

Problem 7   A chemical storeroom has a 90% acid solution and a 40% acid solution. How many centiliters must be taken from each to obtain 25 centiliters of a 50% acid solution?

Answers to Matched Problems
1. 3, 5, 7   **2.** 20 centimeters   **3.** 50 minutes   **4.** 50 miles per hour   **5.** 3.6 hours
6. 2,230 $12 tickets; 2,770 $20 tickets
7. 5 centiliters of 90% solution; 20 centiliters of 40% solution

EXERCISE 2-2

APPLICATIONS   *These problems are not grouped from easy (A) to difficult or theoretical (C). Instead, they are grouped according to subject area. As before, the most difficult problems are marked with two stars (★★), the moderately difficult problems are marked with one star (★), and the easier problems are not marked.*

Numbers

1. Find a number such that 10 less than two-thirds the number is one-fourth the number.

2. Find a number such that 6 more than one-half the number is two-thirds the number.

3. Find four consecutive even integers so that the sum of the first three is 2 more than twice the fourth.

4. Find three consecutive even integers so that the first plus twice the second is twice the third.

Geometry

5. Find the dimensions of a rectangle with a perimeter of 54 meters, if its length is 3 meters less than twice its width.

6. A rectangle 24 meters long has the same area as a square that is 12 meters on a side. What are the dimensions of the rectangle?

7. Find the perimeter of a triangle if one side is 16 feet, another side is two-sevenths the perimeter, and the third side is one-third the perimeter.

8. Find the perimeter of a triangle if one side is 11 centimeters, another side is one-fifth the perimeter, and the third side is one-fourth the perimeter.

Business and Economics

9. The sale price on a camera after a 20% discount is $72. What was the price before the discount?

10. A stereo store marks up each item it sells 60% above wholesale price. What is the wholesale price on a cassette player that retails at $144?

11. It costs a small recording company $17,680 to prepare a record album. This is a one-time fixed cost that covers recording, album design, and so on. Variable costs, including such things as manufacturing, marketing, and royalties, are $4.60 per record. If the album is sold to record shops for $8 each, how many must be sold for the company to break even; that is, so that costs will equal revenues?

12. A videocassette manufacturer has determined that its weekly cost equation is $C = 3,000 + 10x$, where x is the number of cassettes produced and sold each week. If cassettes are sold for $15 each to distributors, how many must be sold each week for the manufacturer to break even? (Refer to Problem 11.)

★**13.** Suppose you have $12,000 to invest. If part is invested at 10% and the rest at 15%, how much should be invested at each rate to yield 12% on the total amount invested?

★**14.** An investor has $20,000 to invest. If part is invested at 8% and the rest at 12%, how much should be invested at each rate to yield 11% on the total amount invested?

Earth Science

★**15.** In 1984, the Soviets led the world in drilling the deepest hole in the Earth's crust—more than 12 kilometers deep. They found that below 3 kilometers the temperature T increased 2.5°C for each additional 100 meters of depth.
(A) If the temperature at 3 kilometers is 30°C and x is the depth of the hole in kilometers, write an equation using x that will give the temperature T in the hole at any depth beyond 3 kilometers.

(B) What would the temperature be at 15 kilometers? (The limit for their drilling equipment was about 300°C.)

(C) At what depth (in kilometers) would they reach a temperature of 280°C?

★ **16.** Because air is not as dense at high altitudes, planes require a higher ground speed to become airborne. A rule of thumb is 3% more ground speed per 1,000 feet of elevation, assuming no wind and no change in air temperature. (Compute numerical answers to 3 significant digits.)

(A) Let

$$V_s = \text{Takeoff ground speed at sea level for a particular plane (in miles per hour)}$$

$$A = \text{Altitude above sea level (in thousands of feet)}$$

$$V = \text{Takeoff ground speed at altitude } A \text{ for the same plane (in miles per hour)}$$

Write a formula relating these three quantities.

(B) What takeoff ground speed would be required at Lake Tahoe airport (6,400 feet), if takeoff ground speed at San Francisco airport (sea level) is 120 miles per hour?

(C) If a landing strip at a Colorado Rockies hunting lodge (8,500 feet) requires a takeoff ground speed of 125 miles per hour, what would be the takeoff ground speed in Los Angeles (sea level)?

(D) If the takeoff ground speed at sea level is 135 miles per hour and the takeoff ground speed at a mountain resort is 155 miles per hour, what is the altitude of the mountain resort in thousands of feet?

★★ **17.** An earthquake emits a primary wave and a secondary wave. Near the surface of the Earth the primary wave travels at about 5 miles per second, and the secondary wave travels at about 3 miles per second. From the time lag between the two waves arriving at a given seismic station, it is possible to estimate the distance to the quake. Suppose a station measures a time difference of 12 seconds between the arrival of the two waves. How far is the earthquake from the station? (The *epicenter* can be located by obtaining distance bearings at three or more stations.)

★★ **18.** A ship using sound-sensing devices above and below water recorded a surface explosion 39 seconds sooner on its underwater device than on its above-water device. If sound travels in air at about 1,100 feet per second and in water at about 5,000 feet per second, how far away was the explosion?

Life Science

19. A naturalist for a fish and game department estimated the total number of trout in a certain lake using the popular capture–mark–recapture technique. She net-

ted, marked, and released 200 trout. A week later, allowing for thorough mixing, she again netted 200 trout and found 8 marked ones among them. Assuming that the ratio of marked trout to the total number in the second sample is the same as the ratio of all marked fish in the first sample to the total trout population in the lake, estimate the total number of fish in the lake.

20. Repeat Problem 19 with a first (marked) sample of 300 and a second sample of 180 with only 6 marked trout.

Chemistry

★ **21.** How many gallons of distilled water must be mixed with 50 gallons of 30% alcohol solution to obtain a 25% solution?

★ **22.** How many gallons of hydrochloric acid must be added to 12 gallons of a 30% solution to obtain a 40% solution?

★ **23.** A chemist has two solutions of sulfuric acid—a 20% solution and an 80% solution. How much of each should be used to obtain 100 liters of a 62% solution?

★ **24.** A fuel oil distributor has 120,000 gallons of fuel with 0.9% sulfur content, which exceeds pollution control standards of 0.8% sulfur content. How many gallons of fuel oil with a 0.3% sulfur content must be added to the 120,000 gallons to obtain fuel oil that will comply with the pollution control standards?

Rate–Time

★ **25.** An old computer can do the weekly payroll in 5 hours. A newer computer can do the same payroll in 3 hours. The old computer starts on the payroll, and after 1 hour the newer computer is brought on-line to work with the older computer until the job is finished. How long will it take both computers working together to finish the job? (Assume the computers operate independently.)

★ **26.** One pump can fill a gasoline storage tank in 8 hours. With a second pump working simultaneously, the tank can be filled in 3 hours. How long would it take the second pump to fill the tank operating alone?

★★ **27.** The cruising speed of an airplane is 150 miles per hour (relative to the ground). You wish to hire the plane for a 3-hour sightseeing trip. You instruct the pilot to fly north as far as he can and still return to the airport at the end of the allotted time.

(A) How far north should the pilot fly if the wind is blowing from the north at 30 miles per hour?

(B) How far north should the pilot fly if there is no wind?

★ **28.** Suppose you are at a river resort and rent a motor boat for 5 hours starting at 7 A.M. You are told that the boat will travel at 8 miles per hour upstream and 12 miles per hour returning. You decide that you would like to go as far up the river as you can and still be back at noon. At what time should you turn back, and how far from the resort will you be at that time?

Music

★ **29.** A major chord in music is composed of notes whose frequencies are in the ratio $4 : 5 : 6$. If the first note of a chord has a frequency of 264 hertz (middle C on the piano), find the frequencies of the other two notes. [*Hint:* Set up two proportions using $4 : 5$ and $4 : 6$.]

★ **30.** A minor chord is composed of notes whose frequencies are in the ratio $10 : 12 : 15$. If the first note of a minor chord is A, with a frequency of 220 hertz, what are the frequencies of the other two notes?

Psychology

31. In an experiment on motivation, Professor Brown trained a group of rats to run down a narrow passage in a cage to receive food in a goal box. He then put a harness on each rat and connected it to an overhead wire attached to a scale. In this way he could place the rat different distances from the food and measure the pull (in grams) of the rat toward the food. He found that the relationship between motivation (pull) and position was given approximately by the equation

$$p = -\tfrac{1}{5}d + 70 \qquad 30 \le d \le 170$$

where pull p is measured in grams and distance d in

centimeters. When the pull registered was 40 grams, how far was the rat from the goal box?

32. Professor Brown performed the same kind of experiment as described in Problem 31, except that he replaced the food in the goal box with a mild electric shock. With the same kind of apparatus, he was able to measure the avoidance strength relative to the distance from the object to be avoided. He found that the avoidance strength a (measured in grams) was related to the distance d that the rat was from the shock (measured in centimeters) approximately by the equation

$$a = -\tfrac{4}{3}d + 230 \qquad 30 \le d \le 170$$

If the same rat were trained as described in this problem and in Problem 31, at what distance (to one decimal place) from the goal box would the approach and avoidance strengths be the same? (What do you think the rat would do at this point?)

Puzzle

33. An oil-drilling rig in the Gulf of Mexico stands so that one-fifth of it is in sand, 20 feet of it is in water, and two-thirds of it is in the air. What is the total height of the rig?

34. During a camping trip in the North Woods in Canada, a couple went one-third of the way by boat, 10 miles by foot, and one-sixth of the way by horse. How long was the trip?

★★ **35.** After exactly 12 o'clock noon, what time will the hands of a clock be together again?

SECTION 2-3　Linear Inequalities

- Inequality Relations and Interval Notation
- Solving Linear Inequalities
- Applications

We now turn to the problem of solving linear inequalities in one variable, such as

$$3(x - 5) \ge 5(x + 7) - 10 \qquad \text{and} \qquad -4 \le 3 - 2x < 7$$

● **Inequality Relations and Interval Notation**

The above mathematical forms involve the **inequality**, or **order**, **relation**—that is, "less than" and "greater than" relations. Just as we use $=$ to replace the words "is equal to," we use the **inequality symbols** $<$ and $>$ to represent "is less than" and "is greater than," respectively.

While it probably seems obvious to you that

$$2 < 4 \qquad 5 > 0 \qquad 25{,}000 > 1$$

are true, it may not seem as obvious that

$$-4 < -2 \qquad 0 > -5 \qquad -25{,}000 < -1$$

To make the inequality relation precise so that we can interpret it relative to all real numbers, we need a precise definition of the concept.

DEFINITION 1

$a < b$ and $b > a$

For a and b real numbers, we say that **a is less than b** or **b is greater than a** and write

$$a < b \qquad \text{or} \qquad b > a$$

if there exists a positive real number p such that $a + p = b$ (or equivalently, $b - a = p$).

We certainly expect that if a positive number is added to *any* real number, the sum is larger than the original. That is essentially what the definition states.

When we write

$$a \leq b$$

we mean $a < b$ or $a = b$ and say **a is less than or equal to b**. When we write

$$a \geq b$$

we mean $a > b$ or $a = b$ and say **a is greater than or equal to b**.

The inequality symbols $<$ and $>$ have a very clear geometric interpretation on the real number line. If $a < b$, then a is to the left of b; if $c > d$, then c is to the right of d (Fig. 1).

FIGURE 1 $a < b, c > d$.

It is an interesting and useful fact that for any two real numbers a and b, either $a < b$, or $a > b$, or $a = b$. This is called the **trichotomy property** of real numbers.

The double inequality $a < x \leq b$ means that $x > a$ *and* $x \leq b$; that is, x is between a and b, including b but not including a. The set of all real numbers x satisfying the inequality $a < x \leq b$ is called an **interval** and is represented by $(a, b]$. Thus,

$$(a, b] = \{x \mid a < x \leq b\}$$

The number a is called the **left endpoint** of the interval, and the symbol "(" indicates that a is not included in the interval. The number b is called the **right endpoint** of the interval, and the symbol "]" indicates that b is included in the interval. Other types of intervals of real numbers are shown in Table 1.

TABLE 1   **Interval Notation**

Interval notation	Inequality notation	Line graph	Type
$[a, b]$	$a \leq x \leq b$		Closed
$[a, b)$	$a \leq x < b$		Half-open
$(a, b]$	$a < x \leq b$		Half-open
(a, b)	$a < x < b$		Open
$[b, \infty)$	$x \geq b$		Closed
(b, ∞)	$x > b$		Open
$(-\infty, a]$	$x \leq a$		Closed
$(-\infty, a)$	$x < a$		Open

Note that the symbol "∞," read "infinity," used in Table 1 is not a numeral. When we write $[b, \infty)$, we are simply referring to the interval starting at b and continuing indefinitely to the right. We would never write $[b, \infty]$ or $b \leq x \leq \infty$, since ∞ cannot be used as an endpoint of an interval. The interval $(-\infty, \infty)$ represents the set of real numbers R, since its graph is the entire real number line.

CAUTION

It is important to note that

$$5 > x \geq -3 \quad \text{is equivalent to } [-3, 5) \text{ and not to } (5, -3]$$

In interval notation, the smaller number is always written to the left. Thus, it may be useful to rewrite the inequality as $-3 \leq x < 5$ before rewriting it in interval notation.

EXAMPLE 1   Graphing Intervals and Inequalities

Write each of the following in inequality notation and graph on a real number line:

(A) $[-2, 3)$ (B) $(-4, 2)$ (C) $[-2, \infty)$ (D) $(-\infty, 3)$

Solutions   (A) $-2 \le x < 3$

(B) $-4 < x < 2$

(C) $x \ge -2$

(D) $x < 3$

Problem 1   Write each of the following in interval notation and graph on a real number line:

(A) $-3 < x \le 3$ (B) $2 \ge x \ge -1$ (C) $x > 1$ (D) $x \le 2$

Since intervals are sets of real numbers, the set operations of *union* and *intersection* are often useful when working with intervals. The **union** of sets A and B, denoted by $A \cup B$, is the set formed by combining all the elements of A and all the elements of B. The **intersection** of sets A and B, denoted by $A \cap B$, is the set of elements of A that are also in B. Symbolically:

DEFINITION 2   **Union and Intersection**

UNION: $A \cup B = \{x | x$ is in A **or** x is in $B\}$
$\{1, 2, 3\} \cup \{2, 3, 4, 5\} = \{1, 2, 3, 4, 5\}$

INTERSECTION: $A \cap B = \{x | x$ is in A **and** x is in $B\}$
$\{1, 2, 3\} \cap \{2, 3, 4, 5\} = \{2, 3\}$

EXAMPLE 2   **Graphing Unions and Intersections of Intervals**

If $A = [-2, 3]$, $B = (1, 6)$, and $C = (4, \infty)$, graph the indicated sets and write as a single interval, if possible:

(A) $A \cup B$ and $A \cap B$ (B) $A \cup C$ and $A \cap C$

Solution   (A)

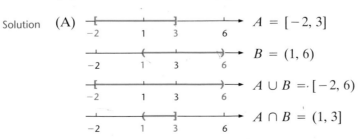

$A = [-2, 3]$

$B = (1, 6)$

$A \cup B = [-2, 6)$

$A \cap B = (1, 3]$

(B)

$A = [-2, 3]$

$C = (4, \infty)$

$A \cup C = [-2, 3] \cup (4, \infty)$

$A \cap C = \varnothing$

Problem 2 If $D = [-4, 1)$, $E = (-1, 3]$, and $F = [2, \infty)$, graph the indicated sets and write as a single interval, if possible:

(A) $D \cup E$ (B) $D \cap E$ (C) $E \cup F$ (D) $E \cap F$

• Solving Linear Inequalities

We now turn to the problem of solving linear inequalities in one variable, such as

$$2(2x + 3) < 6(x - 2) + 10 \quad \text{and} \quad -3 < 2x + 3 \le 9$$

The **solution set** for an inequality is the set of all values of the variable that make the inequality a true statement. Each element of the solution set is called a **solution** of the inequality. To **solve an inequality** is to find its solution set. Two inequalities are **equivalent** if they have the same solution set for a given replacement set. Just as with equations, we perform operations on inequalities that produce simpler equivalent inequalities, and continue the process until an inequality is reached whose solution is obvious. The properties of inequalities given in Theorem 1 can be used to produce equivalent inequalities.

Theorem 1

Inequality Properties

For a, b, and c any real numbers:

1. If $a < b$ and $b < c$, then $a < c$. **Transitive Property**
2. If $a < b$, then $a + c < b + c$. **Addition Property**
   $-2 < 4 \qquad -2 + 3 < 4 + 3$
3. If $a < b$, then $a - c < b - c$. **Subtraction Property**
   $-2 < 4 \qquad -2 - 3 < 4 - 3$
4. If $a < b$ and c is positive, then $ca < cb$.
   $-2 < 4 \qquad\qquad 3(-2) < 3(4)$
5. If $a < b$ and c is negative, then $ca > cb$.
   $-2 < 4 \qquad\qquad (-3)(-2) > (-3)(4)$

Multiplication Property (Note difference between 4 and 5.)

6. If $a < b$ and c is positive, then $\dfrac{a}{c} < \dfrac{b}{c}$.

$-2 < 4$ $\qquad\qquad$ $\dfrac{-2}{2} < \dfrac{4}{2}$

7. If $a < b$ and c is negative, then $\dfrac{a}{c} > \dfrac{b}{c}$.

$-2 < 4$ $\qquad\qquad$ $\dfrac{-2}{-2} > \dfrac{4}{-2}$

Division Property (Note difference between 6 and 7.)

Similar properties hold if each inequality sign is reversed, or if $<$ is replaced with $\le$ and $>$ is replaced with $\ge$. Thus, we find that we can perform essentially the same operations on inequalities that we perform on equations. When working with inequalities, however, we have to be particularly careful of the use of the multiplication and division properties.

The order of the inequality reverses if we multiply or divide both sides of an inequality statement by a negative number.

Now let's see how the inequality properties are used to solve linear inequalities. Several examples will illustrate the process.

EXAMPLE 3   **Solving a Linear Inequality**

Solve and graph: $2(2x + 3) - 10 < 6(x - 2)$

Solution $\qquad$ $2(2x + 3) - 10 < 6(x - 2)$

$\qquad\qquad 4x + 6 - 10 < 6x - 12$ $\qquad$ Simplify left and right sides.

$\qquad\qquad\quad 4x - 4 < 6x - 12$

$\qquad\qquad \boxed{4x - 4 + 4 \le 6x - 12 + 4}$ $\qquad$ Addition property

$\qquad\qquad\quad 4x < 6x - 8$

$\qquad\qquad \boxed{4x - 6x < 6x - 8 - 6x}$ $\qquad$ Subtraction property

$\qquad\qquad\quad -2x < -8$

$\qquad\qquad \boxed{\dfrac{-2x}{-2} > \dfrac{-8}{-2}}$ $\qquad$ Division property—note that order reverses since -2 is negative

$\qquad\qquad\quad x > 4 \qquad$ or $\qquad (4, \infty)$ $\qquad$ Solution set

$\qquad\qquad$ Graph of solution set

Problem 3 $\quad$ Solve and graph: $3(x - 1) \ge 5(x + 2) - 5$

EXAMPLE 4   Solving a Linear Inequality Involving Fractions

Solve and graph: $\dfrac{2x-3}{4} + 6 \geq 2 + \dfrac{4x}{3}$

Solution

$$\dfrac{2x-3}{4} + 6 \geq 2 + \dfrac{4x}{3}$$

$$12 \cdot \dfrac{2x-3}{4} + 12 \cdot 6 \geq 12 \cdot 2 + 12 \cdot \dfrac{4x}{3}$$ Multiply both sides by 12, the LCD.

$$3(2x-3) + 72 \geq 24 + 4(4x)$$

$$6x - 9 + 72 \geq 24 + 16x$$

$$6x + 63 \geq 24 + 16x$$

$$-10x \geq -39$$

$$x \leq 3.9 \qquad \text{or} \qquad (-\infty, 3.9]$$ Order reverses since both sides are divided by -10, a negative number.

Problem 4   Solve and graph: $\dfrac{4x-3}{3} + 8 < 6 + \dfrac{3x}{2}$

EXAMPLE 5   Solving a Double Inequality

Solve and graph: $-3 \leq 4 - 7x < 18$

Solution   We proceed as before, except we try to isolate x in the middle with a coefficient of 1.

$$-3 \leq 4 - 7x < 18$$

$$-3 - 4 \leq 4 - 7x - 4 < 18 - 4$$ Subtract 4 from each member.

$$-7 \leq -7x < 14$$

$$\dfrac{-7}{-7} \geq \dfrac{-7x}{-7} > \dfrac{14}{-7}$$ Divide each member by -7 and reverse each inequality.

$$1 \geq x > -2 \qquad \text{or} \qquad -2 < x \leq 1 \qquad \text{or} \qquad (-2, 1]$$

Problem 5   Solve and graph: $-3 < 7 - 2x \leq 7$

• Applications

EXAMPLE 6   Chemistry

In a chemistry experiment, a solution of hydrochloric acid is to be kept between 30°C and 35°C—that is, $30 \leq C \leq 35$. What is the range in temperature in degrees Fahrenheit if the Celsius/Fahrenheit conversion formula is $C = \frac{5}{9}(F - 32)$?

Solution

$$30 \leq C \leq 35$$

$$30 \leq \frac{5}{9}(F - 32) \leq 35 \qquad \text{Replace } C \text{ with } \frac{5}{9}(F - 32).$$

$$\frac{9}{5} \cdot 30 \leq \frac{9}{5} \cdot \frac{5}{9}(F - 32) \leq \frac{9}{5} \cdot 35 \qquad \text{Multiply each member by } \frac{9}{5}.$$

$$54 \leq F - 32 \leq 63$$

$$54 + 32 \leq F - 32 + 32 \leq 63 + 32 \qquad \text{Add 32 to each member.}$$

$$86 \leq F \leq 95$$

The range of the temperature is from 86°F to 95°F, inclusive.

Problem 6

A film developer is to be kept between 68°F and 77°F—that is, $68 \leq F \leq 77$. What is the range in temperature in degrees Celsius if the Celsius/Fahrenheit conversion formula is $F = \frac{9}{5}C + 32$?

Answers to Matched Problems

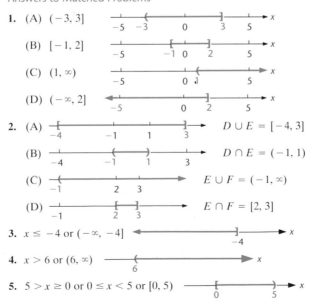

1. (A) $(-3, 3]$

   (B) $[-1, 2]$

   (C) $(1, \infty)$

   (D) $(-\infty, 2]$

2. (A) $D \cup E = [-4, 3]$

   (B) $D \cap E = (-1, 1)$

   (C) $E \cup F = (-1, \infty)$

   (D) $E \cap F = [2, 3]$

3. $x \leq -4$ or $(-\infty, -4]$

4. $x > 6$ or $(6, \infty)$

5. $5 > x \geq 0$ or $0 \leq x < 5$ or $[0, 5)$

6. $20 \leq C \leq 25$; the range in temperature is from 20°C to 25°C

EXERCISE 2-3

A _____

In Problems 1–6, rewrite in inequality notation and graph on a real number line.

1. $[-8, 7]$ **2.** $(-4, 8)$ **3.** $[-6, 6)$ **4.** $(-3, 3]$ **5.** $[-6, \infty)$ **6.** $(-\infty, 7)$

In Problems 7–12, rewrite in interval notation and graph on a real number line.

7. $-2 < x \le 6$ **8.** $-5 \le x \le 5$ **9.** $-7 < x < 8$ **10.** $-4 \le x < 5$

11. $x \le -2$ **12.** $x > 3$

In Problems 13–16, write in interval and inequality notation.

13.

14.

15.

16.

In Problems 17–30, solve and graph.

17. $7x - 8 < 4x + 7$ **18.** $4x + 8 \ge x - 1$ **19.** $3 - x \ge 5(3 - x)$ **20.** $2(x - 3) + 5 < 5 - x$

21. $\dfrac{N}{-2} > 4$ **22.** $\dfrac{M}{-3} \le -2$ **23.** $-5t < -10$ **24.** $-7n \ge 21$

25. $3 - m < 4(m - 3)$ **26.** $2(1 - u) \ge 5u$ **27.** $-2 - \dfrac{B}{4} \le \dfrac{1 + B}{3}$ **28.** $\dfrac{y - 3}{4} - 1 > \dfrac{y}{2}$

29. $-4 < 5t + 6 \le 21$ **30.** $2 \le 3m - 7 < 14$

B _____

In Problems 31–42, graph the indicated set and write as a single interval, if possible.

31. $(-5, 5) \cup [4, 7]$ **32.** $(-5, 5) \cap [4, 7]$ **33.** $[-1, 4) \cap (2, 6]$ **34.** $[-1, 4) \cup (2, 6]$

35. $(-\infty, 1) \cup (-2, \infty)$ **36.** $(-\infty, 1) \cap (2, \infty)$ **37.** $(-\infty, -1) \cup [3, 7)$ **38.** $(1, 6] \cup [9, \infty)$

39. $[2, 3] \cup (1, 5)$ **40.** $[2, 3] \cap (1, 5)$ **41.** $(-\infty, 4) \cup (-1, 6]$ **42.** $(-3, 2) \cup [0, \infty)$

In Problems 43–54, solve and graph.

43. $\dfrac{q}{7} - 3 > \dfrac{q - 4}{3} + 1$ **44.** $\dfrac{p}{3} - \dfrac{p - 2}{2} \le \dfrac{p}{4} - 4$

45. $\dfrac{2x}{5} - \dfrac{1}{2}(x - 3) \le \dfrac{2x}{3} - \dfrac{3}{10}(x + 2)$ **46.** $\dfrac{2}{3}(x + 7) - \dfrac{x}{4} > \dfrac{1}{2}(3 - x) + \dfrac{x}{6}$

47. $-4 \le \frac{9}{5}x + 32 \le 68$ **48.** $-1 \le \frac{2}{3}A + 5 \le 11$ **49.** $-12 < \frac{3}{4}(2 - x) \le 24$ **50.** $24 \le \frac{2}{3}(x - 5) < 36$

51. $16 < 7 - 3x \le 31$ **52.** $-1 \le 9 - 2x < 5$ **53.** $-6 < -\frac{2}{5}(1 - x) \le 4$ **54.** $15 \le 7 - \frac{2}{5}x \le 21$

Use a calculator to solve each of the inequalities in Problems 55–58. Write answers using inequality notation.

55. $5.23(x - 0.172) \le 6.02x - 0.427$ **56.** $72.3x - 4.07 > 9.02(11.7x - 8.22)$

57. $-0.703 < 0.112 - 2.28x < 0.703$ **58.** $-4.26 < 3.88 - 6.07x < 5.66$

∫ *Problems 59–64 are calculus-related. For what real number(s) x does each expression represent a real number?*

59. $\sqrt{1 - x}$ **60.** $\sqrt{x + 5}$ **61.** $\sqrt{3x + 5}$ **62.** $\sqrt{7 - 2x}$ **63.** $\dfrac{1}{\sqrt[4]{2x + 3}}$ **64.** $\dfrac{1}{\sqrt[4]{5 - 6x}}$

65. What can be said about the signs of the numbers a and b in each case?
(A) $ab > 0$ (B) $ab < 0$
(C) $\dfrac{a}{b} > 0$ (D) $\dfrac{a}{b} < 0$

66. What can be said about the signs of the numbers a, b, and c in each case?
(A) $abc > 0$ (B) $\dfrac{ab}{c} < 0$
(C) $\dfrac{a}{bc} > 0$ (D) $\dfrac{a^2}{bc} < 0$

67. Replace each question mark with $<$ or $>$, as appropriate:
(A) If $a - b = 1$, then a ? b.
(B) If $u - v = -2$, then u ? v.

68. For what p and q is $p + q < p - q$?

C

69. If both a and b are negative numbers and b/a is greater than 1, then is $a - b$ positive or negative?

70. If both a and b are positive numbers and b/a is greater than 1, then is $a - b$ positive or negative?

71. Indicate true (T) or false (F):
(A) If $p > q$ and $m > 0$, then $mp < mq$.
(B) If $p < q$ and $m < 0$, then $mp > mq$.
(C) If $p > 0$ and $q < 0$, then $p + q > q$.

72. Assume that $m > n > 0$; then

$$mn > n^2$$
$$mn - m^2 > n^2 - m^2$$
$$m(n - m) > (n + m)(n - m)$$
$$m > n + m$$
$$0 > n$$

But it was assumed that $n > 0$. Find the error.

Prove each inequality property in Problems 73–76, given a, b, and c are arbitrary real numbers.

73. If $a < b$, then $a + c < b + c$.

74. If $a < b$, then $a - c < b - c$.

75. (A) If $a < b$ and c is positive, then $ca < cb$.
(B) If $a < b$ and c is negative, then $ca > cb$.

76. (A) If $a < b$ and c is positive, then $\dfrac{a}{c} < \dfrac{b}{c}$.
(B) If $a < b$ and c is negative, then $\dfrac{a}{c} > \dfrac{b}{c}$.

APPLICATIONS   *Write all answers using inequality notation.*

77. Earth Science. The Soviets, in drilling the world's deepest hole in 1984, found that the temperature x kilometers below the surface of the Earth was given by

$$T = 30 + 25(x - 3) \qquad 3 \le x \le 15$$

where T is temperature in degrees Celsius. At what depth will the temperature be between 200°C and 300°C, inclusive?

78. Earth Science. As dry air moves upward it expands, and in so doing it cools at a rate of about 5.5°F for each

1,000-foot rise up to about 40,000 feet. If the ground temperature is 70°F, then the temperature T at height h is given approximately by $T = 70 - 0.0055h$. For what range in altitude will the temperature be between 26°F and -40°F, inclusive?

79. Business and Economics. For a business to make a profit it is clear that revenue R must be greater than cost C; in short, a profit will result only if $R > C$. If a company manufactures records and its cost equation for a week is $C = 300 + 1.5x$ and its revenue equation is $R = 2x$, where x is the number of records sold in a week, how many records must be sold for the company to realize a profit?

80. Business and Economics. A videocassette manufacturer has determined that its weekly cost equation is $C = 3,000 + 10x$, where x is the number of cassettes produced and sold each week. If cassettes are sold for $15 each to distributors, how many must be sold each week for the company to make a profit?

81. Energy. If the power demands in a 110-volt electric circuit in a home vary between 220 and 2,750 watts, what is the range of current flowing through the circuit? ($W = EI$, where W = Power in watts, E = Pressure in volts, and I = Current in amperes.)

82. Psychology. A person's IQ is given by the formula

$$IQ = \frac{MA}{CA} \, 100$$

where MA is mental age and CA is chronological age. If

$$80 \le IQ \le 140$$

for a group of 12-year-old children, find the range of their mental ages.

★83. Finance. If an individual aged 65–69 continues to work after Social Security benefits start, benefits will be reduced when earnings exceed an earnings limitation. In 1989, benefits were reduced by $1 for every $2 earned in excess of $8,880. Find the range of benefit reductions for individuals earning between $13,000 and $16,000.

★84. Finance. Refer to Problem 83. In 1990 the law was changed so that benefits were reduced by $1 for every $3 earned in excess of $8,880. Find the range of benefit reductions for individuals earning between $13,000 and $16,000.

SECTION 2-4 Absolute Value in Equations and Inequalities

- Absolute Value and Distance
- Absolute Value in Equations and Inequalities
- Absolute Value and Radicals

This section discusses solving absolute value equations and inequalities.

- **Absolute Value and Distance**

We start with a geometric definition of absolute value. If a is the coordinate of a point on a real number line, then the distance from the origin to a is represented by $|a|$ and is referred to as the **absolute value** of a. Thus, $|5| = 5$ since the point with coordinate 5 is five units from the origin, and $|-6| = 6$ since the point with coordinate -6 is six units from the origin (Fig. 1).

FIGURE 1 Absolute value.

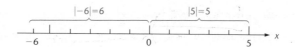

Symbolically, and more formally, we define absolute value as follows:

DEFINITION 1   **Absolute Value**

$$|x| = \begin{cases} x & \text{if } x \geq 0 \\ -x & \text{if } x < 0 \end{cases} \quad \begin{matrix} |4| = 4 \\ |-3| = -(-3) = 3 \end{matrix}$$

[*Note:* $-x$ is positive if x is negative.]

Both the geometric and nongeometric definitions of absolute value are useful, as will be seen in the material that follows. Remember:

The absolute value of a number is never negative.

EXAMPLE 1   Absolute Value of a Real Number

(A) $|\pi - 3| = \pi - 3$ Since $\pi \approx 3.14$, $\pi - 3$ is positive.
(B) $|3 - \pi| = -(3 - \pi) = \pi - 3$ Since $3 - \pi$ is negative

Problem 1   Write without the absolute value sign:

(A) $|8|$ (B) $|\sqrt[3]{9} - 2|$ (C) $|-\sqrt{2}|$ (D) $|2 - \sqrt[3]{9}|$

Following the same reasoning used in Example 1, the next theorem can be proved (see Problem 93 in Exercise 2-4).

Theorem 1   For all real numbers a and b,

$$|b - a| = |a - b|$$

We use this result in defining the distance between two points on a real number line.

DEFINITION 2   **Distance between Points A and B**

Let A and B be two points on a real number line with coordinates a and b, respectively. The **distance between A and B** is given by

$$d(A, B) = |b - a|$$

This distance is also called the **length of the line segment** joining A and B.

EXAMPLE 2   **Distance between Points on a Number Line**

Find the distance between points A and B with coordinates a and b, respectively, as given.

(A) $a = 4,\ \ b = 9$ (B) $a = 9,\ \ b = 4$ (C) $a = 0,\ \ b = 6$

Solutions   (A)

$$d(A,B)=|9-4|=|5|=5$$

(B)

$$d(A,B)=|4-9|=|-5|=5$$

(C)

$$d(A,B)=|6-0|=|6|=6$$

It should be clear, since $|b - a| = |a - b|$, that

$$d(A, B) = d(B, A)$$

Hence, in computing the distance between two points on a real number line, it does not matter how the two points are labeled—point A can be to the left or to the right of point B. Note also that if A is at the origin O, then

$$d(O, B) = |b - 0| = |b|$$

Problem 2   Use the number line below to find the indicated distances.

(A) $d(C, D)$ (B) $d(D, C)$ (C) $d(A, B)$

(D) $d(A, C)$ (E) $d(O, A)$ (F) $d(D, A)$

● **Absolute Value in Equations and Inequalities**

Absolute value is frequently encountered in equations and inequalities. Some of these forms have immediate geometric interpretation.

EXAMPLE 3   **Solving Absolute Value Problems Geometrically**

Solve geometrically and graph. Write solutions in both inequality and interval notation, where appropriate.

(A) $|x - 3| = 5$ (B) $|x - 3| < 5$

(C) $0 < |x - 3| < 5$ (D) $|x - 3| > 5$

Solutions   (A) Geometrically, $|x - 3|$ represents the distance between x and 3. Thus, in $|x - 3| = 5, x$ is a number whose distance from 3 is 5. That is,

$$x = 3 \pm 5 = -2 \quad \text{or} \quad 8$$

The solution set is $\{-2, 8\}$.   This *is not* interval notation.

(B) Geometrically, in $|x - 3| < 5, x$ is a number whose distance from 3 is less than 5; that is,

$$-2 < x < 8$$

The solution set is $(-2, 8)$.   This *is* interval notation.

(C) The form $0 < |x - 3| < 5$ is frequently encountered in calculus and more advanced mathematics. Geometrically, x is a number whose distance from 3 is less than 5, but x cannot equal 3. Thus,

$$-2 < x < 8 \quad x \neq 3 \quad \text{or} \quad (-2, 3) \cup (3, 8)$$

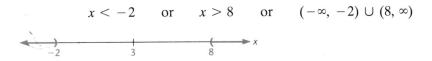

(D) Geometrically, in $|x - 3| > 5, x$ is a number whose distance from 3 is greater than 5; that is,

$$x < -2 \quad \text{or} \quad x > 8 \quad \text{or} \quad (-\infty, -2) \cup (8, \infty)$$

CAUTION   Do not confuse solutions like

$$-2 < x \quad \text{and} \quad x < 8$$

which can also be written as

$$-2 < x < 8 \quad \text{or} \quad (-2, 8)$$

with solutions like

$$x < -2 \quad \text{or} \quad x > 8$$

which cannot be written as a double inequality or as a single interval.

We summarize the preceding results in Table 1.

TABLE 1 **Geometric Interpretation of Absolute Value Equations and Inequalities**

Form ($d > 0$)	Geometric interpretation	Solution	Graph
$\lvert x - c \rvert = d$	Distance between x and c is equal to d.	$\{c - d, c + d\}$	
$\lvert x - c \rvert < d$	Distance between x and c is less than d.	$(c - d, c + d)$	
$0 < \lvert x - c \rvert < d$	Distance between x and c is less than d, but $x \neq c$.	$(c - d, c) \cup (c, c + d)$	
$\lvert x - c \rvert > d$	Distance between x and c is greater than d.	$(-\infty, c - d) \cup (c + d, \infty)$	

Problem 3   Solve geometrically and graph. Write solutions in both inequality and interval notation, where appropriate.

(A) $\lvert x + 2 \rvert = 6$ (B) $\lvert x + 2 \rvert < 6$
(C) $0 < \lvert x + 2 \rvert < 6$ (D) $\lvert x + 2 \rvert > 6$

[*Hint:* $\lvert x + 2 \rvert = \lvert x - (-2) \rvert$.]

Reasoning geometrically as before (noting that $\lvert x \rvert = \lvert x - 0 \rvert$) leads to Theorem 2.

Theorem 2   **Properties of Equations and Inequalities Involving $\lvert x \rvert$**

For $p > 0$:

1. $\lvert x \rvert = p$ is equivalent to $x = p$ or $x = -p$.

2. $\lvert x \rvert < p$ is equivalent to $-p < x < p$.

3. $\lvert x \rvert > p$ is equivalent to $x < -p$ or $x > p$.

If we replace x in Theorem 2 with $ax + b$, we obtain the more general Theorem 3.

Theorem 3	**Properties of Equations and Inequalities Involving $\lvert ax + b\rvert$**

For $p > 0$:

1. $\lvert ax + b\rvert = p$ is equivalent to $ax + b = p$ or $ax + b = -p$.
2. $\lvert ax + b\rvert < p$ is equivalent to $-p < ax + b < p$.
3. $\lvert ax + b\rvert > p$ is equivalent to $ax + b < -p$ or $ax + b > p$.

EXAMPLE 4 Solving Absolute Value Problems

Solve, and write solutions in both inequality and interval notation, where appropriate.

(A) $\lvert 3x + 5\rvert = 4$ (B) $\lvert x\rvert < 5$ (C) $\lvert 2x - 1\rvert < 3$ (D) $\lvert 7 - 3x\rvert \le 2$

Solutions

(A) $\lvert 3x + 5\rvert = 4$

$$3x + 5 = \pm 4$$
$$3x = -5 \pm 4$$
$$x = \frac{-5 \pm 4}{3}$$
$$x = -3, \ -\tfrac{1}{3}$$
$$\text{or} \quad \{-3, \ -\tfrac{1}{3}\}$$

(B) $\lvert x\rvert < 5$

$$-5 < x < 5$$
$$\text{or} \quad (-5, 5)$$

(C) $\lvert 2x - 1\rvert < 3$

$$-3 < 2x - 1 < 3$$
$$-2 < 2x < 4$$
$$-1 < x < 2$$
$$\text{or} \quad (-1, 2)$$

(D) $\lvert 7 - 3x\rvert \le 2$

$$-2 \le 7 - 3x \le 2$$
$$-9 \le -3x \le -5$$
$$3 \ge x \ge \tfrac{5}{3}$$
$$\tfrac{5}{3} \le x \le 3$$
$$\text{or} \ [\tfrac{5}{3}, 3]$$

Problem 4 Solve, and write solutions in both inequality and interval notation, where appropriate.

(A) $\lvert 2x - 1\rvert = 8$ (B) $\lvert x\rvert \le 7$ (C) $\lvert 3x + 3\rvert \le 9$ (D) $\lvert 5 - 2x\rvert < 9$

EXAMPLE 5 Solving Absolute Value Inequalities

Solve, and write solutions in both inequality and interval notation.

(A) $\lvert x\rvert > 3$ (B) $\lvert 2x - 1\rvert \ge 3$ (C) $\lvert 7 - 3x\rvert > 2$

Solutions
(A) $|x| > 3$

$x < -3$ or $x > 3$ Inequality notation

$(-\infty, -3) \cup (3, \infty)$ Interval notation

(B) $|2x - 1| \geq 3$

$2x - 1 \leq -3$ or $2x - 1 \geq 3$

$2x \leq -2$ or $2x \geq 4$

$x \leq -1$ or $x \geq 2$ Inequality notation

$(-\infty, -1] \cup [2, \infty)$ Interval notation

(C) $|7 - 3x| > 2$

$7 - 3x < -2$ or $7 - 3x > 2$

$-3x < -9$ or $-3x > -5$

$x > 3$ or $x < \frac{5}{3}$ Inequality notation

$(-\infty, \frac{5}{3}) \cup (3, \infty)$ Interval notation

Problem 5   Solve, and write solutions in both inequality and interval notation.

(A) $|x| \geq 5$ (B) $|4x - 3| > 5$ (C) $|6 - 5x| > 16$

EXAMPLE 6   **An Absolute Value Problem with Two Cases**

Solve: $|x + 4| = 3x - 8$

Solution   Theorem 3 does not apply directly, since $3x - 8$ is not known to be positive. However, we can use the definition of absolute value and two cases: $x + 4 \geq 0$ and $x + 4 < 0$.

Case 1. $x + 4 \geq 0$ (that is, $x \geq -4$)
For this case, the possible values of x are in the set $\{x \mid x \geq -4\}$.

$|x + 4| = 3x - 8$

$x + 4 = 3x - 8$ $|a| = a$ for $a \geq 0$

$-2x = -12$

$x = 6$ A solution, since 6 is among the possible values of x

The check is left to the reader.

Case 2. $x + 4 < 0$ (that is, $x < -4$)
In this case, the possible values of x are in the set $\{x \mid x < -4\}$.

$|x + 4| = 3x - 8$

$-(x + 4) = 3x - 8$ $|a| = -a$ for $a < 0$

$-x - 4 = 3x - 8$

$-4x = -4$

$x = 1$ Not a solution, since 1 is not among the possible values of x

Combining both cases, we see that the only solution is $x = 6$.

Problem 6 Solve: $|3x - 4| = x + 5$

● **Absolute Value and Radicals**

In Section 1-7, we found that if x is positive or 0, then

$$\sqrt{x^2} = x$$

If x is negative, however, we must write

$$\sqrt{x^2} = -x \quad \sqrt{(-2)^2} = -(-2) = 2$$

Thus, for x any real number,

$$\sqrt{x^2} = \begin{cases} x & \text{if } x \geq 0 \\ -x & \text{if } x < 0 \end{cases}$$

But this is exactly how we defined $|x|$ at the beginning of this section (see Definition 1). Thus, for x any real number,

$$\sqrt{x^2} = |x|$$

Answers to Matched Problems
1. (A) 8   (B) $\sqrt[3]{9} - 2$ (C) $\sqrt{2}$ (D) $\sqrt[3]{9} - 2$
2. (A) 4   (B) 4   (C) 6   (D) 11   (E) 8   (F) 15

3. (A) $x = -8, 4$ or $\{-8, 4\}$

(B) $-8 < x < 4$ or $(-8, 4)$

(C) $-8 < x < 4, x \neq -2$, or $(-8, -2) \cup (-2, 4)$

(D) $x < -8$ or $x > 4$, or $(-\infty, -8) \cup (4, \infty)$

4. (A) $x = -\frac{7}{2}, \frac{9}{2}$ or $\{-\frac{7}{2}, \frac{9}{2}\}$ (B) $-7 \leq x \leq 7$ or $[-7, 7]$ (C) $-4 \leq x \leq 2$ or $[-4, 2]$
(D) $-2 < x < 7$ or $(-2, 7)$
5. (A) $x \leq -5$ or $x \geq 5$, or $(-\infty, -5) \cup (5, \infty)$ (B) $x < -\frac{1}{2}$ or $x > 2$, or $(-\infty, -\frac{1}{2}) \cup (2, \infty)$
(C) $x < -2$ or $x > \frac{22}{5}$, or $(-\infty, -2) \cup (\frac{22}{5}, \infty)$
6. $x = -\frac{1}{4}, \frac{9}{2}$ or $\{-\frac{1}{4}, \frac{9}{2}\}$

EXERCISE 2-4

A ───────────────────────

In Problems 1–8, simplify, and write without absolute value signs. Do not replace radicals with decimal approximations.

1. $|\sqrt{5}|$

2. $|-\frac{3}{4}|$

3. $|(-6) - (-2)|$

4. $|(-2) - (-6)|$

5. $|5 - \sqrt{5}|$

6. $|\sqrt{7} - 2|$

7. $|\sqrt{5} - 5|$

8. $|2 - \sqrt{7}|$

In Problems 9–14, find the distance between points A and B with coordinates a and b respectively, as given.

9. $a = -7, \quad b = 5$
10. $a = 3, \quad b = 12$
11. $a = 5, \quad b = -7$
12. $a = 12, \quad b = 3$
13. $a = -16, \quad b = -25$
14. $a = -9, \quad b = -17$

In Problems 15–20, use the number line below to find the indicated distances.

15. $d(B, O)$
16. $d(A, B)$
17. $d(O, B)$
18. $d(B, A)$
19. $d(B, C)$
20. $d(D, C)$

In Problems 21–38, solve and graph. When applicable, write answers using both inequality notation and interval notation.

21. $|x| = 7$
22. $|x| = 5$
23. $|x| \le 7$
24. $|t| \le 5$
25. $|x| \ge 7$
26. $|x| \ge 5$
27. $|y - 5| = 3$
28. $|t - 3| = 4$
29. $|y - 5| < 3$
30. $|t - 3| < 4$
31. $|y - 5| > 3$
32. $|t - 3| > 4$
33. $|u + 8| = 3$
34. $|x + 1| = 5$
35. $|u + 8| \le 3$
36. $|x + 1| \le 5$
37. $|u + 8| \ge 3$
38. $|x + 1| \ge 5$

B _____

In Problems 39–60, solve each equation or inequality. When applicable, write answers using both inequality notation and interval notation.

39. $|3x + 4| = 8$
40. $|2x - 3| = 5$
41. $|5x - 3| \le 12$
42. $|2x - 3| \le 5$
43. $|2y - 8| > 2$
44. $|3u + 4| > 3$
45. $|5t - 7| = 11$
46. $|6m + 9| = 13$
47. $|9 - 7u| < 14$
48. $|7 - 9M| < 15$
49. $|1 - \frac{2}{3}x| \ge 5$
50. $|\frac{3}{4}x + 3| \ge 9$
51. $|\frac{9}{5}C + 32| < 31$
52. $|\frac{5}{9}(F - 32)| < 40$
53. $\sqrt{x^2} < 2$
54. $\sqrt{u^2} \le 5$
55. $\sqrt{t^2} \ge 4$
56. $\sqrt{m^2} > 3$
57. $\sqrt{(1 - 3t)^2} \le 2$
58. $\sqrt{(3 - 2x)^2} < 5$
59. $\sqrt{(2t - 3)^2} > 3$
60. $\sqrt{(3m + 5)^2} \ge 4$

Interpret Problems 61–68 geometrically.

61. $|m - 2| = 4$
62. $|n - 1| = 2$
63. $|x + 1| = 7$
64. $|y + 3| = 2$
65. $|u - 4| < 5$
66. $|x - 4| \ge 3$
67. $|z - 2| \le 3$
68. $|x - 1| > 4$

Write each of the statements in Problems 69–78 as an absolute value equation or inequality.

69. x is 4 units from 3.
70. y is 3 units from 1.
71. m is 5 units from -2.
72. n is 7 units from -5.
73. x is less than 5 units from 3.
74. z is less than 8 units from -2.
75. p is more than 6 units from -2.
76. c is no greater than 7 units from -3.
77. q is no less than 2 units from 1.
78. d is no more than 4 units from 5.

C _____

∫ *Problems 79–82 are calculus-related. Solve and graph. Write each solution using interval notation.*

79. $0 < |x - 3| < 0.1$ **80.** $0 < |x - 5| < 0.01$ **81.** $0 < |x - c| < d$ **82.** $0 < |x - 4| < d$

In Problems 83–92, for what values of x does each hold?

83. $|x - 5| = x - 5$ **84.** $|x + 7| = x + 7$

85. $|x + 8| = -(x + 8)$ **86.** $|x - 11| = -(x - 11)$

87. $|4x + 3| = 4x + 3$ **88.** $|5x - 9| = (5x - 9)$

89. $|5x - 2| = -(5x - 2)$ **90.** $|3x + 7| = -(3x + 7)$

91. $|3 - x| + 3 = |2 - 3x|$ **92.** $|x - 1| + |x + 3| = 6$

93. Prove that $|b - a| = |a - b|$ for all real numbers a and b.

94. Prove that $|x|^2 = x^2$ for all real numbers x.

95. Prove that the average of two numbers is between the two numbers; that is, if $m < n$, then

$$m < \frac{m + n}{2} < n$$

96. Prove that for $m < n$,

$$d\left(m, \frac{m + n}{2}\right) = d\left(\frac{m + n}{2}, n\right)$$

97. Prove that $|-m| = |m|$.

98. Prove that $|m| = |n|$ if and only if $m = n$ or $m = -n$.

99. Prove that for $n \neq 0$,

$$\left|\frac{m}{n}\right| = \frac{|m|}{|n|}$$

100. Prove that $|mn| = |m||n|$.

101. Prove that $-|m| \leq m \leq |m|$.

102. Prove the **triangle inequality**:

$$|m + n| \leq |m| + |n|$$

Hint: Use Problem 101 to show that

$$-|m| - |n| \leq m + n \leq |m| + |n|$$

103. If a and b are real numbers, prove that the maximum of a and b is given by

$$\max(a, b) = \tfrac{1}{2}[a + b + |a - b|]$$

104. Prove that the minimum of a and b is given by

$$\min(a, b) = \tfrac{1}{2}[a + b - |a - b|]$$

APPLICATIONS

105. Statistics. Inequalities of the form

$$\left|\frac{x - m}{s}\right| < n$$

occur frequently in statistics. If $m = 45.4$, $s = 3.2$, and $n = 1$, solve for x.

106. Statistics. Repeat Problem 105 for $m = 28.6$, $s = 6.5$, and $n = 2$.

★ **107. Business.** The daily production P in an automobile assembly plant is within 20 units of 500 units. Express the daily production as an absolute value inequality.

★ **108. Chemistry.** In a chemical process, the temperature T is to be kept within 10°C of 200°C. Express this restriction as an absolute value inequality.

★**109.** **Significant Digits.** If $N = 2.37$ represents a measurement, then we assume an accuracy of 2.37 ± 0.005. Express the accuracy assumption using an absolute value inequality.

★**110.** **Significant Digits.** If $N = 3.65 \times 10^{-3}$ is a number from a measurement, then we assume an accuracy of $3.65 \times 10^{-3} \pm 5 \times 10^{-6}$. Express the accuracy assumption using an absolute value inequality.

SECTION 2-5 **Complex Numbers**

- Introductory Remarks
- The Complex Number System
- Complex Numbers and Radicals

• Introductory Remarks

The Pythagoreans (500–275 B.C.) found that the simple equation

$$x^2 = 2 \tag{1}$$

had no rational number solutions. If equation (1) were to have a solution, then a new kind of number had to be invented—an irrational number. The irrational numbers $\sqrt{2}$ and $-\sqrt{2}$ are both solutions to equation (1). Irrational numbers were not put on a firm mathematical foundation until the last century. The rational and irrational numbers together constitute the real number system.

Is there any need to consider another number system? Yes, if we want the simple equation

$$x^2 = -1$$

to have a solution. If x is any real number, then $x^2 \geq 0$. Thus, $x^2 = -1$ cannot have any real number solutions. Once again a new type of number must be invented, a number whose square can be negative. These new numbers are among the numbers called *complex numbers*. The complex numbers evolved over a long period of time, but, like the real numbers, it was not until the last century that they were placed on a firm mathematical foundation.

TABLE 1 **Brief History of Complex Numbers**

Approximate date	Person	Event
50	Heron of Alexandria	First recorded encounter of a square root of a negative number
850	Mahavira of India	Said that a negative has no square root, since it is not a square
1545	Cardano of Italy	Solutions to cubic equations involved square roots of negative numbers.
1637	Descartes of France	Introduced the terms *real* and *imaginary*
1748	Euler of Switzerland	Used i for $\sqrt{-1}$
1832	Gauss of Germany	Introduced the term *complex number*

• The Complex Number System   We start the development of the complex number system by defining a complex number and several special types of complex numbers. We then define equality, addition, and multiplication in this system, and from these definitions the important special properties and operational rules for addition, subtraction, multiplication, and division will follow.

DEFINITION 1

Complex Number

A **complex number** is a number of the form

$$a + bi \qquad \textbf{Standard Form}$$

where a and b are real numbers and i is called the **imaginary unit**.

The imaginary unit i introduced in Definition 1 is not a real number. It is a special symbol used in the representation of the elements in this new complex number system.

Some examples of complex numbers are

$$3 - 2i \qquad \tfrac{1}{2} + 5i \qquad 2 - \tfrac{1}{3}i$$
$$0 + 3i \qquad 5 + 0i \qquad 0 + 0i$$

Particular kinds of complex numbers are given special names as follows:

DEFINITION 2

Names for Particular Kinds of Complex Numbers

Imaginary Unit:	i
Complex Number:	$a + bi$ $\quad a$ and b real numbers
Imaginary Number:	$a + bi$ $\quad b \neq 0$
Pure Imaginary Number:	$0 + bi = bi$ $\quad b \neq 0$
Real Number:	$a + 0i = a$
Zero:	$0 + 0i = 0$
Conjugate of $a + bi$:	$a - bi$

EXAMPLE 1   Special Types of Complex Numbers

Given the list of complex numbers:

$$3 - 2i \qquad\qquad \tfrac{1}{2} + 5i \qquad\qquad 2 - \tfrac{1}{3}i$$
$$0 + 3i = 3i \qquad 5 + 0i = 5 \qquad 0 + 0i = 0$$

(A) List all the imaginary numbers, pure imaginary numbers, real numbers, and zero.

(B) Write the conjugate of each.

Solutions

(A) Imaginary numbers: $3 - 2i, \frac{1}{2} + 5i, 2 - \frac{1}{3}i, 3i$
Pure imaginary numbers: $0 + 3i = 3i$
Real numbers: $5 + 0i = 5, 0 + 0i = 0$
Zero: $0 + 0i = 0$

(B) $3 + 2i$ $\qquad \frac{1}{2} - 5i$ $\qquad 2 + \frac{1}{3}i$
$0 - 3i = -3i$ $\quad 5 - 0i = 5$ $\quad 0 - 0i = 0$

Problem 1

Given the list of complex numbers:

$$6 + 7i \qquad \sqrt{2} - \tfrac{1}{3}i \qquad 0 - i = -i$$

$$0 + \tfrac{2}{3}i = \tfrac{2}{3}i \qquad -\sqrt{3} + 0i = -\sqrt{3} \qquad 0 - 0i = 0$$

(A) List all the imaginary numbers, pure imaginary numbers, real numbers, and zero.

(B) Write the conjugate of each.

In Definition 2, notice that we identify a complex number of the form $a + 0i$ with the real number a, a complex number of the form $0 + bi, b \neq 0$, with the **pure imaginary number** bi, and the complex number $0 + 0i$ with the real number 0. Thus, a real number is also a complex number, just as a rational number is also a real number. Any complex number that is not a real number is called an **imaginary number**. If we combine the set of all real numbers with the set of all imaginary numbers, we obtain C, **the set of complex numbers**. The relationship of the complex number system to the other number systems we have studied is shown in Figure 1.

FIGURE 1

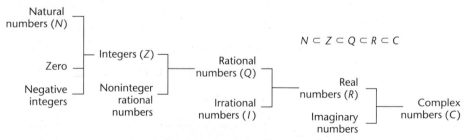

$$N \subset Z \subset Q \subset R \subset C$$

To use complex numbers, we must know how to add, subtract, multiply, and divide them. We start by defining equality, addition, and multiplication.

DEFINITION 3

Equality and Basic Operations

1. **Equality:** $\quad a + bi = c + di$ if and only if $a = c$ and $b = d$
2. **Addition** $\quad (a + bi) + (c + di) = (a + c) + (b + d)i$
3. **Multiplication:** $\quad (a + bi)(c + di) = (ac - bd) + (ad + bc)i$

In Section 1-1 we listed the basic properties of the real number system. Using Definition 3, it can be shown that the complex number system possesses the same properties. That is:

1. Addition and multiplication of complex numbers are commutative and associative operations.

2. There is an additive identity and a multiplicative identity for complex numbers.

3. Every complex number has an additive inverse or negative.

4. Every nonzero complex number has a multiplicative inverse or reciprocal.

5. Multiplication distributes over addition.

As a consequence of these properties, we can manipulate complex number symbols of the form $a + bi$ just like we manipulate binomials of the form $a + bx$, as long as we remember that i is a special symbol for the imaginary unit, not for a real number. Thus, you will not have to memorize the definitions of addition and multiplication of complex numbers. We now discuss these operations and some of their properties. Others will be considered in Exercise 2-5.

EXAMPLE 2   **Addition of Complex Numbers**

Carry out each operation and express the answer in standard form:

(A) $(2 - 3i) + (6 + 2i)$ (B) $(-5 + 4i) + (0 + 0i)$

Solution   (A) We could apply the definition of addition directly, but it is easier to use complex number properties.

$$(2 - 3i) + (6 + 2i) = 2 - 3i + 6 + 2i \qquad \text{Remove parentheses.}$$
$$= (2 + 6) + (-3 + 2)i \qquad \text{Combine like terms.}$$
$$= 8 - i$$

(B) $(-5 + 4i) + (0 + 0i) = -5 + 4i + 0 + 0i$
$$= -5 + 4i$$

Problem 2   Carry out each operation and express the answer in standard form:

(A) $(3 + 2i) + (6 - 4i)$ (B) $(0 + 0i) + (7 - 5i)$

Example 2B and Problem 2B illustrate the following general result: For any complex number $a + bi$,

$$(a + bi) + (0 + 0i) = (0 + 0i) + (a + bi) = a + bi$$

Thus, $0 + 0i$ is the **additive identity** or **zero** for the complex numbers. We anticipated this result in Definition 1 when we identified the complex number $0 + 0i$ with the real number 0.

We could define negatives and subtraction in terms of the additive inverse of a complex number, as we did for real numbers, but once again it is much easier to use the properties of complex numbers.

EXAMPLE 3 Negation and Subtraction

Carry out each operation and express the answer in standard form:

(A) $-(4 - 5i)$ (B) $(7 - 3i) - (6 + 2i)$
(C) $(-2 + 7i) + (2 - 7i)$

Solution (A) $-(4 - 5i) \boxed{= (-1)(4 - 5i)} = -4 + 5i$

(B) $(7 - 3i) - (6 + 2i) = 7 - 3i - 6 - 2i$
$$= 1 - 5i$$

(C) $(-2 + 7i) + (2 - 7i) = -2 + 7i + 2 - 7i = 0$

Problem 3 Carry out each operation and express the answer in standard form:

(A) $-(-3 + 2i)$ (B) $(3 - 5i) - (1 - 3i)$
(C) $(-4 + 9i) + (4 - 9i)$

In general, the **additive inverse** or **negative** of $a + bi$ is $-a - bi$ since

$$(a + bi) + (-a - bi) = (-a - bi) + (a + bi) = 0$$

(see Example 3C and Problem 3C).

Now we turn our attention to multiplication. First, we use the definition of multiplication to see what happens to the complex unit i when it is squared:

$$i^2 = \overset{a}{(0} + \overset{b}{1i})\overset{c}{(0} + \overset{d}{1i})$$

$$= (\overset{a}{0} \cdot \overset{c}{0} - \overset{b}{1} \cdot \overset{d}{1}) + (\overset{a}{0} \cdot \overset{d}{1} + \overset{b}{1} \cdot \overset{c}{0})i$$

$$= -1 + 0i$$

$$= -1$$

Thus, we have proved that

$$i^2 = -1$$

We now have a number whose square is negative and a solution to $x^2 = -1$. Since $i^2 = -1$, we define $\sqrt{-1}$ to be the imaginary unit i. Thus,

$$i = \sqrt{-1} \qquad \text{and} \qquad -i = -\sqrt{-1}$$

Just as was the case with addition and subtraction, multiplication of complex numbers can be carried out by using the properties of complex numbers rather than the definition of multiplication. We just replace i^2 with -1 each time it occurs.

EXAMPLE 4   Multiplying Complex Numbers

Carry out each operation and express the answer in standard form:

(A) $(2 - 3i)(6 + 2i)$ (B) $1(3 - 5i)$
(C) $i(1 + i)$ (D) $(3 + 4i)(3 - 4i)$

Solution   (A) $(2 - 3i)(6 + 2i)\ \boxed{= 2(6 + 2i) - 3i(6 + 2i)}$

$$= 12 + 4i - 18i - 6i^2$$
$$= 12 - 14i - 6(-1) \quad \text{Replace } i^2 \text{ with } -1.$$
$$= 18 - 14i$$

(B) $1(3 - 5i)\ \boxed{= 1 \cdot 3 - 1 \cdot 5i}\ = 3 - 5i$

(C) $i(1 + i) = i + i^2 = i - 1 = -1 + i$

(D) $(3 + 4i)(3 - 4i) = 9 - 12i + 12i - 16i^2$
$$= 9 + 16 = 25$$

Problem 4   Carry out each operation and express the answer in standard form:

(A) $(5 + 2i)(4 - 3i)$ (B) $3(-2 + 6i)$
(C) $i(2 - 3i)$ (D) $(2 + 3i)(2 - 3i)$

For any complex number $a + bi$,

$$1(a + bi) = (a + bi)1 = a + bi$$

(see Example 4B). Thus, 1 is the **multiplicative identity** for complex numbers, just as it is for real numbers.

Earlier we stated that every nonzero complex number has a multiplicative inverse or reciprocal. We will denote this as a fraction, just as we do with real numbers. Thus,

$$\frac{1}{a + bi} \qquad \text{is the \textbf{reciprocal} of} \qquad a + bi \qquad a + bi \neq 0$$

The following important property of the conjugate of a complex number is used to express reciprocals and quotients in standard form.

Theorem 1	**Product of a Complex Number and Its Conjugate**
	$$(a + bi)(a - bi) = a^2 + b^2 \quad \text{A real number}$$

EXAMPLE 5   Reciprocals and Quotients

Carry out each operation and express the answer in standard form:

(A) $\dfrac{1}{2 + 3i}$ (B) $\dfrac{7 - 3i}{1 + i}$

Solution   (A) Multiply numerator and denominator by the conjugate of the denominator:

$$\frac{1}{2 + 3i} = \frac{1}{2 + 3i} \cdot \frac{2 - 3i}{2 - 3i} \left| \begin{array}{c} = \dfrac{2 - 3i}{4 - 9i^2} = \dfrac{2 - 3i}{4 + 9} \end{array} \right.$$

$$= \frac{2 - 3i}{13} = \frac{2}{13} - \frac{3}{13} i$$

This answer can be checked by multiplication:

Check
$$(2 + 3i)\left(\frac{2}{13} - \frac{3}{13} i\right) = \frac{4}{13} - \frac{6}{13} i + \frac{6}{13} i - \frac{9}{13} i^2$$

$$= \frac{4}{13} + \frac{9}{13} = 1$$

(B) $\dfrac{7 - 3i}{1 + i} = \dfrac{7 - 3i}{1 + i} \cdot \dfrac{1 - i}{1 - i} \left| \begin{array}{c} = \dfrac{7 - 7i - 3i + 3i^2}{1 - i^2} \end{array} \right.$

$$= \frac{4 - 10i}{2} = 2 - 5i$$

Check
$$(1 + i)(2 - 5i) = 2 - 5i + 2i - 5i^2 = 7 - 3i$$

Problem 5   Carry out each operation and express the answer in standard form:

(A) $\dfrac{1}{4 + 2i}$ (B) $\dfrac{6 + 7i}{2 - i}$

EXAMPLE 6 **Combined Operations**

Carry out the indicated operations and write each answer in standard form:

(A) $(3 - 2i)^2 - 6(3 - 2i) + 13$ (B) $\dfrac{2 - 3i}{2i}$

Solutions   (A) $(3 - 2i)^2 - 6(3 - 2i) + 13 = 9 - 12i + 4i^2 - 18 + 12i + 13$
$$= 9 - 12i - 4 - 18 + 12i + 13$$
$$= 0$$

(B) If a complex number is divided by a pure imaginary number, we can make the denominator real by multiplying numerator and denominator by i.

$$\frac{2 - 3i}{2i} \cdot \frac{i}{i} = \frac{2i - 3i^2}{2i^2} = \frac{2i + 3}{-2} = -\frac{3}{2} - i$$

Problem 6   Carry out the indicated operations and write each answer in standard form:

(A) $(3 + 2i)^2 - 6(3 + 2i) + 13$ (B) $\dfrac{4 - i}{3i}$

Natural number powers of i take on particularly simple forms. Try generalizing from the pattern in the following first eight powers of i:

i $\qquad\qquad\qquad\qquad\qquad i^5 = i^4 \cdot i = (1)i = i$

$i^2 = -1$ $\qquad\qquad\qquad\quad i^6 = i^4 \cdot i^2 = 1(-1) = -1$

$i^3 = i^2 \cdot i = (-1)i = -i$ $\qquad i^7 = i^4 \cdot i^3 = 1(-i) = -i$

$i^4 = i^2 \cdot i^2 = (-1)(-1) = 1$ $\qquad i^8 = i^4 \cdot i^4 = 1 \cdot 1 = 1$

In general, it can be shown that for n a natural number, i^n must be -1, 1, $-i$, or i. To compute i^n for n larger than 4, we take advantage of the equality $i^4 = 1$. Consider the following example:

Largest integer in 27 exactly divisible by 4

$$i^{27} = i^{24} \cdot i^3 \boxed{= (i^4)^6 \cdot i^2 \cdot i = 1^6(-1)i} = -i$$

● **Complex Numbers and Radicals**   Recall that we say that a is a square root of b if $a^2 = b$. If x is a positive real number, then x has two square roots, the principal square root, denoted by $\sqrt{x}$, and its negative, $-\sqrt{x}$ (Section 1-7). If x is a negative real number, then x still has two square roots, but now these square roots are imaginary numbers.

DEFINITION 4 | **Principal Square Root of a Negative Real Number**

The **principal square root of a negative real number**, denoted by $\sqrt{-a}$ where a is positive, is defined by

$$\sqrt{-a} = i\sqrt{a} \qquad \sqrt{-3} = i\sqrt{3} \qquad \sqrt{-9} = i\sqrt{9} = 3i$$

The other square root of $-a$, $a > 0$, is $-\sqrt{-a} = -i\sqrt{a}$.

Note in Definition 4 that we wrote $i\sqrt{a}$ and $i\sqrt{3}$ in place of the standard forms $\sqrt{a}\,i$ and $\sqrt{3}\,i$. We follow this convention whenever it appears that i might accidentally slip under a radical sign ($\sqrt{a i} \neq \sqrt{a}\,i$, but $\sqrt{a}\,i = i\sqrt{a}$). Definition 4 is motivated by the fact that

$$(i\sqrt{a})^2 = i^2 a = -a$$

EXAMPLE 7 Complex Numbers and Radicals

Write in standard form:

(A) $\sqrt{-4}$ (B) $4 + \sqrt{-5}$ (C) $\dfrac{-3 - \sqrt{-5}}{2}$ (D) $\dfrac{1}{1 - \sqrt{-9}}$

Solutions   (A) $\sqrt{-4} = i\sqrt{4} = 2i$ (B) $4 + \sqrt{-5} = 4 + i\sqrt{5}$

(C) $\dfrac{-3 - \sqrt{-5}}{2} = \dfrac{-3 - i\sqrt{5}}{2} = -\dfrac{3}{2} - \dfrac{\sqrt{5}}{2}i$

(D) $\dfrac{1}{1 - \sqrt{-9}} = \dfrac{1}{1 - 3i} = \dfrac{1 \cdot (1 + 3i)}{(1 - 3i) \cdot (1 + 3i)}$

$$= \dfrac{1 + 3i}{1 - 9i^2} = \dfrac{1 + 3i}{10} = \dfrac{1}{10} + \dfrac{3}{10}i$$

Problem 7 Write in standard form:

(A) $\sqrt{-16}$ (B) $5 + \sqrt{-7}$ (C) $\dfrac{-5 - \sqrt{-2}}{2}$ (D) $\dfrac{1}{3 - \sqrt{-4}}$

CAUTION   Always write numbers such as $1 - \sqrt{-9}$ in the form $a + bi$ before performing any operations. In Example 7D, note what happens if we do not do this and try to proceed using properties of radicals for real numbers.

$$\frac{1}{1 - \sqrt{-9}} = \frac{1}{1 - \sqrt{-9}} \cdot \frac{1 + \sqrt{-9}}{1 + \sqrt{-9}}$$

$$= \frac{1 + \sqrt{-9}}{1 - \sqrt{-9}\sqrt{-9}} \neq \frac{1 + \sqrt{-9}}{1 - \sqrt{(-9)(-9)}}$$

The reason for $\neq$ in the last step is that

$$\sqrt{-9}\sqrt{-9} \neq \sqrt{(-9)(-9)}$$

They are not equal because the left side equals $3i \cdot 3i = -9$ and the right side equals $\sqrt{81} = 9$. From Theorem 1 in Section 1-7 we know that if a and b are positive real numbers, then

$$\sqrt{a}\sqrt{b} = \sqrt{ab} \quad \text{and} \quad \frac{\sqrt{a}}{\sqrt{b}} = \sqrt{\frac{a}{b}}$$

If a and b are both negative, then neither property holds.

Early resistance to these new numbers is suggested by the words used to name them: *complex* and *imaginary*. In spite of this early resistance, complex numbers have come into widespread use in both pure and applied mathematics. They are used extensively, for example, in electrical engineering, physics, chemistry, statistics, and aeronautical engineering. Our first use of them will be in connection with solutions of second-degree equations in the next section.

Answers to Matched Problems
1. (A) Imaginary numbers: $6 + 7i, \sqrt{2} - \frac{1}{3}i, 0 - i = -i, 0 + \frac{2}{3}i = \frac{2}{3}i$
      Pure imaginary numbers: $0 - i = -i, 0 + \frac{2}{3}i = \frac{2}{3}i$
      Real numbers: $-\sqrt{3} + 0i = -\sqrt{3}, 0 - 0i = 0$
      Zero: $0 - 0i = 0$
   (B) $6 - 7i, \sqrt{2} + \frac{1}{3}i, 0 + i = i, 0 - \frac{2}{3}i = -\frac{2}{3}i, -\sqrt{3} - 0i = -\sqrt{3}, 0 + 0i = 0$
2. (A) $9 - 2i$ (B) $7 - 5i$
3. (A) $3 - 2i$ (B) $2 - 2i$ (C) 0
4. (A) $26 - 7i$ (B) $-6 + 18i$ (C) $3 + 2i$ (D) 13
5. (A) $\frac{1}{5} - \frac{1}{10}i$ (B) $1 + 4i$
6. (A) 0 (B) $-\frac{1}{3} - \frac{4}{3}i$
7. (A) $4i$ (B) $5 + i\sqrt{7}$ (C) $-\frac{5}{2} - (\sqrt{2}/2)i$ (D) $\frac{3}{13} + \frac{2}{13}i$

EXERCISE 2-5

A ⎯⎯⎯⎯⎯⎯⎯⎯⎯⎯⎯⎯⎯⎯⎯⎯⎯⎯⎯⎯⎯⎯⎯⎯⎯⎯⎯⎯⎯⎯⎯⎯⎯

In Problems 1–26, perform the indicated operations and write each answer in standard form.

1. $(2 + 4i) + (5 + i)$ 2. $(3 + i) + (4 + 2i)$ 3. $(-2 + 6i) + (7 - 3i)$ 4. $(6 - 2i) + (8 - 3i)$

5. $(6 + 7i) - (4 + 3i)$ **6.** $(9 + 8i) - (5 + 6i)$ **7.** $(3 + 5i) - (-2 - 4i)$ **8.** $(8 - 4i) - (11 - 2i)$

9. $(4 - 5i) + 2i$ **10.** $6 + (3 - 4i)$ **11.** $(4i)(6i)$ **12.** $(3i)(8i)$

13. $-3i(2 - 4i)$ **14.** $-2i(5 - 3i)$ **15.** $(3 + 3i)(2 - 3i)$ **16.** $(-2 - 3i)(3 - 5i)$

17. $(2 - 3i)(7 - 6i)$ **18.** $(3 + 2i)(2 - i)$ **19.** $(7 + 4i)(7 - 4i)$ **20.** $(5 + 3i)(5 - 3i)$

21. $\dfrac{1}{2 + i}$ **22.** $\dfrac{1}{3 - i}$ **23.** $\dfrac{3 + i}{2 - 3i}$ **24.** $\dfrac{2 - i}{3 + 2i}$

25. $\dfrac{13 + i}{2 - i}$ **26.** $\dfrac{15 - 3i}{2 - 3i}$

B _____

In Problems 27–36, convert imaginary numbers to standard form, perform the indicated operations, and express answers in standard form.

27. $(2 - \sqrt{-4}) + (5 - \sqrt{-9})$ **28.** $(3 - \sqrt{-4}) + (-8 + \sqrt{-25})$

29. $(9 - \sqrt{-9}) - (12 - \sqrt{-25})$ **30.** $(-2 - \sqrt{-36}) - (4 + \sqrt{-49})$

31. $(3 - \sqrt{-4})(-2 + \sqrt{-49})$ **32.** $(2 - \sqrt{-1})(5 + \sqrt{-9})$

33. $\dfrac{5 - \sqrt{-4}}{7}$ **34.** $\dfrac{6 - \sqrt{-64}}{2}$

35. $\dfrac{1}{2 - \sqrt{-9}}$ **36.** $\dfrac{1}{3 - \sqrt{-16}}$

Write Problems 37–42 in standard form.

37. $\dfrac{2}{5i}$ **38.** $\dfrac{1}{3i}$ **39.** $\dfrac{1 + 3i}{2i}$ **40.** $\dfrac{2 - i}{3i}$

41. $(2 - 3i)^2 - 2(2 - 3i) + 9$ **42.** $(2 - i)^2 + 3(2 - i) - 5$

43. Evaluate $x^2 - 2x + 2$ for $x = 1 - i$.   **44.** Evaluate $x^2 - 2x + 2$ for $x = 1 + i$.

45. Simplify: i^{18}, i^{32}, and i^{67} **46.** Simplify: i^{21}, i^{43}, and i^{52}

47. For what real values of x and y will the following equation be a true statement?

$$(2x - 1) + (3y + 2)i = 5 - 4i$$

48. For what real values of x and y will the following equation be a true statement?

$$3x + (y - 2)i = (5 - 2x) + (3y - 8)i$$

In Problems 49–52, for what real values of x does each expression represent an imaginary number?

49. $\sqrt{3 - x}$ **50.** $\sqrt{5 + x}$ **51.** $\sqrt{2 - 3x}$ **52.** $\sqrt{3 + 2x}$

Use a calculator to compute Problems 53–56. Write in standard form a + bi, where a and b are computed to two decimal places.

53. $(3.17 - 4.08i)(7.14 + 2.76i)$ **54.** $(6.12 + 4.92i)(1.82 - 5.05i)$

55. $\dfrac{8.14 + 2.63i}{3.04 + 6.27i}$ **56.** $\dfrac{7.66 + 3.33i}{4.72 - 2.68i}$

C
_____

In Problems 57–62, perform the indicated operations, and write each answer in standard form.

57. $(a + bi) + (c + di)$

58. $(a + bi) - (c + di)$

59. $(a + bi)(a - bi)$

60. $(u - vi)(u + vi)$

61. $(a + bi)(c + di)$

62. $\dfrac{a + bi}{c + di}$

63. Show that $i^{4k} = 1$, k a natural number

64. Show that $i^{4k+1} = i$, k a natural number

Supply the reasons in the proofs for the theorems stated in Problems 65 and 66.

65. *Theorem:* The complex numbers are commutative under addition.

    Proof: Let $a + bi$ and $c + di$ be two arbitrary complex numbers; then:

Statement
1. $(a + bi) + (c + di) = (a + c) + (b + d)i$
2. $\qquad\qquad\qquad\quad = (c + a) + (d + b)i$
3. $\qquad\qquad\qquad\quad = (c + di) + (a + bi)$

Reason
1.
2.
3.

66. *Theorem:* The complex numbers are commutative under multiplication.

    Proof: Let $a + bi$ and $c + di$ be two arbitrary complex numbers; then:

Statement
1. $(a + bi) \cdot (c + di) = (ac - bd) + (ad + bc)i$
2. $\qquad\qquad\qquad\quad = (ca - db) + (da + cb)i$
3. $\qquad\qquad\qquad\quad = (c + di)(a + bi)$

Reason
1.
2.
3.

Letters z and w are often used as complex variables, where $z = x + yi$, $w = u + vi$, and x, y, u, v are real numbers. The conjugates of z and w, denoted by $\bar{z}$ and $\bar{w}$, respectively, are given by $\bar{z} = x - yi$ and $\bar{w} = u - vi$. Prove the properties of conjugates stated in Problems 67–74.

67. $z\bar{z}$ is a real number.

68. $z + \bar{z}$ is a real number.

69. $\bar{z} = z$ if and only if z is real.

70. $\bar{\bar{z}} = z$

71. $\overline{z + w} = \bar{z} + \bar{w}$

72. $\overline{z - w} = \bar{z} - \bar{w}$

73. $\overline{zw} = \bar{z} \cdot \bar{w}$

74. $\overline{z/w} = \bar{z}/\bar{w}$

SECTION 2-6   Quadratic Equations

- Solution by Factoring
- Solution by Square Root
- Solution by Completing the Square
- Solution by Quadratic Formula
- Applications

The next class of equations we consider are the second-degree polynomial equations in one variable, called *quadratic equations*.

DEFINITION 1   **Quadratic Equation**

A **quadratic equation** in one variable is any equation that can be written in the form

$$ax^2 + bx + c = 0 \qquad a \neq 0 \qquad \textbf{Standard Form}$$

where x is a variable and a, b, and c are constants.

Now that we have discussed the complex number system, we will use complex numbers when solving equations. Recall that a solution of an equation is also called a *root* of the equation. A real number solution of an equation is called a **real root**, and an imaginary number solution is called an **imaginary root**. In this section we develop methods for finding all real and imaginary roots of a quadratic equation.

● **Solution by Factoring**

If $ax^2 + bx + c$ can be written as the product of two first-degree factors, then the quadratic equation can be quickly and easily solved. The method of solution by factoring rests on the zero property of complex numbers, which is a generalization of the zero property of real numbers introduced in Section 1-1.

Zero Property

If m and n are complex numbers, then

$$m \cdot n = 0 \qquad \text{if and only if} \qquad m = 0 \text{ or } n = 0 \text{ (or both)}$$

EXAMPLE 1   **Solving Quadratic Equations by Factoring**

Solve by factoring:

(A) $6x^2 - 19x - 7 = 0$ (B) $x^2 - 6x + 5 = -4$ (C) $2x^2 = 3x$

Solutions   (A) $6x^2 - 19x - 7 = 0$

$(2x - 7)(3x + 1) = 0$ Factor left side.

$2x - 7 = 0 \qquad \text{or} \qquad 3x + 1 = 0$

$x = \frac{7}{2} \qquad\qquad\qquad x = -\frac{1}{3}$

The solution set is $\{-\frac{1}{3}, \frac{7}{2}\}$.

(B) $x^2 - 6x + 5 = -4$

$x^2 - 6x + 9 = 0$ Write in standard form.

$(x - 3)^2 = 0$ Factor left side.

$x = 3$

The solution set is {3}. The equation has one root, 3. But since it came from two factors, we call 3 a **double root**.

(C)
$$2x^2 = 3x$$
$$2x^2 - 3x = 0$$
$$x(2x - 3) = 0$$
$$x = 0 \quad \text{or} \quad 2x - 3 = 0$$
$$x = \tfrac{3}{2}$$

Solution set: $\{0, \tfrac{3}{2}\}$

Problem 1 Solve by factoring:

(A) $3x^2 + 7x - 20 = 0$ (B) $4x^2 + 12x + 9 = 0$ (C) $4x^2 = 5x$

CAUTION

1. One side of an equation must be 0 before the zero property can be applied. Thus

$$x^2 - 6x + 5 = -4$$
$$(x - 1)(x - 5) = -4$$

does not imply that $x - 1 = -4$ or $x - 5 = -4$. See Example 1B for the correct solution of this equation.

2. The equations

$$2x^2 = 3x \quad \text{and} \quad 2x = 3$$

are not equivalent. The first has solution set $\{0, \tfrac{3}{2}\}$, while the second has solution set $\{\tfrac{3}{2}\}$. The root $x = 0$ is lost when each member of the first equation is divided by the variable x. See Example 1C for the correct solution of this equation.

Do not divide both members of an equation by an expression containing the variable for which you are solving. You may be dividing by 0.

• **Solution by Square Root**

We now turn our attention to quadratic equations that do not have the first-degree term—that is, equations of the special form

$$ax^2 + c = 0 \quad a \neq 0$$

The method of solution of this special form makes direct use of the square root property:

> **Square Root Property**
>
> If $A^2 = C$, then $A = \pm\sqrt{C}$.

The use of this property is illustrated in the next example.

Note: It is common practice to represent solutions of quadratic equations informally by the last equation rather than by writing a solution set using set notation. From now on, we will follow this practice unless a particular emphasis is desired.

EXAMPLE 2   **Using the Square Root Property**

Solve using the square root property:

(A) $2x^2 - 3 = 0$ (B) $3x^2 + 27 = 0$ (C) $(x + \tfrac{1}{2})^2 = \tfrac{5}{4}$

Solutions

(A) $2x^2 - 3 = 0$

$x^2 = \tfrac{3}{2}$

$x = \pm\sqrt{\tfrac{3}{2}}$ or $\pm\dfrac{\sqrt{6}}{2}$ Solution set: $\left\{\dfrac{-\sqrt{6}}{2}, \dfrac{\sqrt{6}}{2}\right\}$

(B) $3x^2 + 27 = 0$

$x^2 = -9$

$x = \pm\sqrt{-9}$ or $\pm 3i$ Solution set: $\{-3i, 3i\}$

(C) $(x + \tfrac{1}{2})^2 = \tfrac{5}{4}$

$x + \tfrac{1}{2} = \pm\sqrt{\tfrac{5}{4}}$

$x = -\dfrac{1}{2} \pm \dfrac{\sqrt{5}}{2}$

$\quad = \dfrac{-1 \pm \sqrt{5}}{2}$

Problem 2   Solve using the square root property:

(A) $3x^2 - 5 = 0$ (B) $2x^2 + 8 = 0$ (C) $(x + \tfrac{1}{3})^2 = \tfrac{2}{9}$

● **Solution by Completing the Square**

The methods of square root and factoring are generally fast when they apply; however, there are equations, such as $x^2 + 6x - 2 = 0$ (see Example 4A), that cannot be solved directly by these methods. A more general procedure must be developed to take care of this type of equation—for example, the method of completing the square. This method is based on the process of transforming the standard quadratic equation

$$ax^2 + bx + c = 0$$

into the form

$$(x + A)^2 = B$$

where A and B are constants. The last equation can easily be solved by using the square root property. But how do we transform the first equation into the second? The following brief discussion provides the key to the process.

What number must be added to $x^2 + bx$ so that the result is the square of a first-degree polynomial? There is a simple mechanical rule for finding this number, based on the square of the following binomials:

$$(x + m)^2 = x^2 + 2mx + m^2$$
$$(x - m)^2 = x^2 - 2mx + m^2$$

In either case, we see that the third term on the right is the square of one-half of the coefficient of x in the second term on the right. This observation leads directly to the rule for completing the square.

Completing the Square

To complete the square of a quadratic of the form $x^2 + bx$, add the square of one-half the coefficient of x; that is, add $(b/2)^2$. Thus,

$x^2 + bx$

$$x^2 + bx + \left(\frac{b}{2}\right)^2 = \left(x + \frac{b}{2}\right)^2$$

$x^2 + 5x$

$$x^2 + 5x + \left(\frac{5}{2}\right)^2 = \left(x + \frac{5}{2}\right)^2$$

EXAMPLE 3   Completing the Square

Complete the square for each of the following:

(A) $x^2 - 3x$ (B) $x^2 - bx$

Solutions   (A) $x^2 - 3x$

$$x^2 - 3x + \tfrac{9}{4} = (x - \tfrac{3}{2})^2 \quad \text{Add } \left(\frac{-3}{2}\right)^2; \text{ that is, } \frac{9}{4}.$$

(B) $x^2 - bx$

$$x^2 - bx + \frac{b^2}{4} = \left(x - \frac{b}{2}\right)^2 \quad \text{Add } \left(\frac{-b}{2}\right)^2; \text{ that is, } \frac{b^2}{4}.$$

Problem 3 Complete the square for each of the following:

(A) $x^2 - 5x$ (B) $x^2 + mx$

It is important to note that the rule for completing the square applies only to quadratic forms in which the coefficient of the second-degree term is 1. This causes little trouble, however, as you will see. We now solve two equations by the method of completing the square.

EXAMPLE 4 **Solution by Completing the Square**

Solve by completing the square:

(A) $x^2 + 6x - 2 = 0$ (B) $2x^2 - 4x + 3 = 0$

Solutions (A) $x^2 + 6x - 2 = 0$

$$x^2 + 6x = 2$$

$$x^2 + 6x + 9 = 2 + 9 \qquad \text{Complete the square on the left side, and add the same number to the right side.}$$

$$(x + 3)^2 = 11$$

$$x + 3 = \pm\sqrt{11}$$

$$x = -3 \pm \sqrt{11}$$

(B) $2x^2 - 4x + 3 = 0$

$$x^2 - 2x + \tfrac{3}{2} = 0 \qquad \text{Make the leading coefficient 1 by dividing by 2.}$$

$$x^2 - 2x = -\tfrac{3}{2}$$

$$x^2 - 2x + 1 = -\tfrac{3}{2} + 1 \qquad \text{Complete the square on the left side and add the same number to the right side.}$$

$$(x - 1)^2 = -\tfrac{1}{2} \qquad \text{Factor the left side.}$$

$$x - 1 = \pm\sqrt{-\tfrac{1}{2}}$$

$$x = 1 \pm i\sqrt{\tfrac{1}{2}}$$

$$= 1 \pm \frac{\sqrt{2}}{2} i \qquad \text{Answer in } a + bi \text{ form.}$$

Problem 4 Solve by completing the square:

(A) $x^2 + 8x - 3 = 0$ (B) $3x^2 - 12x + 13 = 0$

● **Solution by Quadratic Formula**

Now consider the general quadratic equation with unspecified coefficients:

$$ax^2 + bx + c = 0 \qquad a \neq 0$$

We can solve it by completing the square exactly as we did in Example 4B. To make the leading coefficient 1, we must multiply both sides of the equation by $1/a$. Thus,

$$x^2 + \frac{b}{a}x + \frac{c}{a} = 0$$

Adding $-c/a$ to both sides of the equation and then completing the square of the left side, we have

$$x^2 + \frac{b}{a}x + \frac{b^2}{4a^2} = \frac{b^2}{4a^2} - \frac{c}{a}$$

We now factor the left side and solve using the square root property:

$$\left(x + \frac{b}{2a}\right)^2 = \frac{b^2 - 4ac}{4a^2}$$

$$x + \frac{b}{2a} = \pm\sqrt{\frac{b^2 - 4ac}{4a^2}}$$

$$x = -\frac{b}{2a} \pm \frac{\sqrt{b^2 - 4ac}}{2a} \qquad \text{See Problem 75 in Exercise 2-6.}$$

$$= \frac{-b \pm \sqrt{b^2 - 4ac}}{2a}$$

We have thus derived the well-known and widely used **quadratic formula**:

Theorem 1   **Quadratic Formula**

If $ax^2 + bx + c = 0, a \neq 0$, then

$$x = \frac{-b \pm \sqrt{b^2 - 4ac}}{2a}$$

The quadratic formula should be memorized and used to solve quadratic equations when other methods fail or are more difficult to apply.

EXAMPLE 5   **Using the Quadratic Formula**

Solve $2x + \frac{3}{2} = x^2$ by use of the quadratic formula. Leave the answer in simplest radical form.

Solution

$$2x + \tfrac{3}{2} = x^2$$

$$4x + 3 = 2x^2 \qquad \text{Multiply both sides by 2.}$$

$$2x^2 - 4x - 3 = 0 \qquad \text{Write in standard form.}$$

$$x = \frac{-b \pm \sqrt{b^2 - 4ac}}{2a} \qquad a = 2,\, b = -4,\, c = -3$$

$$= \frac{-(-4) \pm \sqrt{(-4)^2 - 4(2)(-3)}}{2(2)}$$

$$= \frac{4 \pm \sqrt{40}}{4} = \frac{4 \pm 2\sqrt{10}}{4} = \frac{2 \pm \sqrt{10}}{2}$$

CAUTION

1. $-4^2 \neq (-4)^2 \qquad -4^2 = -16 \text{ and } (-4)^2 = 16$

2. $2 + \dfrac{\sqrt{10}}{2} \neq \dfrac{2 + \sqrt{10}}{2} \qquad 2 + \dfrac{\sqrt{10}}{2} = \dfrac{4 + \sqrt{10}}{2}$

3. $\dfrac{\cancel{4} \pm 2\sqrt{10}}{\cancel{4}} \neq \pm 2\sqrt{10} \qquad \dfrac{4 \pm 2\sqrt{10}}{4} = \dfrac{2(2 \pm \sqrt{10})}{4} = \dfrac{2 \pm \sqrt{10}}{2}$

Problem 5

Solve $x^2 - \tfrac{5}{2} = -3x$ by use of the quadratic formula. Leave the answer in simplest radical form.

EXAMPLE 6   **Using the Quadratic Formula with a Calculator**

Solve $5.37x^2 - 6.03x + 1.17 = 0$ to two decimal places using a calculator.

Solution

$$5.37x^2 - 6.03x + 1.17 = 0$$

$$x = \frac{6.03 \pm \sqrt{(-6.03)^2 - 4(5.37)(1.17)}}{2(5.37)}$$

$$= 0.25,\ 0.87$$

Calculator Steps for the Solution Involving the Negative Radical

A: $(\ 6.03\ ][\ -\ ][(\ 6.03\ ][\ x^2\ ][\ -\ ][\ 4\ ][\ \times\ ][\ 5.37\ ][\ \times\ ][\ 1.17\ ][\ )\ ][\ \sqrt{x}\ ][\ )\ ][\ \div\ ]$
$[(\ 2\ ][\ \times\ ][\ 5.37\ ][\ )\ ][\ =\ ]$

P: $[\ 6.03\ ][\ \text{Enter}\ ][\ 6.03\ ][\ x^2\ ][\ 4\ ][\ \text{Enter}\ ][\ 5.37\ ][\ \times\ ][\ 1.17\ ][\ \times\ ]$
$[\ -\ ][\ \sqrt{x}\ ][\ -\ ][\ 2\ ][\ \text{Enter}\ ][\ 5.37\ ][\ \times\ ][\ \div\ ]$

Problem 6

Solve $2.79x^2 + 5.07x - 7.69 = 0$ to two decimal places using a calculator.

We conclude this part of the discussion by noting that $b^2 - 4ac$ in the quadratic formula is called the **discriminant** and gives us useful information about the corresponding roots as shown in Table 1.

TABLE 1 **Discriminant and Roots**

Discriminant $b^2 - 4ac$	Roots of $ax^2 + bx + c = 0$ a, b, and c real numbers, $a \neq 0$
Positive	Two distinct real roots
0	One real root (a double root)
Negative	Two imaginary roots, one the conjugate of the other

For example:

(A) $2x^2 - 3x - 4 = 0$ has two real roots, since

$$b^2 - 4ac = (-3)^2 - 4(2)(-4) = 41 > 0$$

(B) $4x^2 - 4x + 1 = 0$ has one real (double) root, since

$$b^2 - 4ac = (-4)^2 - 4(4)(1) = 0$$

(C) $2x^2 - 3x + 4 = 0$ has two imaginary roots, since

$$b^2 - 4ac = (-3)^2 - 4(2)(4) = -23 < 0$$

● **Applications**   We now consider several applications that make use of quadratic equations. First, the strategy for solving word problems, presented earlier in Section 2-2, is repeated below.

Strategy for Solving Word Problems

1. Read the problem carefully—several times if necessary; that is, until you understand the problem, know what is to be found, and know what is given.

2. Let one of the unknown quantities be represented by a variable, say x, and try to represent all other unknown quantities in terms of x. This is an important step and must be done carefully.

3. If appropriate, draw figures or diagrams and label known and unknown parts.

4. Look for formulas connecting the known quantities to the unknown quantities.

5. Form an equation relating the unknown quantities to the known quantities.

6. Solve the equation and write answers to *all* questions asked in the problem.

7. Check and interpret all solutions in terms of the original problem—not just the equation found in step 5—since a mistake may have been made in setting up the equation in step 5.

EXAMPLE 7   Setting Up and Solving a Word Problem

The sum of a number and its reciprocal is $\frac{13}{6}$. Find all such numbers.

Solution   Let $x =$ the number; then:

$$x + \frac{1}{x} = \frac{13}{6}$$

$$(6x)x + (6x)\frac{1}{x} = (6x)\frac{13}{6}$$ Multiply both sides by $6x$. [*Note:* $x \neq 0$.]

$$6x^2 + 6 = 13x$$ A quadratic equation

$$6x^2 - 13x + 6 = 0$$

$$(2x - 3)(3x - 2) = 0$$

$$2x - 3 = 0 \quad \text{or} \quad 3x - 2 = 0$$

$$x = \tfrac{3}{2} \qquad\qquad x = \tfrac{2}{3}$$

Thus, two such numbers are $\frac{3}{2}$ and $\frac{2}{3}$.

Check $\tfrac{3}{2} + \tfrac{2}{3} = \tfrac{13}{6} \qquad \tfrac{2}{3} + \tfrac{3}{2} = \tfrac{13}{6}$

Problem 7   The sum of two numbers is 23 and their product is 132. Find the two numbers. [*Hint:* If one number is x, then the other number is $23 - x$.]

EXAMPLE 8   A Distance–Rate–Time Problem

An excursion boat takes 1.6 hours longer to go 36 miles up a river than to return. If the rate of the current is 4 miles per hour, what is the rate of the boat in still water?

Solution   Let

$$x = \text{Rate of boat in still water}$$

$$x + 4 = \text{Rate downstream}$$

$$x - 4 = \text{Rate upstream}$$

$$\left(\begin{array}{c}\text{Time}\\\text{upstream}\end{array}\right) - \left(\begin{array}{c}\text{Time}\\\text{downstream}\end{array}\right) = 1.6$$

$$\frac{36}{x - 4} - \frac{36}{x + 4} = 1.6 \qquad\qquad T = \frac{D}{R}$$

$$36(x + 4) - 36(x - 4) = 1.6(x - 4)(x + 4)$$

$$36x + 144 - 36x + 144 = 1.6x^2 - 25.6$$

$$1.6x^2 = 313.6$$

$$x^2 = 196$$

$$x = \sqrt{196} = 14$$

The rate in still water is 14 miles per hour.

[*Note:* $-\sqrt{196} = -14$ must be discarded, since it doesn't make sense in the problem to have a negative rate.]

Check

$$\text{Time upstream} = \frac{D}{R} = \frac{36}{14 - 4} = 3.6$$

$$\text{Time downstream} = \frac{D}{R} = \frac{36}{14 + 4} = 2$$

$$\overline{1.6} \quad \text{Difference of times}$$

Problem 8   Two boats travel at right angles to each other after leaving a dock at the same time. One hour later they are 25 miles apart. If one boat travels 5 miles per hour faster than the other, what is the rate of each? [*Hint:* Use the Pythagorean theorem,* remembering that distance equals rate times time.]

EXAMPLE 9   A Quantity–Rate–Time Problem

A payroll can be completed in 4 hours by two computers working simultaneously. How many hours are required for each computer to complete the payroll alone if the older model requires 3 hours longer than the newer model? Compute answers to two decimal places.

Solution   Let

$$x = \text{Time for new model to complete the payroll alone}$$

$$x + 3 = \text{Time for old model to complete the payroll alone}$$

$$4 = \text{Time for both computers to complete the payroll together}$$

Then,

$$\frac{1}{x} = \text{Rate for new model} \qquad \text{Completes } \frac{1}{x} \text{ of the payroll per hour}$$

$$\frac{1}{x + 3} = \text{Rate for old model} \qquad \text{Completes } \frac{1}{x + 3} \text{ of the payroll per hour}$$

$$\begin{pmatrix} \text{Part of job} \\ \text{completed by} \\ \text{new model in} \\ \text{4 hours} \end{pmatrix} + \begin{pmatrix} \text{Part of job} \\ \text{completed by} \\ \text{old model in} \\ \text{4 hours} \end{pmatrix} = 1 \text{ whole job}$$

$$\frac{1}{x}(4) \quad + \quad \frac{1}{x + 3}(4) \quad = 1$$

$$\frac{4}{x} \quad + \quad \frac{4}{x + 3} \quad = 1$$

Pythagorean theorem: A triangle is a right triangle if and only if the square of the length of the longest side is equal to the sum of the squares of the lengths of the two shorter sides: $c^2 = a^2 + b^2$.

$$4(x + 3) + 4x = x(x + 3) \quad \text{Multiply both sides by } x(x + 3).$$

$$4x + 12 + 4x = x^2 + 3x$$

$$x^2 - 5x - 12 = 0$$

$$x = \frac{5 \pm \sqrt{73}}{2}$$

$$x = \frac{5 + \sqrt{73}}{2} \approx 6.77 \qquad \frac{5 - \sqrt{73}}{2} \approx -1.77 \text{ is}$$

discarded since x cannot be negative.

$$x + 3 = 9.77$$

The new model would complete the payroll in 6.77 hours working alone, and the old model would complete the payroll in 9.77 hours working alone.

Check

$$\frac{1}{6.77}(4) + \frac{1}{9.77}(4) \stackrel{?}{=} 1$$

$$1.000\ 259 \stackrel{\checkmark}{\approx} 1$$

[*Note:* We do not expect the check to be exact, since we rounded the answers to two decimal places. An exact check would be produced by using $x = (5 + \sqrt{73})/2$. The latter is left to the reader.]

Problem 9 Two technicians can complete a mailing in 3 hours when working together. Alone, one can complete the mailing 2 hours faster than the other. How long will it take each person to complete the mailing alone? Compute the answers to two decimal places.

Answers to Matched Problems
1. (A) $x = -4, \frac{5}{3}$ (B) $x = -\frac{3}{2}$ (a double root) (C) $x = 0, \frac{5}{4}$
2. (A) $x = \pm\sqrt{\frac{5}{3}}$ or $\pm\sqrt{15}/3$ (B) $x = \pm 2i$ (C) $x = (-1 \pm \sqrt{2})/3$
3. (A) $x^2 - 5x + \frac{25}{4} = (x - \frac{5}{2})^2$ (B) $x^2 + mx + (m^2/4) = [x + (m/2)]^2$
4. (A) $x = -4 \pm \sqrt{19}$ (B) $x = (6 \pm i\sqrt{3})/3$ or $2 \pm (\sqrt{3}/3)i$ 5. $x = (-3 \pm \sqrt{19})/2$
6. $x = -2.80, 0.98$ 7. 11 and 12 8. 15 and 20 miles per hour 9. 5.16 and 7.16 hours

EXERCISE 2-6 *Leave all answers involving radicals in simplified radical form unless otherwise stated.*

A

In Problems 1–6, solve by factoring.

1. $4u^2 = 8u$ 2. $3A^2 = -12A$ 3. $9y^2 = 12y - 4$ 4. $16x^2 + 8x = -1$
5. $11x = 2x^2 + 12$ 6. $8 - 10x = 3x^2$

In Problems 7–18, solve by using the square root property.

7. $m^2 - 12 = 0$ **8.** $y^2 - 45 = 0$ **9.** $x^2 + 25 = 0$ **10.** $x^2 + 16 = 0$

11. $9y^2 - 16 = 0$ **12.** $4x^2 - 9 = 0$ **13.** $4x^2 + 25 = 0$ **14.** $16a^2 + 9 = 0$

15. $(n + 5)^2 = 9$ **16.** $(m - 3)^2 = 25$ **17.** $(d - 3)^2 = -4$ **18.** $(t + 1)^2 = -9$

In Problems 19–26, solve using the quadratic formula.

19. $x^2 - 10x - 3 = 0$ **20.** $x^2 - 6x - 3 = 0$ **21.** $x^2 + 8 = 4x$ **22.** $y^2 + 3 = 2y$

23. $2x^2 + 1 = 4x$ **24.** $2m^2 + 3 = 6m$ **25.** $5x^2 + 2 = 2x$ **26.** $7x^2 + 6x + 4 = 0$

B

In Problems 27–34, solve by completing the square.

27. $x^2 - 6x - 3 = 0$ **28.** $y^2 - 10y - 3 = 0$

29. $2y^2 - 6y + 3 = 0$ **30.** $2d^2 - 4d + 1 = 0$

31. $3x^2 - 2x - 2 = 0$ **32.** $3x^2 + 5x - 4 = 0$

33. $x^2 + mx + n = 0$ **34.** $ax^2 + bx + c = 0, a \neq 0$

In Problems 35–52, solve by any method.

35. $12x^2 + 7x = 10$ **36.** $9x^2 + 9x = 4$

37. $(2y - 3)^2 = 5$ **38.** $(3m + 2)^2 = -4$

39. $x^2 = 3x + 1$ **40.** $x^2 + 2x = 2$

41. $7n^2 = -4n$ **42.** $8u^2 + 3u = 0$

43. $1 + \dfrac{8}{x^2} = \dfrac{4}{x}$ **44.** $\dfrac{2}{u} = \dfrac{3}{u^2} + 1$

45. $\dfrac{24}{10 + m} + 1 = \dfrac{24}{10 - m}$ **46.** $\dfrac{1.2}{y - 1} + \dfrac{1.2}{y} = 1$

47. $\dfrac{2}{x - 2} = \dfrac{4}{x - 3} - \dfrac{1}{x + 1}$ **48.** $\dfrac{3}{x - 1} - \dfrac{2}{x + 3} = \dfrac{4}{x - 2}$

49. $\dfrac{x + 2}{x + 3} - \dfrac{x^2}{x^2 - 9} = 1 - \dfrac{x - 1}{3 - x}$ **50.** $\dfrac{11}{x^2 - 4} + \dfrac{x + 3}{2 - x} = \dfrac{2x - 3}{x + 2}$

51. $|3u - 2| = u^2$ **52.** $|12 + 7x| = x^2$

In Problems 53–56, solve for the indicated variable in terms of the other variables. Use positive square roots only.

53. $s = \frac{1}{2}gt^2$ for t **54.** $a^2 + b^2 = c^2$ for a

55. $P = EI - RI^2$ for I **56.** $A = P(1 + r)^2$ for r

Solve Problems 57–60 to two decimal places using a calculator.

57. $2.07x^2 - 3.79x + 1.34 = 0$ **58.** $0.61x^2 - 4.28x + 2.93 = 0$

59. $4.83x^2 + 2.04x - 3.18 = 0$ **60.** $5.13x^2 + 7.27x - 4.32 = 0$

Use the discriminant to determine whether the equations in Problems 61–64 have real solutions.

61. $0.0134x^2 + 0.0414x + 0.0304 = 0$

62. $0.543x^2 - 0.182x + 0.003\ 12 = 0$

63. $0.0134x^2 + 0.0214x + 0.0304 = 0$

64. $0.543x^2 - 0.182x + 0.0312 = 0$

C

Solve Problems 65–68 and leave answers in simplified radical form (i is the imaginary unit).

65. $\sqrt{3}x^2 = 8\sqrt{2}x - 4\sqrt{3}$

66. $2\sqrt{2}x + \sqrt{3} = \sqrt{3}x^2$

67. $x^2 + 2ix = 3$

68. $x^2 = 2ix - 3$

In Problems 69 and 70, find all solutions.

69. $x^3 - 1 = 0$

70. $x^4 - 1 = 0$

71. Indicate true (T) or false (F):
A quadratic equation with rational coefficients can have one rational root and one irrational root.

72. Indicate true (T) or false (F):
A quadratic equation with real coefficients can have one real root and one imaginary root.

73. Show that if r_1 and r_2 are the two roots of $ax^2 + bx + c = 0$, then $r_1r_2 = c/a$.

74. For r_1 and r_2 in Problem 73, show that $r_1 + r_2 = -b/a$.

75. In one stage of the derivation of the quadratic formula, we replaced the expression

$$\pm\sqrt{(b^2 - 4ac)/4a^2}$$

with

$$\pm\sqrt{b^2 - 4ac}/2a$$

What justifies using $2a$ in place of $|2a|$?

76. Find the error in the following "proof" that two arbitrary numbers are equal to each other: Let a and b be arbitrary numbers such that $a \neq b$. Then

$$(a - b)^2 = a^2 - 2ab + b^2 = b^2 - 2ab + a^2$$
$$(a - b)^2 = (b - a)^2$$
$$a - b = b - a$$
$$2a = 2b$$
$$a = b$$

APPLICATIONS

77. Numbers. Find two numbers such that their sum is 21 and their product is 104.

78. Numbers. Find all numbers with the property that when the number is added to itself the sum is the same as when the number is multiplied by itself.

79. Numbers. Find two consecutive positive even integers whose product is 168.

80. Numbers. The sum of a number and its reciprocal is $\frac{10}{3}$. Find the number.

81. Geometry. If the length and width of a 4- by 2-inch rectangle are each increased by the same amount, the area of the new rectangle will be twice that of the original. What are the dimensions of the new rectangle (to two decimal places)?

82. Geometry. Find the base b and height h of a triangle with an area of 2 square feet if its base is 3 feet longer than its height and the formula for area is $A = \frac{1}{2}bh$.

83. Business. If $\$P$ are invested at an interest rate r compounded annually, at the end of 2 years the amount will be $A = P(1 + r)^2$. At what interest rate will $\$1,000$ increase to $\$1,440$ in 2 years? [*Note:* $A = \$1,440$ and $P = \$1,000$.]

★84. Economics. In a certain city the demand equation for CDs is $q_d = 75,000/p$, where q_d is the quantity of CDs demanded on a given day if the selling price is $\$p$ per CD. On the other hand, the supply equation is $q_s = 2,000p - 25,000$, where q_s is the quantity of CDs a supplier is willing to supply at $\$p$ per CD. At what price will supply equal demand? That is, at what price will

$q_d = q_s$? In economic theory the price at which supply equals demand is called the **equilibrium point**, the point at which the price ceases to change.

85. Puzzle. Two planes travel at right angles to each other after leaving the same airport at the same time. One hour later they are 260 miles apart. If one travels 140 miles per hour faster than the other, what is the rate of each?

86. Navigation. A speedboat takes 1 hour longer to go 24 miles up a river than to return. If the boat cruises at 10 miles per hour in still water, what is the rate of the current?

★**87. Engineering.** One pipe can fill a tank in 5 hours less than another. Together they can fill the tank in 5 hours. How long would it take each alone to fill the tank? Compute the answer to two decimal places.

★★**88. Engineering.** Two gears rotate so that one completes 1 more revolution per minute than the other. If it takes the smaller gear 1 second less than the larger gear to complete $\frac{1}{5}$ revolution, how many revolutions does each gear make in 1 minute?

★**89. Physics—Engineering.** For a car traveling at a speed of v miles per hour, under the best possible conditions the shortest distance d necessary to stop it (including reaction time) is given by the empirical formula $d = 0.044v^2 + 1.1v$, where d is measured in feet. Estimate the speed of a car that requires 165 feet to stop in an emergency.

★**90. Physics—Engineering.** If a projectile is shot vertically into the air (from the ground) with an initial velocity of 176 feet per second, its distance y (in feet) above the ground t seconds after it is shot is given by $y = 176t - 16t^2$ (neglecting air resistance).
   (A) Find the times when y is 0, and interpret the results physically.
   (B) Find the times when the projectile is 16 feet off the ground. Compute answers to two decimal places.

★**91. Construction.** A developer wants to erect a rectangular building on a triangular-shaped piece of property that is 200 feet wide and 400 feet long (see the figure at the top of the next column). Find the dimensions of the building if its cross-sectional area is 15,000 square feet. [*Hint:* Use Euclid's theorem* to find a relationship between the length and width of the building.]

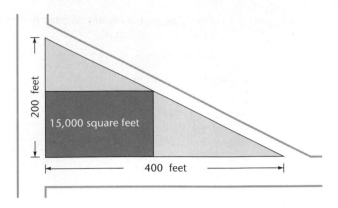

★**92. Architecture.** An architect is designing a small A-frame cottage for a resort area. A cross section of the cottage is an isosceles triangle with an area of 98 square feet. The front wall of the cottage must accommodate a sliding door that is 6 feet wide and 8 feet high (see the figure). Find the width and height of the cross section of the cottage. [*Recall:* The area of a triangle with base b and altitude h is $bh/2$.]

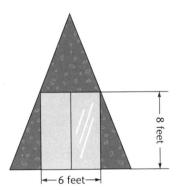

93. Transportation. A delivery truck leaves a warehouse and travels north to factory A. From factory A the truck travels east to factory B and then returns directly to the warehouse (see the figure). The driver recorded the truck's odometer reading at the warehouse at both the beginning and the end of the trip and also at factory B, but forgot to record it at factory A (see the table). The driver does recall that it was further from the warehouse to factory A than it was from factory A to factory B. Since delivery charges are based on distance from the

Euclid's theorem: If two triangles are similar, their corresponding sides are proportional:

$$\frac{a}{a'} = \frac{b}{b'} = \frac{c}{c'}$$

warehouse, the driver needs to know how far factory *A* is from the warehouse. Find this distance.

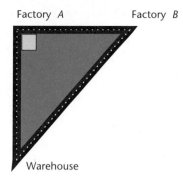

Factory *A*   Factory *B*

Warehouse

	Odometer Readings
Warehouse	5 2 8 4 6
Factory *A*	5 2 ? ? ?
Factory *B*	5 2 9 3 7
Warehouse	5 3 0 0 2

★★ **94. Construction.** A $\frac{1}{4}$-mile track for racing stock cars consists of two semicircles connected by parallel straightaways (see the figure). In order to provide sufficient room for pit crews, emergency vehicles, and spectator parking, the track must enclose an area of 100,000 square feet. Find the length of the straightaways and the diameter of the semicircles to the nearest foot. [*Recall:* The area *A* and circumference *C* of a circle of diameter *d* are given by $A = \pi d^2/4$ and $C = \pi d$.]

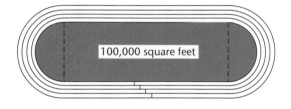

100,000 square feet

SECTION $2\text{-}7$ **Equations Reducible to Quadratic Form**

- Equations Involving Radicals
- Equations Involving Rational Exponents

• Equations Involving Radicals

In solving an equation involving a radical like

$$x = \sqrt{x + 2}$$

it appears that we can remove the radical by squaring each side and then proceed to solve the resulting quadratic equation. Thus,

$$x^2 = (\sqrt{x + 2})^2$$
$$x^2 = x + 2$$
$$x^2 - x - 2 = 0$$
$$(x - 2)(x + 1) = 0$$
$$x = 2, -1$$

Now we check these results in the original equation.

$$\text{Check: } x = 2 \qquad \text{Check: } x = -1$$

$$x = \sqrt{x + 2} \qquad\qquad x = \sqrt{x + 2}$$

$$2 \overset{?}{=} \sqrt{2 + 2} \qquad\qquad -1 \overset{?}{=} \sqrt{-1 + 2}$$

$$2 \overset{?}{=} \sqrt{4} \qquad\qquad -1 \overset{?}{=} \sqrt{1}$$

$$2 \overset{\checkmark}{=} 2 \qquad\qquad -1 \neq 1$$

Thus, 2 is a solution, but -1 is not. These results are a special case of Theorem 1.

Theorem 1

Squaring Operation on Equations

If both sides of an equation are squared, then the solution set of the original equation is a subset of the solution set of the new equation.

Equation	Solution Set
$x = 3$	$\{3\}$
$x^2 = 9$	$\{-3, 3\}$

This theorem provides us with a method of solving some equations involving radicals. It is important to remember that any new equation obtained by raising both members of an equation to the same power may have solutions, called **extraneous solutions**, that are not solutions of the original equation. On the other hand, any solution of the original equation must be among those of the new equation.

> **Every solution of the new equation must be checked in the original equation to eliminate extraneous solutions.**

CAUTION   Remember that $\sqrt{9}$ represents the *positive* square root of 9 and $-\sqrt{9}$ represents the *negative* square root of 9. It is correct to use the symbol $\pm$ to combine these two roots when solving an equation:

$$x^2 = 9 \quad \text{implies} \quad x = \pm\sqrt{9} = \pm 3$$

But it is incorrect to use $\pm$ when evaluating the positive square root of a number:

$$\sqrt{9} \neq \pm 3 \quad \sqrt{9} = 3$$

EXAMPLE 1   Solving Equations Involving Radicals

Solve:

(A) $x + \sqrt{x - 4} = 4$ (B) $\sqrt{2x + 3} - \sqrt{x - 2} = 2$

Solutions   (A) $x + \sqrt{x - 4} = 4$

$\qquad\qquad \sqrt{x - 4} = 4 - x \qquad\qquad$ Isolate radical on one side.

$\qquad\qquad\quad x - 4 = 16 - 8x + x^2 \quad$ Square both sides.

$\qquad\quad x^2 - 9x + 20 = 0$

$\qquad\quad (x - 5)(x - 4) = 0$

$\qquad\qquad\qquad\qquad x = 5, 4$

Check
$$x = 5 \qquad\qquad\qquad x = 4$$
$$x + \sqrt{x - 4} = 4 \qquad x + \sqrt{x - 4} = 4$$
$$5 + \sqrt{5 - 4} \overset{?}{=} 4 \qquad 4 + \sqrt{4 - 4} \overset{?}{=} 4$$
$$6 \neq 4 \qquad\qquad\qquad 4 \overset{\checkmark}{=} 4$$

This shows that 4 is a solution to the original equation and 5 is extraneous. Thus

$$x = 4 \quad \text{Only one solution}$$

(B) To solve an equation that contains more than one radical, isolate one radical at a time and square both sides to eliminate it. Repeat this process until all the radicals are eliminated.

$$\sqrt{2x + 3} - \sqrt{x - 2} = 2$$

$\qquad\qquad \sqrt{2x + 3} = \sqrt{x - 2} + 2 \qquad\qquad$ Isolate one of the radicals.

$\qquad\qquad\quad 2x + 3 = x - 2 + 4\sqrt{x - 2} + 4 \quad$ Square both sides.

$\qquad\qquad\qquad x + 1 = 4\sqrt{x - 2} \qquad\qquad$ Isolate the remaining radical.

$\qquad\quad x^2 + 2x + 1 = 16(x - 2) \qquad\qquad$ Square both sides.

$\qquad\quad x^2 - 14x + 33 = 0$

$\qquad\quad (x - 3)(x - 11) = 0$

$\qquad\qquad\qquad\qquad x = 3, 11$

Check
$$x = 3 \qquad\qquad\qquad\qquad x = 11$$
$$\sqrt{2x + 3} - \sqrt{x - 2} = 2 \qquad \sqrt{2x + 3} - \sqrt{x - 2} = 2$$
$$\sqrt{2(3) + 3} - \sqrt{3 - 2} \overset{?}{=} 2 \qquad \sqrt{2(11) + 3} - \sqrt{11 - 2} \overset{?}{=} 2$$
$$2 \overset{\checkmark}{=} 2 \qquad\qquad\qquad\qquad 2 \overset{\checkmark}{=} 2$$

Both solutions check. Thus,

$$x = 3, 11 \quad \text{Two solutions}$$

Problem 1   Solve:

(A) $x - 5 = \sqrt{x - 3}$ \qquad (B) $\sqrt{2x + 5} + \sqrt{x + 2} = 5$

CAUTION

When squaring an expression like $\sqrt{x-2}+2$, be certain to correctly apply the formula for squaring the sum of two terms (see Section 1-2):

$$(u + v)^2 = u^2 + 2uv + v^2$$
$$(\sqrt{x-2}+2)^2 = (\sqrt{x-2})^2 + 2(\sqrt{x-2})(2) + (2)^2$$
$$= x - 2 + 4\sqrt{x-2} + 4$$

Do not omit the middle term in this product:

$$(\sqrt{x-2}+2)^2 \neq x - 2 + 4$$

● Equations Involving Rational Exponents

To solve the equation

$$x^{2/3} - x^{1/3} - 6 = 0$$

write it in the form

$$(x^{1/3})^2 - x^{1/3} - 6 = 0$$

You can now recognize that the equation is quadratic in $x^{1/3}$. So, we solve for $x^{1/3}$ first, and then solve for x. We can solve the equation directly or make the substitution $u = x^{1/3}$, solve for u, and then solve for x. Both methods of solution are shown below.

Method I. Direct solution:

$$(x^{1/3})^2 - x^{1/3} - 6 = 0$$
$$(x^{1/3} - 3)(x^{1/3} + 2) = 0 \qquad \text{Factor left side.}$$
$$x^{1/3} = 3 \qquad \text{or} \qquad x^{1/3} = -2$$
$$(x^{1/3})^3 = 3^3 \qquad\qquad (x^{1/3})^3 = (-2)^3 \qquad \text{Cube both sides.}$$
$$x = 27 \qquad\qquad\qquad x = -8$$

Solution set: $\{-8, 27\}$

Method II. Using substitution:
Let $u = x^{1/3}$, solve for u, and then solve for x.

$$u^2 - u - 6 = 0$$
$$(u - 3)(u + 2) = 0$$
$$u = 3, -2$$

Replacing u with $x^{1/3}$, we obtain

$$x^{1/3} = 3 \qquad \text{or} \qquad x^{1/3} = -2$$
$$x = 27 \qquad\qquad x = -8$$

Solution set: $\{-8, 27\}$

In general, if an equation that is not quadratic can be transformed to the form

$$au^2 + bu + c = 0$$

where u is an expression in some other variable, then the equation is called a **quadratic form**. Once recognized as a quadratic form, an equation often can be solved using quadratic methods.

EXAMPLE 2   Solving Quadratic Forms

Solve as far as possible using techniques we have developed up to this point. (Some equations may have additional imaginary solutions that you will not be able to find without further study in the theory of equations.)

(A) $x^{10} + 6x^5 - 16 = 0$
(B) $3x^{-4} + 6x^{-2} - 1 = 0$ (Find all real solutions.)

Solutions   (A) For $x^{10} + 6x^5 - 16 = 0$, let $u = x^5$ and solve:

$$u^2 + 6u - 16 = 0$$
$$(u + 8)(u - 2) = 0$$
$$u = -8, 2$$

Thus,

$$x^5 = -8 \qquad \text{or} \qquad x^5 = 2$$
$$x = \sqrt[5]{-8} = -\sqrt[5]{8} \qquad\qquad x = \sqrt[5]{2} \quad \textit{Note: } x \neq (-8)^5 \text{ and } x \neq 2^5.$$

Two solutions are $-\sqrt[5]{8}$ and $\sqrt[5]{2}$. In Chapter 4 we will see that there are eight other solutions.

(B) The equation $3x^{-4} + 6x^{-2} - 1 = 0$ is quadratic in x^{-2}, so we could solve for x^{-2} first and then for x. However, we prefer first to convert the equation to one involving positive exponents. We do this by multiplying both sides by $x^4, x \neq 0$:

$$3x^{-4} + 6x^{-2} - 1 = 0$$
$$3 + 6x^2 - x^4 = 0 \quad \text{Multiply both sides by } x^4, x \neq 0$$
$$(x^2)^2 - 6x^2 - 3 = 0 \quad \text{Quadratic in } x^2$$

The left side doesn't factor using integer coefficients, so we solve for x^2 using the quadratic formula:

$$x^2 = \frac{6 \pm \sqrt{36 - 4(1)(-3)}}{2}$$
$$x^2 = 3 \pm 2\sqrt{3}$$
$$x = \pm\sqrt{3 \pm 2\sqrt{3}}$$

But since $3 - 2\sqrt{3}$ is negative and leads to imaginary roots, it must be discarded (we were to find all real roots). Thus,

$$x = \pm\sqrt{3 + 2\sqrt{3}} \quad \text{Two real roots}$$

Problem 2   Solve as far as possible using techniques we have developed up to this point.

(A) $x^{2/3} - x^{1/3} - 12 = 0$ (B) $x^4 - 5x^2 + 4 = 0$
(C) $3x^{-4} = 1 - 10x^{-2}$ (Find all real solutions.)

EXAMPLE 3   **Setting Up and Solving a Word Problem**

The diagonal of a rectangle is 10 inches, and the area is 45 square inches. Find the dimensions of the rectangle correct to one decimal place.

Solution   Draw a rectangle and label the dimensions as shown in the margin. From the Pythagorean theorem,

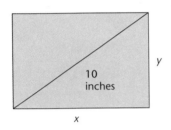

$$x^2 + y^2 = 10^2$$

Thus,

$$y = \sqrt{100 - x^2}$$

Since the area of the rectangle is given by xy, we have

$$x\sqrt{100 - x^2} = 45 \qquad \text{Area of the rectangle}$$
$$x^2(100 - x^2) = 2{,}025 \quad \text{Square both sides.}$$
$$100x^2 - x^4 = 2{,}025$$
$$(x^2)^2 - 100x^2 + 2{,}025 = 0 \qquad \text{Quadratic in } x^2$$
$$x^2 = \frac{100 \pm \sqrt{100^2 - 4(1)(2{,}025)}}{2}$$
$$x^2 = 50 \pm 5\sqrt{19}$$
$$x = \sqrt{50 \pm 5\sqrt{19}} \qquad \text{Discard the negative solutions since } x > 0.$$

If $x = \sqrt{50 + 5\sqrt{19}} \approx 8.5$, then

$$y = \sqrt{100 - x^2}$$
$$= \sqrt{100 - (50 + 5\sqrt{19})}$$
$$= \sqrt{50 - 5\sqrt{19}} \approx 5.3$$

Thus, the dimensions of the rectangle to one decimal place are 8.5 inches by 5.3 inches. Notice that if $x = \sqrt{50 - 5\sqrt{19}}$, then $y = \sqrt{50 + 5\sqrt{19}}$, and the dimensions are still 8.5 inches by 5.3 inches.

Check   Area: $(8.5)(5.3) = 45.05 \approx 45$

Diagonal: $\sqrt{8.5^2 + 5.3^2} = \sqrt{100.34} \approx 10$

Note: An exact check can be obtained by using $\sqrt{50 - 5\sqrt{19}}$ and $\sqrt{50 + 5\sqrt{19}}$ in place of these decimal approximations. This is left to the reader.

Problem 3   If the area of a right triangle is 24 square inches and the hypotenuse is 12 inches, find the lengths of the legs of the triangle correct to one decimal place.

Answers to Matched Problems
1. (A) $x = 7$ (B) $x = 2$
2. (A) $x = 64, -27$ (B) $x = \pm 1, \pm 2$ (C) $x = \pm \sqrt{5 + 2\sqrt{7}}$ (two real roots)
3. 11.2 inches by 4.3 inches

EXERCISE 2-7   *Unless stated to the contrary, find all solutions possible by the techniques that have been developed so far.*

A

1. $\sqrt[3]{x + 5} = 3$
2. $\sqrt[4]{x - 3} = 2$
3. $\sqrt{5n + 9} = n - 1$
4. $m - 13 = \sqrt{m + 7}$
5. $\sqrt{x + 5} + 7 = 0$
6. $3 + \sqrt{2x - 1} = 0$
7. $\sqrt{3x + 4} = 2 + \sqrt{x}$
8. $\sqrt{3w - 2} - \sqrt{w} = 2$
9. $y^4 - 2y^2 - 8 = 0$
10. $x^4 - 7x^2 - 18 = 0$
11. $x^{10} + 3x^5 - 10 = 0$
12. $x^{10} - 7x^5 - 8 = 0$
13. $2x^{2/3} + 3x^{1/3} - 2 = 0$
14. $x^{2/3} - 3x^{1/3} - 10 = 0$
15. $(m^2 - m)^2 - 4(m^2 - m) = 12$
16. $(x^2 + 2x)^2 - (x^2 + 2x) = 6$

B

17. $\sqrt{u - 2} = 2 + \sqrt{2u + 3}$
18. $\sqrt{3t + 4} + \sqrt{t} = -3$
19. $\sqrt{3y - 2} = 3 - \sqrt{3y + 1}$
20. $\sqrt{2x - 1} - \sqrt{x - 4} = 2$
21. $\sqrt{7x - 2} - \sqrt{x + 1} = \sqrt{3}$
22. $\sqrt{3x + 6} - \sqrt{x + 4} = \sqrt{2}$
23. $3n^{-2} - 11n^{-1} - 20 = 0$
24. $6x^{-2} - 5x^{-1} - 6 = 0$
25. $9y^{-4} - 10y^{-2} + 1 = 0$
26. $4x^{-4} - 17x^{-2} + 4 = 0$
27. $y^{1/2} - 3y^{1/4} + 2 = 0$
28. $4x^{-1} - 9x^{-1/2} + 2 = 0$
29. $(m - 5)^4 + 36 = 13(m - 5)^2$
30. $(x - 3)^4 + 3(x - 3)^2 = 4$

C

31. $\sqrt{5 - 2x} - \sqrt{x + 6} = \sqrt{x + 3}$
32. $\sqrt{2x + 3} - \sqrt{x - 2} = \sqrt{x + 1}$
33. $2 + 3y^{-4} = 6y^{-2}$ (Find all real roots.)
34. $4m^{-2} = 2 + m^{-4}$ (Find all real roots.)

Solve Problems 35–38 two ways: by squaring and by substitution.

35. $m - 7\sqrt{m} + 12 = 0$

36. $y - 6 + \sqrt{y} = 0$

37. $t - 11\sqrt{t} + 18 = 0$

38. $x = 15 - 2\sqrt{x}$

APPLICATIONS

39. Manufacturing. A lumber mill cuts rectangular beams from circular logs (see the figure). If the diameter of the log is 16 inches and the cross-sectional area of the beam is 120 square inches, find the dimensions of the cross section of the beam correct to one decimal place.

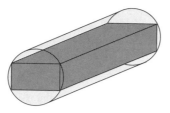

40. Design. A food-processing company packages an assortment of their products in circular metal tins 12 inches in diameter. Four identically sized rectangular boxes are used to divide the tin into eight compartments (see the figure). If the cross-sectional area of each box is 15 square inches, find the dimensions of the boxes correct to one decimal place.

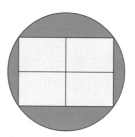

★**41. Construction.** A water trough is constructed by bending a 4- by 6-foot rectangular sheet of metal down the middle and attaching triangular ends (see the figure). If the volume of the trough is 9 cubic feet, find the width correct to two decimal places.

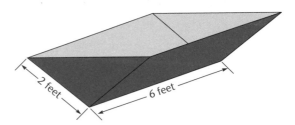

★**42. Design.** A paper drinking cup in the shape of a right circular cone is constructed from 125 square centimeters of paper (see the figure). If the height of the cone is 10 centimeters, find the radius correct to two decimal places.

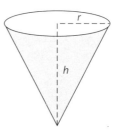

Lateral surface area:

$$S = \pi r \sqrt{r^2 + h^2}$$

SECTION 2-8 Polynomial and Rational Inequalities

- Polynomial Inequalities
- Rational Inequalities

In this section we solve fairly simple polynomial and rational inequalities of the form

$$2x^2 - 3x - 4 < 0 \quad \text{and} \quad \frac{x - 2}{x^2 - x - 3} \geq 0$$

Even though we limit the discussion to quadratic inequalities and rational inequalities with numerators and denominators of degree 2 or less (you will see why below), the theory presented applies to polynomial and rational inequalities in general. In Chapter 4, with additional theory, we will be able to use the methods developed here to solve polynomial and rational inequalities of a more general nature. Also, the process—with only slight modification of key theorems—applies to other forms encountered in calculus.

Why so much interest in solving inequalities? Most significant applications of mathematics involve more use of inequalities than equalities. In the real world few things are exact.

● **Polynomial Inequalities**

We know how to solve linear inequalities such as

$$3x - 7 \geq 5(x - 2) + 3$$

But how do we solve quadratic (or higher-degree polynomial) inequalities such as the one given below?

$$x^2 + 2x < 8 \tag{1}$$

We first write the inequality in **standard form**; that is, we transfer all nonzero terms to the left side, leaving only 0 on the right side:

$$x^2 + 2x - 8 < 0 \quad \text{Standard form} \tag{2}$$

In this example, we are looking for values of x that will make the quadratic on the left side less than 0—that is, negative.

The following theorem provides the basis for an effective way of solving this problem. Theorem 1 makes direct use of the *real zeros* of the polynomial on the left side of inequality (2). **Real zeros** of a polynomial are those real numbers that make the polynomial equal to 0—that is, the real roots of the corresponding polynomial equation. If a polynomial has one or more real zeros, then plotting these zeros on a real number line divides the line into two or more intervals.

Theorem 1   **Sign of a Polynomial over a Real Number Line**

A nonzero polynomial will have a constant sign (either always positive or always negative) within each interval determined by its real zeros plotted on a number line. If a polynomial has no real zeros, then the polynomial is either positive over the whole real number line or negative over the whole real number line.

We now complete the solution of inequality (1) using Theorem 1. After writing (1) in standard form, as we did in inequality (2), we find the real zeros of the polynomial on the left side by solving the corresponding polynomial equation:

$$x^2 + 2x - 8 = 0 \quad \text{Can be solved by factoring}$$
$$(x - 2)(x + 4) = 0$$
$$x = -4, 2 \quad \text{Real zeros of the polynomial } x^2 + 2x - 8$$

FIGURE 1 Real zeros of $x^2 + 2x - 8$.

Next, we plot the real zeros, -4 and 2, on a number line (Fig. 1) and note that three intervals are determined: $(-\infty, -4)$, $(-4, 2)$, and $(2, \infty)$.

From Theorem 1 we know that the polynomial has constant sign on each of these three intervals. If we select a **test number** in each interval and evaluate the polynomial at that number, then the sign of the polynomial at this test number must be the sign for the whole interval. Since any number within an interval can be used as a test number, we generally choose test numbers that result in easy computations. In this example, we choose -5, 0, and 3. Table 1 shows the computations.

TABLE 1   Polynomial: $x^2 + 2x - 8 = (x - 2)(x + 4)$

Test number	-5	0	3
Value of polynomial for test number	7	-8	7
Sign of polynomial in interval containing test number	$+$	$-$	$+$
Interval containing test number	$(-\infty, -4)$	$(-4, 2)$	$(2, \infty)$

FIGURE 2 Sign chart for $x^2 + 2x - 8$.

Using the information in Table 1, we construct a **sign chart** for the polynomial, as shown in Figure 2.

Thus, $x^2 + 2x - 8$ is negative within the interval $(-4, 2)$, and we have solved the inequality. The solution and graph are given below and in Figure 3:

$$-4 < x < 2 \quad \text{Inequality notation}$$

$$(-4, 2) \quad \text{Interval notation}$$

FIGURE 3 Solution of $x^2 + 2x < 8$.

Note: If $<$ in the original problem had been $\leq$ instead, then we would have included the zeros of the polynomial in the solution set.

The steps in the above process are summarized in the following box:

Key Steps in Solving Polynomial Inequalities

Step 1. Write the polynomial inequality in standard form (a form where the right-hand side is 0).

Step 2. Find all real zeros of the polynomial (the left side of the standard form).

Step 3. Plot the real zeros on a number line, dividing the number line into intervals.

Step 4. Choose a test number (that is easy to compute with) in each interval, and evaluate the polynomial for each number (a small table is useful).

Step 5. Use the results of step 4 to construct a sign chart, showing the sign of the polynomial in each interval.

Step 6. From the sign chart, write down the solution of the original polynomial inequality (and draw the graph, if required).

With a little experience, many of the above steps can be combined and the process streamlined to two or three key operational steps. The critical part of the method is step 2, finding all real zeros of the polynomial. At this point we can find all real zeros of any quadratic polynomial (see Section 2-6). Finding real zeros of higher-degree polynomials is more difficult, and the process is considered in detail in Chapter 4.

We now turn to a significant application that involves a polynomial inequality.

EXAMPLE 1   **Profit and Loss Analysis**

A company manufactures and sells flashlights. For a particular model, the marketing research and financial departments estimate that at a price of $$p$ per unit, the weekly cost C and revenue R (in thousands of dollars) will be given by the equations

$$C = 7 - p \qquad \text{Cost equation}$$

$$R = 5p - p^2 \qquad \text{Revenue equation}$$

Find prices (including a graph) for which the company will realize:

(A) A profit   (B) A loss

Solutions   (A) A profit will result if cost is less than revenue, that is, if

$$C < R$$

$$7 - p < 5p - p^2$$

We solve this inequality following the steps outlined above.

Step 1. Write the polynomial inequality in standard form.

$$p^2 - 6p + 7 < 0 \qquad \text{Standard form}$$

Step 2. Find all real zeros of the polynomial.

$$p^2 - 6p + 7 = 0$$

$$p = \frac{6 \pm \sqrt{36 - 28}}{2} \qquad \text{Solve, using the quadratic formula.}$$

$$= 3 \pm \sqrt{2}$$

$$\approx \$1.59, \$4.41 \qquad \text{Real zeros of the polynomial rounded to the nearest cent}$$

Step 3. Plot the real zeros on a number line.

The two real zeros determine three intervals: $(-\infty, 1.59)$, $(1.59, 4.41)$, and $(4.41, \infty)$.

Step 4. Choose a test number in each interval, and construct a table.

Polynomial: $p^2 - 6p + 7$

Test number	1	2	5
Value of polynomial for test number	2	-1	2
Sign of polynomial in interval containing test number	$+$	$-$	$+$
Interval containing test number	$(-\infty, 1.59)$	$(1.59, 4.41)$	$(4.41, \infty)$

Step 5. Construct a sign chart.

$$+\ +\ +\ |\ -\ -\ -\ -\ |\ +\ +\ +$$

Sign chart for $p^2 - 6p + 7$

at $\$1.59$ and $\$4.41$

Step 6. Write the solution and draw the graph.

Referring to the sign chart for the polynomial $p^2 - 6p + 7$ in step 5, we see that $p^2 - 6p + 7 < 0$, and a profit will occur ($C < R$), for

$$\$1.59 < p < \$4.41 \qquad \qquad p \quad \text{Profit}$$

(B) A loss will result if cost is greater than revenue; that is, if

$$C > R$$
$$7 - p > 5p - p^2$$

Writing this polynomial inequality in standard form, we obtain the same inequality that was obtained in step 1 of part (A), except the order of the inequality is reversed:

$$p^2 - 6p + 7 > 0 \qquad \text{Standard form}$$

Referring to the sign chart for the polynomial $p^2 - 6p + 7$ in step 5 of part (A), we see that $p^2 - 6p + 7 > 0$, and a loss will occur ($C > R$), for

$$p < \$1.59 \qquad \text{or} \qquad p > \$4.41 \qquad \qquad p$$

Since a negative price doesn't make sense, we must modify this result by deleting any number to the left of 0. Thus, a loss will occur for the following prices:

$$\$0 \leq p < \$1.59 \qquad \text{or} \qquad p > \$4.41 \qquad \qquad p$$

The real zeros are not included, because they are the values for which $R = C$, the **break-even** values for the company.

Problem 1 A company manufactures and sells computer printer ribbons. For a particular ribbon, the marketing research and financial departments estimate that at a price of $\$p$ per unit, the weekly cost C and revenue R (in thousands of dollars) will be given by the equations

$$C = 13 - p \qquad \text{Cost equation}$$

$$R = 7p - p^2 \qquad \text{Revenue equation}$$

Find prices (including a graph) for which the company will realize:

(A) A profit   (B) A loss

● Rational Inequalities

The steps for solving polynomial inequalities can, with slight modification, be used to solve rational inequalities such as

$$\frac{x - 3}{x + 5} > 0 \qquad \text{and} \qquad \frac{x^2 + 5x - 6}{5 - x} \le 3$$

If, after suitable operations on an inequality, the right side is 0 and the left side is of the form P/Q, where P and Q are nonzero polynomials, then the inequality is said to be a **rational inequality in standard form**. When the real zeros (if they exist) of the polynomials P and Q are plotted on a number line, they divide the line into two or more intervals. The following theorem, which includes Theorem 1 as a special case, provides a basis for solving rational inequalities in standard form.

Theorem 2

Sign of a Rational Expression over a Real Number Line

The rational expression P/Q, where P and Q are nonzero polynomials, will have a constant sign (either always positive or always negative) within each interval determined by the real zeros of P and Q plotted on a number line. If neither P nor Q have real zeros, then the rational expression P/Q is either positive over the whole real number line or negative over the whole real number line.

We will illustrate the use of Theorem 2 through an example.

EXAMPLE 2 Solving a Rational Inequality

Solve and graph: $\dfrac{x^2 - 3x - 10}{1 - x} \ge 2$

Solution   We might be tempted to start by multiplying both sides by $1 - x$ (as we would do if the inequality were an equation). However, since we don't know whether $1 - x$ is positive or negative, we don't know whether the order of the inequality is to be changed. We proceed instead as follows (modifying the steps for solving polynomial inequalities as needed):

Step 1. Write the inequality in standard form.

$$\frac{x^2 - 3x - 10}{1 - x} \geq 2$$

$$\frac{x^2 - 3x - 10}{1 - x} - 2 \geq 0 \qquad \text{Subtract 2 from both sides.}$$

$$\frac{x^2 - 3x - 10 - 2(1 - x)}{1 - x} \geq 0 \qquad \text{Combine left side into a single fraction.}$$

$$\frac{x^2 - x - 12}{1 - x} \geq 0 \qquad \text{Standard form: } \frac{P}{Q} \geq 0$$

The left side of the last inequality is a rational expression of the form P/Q, where $P = x^2 - x - 12$ and $Q = 1 - x$. Our problem now is to find all values of x so that $P/Q \geq 0$; that is, so that P/Q is positive or 0.

Step 2. Find all real zeros for polynomials P and Q.

$$x^2 - x - 12 = 0$$

$$(x + 3)(x - 4) = 0$$

$$x = -3, 4 \qquad \text{Real zeros for } P$$

$$1 - x = 0$$

$$x = 1 \qquad \text{Real zero for } Q$$

[*Note:* The real zeros for P make P/Q equal to 0; thus, the equality part of the original inequality is satisfied for these zeros and they must be included in the final solution set. On the other hand, since division by 0 is never allowed, P/Q is not defined at the zeros of Q. Thus, the real zeros of Q must *not* be included in the solution set.]

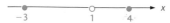

Step 3. Plot the real zeros for P and Q on a number line.

The three zeros of P and Q determine four intervals: $(-\infty, -3)$, $(-3, 1)$, $(1, 4)$, and $(4, \infty)$. Note that we use solid dots at -3 and 4 to indicate that these zeros of P are part of the solution set. However, we use an open dot at 1 to indicate that this zero of Q is not part of the solution set. Remember, P/Q is not defined at the zeros of Q.

Step 4. Choose a test number in each interval, and construct a table.

Rational Expression: $\dfrac{x^2 - x - 12}{1 - x} = \dfrac{(x + 3)(x - 4)}{1 - x}$

Test number	-4	0	2	5
Value of P/Q	$\frac{8}{5}$	-12	10	-2
Sign of P/Q	$+$	$-$	$+$	$-$
Interval	$(-\infty, -3)$	$(-3, 1)$	$(1, 4)$	$(4, \infty)$

Step 5. Construct a sign chart.

$$\begin{array}{c}
+\,+ \quad -\ -\ -\ -\ -\quad +\,+\,+\quad -\ - \\
\xrightarrow{\;\;\;\;\;\;\bullet\;\;\;\;\;\;\;\;\;\;\;\;\circ\;\;\;\;\;\;\bullet\;\;\;\;\;\;}\; x \qquad \text{Sign chart for } \dfrac{x^2 - x - 12}{1 - x}\\
\quad\;\; -3 \qquad\quad 1 \qquad 4
\end{array}$$

Step 6. Write the solution and draw the graph.

From the sign chart, we see that

$$\frac{x^2 - x - 12}{1 - x} \geq 0 \qquad \text{and} \qquad \frac{x^2 - 3x - 10}{1 - x} \geq 2$$

for

$$x \leq -3 \qquad \text{or} \qquad 1 < x \leq 4 \qquad \text{Inequality notation}$$

$$(-\infty, -3] \cup (1, 4] \qquad\qquad \text{Interval notation}$$

Problem 2   Solve and graph: $\dfrac{3}{2 - x} \leq \dfrac{1}{x + 4}$

Answers to Matched Problems

1. (A) Profit: $\$2.27 < p < \$5.73$ (B) Loss: $\$0 \leq p < \$2.27$ or $p > \$5.73$

$\xrightarrow{\qquad(\qquad\quad)\qquad\qquad}\, p$
$\qquad\;\;\$2.27\qquad\;\$5.73$

$\xrightarrow{[\qquad)\qquad\qquad(\qquad}\, p$
$\;\;\$0\quad\$2.27\qquad\;\;\$5.73$

2. $-4 < x \leq -\frac{5}{2}$ or $x > 2$
$(-4, -\frac{5}{2}] \cup (2, \infty)$

$\xrightarrow{\;\;(\qquad]\qquad\qquad(\qquad}\, x$
$\;\;-4\;\;-\frac{5}{2}\qquad\quad 2$

EXERCISE **2-8**

A

In Problems 1–14, solve and graph. Express answers in both inequality and interval notation.

1. $x^2 < 10 - 3x$ **2.** $x^2 + x < 12$ **3.** $x^2 + 21 > 10x$ **4.** $x^2 + 7x + 10 > 0$

5. $x^2 \leq 8x$ **6.** $x^2 + 6x \geq 0$ **7.** $x^2 + 5x \leq 0$ **8.** $x^2 \leq 4x$

9. $x^2 > 4$

10. $x^2 \le 9$

11. $\dfrac{x - 2}{x + 4} \le 0$

12. $\dfrac{x + 3}{x - 1} \ge 0$

13. $\dfrac{x + 4}{1 - x} \le 0$

14. $\dfrac{3 - x}{x + 5} \le 0$

$\dfrac{x^2 + 1}{x^2 - 1} \ge 0$

B

In Problems 15–26, solve and graph. Express answers in both inequality and interval notation.

15. $\dfrac{x^2 + 5x}{x - 3} \ge 0$

16. $\dfrac{x - 4}{x^2 + 2x} \le 0$

17. $\dfrac{(x + 1)^2}{x^2 + 2x - 3} \le 0$

18. $\dfrac{x^2 - x - 12}{x^2 + 4} \le 0$

19. $\dfrac{1}{x} < 4$

20. $\dfrac{5}{x} > 3$

21. $\dfrac{3x + 1}{x + 4} \le 1$

22. $\dfrac{5x - 8}{x - 5} \ge 2$

23. $\dfrac{2}{x + 1} \ge \dfrac{1}{x - 2}$

24. $\dfrac{3}{x - 3} \le \dfrac{2}{x + 2}$

25. $x^3 + 2x^2 \le 8x$

26. $2x^3 + x^2 > 6x$

 For what real values of x will each expression in Problems 27–32 represent a real number? Write answers using inequality notation.

27. $\sqrt{x^2 - 9}$

28. $\sqrt{4 - x^2}$

29. $\sqrt{2x^2 + x - 6}$

30. $\sqrt{3x^2 - 7x - 6}$

31. $\sqrt{\dfrac{x + 7}{3 - x}}$

32. $\sqrt{\dfrac{x - 1}{x + 3}}$

C

In Problems 33–44, solve and graph. Express answers in both inequality and interval notation.

33. $x^2 + 1 < 2x$

34. $x^2 + 25 < 10x$

35. $x^2 < 3x - 3$

36. $x^2 + 3 > 2x$

37. $x^2 - 1 \ge 4x$

38. $2x + 2 > x^2$

39. $x^3 > 2x^2 + x$

40. $x^3 \le 4x^2 + 3x$

41. $4x^4 + 4 \le 17x^2$

42. $x^4 + 36 \ge 13x^2$

43. $|x^2 - 1| \le 3$

44. $\left| \dfrac{x + 1}{x} \right| > 2$

APPLICATIONS

45. Profit and Loss Analysis. At a price of $\$p$ per unit, the marketing department in a company estimates that the weekly cost C and revenue R (in thousands of dollars) will be given by the equations

$$C = 28 - 2p \quad \text{Cost equation}$$
$$R = 9p - p^2 \quad \text{Revenue equation}$$

Find the prices for which the company has
(A) A profit   (B) A loss

46. Profit and Loss Analysis. At a price of $\$p$ per unit, the marketing department in a company estimates that the weekly cost C and revenue R (in thousands of dollars) will be given by the equations

$$C = 27 - 2p \quad \text{Cost equation}$$
$$R = 10p - p^2 \quad \text{Revenue equation}$$

Find the prices for which the company has
(A) A profit   (B) A loss

47. Physics. If an object is shot straight up from the ground with an initial velocity of 112 feet per second, its distance d (in feet) above the ground at the end of t seconds (neglecting air resistance) is given by $d = 112t - 16t^2$. Find the interval of time for which the object is 160 feet above the ground or higher.

48. Physics. In Problem 47, find the interval of time for which the object is above the ground.

★**49. Safety Research.** It is of considerable importance to know the shortest distance d (in feet) in which a car can be stopped, including reaction time of the driver, at various speeds v (in miles per hour). Safety research has produced the formula $d = 0.044v^2 + 1.1v$ for a given car. At what speeds will it take the car more than 330 feet to stop?

★**50. Safety Research.** Using the information in Problem 49, at what speeds will it take a car less than 220 feet to stop?

★★**51. Marketing.** When successful new software is first introduced, the weekly sales generally increase rapidly for a period of time and then begin to decrease. Suppose that the weekly sales S (in thousands of units) t weeks after the software is introduced are given by

$$S = \frac{200t}{t^2 + 100}$$

When will sales be 8 thousand units per week or more?

★★**52. Medicine.** A drug is injected into the bloodstream of a patient through her right arm. The concentration (in milligrams per milliliter) of the drug in the bloodstream of the left arm t hours after the injection is given approximately by

$$C = \frac{0.12t}{t^2 + 2}$$

When will the concentration of the drug in the left arm be 0.04 milligram per milliliter or greater?

Chapter 2 Review

2-1   LINEAR EQUATIONS

A **solution** or **root** of an equation is a number in the **domain** or **replacement set** of the variable that when substituted for the variable makes the equation a true statement. An equation is an **identity** if it is true for all values from the domain of the variable and a **conditional equation** if it is true for some domain values and false for others. Two equations are **equivalent** if they have the same **solution set**. The **properties of equality** are used to solve equations:

1. If $a = b$, then $a + c = b + c$. Addition Property

2. If $a = b$, then $a - c = b - c$. Subtraction Property

3. If $a = b$, then $ca = cb$, $c \neq 0$. Multiplication Property

4. If $a = b$, then $\dfrac{a}{c} = \dfrac{b}{c}$, $c \neq 0$. Division Property

5. If $a = b$, then either may replace the other in any statement without changing the truth or falsity of statement. Substitution Property

An equation that can be written in the **standard form** $ax + b = 0$, $a \neq 0$, is a **linear** or **first-degree equation**.

2-2   APPLICATIONS INVOLVING LINEAR EQUATIONS

Strategy for Solving Word Problems

1. Read the problem carefully—several times if necessary; that is, until you understand the problem, know what is to be found, and know what is given.

2. Let one of the unknown quantities be represented by a variable, say x, and try to represent all other unknown quantities in terms of x. This is an important step and must be done carefully.

3. If appropriate, draw figures or diagrams and label known and unknown parts.

4. Look for formulas connecting the known quantities to the unknown quantities.

5. Form an equation relating the unknown quantities to the known quantities.

6. Solve the equation and write answers to *all* questions asked in the problem.

7. Check and interpret all solutions in terms of the original problem—not just the equation found in step 5—since a mistake may have been made in setting up the equation in step 5.

If Q is **quantity** produced or **distance** traveled at an average or uniform **rate** R in T units of **time**, then the **quantity–rate–time formulas** are

$$R = \frac{Q}{T} \qquad Q = RT \qquad T = \frac{Q}{R}$$

2-3   LINEAR INEQUALITIES

The inequality symbols $<$, $>$, $\leq$, $\geq$ are used to express **inequality relations**. **Line graphs**, **interval notation**, and the set operations of **union** and **intersection** are used to describe inequality relations. A **solution** of a linear inequality in one variable is a value of the variable that makes the inequality a true statement. Two inequalities are **equivalent** if they have the same **solution set**. **Inequality properties** are used to solve inequalities:

1. If $a < b$ and $b < c$, then $a < c$.   Transitive Property

2. If $a < b$, then $a + c < b + c$.   Addition Property

3. If $a < b$, then $a - c < b - c$.   Subtraction Property

4. If $a < b$ and $c > 0$, then $ca < cb$.
5. If $a < b$ and $c < 0$, then $ca > cb$.   Multiplication Property

6. If $a < b$ and $c > 0$, then $\frac{a}{c} < \frac{b}{c}$.
7. If $a < b$ and $c < 0$, then $\frac{a}{c} > \frac{b}{c}$.   Division Property

The order of an inequality reverses if we multiply or divide both sides of an inequality statement by a negative number.

2-4   ABSOLUTE VALUE IN EQUATIONS AND INEQUALITIES

The **absolute value** of a number x is the distance on a real number line from the origin to the point with coordinate x and is given by

$$|x| = \begin{cases} x & \text{if } x \geq 0 \\ -x & \text{if } x < 0 \end{cases}$$

The **distance between points A and B** with coordinates a and b, respectively, is $d(A, B) = |b - a|$, which has the following **geometric interpretations**:

$$|x - c| = d \qquad \text{Distance between } x \text{ and } c \text{ is equal to } d.$$

$$|x - c| < d \qquad \text{Distance between } x \text{ and } c \text{ is less than } d.$$

$$0 < |x - c| < d \qquad \text{Distance between } x \text{ and } c \text{ is less than } d, \text{ but } x \neq c.$$

$$|x - c| > d \qquad \text{Distance between } x \text{ and } c \text{ is greater than } d.$$

Equations and inequalities involving absolute values are solved using the following relationships for $p > 0$:

1. $|x| = p$ is equivalent to $x = p$ or $x = -p$.

2. $|x| < p$ is equivalent to $-p < x < p$.

3. $|x| > p$ is equivalent to $x < -p$ or $x > p$.

These relationships also hold if x is replaced with $ax + b$. For x any real number, $\sqrt{x^2} = |x|$.

2-5   COMPLEX NUMBERS

A **complex number** in **standard form** is a number in the form $a + bi$, where a and b are real numbers and i is the **imaginary unit**. If $b \neq 0$, then $a + bi$ is also called an **imaginary number**. If $a = 0$, then $0 + bi = bi$ is also called a **pure imaginary number**. If $b = 0$, then $a + 0i = a$ is a **real number**. The complex **zero** is $0 + 0i = 0$. The **conjugate** of $a + bi$ is $a - bi$. **Equality**, **addition**, and **multiplication** are defined as follows:

1. $a + bi = c + di$ if and only if $a = c$ and $b = d$

2. $(a + bi) + (c + di) = (a + c) + (b + d)i$

3. $(a + bi)(c + di) = (ac - bd) + (ad + bc)i$

Since complex numbers obey the same commutative, associative, and distributive properties as real numbers, most operations with complex numbers are performed by using these properties and the fact that $i^2 = -1$. The **property of conjugates**,

$$(a + bi)(a - bi) = a^2 + b^2$$

can be used to find **reciprocals** and **quotients**. If $a > 0$, then the **principal square root of the negative real number** $-a$ is $\sqrt{-a} = i\sqrt{a}$.

2-6   QUADRATIC EQUATIONS

A **quadratic equation** in **standard form** is an equation that can be written in the form

$$ax^2 + bx + c = 0 \qquad a \neq 0$$

where x is a variable and a, b, and c are constants. Methods of solution include:

1. Factoring and using the **zero property**:

$$m \cdot n = 0 \qquad \text{if and only if} \qquad m = 0 \text{ or } n = 0 \text{ (or both)}$$

2. Using the **square root property**:

$$\text{If } A^2 = C, \text{ then } A = \pm\sqrt{C}$$

3. **Completing the square**:

$$x^2 + bx + \left(\frac{b}{2}\right)^2 = \left(x + \frac{b}{2}\right)^2$$

4. Using the **quadratic formula**:

$$x = \frac{-b \pm \sqrt{b^2 - 4ac}}{2a}$$

If the **discriminant $b^2 - 4ac$** is positive, the equation has two distinct **real roots**; if the discriminant is 0, the equation has one real **double root**; and if the discriminant is negative, the equation has two **imaginary roots**, each the conjugate of the other.

2-7 EQUATIONS REDUCIBLE TO QUADRATIC FORM

A **square root radical** can be eliminated from an equation by isolating the radical on one side of the equation and squaring both sides of the equation. The new equation formed by squaring both sides may have **extraneous solutions**. Consequently, **every solution of the new equation must be checked in the original equation to eliminate extraneous solutions**. If an equation contains more than one radical, then the process of isolating a radical and squaring both sides can be repeated until all radicals are eliminated. If a substitution transforms an equation into the form $au^2 + bu + c = 0$, where u is an expression in some other variable, then the equation is a **quadratic form** that can be solved by quadratic methods.

2-8 POLYNOMIAL AND RATIONAL INEQUALITIES

An inequality is in **standard form** if the right side is 0. If the left side is a **polynomial**, then the **real zeros** of this polynomial divide the real number line into intervals with the property that the polynomial has constant sign over each interval. Selecting a **test number** in each interval and constructing a **sign chart** produces the solution to the inequality. If the left side of an inequality is a **rational expression** of the form P/Q, where P and Q are polynomials, then the real zeros of both polynomials are used to divide the real number line into intervals over which P/Q has constant sign. Since **division by zero is never allowed**, the real zeros of Q must always be excluded from the solution set.

Chapter 2 Review Exercise

Work through all the problems in this chapter review and check answers in the back of the book. Answers to all review problems are there, and following each answer is a number in italics indicating the section in which that type of problem is discussed. Where weaknesses show up, review appropriate sections in the text.

A

Solve Problems 1 and 2.

1. $0.05x + 0.25(30 - x) = 3.3$

2. $\dfrac{5x}{3} - \dfrac{4 + x}{2} = \dfrac{x - 2}{4} + 1$

Solve and graph Problems 3–7.

3. $3(2 - x) - 2 \leq 2x - 1$

4. $|y + 9| < 5$

5. $|3 - 2x| \leq 5$

6. $x^2 + x < 20$

7. $x^2 \geq 4x + 21$

8. Perform the indicated operations and write the answers in standard form:
  (A) $(-3 + 2i) + (6 - 8i)$
  (B) $(3 - 3i)(2 + 3i)$
  (C) $\dfrac{13 - i}{5 - 3i}$

Solve Problems 9–14.

9. $2x^2 - 7 = 0$

10. $2x^2 = 4x$

11. $2x^2 = 7x - 3$

12. $m^2 + m + 1 = 0$

13. $y^2 = \frac{3}{2}(y + 1)$

14. $\sqrt{5x - 6} - x = 0$

15. For what values of x does $\sqrt{3 - 5x}$ represent a real number?

B

Solve Problems 16 and 17.

16. $\dfrac{7}{2 - x} = \dfrac{10 - 4x}{x^2 + 3x - 10}$

17. $\dfrac{u - 3}{2u - 2} = \dfrac{1}{6} - \dfrac{1 - u}{3u - 3}$

Solve and graph Problems 18–22.

18. $\dfrac{x + 3}{8} \leq 5 - \dfrac{2 - x}{3}$

19. $|3x - 8| > 2$

20. $\dfrac{1}{x} < 2$

21. $\dfrac{3}{x - 4} \leq \dfrac{2}{x - 3}$

22. $\sqrt{(1 - 2m)^2} \leq 3$

23. For what real values of x does the expression below represent a real number?

$$\sqrt{\dfrac{x + 4}{2 - x}}$$

24. If the coordinates of A and B on a real number line are -8 and -2, respectively, find:
  (A) $d(A, B)$ (B) $d(B, A)$

25. Perform the indicated operations and write the final answers in standard form:
  (A) $(3 + i)^2 - 2(3 + i) + 3$ (B) i^{27}

26. Convert to $a + bi$ forms, perform the indicated operations, and write the final answers in standard form:
  (A) $(2 - \sqrt{-4}) - (3 - \sqrt{-9})$
  (B) $\dfrac{2 - \sqrt{-1}}{3 + \sqrt{-4}}$
  (C) $\dfrac{4 + \sqrt{-25}}{\sqrt{-4}}$

In Problems 27–32, find all solutions possible using techniques we have developed so far.

27. $\left(u + \dfrac{5}{2}\right)^2 = \dfrac{5}{4}$

28. $1 + \dfrac{3}{u^2} = \dfrac{2}{u}$

29. $\dfrac{x}{x^2 - x - 6} - \dfrac{2}{x - 3} = 3$

30. $2x^{2/3} - 5x^{1/3} - 12 = 0$

31. $m^4 + 5m^2 - 36 = 0$

32. $\sqrt{y - 2} - \sqrt{5y + 1} = -3$

Use a calculator to solve Problems 33–36, and compute to two decimal places.

33. $2.15x - 3.73(x - 0.93) = 6.11x$

34. $-1.52 \leq 0.77 - 2.04x \leq 5.33$

35. $\dfrac{3.77 - 8.47i}{6.82 - 7.06i}$

36. $6.09x^2 + 4.57x - 8.86 = 0$

Solve Problems 37–39 for the indicated variable in terms of the other variables.

37. $P = M - Mdt$ for M (mathematics of finance)

38. $P = EI - RI^2$ for I (electrical engineering)

39. $x = \dfrac{4y + 5}{2y + 1}$ for y

C

40. For what values of a and b is the inequality $a + b < b - a$ true?

41. If a and b are negative numbers and $a > b$, then is a/b greater than 1 or less than 1?

42. Solve for x in terms of y: $y = \dfrac{1}{1 - \dfrac{1}{1 - x}}$

43. Solve and graph: $0 < |x - 6| < d$

Solve Problems 44 and 45.

44. $2x^2 = \sqrt{3}x - \frac{1}{2}$ (Find all roots.)

45. $4 = 8x^{-2} + x^{-4}$ (Find all real roots.)

46. Evaluate: $(a + bi)\left(\dfrac{a}{a^2 + b^2} - \dfrac{b}{a^2 + b^2}i\right)$, $a, b \neq 0$

Solve Problems 47–49.

47. $2x > \dfrac{x^2}{5} + 5$

48. $\dfrac{x^2}{4} + 4 \geq 2x$

49. $\left| x - \dfrac{8}{x} \right| \geq 2$

APPLICATIONS

50. Numbers. Find a number such that subtracting its reciprocal from the number gives $\frac{16}{15}$.

51. Sports Medicine. The following quotation was found in a sports medicine handout: "The idea is to raise and sustain your heart rate to 70% of its maximum safe rate for your age. One way to determine this is to subtract your age from 220 and multiply by 0.7."
(A) If H is the maximum safe sustained heart rate (in beats per minute) for a person of age A (in years), write a formula relating H and A.
(B) What is the maximum safe sustained heart rate for a 20-year-old?

(C) If the maximum safe sustained heart rate for a person is 126 beats per minute, how old is the person?

★**52. Chemistry.** A chemical storeroom has an 80% alcohol solution and a 30% alcohol solution. How many milliliters of each should be used to obtain 50 milliliters of a 60% solution?

★**53. Rate–Time.** An excursion boat takes 2 hours longer to go 45 miles up a river than to return. If the boat's speed in still water is 12 miles per hour, what is the rate of the current?

54. Cost Analysis. Cost equations for manufacturing companies are often quadratic in nature. If the cost equation for manufacturing inexpensive calculators is $C = x^2 - 10x + 31$, where C is the cost of manufacturing x units per week (both in thousands), find:

(A) The output for a $15 thousand weekly cost

(B) The output for a $6 thousand weekly cost

55. **Break-Even Analysis.** The manufacturing company in Problem 54 sells its calculators to wholesalers for $3 each. Thus, its revenue equation is $R = 3x$, where R is revenue and x is the number of units sold per week (both in thousands). Find the break-even point(s) for the company—that is, the output at which revenue equals cost.

★56. **Profit Analysis.** Referring to Problems 54 and 55, find all output levels for which a profit will result—that is, for which $R > C$.

★57. **Chemistry.** If the temperature T of a solution must be kept within 5°C of 110°C, express this restriction as an absolute value inequality.

★58. **Design.** The pages of a textbook have uniform margins of 2 centimeters on all four sides (see the figure). If the area of the entire page is 480 square centimeters and the area of the printed portion is 320 square centimeters, find the dimensions of the page.

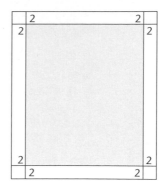

★59. **Design.** A landscape designer uses 8-foot timbers to form a pattern of isosceles triangles along the wall of a building (see the figure). If the area of each triangle is 24 square feet, find the base correct to two decimal places.

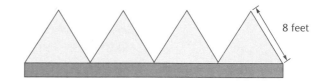

8 feet

Cumulative Review Exercise
Chapters 1 and 2

Work through all the problems in this cumulative review and check answers in the back of the book. Answers to all review problems are there, and following each answer is a number in italics indicating the section in which that type of problem is discussed. Where weaknesses show up, review appropriate sections in the text.

A

1. Solve for x: $\dfrac{7x}{5} - \dfrac{3 + 2x}{2} = \dfrac{x - 10}{3} + 2$

2. Replace each question mark with an appropriate expression that will illustrate the use of the indicated real number property:

   (A) Distributive: $c(a + b) = ?$

   (B) Associative ($+$): $(a + b) + c = ?$

   (C) Commutative ($\cdot$): $(a + b)c = ?$

In Problems 3–7, perform the indicated operation and simplify.

3. $4x(x - 3) + 2(3x + 5)$

4. $3x(x - 7) - (5x - 9)$

5. $(2x - 5y)(3x + 4y)$

6. $(2a - 3b)(2a + 3b)$

7. $(5m + 2n)^2$

Solve and graph Problems 8–10.

8. $2(3 - y) + 4 \leq 5 - y$

9. $|x - 2| < 7$

10. $x^2 + 3x \geq 10$

In Problems 11 and 12, write each polynomial in a completely factored form relative to the integers. If the polynomial is prime relative to the integers, say so.

11. $x^2 - 3x + 10$

12. $6t^2 + 7t - 5$

In Problems 13 and 14, perform the indicated operations and reduce to lowest terms. Represent all compound fractions as simple fractions reduced to lowest terms.

13. $\dfrac{6}{x^2 - 3x} - \dfrac{4}{x^2 - 2x}$

14. $\dfrac{\dfrac{4}{y} - y}{\dfrac{2}{y^2} - \dfrac{1}{2}}$

15. Perform the indicated operations and write the answer in standard form:
(A) $(2 - 3i) - (-5 + 7i)$
(B) $(1 + 4i)(3 - 5i)$
(C) $\dfrac{5 + i}{2 + 3i}$

Simplify Problems 16–18 and write answers using positive exponents only. All variables represent positive real numbers.

16. $3x^2(xy^2)^5$

17. $\dfrac{(2xy^2)^3}{4x^4y^4}$

18. $(a^4b^{-6})^{3/2}$

Solve Problems 19–22.

19. $3x^2 = -12x$

20. $4x^2 - 20 = 0$

21. $x^2 - 6x + 2 = 0$

22. $x - \sqrt{12 - x} = 0$

23. Change to radical form: $5a^{3/4}$

24. Change to rational exponent form: $2\sqrt[5]{x^2y^3}$

Simplify Problems 25–27.

25. $xy^2\sqrt[3]{x^4y^8}$

26. $\sqrt{2xy^3}\sqrt{6x^2y}$

27. $\dfrac{3 + \sqrt{2}}{2 + 3\sqrt{2}}$

28. For what values of x does $\sqrt{2 + 3x}$ represent a real number?

B

29. Write using the listing method:

$\{x\,|\,x \text{ is a prime factor of } 60\}$

30. Indicate true (T) or false (F):
(A) If $ab = 1$, then either $a = 1$ or $b = 1$ or both.
(B) If $ab = 0$, then either $a = 0$ or $b = 0$ or both.
(C) A real number with a repeating decimal expansion is an irrational number.

Solve Problems 31–33.

31. $\dfrac{x + 3}{2x + 2} + \dfrac{5x + 2}{3x + 3} = \dfrac{5}{6}$

32. $\dfrac{3}{x} = \dfrac{6}{x + 1} - \dfrac{1}{x - 1}$

33. $2x + 1 = 3\sqrt{2x - 1}$

34. Indicate which of the following algebraic expressions are polynomials. State the degree of each polynomial.

(A) $5x^2 - \sqrt{3}x + 7$
(B) $4y^3 + \sqrt{5y} - 11$
(C) $3x^3 + 2x^2y^2 + 5y^3$
(D) $2x^{-2} + 3x^{-1} - 4$

In Problems 35–37, perform the indicated operations and simplify.

35. $(a + 2b)(2a + b) - (a - 2b)(2a - b)$

36. $3(x + h)^2 - 4(x + h) - (3x^2 - 4x)$

37. $(4m + 3n)^3$

In Problems 38–40, write in a completely factored form relative to the integers.

38. $(2y + 4)^2 - y^2$

39. $a^3 + 3a^2 - 4a - 12$

40. $3x^4(x + 1)^2 + 4x^3(x + 1)^3$

In Problems 41–43, perform the indicated operations and reduce to lowest terms. Represent all compound fractions as simple fractions reduced to lowest terms.

41. $\dfrac{-3x^4(x + 1)^2 + 4x^3(x + 1)^3}{(x + 1)^6}$

42. $\dfrac{2a + 4b}{(a^2 - b^2)} + \dfrac{3a}{a^2 - 3ab + 2b^2} - \dfrac{3b}{a^2 - ab - 2b^2}$

43. $\dfrac{2 - \dfrac{4}{2 - \dfrac{x}{y}}}{2 - \dfrac{4}{2 + \dfrac{x}{y}}}$

Solve and graph Problems 44–46.

44. $|4x - 9| > 3$

45. $\sqrt{(3m - 4)^2} \leq 2$

46. $\dfrac{2}{x + 1} \geq \dfrac{1}{x - 2}$

47. For what real values of x does the expression below represent a real number?

$$\frac{\sqrt{x - 2}}{x - 4}$$

48. Perform the indicated operations and write the final answers in standard form:

(A) $(2 - 3i)^2 - (4 - 5i)(2 - 3i) - (2 + 10i)$
(B) $\frac{3}{5} + \frac{4}{5}i + \dfrac{1}{\frac{3}{5} + \frac{4}{5}i}$ (C) i^{35}

49. Convert to $a + bi$ forms, perform the indicated operations, and write the final answers in standard form:

(A) $(5 + 2\sqrt{-9}) - (2 - 3\sqrt{-16})$
(B) $\dfrac{2 + 7\sqrt{-25}}{3 - \sqrt{-1}}$ (C) $\dfrac{12 - \sqrt{-64}}{\sqrt{-4}}$

50. Evaluate to four significant digits using a calculator:

(A) 0.9274^{23}
(B) $\sqrt[7]{12.47}$
(C) $\sqrt[5]{\sqrt[3]{9} + 4}$
(D) $\dfrac{(52{,}180{,}000{,}000)(0.000\ 000\ 002\ 973)}{0.000\ 000\ 000\ 000\ 271}$

In Problems 51–54, perform the indicated operations and express answers in simplified form. All radicands represent positive real numbers.

51. $3xy^2\sqrt[3]{16x^5y^7}$

52. $\sqrt[5]{\dfrac{7a^3}{16b^4}}$

53. $\sqrt[3]{\sqrt{8y^9}}$

54. $\dfrac{t}{\sqrt{t + 9} - 3}$

55. Write in the form $ax^p + bx^q$, where a and b are real numbers and p and q are rational numbers:

(A) $\dfrac{2x^3 + 10}{5x^2}$ (B) $\dfrac{2\sqrt{x} - 5}{4\sqrt{x}}$

Solve Problems 56–59.

56. $1 + \dfrac{14}{y^2} = \dfrac{6}{y}$

57. $4x^{2/3} - 4x^{1/3} - 3 = 0$

58. $u^4 + u^2 - 12 = 0$

59. $\sqrt{8t - 2} - 2\sqrt{t} = 1$

Use a calculator to solve Problems 60 and 61. Compute answers to two decimal places.

60. $-3.45 < 1.86 - 0.33x \le 7.92$

61. $2.35x^2 + 10.44x - 16.47 = 0$

62. Solve for y in terms of x:

$$\frac{x - 2}{x + 1} = \frac{2y + 1}{y - 2}$$

C

63. Evaluate $x^2 - x - 1$ for $x = \dfrac{1}{2} - \dfrac{\sqrt{5}}{2}$.

64. Evaluate $x^2 - x + 2$ for $x = \dfrac{1}{2} - \dfrac{i}{2}\sqrt{7}$.

65. Simplify $(x + 1)^3 - (x - 1)^3$.

66. For what values of a and b is the inequality $a - b < b - a$ true?

67. Factor completely relative to the integers:

$$9b^4 - 16b^2(a^2 - 2a + 1)$$

68. Simplify:

$$\frac{1 - \dfrac{1}{1 + \dfrac{1}{1 + m}}}{\dfrac{m}{2 - m} + \dfrac{m}{2 + m}}$$

69. Solve for y in terms of x:

$$\frac{x + y}{y - \dfrac{x + y}{x - y}} = 1$$

70. Find all roots: $3x^2 = 2\sqrt{2}x - 1$.

71. Find all real roots: $1 = 6x^{-2} + 9x^{-4}$.

72. Simplify and express answers using positive exponents only:

(A) $(a^{1/4} - b^{1/4})(a^{1/4} + b^{1/4})(a^{1/2} + b^{1/2})$

(B) $\left(\dfrac{x^{-1}y + xy^{-1}}{xy^{-1} - x^{-1}y}\right)^{-1}$

73. Rationalize the numerator: $\dfrac{\sqrt[3]{8 + h} - 2}{h}$.

74. Write in standard form: $\dfrac{a + bi}{a - bi}, a, b \ne 0$.

75. Solve: $\left|\dfrac{x + 4}{x}\right| < 3$.

76. Write in simplified form: $\sqrt[n]{\left(\dfrac{x^{2n^2}}{x^{4n}}\right)^{1/(n-2)}}, n > 2$.

APPLICATIONS

77. **Numbers.** Find a number such that the number exceeds its reciprocal by $\frac{3}{2}$.

78. **Rate–Time.** An older conveyor belt working alone can fill a railroad car with ore in 14 minutes. The older belt and a newer belt operating together can fill the car in 6 minutes. How long will it take the newer belt working alone to fill the car with ore?

★ 79. **Rate–Time.** A boat travels upstream for 35 miles and then returns to its starting point. If the round-trip took 4.8 hours and the boat's speed in still water is 15 miles per hour, what is the speed of the current?

★ 80. **Chemistry.** How many gallons of distilled water must be mixed with 24 gallons of a 90% sulfuric acid solution to obtain a 60% solution?

81. **Break-Even Analysis.** The publisher's fixed costs for the production of a new cookbook are $41,800. Variable costs are $4.90 per book. If the book is sold to bookstores for $9.65, how many must be sold for the publisher to break even?

82. **Finance.** An investor instructs a broker to purchase a certain stock whenever the price per share p of the stock is within $10 of $200. Express this instruction as an absolute value inequality.

83. **Construction.** A square sandbox with no bottom is constructed out of concrete (see the figure). The walls of the sandbox are 6 inches thick and 1 foot high. If the outer dimension of the sandbox is x feet, write an al-

gebraic expression in terms of x that represents the volume of the concrete used to construct the sandbox. Simplify the expression.

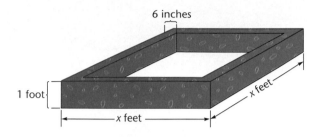

6 inches

1 foot

x feet

x feet

84. **Profit and Loss Analysis.** At a price of $p per unit, the marketing department in a company estimates that the weekly cost C and the weekly revenue R, in thousands of dollars, will be given by the equations

$$C = 88 - 12p \qquad \text{Cost equation}$$
$$R = 15p - 2p^2 \qquad \text{Revenue equation}$$

Find the prices for which the company has:
(A) A profit   (B) A loss

★ 85. **Shipping.** A ship leaves port A, sails east to port B, and then north to port C, a total distance of 115 miles. The next day the ship sails directly from port C back to port A, a distance of 85 miles. Find the distance between ports A and B and between ports B and C.

Graphs and Functions

3

The function concept is one of the most important ideas in mathematics. The study of either the theory or the applications of mathematics beyond the most elementary level requires a firm understanding of functions and their graphs. In the first two sections of this chapter we consider some basic geometric concepts, including the graphs of circles and straight lines. In the remaining sections we introduce the important concept of a function, discuss basic properties of functions, examine specific types of functions, and consider operations that can be performed with functions. Much of the remainder of this book is concerned with applying the ideas introduced in this chapter to a variety of different types of functions. Notice the chapter titles following this chapter. Effort made to understand and use the function concept correctly from the beginning will be rewarded many times in this course and in most future courses that involve mathematics and its applications.

SECTION 3-1 Basic Tools; Circles

- Cartesian Coordinate System
- Graphing: Point by Point
- Symmetry
- Distance between Two Points
- Circles

In this section we develop some of the basic tools used in analytic geometry and apply these tools to the graphing of equations and to the derivation of the equation of a circle.

• Cartesian Coordinate System

Just as we formed a real number line by establishing a one-to-one correspondence between the points on a line and the elements in the set of real numbers, we can form a **real plane** by establishing a one-to-one correspondence between the points in a plane and elements in the set of all ordered pairs of real numbers. This can be done by means of a Cartesian coordinate system.

Recall that to form a **Cartesian** or **rectangular coordinate system**, we select two real number lines, one horizontal and one vertical, and let them cross through their origins as indicated in Figure 1. Up and to the right are the usual choices for the positive directions. These two number lines are called the **horizontal axis** and the **vertical axis**, or, together, the **coordinate axes**. The horizontal axis is usually referred to as the **x axis** and the vertical axis as the **y axis**, and each is labeled accordingly. Other labels may be used in certain situations. The coordinate axes divide the plane into four parts called **quadrants**, which are numbered counterclockwise from I to IV (see Figure 1).

Now we want to assign *coordinates* to each point in the plane. Given an arbitrary point P in the plane, pass horizontal and vertical lines through the point (Figure 2). The vertical line will intersect the horizontal axis at a point with coordinate a, and the horizontal line will intersect the vertical axis at a point with

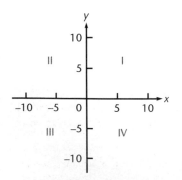

FIGURE 1 Cartesian coordinate system.

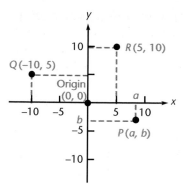

FIGURE 2 Coordinates in a plane.

coordinate b. These two numbers written as the ordered pair* (a, b) form the **coordinates** of the point P. The first coordinate a is called the **abscissa** of P; the second coordinate b is called the **ordinate** of P. The abscissa of Q in Figure 2 is -10, and the ordinate of Q is 5. The coordinates of a point can also be referenced in terms of the axis labels. The **x coordinate** of R in Figure 2 is 5, and the **y coordinate** of R is 10. The point with coordinates $(0, 0)$ is called the **origin**.

The procedure we have just described assigns to each point P in the plane a unique pair of real numbers (a, b). Conversely, if we are given an ordered pair of real numbers (a, b), then, reversing this procedure, we can determine a unique point P in the plane. Thus:

> There is a one-to-one correspondence between the points in a plane and the elements in the set of all ordered pairs of real numbers.

This is often referred to as the **fundamental theorem of analytic geometry**.

● **Graphing: Point by Point**

The fundamental theorem of analytic geometry allows us to look at algebraic forms geometrically and to look at geometric forms algebraically. We begin by considering an algebraic form, an equation in two variables:

$$y = x^2 - 4 \tag{1}$$

A **solution** to equation (1) is an ordered pair of real numbers (a, b) such that

$$b = a^2 - 4$$

The **solution set** of equation (1) is the set of all its solutions. More formally,

Solution set of equation (1): $\{(x, y) | y = x^2 - 4\}$

To find a solution to equation (1) we simply replace one of the variables with a number and solve for the other variable. For example, if $x = 2$, then $y = 2^2 - 4 = 0$, and the ordered pair $(2, 0)$ is a solution. Similarly, if $y = 5$, then $5 = x^2 - 4, x^2 = 9, x = \pm 3$, and the ordered pairs $(3, 5)$ and $(-3, 5)$ are solutions.

Sometimes replacing one variable with a number and solving for the other variable will introduce imaginary numbers. For example, if $y = -5$ in equation (1), then

$$-5 = x^2 - 4$$
$$x^2 = -1$$
$$x = \pm\sqrt{-1} = \pm i$$

Thus, $(-i, 5)$ and $(i, 5)$ are solutions to $y = x^2 - 4$. However, the coordinates of a point in a rectangular coordinate system must be real numbers. Therefore, **when graphing an equation, we only consider those values of the variables that produce real solutions to the equation.**

*An **ordered pair** of real numbers is a pair of numbers in which the order is specified. We now use (a, b) as the coordinates of a point in a plane. In Chapter 2 we used (a, b) to represent an interval on a real number line. These concepts are not the same. You must always interpret the symbol (a, b) in terms of the context in which it is used.

The **graph of an equation in two variables** is the graph of its solution set. In equation (1), we find that its solution set will have infinitely many elements and its graph will extend off any paper we might choose, no matter how large. Thus, **to sketch the graph of an equation**, we include enough points from its solution set so that the total graph is apparent.

EXAMPLE 1   **Graphing an Equation Using Point-by-Point Plotting**

Sketch a graph of $y = x^2 - 4$.

Solution   We make up a table of solutions—ordered pairs of real numbers that satisfy the given equation.

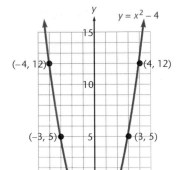

x	-4	-3	-2	-1	0	1	2	3	4
y	12	5	0	-3	-4	-3	0	5	12

After plotting these solutions, if there are any portions of the graph that are unclear, we plot additional points until the shape of the graph is apparent. Then we join all these plotted points with a smooth curve, as shown in the figure. Arrowheads are used to indicate that the graph continues beyond the portion shown here with no significant changes in shape.

The resulting figure is called a *parabola*. Notice that if we fold the paper along the y axis, the right side will match the left side. We say that the graph is *symmetric with respect to the y axis* and call the y axis the *axis of the parabola*. More will be said about parabolas later in the text.

Problem 1   Sketch a graph of $y^2 = x$.

The use of graphic calculators and computers to draw graphs has become commonplace, and some of the exercise sets in this book contain problems that are designed to be solved with the aid of one of these electronic devices. These problems are clearly marked and easily omitted if no such device is available. In any case, whether graphs are being sketched by hand with pencil and paper or electronically with a calculator or a computer, a sound understanding of the underlying mathematical concepts is essential.

The procedure used to sketch the graph of $y = x^2 - 4$ in Example 1 is called **point-by-point plotting**. An important aspect of this course, and later in calculus, is the development of concepts that can be used as graphing aids to minimize the amount of point-by-point plotting required to draw a graph. A particularly useful graphing aid is *symmetry*, which we now discuss.

● **Symmetry**   We noticed that the graph of $y = x^2 - 4$ in Example 1 is *symmetric with respect to the y axis*; that is, the two parts of the graph coincide if the paper is folded along the y axis. Similarly, we say that a graph is *symmetric with respect to the x axis* if the parts above and below the x axis coincide when the paper is folded along the x axis. In general, we define symmetry with respect to the y axis, x axis, and origin as follows:

DEFINITION 1   **Symmetry**

A graph is **symmetric with respect to**:

1. **The y axis** if $(-a, b)$ is on the graph whenever (a, b) is on the graph—the two points are equidistant from the y axis.

2. **The x axis** if $(a, -b)$ is on the graph whenever (a, b) is on the graph—the two points are equidistant from the x axis.

3. **The origin** if $(-a, -b)$ is on the graph whenever (a, b) is on the graph—the two points are equidistant from the origin on a line through the origin.

Figure 3 illustrates these three types of symmetry.

FIGURE 3   Symmetry.

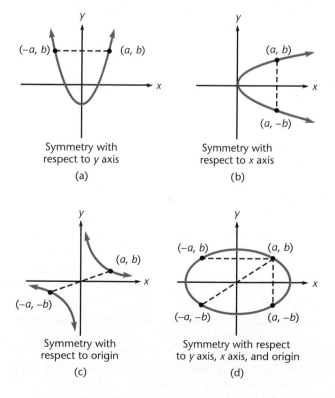

Symmetry with
respect to y axis
(a)

Symmetry with
respect to x axis
(b)

Symmetry with
respect to origin
(c)

Symmetry with respect
to y axis, x axis, and origin
(d)

Given an equation, if we can determine the symmetry properties of its graph ahead of time, we can save a lot of time and energy in sketching the graph. For example, we know that the graph of $y = x^2 - 4$ in Example 1 is symmetric with respect to the y axis, so we can carefully sketch only the right side of the graph;

then reflect the result across the y axis to obtain the whole sketch—the point-by-point plotting is cut in half!

The tests for symmetry are given in Theorem 1. These tests are easily applied and are very helpful aids to graphing. Recall, two equations are equivalent if they have the same solution set.

Theorem 1   **Tests for Symmetry**

Symmetry with respect to the:	Equation is equivalent when:
y axis	x is replaced with $-x$
x axis	y is replaced with $-y$
Origin	x and y are replaced with $-x$ and $-y$

Let's apply the tests for symmetry to $y = x^2 - 4$ from Example 1.

Test y axis symmetry. Replace x with $-x$:

$$y = (-x)^2 - 4$$
$$y = x^2 - 4$$

The equation remains equivalent, so the graph is symmetric with respect to the y axis.

Test x axis symmetry. Replace y with $-y$:

$$-y = x^2 - 4 \qquad \text{or} \qquad y = -x^2 + 4$$

which is not equivalent to $y = x^2 - 4$. Thus, the graph is not symmetric with respect to the x axis.

Test origin symmetry. Replace x with $-x$ and y with $-y$:

$$-y = (-x)^2 - 4$$
$$-y = x^2 - 4 \qquad \text{or} \qquad y = -x^2 + 4$$

which is not equivalent to $y = x^2 - 4$. The graph is not symmetric with respect to the origin.

The only symmetry we have uncovered is symmetry with respect to the y axis (see the figure for Example 1).

EXAMPLE 2   **Using Symmetry as an Aid to Graphing**

Test for symmetry and graph:

$$(A) \ y = x^3 \qquad (B) \ y = |x| \qquad (C) \ x^2 + y^2 = 36$$

Solution   (A) **Symmetry tests for $y = x^3$.**

Test y Axis	**Test x Axis**	**Test Origin**
Replace x with $-x$:	Replace y with $-y$:	Replace x with $-x$ and y with $-y$:

$$
\begin{array}{lll}
y = (-x)^3 & -y = x^3 & -y = (-x)^3 \\
y = -x^3 & y = -x^3 & -y = -x^3 \\
 & & y = x^3
\end{array}
$$

The only test that produces an equivalent equation is replacing x with $-x$ and y with $-y$. Thus, the only symmetry property for the graph of $y = x^3$ is symmetry with respect to the origin.

Graph. Note that positive values of x produce positive values for y and negative values of x produce negative values for y. Therefore, the graph occurs in the first and third quadrants. We make a careful sketch in the first quadrant; then reflect these points through the origin to obtain the complete sketch shown in the figure.

x	0	1	2
y	0	1	8

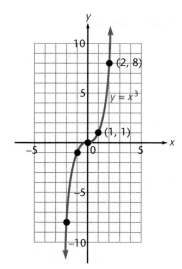

At a glance, the graph shows us how y varies as x varies. A graph is a visual aid and should be constructed to impart the maximum amount of information using the least amount of effort on the part of the observer. Label coordinate axes and indicate scales on both axes.

(B) **Symmetry tests for $y = |x|$.**

Text y Axis	**Test x Axis**	**Test Origin**
Replace x with $-x$:	Replace y with $-y$:	Replace x with $-x$ and y with $-y$:

$$
\begin{array}{lll}
y = |-x| & -y = |x| & -y = |-x| \\
y = |x| & y = -|x| & -y = |x| \\
 & & y = -|x|
\end{array}
$$

Thus, the only symmetry property for the graph of $y = |x|$ is symmetry with respect to the y axis.

Graph. Since $|x|$ is never negative, this graph occurs in the first and second quadrants. We make a careful sketch in the first quadrant, then reflect this graph across the y axis to obtain the complete sketch shown in the figure.

x	0	2	4
y	0	2	4

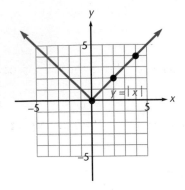

(C) Since both x and y occur to only even powers in $x^2 + 4y^2 = 36$, the equation will remain equivalent if x is replaced with $-x$ or if y is replaced with $-y$. Consequently, the graph is symmetric with respect to the y axis, x axis, and origin. We need to make a careful sketch in only the first quadrant, reflect this graph across the y axis, and then reflect everything across the x axis. To find quadrant I solutions, we solve the equation for either y in terms of x or x in terms of y. We choose the latter because the result is simpler to work with.

$$x^2 + 4y^2 = 36$$
$$x^2 = 36 - 4y^2$$
$$x = \pm\sqrt{36 - 4y^2}$$

To obtain the quadrant I portion of the graph, we sketch $x = \sqrt{36 - 4y^2}$ for $0 \le y \le 3$. Note that $36 - 4y^2 < 0$ for $y > 3$, so there are no real solutions for $y > 3$.

x	6	$\sqrt{32} \approx 5.7$	$\sqrt{20} \approx 4.5$	0
y	0	1	2	3

Choose values for y and solve for x.

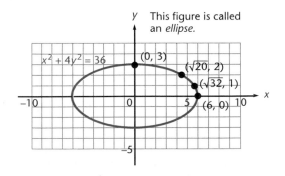

This figure is called an *ellipse*.

Problem 2 Test for symmetry and graph:

(A) $y = x$ (B) $y = -|x|$ (C) $9x^2 + y^2 = 36$

• Distance between Two Points

Analytic geometry is concerned with two basic problems:

1. Given an equation, find its graph.

2. Given a figure (line, circle, parabola, ellipse, etc.) in a coordinate system, find its equation.

So far we have concentrated on the first problem. We now introduce a basic tool that is used extensively in solving the second problem. This basic tool is the *distance-between-two-points formula*, which is easily derived using the Pythagorean theorem (see footnote, Section 2-6). Let $P_1(x_1, y_1)$ and $P_2(x_2, y_2)$ be two points in a rectangular coordinate system (the scale on each axis is assumed to be the same). Then referring to Figure 4, we see that

$$[d(P_1, P_2)]^2 = |x_2 - x_1|^2 + |y_2 - y_1|^2$$
$$= (x_2 - x_1)^2 + (y_2 - y_1)^2 \quad \text{Since } |N|^2 = N^2.$$

FIGURE 4 Distance between two points.

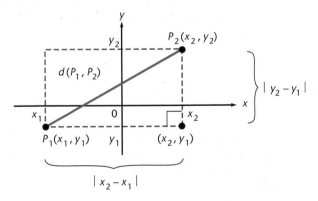

Thus:

Theorem 2	**Distance between $P_1(x_1, y_1)$ and $P_2(x_2, y_2)$**
	$$d(P_1, P_2) = \sqrt{(x_2 - x_1)^2 + (y_2 - y_1)^2}$$

EXAMPLE 3 Using the Distance-between-Two-Points Formula

Find the distance between the points $(-3, 5)$ and $(-2, -8)$.*

---

*We often speak of the point (a, b) when we are referring to the point with coordinates (a, b). This shorthand, though not accurate, causes little trouble, and we will continue the practice.

Solution   It doesn't matter which point we designate as P_1 or P_2 because of the squaring in the formula. Let $(x_1, y_1) = (-3, 5)$ and $(x_2, y_2) = (-2, -8)$. Then

$$d = \sqrt{[(-2) - (-3)]^2 + [(-8) - 5]^2}$$
$$= \sqrt{(-2 + 3)^2 + (-8 - 5)^2} = \sqrt{1^2 + (-13)^2} = \sqrt{1 + 169} = \sqrt{170}$$

Problem 3   Find the distance between the points $(6, -3)$ and $(-7, -5)$.

● Circles   The distance-between-two-points formula would still be helpful if its only use were to find actual distances between points, such as in Example 3. However, its more important use is in finding equations of figures in a rectangular coordinate system. We will use it to derive the standard equation of a circle. We start with a coordinate-free definition of a circle.

DEFINITION 2   **Circle**

A **circle** is the set of all points in a plane equidistant from a fixed point. The fixed distance is called the **radius**, and the fixed point is called the **center**.

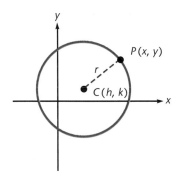

Let's find the equation of a circle with radius r ($r > 0$) and center C at (h, k) in a rectangular coordinate system (Figure 5). The point $P(x, y)$ is on the circle if and only if $d(P, C) = r$; that is, if and only if

$$\sqrt{(x - h)^2 + (y - k)^2} = r \qquad r > 0$$

or, equivalently,

$$(x - h)^2 + (y - k)^2 = r^2 \qquad r > 0$$

FIGURE 5   Circle.

Theorem 3   **Standard Equation of a Circle**

1. Circle with radius r and center at (h, k):

$$(x - h)^2 + (y - k)^2 = r^2 \qquad r > 0$$

2. Circle with radius r and center at $(0, 0)$:

$$x^2 + y^2 = r^2 \qquad r > 0$$

EXAMPLE 4   Equations and Graphs of Circles

Find the equation of a circle with radius 4 and center at:

(A) $(-3, 6)$ (B) $(0, 0)$

Graph each equation.

Solutions   (A) $(h, k) = (-3, 6)$ and $r = 4$:   (B) $(h, k) = (0, 0)$ and $r = 4$:

$$(x - h)^2 + (y - k)^2 = r^2$$ $$x^2 + y^2 = r^2$$

$$[x - (-3)]^2 + (y - 6)^2 = 4^2$$ $$x^2 + y^2 = 4^2$$

$$(x + 3)^2 + (y - 6)^2 = 16$$ $$x^2 + y^2 = 16$$

To graph the equation, locate the   To graph the equation, locate the
center $C(-3, 6)$ and draw a   center at the origin and draw a
circle of radius 4.   circle of radius 4.

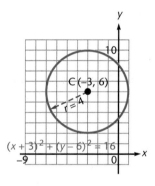

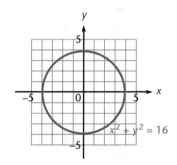

Problem 4   Find the equation of a circle with radius 3 and center at:

(A) $(3, -2)$ (B) $(0, 0)$

Graph each equation.

EXAMPLE 5   Finding the Center and Radius of a Circle

Find the center and radius of the circle with equation $x^2 + y^2 + 6x - 4y = 23$.

Solution   We transform the equation into the form $(x - h)^2 + (y - k)^2 = r^2$ by completing the square relative to x and relative to y (see Section 2-6). From this standard form we can determine the center and radius.

$$x^2 + y^2 + 6x - 4y = 23$$

$$(x^2 + 6x \quad) + (y^2 - 4y \quad) = 23$$

$$(x^2 + 6x + 9) + (y^2 - 4y + 4) = 23 + 9 + 4 \quad \text{Complete the squares.}$$

$$(x + 3)^2 + (y - 2)^2 = 36$$

$$[x - (-3)]^2 + (y - 2)^2 = 6^2$$

Center: $C(h, k) = C(-3, 2)$

Radius: $r = \sqrt{36} = 6$

Problem 5   Find the center and radius of the circle with equation $x^2 + y^2 - 8x + 10y = -25$.

Answers to Matched Problems

1.

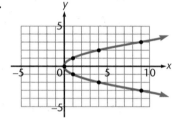

2. (A) Symmetric with respect to the origin

(B) Symmetric with respect to the y axis

(C) Symmetric with respect to the x axis, y axis, and origin

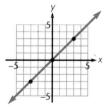   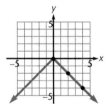   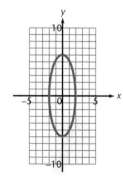

3. $d = \sqrt{173}$ **4.** (A) $(x - 3)^2 + (y + 2)^2 = 9$ (B) $x^2 + y^2 = 9$

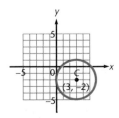   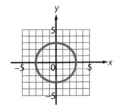

5. $(x - 4)^2 + (y + 5)^2 = 16$; radius: 4, center: $(4, -5)$

75. Architecture. An arched doorway is formed by placing a circular arc on top of a rectangle (see the figure). If the doorway is 4 feet wide and the height of the arc above its ends is 1 foot, what is the radius of the circle containing the arc? [*Hint:* Note that $(2, r - 1)$ must satisfy $x^2 + y^2 = r^2$.]

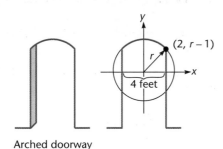

Arched doorway

76. Engineering. The cross section of a rivet has a top that is an arc of a circle (see the figure). If the ends of the arc are 12 millimeters apart and the top is 4 millimeters above the ends, what is the radius of the circle containing the arc?

Rivet

★77. Construction. Town B is located 36 miles east and 15 miles north of town A (see the figure). A local telephone company wants to position a relay tower so that the distance from the tower to town B is twice the distance from the tower to town A.
   (A) Show that the tower must lie on a circle, find the center and radius of this circle, and graph.
   (B) If the company decides to position the tower on this circle at a point directly east of town A, how far from town A should they place the tower? Compute answer to one decimal place.

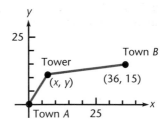

★78. Construction. Repeat Problem 77 if the distance from the tower to town A is twice the distance from the tower to town B.

SECTION 3-2 Straight Lines

- Graphs of First-Degree Equations in Two Variables
- Slope of a Line
- Equations of a Line—Special Forms
- Parallel and Perpendicular Lines

In this section we investigate some standard equations whose graphs are straight lines and determine how to find an equation of a straight line, given information about the line.

• Graphs of First-Degree Equations in Two Variables

With your past experience in graphing equations in two variables, you probably remember that first-degree equations in two variables, such as

$$y = -3x + 5 \qquad 3x - 4y = 9 \qquad y = -\tfrac{2}{3}x$$

have graphs that are straight lines. This fact is stated in Theorem 1. For a partial proof of this theorem, see Problem 78 in Exercise 3-2.

Theorem 1	**The Equation of a Straight Line**

If A, B, and C are constants, with A and B not both 0, and x and y are variables, then the graph of the equation

$$Ax + By = C \qquad \textbf{Standard Form} \qquad (1)$$

is a straight line. Any straight line in a rectangular coordinate system has an equation of this form.

Also, the graph of any equation of the form

$$y = mx + b \qquad\qquad\qquad (2)$$

where m and b are constants, is a straight line. Form (2), which we will discuss in detail later, is simply a special case of form (1) for $B \neq 0$. This can be seen by solving form (1) for y in terms of x:

$$y = -\frac{A}{B}x + \frac{C}{B} \qquad B \neq 0$$

To graph either equation (1) or (2), we plot any two points from the solution set and use a straightedge to draw a line through these two points. The points where the line crosses the axes are convenient to use and easy to find. The **y intercept*** is the ordinate of the point where the graph crosses the y axis, and the **x intercept** is the abscissa of the point where the graph crosses the x axis. To find the y intercept, let $x = 0$ and solve for y; to find the x intercept, let $y = 0$ and solve for x. It is often advisable to find a third point as a check point. All three points must lie on the same straight line or a mistake has been made.

EXAMPLE 1   Using Intercepts to Graph a Straight Line

Graph the equation $3x - 4y = 12$.

Solution   Find intercepts, a third check point (optional), and draw a line through the two (three) points.

x	0	4	8
y	-3	0	3

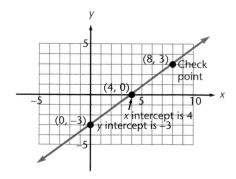

---

*If the x intercept is a and the y intercept is b, then the graph of the line passes through the points $(a, 0)$ and $(0, b)$. It is common practice to refer to both the numbers a and b and the points $(a, 0)$ and $(0, b)$ as the x and y intercepts of the line.

Problem 1 Graph the equation $4x + 3y = 12$.

• **Slope of a Line** If we take two points $P_1(x_1, y_1)$ and $P_2(x_2, y_2)$ on a line, then the ratio of the change in y to the change in x as we move from point P_1 to point P_2 is called the **slope** of the line. Roughly speaking, slope is a measure of the "steepness" of a line. Sometimes the change in x is called the **run** and the change in y is called the **rise**.

DEFINITION 1 **Slope of a Line**

If a line passes through two distinct points $P_1(x_1, y_1)$ and $P_2(x_2, y_2)$, then its slope m is given by the formula

$$m = \frac{y_2 - y_1}{x_2 - x_1} \qquad x_1 \neq x_2$$

$$= \frac{\text{Vertical change (rise)}}{\text{Horizontal change (run)}}$$

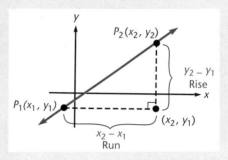

For a horizontal line, y doesn't change as x changes; hence, its slope is 0. On the other hand, for a vertical line, x doesn't change as y changes; hence, $x_1 = x_2$ and its slope is not defined:

$$\frac{y_2 - y_1}{x_2 - x_1} = \frac{y_2 - y_1}{0} \qquad \text{For a vertical line, slope is not defined.}$$

In general, the slope of a line may be positive, negative, 0, or not defined. Each of these cases is interpreted geometrically as shown in Table 1.

TABLE 1 **Geometric Interpretation of Slope**

Line	Slope	Example
Rising as x moves from left to right	Positive	
Falling as x moves from left to right	Negative	
Horizontal	0	
Vertical	Not defined	

In using the formula to find the slope of the line through two points, it doesn't matter which point is labeled P_1 or P_2, since changing the labeling will change the sign in both the numerator and denominator of the slope formula:

$$\frac{y_2 - y_1}{x_2 - x_1} = \frac{y_1 - y_2}{x_1 - x_2} \qquad \text{For example:} \quad \frac{5 - 2}{7 - 3} = \frac{2 - 5}{3 - 7}$$

In addition, it is important to note that the definition of slope doesn't depend on the two points chosen on the line as long as they are distinct. This follows from the fact that the ratios of corresponding sides of similar triangles are equal.

EXAMPLE 2   Finding Slopes

Sketch a line through each pair of points and find the slope of each line.

(A) $(-3, -4), (3, 2)$ (B) $(-2, 3), (1, -3)$
(C) $(-4, 2), (3, 2)$ (D) $(2, 4), (2, -3)$

Solutions   (A)

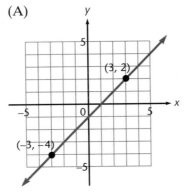

$$m = \frac{2 - (-4)}{3 - (-3)} = \frac{6}{6} = 1$$

(B)

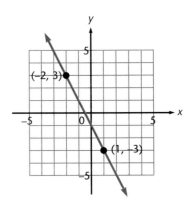

$$m = \frac{-3 - 3}{1 - (-2)} = \frac{-6}{3} = -2$$

(C)

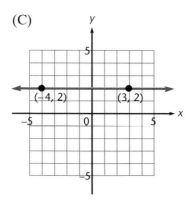

$$m = \frac{2 - 2}{3 - (-4)} = \frac{0}{7} = 0$$

(D)

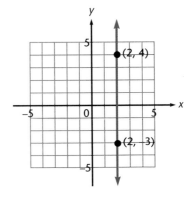

$$m = \frac{-3 - 4}{2 - 2} = \frac{-7}{0};$$

slope is not defined

Problem 2   Find the slope of the line through each pair of points. Do not graph.

(A) $(-3, -3), (2, -3)$ (B) $(-2, -1), (1, 2)$
(C) $(0, 4), (2, -4)$ (D) $(-3, 2), (-3, -1)$

● **Equations of a Line—Special Forms**

The constants m and b in the equation

$$y = mx + b \qquad (3)$$

have special geometric significance.

If we let $x = 0$, then $y = b$, and we observe that the graph of equation (3) crosses the y axis at $(0, b)$. The constant b is the y *intercept*. For example, the y intercept of the graph of $y = -3x - 2$ is -2.

To determine the geometric significance of m, we proceed as follows: If $y = mx + b$, then by setting $x = 0$ and $x = 1$, we conclude that both $(0, b)$ and $(1, m + b)$ lie on the graph, which is a line. Hence, the slope of this line is given by:

$$\text{Slope} = \frac{y_2 - y_1}{x_2 - x_1} = \frac{(m + b) - b}{1 - 0} = m$$

Thus, m is the slope of the line given by $y = mx + b$. Now we know why equation (3) is called the **slope–intercept form** of an equation of a line.

Theorem 2   **Slope–Intercept Form**

$$y = mx + b$$

$$m = \frac{\text{Rise}}{\text{Run}} = \text{Slope}$$

$$b = y \text{ intercept}$$

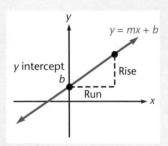

EXAMPLE 3   **Using the Slope–Intercept Form**

(A) Write the slope–intercept equation of a line with slope $\frac{2}{3}$ and y intercept -5.
(B) Find the slope and y intercept, and graph $y = -\frac{3}{4}x - 2$.

Solutions   (A) Substitute $m = \frac{2}{3}$ and $b = -5$ in $y = mx + b$ to obtain

$$y = \tfrac{2}{3}x - 5$$

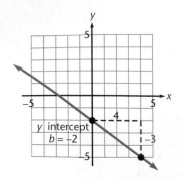

(B) The y intercept of $y = -\frac{3}{4}x - 2$ is -2, so the point $(0, -2)$ is on the graph. The slope of the line is $-\frac{3}{4}$, so when the x coordinate of $(0, -2)$ increases (runs) by 4 units, the y coordinate changes (rises) by -3. The resulting point $(4, -5)$ is easily plotted, and the two points yield the graph of the line. In short, we start at the y intercept -2, and move 4 units to the right and 3 units down to obtain a second point. We then draw a line through the two points, as shown in the figure.

Problem 3   Write the slope–intercept equation of the line with slope $\frac{2}{3}$ and y intercept 1. Graph the equation.

In Example 3 we found the equation of a line with a given slope and y intercept. It is also possible to find the equation of a line passing through a given point with a given slope or to find the equation of a line containing two given points.

Suppose a line has slope m and passes through a fixed point (x_1, y_1). If the point (x, y) is any other point on the line, then

$$\frac{y - y_1}{x - x_1} = m \qquad x \neq x_1$$

that is,

$$y - y_1 = m(x - x_1) \tag{4}$$

We now observe that (x_1, y_1) also satisfies equation (4) and conclude that (4) is an equation of a line with slope m that passes through (x_1, y_1).

We have just obtained the **point–slope form** of the equation of a line.

Theorem 3   **Point–Slope Form**

An equation of a line through a point $P_1(x_1, y_1)$ with slope m is

$$y - y_1 = m(x - x_1)$$

Remember that $P(x, y)$ is a variable point and $P_1(x_1, y_1)$ is fixed.

The point–slope form is extremely useful, since it enables us to find an equation for a line if we know its slope and the coordinates of a point on the line or if we know the coordinates of two points on the line. In the latter case, we find the slope first using the coordinates of the two points; then we use the point–slope form with either of the two given points.

EXAMPLE 4 **Using the Point–Slope Form**

(A) Find an equation for the line that has slope $\frac{2}{3}$ and passes through the point $(-2, 1)$. Write the final answer in the standard form $Ax + By = C$.

(B) Find an equation for the line that passes through the two points $(4, -1)$ and $(-8, 5)$. Write the final answer in the slope–intercept form $y = mx + b$.

Solutions (A) Let $m = \frac{2}{3}$ and $(x_1, y_1) = (-2, 1)$. Then

$$y - y_1 = m(x - x_1)$$
$$y - 1 = \tfrac{2}{3}[x - (-2)]$$
$$y - 1 = \tfrac{2}{3}(x + 2)$$
$$3y - 3 = 2x + 4$$
$$-2x + 3y = 7 \quad \text{or} \quad 2x - 3y = -7$$

(B) First, find the slope of the line by using the slope formula:

$$m = \frac{y_2 - y_1}{x_2 - x_1} = \frac{5 - (-1)}{-8 - 4} = \frac{6}{-12} = -\frac{1}{2}$$

Now let (x_1, y_1) be either of the two given points and proceed as in part (A)—we choose $(x_1, y_1) = (4, -1)$:

$$y - y_1 = m(x - x_1)$$
$$y - (-1) = -\tfrac{1}{2}(x - 4)$$
$$y + 1 = -\tfrac{1}{2}(x - 4)$$
$$y + 1 = -\tfrac{1}{2}x + 2$$
$$y = -\tfrac{1}{2}x + 1$$

You should verify that using $(-8, 5)$, the other given point, produces the same equation.

Problem 4 (A) Find an equation for the line that has slope $-\frac{2}{5}$ and passes through the point $(3, -2)$. Write the final answer in the standard form $Ax + By = C$.

(B) Find an equation for the line that passes through the two points $(-3, 1)$ and $(7, -3)$. Write the final answer in the slope–intercept form $y = mx + b$.

EXAMPLE 5   **Business Markup Policy**

A sporting goods store sells a fishing rod that cost $60 for $82 and a pair of cross-country ski boots that cost $80 for $106.

(A) If the markup policy of the store for items that cost more than $30 is assumed to be linear and is reflected in the pricing of these two items, write an equation that relates retail price R to cost C.

(B) Use the equation to find the retail price for a pair of running shoes that cost $40.

Solutions   (A) If the retail price R is assumed to be linearly related to cost C, then we are looking for an equation whose graph passes through $(C_1, R_1) = (60, 82)$ and $(C_2, R_2) = (80, 106)$. We find the slope, and then use the point–slope form to find the equation.

$$m \boxed{= \frac{R_2 - R_1}{C_2 - C_1}} = \frac{106 - 82}{80 - 60} = \frac{24}{20} = 1.2$$

$$R - R_1 = m(C - C_1)$$

$$R - 82 = 1.2(C - 60)$$

$$R - 82 = 1.2C - 72$$

$$R = 1.2C + 10$$

(B) $R = 1.2(40) + 10 = \$58$

Problem 5   The management of a company that manufactures ballpoint pens estimates costs for running the company to be $200 per day at zero output and $700 per day at an output of 1,000 pens.

(A) Assuming total cost per day C is linearly related to total output per day x, write an equation relating these two quantities.

(B) What is the total cost per day for an output of 5,000 pens?

The simplest equations of lines are those for horizontal and vertical lines. Consider the following two equations:

$$x + 0y = a \quad \text{or} \quad x = a \qquad (5)$$

$$0x + y = b \quad \text{or} \quad y = b \qquad (6)$$

In equation (5), y can be any number as long as $x = a$. Thus, the graph of $x = a$ is a vertical line crossing the x axis at $(a, 0)$. In equation (6), x can be any number as long as $y = b$. Thus, the graph of $y = b$ is a horizontal line crossing the y axis at $(0, b)$. We summarize these results as follows:

Theorem 4	**Vertical and Horizontal Lines**

Equation | Graph
$x = a$ (short for $x + 0y = a$) | Vertical line through $(a, 0)$
 | (Slope is undefined.)
$y = b$ (short for $0x + y = b$) | Horizontal line through $(0, b)$
 | (Slope is 0.)

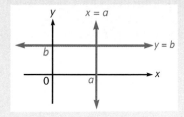

EXAMPLE 6   **Graphing Horizontal and Vertical Lines**

Graph the line $x = -2$ and the line $y = 3$.

Solution

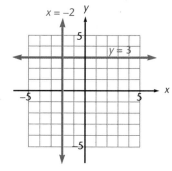

Problem 6   Graph the line $x = 4$ and the line $y = -2$.

The various forms of the equation of a line that we have discussed are summarized in Table 2 for convenient reference.

TABLE 2	**Equations of a Line**		

Standard form	$Ax + By = C$	A and B not both 0
Slope–intercept form	$y = mx + b$	Slope: m; y intercept: b
Point–slope form	$y - y_1 = m(x - x_1)$	Slope: m; Point: (x_1, y_1)
Horizontal line	$y = b$	Slope: 0
Vertical line	$x = a$	Slope: Undefined

● **Parallel and Perpendicular Lines**

From geometry, we know that two vertical lines are parallel to each other and that a horizontal line and a vertical line are perpendicular to each other. How can we tell when two nonvertical lines are parallel or perpendicular to each other? Theorem 5, which we state without proof, provides a convenient test.

Theorem 5

Parallel and Perpendicular Lines

Given two nonvertical lines L_1 and L_2 with slopes m_1 and m_2, respectively, then

$$L_1 \parallel L_2 \qquad \text{if and only if} \qquad m_1 = m_2$$
$$L_1 \perp L_2 \qquad \text{if and only if} \qquad m_1 m_2 = -1$$

The symbols $\parallel$ and $\perp$ mean, respectively, "is parallel to" and "is perpendicular to." In the case of perpendicularity, the condition $m_1 m_2 = -1$ also can be written as

$$m_2 = -\frac{1}{m_1} \qquad \text{or} \qquad m_1 = -\frac{1}{m_2}$$

Thus:

Two nonvertical lines are perpendicular if and only if their slopes are the negative reciprocals of each other.

EXAMPLE 7   **Parallel and Perpendicular Lines**

Given the line $L: 3x - 2y = 5$ and the point $P(-3, 5)$, find an equation of a line through P that is:

(A) Parallel to L (B) Perpendicular to L

Write the final answers in the slope–intercept form $y = mx + b$.

Solutions   First, find the slope of L by writing $3x - 2y = 5$ in the equivalent slope–intercept form $y = mx + b$:

$$3x - 2y = 5$$
$$-2y = -3x + 5$$
$$y = \tfrac{3}{2}x - \tfrac{5}{2}$$

Thus, the slope of L is $\tfrac{3}{2}$. The slope of a line parallel to L is the same, $\tfrac{3}{2}$, and the slope of a line perpendicular to L is $-\tfrac{2}{3}$. We now can find the equations of the two lines in parts (A) and (B) using the point–slope form.

(A) Parallel ($m = \frac{3}{2}$):

$$y - y_1 = m(x - x_1)$$
$$y - 5 = \tfrac{3}{2}(x + 3)$$
$$y - 5 = \tfrac{3}{2}x + \tfrac{9}{2}$$
$$y = \tfrac{3}{2}x + \tfrac{19}{2}$$

(B) Perpendicular ($m = -\frac{2}{3}$):

$$y - y_1 = m(x - x_1)$$
$$y - 5 = -\tfrac{2}{3}(x + 3)$$
$$y - 5 = -\tfrac{2}{3}x - 2$$
$$y = -\tfrac{2}{3}x + 3$$

Problem 7 Given the line L: $4x + 2y = 3$ and the point $P(2, -3)$, find an equation of a line through P that is:

(A) Parallel to L (B) Perpendicular to L

Write the final answers in the slope–intercept form $y = mx + b$.

Answers to Matched Problems

1.

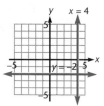

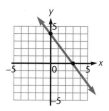

2. (A) $m = 0$ (B) $m = 1$
   (C) $m = -4$ (D) m is not defined

3. $y = \tfrac{2}{3}x + 1$

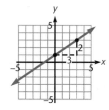

4. (A) $2x + 5y = -4$ (B) $y = -\tfrac{2}{5}x - \tfrac{1}{5}$

5. (A) $C = 0.5x + 200$ (B) $\$2{,}700$

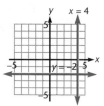

7. (A) $y = -2x + 1$ (B) $y = \tfrac{1}{2}x - 4$

EXERCISE 3-2

A

In Problems 1–6, use the graph of each line to find the x intercept, y intercept, and slope.

1.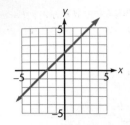

2.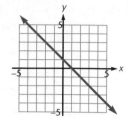

3.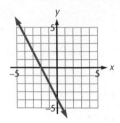

4.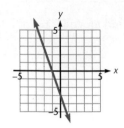

5.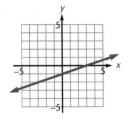

6.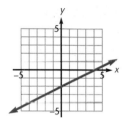

Graph each equation in Problems 7–20, and indicate the slope, if it exists.

7. $y = -\frac{3}{5}x + 4$ **8.** $y = -\frac{3}{2}x + 6$ **9.** $y = -\frac{3}{4}x$ **10.** $y = \frac{2}{3}x - 3$

11. $2x - 3y = 15$ **12.** $4x + 3y = 24$ **13.** $4x - 5y = -24$ **14.** $6x - 7y = -49$

15. $\dfrac{y}{8} - \dfrac{x}{4} = 1$ **16.** $\dfrac{y}{6} - \dfrac{x}{5} = 1$ **17.** $x = -3$ **18.** $y = -2$

19. $y = 3.5$ **20.** $x = 2.5$

In Problems 21–24, write the slope–intercept form of the equation of the line with indicated slope and y intercept.

21. Slope $= 1$; y intercept $= 0$ **22.** Slope $= -1$; y intercept $= 7$

23. Slope $= -\frac{2}{3}$; y intercept $= -4$ **24.** Slope $= \frac{5}{3}$; y intercept $= 6$

B _____

In Problems 25–42, write an equation of the line that contains the indicated point(s) and/or has the indicated slope and/or has the indicated intercepts. Write the final equation in the slope–intercept form $y = mx + b$ or in the form $x = c$.

25. $(0, 4)$; $m = -3$ **26.** $(2, 0)$; $m = 2$

27. $(-5, 4)$; $m = -\frac{2}{5}$ **28.** $(3, -3)$; $m = -\frac{1}{3}$

29. $(5, 5)$; $m = 0$ **30.** $(-4, -2)$; $m = \frac{1}{2}$

31. $(1, 6)$; $(5, -2)$ **32.** $(-3, 4)$; $(6, 1)$

33. $(-4, 8)$; $(2, 0)$ **34.** $(2, -1)$; $(10, 5)$

35. $(-3, 4)$; $(5, 4)$ **36.** $(0, -2)$; $(4, -2)$

37. $(4, 6)$; $(4, -3)$ **38.** $(-3, 1)$; $(-3, -4)$

39. x intercept 6; y intercept 2   **40.** x intercept 3; y intercept 4

41. x intercept -4; y intercept 3   **42.** x intercept -4; y intercept -5

In Problems 43–54, write an equation of the line that contains the indicated point and meets the indicated condition(s). Write the final answer in standard form $Ax + By = C, A \geq 0$.

43. $(-3, 4)$; parallel to $y = 3x - 5$

44. $(-4, 0)$; parallel to $y = -2x + 1$

45. $(2, -3)$; perpendicular to $y = -\frac{1}{3}x$

46. $(-2, -4)$; perpendicular to $y = \frac{2}{3}x - 5$

47. $(2, 5)$; parallel to y axis

48. $(7, 3)$; parallel to x axis

49. $(3, -2)$; vertical

50. $(-2, -3)$; horizontal

51. $(5, 0)$; parallel to $3x - 2y = 4$

52. $(3, 5)$; parallel to $3x + 4y = 8$

53. $(0, -4)$; perpendicular to $x + 3y = 9$

54. $(-2, 4)$; perpendicular to $4x + 5y = 0$

55. Graph $y = mx + 2$ for $m = 2$, $m = \frac{1}{2}$, $m = 0$, $m = -\frac{1}{2}$, and $m = -2$, all on the same coordinate system.

56. Graph $y = -\frac{1}{2}x + b$ for $b = -3$, $b = 0$, and $b = 3$, all on the same coordinate system.

Problems 57–62 refer to the quadrilateral with vertices $A(0, 2)$, $B(4, -1)$, $C(1, -5)$, and $D(-3, -2)$.

57. Show that $AB \parallel DC$.

58. Show that $DA \parallel CB$.

59. Show that $AB \perp BC$.

60. Show that $AD \perp DC$.

61. Find an equation of the perpendicular bisector of AD. [*Hint:* First find the midpoint of AD.]

62. Find an equation of the perpendicular bisector of AB.

Problems 63–68 are calculus-related. Recall that a line tangent to a circle at a point is perpendicular to the radius drawn to that point (see the figure). Find the equation of the line tangent to the circle at the indicated point. Write the final answer in the standard form $Ax + By = C, A \geq 0$. Graph the circle and the tangent line on the same coordinate system.

63. $x^2 + y^2 = 25$, $(3, 4)$

64. $x^2 + y^2 = 100$, $(-8, 6)$

65. $x^2 + y^2 = 50$, $(5, -5)$

66. $x^2 + y^2 = 80$, $(-4, -8)$

67. $(x - 3)^2 + (y + 4)^2 = 169$, $(8, -16)$

68. $(x + 5)^2 + (y - 9)^2 = 289$, $(-13, -6)$

C

Sketch the graphs of the equations in Problems 69–74.

69. $y = |\frac{1}{2}x|$

70. $y = |2x|$

71. $y = |x - 1|$

72. $y = |x + 2|$

73. $x^2 - y^2 = 0$

74. $4y^2 - 9x^2 = 0$

75. Prove that the lines $Ax + By = C$ and $Ax + By = D$, $C \neq D$, are parallel. (Consider two cases, $B \neq 0$, and $B = 0$.)

76. Prove that the lines $Ax + By = C$ and $Bx - Ay = C$ are perpendicular. (Consider two cases, $B \neq 0$ and $B = 0$.)

77. Prove that if a line L has x intercept $(a, 0)$ and y intercept $(0, b)$, then the equation of L can be written in the **intercept form**

$$\frac{x}{a} + \frac{y}{b} = 1 \qquad a, b \neq 0$$

78. Let

$$P_1(x_1, y_1) = P_1(x_1, mx_1 + b)$$
$$P_2(x_2, y_2) = P_2(x_2, mx_2 + b)$$
$$P_3(x_3, y_3) = P_3(x_3, mx_3 + b)$$

be three arbitrary points that satisfy $y = mx + b$ with $x_1 < x_2 < x_3$. Show that $P_1, P_2,$ and P_3 are **collinear**; that is, they lie on the same line. [*Hint:* Use the distance formula and show that $d(P_1, P_2) + d(P_2, P_3) = d(P_1, P_3)$.] This proves that the graph of $y = mx + b$ is a straight line.

APPLICATIONS

79. Physics. The two temperature scales Fahrenheit (F) and Celsius (C) are linearly related. It is known that water freezes at 32°F or 0°C and boils at 212°F or 100°C.
(A) Find a linear equation that expresses F in terms of C.
(B) If a European family sets its house thermostat at 20°C, what is the setting in degrees Fahrenheit? If the outside temperature in Milwaukee is 86°F, what is the temperature in degrees Celsius?
(C) What is the slope of the graph of the linear equation found in part (A)? (The slope indicates the change in Fahrenheit degrees per unit change in Celsius degrees.)

80. Physics. Hooke's law states that the relationship between the stretch s of a spring and the weight w causing the stretch is linear (a principle upon which all spring scales are constructed). For a particular spring, a 5-pound weight causes a stretch of 2 inches, while with no weight the stretch of the spring is 0.
(A) Find a linear equation that expresses s in terms of w.
(B) What weight will cause a stretch of 3.6 inches?
(C) What is the slope of the graph of the equation?

81. Business—Depreciation. A copy machine was purchased by a law firm for $8,000 and is assumed to have a depreciated value of $0 after 5 years. The firm takes straight-line depreciation over the 5-year period.
(A) Find a linear equation that expresses value V in dollars in terms of time t in years.
(B) What is the depreciated value after 3 years?
(C) What is the slope of the graph of the equation found in part (A)? (The slope indicates the decrease in value per year.)

82. Business—Markup Policy. A clothing store sells a shirt costing $20 for $33 and a jacket costing $60 for $93.
(A) If the markup policy of the store for items costing over $10 is assumed to be linear, write an equation that expresses retail price R in terms of cost C (wholesale price).
(B) What does a store pay for a suit that retails for $240?
(C) What is the slope of the graph of the equation found in part (A)? (The slope indicates the change in retail price per unit change in cost.)

83. Flight Conditions. In stable air, the air temperature drops about 5°F for each 1,000-foot rise in altitude.
(A) If the temperature at sea level is 70°F and a commercial pilot reports a temperature of −20°F at 18,000 feet, write a linear equation that expresses temperature T in terms of altitude A (in thousands of feet).
(B) How high is the aircraft if the temperature is 0°F?
(C) What is the slope of the graph of the equation found in part (A)? Interpret.

★**84. Flight Navigation.** An airspeed indicator on some aircraft is affected by the changes in atmospheric pressure at different altitudes. A pilot can estimate the true airspeed by observing the indicated airspeed and adding to it about 2% for every 1,000 feet of altitude.
(A) If a pilot maintains a constant reading of 200 miles per hour on the airspeed indicator as the aircraft climbs from sea level to an altitude of 10,000 feet, write a linear equation that expresses true airspeed T (miles per hour) in terms of altitude A (thousands of feet).
(B) What would be the true airspeed of the aircraft at 6,500 feet?
(C) What is the slope of the graph of the equation found in part (A)? Interpret.

★**85. Oceanography.** After about 9 hours of a steady wind, the height of waves in the ocean is approximately linearly related to the duration of time the wind has been blowing. During a storm with 50-knot winds, the wave height after 9 hours was found to be 23 feet, and after 24 hours it was 40 feet.

(A) If t is time after the 50-knot wind started to blow and h is the wave height in feet, write a linear equation that expresses height h in terms of time t.

(B) How long will the wind have been blowing for the waves to be 50 feet high?

Express all calculated quantities to three significant digits.

86. **Oceanography.** As a diver descends into the ocean, pressure increases linearly with depth. The pressure is 15 pounds per square inch on the surface and 30 pounds per square inch 33 feet below the surface.

(A) If p is the pressure in pounds per square inch and d is the depth below the surface in feet, write an equation that expresses p in terms of d.

(B) How deep can a scuba diver go if the safe pressure for his equipment and experience is 40 pounds per square inch?

★87. **Medicine.** Cardiovascular research has shown that above the 210 cholesterol level, each 1% increase in cholesterol level increases coronary risk 2%. For a par-

ticular age group, the coronary risk at a 210 cholesterol level is found to be 0.160 and at a level of 231 the risk is found to be 0.192.

(A) Find a linear equation that expresses risk R in terms of cholesterol level C.

(B) What is the risk for a cholesterol level of 260?

(C) What is the slope of the graph of the equation found in part (A)? Interpret.

Express all calculated quantities to three significant digits.

★88. **Demographics.** The average number of persons per household in the United States has been shrinking steadily for as long as statistics have been kept and is approximately linear with respect to time. In 1900, there were about 4.76 persons per household and in 1990, about 2.5.

(A) If N represents the average number of persons per household and t represents the number of years since 1900, write a linear equation that expresses N in terms of t.

(B) What is the predicted household size in the year 2000?

Express all calculated quantities to three significant digits.

SECTION 3-3 Functions

- Definition of a Function
- Functions Defined by Equations
- Function Notation
- Application
- A Brief History of the Function Concept

The idea of correspondence plays a central role in the formulation of the function concept. You have already had experiences with correspondences in everyday life. For example:

To each person there corresponds an age.

To each item in a store there corresponds a price.

To each automobile there corresponds a license number.

To each circle there corresponds an area.

To each number there corresponds its cube.

One of the most important aspects of any science (managerial, life, social, physical, computer, etc.) is the establishment of correspondences among various types of phenomena. Once a correspondence is known, predictions can be made. A chemist can use a gas law to predict the pressure of an enclosed gas, given its temperature. An engineer can use a formula to predict the deflections of a beam subject to different loads. A computer scientist can use formulas to compare the efficiency of algorithms for sorting data stored in a computer. An economist would like to be able to predict interest rates, given the rate of change of the money supply. And so on.

● **Definition of a Function**

What do all the preceding examples have in common? Each describes the matching of elements from one set with the elements in a second set. Consider Tables 1–3, which list values for the cube, square, and square root, respectively.

TABLE 1	
Domain (number)	Range (cube)
−2 ⟶ −8
−1 ⟶ −1
0 ⟶ 0
1 ⟶ 1
2 ⟶ 8

TABLE 2

Domain (number)   Range (square)

−2
−1   4
0   1
1   0
2

TABLE 3

Domain (number)   Range (square root)

0 ⟶ 0
1   1
         −1
4   2
         −2
9   3
         −3

Tables 1 and 2 define functions, since to each domain value there corresponds exactly one range value. For example, the cube of -2 is -8 and no other number. On the other hand, Table 3 does not define a function, since to at least one domain value there corresponds more than one range value. For example, to the domain value 9 there corresponds the range values -3 and 3, both square roots of 9. The very important term *function* is now defined.

DEFINITION 1   **Rule Form of the Definition of a Function**

A **function** is a rule that produces a correspondence between two sets of elements such that to each element in the first set there corresponds *one and only one* element in the second set.

The first set is called the **domain**, and the second set is called the **range**.

Since a function is a rule that pairs each element in the domain with a corresponding element in the range, this correspondence can be illustrated by using ordered pairs of elements, where the first component represents a domain element and the second component represents the corresponding range element. Thus, the functions defined in Tables 1 and 2 can be written as follows:

$$\text{Function } 1 = \{(-2, -8), (-1, -1), (0, 0), (1, 1), (2, 8)\}$$
$$\text{Function } 2 = \{(-2, 4), (-1, 1), (0, 0), (1, 1), (2, 4)\}$$

In both cases, notice that no two ordered pairs have the same first component and different second components. On the other hand, if we list the set A of ordered pairs determined by Table 3, we have

$$A = \{(0, 0), (1, 1), (1, -1), (4, 2), (4, -2), (9, 3), (9, -3)\}$$

In this case, there are ordered pairs with the same first component and different second components; for example, $(1, 1)$ and $(1, -1)$ both belong to the set A. Once again, we see that Table 3 does not define a function.

This suggests an alternative but equivalent way of defining functions that produces additional insight into this concept.

DEFINITION 2   **Set Form of the Definition of a Function**

A **function** is a set of ordered pairs with the property that no two ordered pairs have the same first component and different second components.

The set of all first components in a function is called the **domain** of the function, and the set of all second components is called the **range**.

EXAMPLE 1   Functions Defined as Sets of Ordered Pairs

(A) The set $S = \{(1, 4), (2, 3), (3, 2), (4, 3), (5, 4)\}$ defines a function since no two ordered pairs have the same first component and different second components. The domain and range are

$$\text{Domain} = \{1, 2, 3, 4, 5\} \quad \text{Set of first components}$$
$$\text{Range} = \{2, 3, 4\} \quad \text{Set of second components}$$

(B) The set $T = \{(1, 4), (2, 3), (3, 2), (2, 4), (1, 5)\}$ does not define a function since there are ordered pairs with the same first component and different second components [for example, $(1, 4)$ and $(1, 5)$].

Problem 1   Determine whether each set defines a function. If it does, then state the domain and range.

(A) $S = \{(-2, 1), (-1, 2), (0, 0), (-1, 1), (-2, 2)\}$
(B) $T = \{(-2, 1), (-1, 2), (0, 0), (1, 2), (2, 1)\}$

• Functions Defined by Equations

Both versions of the definition of a function are quite general, with no restrictions on the type of elements that make up the domain or range. Points in the plane and complex numbers are two examples of domain and range elements that are used in more advanced courses. In this text, unless otherwise indicated, **the domain and range of a function will be sets of real numbers**.

Defining a function by displaying the rule of correspondence in a table or listing all the ordered pairs in the function only works if the domain and range are finite sets. Functions with finite domains and ranges are used extensively in certain specialized areas, such as computer science, but most applications of functions involve infinite domains and ranges. If the domain and range of a function are infinite sets, then the rule of correspondence cannot be displayed in a table, and it is not possible to actually list all the ordered pairs belonging to the function. For most functions, we use an equation in two variables to specify both the rule of correspondence and the set of ordered pairs.

Consider the equation

$$y = x^2 + 2x \qquad x \text{ any real number} \qquad\qquad (1)$$

This equation assigns to each domain value x exactly one range value y. For example,

$$\text{If } x = 4 \qquad \text{then} \qquad y = (4)^2 + 2(4) = 24$$
$$\text{If } x = -\tfrac{1}{3} \qquad \text{then} \qquad y = (-\tfrac{1}{3})^2 + 2(-\tfrac{1}{3}) = -\tfrac{5}{9}$$

Thus, we can view equation (1) as a function with rule of correspondence

$$y = x^2 + 2x \quad x^2 + 2x \text{ corresponds to } x$$

or, equivalently, as a function with set of ordered pairs

$$\{(x, y) | y = x^2 + 2x, x \text{ a real number}\}$$

The variable x is called an *independent variable*, indicating that values can be assigned "independently" to x from the domain. The variable y is called a *dependent variable*, indicating that the value of y "depends" on the value assigned to x and on the given equation. In general, any variable used as a placeholder for domain values is called an **independent variable**; any variable that is used as a placeholder for range values is called a **dependent variable**.

Which equations can be used to define functions?

Functions Defined by Equations

In an equation in two variables, if to each value of the independent variable there corresponds exactly one value of the dependent variable, then the equation defines a function.

If there is any value of the independent variable to which there corresponds more than one value of the dependent variable, then the equation does not define a function.

EXAMPLE 2   Determining if an Equation Defines a Function

Determine which of the following equations define functions with independent variable x and domain all real numbers:

(A) $y^3 - x = 1$ (B) $y^2 - x^2 = 9$

Solutions   (A) Solving for the dependent variable y, we have

$$y^3 - x = 1 \qquad (2)$$
$$y^3 = 1 + x$$
$$y = \sqrt[3]{1 + x}$$

Since $1 + x$ is a real number for each real number x and since each real number has exactly one real cube root, equation (2) assigns exactly one value of the dependent variable, $y = \sqrt[3]{1 + x}$, to each value of the independent variable x. Thus, equation (2) defines a function.

(B) Solving for the dependent variable y, we have

$$y^2 - x^2 = 9 \qquad (3)$$
$$y^2 = 9 + x^2$$
$$y = \pm\sqrt{9 + x^2}$$

Since $9 + x^2$ is always a positive real number and since each positive real number has two real square roots, each value of the independent variable x corresponds to two values of the dependent variable, $y = -\sqrt{9 + x^2}$ and $y = \sqrt{9 + x^2}$. Thus, equation (3) does not define a function.

Problem 2   Determine which of the following equations define functions with independent variable x and domain all real numbers:

(A) $y^2 - x^4 = 4$ (B) $y^3 - x^3 = 3$

Notice that we have used the phrase "an equation defines a function" rather than "an equation is a function." This is a somewhat technical distinction, but it is employed consistently in mathematical literature and we will adhere to it in this text.

It is very easy to determine whether an equation defines a function if you have the graph of the equation. The two equations we considered in Example 2 are graphed in Figure 1 at the top of the next page.

FIGURE 1 Graphs of equations and the vertical line test.

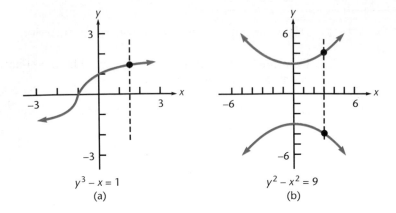

$y^3 - x = 1$
(a)

$y^2 - x^2 = 9$
(b)

In Figure 1(a), each vertical line intersects the graph of the equation $y^3 - x = 1$ in exactly one point. This shows that each value of the independent variable x corresponds to exactly one value of the dependent variable y and confirms our conclusion that this equation defines a function. On the other hand, Figure 1(b) shows that there exist vertical lines that intersect the graph of $y^2 - x^2 = 9$ in two points. This indicates that there exist values of the independent variable x that correspond to two different values of the dependent variable y, which confirms our conclusion that this equation does not define a function. These observations are generalized in Theorem 1.

Theorem 1

Vertical Line Test for a Function

An equation defines a function if each vertical line in the rectangular coordinate system passes through at most one point on the graph of the equation.

If any vertical line passes through two or more points on the graph of an equation, then the equation does not define a function.

In Example 2, the domain was given in the statement of the problem. In many cases, this will not be done. Unless stated to the contrary, we will adhere to the following convention regarding domains and ranges for functions defined by equations:

Agreement on Domains and Ranges

If a function is defined by an equation and the domain is not indicated, then we assume that the domain is the set of all real number replacements of the independent variable that produce *real values* for the dependent variable. The range is the set of all values of the dependent variable corresponding to these domain values.

EXAMPLE 3   Finding the Domain of a Function

Find the domain of the function defined by the equation $y = \sqrt{x - 3}$, assuming x is the independent variable.

Solution   For y to be real, $x - 3$ must be greater than or equal to 0. That is,

$$x - 3 \geq 0 \qquad \text{or} \qquad x \geq 3$$

Thus,

$$\text{Domain: } \{x | x \geq 3\} \text{ or } [3, \infty)$$

Note that in many cases we will dispense with the use of set notation and simply write $x \geq 3$ instead of $\{x | x \geq 3\}$.

Problem 3   Find the domain of the function defined by the equation $y = \sqrt{x + 5}$, assuming x is the independent variable.

● **Function Notation**   We will use letters to name functions and to provide a very important and convenient notation for defining functions. For example, if f is the name of the function defined by the equation $y = 2x + 1$, then instead of the more formal representations

$$f: y = 2x + 1 \qquad \text{Rule of correspondence}$$

or

$$f: \{(x, y) | y = 2x + 1\} \qquad \text{Set of ordered pairs}$$

we simply write

$$f(x) = 2x + 1 \qquad \text{Function notation}$$

The symbol $f(x)$ is read "f of x," "f at x," or "the value of f at x" and represents the number in the range of the function f to which the domain value x is paired. Thus, $f(3)$ is the range value for the function f associated with the domain value 3. We find this range value by replacing x with 3 wherever x occurs in the function definition

$$f(x) = 2x + 1$$

and evaluating the right side,

$$f(3) = 2 \cdot 3 + 1$$
$$= 6 + 1$$
$$= 7$$

The statement $f(3) = 7$ indicates in a concise way that the function f assigns the range value 7 to the domain value 3 or, equivalently, that the ordered pair (3, 7) belongs to f.

The symbol $f{:}x \rightarrow f(x)$, read "f maps x into $f(x)$," is also used to denote the relationship between the domain value x and the range value $f(x)$ (see Figure 2). Whenever we write $y = f(x)$, we assume that x is an independent variable and that y and $f(x)$ both represent the dependent variable.

FIGURE 2 Function notation.

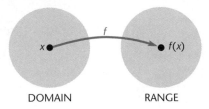

DOMAIN   RANGE

The function f "maps" the domain value x into the range value $f(x)$.

Letters other than f and x can be used to represent functions and independent variables. For example,

$$g(t) = t^2 - 3t + 7$$

defines g as a function of the independent variable t. To find $g(-2)$, we replace t by -2 wherever t occurs in

$$g(t) = t^2 - 3t + 7$$

and evaluate the right side:

$$\begin{aligned} g(-2) &= (-2)^2 - 3(-2) + 7 \\ &= 4 + 6 + 7 \\ &= 17 \end{aligned}$$

Thus, the function g assigns the range value 17 to the domain value -2; the ordered pair $(-2, 17)$ belongs to g.

It is important to understand and remember the definition of the symbol $f(x)$:

DEFINITION 3   **The Symbol $f(x)$**

The symbol $f(x)$ represents the real number in the range of the function f corresponding to the domain value x. Symbolically, $f{:}x \rightarrow f(x)$. The ordered pair $(x, f(x))$ belongs to the function f. If x is a real number that is not in the domain of f, then f is **not defined** at x and $f(x)$ **does not exist**.

EXAMPLE 4   Evaluating Functions

For

$$f(x) = \frac{15}{x - 3} \qquad g(x) = 16 + 3x - x^2 \qquad h(x) = \sqrt{25 - x^2}$$

find:

(A) $f(6)$ (B) $g(-7)$ (C) $h(10)$ (D) $f(0) + g(4) - h(-3)$

Solution (A) $f(6) \boxed{= \dfrac{15}{6-3}} = \dfrac{15}{3} = 5$

(B) $g(-7) \boxed{= 16 + 3(-7) - (-7)^2} = 16 - 21 - 49 = -54$

(C) $h(10) \boxed{= \sqrt{25 - 10^2}} = \sqrt{25 - 100} = \sqrt{-75}$

But $\sqrt{-75}$ is not a real number. Since we have agreed to restrict the domain of a function to values of x that produce real values for the function, 10 is not in the domain of h and $h(10)$ is not defined.

(D) $f(0) + g(4) - h(-3)$

$$\boxed{= \frac{15}{0-3} + [16 + 3(4) - 4^2] - \sqrt{25 - (-3)^2}}$$

$$= \frac{15}{-3} + 12 - \sqrt{16}$$

$$= -5 + 12 - 4 = 3$$

Problem 4 Use the functions in Example 4 to find:

(A) $f(-2)$ (B) $g(6)$ (C) $h(-8)$ (D) $\dfrac{f(8)}{h(4)}$

EXAMPLE 5 **Finding Domains of Functions**

Find the domains of functions f, g, and h:

$$f(x) = \frac{15}{x-3} \qquad g(x) = 16 + 3x - x^2 \qquad h(x) = \sqrt{25 - x^2}$$

Solution **Domain of f**
The fraction $15/(x-3)$ represents a real number for all replacements of x by real numbers except $x = 3$, since division by 0 is not defined. Thus, $f(3)$ does not exist, and the domain of f is the set of all real numbers except 3. We often indicate this by writing

$$f(x) = \frac{15}{x-3} \qquad x \neq 3$$

Domain of g
The domain is R, the set of all real numbers, since $16 + 3x - x^2$ represents a real number for all replacements of x by real numbers.

Domain of *h*

The domain is the set of all real numbers x such that $\sqrt{25 - x^2}$ is a real number— that is, such that $25 - x^2 \geq 0$. Solving $25 - x^2 = (5 - x)(5 + x) \geq 0$ by methods discussed in Section 2-7, we find

$$\text{Domain:}\quad -5 \leq x \leq 5 \quad \text{or} \quad [-5, 5]$$

Problem 5   Find the domains of functions F, G, and H:

$$F(x) = x^2 + 5x - 2 \qquad G(x) = \sqrt{\frac{x - 2}{x + 3}} \qquad H(x) = \frac{4}{x - 2}$$

In addition to evaluating functions at specific numbers, it is important to be able to evaluate functions at expressions that involve one or more variables. For example, the **difference quotient**

$$\frac{f(x + h) - f(x)}{h} \qquad x \text{ and } x + h \text{ in the domain of } f, h \neq 0$$

is studied extensively in a calculus course.

EXAMPLE 6   Evaluating and Simplifying a Difference Quotient

For $f(x) = x^2 + 4x + 5$, find and simplify:

(A) $f(x + h)$ (B) $\dfrac{f(x + h) - f(x)}{h}, h \neq 0$

Solution   (A) To find $f(x + h)$, we replace x with $x + h$ everywhere it appears in the equation that defines f and simplify:

$$\begin{aligned} f(x + h) &= (x + h)^2 + 4(x + h) + 5 \\ &= x^2 + 2xh + h^2 + 4x + 4h + 5 \end{aligned}$$

(B) Using the result of part (A), we get

$$\begin{aligned} \frac{f(x + h) - f(x)}{h} &= \frac{x^2 + 2xh + h^2 + 4x + 4h + 5 - (x^2 + 4x + 5)}{h} \\ &= \frac{x^2 + 2xh + h^2 + 4x + 4h + 5 - x^2 - 4x - 5}{h} \\ &= \frac{2xh + h^2 + 4h}{h} = \frac{h(2x + h + 4)}{h} = 2x + h + 4 \end{aligned}$$

Problem 6   Repeat Example 6 for $f(x) = x^2 + 3x + 7$.

CAUTION

1. If f is a function, then the symbol $f(x + h)$ represents the value of f at the number $x + h$ and must be evaluated by replacing the independent variable in the equation that defines f with the expression $x + h$, as we did in Example 6. Do not confuse this notation with the familiar algebraic notation for multiplication:

$$f(x + h) \neq fx + fh \qquad f(x + h) \text{ is function notation.}$$

$$4(x + h) = 4x + 4h \qquad 4(x + h) \text{ is algebraic multiplication notation.}$$

2. There is another common incorrect interpretation of the symbol $f(x + h)$. If f is an arbitrary function, then

$$f(x + h) \neq f(x) + f(h)$$

It is possible to find some particular functions for which $f(x + h) = f(x) + f(h)$ is a true statement, but in general these two expressions are not equal.

• Application

EXAMPLE 7   Construction

A rectangular feeding pen for cattle is to be made with 100 meters of fencing.

(A) If x represents the width of the pen, express its area $A(x)$ in terms of x.
(B) What is the domain of the function A (determined by the physical restrictions)?

Solutions   (A) Draw a figure and label the sides.

Perimeter = 100 meters of fencing.
Half the perimeter = 50.
If x = Width, then $50 - x$ = Length.

x (Width)

$50 - x$ (Length)

$$A(x) = (\text{Width})(\text{Length}) = x(50 - x)$$

(B) To have a pen, x must be positive, but x must also be less than 50 (or the length will not exist). Thus,

Domain: $0 < x < 50$ Inequality notation

$(0, 50)$ Interval notation

Problem 7   Rework Example 7 with the added assumption that a large barn is to be used as one side of the pen.

● **A Brief History of the Function Concept**   The history of the use of functions in mathematics illustrates the tendency of mathematicians to extend and generalize each concept. The word "function" appears to have been first used by Leibniz in 1694 to stand for any quantity associated with a curve. By 1718, Johann Bernoulli considered a function any expression made up of constants and a variable. Later in the same century, Euler came to regard a function as any equation made up of constants and variables. Euler made extensive use of the extremely important notation $f(x)$, although its origin is generally attributed to Clairaut (1734).

The form of the definition of function that has been used until well into this century (many texts still contain this definition) was formulated by Dirichlet (1805–1859). He stated that, if two variables x and y are so related that for each value of x there corresponds exactly one value of y, then y is said to be a (single-valued) function of x. He called x, the variable to which values are assigned at will, the independent variable, and y, the variable whose values depend on the values assigned to x, the dependent variable. He called the values assumed by x the domain of the function, and the corresponding values assumed by y the range of the function.

Now, since set concepts permeate almost all mathematics, we have the more general definition of function presented in this section in terms of sets of ordered pairs of elements.

Answers to Matched Problems

1. (A) S does not define a function
   (B) T defines a function with domain $\{-2, -1, 0, 1, 2\}$ and range $\{0, 1, 2\}$
2. (A) Does not define a function   (B) Defines a function
3. $x \geq -5$ Inequality notation
   $[-5, \infty)$ Interval notation
4. (A) -3 (B) -2 (C) Does not exist   (D) 1
5. Domain of F: all real numbers
   Domain of G: $x < -3$ or $x \geq 2$ Inequality notation
                  $(-\infty, -3) \cup [2, \infty)$ Interval notation
   Domain of H: All real numbers except 2
6. (A) $x^2 + 2xh + h^2 + 3x + 3h + 7$ (B) $2x + h + 3$
7. (A) $A(x) = x(100 - 2x)$ (B) Domain: $0 < x < 50$ Inequality notation
                                              $(0, 50)$ Interval notation

EXERCISE 3-3

A

Indicate whether each table in Problems 1–6 defines a function.

1. Domain   Range

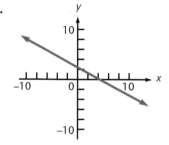

2. Domain   Range

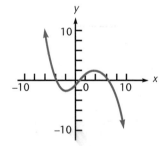

3. Domain   Range

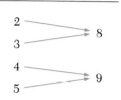

4. Domain   Range

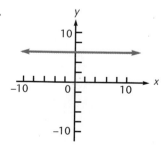

5. Domain   Range

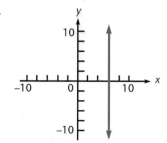

6. Domain   Range

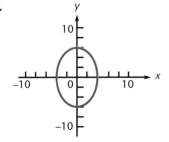

Indicate whether each set in Problems 7–12 defines a function. Find the domain and range of each function.

7. $\{(2, 4), (3, 6), (4, 8), (5, 10)\}$

8. $\{(-1, 4), (0, 3), (1, 2), (2, 1)\}$

9. $\{(10, -10), (5, -5), (0, 0), (5, 5), (10, 10)\}$

10. $\{(-10, 10), (-5, 5), (0, 0), (5, 5), (10, 10)\}$

11. $\{(0, 1), (1, 1), (2, 1), (3, 2), (4, 2), (5, 2)\}$

12. $\{(1, 1), (2, 1), (3, 1), (1, 2), (2, 2), (3, 2)\}$

Indicate whether each graph in Problems 13–18 is the graph of a function.

13.

14.

15.

16.

17.

18.

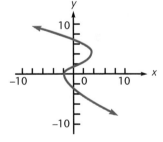
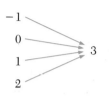

Problems 19–30 refer to the functions

$$f(x) = 3x - 5 \qquad g(t) = 4 - t$$
$$F(m) = 3m^2 + 2m - 4 \qquad G(u) = u - u^2$$

Evaluate as indicated.

19. $f(-1)$ **20.** $g(6)$ **21.** $G(-2)$ **22.** $F(-3)$

23. $F(-1) + f(3)$ **24.** $G(2) - g(-3)$ **25.** $2F(-2) - G(-1)$ **26.** $3G(-2) + 2F(-1)$

27. $\dfrac{f(0) \cdot g(-2)}{F(-3)}$ **28.** $\dfrac{g(4) \cdot f(2)}{G(1)}$ **29.** $\dfrac{f(4) - f(2)}{2}$ **30.** $\dfrac{g(5) - g(3)}{2}$

B _____

Which equations in Problems 31–48 define a function, given that x is an independent variable?

31. $y = 3 - x$ **32.** $y = 2x + 3$ **33.** $y = 2x^2 - 3x + 5$ **34.** $y = (2 - x)^3$

35. $y^2 - x = 2$ **36.** $y - x^2 = 2$ **37.** $y = |x - 2|$ **38.** $|y| = x + 2$

39. $x^2 + y^2 = 81$ **40.** $16x^2 + y^2 = 16$ **41.** $y^3 = x$ **42.** $y^5 = x$

43. $y = \dfrac{x + 3}{x - 2}$ **44.** $y = \dfrac{x - 4}{2x^2 + 7x - 4}$ **45.** $y = \sqrt{x + 1}$ **46.** $y^2 = x + 1$

47. $y = \sqrt{\dfrac{x + 3}{x - 2}}$ **48.** $y = \sqrt{x^2 + x - 12}$

In Problems 49–64, find the domain of the indicated function.

49. $f(x) = 3x + 8$ **50.** $g(x) = -2x + 11$

51. $h(x) = \sqrt{x + 2}$ **52.** $k(x) = \sqrt{4 - x}$

53. $s(x) = \dfrac{2 + 3x}{4 - x}$ **54.** $m(x) = \dfrac{3 - 5x}{7 + x}$

55. $n(x) = \dfrac{x^2 - 2x + 9}{x^2 - 2x - 8}$ **56.** $p(x) = \dfrac{x^2 + 11}{x^2 + 3x - 10}$

57. $F(x) = \sqrt{4 - x^2}$ **58.** $G(x) = \sqrt{x^2 - 9}$

59. $H(x) = \sqrt{x^2 - 3x - 4}$ **60.** $K(x) = \sqrt{3 - 2x - x^2}$

61. $L(x) = \sqrt{\dfrac{5 - x}{x - 2}}$ **62.** $M(x) = \sqrt{\dfrac{x - 1}{6 - x}}$

63. $N(x) = \dfrac{1}{\sqrt[3]{x^2 - 1}}$ **64.** $P(x) = \dfrac{1}{\sqrt[3]{8 - x^3}}$

65. If $F(s) = 3s + 15$, find: $\dfrac{F(2 + h) - F(2)}{h}$ **66.** If $K(r) = 7 - 4r$, find: $\dfrac{K(1 + h) - K(1)}{h}$

67. If $g(x) = 2 - x^2$, find: $\dfrac{g(3 + h) - g(3)}{h}$ **68.** If $P(m) = 2m^2 + 3$, find: $\dfrac{P(2 + h) - P(2)}{h}$

69. If $Q(t) = t^2 - 2t + 1$, find: $\dfrac{Q(1 + h) - Q(1)}{h}$ **70.** If $S(u) = 3u^2 + 5u - 7$, find: $\dfrac{S(3 + h) - S(3)}{h}$

71. If $L(w) = -2w^2 + 3w - 1$, find:

$$\dfrac{L(-2 + h) - L(-2)}{h}$$

72. If $D(p) = -3p^2 - 4p + 9$, find:

$$\dfrac{D(-1 + h) - D(-1)}{h}$$

C

 In Problems 73–80, find and simplify:

$$\text{(A)} \quad \frac{f(x + h) - f(x)}{h} \qquad \text{(B)} \quad \frac{f(x) - f(a)}{x - a}$$

73. $f(x) = 3x - 4$

74. $f(x) = -2x + 5$

75. $f(x) = x^2 - 1$

76. $f(x) = x^2 + x - 1$

77. $f(x) = -3x^2 + 9x - 12$

78. $f(x) = -x^2 - 2x - 4$

79. $f(x) = x^3$

80. $f(x) = x^3 + x$

81. The area of a rectangle is 64 square inches. Express the perimeter $P(w)$ as a function of the width w.

82. The perimeter of a rectangle is 50 inches. Express the area $A(w)$ as a function of the width w.

83. The altitude of a right triangle is 5 meters. Express the hypotenuse $h(b)$ as a function of the base b.

84. The altitude of a right triangle is 4 meters. Express the base $b(h)$ as a function of the hypotenuse h.

APPLICATIONS 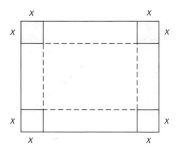 🌐 *Most of the applications in this section are calculus-related. That is, similar problems will appear in a calculus course, but additional analysis of the functions will be required.*

85. Cost Function. The fixed costs per day for a doughnut shop are $300, and the variable costs are $1.75 per dozen doughnuts produced. If x dozen doughnuts are produced daily, express the daily cost $C(x)$ as a function of x.

86. Cost Function. The fixed costs per day for a ski manufacturer are $3,750, and the variable costs are $68 per pair of skis produced. If x pairs of skis are produced daily, express the daily cost $C(x)$ as a function of x.

87. Physics—Rate. The distance in feet that an object falls in a vacuum is given by $s(t) = 16t^2$, where t is time in seconds. Find:
(A) $s(0), s(1), s(2), s(3)$
(B) $\dfrac{s(2 + h) - s(2)}{h}$
(C) What happens in part (B) when h tends to 0? Interpret physically.

88. Physics—Rate. An automobile starts from rest and travels along a straight and level road. The distance in feet traveled by the automobile is given by $s(t) = 10t^2$, where t is time in seconds. Find:
(A) $s(8), s(9), s(10), s(11)$
(B) $\dfrac{s(11 + h) - s(11)}{h}$
(C) What happens in part (B) as h tends to 0? Interpret physically.

89. Manufacturing. A candy box is to be made out of a piece of cardboard that measures 8 by 12 inches.

Squares, x inches on a side, will be cut from each corner, and then the ends and sides will be folded up (see the figure below). Find a formula for the volume of the box $V(x)$ in terms of x. From practical considerations, what is the domain of the function V?

90. Construction. A rancher has 20 miles of fencing to fence a rectangular piece of grazing land along a straight river. If no fence is required along the river and the sides perpendicular to the river are x miles long, find a formula for the area $A(x)$ of the rectangle in terms of x. From practical considerations, what is the domain of the function A?

★ **91. Construction.** The manager of an animal clinic wants to construct a kennel with four individual pens, as indicated in the figure. State law requires that each pen have a gate 3 feet wide and an area of 50 square feet. If x is the width of one pen, express the total amount of fencing $F(x)$ (excluding the gates) required for the construction of the kennel as a function of x. Complete the following table [round values of $F(x)$ to one decimal place]:

x	4	5	6	7
$F(x)$				

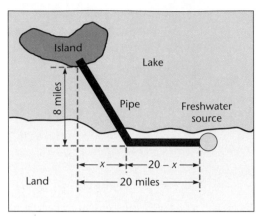

★**92. Architecture.** An architect wants to design a window with an area of 24 square feet in the shape of a rectangle surmounted by a semicircle, as indicated in the figure. If x is the width of the window, express the perimeter $P(x)$ of the window as a function of x. Complete the table below [round each value of $P(x)$ to one decimal place]:

x	4	5	6	7
$P(x)$				

★**93. Construction.** A freshwater pipeline is to be run from a source on the edge of a lake to a small resort community on an island 8 miles offshore, as indicated in the figure. It costs \$10,000 per mile to lay the pipe on land and \$15,000 per mile to lay the pipe in the lake. Express the total cost $C(x)$ of constructing the pipeline as a function of x. From practical considerations, what is the domain of the function C?

★**94. Weather.** An observation balloon is released at a point 10 miles from the station that receives its signal and rises vertically, as indicated in the figure. Express the distance $d(h)$ between the balloon and the receiving station as a function of the altitude h of the balloon.

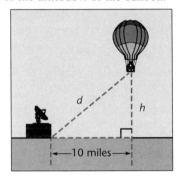

★★**95. Operational Costs.** The cost per hour for fuel for running a train is $v^2/5$ dollars, where v is the speed in miles per hour. (Note that cost goes up as the square of the speed.) Other costs, including labor, are \$400 per hour. Express the total cost of a 500-mile trip as a function of the speed v.

★★**96. Operational Costs.** Refer to Problem 95. If it takes t hours for the train to complete the 500-mile trip, express the total cost as a function of t.

SECTION 3-4 Graphing Functions

- Basic Concepts
- Linear Functions
- Quadratic Functions
- Piecewise-Defined Functions
- The Greatest Integer Function

In this section we take another look at the graphs of linear equations, this time using the function concepts introduced in the preceding section. We also develop procedures for graphing functions defined by quadratic equations and functions formed by "piecing" together two or more other functions. We begin by discussing some general concepts related to the graphs of functions.

● Basic Concepts

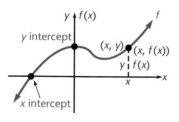

FIGURE 1 Graph of a function.

Each function that has a real number domain and range has a graph—the graph of the ordered pairs of real numbers that constitute the function. When functions are graphed, domain values usually are associated with the horizontal axis and range values with the vertical axis. Thus, the **graph of a function f** is the same as the graph of the equation

$$y = f(x)$$

where x is the independent variable and the abscissa of a point on the graph of f. The variables y and $f(x)$ are dependent variables, and either is the ordinate of a point on the graph of f (see Figure 1).

The abscissa of a point where the graph of a function crosses the x axis is called an ***x* intercept** of the function. The x intercepts are determined by finding the real solutions of the equation $f(x) = 0$, if any exist. The ordinate of a point where the graph of a function crosses the y axis is called the ***y* intercept** of the function. The y intercept is given by $f(0)$, provided 0 is in the domain of f. Note that a function can have more than one x intercept but can never have more than one y intercept—a consequence of the vertical line test discussed in the preceding section.

We now take a look at increasing and decreasing properties of functions. Intuitively, a function is increasing over an interval I in its domain if its graph rises as the independent variable increases over I. A function is decreasing over I if its graph falls as the independent variable increases over I (Figure 2).

FIGURE 2 Increasing, decreasing, and constant functions.

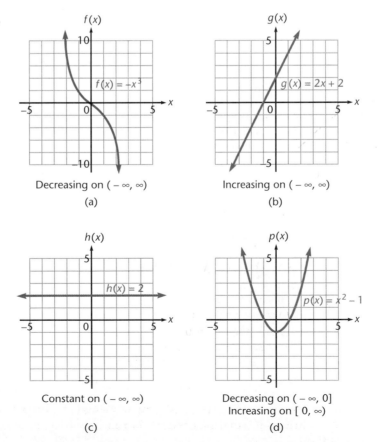

More formally, we define increasing, decreasing, and constant functions as follows:

DEFINITION 1   **Increasing, Decreasing, and Constant Functions**

Let I be an interval in the domain of a function f. Then:

1. f is **increasing** on I if $f(b) > f(a)$ whenever $b > a$ in I.
2. f is **decreasing** on I if $f(b) < f(a)$ whenever $b > a$ in I.
3. f is **constant** on I if $f(a) = f(b)$ for all a and b in I.

• **Linear Functions**   We now apply the general concepts discussed above to a specific class of functions known as *linear functions*.

DEFINITION 2   **Linear Function**

A function f is a **linear function** if

$$f(x) = mx + b \qquad m \neq 0$$

where m and b are real numbers.

Graphing a linear function is equivalent to graphing the equation

$$y = mx + b$$

which we recognize as the equation of a line with slope m and y intercept b. Since the expression $mx + b$ represents a real number for all real number replacements of x, the domain of a linear function is the set of all real numbers. The restriction $m \neq 0$ in the definition of a linear function implies that the graph is not a horizontal line. Hence, the range of a linear function is also the set of all real numbers.

Graph of $f(x) = mx + b$, $m \neq 0$

The graph of a linear function f is a nonvertical and nonhorizontal straight line with slope m and y intercept b.

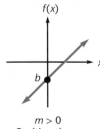
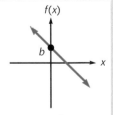

$m > 0$
Positive slope
Increasing on $(-\infty, \infty)$

$m < 0$
Negative slope
Decreasing on $(-\infty, \infty)$

Domain: All real numbers   Range: All real numbers

Notice that two types of lines are not the graphs of linear functions. A vertical line with equation $x = a$ does not pass the vertical line test and cannot define a function. A horizontal line with equation $y = b$ does pass the vertical line test and does define a function. However, a function of the form

$$f(x) = b \quad \text{Constant function}$$

is called a **constant function**, not a linear function.

EXAMPLE 1   Graphing a Linear Function

Find the slope and intercepts, and then sketch the graph of the linear function defined by

$$f(x) = -\tfrac{2}{3}x + 4$$

Solution   The y intercept is $f(0) = 4$, and the slope is $-\tfrac{2}{3}$. To find the x intercept, we solve the equation $f(x) = 0$ for x:

$$f(x) = 0$$
$$-\tfrac{2}{3}x + 4 = 0$$
$$-\tfrac{2}{3}x = -4$$
$$x = (-\tfrac{3}{2})(-4) = 6 \quad x \text{ intercept}$$

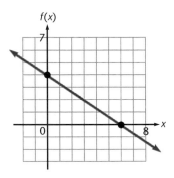

Problem 1   Find the slope and intercepts, and then sketch the graph of the linear function defined by

$$f(x) = \tfrac{3}{2}x - 6$$

● **Quadratic Functions**   Just as we used the first-degree polynomial $mx + b$, $m \neq 0$, to define a linear function, we use the second-degree polynomial $ax^2 + bx + c$, $a \neq 0$, to define a *quadratic function*.

DEFINITION 3

Quadratic Function

A function f is a **quadratic function** if

$$f(x) = ax^2 + bx + c \qquad a \neq 0 \qquad (1)$$

where a, b, and c are real numbers.

Since the expression $ax^2 + bx + c$ represents a real number for all real number replacements of x,

the domain of a quadratic function is the set of all real numbers.

The range of a quadratic function and many important features of its graph can be determined by first transforming equation (1) by completing the square into the form

$$f(x) = a(x - h)^2 + k \qquad \text{minimum value} \qquad (2)$$

A brief review of completing the square, which we discussed in Section 2-6, might prove helpful at this point. We illustrate this method through an example and then generalize the results.

Consider the quadratic function given by

$$f(x) = 2x^2 - 8x + 4$$

We start by transforming this function into form (2) by completing the square as follows:

$$
\begin{aligned}
f(x) &= 2x^2 - 8x + 4 \\
&= 2(x^2 - 4x) + 4 && \text{Factor the coefficient of } x^2 \text{ out of the first two terms.} \\
&= 2(x^2 - 4x + ?) + 4 \\
&= 2(x^2 - 4x + 4) + 4 - 8 && \text{We add 4 to complete the square inside the parentheses. But because of the 2 outside the parentheses, we have actually added 8, so we must subtract 8.} \\
&= 2(x - 2)^2 - 4 && \text{The transformation is complete.}
\end{aligned}
$$

Thus,

$$f(x) = 2(x - 2)^2 - 4 \qquad (3)$$

If $x = 2$, then $2(x - 2)^2 = 0$ and $f(2) = -4$. For any other value of x, the positive number $2(x - 2)^2$ is added to -4, thus making $f(x)$ larger. Therefore,

$$f(2) = -4$$

is the *minimum value* of $f(x)$ for all x—a very important result! Furthermore, if we choose any two x values that are equidistant from the vertical line $x = 2$, we will obtain the same value for the function, as Table 1 illustrates.

TABLE 1   **Values of a Quadratic Function**

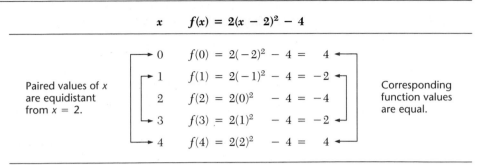

	x	$f(x) = 2(x - 2)^2 - 4$

Paired values of x are equidistant from $x = 2$.

$0 \quad f(0) = 2(-2)^2 - 4 = \quad 4$
$1 \quad f(1) = 2(-1)^2 - 4 = -2$
$2 \quad f(2) = 2(0)^2 \quad - 4 = -4$
$3 \quad f(3) = 2(1)^2 \quad - 4 = -2$
$4 \quad f(4) = 2(2)^2 \quad - 4 = \quad 4$

Corresponding function values are equal.

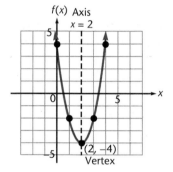

FIGURE 3 Graph of $f(x) = 2(x - 2)^2 + 4$.

Plotting the points from Table 1 and connecting them with a smooth curve, we obtain the graph shown in Figure 3. This graph should look familiar to you. It is a parabola. The vertical line $x = 2$ is a line of symmetry.

In general, it can be shown that the graph of a quadratic function is always a parabola with line of symmetry parallel to the vertical axis. The lowest or highest point on the parabola, whichever exists, is called the **vertex**. The vertical line of symmetry through the vertex is called the **axis** of the parabola. Thus, for this example, the vertical line $x = 2$ is the axis, and the vertex is $(2, -4)$.

Examining the graph in Figure 3, we see that $f(x)$ is decreasing on $(-\infty, 2]$ and increasing on $[2, \infty)$. Furthermore, $f(x)$ can assume any value greater than or equal to -4, but no values less than -4. Thus, the range of f is

$$\text{Range of } f: y \geq -4 \quad \text{or} \quad [-4, \infty)$$

We also note that this graph has two x intercepts, which are the solutions of the equation $f(x) = 0$. Since we have already transformed $f(x)$ into form (3), we can use this form to find the x intercepts:

$$f(x) = 0$$
$$2(x - 2)^2 - 4 = 0$$
$$2(x - 2)^2 = 4$$
$$(x - 2)^2 = 2$$
$$x - 2 = \pm\sqrt{2}$$
$$x = 2 \pm \sqrt{2}$$

Thus, the x intercepts are $x = 2 + \sqrt{2}$ and $x = 2 - \sqrt{2}$.

For completeness, we note that the y intercept is $f(0) = 4$, as shown in Table 1.

Note the important results we have obtained from transforming $f(x)$ into form (3):

The axis of the parabola

The vertex of the parabola

The minimum value of $f(x)$

The graph of $y = f(x)$

The range of f

The x intercepts

The y intercept

Starting with the general quadratic function defined by $f(x) = ax^2 + bx + c$, $a \neq 0$, and completing the square (see Section 2-6), we obtain

$$f(x) = a\left(x + \frac{b}{2a}\right)^2 + \frac{4ac - b^2}{4a}$$

Using the same reasoning as in the preceding example, we obtain the following general result:

Properties of $f(x) = ax^2 + bx + c$, $a \neq 0$, and Its Graph

1. **Axis (of symmetry):** $x = -\dfrac{b}{2a}$

2. **Vertex:** $\left(-\dfrac{b}{2a}, f\left(-\dfrac{b}{2a}\right)\right)$

3. **Maximum or minimum value of $f(x)$:**

$$f\left(-\frac{b}{2a}\right) = \begin{cases} \text{Minimum} & \text{if } a > 0 \\ \text{Maximum} & \text{if } a < 0 \end{cases}$$

4. **Graph (a parabola):**

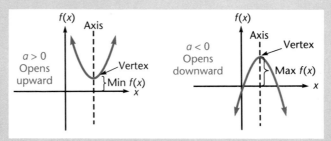

5. **Domain:** All real numbers
   Range: Determine from graph

To graph a quadratic function using the method of completing the square, we can either actually complete the square as in the preceding example or use the properties in the box. Some of you can probably more readily remember a formula, others a process. We use the properties in the box in the next example.

EXAMPLE 2   **Graph of a Quadratic Function**

Graph, finding the axis, vertex, maximum or minimum of $f(x)$, intervals where f is increasing and decreasing, range, and intercepts:

$$f(x) = -x^2 - 2x - 3$$

Solution   Note that $a = -1, b = -2,$ and $c = -3.$

Axis of Symmetry

$$x = -\frac{b}{2a} = -\frac{-2}{2(-1)} = -1$$

Vertex

$$\left(-\frac{b}{2a}, f\left(-\frac{b}{2a}\right)\right) = (-1, f(-1)) = (-1, -2)$$

Maximum Value of $f(x)$
Since $a = -1 < 0,$

$$\text{Max } f(x) = f\left(-\frac{b}{2a}\right) = f(-1) = -2$$

Graph
To graph f, locate the axis and vertex; then plot several points on either side of the axis.

x	-3	-2	-1	0	1
$f(x)$	-6	-3	-2	-3	-6

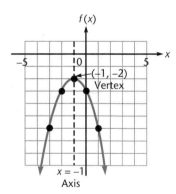

Increasing and Decreasing
From the graph, we see that f is increasing on $(-\infty, -1]$ and decreasing on $[-1, \infty)$.

Range
From the graph, we see that $y = f(x)$ can be any number less than or equal to -2. Thus, the range of f is $y \leq -2$ or $(-\infty, -2]$.

y Intercept

$$f(0) = -(0)^2 - 2(0) - 3 = -3$$

x Intercepts

Since the graph of f does not cross the x axis, there are no x intercepts. The solutions of the equation $f(x) = 0$ are the imaginary numbers $-1 + i\sqrt{2}$ and $-1 - i\sqrt{2}$ (verify this).

Problem 2   Graph, finding the axis, vertex, maximum or minimum of $f(x)$, intervals where f is increasing and decreasing, range, and intercepts:

$$f(x) = -x^2 + 4x - 4$$

● **Piecewise-Defined Functions**

The **absolute value function** can be defined using the definition of absolute value from Section 2-4:

$$f(x) = |x| = \begin{cases} -x & \text{if } x < 0 \\ x & \text{if } x \geq 0 \end{cases}$$

Notice that this function is defined by different formulas for different parts of its domain. Functions whose definitions involve more than one formula are called **piecewise-defined functions**. As the next example illustrates, piecewise-defined functions occur naturally in many applications.

EXAMPLE 3   Rental Charges

A car rental agency charges $0.25 per mile if the total mileage does not exceed 100. If the total mileage exceeds 100, the agency charges $0.25 per mile for the first 100 miles plus $0.15 per mile for the additional mileage. If x represents the number of miles a rented vehicle is driven, express the mileage charge $C(x)$ as a function of x. Find $C(50)$ and $C(150)$, and graph C.

Solution   If $0 \leq x \leq 100$, then

$$C(x) = 0.25x$$

If $x > 100$, then

	Charge for the first 100 miles		Charge for the additional mileage
$C(x) =$	$0.25(100)$	$+$	$0.15(x - 100)$
$=$	25	$+$	$0.15x - 15$
$=$	$10 + 0.15x$		

Thus, we see that C is a piecewise-defined function:

$$C(x) = \begin{cases} 0.25x & \text{if } 0 \leq x \leq 100 \\ 10 + 0.15x & \text{if } x > 100 \end{cases}$$

Piecewise-defined functions are evaluated by first determining which rule applies and then using the appropriate rule to find the value of the function. For example, to evaluate $C(50)$, we use the first rule and obtain

$$C(50) = 0.25(50) = \$12.50 \quad x = 50 \text{ satisfies } 0 \le x \le 100$$

To evaluate $C(150)$, we use the second rule and obtain

$$C(150) = 10 + 0.15(150) = \$32.50 \quad x = 150 \text{ satisfies } x > 100$$

To graph C, we graph each rule in the definition for the indicated values of x.

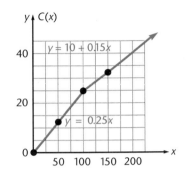

x	$y = 0.25x$
50	12.5
100	25

x	$y = 10 + 0.15x$
100	25
150	32.5

Notice that the two formulas produce the same value at $x = 100$ and that the graph of C contains no breaks. Informally, a graph (or portion of a graph) is said to be **continuous** if it contains no breaks or gaps. (A formal presentation of continuity may be found in calculus texts.)

Problem 3 Refer to Example 3. Find $C(x)$ if the agency charges $0.30 per mile when the total mileage does not exceed 75, and $0.30 per mile for the first 75 miles plus $0.20 per mile for additional mileage when the total mileage exceeds 75. Find $C(50)$ and $C(100)$, and graph C.

EXAMPLE 4 **Graphing a Function Involving Absolute Value**

Graph the function f given by

$$f(x) = x + \frac{x}{|x|}$$

and find its domain and range.

Solution We use the piecewise definition of $|x|$ to find a piecewise definition of f that does not involve $|x|$.

If $x < 0$, then $|x| = -x$ and

$$f(x) = x + \frac{x}{|x|} = x + \frac{x}{-x} = x - 1$$

If $x = 0$, then f is not defined, since division by 0 is not permissible.
If $x > 0$, then $|x| = x$ and

$$f(x) = x + \frac{x}{|x|} = x + \frac{x}{x} = x + 1$$

Thus, a piecewise definition for f is

$$f(x) = \begin{cases} x - 1 & \text{if } x < 0 \\ x + 1 & \text{if } x > 0 \end{cases}$$

Domain: $x \neq 0$ or $(-\infty, 0) \cup (0, \infty)$

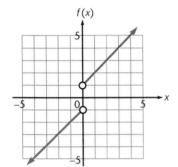

We use this definition to graph f as shown in the margin. Examining this graph, we see that $y = f(x)$ can be any number less than -1 or any number greater than 1. Thus,

Range: $y < -1$ or $y > 1$ or $(-\infty, -1) \cup (1, \infty)$

Notice that we used open dots in the figure at $(0, -1)$ and $(0, 1)$ to indicate that these points do not belong to the graph of f. Because of the break in the graph at $x = 0$, we say that f is **discontinuous** at $x = 0$.

Problem 4   Graph the function f given by

$$f(x) = -\frac{2x}{|x|} - x$$

and find its domain and range.

• The Greatest Integer Function

We conclude this section with a discussion of an interesting and useful function called the *greatest integer function*.

The **greatest integer** of a real number x, denoted by $[\![x]\!]$, is the integer n such that $n \leq x < n + 1$; that is, $[\![x]\!]$ is the largest integer less than or equal to x. For example,

$$[\![3.45]\!] = 3 \qquad [\![-2.13]\!] = -3 \quad \text{Not } -2$$

$$[\![7]\!] = 7 \qquad [\![-8]\!] = -8$$

$$[\![0]\!] = 0$$

The **greatest integer function** f is defined by the equation $f(x) = [\![x]\!]$. A piecewise definition of f for $-2 \le x < 3$ and a sketch of the graph of f for $-5 \le x \le 5$ are shown in Figure 4. Since the domain of f is all real numbers, the piecewise definition continues indefinitely in both directions, as does the stairstep pattern in the figure. Thus, the range of f is the set of all integers. The greatest integer function is an example of a more general class of functions called **step functions**.

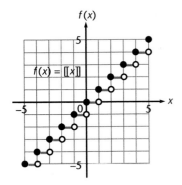

$$f(x) = [\![x]\!] = \begin{cases} \vdots \\ -2 & \text{if } -2 \le x < -1 \\ -1 & \text{if } -1 \le x < 0 \\ 0 & \text{if } 0 \le x < 1 \\ 1 & \text{if } 1 \le x < 2 \\ 2 & \text{if } 2 \le x < 3 \\ \vdots \end{cases}$$

FIGURE 4 Greatest integer function.

Notice in Figure 4 that at each integer value of x there is a break in the graph, and between integer values of x there is no break. Thus, the greatest integer function is discontinuous at each integer n and continuous on each interval of the form $[n, n + 1)$.

Most computer programming languages contain a greatest integer function, usually denoted by int(x), although not all languages define it the same way we have for negative values of x. If you use the greatest integer function in a programming language such as BASIC, Pascal, or FORTRAN, be certain to find out how that language defines this function. The next example illustrates how the greatest integer function can be used to round numbers to a specified number of decimal places, a common operation in computer programming.

EXAMPLE 5 Computer Science

Let

$$f(x) = \frac{[\![10x + 0.5]\!]}{10}$$

Find:

(A) $f(6)$ (B) $f(1.8)$ (C) $f(3.24)$ (D) $f(4.582)$ (E) $f(-2.68)$

What operation does this function perform?

Solutions

(A) $f(6) = \dfrac{[\![60.5]\!]}{10} = \dfrac{60}{10} = 6$

(B) $f(1.8) = \dfrac{[\![18.5]\!]}{10} = \dfrac{18}{10} = 1.8$

(C) $f(3.24) = \dfrac{[\![32.9]\!]}{10} = \dfrac{32}{10} = 3.2$

x	$f(x)$
6	6
1.8	1.8
3.24	3.2
4.582	4.6
-2.68	-2.7

(D) $f(4.582) = \dfrac{[\![46.32]\!]}{10} = \dfrac{46}{10} = 4.6$

(E) $f(-2.68) = \dfrac{[\![-26.3]\!]}{10} = \dfrac{-27}{10} = -2.7$

Comparing the values of x and $f(x)$ in the table in the margin, we conclude that this function rounds decimal fractions to the nearest tenth.

Problem 5 Let $f(x) = [\![x + 0.5]\!]$. Find:

(A) $f(6)$ (B) $f(1.8)$ (C) $f(3.24)$ (D) $f(-4.3)$ (E) $f(-2.69)$

What operation does this function perform?

Answers to Matched Problems

1. y intercept: $f(0) = -6$
   x intercept: 4
   Slope: $\frac{3}{2}$

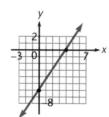

2. Axis: $x = 2$
   Vertex: $(2, f(2)) = (2, 0)$
   Max $f(x)$: $f(2) = 0$
   Increasing: $(-\infty, 2]$
   Decreasing: $[2, \infty)$
   Range: $(-\infty, f(2)] = (-\infty, 0]$
   y intercept: $f(0) = -4$
   x intercept: $x = 2$

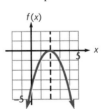

3. $C(x) = \begin{cases} 0.3x & \text{if } 0 \le x \le 75 \\ 7.5 + 0.2x & \text{if } x < 75 \end{cases}$

   $C(50) = \$15; \ C(100) = \$27.50$

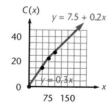

4. $f(x) = \begin{cases} 2 - x & \text{if } x < 0 \\ -2 - x & \text{if } x > 0 \end{cases}$

   Domain: $x \ne 0$ or $(-\infty, 0) \cup (0, \infty)$
   Range: $(-\infty, -2) \cup (2, \infty)$

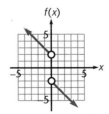

5. (A) 6   (B) 2   (C) 3   (D) -4 (E) -3;
   f rounds decimal fractions to the nearest integer.

EXERCISE 3-4

A _____

Problems 1–4 refer to functions f, g, p, and q given by the following graphs. (Assume the graphs continue as indicated beyond the parts shown.)

1. For the function f at the right, find:
 (A) Domain
 (B) Range
 (C) x intercepts
 (D) y intercept
 (E) Intervals over which f is increasing
 (F) Intervals over which f is decreasing
 (G) Intervals over which f is constant
 (H) Any points of discontinuity

2. Repeat Problem 1 for the function g.

3. Repeat Problem 1 for the function p.

4. Repeat Problem 1 for the function q.

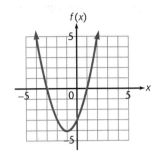
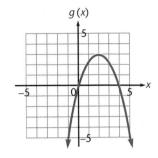
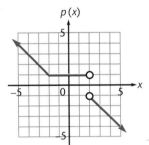
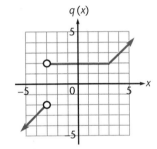

In Problems 5–8, find the slope and intercepts, and then sketch the graph.

5. $f(x) = 2x + 4$ **6.** $f(x) = 3x - 3$ **7.** $f(x) = -\frac{1}{2}x - \frac{5}{3}$ **8.** $f(x) = -\frac{3}{4}x + \frac{6}{5}$

In Problems 9 and 10, find a linear function with the indicated slope and y intercept.

9. Slope -3; y intercept 5   **10.** Slope 4; y intercept -7

In Problems 11 and 12, find a linear function whose graph passes through the indicated points.

11. $(-2, 5); (4, 2)$ **12.** $(3, -2); (9, 6)$

In Problems 13 and 14, find a linear function f satisfying the given conditions.

13. $f(-2) = 7$ and $f(4) = -2$ **14.** $f(-3) = -2$ and $f(5) = 4$

B _____

In Problems 15–18, graph, finding the axis, vertex, maximum or minimum, and range.

15. $f(x) = (x - 3)^2 + 2$ **16.** $f(x) = \frac{1}{2}(x + 2)^2 - 4$

17. $f(x) = -(x + 3)^2 - 2$ **18.** $f(x) = -(x - 2)^2 + 4$

In Problems 19–22, graph, finding the axis, vertex, x intercepts, and y intercept.

19. $f(x) = x^2 - 4x - 5$

20. $f(x) = x^2 - 6x + 5$

21. $f(x) = -x^2 + 6x$

22. $f(x) = -x^2 + 2x + 8$

In Problems 23–26, graph, finding the axis, vertex, intervals over which f is increasing, and intervals over which f is decreasing.

23. $f(x) = x^2 + 6x + 11$

24. $f(x) = x^2 - 8x + 14$

25. $f(x) = -x^2 + 6x - 6$

26. $f(x) = -x^2 - 10x - 24$

In Problems 27–38, graph, finding the domain, range, and any points of discontinuity.

27. $f(x) = \begin{cases} x + 1 & \text{if } -1 \le x < 0 \\ -x + 1 & \text{if } 0 \le x \le 1 \end{cases}$

28. $f(x) = \begin{cases} x & \text{if } -2 \le x < 1 \\ -x + 2 & \text{if } 1 \le x \le 2 \end{cases}$

29. $f(x) = \begin{cases} -2 & \text{if } -3 \le x < -1 \\ 4 & \text{if } -1 < x \le 2 \end{cases}$

30. $f(x) = \begin{cases} 1 & \text{if } -2 \le x < 2 \\ -3 & \text{if } 2 < x \le 5 \end{cases}$

31. $f(x) = \begin{cases} x + 2 & \text{if } x < -1 \\ x - 2 & \text{if } x \ge -1 \end{cases}$

32. $f(x) = \begin{cases} -1 - x & \text{if } x \le 2 \\ 5 - x & \text{if } x > 2 \end{cases}$

33. $g(x) = \begin{cases} x^2 + 1 & \text{if } x < 0 \\ -x^2 - 1 & \text{if } x > 0 \end{cases}$

34. $h(x) = \begin{cases} -x^2 - 2 & \text{if } x < 0 \\ x^2 + 2 & \text{if } x > 0 \end{cases}$

35. $k(x) = \begin{cases} 1 & \text{if } x < 0 \\ x + 1 & \text{if } 0 \le x < 2 \\ 2 & \text{if } x \ge 2 \end{cases}$

36. $g(x) = \begin{cases} -x & \text{if } x < 0 \\ 2 & \text{if } 0 \le x < 2 \\ x - 2 & \text{if } x \ge 2 \end{cases}$

37. $p(x) = \begin{cases} 0 & \text{if } x < 0 \\ 4 - x^2 & \text{if } 0 \le x < 2 \\ 0 & \text{if } x \ge 2 \end{cases}$

38. $T(x) = \begin{cases} 0 & \text{if } x < 0 \\ x^2 & \text{if } 0 \le x < 2 \\ 0 & \text{if } x \ge 2 \end{cases}$

C ──────────────────────────────

In Problems 39–44, graph, finding the axis, vertex, maximum or minimum of f(x), range, intercepts, intervals over which f is increasing, and intervals over which f is decreasing.

39. $f(x) = \frac{1}{2}x^2 + 2x + 3$ $f(x) = x^2 + x + 1$

40. $f(x) = 2x^2 - 12x + 14$

41. $f(x) = 4x^2 - 12x + 9$

42. $f(x) = -\frac{1}{2}x^2 + 4x - 10$

43. $f(x) = -2x^2 - 8x - 2$

44. $f(x) = -4x^2 - 4x - 1$

In Problems 45–50, find a piecewise definition of f that does not involve the absolute value function (see Example 4). Sketch the graph, and find the domain, range, and any points of discontinuity.

45. $f(x) = \dfrac{|x|}{x}$

46. $f(x) = x|x|$

47. $f(x) = x + \dfrac{|x - 1|}{x - 1}$

48. $f(x) = x + 2\dfrac{|x + 1|}{x + 1}$

49. $f(x) = |x| + |x - 2|$

50. $f(x) = |x| - |x - 3|$

where v is the velocity of the cycle in feet per second as it leaves the ramp.

(A) How fast must the cycle be traveling when it leaves the ramp in order to follow the trajectory illustrated in the figure?

(B) What is the maximum height of the cycle above the ground as it follows this trajectory?

★★**84.** **Physics.** The trajectory of a circus performer shot from a cannon is given by the graph of the function

$$f(x) = x - \frac{1}{100}x^2$$

Both the cannon and the net are 10 feet high (see the figure).

(A) How far from the muzzle of the cannon should the center of the net be placed so that the performer lands in the center of the net?

(B) What is the maximum height of the performer above the ground?

SECTION 3-5 Aids to Graphing Functions

* Graphs of Basic Functions
* Graphing Aids

The functions f and g given by

$$f(x) = x^2 \qquad \text{and} \qquad g(x) = 2x^2 + 3$$

satisfy the algebraic relationship

$$g(x) = 2f(x) + 3$$

In this section we see how this algebraic relationship and a knowledge of the graph of f can be used as an aid to graphing g. The techniques developed in this section provide an efficient method for graphing many frequently encountered functions.

• Graphs of Basic Functions

To make efficient use of the techniques discussed in this section, you need to be familiar with the graphs of the six basic functions shown in Figure 1 on the next page. The graphs of these basic functions can then be used to form the graphs of more complicated functions. Study the graph of each function shown in Figure 1 carefully, until you can draw it without reference to the figure (a calculator might be helpful in some cases).

Notice in Figure 1 that the graphs of the square and absolute value functions are symmetric with respect to the vertical axis, while the graphs of the identity, cube, and cube root functions are symmetric with respect to the origin. The terms *even* and *odd*, as defined below, are commonly used to describe the symmetry properties of graphs of functions.

FIGURE 1 Graphs of basic functions.

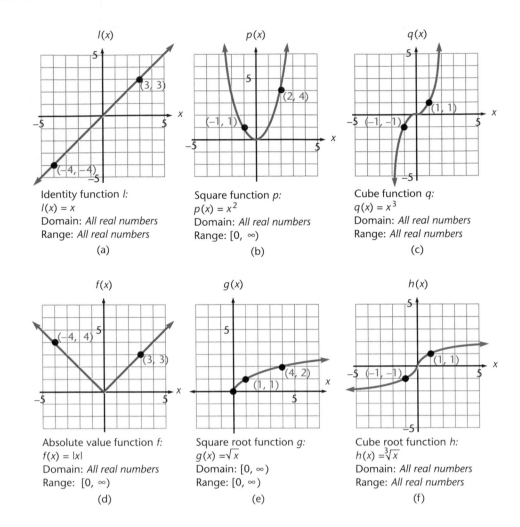

Identity function *l*:
$l(x) = x$
Domain: *All real numbers*
Range: *All real numbers*
(a)

Square function *p*:
$p(x) = x^2$
Domain: *All real numbers*
Range: $[0, \infty)$
(b)

Cube function *q*:
$q(x) = x^3$
Domain: *All real numbers*
Range: *All real numbers*
(c)

Absolute value function *f*:
$f(x) = |x|$
Domain: *All real numbers*
Range: $[0, \infty)$
(d)

Square root function *g*:
$g(x) = \sqrt{x}$
Domain: $[0, \infty)$
Range: $[0, \infty)$
(e)

Cube root function *h*:
$h(x) = \sqrt[3]{x}$
Domain: *All real numbers*
Range: *All real numbers*
(f)

DEFINITION 1   **Even and Odd Functions**

A function is called an **even function** if its graph is symmetric with respect to the vertical axis and an **odd function** if its graph is symmetric with respect to the origin.

Thus, the square and absolute value functions are even functions, and the identity, cube, and cube root functions are odd functions. Notice in Figure 1(e) that the square root function is not symmetric with respect to the vertical axis or the origin. Consequently, the square root function is neither even nor odd.

If we can determine whether a function is even or odd before we draw its graph, then we can use this information as an aid to the graphing process. As a consequence of the tests for symmetry discussed in Section 3-1, we have the following tests for even and odd functions:

Theorem 1 **Tests for Even and Odd Functions**

If $f(-x) = f(x)$, then f is an **even function**.

If $f(-x) = -f(x)$, then f is an **odd function**.

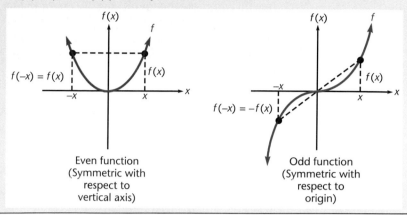

Even function   Odd function
(Symmetric with   (Symmetric with
respect to   respect to
vertical axis)   origin)

EXAMPLE 1 **Testing for Even and Odd Functions**

Without graphing, determine whether the functions f, g, and h are even, odd, or neither:

(A) $f(x) = x^4 + 1$ (B) $g(x) = x^3 + 1$ (C) $h(x) = x^5 + x$

Solutions   (A) $f(x) = x^4 + 1$

$f(-x) = (-x)^4 + 1$

$\qquad = x^4 + 1$ $(-x)^4 = [(-1)x]^4 = (-1)^4x^4 = x^4$, since $(-1)^4 = 1$.

$\qquad = f(x)$

Therefore, f is even.

(B) $g(x) = x^3 + 1$

$g(-x) = (-x)^3 + 1$

$\qquad = -x^3 + 1$ $(-x)^3 = (-1)^3x^3 = -x^3$, since $(-1)^3 = -1$.

$-g(x) = -(x^3 + 1)$

$\qquad = -x^3 - 1$

Since $g(-x) \neq g(x)$ and $g(-x) \neq -g(x)$, g is neither even nor odd.

(C) $h(x) = x^5 + x$

$h(-x) = (-x)^5 + (-x) = -x^5 - x = -(x^5 + x) = -h(x)$

Therefore, h is odd.

Problem 1 Without graphing, determine whether the functions F, G, and H are even, odd, or neither:

(A) $F(x) = x^3 - 2x$ (B) $G(x) = x^2 + 1$ (C) $H(x) = 2x + 4$

In the solution of Example 1 notice that we used the fact that

$$(-x)^n = \begin{cases} x^n & \text{if } n \text{ is an even integer} \\ -x^n & \text{if } n \text{ is an odd integer} \end{cases}$$

It is this property that motivates the use of the terms *even* and *odd* when describing symmetry properties of the graphs of functions. In addition to being an aid to graphing, certain problems and developments in calculus and more advanced mathematics are simplified if we recognize the presence of either an even or an odd function.

● **Graphing Aids**   We now turn to the central idea of this section: using the known graphs of basic functions to graph related functions. This involves relationships of the form

$$F(x) = f(x) + k \qquad G(x) = f(x + k) \qquad \text{and} \qquad H(x) = kf(x)$$

where f is a function whose graph is known and k is a constant. These graphing aids will be illustrated through examples.

EXAMPLE 2   Vertical Shifts

Graph $f(x) = x^2$, $F(x) = x^2 + 3$, and $G(x) = x^2 - 2$.

Solution   We already know that the graph of $y = x^2$ is a parabola opening upward; it has its vertex at the origin and has as its axis the y axis, as shown in figure (a). To graph $y = x^2 + 3$, we add 3 to each ordinate value for the graph of $y = x^2$, as shown in figure (b). To graph $y = x^2 - 2$, we subtract 2 from each ordinate value, as shown in figure (c). The net result is that the graph of $y = x^2 + 3$ is just the graph of $y = x^2$ shifted upward 3 units, and the graph of $y = x^2 - 2$ is just the graph of $y = x^2$ shifted downward 2 units.

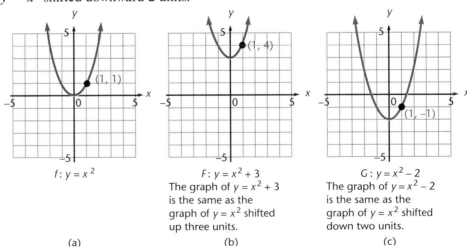

The functions F and G in Example 2 are related to the function f by the equations $F(x) = f(x) + 3$ and $G(x) = f(x) - 2$ and their graphs are formed by shifting the graph of f up 3 units and down 2 units, respectively. These results are generalized in Theorem 2.

Theorem 2	**Vertical Shifting (Translation)**

To Graph:	Shift the Graph of $y = f(x)$:
1. $y = f(x) + k, k > 0$	**1.** Up k units
2. $y = f(x) - k, k > 0$	**2.** Down k units

Problem 2   Graph $f(x) = |x| + k$ for $k = 0, k = 2$, and $k = -3$.

We now turn to horizontal shifting.

EXAMPLE 3   **Horizontal Shifts**

Graph $f(x) = x^2$, $P(x) = (x + 2)^2$, and $Q(x) = (x - 3)^2$.

Solution   Observe the following:

$$f(x) = x^2 \qquad P(x) = (x + 2)^2 \qquad Q(x) = (x - 3)^2$$
$$f(a) = a^2 \qquad P(a - 2) = (a - 2 + 2)^2 \qquad Q(a + 3) = (a + 3 - 3)^2$$
$$(a, a^2) \in f \qquad\qquad = a^2 \qquad\qquad = a^2$$
$$(a - 2, a^2) \in P \qquad\qquad (a + 3, a^2) \in Q$$

Thus, the point with abscissa a on the graph of $y = f(x) = x^2$ has the same ordinate value, a^2, as the point with abscissa $a - 2$ on the graph of $y = P(x) = (x + 2)^2$ and the point with abscissa $a + 3$ on the graph of $y = Q(x) = (x - 3)^2$. We conclude that the graph of $y = (x + 2)^2$ is the same as the graph of $y = x^2$ shifted to the left 2 units, as shown in figure (b). And the graph of $y = (x - 3)^2$ is the same as the graph of $y = x^2$ shifted to the right 3 units, as shown in figure (c).

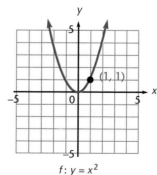

$f: y = x^2$

(a)

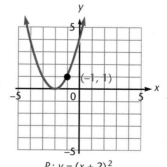

$P: y = (x + 2)^2$
The graph of $y = (x + 2)^2$ is the same as the graph of $y = x^2$ shifted to the left two units.

(b)

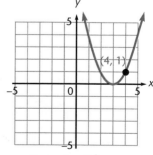

$Q: y = (x - 3)^2$
The graph of $y = (x - 3)^2$ is the same as the graph of $y = x^2$ shifted to the right three units.

(c)

The functions P and Q in Example 3 are related to the function f by the equations

$$P(x) = f(x + 2) \quad \text{and} \quad Q(x) = f(x - 3)$$

and their graphs are formed by shifting the graph of f to the left 2 units and to the right 3 units, respectively. These results are generalized in Theorem 3.

Theorem 3

$f(x) = x^2$

$f(x+h) = (x+h)^2$

Horizontal Shifting (Translation)

To Graph:	Shift the Graph of $y = f(x)$:
1. $y - f(x + h), h > 0$	1. To the left h units
2. $y = f(x - h), h > 0$	2. To the right h units

CAUTION   Do not confuse the direction of a horizontal shift with that of a vertical shift. The vertical shift $y = f(x) + 2$ shifts the graph of f upward, which is the positive direction on the y axis. However, the horizontal shift $y = f(x + 2)$ shifts the graph of f to the left, which is the negative direction on the x axis. Also, the horizontal shift $y = f(x - 2)$ shifts the graph of f to the right, whereas the vertical shift $y = f(x) - 2$ shifts the graph of f downward.

Problem 3   Graph $f(x) = |x + h|$ for $h = 0, h = 2$, and $h = -3$.

We consider reflections, expansions, and contractions of graphs in the next example.

EXAMPLE 4   **Reflections, Expansions, and Contractions**

Graph $f(x) = x^2$, $R(x) = -x^2$, $S(x) = 2x^2$, and $T(x) = \frac{1}{2}x^2$.

Solution   If we take the negative of each ordinate value on the graph of $y = x^2$ [figure (a)], we obtain the graph of $y = -x^2$, as shown in figure (b). If we double each ordinate value on the graph of $y = x^2$, we obtain the graph of $y = 2x^2$, as shown in figure (c). And if we cut each ordinate value in half on the graph of $y = x^2$, we obtain the graph of $y = \frac{1}{2}x^2$, as shown in figure (d).

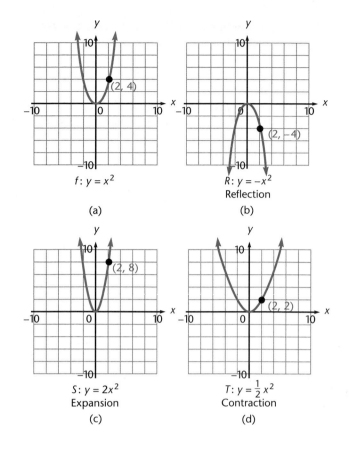

(a) $f: y = x^2$

(b) $R: y = -x^2$ Reflection

(c) $S: y = 2x^2$ Expansion

(d) $T: y = \frac{1}{2} x^2$ Contraction

Following the same line of reasoning as in Example 4, the results can be generalized as given in Theorem 4.

Theorem 4

Reflection, Expansion, and Contraction

To Graph:

1. $y = -f(x)$
     1. Reflect the graph of $y = f(x)$ across the x axis.

2. $y = Cf(x), \quad C > 1$
     2. Expand the graph of $y = f(x)$ by multiplying each ordinate value by C.

3. $y = Cf(x), \quad 0 < C < 1$
     3. Contract the graph of $y = f(x)$ by multiplying each ordinate value by C.

Problem 4   Graph $f(x) = C|x|$ for $C = 1$, $C = -1$, $C = 2$, and $C = \frac{1}{2}$.

All the graphing aids for functions discussed in this section are summarized in the next box for convenient reference.

Graphing Aids for Functions

Symmetry Tests:

$f(-x) = f(x)$ Graph of $y = f(x)$ is symmetric with respect to the vertical axis.

$f(-x) = -f(x)$ Graph of $y = f(x)$ is symmetric with respect to the origin.

Vertical Translation:

$y = f(x) + k, k > 0$ Shift the graph of $y = f(x)$ upward k units.
$y = f(x) - k, k > 0$ Shift the graph of $y = f(x)$ downward k units.

Horizontal Translation:

$y = f(x - h), h > 0$ Shift the graph of $y = f(x)$ to the right h units.

$y = f(x + h), h > 0$ Shift the graph of $y = f(x)$ to the left h units.

Reflection:

$y = -f(x)$ Reflect the graph of $y = f(x)$ in the x axis.

Expansion and Contraction:

$y = Cf(x), C > 1$ Expand the graph of $y = f(x)$ by multiplying each ordinate value by C.
$y = Cf(x), 0 < C < 1$ Contract the graph of $y = f(x)$ by multiplying each ordinate value by C.

More complicated functions often can be graphed by applying several of these graphing aids. Example 5 illustrates this process.

EXAMPLE 5   Combining Graphing Aids

Graph:

(A) $f(x) = 3 - x^3$ (B) $g(x) = 2\sqrt{x + 2}$

Solutions   (A) To graph $y = 3 - x^3$, we first reflect the graph of $y = x^3$ [figure (a)] across the x axis, as shown in figure (b), and then we shift the graph of $y = -x^3$ up 3 units, as shown in figure (c).

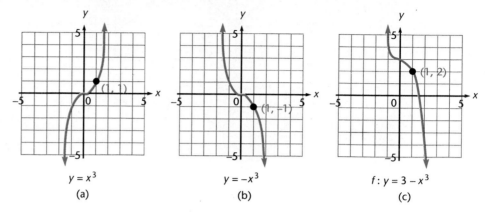

(B) To form the graph of $y = 2\sqrt{x + 2}$, we first shift the graph of $y = \sqrt{x}$ [figure (a)] to the left 2 units, as shown in figure (b), and then we multiply the ordinate values by 2, as shown in figure (c).

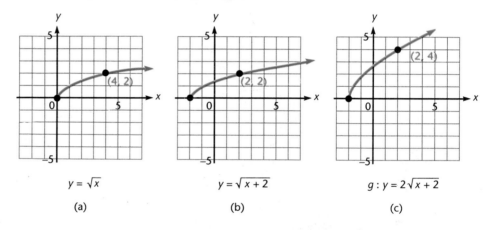

Problem 5   Graph:

(A) $f(x) = 4 - \sqrt{x}$ (B) $g(x) = \frac{1}{4}(x - 1)^3$

Answers to Matched Problems
1. (A) Odd   (B) Even   (C) Neither
2.

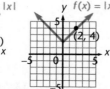
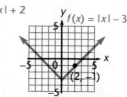

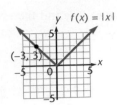

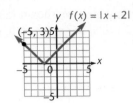

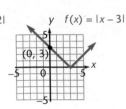

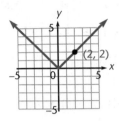

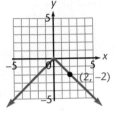

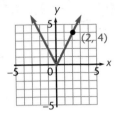

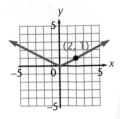

$$f(x) = |x|$$

$$f(x) = -|x|$$

$$f(x) = 2|x|$$

$$f(x) = \frac{1}{2}|x|$$

EXERCISE 3-5

A _____

Without graphing, indicate whether each function in Problems 1–10 is even, odd, or neither.

1. $g(x) = x^3 + x$

2. $f(x) = x^5 - x$

3. $m(x) = x^4 + 3x^2$

4. $h(x) = x^4 - x^2$

5. $F(x) = x^5 + 1$

6. $f(x) = x^5 - 3$

7. $G(x) = x^4 + 2$

8. $P(x) = x^4 - 4$

9. $q(x) = x^2 + x - 3$

10. $n(x) = 2x - 3$

Problems 11–22 refer to the functions f and g given by the graphs below (the domain of each function is $[-2, 2]$).

Use the graph of f or g, as required, to graph each given function.

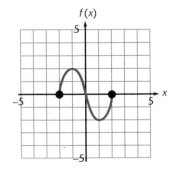

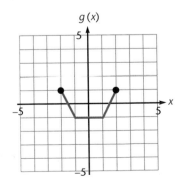

11. $f(x) + 2$

12. $g(x) - 1$

13. $g(x) + 2$

14. $f(x) - 1$

15. $f(x - 2)$ **16.** $g(x - 1)$ **17.** $g(x + 2)$ **18.** $f(x - 1)$

19. $-f(x)$ **20.** $-g(x)$ **21.** $2g(x)$ **22.** $\frac{1}{2}f(x)$

B _____

Graph the functions in Problems 23–28.

23. $f(x) = \sqrt{x} + k$ for $k = 0, k = 2, k = -3$ **24.** $g(x) = -x^2 + k$ for $k = 0, k = 4, k = -1$

25. $F(x) = \sqrt{x + h}$ for $h = 0, h = 4, h = -1$ **26.** $G(x) = -(x + h)^2$ for $h = 0, h = 1, h = -2$

27. $p(x) = C\sqrt{x}$ for $C = 1, C = -1, C = 2, C = \frac{1}{2}$ **28.** $Q(x) = Cx$ for $C = 1, C = -1, C = 2, C = \frac{1}{2}$

Indicate how the graph of each function in Problems 29–34 is related to the graph of $y = x^2$, $y = |x|$, or $y = \sqrt{x}$. Graph each function.

29. $g(x) = -(x + 2)^2$ **30.** $h(x) = -(x - 3)^2$

31. $G(x) = |x + 2|$ **32.** $H(x) = -|x - 3|$

33. $P(x) = -\sqrt{x - 1}$ **34.** $Q(x) = -\sqrt{x + 2}$

The graph of $f(x) = \sqrt{1 - x^2}$ is the upper half of the circle with radius 1 and center (0, 0) (see the figure). Use the graph of f to graph each function in Problems 35–38.

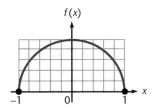

35. $g(x) = -\sqrt{1 - x^2}$ **36.** $h(x) = 1 + \sqrt{1 - x^2}$

37. $G(x) = \sqrt{1 - (x - 1)^2}$ **38.** $H(x) = \sqrt{1 - (x + 2)^2}$

Graph each function in Problems 39–48 using the aids to graphing discussed in this section. Indicate whether a function is even or odd, and show any points of discontinuity.

39. $g(x) = -4$ **40.** $f(x) = -2$

41. $S(x) = -2|x + 2|$ **42.** $f(x) = -\frac{1}{2}(x - 2)^2$

43. $g(x) = \frac{1}{2}(x + 1)^2 - 2$ **44.** $h(x) = -2|x - 2| + 2$

45. $p(x) = -\frac{1}{4}x^3$ **46.** $T(x) = -3x^{1/3}$

47. $f(x) = 2x^{1/3} + 4$ **48.** $h(x) = \frac{1}{2}x^3 - 4$

error in book

$f(x) = C(x - h)^2 + k$

In Problems 49–52, use completing the square to transform each quadratic function f into the form $f(x) = C(x + h)^2 + k$ where C, h, and k are constants. Indicate how the graph of f is related to the graph of the function $p(x) = x^2$. Graph $y = f(x)$.

49. $f(x) = 2x^2 - 8x + 4$ **50.** $f(x) = 2x^2 + 4x - 1$

51. $f(x) = -\frac{1}{2}x^2 + 2x + 1$ **52.** $f(x) = -\frac{1}{4}x^2 - x + 4$

C _____

Graph each function in Problems 53–58 using the aids to graphing discussed in this section. Indicate whether a function is even or odd, and show any points of discontinuity.

53. $p(x) = -[\![x]\!]$

54. $q(x) = [\![x]\!] + 2$

55. $m(x) = -[\![x - 1]\!]$

56. $g(x) = [\![x + 3]\!]$

57. $r(x) = 2[\![x - 3]\!] + 1$

58. $S(x) = 3 - \frac{1}{2}[\![x + 1]\!]$

59. Let f be any function with the property that $-x$ is in the domain of f whenever x is in the domain of f, and let E and O be the functions defined by

$$E(x) = \tfrac{1}{2}[f(x) + f(-x)]$$

and

$$O(x) = \tfrac{1}{2}[f(x) - f(-x)]$$

(A) Show that E is always even.
(B) Show that O is always odd.
(C) Show that $f(x) = E(x) + O(x)$. What is your conclusion?

60. Let f be any function with the property that $-x$ is in the domain of f whenever x is in the domain of f, and let $g(x) = xf(x)$.
(A) If f is even, is g even, odd, or neither?
(B) If f is odd, is g even, odd, or neither?

 Problems 61–72 require the use of a graphic calculator or a computer.

In Problems 61–64, graph $f(x)$, $f(|x|)$, and $f(-|x|)$ in separate viewing rectangles.

61. $f(x) = (x - 1)^2 - 1$

62. $f(x) = 8 - (x + 2)^2$

63. $f(x) = -0.2(x + 2)^3 + 4$

64. $f(x) = 0.1(x - 1)^3 - 2$

65. Describe the relationship between the graphs of $f(x)$ and $f(|x|)$ in Problems 61–64.

66. Describe the relationship between the graphs of $f(x)$ and $f(-|x|)$ in Problems 61–64.

In Problems 67–70, graph $f(x)$, $|f(x)|$, and $-|f(x)|$ in separate viewing rectangles.

67. $f(x) = 0.2x^2 - 5$

68. $f(x) = 4 - 0.25x^2$

69. $f(x) = 4 - 0.1(x + 2)^3$

70. $f(x) = 0.25(x - 1)^3 - 1$

71. Describe the relationship between the graphs of $f(x)$ and $|f(x)|$ in Problems 67–70.

72. Describe the relationship between the graphs of $f(x)$ and $-|f(x)|$ in Problems 67–70.

_____

APPLICATIONS

73. Production Costs. Total production costs for a product can be broken down into fixed costs, which do not depend on the number of units produced, and variable costs, which do depend on the number of units pro-

duced. Thus, the total cost of producing x units of the product can be expressed in the form

$$C(x) = K + f(x)$$

where K is a constant that represents the fixed costs and $f(x)$ is a function that represents the variable costs. Use the graph of the variable-cost function $f(x)$ shown in the figure to graph the total cost function if the fixed costs are $30,000.

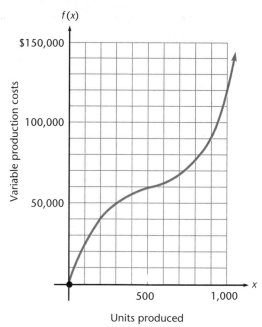

Units produced

★ **74. Cost Function.** Refer to the variable-cost function $f(x)$ in Problem 73. Suppose construction of a new production facility results in a 25% decrease in the variable cost at all levels of output. If F is the new variable-cost function, use the graph of f to graph $y = F(x)$.

75. Timber Harvesting. To determine when a forest should be harvested, forest managers often use formulas to estimate the number of board feet a tree will produce. A board foot equals 1 square foot of wood, 1 inch thick. Suppose that the number of board feet y yielded by a tree can be estimated by

$$y = f(x) = 15 + 0.004(x - 10)^3$$

where x is the diameter of the tree in inches measured at a height of 4 feet above the ground. Graph $y = f(x)$ for $10 \leq x \leq 25$.

76. Safety Research. If a person driving a vehicle slams on the brakes and skids to a stop, the speed v in miles per hour of the vehicle at the time the brakes are applied is given approximately by

$$v = f(x) = C\sqrt{x}$$

where x is the length in feet of the skid marks and C is a constant that depends on the road conditions and the weight of the vehicle. On the same set of axes, graph $v = f(x), 0 \leq x \leq 100$, for $C = 3, 4$, and 5.

77. Family of Curves. In calculus, solutions to certain types of problems often involve an unspecified constant. For example, consider the equation

$$y = \frac{1}{C}x^2 - C$$

where C is a positive constant. The collection of graphs of this equation for all permissible values of C is called a **family of curves.** On the same axes, graph the members of this family corresponding to $C = 1, 2, 3$, and 4.

78. Family of Curves. A family of curves is defined by the equation

$$y = 2C - \frac{2}{C}x^2$$

where C is a positive constant. On the same axes, graph the members of this family corresponding to $C = 1, 2, 3$, and 4.

79. Fluid Flow. A cubic tank is 4 feet on a side and is initially full of water. Water flows out an opening in the bottom of the tank at a rate proportional to the square root of the depth (see the figure). Using advanced concepts from mathematics and physics, it can be shown that the volume of the water in the tank t minutes after the water begins to flow is given by

$$V(t) = \frac{64}{C^2}(C - t)^2 \qquad 0 \leq t \leq C$$

where C is a constant that depends on the size of the opening. Graph $V(t)$ for $C = 1, C = 2, C = 4$, and $C = 8$.

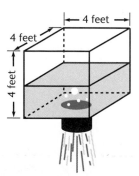

80. Evaporation. A water trough with triangular ends is 9 feet long, 4 feet wide, and 2 feet deep (see the figure on the next page). Initially, the trough is full of water, but due to evaporation, the volume of the water in the trough decreases at a rate proportional to the square root of the volume. Using advanced concepts from mathematics and physics, it can be shown that the volume after t hours is given by

$$V(t) = \frac{1}{C^2}(t + 6C)^2 \qquad 0 \le t \le 6|C|$$

where C is a constant. Graph $V(t)$ for $C = -4$, $C = -5$, and $C = -6$.

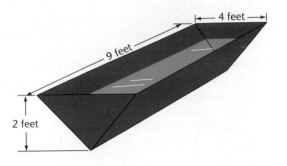

SECTION 3-6   Rational Functions

- Rational Functions
- Vertical and Horizontal Asymptotes
- Graphing Rational Functions

• **Rational Functions**   Linear and quadratic functions are examples of *polynomial functions*. In general, an **nth-degree polynomial function** is defined by an equation of the form

$$P(x) = a_n x^n + a_{n-1} x^{n-1} + \cdots + a_1 x + a_0 \qquad a_n \ne 0$$

where the coefficients a_i are real numbers and n is an integer, $n \ge 0$.

Just as rational numbers are defined in terms of quotients of integers, rational functions are defined in terms of quotients of polynomials. The following equations define rational functions:

$$f(x) = \frac{x - 1}{x^2 - x - 6} \qquad g(x) = \frac{1}{x} \qquad h(x) = \frac{x^3 - 1}{x}$$

$$p(x) = 2x^2 - 3 \qquad q(x) = 3 \qquad r(x) = 0$$

In general, a function f is a **rational function** if

$$f(x) = \frac{n(x)}{d(x)} \qquad d(x) \ne 0$$

where $n(x)$ and $d(x)$ are polynomials. The **domain of f** is the set of all real numbers x such that $d(x) \ne 0$.

If $x = a$ and $d(a) = 0$, then f is not defined at $x = a$ and there can be no point on the graph of f with abscissa $x = a$. Remember, division by 0 is never allowed. It can be shown that:

If $f(x) = n(x)/d(x)$ and $d(a) = 0$, then f is discontinuous at $x = a$ and the graph of f has a hole or break at $x = a$.

If $x = a$ is in the domain of f(x) and $n(a) = 0$, then the graph of f crosses the x axis at $x = a$. Thus:

If $f(x) = n(x)/d(x)$, $n(a) = 0$, and $d(a) \neq 0$, then $x = a$ is an *x* intercept for the graph of f.

What happens if both $n(a) = 0$ and $d(a) = 0$? In this case, it can be shown that $x - a$ is a factor of both $n(x)$ and $d(x)$, and thus, $f(x)$ is not in lowest terms (see Section 1-4).

Unless specifically stated to the contrary, we assume that all the rational functions we consider are reduced to lowest terms.

EXAMPLE 1   **Finding the Domain and *x* Intercepts for a Rational Function**

Find the domain and *x* intercepts for: $f(x) = \dfrac{2x^2 - 2x - 4}{x^2 - 9}$

Solution

$$f(x) = \frac{n(x)}{d(x)} = \frac{2x^2 - 2x - 4}{x^2 - 9} = \frac{2(x - 2)(x + 1)}{(x - 3)(x + 3)}$$

Since $d(3) = 0$ and $d(-3) = 0$, the domain of f is

$$x \neq \pm 3 \qquad \text{or} \qquad (-\infty, -3) \cup (-3, 3) \cup (3, \infty)$$

Since $n(2) = 0$ and $n(-1) = 0$, the graph of f crosses the *x* axis at $x = 2$ and $x = -1$.

Problem 1   Find the domain and *x* intercepts for: $f(x) = \dfrac{3x^2 - 12}{x^2 + 2x - 3}$

● **Vertical and Horizontal Asymptotes**

Even though a rational function f may be discontinuous at $x = a$ (no graph for $x = a$), it is still useful to know what happens to the graph of f when x is close to a. For example, consider the very simple rational function f defined by

$$f(x) = \frac{1}{x}$$

It is clear that the function f is discontinuous at $x = 0$. But what happens to $f(x)$ when x approaches 0 from either side of 0? Some tables will give us an idea of what happens to $f(x)$ when x gets close to 0. From Table 1 on the next page, we see that as x approaches 0 from the right, $1/x$ gets larger and larger—that is, $1/x$ increases without bound. We write this symbolically* as

$$\frac{1}{x} \to \infty \qquad \text{as} \qquad x \to 0^+$$

---

*Remember, the symbol ∞ does not represent a real number. In this context, ∞ is used to indicate that the values of $1/x$ increase without bound. That is, $1/x$ exceeds any given number N no matter how large N is chosen.

TABLE 1   **Behavior of $1/x$ as $x \to 0^+$**

x	1	0.1	0.01	0.001	0.0001	0.000 01	0.000 001	...	x approaches 0 from the right ($x \to 0^+$)
$1/x$	1	10	100	1,000	10,000	100,000	1,000,000	...	$1/x$ increases without bound ($1/x \to \infty$)

If x approaches 0 from the left, then x and $1/x$ are both negative and the values of $1/x$ decrease without bound (see Table 2). This is denoted as

$$\frac{1}{x} \to -\infty \quad \text{as} \quad x \to 0^-$$

TABLE 2   **Behavior of $1/x$ as $x \to 0^-$**

x	-1	-0.1	-0.01	-0.001	-0.0001	$-0.000\ 01$	$-0.000\ 001$	...	x approaches 0 from the left $(x \to 0^-)$
$1/x$	-1	-10	-100	$-1,000$	$-10,000$	$-100,000$	$-1,000,000$	...	$1/x$ decreases without bound $(1/x \to -\infty)$

The graph of $f(x) = 1/x$ for $-1 \leq x \leq 1$, $x \neq 0$, is shown in Figure 1. The behavior of f as x approaches 0 from the right is illustrated on the graph by drawing a curve that becomes almost vertical and placing an arrow on the curve to indicate that the values of $1/x$ continue to increase without bound as x approaches 0 from the right. The behavior as x approaches 0 from the left is illustrated in a similar manner.

FIGURE 1 $f(x) = \dfrac{1}{x}$ near $x = 0$.

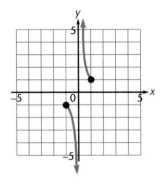

Now we look at the behavior of $f(x) = 1/x$ as $|x|$ gets very large—that is, as $x \to \infty$ and as $x \to -\infty$. Consider Tables 3 and 4. As x increases without bound, $1/x$ is positive and approaches 0 from above. As x decreases without bound, $1/x$ is negative and approaches 0 from below. For our purposes, it is not necessary to distinguish between $1/x$ approaching 0 from above and from below. Thus, we will describe this behavior by writing

$$\frac{1}{x} \to 0 \quad \text{as} \quad x \to \infty \text{ and as } x \to -\infty$$

TABLE 3   **Behavior of $1/x$ as $x \to \infty$**

x	1	10	100	1,000	10,000	100,000	1,000,000	. . .	x increases without bound $(x \to \infty)$
$1/x$	1	0.1	0.01	0.001	0.0001	0.000 01	0.000 001	. . .	$1/x$ approaches 0 $(1/x \to 0)$

TABLE 4   **Behavior of $1/x$ as $x \to -\infty$**

x	-1	-10	-100	$-1,000$	$-10,000$	$-100,000$	$-1,000,000$	. . .	x decreases without bound $(x \to -\infty)$
$1/x$	-1	-0.1	-0.01	-0.001	-0.0001	$-0.000\ 01$	$-0.000\ 001$	. . .	$1/x$ approaches 0 $(1/x \to 0)$

The completed graph of $f(x) = 1/x$ is shown in Figure 2. Notice that the behavior as $x \to \infty$ and as $x \to -\infty$ is illustrated by drawing a curve that is almost horizontal and adding arrows at the ends. The curve in Figure 2 is an example of a plane curve called a **hyperbola**, and the coordinate axes are called **asymptotes** for this curve. The y axis is a *vertical asymptote* for $1/x$, and the x axis is a *horizontal asymptote* for $1/x$.

FIGURE 2 $f(x) = \dfrac{1}{x}$, $x \neq 0$.

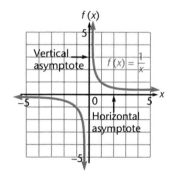

DEFINITION 1

Horizontal and Vertical Asymptotes

The line $x = a$ is a **vertical asymptote** for the graph of $y = f(x)$ if $f(x)$ either increases or decreases without bound as x approaches a from the right or from the left. Symbolically,

$$f(x) \to \infty \quad \text{or} \quad f(x) \to -\infty \quad \text{as} \quad x \to a^+ \quad \text{or} \quad x \to a^-$$

The line $y = b$ is a **horizontal asymptote** for the graph of $y = f(x)$ if $f(x)$ approaches b as x increases without bound or as x decreases without bound. Symbolically,

$$f(x) \to b \quad \text{as} \quad x \to \infty \quad \text{or} \quad x \to -\infty$$

Graphing rational functions is considerably aided by locating vertical and horizontal asymptotes first, if they exist. Using the same kind of reasoning as in the preceding discussion, we state the following general method of locating vertical asymptotes for rational functions.

| Theorem 1 | **Vertical Asymptotes and Rational Functions** |

Let f be a rational function defined by

$$f(x) = \frac{n(x)}{d(x)}$$

where $n(x)$ and $d(x)$ are polynomials. If a is a real number such that $d(a) = 0$ and $n(a) \neq 0$, then the line $x = a$ is a vertical asymptote of the graph of $y = f(x)$.

To gain added insight in locating horizontal asymptotes for the graph of $f(x) = n(x)/d(x)$, we now consider the following three examples:

1. **Degree of numerator less than degree of denominator**

$$f(x) = \frac{2x - 1}{3x^2 - 2x}$$

$$= \frac{\dfrac{2x}{x^2} - \dfrac{1}{x^2}}{\dfrac{3x^2}{x^2} - \dfrac{2x}{x^2}}$$ Divide each term in numerator and denominator by x^2, the highest power of x that appears in the numerator and denominator.

$$= \frac{\dfrac{2}{x} - \dfrac{1}{x^2}}{3 - \dfrac{2}{x}} \qquad \rightarrow \frac{0 - 0}{3 - 0} = 0$$

As $x \to \infty$ or $-\infty$, $2/x \to 0$ and $1/x^2 \to 0$; hence, $f(x) \to 0$. Thus, the line $y = 0$ is a horizontal asymptote.

2. **Degree of numerator equal to degree of denominator**

$$g(x) = \frac{2x^2 - 1}{3x^2 - 2x}$$

$$= \frac{\dfrac{2x^2}{x^2} - \dfrac{1}{x^2}}{\dfrac{3x^2}{x^2} - \dfrac{2x}{x^2}}$$ Divide each term in numerator and denominator by x^2, the highest power of x that appears in the numerator and denominator.

$$= \frac{2 - \dfrac{1}{x^2}}{3 - \dfrac{2}{x}} \qquad \rightarrow \frac{2 - 0}{3 - 0} = \frac{2}{3}$$

As $x \to \infty$ or $-\infty$, $1/x^2 \to 0$ and $2/x \to 0$; hence, $g(x) \to \frac{2}{3}$. Thus, $y = \frac{2}{3}$ is a horizontal asymptote.

3. Degree of numerator greater than degree of denominator:

$$h(x) = \frac{2x^3 - 1}{3x^2 - 2x}$$

$$= \frac{\dfrac{2x^3}{x^3} - \dfrac{1}{x^3}}{\dfrac{3x^2}{x^3} - \dfrac{2x}{x^3}}$$

Divide each term in numerator and denominator by x^3, the highest power of x that appears in the numerator and denominator.

$$= \frac{2 - \dfrac{1}{x^3}}{\dfrac{3}{x} - \dfrac{2}{x^2}} \qquad \to \frac{2 - 0}{0 - 0}; \text{ undefined}$$

As $x \to \infty$ or $-\infty$, the numerator approaches 2 and the denominator approaches 0. Thus, $|h(x)|$ increases without bound and there is no horizontal asymptote.

Reasoning in the same way as in these examples, we can establish the general method of locating horizontal asymptotes stated in Theorem 2.

Theorem 2	**Horizontal Asymptotes and Rational Functions**

Let f be a rational function defined by the quotient of two polynomials as follows:

$$f(x) = \frac{a_m x^m + \cdots + a_1 x + a_0}{b_n x^n + \cdots + b_1 x + b_0} \qquad a_m, b_n \neq 0$$

1. For $m < n$, the x axis ($y = 0$) is a horizontal asymptote.

2. For $m = n$, the line $y = a_m/b_n$ is a horizontal asymptote.

3. For $m > n$, there are no horizontal asymptotes.

EXAMPLE 2   **Finding Vertical and Horizontal Asymptotes for a Rational Function**

Find all vertical and horizontal asymptotes for

$$f(x) = \frac{n(x)}{d(x)} = \frac{2x^2 - 2x - 4}{x^2 - 9}$$

Solution   Since $d(x) = x^2 - 9 = (x - 3)(x + 3)$, the graph of $f(x)$ has vertical asymptotes at $x = 3$ and $x = -3$ (Theorem 1). Since $n(x)$ and $d(x)$ have the same degree, the line

$$y \boxed{= \frac{a_2}{b_2}} = \frac{2}{1} = 2 \qquad a_2 = 2, \, b_2 = 1$$

is a horizontal asymptote (Theorem 2, part 2).

Problem 2   Find all vertical and horizontal asymptotes for

$$f(x) = \frac{3x^2 - 12}{x^2 + 2x - 3}$$

● **Graphing Rational Functions**   We now use the techniques for locating asymptotes, along with other graphing aids discussed in the text, to graph several rational functions. First, we outline a systematic approach to the problem of graphing rational functions:

Graphing a Rational Function: $f(x) = n(x)/d(x)$

Step 1. Intercepts. Find the real solutions of the equation $n(x) = 0$ and use these solutions to plot any x intercepts of the graph of f. Evaluate $f(0)$, if it exists, and plot the y intercept.

Step 2. Vertical Asymptotes. Find the real solutions of the equation $d(x) = 0$ and use these solutions to determine the domain of f, the points of discontinuity, and the vertical asymptotes. Sketch any vertical asymptotes as dashed lines.

Step 3. Sign Chart. Construct a sign chart for f and use it to determine the behavior of the graph near each vertical asymptote.

Step 4. Horizontal Asymptotes. Determine whether there is a horizontal asymptote and if so, sketch it as a dashed line.

Step 5. Symmetry. Determine symmetry with respect to the vertical axis and the origin.

Step 6. Complete the Sketch. Complete the sketch of the graph by plotting additional points and joining these points with a smooth continuous curve over each interval in the domain of f. Do not cross any points of discontinuity.

EXAMPLE 3   Graphing a Rational Function

Graph: $y = f(x) = \dfrac{2x}{x - 3}$

Solution

$$f(x) = \frac{2x}{x - 3} = \frac{n(x)}{d(x)}$$

Step 1. Intercepts. Find real zeros of $n(x) = 2x$ and find $f(0)$:

$$2x = 0$$

$$x = 0 \quad x \text{ intercept}$$

$$f(0) = 0 \quad y \text{ intercept}$$

The graph crosses the coordinate axes only at the origin. Plot this intercept, as shown in figure (a).

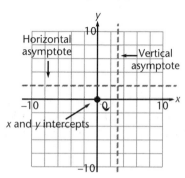

Intercepts and asymptotes

(a)

Step 2. Vertical Asymptotes. Find real zeros of $d(x) = x - 3$:

$$x - 3 = 0$$

$$x = 3$$

The domain of f is $(-\infty, 3) \cup (3, \infty)$, f is discontinuous at $x = 3$, and the graph has a vertical asymptote at $x = 3$. Sketch this asymptote, as shown in figure (a).

Test number	-1	1	4
Value of f	$\frac{1}{2}$	-1	8
Sign of f	$+$	$-$	$+$

$$+ + + \mid - - - - \mid + + + \atop 0 \qquad 3 \qquad\quad x$$

Step 3. Sign Chart. Construct a sign chart for $f(x)$ (review Section 2-8), as shown in the margin. Since $x = 3$ is a vertical asymptote and $f(x) < 0$ for $0 < x < 3$,

$$f(x) \to -\infty \quad \text{as} \quad x \to 3^{-}$$

Since $x = 3$ is a vertical asymptote and $f(x) > 0$ for $x > 3$,

$$f(x) \to \infty \quad \text{as} \quad x \to 3^{+}$$

Notice how much information is contained in the sign chart for f. The solid dot determines the x intercept; the open dot determines the domain, the point of discontinuity, and the vertical asymptote; and the signs of $f(x)$ determine the behavior of the graph at the vertical asymptote.

Step 4. Horizontal Asymptote. Since $n(x)$ and $d(x)$ have the same degree, the line $y = 2$ is a horizontal asymptote. Sketch this asymptote, as shown in figure (a).

Step 5. Symmetry.

$$f(-x) = \frac{-2x}{-x-3} = \frac{2x}{x+3}$$

Since $f(-x) \neq f(x)$ and $f(-x) \neq -f(x)$, the graph is not symmetric with respect to the y axis or the origin.

Step 6. Complete the Sketch. By plotting a few additional points, we obtain the graph in figure (b). Notice that the graph is a smooth continuous curve over the interval $(-\infty, 3)$ and over the interval $(3, \infty)$. As expected, there is a break in the graph at $x = 3$.

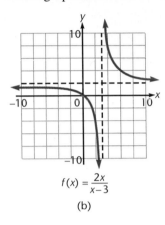

$$f(x) = \frac{2x}{x-3}$$

(b)

As you gain experience in graphing, many of the steps in Example 3 can be done mentally (or on scratch paper) and the process can be speeded up considerably.

Problem 3   Proceed as in Example 3 and graph: $y = f(x) = \dfrac{3x}{x+2}$

EXAMPLE 4   **Graphing a Rational Function**

Graph: $y = f(x) = \dfrac{8}{4-x^2}$

Solution
$$f(x) = \frac{8}{4-x^2} = \frac{n(x)}{d(x)}$$

Step 1. Intercepts. Find real zeros of $n(x) = 8$:

$$n(x) = 8 \neq 0 \qquad \text{for all real numbers } x$$

There are no x intercepts.

$$f(0) = \tfrac{8}{4} = 2 \quad y \text{ intercept}$$

Plot this intercept, as shown in figure (a).

Step 2. Vertical Asymptotes. Find real zeros of $d(x) = 4 - x^2$:

$$d(x) = 4 - x^2 = (2 - x)(2 + x) = 0$$

$$x = -2 \quad \text{or} \quad x = 2$$

The domain of f is $(-\infty, -2) \cup (-2, 2) \cup (2, \infty)$, f is discontinuous at $x = -2$ and $x = 2$, and the graph has vertical asymptotes at $x = -2$ and $x = 2$. Sketch these asymptotes, as shown in figure (a).

Step 3. Sign Chart. Construct a sign chart for $f(x)$, as shown in the margin.

Test number	-3	0	3
Value of f	$-\frac{8}{5}$	2	$-\frac{8}{5}$
Sign of f	$-$	$+$	$-$

$$- - - \mid + + + \mid - - -$$
$$\underset{-2}{\circ} \qquad \underset{2}{\circ} \quad\longrightarrow x$$

$$
\left.
\begin{aligned}
f(x) &\to -\infty & \text{as} && x &\to -2^- \\
f(x) &\to \infty & \text{as} && x &\to -2^+ \\
f(x) &\to \infty & \text{as} && x &\to 2^- \\
f(x) &\to -\infty & \text{as} && x &\to 2^+
\end{aligned}
\right\}
\begin{aligned}
&\text{Behavior of graph} \\
&\text{near asymptotes}
\end{aligned}
$$

Step 4. Horizontal Asymptote. Since $n(x) = 8$ is a polynomial of degree 0 and $d(x) = 4 - x^2$ is a polynomial of degree 2, the x axis is a horizontal asymptote (Theorem 2, part 1), as indicated in figure (a).

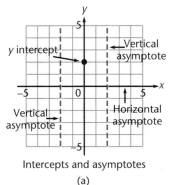

Intercepts and asymptotes

(a)

Step 5. Symmetry.

$$f(-x) = \frac{8}{4 - (-x)^2} = \frac{8}{4 - x^2} = f(x)$$

The graph of $y = f(x)$ is symmetric with respect to the y axis.

Step 6. Complete the Sketch. By plotting additional points, we obtain the graph in figure (b).

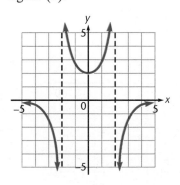

$$f(x) = \frac{8}{4 - x^2}$$

(b)

Problem 4   Graph: $y = f(x) = \dfrac{x^2}{x^2 - 1}$

In the remaining examples we will just list the results of each step in the graphing strategy and omit the computational details.

EXAMPLE 5   **Graphing a Rational Function**

Graph: $y = f(x) = \dfrac{x^2 - 6x + 9}{x^2 + x - 2}$

Solution

$$f(x) = \frac{x^2 - 6x + 9}{x^2 + x - 2} = \frac{(x - 3)^2}{(x + 2)(x - 1)}$$

x intercept: $x = 3$

y intercept: $y = f(0) = -\frac{9}{2} = -4.5$

Domain: $(-\infty, -2) \cup (-2, 1) \cup (1, \infty)$

Points of discontinuity: $x = -2$ and $x = 1$

Vertical asymptotes: $x = -2$ and $x = 1$

$f(x) \to \infty$	as	$x \to -2^-$
$f(x) \to -\infty$	as	$x \to -2^+$
$f(x) \to -\infty$	as	$x \to 1^-$
$f(x) \to \infty$	as	$x \to 2^+$

Horizontal asymptote: $y = 1$

No symmetry

Test number	-3	0	2	4
Value of f	9	$-\frac{9}{2}$	$\frac{1}{4}$	$\frac{1}{18}$
Sign of f	$+$	$-$	$+$	$+$

```
+ + +| - - -| + + +| + + +
─────○───────○───────●──────→ x
    -2       1       3
```

Sketch in the intercepts and asymptotes [figure (a)], then sketch the graph of f [figure (b)].

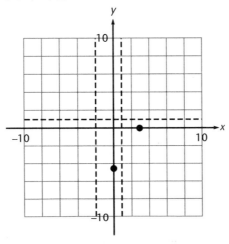

Intercepts and asymptotes

(a)

$f(x) = \dfrac{x^2 - 6x + 9}{x^2 + x - 2}$

(b)

Problem 5 Graph: $y = f(x) = \dfrac{x^2}{x^2 - 7x + 10}$

CAUTION

The graph of a function cannot cross a vertical asymptote, but the same statement is not true for horizontal asymptotes. The graph in Example 5 clearly shows that **the graph of a function can cross a horizontal asymptote**. The definition of a horizontal asymptote requires $f(x)$ to approach b as x increases or decreases without bound, but it does not preclude the possibility that $f(x) = b$ for one or more values of x. In fact, using the cosine function from trigonometry, it is possible to construct a function whose graph crosses a horizontal asymptote an infinite number of times (see Figure 3).

FIGURE 3 Multiple intersections of a graph and a horizontal asymptote.

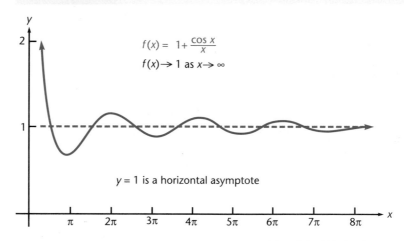

$f(x) = 1 + \dfrac{\cos x}{x}$

$f(x) \rightarrow 1$ as $x \rightarrow \infty$

$y = 1$ is a horizontal asymptote

EXAMPLE 6 **Graphing a Rational Function**

Graph: $y = f(x) = \dfrac{x^2 - 3x - 4}{x - 2}$

Solution

$$f(x) = \frac{x^2 - 3x - 4}{x - 2} = \frac{(x + 1)(x - 4)}{x - 2}$$

x intercepts: $x = -1$ and $x = 4$

y intercept: $y = f(0) = 2$

Domain: $(-\infty, 2) \cup (2, \infty)$

Points of discontinuity: $x = 2$

Vertical asymptote: $x = 2$

$f(x) \rightarrow \infty$ as $x \rightarrow 2^-$

$f(x) \rightarrow -\infty$ as $x \rightarrow 2^+$

No horizontal asymptote

No symmetry

Test number	-2	0	3	5
Value of f	$-\frac{3}{2}$	2	-4	2
Sign of f	$-$	$+$	$-$	$+$

$$- - -\,|+\,+\,+\,|- - -\,|+\,+\,+$$
$${-1}\qquad 2\qquad 4 \qquad x$$

Even though the graph of f does not have a horizontal asymptote, we can still gain some useful information about the behavior of the graph as $x \rightarrow -\infty$ and as $x \rightarrow \infty$ if we first perform a long division:

$$\begin{array}{r}
x - 1 \qquad \text{Quotient} \\
x - 2 \overline{)x^2 - 3x - 4} \\
\underline{x^2 - 2x} \\
-x - 4 \\
\underline{-x + 2} \\
-6 \qquad \text{Remainder}
\end{array}$$

Thus,

$$f(x) = \frac{x^2 - 3x - 4}{x - 2} = x - 1 - \frac{6}{x - 2}$$

As $x \to -\infty$ or $x \to \infty$, $6/(x - 2) \to 0$ and the graph of f approaches the line $y = x - 1$. This line is called an **oblique asymptote** for the graph of f. The asymptotes and intercepts are sketched in figure (a), and the graph of f is sketched in figure (b).

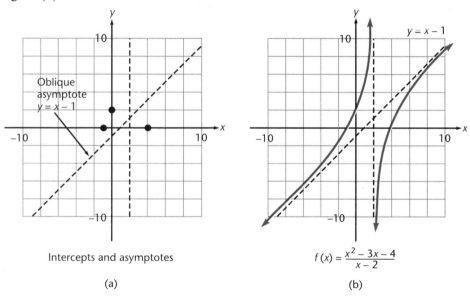

Intercepts and asymptotes

(a)

$f(x) = \dfrac{x^2 - 3x - 4}{x - 2}$

(b)

Generalizing the results of Example 6, we have Theorem 3.

Theorem 3   **Oblique Asymptotes and Rational Functions**

If $f(x) = n(x)/d(x)$, where $n(x)$ and $d(x)$ are polynomials and the degree of $n(x)$ is 1 more than the degree of $d(x)$, then $f(x)$ can be expressed in the form

$$f(x) = mx + b + \frac{r(x)}{d(x)}$$

where the degree of $r(x)$ is less than the degree of $d(x)$. The line

$$y = mx + b$$

is an oblique asymptote for the graph of f. That is,

$$[f(x) - (mx + b)] \to 0 \qquad \text{as} \qquad x \to -\infty \quad \text{or} \quad x \to \infty$$

Problem 6 Graph, including any oblique asymptotes: $y = f(x) = \dfrac{x^2 + 5}{x + 1}$

Answers to Matched Problems
1. Domain: $(-\infty, -3) \cup (-3, 1) \cup (1, \infty)$; x intercepts: $x = -2, x = 2$
2. Vertical asymptotes: $x = -3, x = 1$; horizontal asymptote: $y = 3$
3.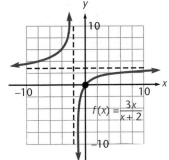
4.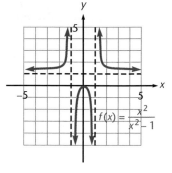
5.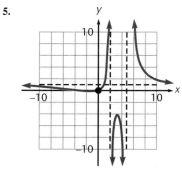

$$f(x) = \dfrac{x^2}{x^2 - 7x + 10}$$

6.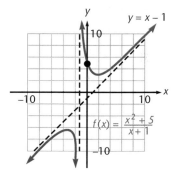

EXERCISE 3-6

A

In Problems 1–8, find the domain and x intercepts. Do not graph.

1. $f(x) = \dfrac{2x - 4}{x + 1}$

2. $g(x) = \dfrac{3x + 6}{x - 1}$

3. $h(x) = \dfrac{x^2 - 1}{x^2 - 16}$

4. $k(x) = \dfrac{x^2 - 36}{x^2 - 25}$

5. $r(x) = \dfrac{x^2 - x - 6}{x^2 - x - 12}$

6. $s(x) = \dfrac{x^2 + x - 12}{x^2 + x - 6}$

7. $F(x) = \dfrac{x}{x^2 + 4}$

8. $G(x) = \dfrac{x^2}{x^2 + 16}$

In Problems 9–16, find all vertical and horizontal asymptotes. Do not graph.

9. $f(x) = \dfrac{2x}{x - 4}$

10. $h(x) = \dfrac{3x}{x + 5}$

11. $s(x) = \dfrac{2x^2 + 3}{3x^2 - 48}$

12. $r(x) = \dfrac{5x^2 - 7x}{2x^2 - 50}$

13. $p(x) = \dfrac{2x}{x^4 + 1}$

14. $q(x) = \dfrac{5x^4}{2x^2 + 3x - 2}$

15. $t(x) = \dfrac{6x^4}{3x^2 - 2x - 5}$

16. $g(x) = \dfrac{3x}{x^4 + 2x^2 + 1}$

B

In Problems 17–46, use the graphing strategy outlined in the text to sketch the graph of each function.

17. $f(x) = \dfrac{1}{x - 4}$ **18.** $g(x) = \dfrac{1}{x + 3}$ **19.** $p(x) = \dfrac{-1}{x - 4}$ **20.** $m(x) = \dfrac{-1}{x + 3}$

21. $f(x) = \dfrac{x}{x + 1}$ **22.** $f(x) = \dfrac{3x}{x - 3}$ **23.** $q(x) = \dfrac{2x - 1}{x}$ **24.** $r(x) = \dfrac{1 - x}{x}$

25. $h(x) = \dfrac{x}{2x - 2}$ **26.** $p(x) = \dfrac{3x}{4x + 4}$ **27.** $f(x) = \dfrac{2x - 4}{x + 3}$ **28.** $f(x) = \dfrac{3x + 3}{2 - x}$

29. $f(x) = \dfrac{4x - 5}{x - 3}$ **30.** $f(x) = \dfrac{-3x - 8}{x + 5}$ **31.** $g(x) = \dfrac{1 - x^2}{x^2}$ **32.** $f(x) = \dfrac{x^2 + 1}{x^2}$

33. $f(x) = \dfrac{9}{x^2 - 9}$ **34.** $g(x) = \dfrac{6}{x^2 - x - 6}$ **35.** $f(x) = \dfrac{x}{x^2 - 1}$ **36.** $p(x) = \dfrac{x}{1 - x^2}$

37. $g(x) = \dfrac{2}{x^2 + 1}$ **38.** $f(x) = \dfrac{x}{x^2 + 1}$ **39.** $h(x) = \dfrac{2x^2}{x^2 + 1}$ **40.** $f(x) = \dfrac{-2x^4}{x^4 + 1}$

41. $f(x) = \dfrac{12x^2}{(3x + 5)^2}$ **42.** $f(x) = \dfrac{7x^2}{(2x - 3)^2}$

43. $f(x) = \dfrac{x^2 - 1}{x^2 + 7x + 10}$ **44.** $f(x) = \dfrac{x^2 + 6x + 8}{x^2 - x - 2}$

45. $f(x) = \dfrac{-2x^2 - 3x + 2}{(x - 2)^2}$ **46.** $f(x) = \dfrac{-2x^2 + 9x - 4}{(x + 1)^2}$

In Problems 47–54, find all vertical, horizontal, and oblique asymptotes. Do not graph.

47. $f(x) = \dfrac{2x^2}{x - 1}$ **48.** $g(x) = \dfrac{3x^2}{x + 2}$

49. $h(x) = \dfrac{x^2 + 1}{x^3 + 1}$ **50.** $k(x) = \dfrac{x^4}{x^2 + 1}$

51. $p(x) = \dfrac{x^3}{x^2 + 1}$ **52.** $q(x) = \dfrac{x^5}{x^3 - 8}$

53. $r(x) = \dfrac{2x^2 - 3x + 5}{x}$ **54.** $s(x) = \dfrac{-3x^2 + 5x + 9}{x}$

C

In Problems 55–60, use the graphing strategy outlined in the text to sketch the graph of each function. Include any oblique asymptotes.

55. $f(x) = \dfrac{x^2 + 1}{x}$ **56.** $g(x) = \dfrac{x^2 - 1}{x}$ **57.** $k(x) = \dfrac{x^2 - 4x + 3}{2x - 4}$ **58.** $h(x) = \dfrac{x^2 + x - 2}{2x - 4}$

59. $F(x) = \dfrac{8 - x^3}{4x^2}$ **60.** $G(x) = \dfrac{x^4 + 1}{x^3}$

$\displaystyle\int$ *In calculus, it is often necessary to consider rational functions that are not in lowest terms, such as the functions given in Problems 61–64. For each function, state the domain, reduce the function to lowest terms, and sketch its graph. Remember to exclude from the graph any points with x values that are not in the domain.*

61. $f(x) = \dfrac{x^2 - 4}{x - 2}$ **62.** $g(x) = \dfrac{x^2 - 1}{x + 1}$ **63.** $r(x) = \dfrac{x + 2}{x^2 - 4}$ **64.** $s(x) = \dfrac{x - 1}{x^2 - 1}$

APPLICATIONS

65. Employee Training. A company producing electronic components used in television sets has established that on the average, a new employee can assemble $N(t)$ components per day after t days of on-the-job training, as given by

$$N(t) = \frac{50t}{t + 4} \qquad t \geq 0$$

Sketch the graph of N, including any vertical or horizontal asymptotes. What does N approach as $t \to \infty$?

66. Physiology. In a study on the speed of muscle contraction in frogs under various loads, researchers W. O. Fems and J. Marsh found that the speed of contraction decreases with increasing loads. More precisely, they found that the relationship between speed of contraction S (in centimeters per second) and load w (in grams) is given approximately by

$$S(w) = \frac{26 + 0.06w}{w} \qquad w \geq 5$$

Sketch the graph of S, including any vertical or horizontal asymptotes. What does S approach as $w \to \infty$?

67. Retention. An experiment on retention is conducted in a psychology class. Each student in the class is given 1 day to memorize the same list of 40 special characters. The lists are turned in at the end of the day, and for each succeeding day for 20 days each student is asked to turn in a list of as many of the symbols as can be recalled. Averages are taken, and it is found that a good approximation of the average number of symbols, $N(t)$, retained after t days is given by

$$N(t) = \frac{5t + 30}{t} \qquad t \geq 1$$

Sketch the graph of N, including any vertical or horizontal asymptotes. What does N approach as $t \to \infty$?

68. Learning Theory. In 1917, L. L. Thurstone, a pioneer in quantitative learning theory, proposed the function

$$f(x) = \frac{a(x + c)}{(x + c) + b}$$

to describe the number of successful acts per unit time that a person could accomplish after x practice sessions. Suppose that for a particular person enrolling in a typing class,

$$f(x) = \frac{50(x + 1)}{(x + 5)} \qquad x \geq 0$$

where $f(x)$ is the number of words per minute the person is able to type after x weeks of lessons. Sketch the graph of f, including any vertical or horizontal asymptotes. What does f approach as $x \to \infty$?

> Recall that the minimum value of a quadratic function of the form
>
> $$f(x) = ax^2 + bx + c \qquad a > 0$$
>
> is min $f(x) = f(-b/2a)$. Using calculus techniques, it can be shown that the minimum value of a function of the form
>
> $$g(x) = ax + b + \frac{c}{x} \qquad a > 0, c > 0$$
>
> is min $g(x) = g(\sqrt{c/a})$. Use this fact in Problems 69–72.

★69. Replacement Time. A desktop office copier has an initial price of $2,500. A maintenance/service contract costs $200 for the first year and increases $50 per year thereafter. It can be shown that the total cost of the copier after n years is given by

$$C(n) = 2,500 + 175n + 25n^2$$

The average cost per year for n years is $\overline{C}(n) = C(n)/n$.
(A) Find the rational function $\overline{C}$.
(B) When is the average cost per year minimum? (This

is frequently referred to as the *replacement time* for this piece of equipment.)
(C) Sketch the graph of $\overline{C}$, including any asymptotes.

★70. Average Cost. The total cost of producing x units of a certain product is given by

$$C(x) = \tfrac{1}{5}x^2 + 2x + 2,000$$

The average cost per unit for producing x units is $\overline{C}(x) = C(x)/x$.

(A) Find the rational function $\overline{C}$.
(B) At what production level will the average cost per unit be minimal?
(C) Sketch the graph of $\overline{C}$, including any asymptotes.

★71. **Construction.** A rectangular dog pen is to be made to enclose an area of 225 square feet.
(A) If x represents the width of the pen, express the length $L(x)$ of the fence required for the pen in terms of x.
(B) Considering the physical limitations, what is the domain of the function L?

(C) Find the dimensions of the pen that will require the least amount of fencing material.
(D) Graph the function L, including any asymptotes.

★72. **Construction.** Rework Problem 71 with the added assumption that the pen is to be divided into two sections, as shown in the figure.

SECTION **3-7**   **Operations on Functions; Composition**

* Operations on Functions
* Composition
* Applications

If two functions f and g are both defined at a number x, then $f(x)$ and $g(x)$ are both real numbers. Thus, it is possible to perform real number operations such as addition, subtraction, multiplication, or division with $f(x)$ and $g(x)$. Furthermore, if $g(x)$ is a number in the domain of f, then it is also possible to evaluate f at $g(x)$. In this section we see how operations on the values of functions can be used to define operations on the functions themselves. Some of these operations have important applications in calculus and more advanced mathematics courses, as well as in science and business.

* **Operations on Functions**

The functions f and g given by

$$f(x) = 2x + 3 \quad \text{and} \quad g(x) = x^2 - 4$$

are defined for all real numbers. Thus, for any real x we can perform the following operations:

$$f(x) + g(x) = 2x + 3 + x^2 - 4 = x^2 + 2x - 1$$
$$f(x) - g(x) = 2x + 3 - (x^2 - 4) = -x^2 + 2x + 7$$
$$f(x)g(x) = (2x + 3)(x^2 - 4) = 2x^3 + 3x^2 - 8x - 12$$

For $x \neq \pm 2$ we can also form the quotient

$$\frac{f(x)}{g(x)} = \frac{2x + 3}{x^2 - 4} \qquad x \neq \pm 2$$

Notice that the result of each operation is a new function. Thus, we have

$$(f + g)(x) = f(x) + g(x) = x^2 + 2x - 1 \qquad \text{Sum}$$

$$(f - g)(x) = f(x) - g(x) = -x^2 + 2x + 7 \qquad \text{Difference}$$

$$(fg)(x) = f(x)g(x) = 2x^3 + 3x^2 - 8x - 12 \qquad \text{Product}$$

$$\left(\frac{f}{g}\right)(x) = \frac{f(x)}{g(x)} = \frac{2x + 3}{x^2 - 4} \qquad x \neq \pm 2 \qquad \text{Quotient}$$

Notice that the sum, difference, and product functions are defined for all values of x, as were f and g, but the domain of the quotient function must be restricted to exclude those values where $g(x) = 0$.

DEFINITION 1   **Operations on Functions**

The **sum**, **difference**, **product**, and **quotient** of the functions f and g are the functions defined by

$$(f + g)(x) = f(x) + g(x) \qquad \text{Sum function}$$

$$(f - g)(x) = f(x) - g(x) \qquad \text{Difference function}$$

$$(fg)(x) = f(x)g(x) \qquad \text{Product function}$$

$$\left(\frac{f}{g}\right)(x) = \frac{f(x)}{g(x)} \qquad g(x) \neq 0 \qquad \text{Quotient function}$$

Each function is defined on the intersection of the domains of f and g, with the exception that the values of x where $g(x) = 0$ must be excluded from the domain of the quotient function.

EXAMPLE 1   **Evaluating Sums, Differences, Products, and Quotients of Functions**

For $f(x) = x^3$ and $g(x) = |x - 3|$, find:

(A) $(f + g)(1)$ (B) $(f - g)(-1)$ (C) $(fg)(2)$ (D) $\left(\dfrac{f}{g}\right)(3)$

Solutions (A) $(f + g)(x) = f(x) + g(x) = x^3 + |x - 3|$
$\qquad (f + g)(1) = f(1) + g(1) = 1^3 + |1 - 3|$
$\qquad\qquad\qquad = 1 + 2 = 3$

(B) $\quad (f - g)(x) = f(x) - g(x) = x^3 - |x - 3|$
$\qquad (f - g)(-1) = f(-1) - g(-1) = (-1)^3 - |(-1) - 3|$
$\qquad\qquad\qquad\quad = -1 - 4 = -5$

(C) $(fg)(x) = f(x)g(x) = x^3|x - 3|$
$\qquad (fg)(2) = f(2)g(2) = 2^3|2 - 3| = 8$

(D) $\left(\dfrac{f}{g}\right)(x) = \dfrac{f(x)}{g(x)} = \dfrac{x^3}{|x - 3|}$

$\left(\dfrac{f}{g}\right)(3) = \dfrac{f(3)}{g(3)} = \dfrac{3^3}{|3 - 3|} = \dfrac{27}{0}$ Does not exist.

Since $g(3) = 0$, 3 is not in the domain of f/g and $(f/g)(3)$ does not exist.

Problem 1   For $f(x) = x^{1/3}$ and $g(x) = x - 1$, find:

(A) $(f + g)(-1)$ (B) $(f - g)(8)$ (C) $(fg)(1)$ (D) $\left(\dfrac{f}{g}\right)(0)$

EXAMPLE 2   **Finding the Sum, Difference, Product, and Quotient Functions**

Let $f(x) = \sqrt{4 - x}$ and $g(x) = \sqrt{3 + x}$. Find the functions $f + g$, $f - g$, fg, and f/g, and find their domains.

Solution

$$(f + g)(x) = f(x) + g(x) = \sqrt{4 - x} + \sqrt{3 + x}$$
$$(f - g)(x) = f(x) - g(x) = \sqrt{4 - x} - \sqrt{3 + x}$$
$$(fg)(x) = f(x)g(x) = \sqrt{4 - x}\,\sqrt{3 + x}$$
$$= \sqrt{(4 - x)(3 + x)}$$
$$= \sqrt{12 + x - x^2}$$
$$\left(\dfrac{f}{g}\right)(x) = \dfrac{f(x)}{g(x)} = \dfrac{\sqrt{4 - x}}{\sqrt{3 + x}} = \sqrt{\dfrac{4 - x}{3 + x}}$$

Domain of f

Domain of g

The domains of f and g are

Domain of f: $x \le 4$ or $(-\infty, 4]$

Domain of g: $x \ge -3$ or $[-3, \infty)$

The intersections of these domains is

$$(-\infty, 4] \cap [-3, \infty) = [-3, 4]$$

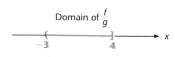
Domain of $f + g$,
$f - g$, and fg

This is the domain of the functions $f + g$, $f - g$, and fg. Since $g(-3) = 0$, $x = -3$ must be excluded from the domain of the quotient function. Thus,

Domain of f
Domain of $\dfrac{f}{g}$

$$\text{Domain of } \dfrac{f}{g}: (-3, 4]$$

Problem 2   Let $f(x) = \sqrt{x}$ and $g(x) = \sqrt{10 - x}$. Find the functions $f + g$, $f - g$, fg, and f/g, and find their domains.

● **Composition**   Consider the function h given by the equation

$$h(x) = \sqrt{2x + 1}$$

Inside the radical is a first-degree polynomial that defines a linear function. So the function h is really a combination of a square root function and a linear function. We can see this more clearly as follows. Let

$$u = 2x + 1 = g(x)$$
$$y = \sqrt{u} = f(u)$$

Then

$$h(x) = f[g(x)]$$

The function h is said to be the *composite* of the two functions f and g. (Loosely speaking, we can think of h as a function of a function.) What can we say about the domain of h given the domains of f and g? In forming the composite $h(x) = f[g(x)]$: **x must be restricted so that x is in the domain of g and $g(x)$ is in the domain of f.** Since the domain of f, where $f(u) = \sqrt{u}$, is the set of nonnegative real numbers, we see that $g(x)$ must be nonnegative; that is,

$$g(x) \geq 0$$
$$2x + 1 \geq 0$$
$$x \geq -\tfrac{1}{2}$$

Thus, the domain of h is this restricted domain of g.

A special function symbol is often used to represent the *composite of two functions*, which we define in general terms below.

DEFINITION 2   **Composite Functions**

Given functions f and g, then $f \circ g$ is called their **composite** and is defined by the equation

$$(f \circ g)(x) = f[g(x)]$$

The domain of $f \circ g$ is the set of all real numbers x in the domain of g where $g(x)$ is in the domain of f.

As an immediate consequence of Definition 2, we have (see Figure 1):

The domain of $f \circ g$ is always a subset of the domain of g, and the range of $f \circ g$ is always a subset of the range of f.

FIGURE 1 Composite functions.

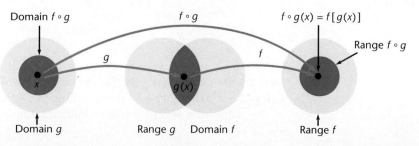

Domain $f \circ g$ $f \circ g$ $f \circ g(x) = f[g(x)]$

g f Range $f \circ g$

x $g(x)$

Domain g Range g Domain f Range f

EXAMPLE 3 · **Finding the Composition of Two Functions**

Find $(f \circ g)(x)$ and $(g \circ f)(x)$ and their domains for $f(x) = x^{10}$ and $g(x) = 3x^4 - 1$.

Solution

$$(f \circ g)(x) = f[g(x)] = f(3x^4 - 1) = (3x^4 - 1)^{10}$$
$$(g \circ f)(x) = g[f(x)] = g(x^{10}) = 3(x^{10})^4 - 1 = 3x^{40} - 1$$

The functions f and g are both defined for all real numbers. If x is any real number, then x is in the domain of g, $g(x)$ is in the domain of f, and, consequently, x is in the domain of $f \circ g$. Thus, the domain of $f \circ g$ is the set of all real numbers. Using similar reasoning, the domain of $g \circ f$ also is the set of all real numbers.

Problem 3   Find $(f \circ g)(x)$ and $(g \circ f)(x)$ and their domains for $f(x) = 2x + 1$ and $g(x) = (x - 1)/2$.

If two functions are both defined for all real numbers, then so is their composition.

If either function is not defined for some real numbers, then, as Example 4 illustrates, the domain of the composition may not be what you first think it should be.

EXAMPLE 4   **Finding the Composition of Two Functions**

Find $(f \circ g)(x)$ and its domain for $f(x) = \dfrac{1}{1 - 2x}$ and $g(x) = \dfrac{1}{x}$.

Solution

$$(f \circ g)(x) = f[g(x)]$$

$$= f\left(\frac{1}{x}\right)$$

$$= \frac{1}{1 - 2\left(\dfrac{1}{x}\right)} \cdot \frac{x}{x} \qquad \text{Multiply numerator and denominator by } x, \ x \neq 0.$$

$$= \frac{x}{x - 2}$$

Since the expression $x/(x - 2)$ is not defined at $x = 2$, we must exclude $x = 2$ from the domain of $f \circ g$. Is this the only number that must be excluded? Notice that g is not defined at $x = 0$. According to Definition 2, $f \circ g$ cannot be defined at any value of x where g is not defined. Hence, 0 must be excluded from the domain of $f \circ g$, even though $x/(x - 2)$ is defined at $x = 0$! Thus, the domain of $f \circ g$ is all real numbers except 0 and 2. Informally, we would write

$$(f \circ g)(x) = \frac{x}{x - 2} \qquad x \neq 0, 2$$

Problem 4   Find $(f \circ g)(x)$ and its domain for $f(x) = \dfrac{1}{1 + 3x}$ and $g(x) = \dfrac{1}{x - 1}$.

CAUTION   The domain of $f \circ g$ cannot always be determined simply by examining the final form of $(f \circ g)(x)$. Any numbers that are excluded from the domain of g must also be excluded from the domain of $f \circ g$.

EXAMPLE 5   **Finding the Composition of Two Functions**

Find $(f \circ g)(x)$ and its domain for $f(x) = \sqrt{4 - x^2}$ and $g(x) = \sqrt{3 - x}$.

Solution   We begin by stating the domains of f and g, a good practice in any composition problem:

$$\text{Domain } f: -2 \leq x \leq 2 \qquad \text{or} \qquad [-2, 2]$$
$$\text{Domain } g: x \leq 3 \qquad \text{or} \qquad (-\infty, 3]$$

Next we find the composition:

$$
\begin{aligned}
(f \circ g)(x) &= f[g(x)] = f(\sqrt{3 - x}) \\
&= \sqrt{4 - (\sqrt{3 - x})^2} \\
&= \sqrt{4 - (3 - x)} \qquad (\sqrt{t})^2 = t, \, t \geq 0 \\
&= \sqrt{1 + x}
\end{aligned}
$$

Even though $\sqrt{1 + x}$ is defined for all $x \geq -1$, we must restrict the domain of $f \circ g$ to those values that also are in the domain of g. Thus,

$$\text{Domain } f \circ g: x \geq -1 \text{ and } x \leq 3 \qquad \text{or} \qquad [-1, 3]$$

Problem 5   Find $(f \circ g)(x)$ and its domain for $f(x) = \sqrt{9 - x^2}$ and $g(x) = \sqrt{x - 1}$.

• **Applications**   In calculus, it is not only important to be able to find the composition of two functions, but also to recognize when a given function is the composition of two simpler functions.

EXAMPLE 6   Recognizing Composition Forms

∫ Express h as a composition of two simpler functions for

$$h(x) = (3x + 5)^5$$

Solution   If we let $f(x) = x^5$ and $g(x) = 3x + 5$, then

$$h(x) = (3x + 5)^5 = f(3x + 5) = f[g(x)] = (f \circ g)(x)$$

and we have expressed h as the composition of f and g.

Problem 6   Express h as a composition of the square root function and a linear function for $h(x) = \sqrt{4x - 7}$.

You will encounter the operations discussed in this section in many different situations. The next example shows how these operations are used in economic analysis.

EXAMPLE 7   Market Research

The research department for an electronics firm estimates that the weekly demand for a certain brand of audiocassette players is given by

$$x = f(p) = 20{,}000 - 1{,}000p \quad \text{Demand function}$$

where x is the number of cassette players retailers are likely to buy per week at $\$p$ per player. The research department also has determined that the total cost (in dollars) of producing x cassette players per week is given by

$$C(x) = 75{,}000 + 4x \quad \text{Cost function}$$

and the total weekly revenue (in dollars) obtained from the sale of these cassette players is given by

$$R(x) = 20x - \frac{1}{1{,}000}x^2 \quad \text{Revenue function}$$

Express the firm's weekly profit as a function of the price p.

Solution   Since profit is revenue minus cost, the profit function is the difference of the revenue and cost functions, $P = R - C$. Since R and C are given as functions of x, we first express P as a function of x:

$$P(x) = (R - C)(x)$$

$$= R(x) - C(x)$$

$$= 20x - \frac{1}{1,000} x^2 - (75,000 + 4x)$$

$$= 16x - \frac{1}{1,000} x^2 - 75,000$$

Next, we use composition to express P as a function of the price p:

$$(P \circ f)(p) = P[f(p)] = P(20,000 - 1,000p)$$

$$= 16(20,000 - 1,000p) - \frac{1}{1,000} (20,000 - 1,000p)^2 - 75,000$$

$$= 320,000 - 16,000p - 400,000 + 40,000p - 1,000p^2 - 75,000$$

$$= -155,000 + 24,000p - 1,000p^2$$

Technically, $P \circ f$ and P are different functions since the first has independent variable p and the second has independent variable x. However, since both functions represent the same quantity, it is customary to use the same symbol to name each function. Thus,

$$P(p) = -155,000 + 24,000p - 1,000p^2$$

expresses the weekly profit P as a function of price p.

Problem 7   Repeat Example 7 for the functions

$$x = f(p) = 10,000 - 1,000p$$

$$C(x) = 90,000 + 5x \qquad R(x) = 10x - \frac{1}{1,000} x^2$$

Answers to Matched Problems
1. (A) -3 (B) -5 (C) 0   (D) 0
2. $(f + g)(x) = \sqrt{x} + \sqrt{10 - x}$, $(f - g)(x) = \sqrt{x} - \sqrt{10 - x}$, $(fg)(x) = \sqrt{10x - x^2}$, $(f/g)(x) = \sqrt{x}/(10 - x)$; the functions $f + g$, $f - g$, and fg have domain $[0, 10]$, the domain of f/g is $[0, 10)$
3. $(f \circ g)(x) = x$, domain $= (-\infty, \infty)$
   $(g \circ f)(x) = x$, domain $= (-\infty, \infty)$
4. $(f \circ g)(x) = (x - 1)/(x + 2)$; domain: $x \neq -2$ and $x \neq 1$ or $(-\infty, -2) \cup (-2, 1) \cup (1, \infty)$
5. $(f \circ g)(x) = \sqrt{10 - x}$; domain: $x \geq 1$ and $x \leq 10$ or $[1, 10]$
6. $h(x) = (f \circ g)(x)$, where $f(x) = \sqrt{x}$ and $g(x) = 4x - 7$
7. $P(p) = -140,000 + 15,000p - 1,000p^2$

EXERCISE 3-7

A

In Problems 1–6, for the indicated functions f and g, find the functions f + g, f − g, fg, and f/g, and find their domains.

1. $f(x) = 4x; \quad g(x) = x + 1$

2. $f(x) = 3x; \quad g(x) = x - 2$

3. $f(x) = 2x^2; \quad g(x) = x^2 + 1$

4. $f(x) = 3x; \quad g(x) = x^2 + 4$

5. $f(x) = 3x + 5; \quad g(x) = x^2 - 1$

6. $f(x) = 2x - 7; \quad g(x) = 9 - x^2$

In Problems 7–12, for the indicated functions f and g, find the functions f ∘ g and g ∘ f, and find their domains.

7. $f(x) = x^3; \quad g(x) = x^2 - x + 1$

8. $f(x) = x^2; \quad g(x) = x^3 + 2x + 4$

9. $f(x) = |x + 1|; \quad g(x) = 2x + 3$

10. $f(x) = |x - 4|; \quad g(x) - 3x + 2$

11. $f(x) = x^{1/3}; \quad g(x) = 2x^3 + 4$

12. $f(x) = x^{2/3}; \quad g(x) = 8 - x^3$

B

In Problems 13–18, for the indicated functions f and g, find the functions f + g, f − g, fg, and f/g, and find their domains.

13. $f(x) = \sqrt{4 - x^2}; \quad g(x) = \sqrt{4 + x^2}$

14. $f(x) = \sqrt{1 - x^2}; \quad g(x) = \sqrt{1 + x^2}$

15. $f(x) = \sqrt{2 - x}; \quad g(x) = \sqrt{x + 3}$

16. $f(x) = \sqrt{x + 4}; \quad g(x) = \sqrt{3 - x}$

17. $f(x) = \sqrt{x} + 2; \quad g(x) = \sqrt{x} - 4$

18. $f(x) = 1 - \sqrt{x}; \quad g(x) = 2 - \sqrt{x}$

19. $f(x) = \sqrt{x^2 + x - 6}; \quad g(x) = \sqrt{7 + 6x - x^2}$

20. $f(x) = \sqrt{8 + 2x - x^2}; \quad g(x) = \sqrt{x^2 - 7x + 10}$

In Problems 21–28, for the indicated functions f and g, find the functions f ∘ g and g ∘ f, and find their domains.

21. $f(x) = (x - 1)^3; \quad g(x) = 1 + x^{1/3}$

22. $f(x) = (x + 2)^{1/3}; \quad g(x) = x^3 - 2$

23. $f(x) = \sqrt{x}; \quad g(x) = x - 4$

24. $f(x) = \sqrt{x}; \quad g(x) = 2x + 5$

25. $f(x) = x + 2; \quad g(x) = \dfrac{1}{x}$

26. $f(x) = x - 3; \quad g(x) = \dfrac{1}{x^2}$

27. $f(x) = |x|; \quad g(x) = \dfrac{1}{x - 1}$

28. $f(x) = |x - 1|; \quad g(x) = \dfrac{1}{x}$

In Problems 29–36, express h as a composition of two simpler functions f and g of the form $f(x) = x^n$ and $g(x) = ax + b$, where n is a rational number and a and b are integers.

29. $h(x) = (2x - 7)^4$

30. $h(x) = (3 - 5x)^7$

31. $h(x) = \sqrt{4 + 2x}$

32. $h(x) = \sqrt{3x - 11}$

33. $h(x) = 3x^7 - 5$

34. $h(x) = 5x^6 + 3$

35. $h(x) = \dfrac{4}{\sqrt{x}} + 3$

36. $h(x) = -\dfrac{2}{\sqrt{x}} + 1$

C

In Problems 37–42, for the indicated functions f and g, find the functions f + g, f − g, fg, and f/g, and find their domains.

37. $f(x) = x + \dfrac{1}{x}; \quad g(x) = x - \dfrac{1}{x}$

38. $f(x) = x - 1; \quad g(x) = x - \dfrac{6}{x - 1}$

39. $f(x) = x + \dfrac{8}{\sqrt{x}};\quad g(x) = x - \dfrac{8}{\sqrt{x}}$

40. $f(x) = x + \sqrt{x};\quad g(x) = x - \sqrt{x}$

41. $f(x) = 1 - \dfrac{x}{|x|};\quad g(x) = 1 + \dfrac{x}{|x|}$

42. $f(x) = x + |x|;\quad g(x) = x - |x|$

In Problems 43–52, for the indicated functions f and g, find the functions f ∘ g and g ∘ f, and find their domains.

43. $f(x) = \sqrt{4 - x};\quad g(x) = x^2$

44. $f(x) = \sqrt{x - 1};\quad g(x) = x^2$

45. $f(x) = \sqrt{x};\quad g(x) = x^2$

46. $f(x) = \dfrac{1}{\sqrt{x}};\quad g(x) = \dfrac{1}{x^2}$

47. $f(x) = \dfrac{x + 5}{x};\quad g(x) = \dfrac{x}{x - 2}$

48. $f(x) = \dfrac{x}{x - 1};\quad g(x) = \dfrac{2x - 4}{x}$

49. $f(x) = \sqrt{4 - x^2};\quad g(x) = \sqrt{x}$

50. $f(x) = \sqrt{x};\quad g(x) = \sqrt{9 - x^2}$

51. $f(x) = \sqrt{25 - x^2};\quad g(x) = \sqrt{9 + x^2}$

52. $f(x) = \sqrt{x^2 - 9};\quad g(x) = \sqrt{x^2 + 25}$

Problems 53–58 require the use of a graphic calculator or a computer. Enter the given expression for $(f \circ g)(x)$ exactly as it is written and graph for $-10 \le x \le 10$. Then simplify the expression, enter the result, and graph in a new viewing rectangle, again for $-10 \le x \le 10$. Find the domain of $f \circ g$. Which is the correct graph of $f \circ g$?

53. $f(x) = \sqrt{5 - x^2}, g(x) = \sqrt{3 - x},$
$(f \circ g)(x) = \sqrt{5 - (\sqrt{3 - x})^2}$

54. $f(x) = \sqrt{6 - x^2}, g(x) = \sqrt{x - 1},$
$(f \circ g)(x) = \sqrt{6 - (\sqrt{x - 1})^2}$

55. $f(x) = \sqrt{x^2 + 5}, g(x) = \sqrt{x^2 - 4},$
$(f \circ g)(x) = \sqrt{(\sqrt{x^2 - 4})^2 + 5}$

56. $f(x) = \sqrt{x^2 + 5}, g(x) = \sqrt{4 - x^2},$
$(f \circ g)(x) = \sqrt{(\sqrt{4 - x^2})^2 + 5}$

57. $f(x) = \sqrt{x^2 + 7}, g(x) = \sqrt{9 - x^2},$
$(f \circ g)(x) = \sqrt{(\sqrt{9 - x^2})^2 + 7}$

58. $f(x) = \sqrt{x^2 + 7}, g(x) = \sqrt{x^2 - 9},$
$(f \circ g)(x) = \sqrt{(\sqrt{x^2 - 9})^2 + 7}$

APPLICATIONS

59. Market Research. The demand x and the price p (in dollars) for a certain product are related by

$$x = f(p) = 4{,}000 - 200p$$

The revenue (in dollars) from the sale of x units is given by

$$R(x) = 20x - \dfrac{1}{200}x^2$$

and the cost (in dollars) of producing x units is given by

$$C(x) = 10x + 30{,}000$$

Express the profit as a function of the price p.

60. Market Research. The demand x and the price p (in dollars) for a certain product are related by

$$x = f(p) = 5{,}000 - 100p$$

The revenue (in dollars) from the sale of x units and the cost (in dollars) of producing x units are given, respectively, by

$$R(x) = 50x - \dfrac{1}{100}x^2 \quad \text{and} \quad C(x) = 20x + 40{,}000$$

Express the profit as a function of the price p.

61. Pollution. An oil tanker aground on a reef is leaking oil that forms a circular oil slick about 0.1 foot thick (see the figure). The radius of the slick (in feet) t minutes after the leak first occurred is given by

$$r(t) = 0.4t^{1/3}$$

Express the volume of the oil slick as a function of t.

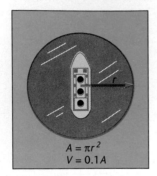

$A = \pi r^2$
$V = 0.1A$

62. Pollution. Repeat Problem 61 if

$$r(t) = \frac{0.3t}{1 + t}$$

★**63. Fluid Flow.** A conical paper cup with diameter 4 inches and height 4 inches is initially full of water. A small hole is made in the bottom of the cup and the water begins to flow out of the cup. Let h and r be the height and radius, respectively, of the water in the cup t minutes after the water begins to flow.

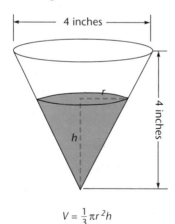

4 inches

4 inches

$V = \frac{1}{3}\pi r^2 h$

(A) Express r as a function of h.
(B) Express the volume V as a function of h.
(C) If the height of the water after t minutes is given by

$$h(t) = 0.5\sqrt{t}$$

express V as a function of t.

★**64. Evaporation.** A water trough with triangular ends is 6 feet long, 4 feet wide, and 2 feet deep. Initially, the trough is full of water, but due to evaporation, the volume of the water is decreasing. Let h and w be the height and width, respectively, of the water in the tank t hours after it began to evaporate.

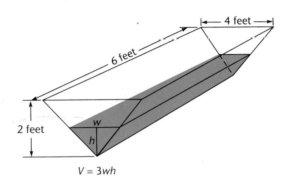

4 feet

6 feet

2 feet

w

h

$V = 3wh$

(A) Express w as a function of h.
(B) Express V as a function of h.
(C) If the height of the water after t hours is given by

$$h(t) = 2 - 0.2\sqrt{t}$$

express V as a function of t.

SECTION 3-8 **Inverse Functions**

- One-to-One Functions
- Inverse Functions

Many important mathematical relationships can be expressed in terms of functions. For example,

$$C = \pi d = f(d)$$ The circumference of a circle is a function of the diameter d.

$$V = s^3 = g(s)$$ The volume of a cube is a function of the edge s.

$$d = 1{,}000 - 100p = h(p)$$ The demand for a product is a function of the price p.

In many cases, we are interested in *reversing* the correspondence determined by a function. Thus,

$$d = \frac{C}{\pi} = m(C)$$ The diameter of a circle is a function of the circumference C.

$$s = \sqrt[3]{V} = n(V)$$ The edge of a cube is a function of the volume V.

$$p = 10 - \frac{1}{100}d = r(d)$$ The price of a product is a function of the demand d.

As these examples illustrate, reversing the relationship between two quantities often produces a new function. This new function is called the *inverse* of the original function. Later in this text we will see that many important functions (for example, logarithmic functions) are actually defined as the inverses of other functions.

In this section, we develop techniques for determining whether the inverse function exists, some general properties of inverse functions, and methods for finding the rule of correspondence that defines the inverse function. A review of Section 3-3 will prove very helpful at this point.

● One-to-One Functions

Recall the set form of the definition of a function:

A function is a set of ordered pairs with the property that no two ordered pairs have the same first component and different second components.

However, it is possible that two ordered pairs in a function could have the same second component and different first components. If this does not happen, then we call the function a *one-to-one function*. It turns out that one-to-one functions are the only functions that have inverse functions.

DEFINITION 1 **One-to-One Function**

A function is one-to-one if no two ordered pairs in the function have the same second component and different first components.

To illustrate this concept, consider the following three sets of ordered pairs:

$$f = \{(0, 3), (0, 5), (4, 7)\}$$ function

$$g = \{(0, 3), (2, 3), (4, 7)\}$$ not 1 to 1

$$h = \{(0, 3), (2, 5), (4, 7)\}$$

Set f is not a function because the ordered pairs $(0, 3)$ and $(0, 5)$ have the same first component and different second components. Set g is a function, but it is not a one-to-one function because the ordered pairs $(0, 3)$ and $(2, 3)$ have the same second component and different first components. But set h is a function and it is one-to-one. Representing these three sets of ordered pairs as rules of correspondence provides some additional insight into this concept.

	f	
Domain		Range
0		3
		5
4		7

f is not a function.

	g	
Domain		Range
0		3
2		
4		7

g is a function, but is not one-to-one.

	h	
Domain		Range
0		3
2		5
4		7

h is a one-to-one function.

EXAMPLE 1   Determining Whether a Function Is One-To-One

Determine whether f is a one-to-one function for:

(A) $f(x) = x^2$ (B) $f(x) = 2x - 1$

Solutions   (A) To show that a function is not one-to-one, all we have to do is find two different ordered pairs in the function with the same second component and different first components. Since

$$f(2) = 2^2 = 4 \quad \text{and} \quad f(-2) = (-2)^2 = 4$$

the ordered pairs $(2, 4)$ and $(-2, 4)$ both belong to f and f is not one-to-one.

(B) To show that a function is one-to-one, we have to show that no two ordered pairs have the same second component and different first components. To do this, we assume there are two ordered pairs $(a, f(a))$ and $(b, f(b))$ in f with the same second components and then show that the first components must also be the same. That is, we show that $f(a) = f(b)$ implies $a = b$. We proceed as follows:

$f(a) = f(b)$ Assume second components are equal.
$2a - 1 = 2b - 1$ Evaluate $f(a)$ and $f(b)$.
$2a = 2b$ Simplify.
$a = b$ Conclusion: f is one-to-one

Thus, by Definition 1, f is a one-to-one function.

Problem 1   Determine whether f is a one-to-one function for:

(A) $f(x) = 4 - x^2$ (B) $f(x) = 4 - 2x$
          Not   is

The methods used in the solution of Example 1 can be stated as a theorem.

Theorem 1 **One-to-One Functions**

1. If $f(a) = f(b)$ for at least one pair of domain values a and b, $a \neq b$, then f is not one-to-one.

2. If the assumption $f(a) = f(b)$ always implies that the domain values a and b are equal, then f is one-to-one.

Applying Theorem 1 is not always easy—try testing $f(x) = x^3 + 2x + 3$, for example. However, if we are given the graph of a function, then there is a simple graphical procedure for determining if the function is one-to-one. If a horizontal line intersects the graph of a function in more than one point, then the function is not one-to-one, as shown in Figure 1(a). However, if each horizontal line intersects the graph in one point, or not at all, then the function is one-to-one, as shown in Figure 1(b). These observations form the basis for the *horizontal line test*.

FIGURE 1 Intersections of graphs and horizontal lines.

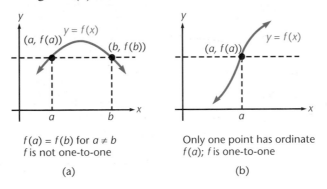

$f(a) = f(b)$ for $a \neq b$
f is not one-to-one

(a)

Only one point has ordinate
$f(a)$; f is one-to-one

(b)

Theorem 2 **Horizontal Line Test**

A function is one-to-one if and only if each horizontal line intersects the graph of the function in at most one point.

The graphs of the functions considered in Example 1 are shown in Figure 2. Applying the horizontal line test to each graph confirms the results we obtained in Example 1.

FIGURE 2 Applying the horizontal line test.

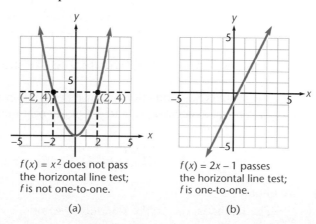

$f(x) = x^2$ does not pass
the horizontal line test;
f is not one-to-one.

(a)

$f(x) = 2x - 1$ passes
the horizontal line test;
f is one-to-one.

(b)

A function that is increasing throughout its domain or decreasing throughout its domain will always pass the horizontal line test [see Figures 3(a) and 3(b). Thus, we have the following theorem.

Theorem 3   **Increasing and Decreasing Functions**

If a function f is increasing throughout its domain or decreasing throughout its domain, then f is a one-to-one function.

FIGURE 3 Increasing, decreasing, and one-to-one functions.

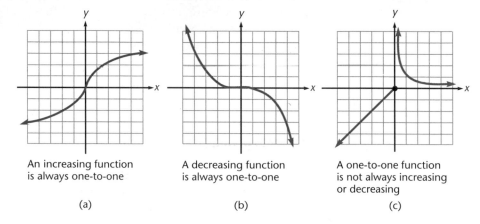

An increasing function   A decreasing function   A one-to-one function
is always one-to-one   is always one-to-one   is not always increasing
                                                               or decreasing

(a)   (b)   (c)

The converse of Theorem 3 is false. To see this, consider the function graphed in Figure 3(c). This function is increasing on $(-\infty, 0]$ and decreasing on $(0, \infty)$, yet the graph passes the horizontal line test. Thus, this is a one-to-one function that is neither an increasing function nor a decreasing function.

● **Inverse Functions**   Now we want to see how we can form a new function by reversing the correspondence determined by a given function. Let g be the function defined as follows:

$$g = \{(-3, 9), (0, 0), (3, 9)\} \quad \text{g is not one-to-one.}$$

Notice that g is not one-to-one because the domain elements -3 and 3 both correspond to the range element 9. We can reverse the correspondence determined by function g simply by reversing the components in each ordered pair in g, producing the following set:

$$G = \{(9, -3), (0, 0), (9, 3)\} \quad \text{G is not a function.}$$

But the result is not a function because the domain element 9 corresponds to two different range elements, -3 and 3. On the other hand, if we reverse the ordered pairs in the function

$$f = \{(1, 2), (2, 4), (3, 9)\} \quad \text{f is one-to-one.}$$

we obtain

$$F = \{(2, 1), (4, 2), (9, 3)\} \quad \textit{F is a function.}$$

This time f is a one-to-one function, and the set F turns out to be a function also. This new function F, formed by reversing all the ordered pairs in f, is called the *inverse* of f and is usually denoted* by f^{-1}. Thus,

$$f^{-1} = \{(2, 1), (4, 2), (9, 3)\} \quad \textit{The inverse of f}$$

Notice that f^{-1} is also a one-to-one function and that the following relationships hold:

$$\text{Domain of } f^{-1} = \{2, 4, 9\} = \text{Range of } f$$
$$\text{Range of } f^{-1} = \{1, 2, 3\} = \text{Domain of } f$$

Thus, reversing all the ordered pairs in a one-to-one function forms a new one-to-one function and reverses the domain and range in the process. We are now ready to present a formal definition of the inverse of a function.

DEFINITION 2   **Inverse of a Function**

If f is a one-to-one function, then the **inverse** of f, denoted f^{-1}, is the function formed by reversing all the ordered pairs in f. Thus,

$$f^{-1} = \{(y, x) \mid (x, y) \text{ is in } f\}$$

If f is not one-to-one, then f **does not have an inverse** and f^{-1} **does not exist.**

The following properties of inverse functions follow directly from the definition.

Theorem 4   **Properties of Inverse Functions**

If f^{-1} exists, then

1. f^{-1} is a one-to-one function.

2. Domain of f^{-1} = Range of f

3. Range of f^{-1} = Domain of f

---

*f^{-1}, read "f inverse," is a special symbol used here to represent the inverse of the function f. It does *not* mean $1/f$.

Finding the inverse of a function defined by a finite set of ordered pairs is easy; just reverse each ordered pair. But how do we find the inverse of a function defined by an equation? Consider the one-to-one function f defined by

$$f(x) = 2x - 1$$

To find f^{-1}, we let $y = f(x)$ and solve for x:

$$y = 2x - 1$$
$$y + 1 = 2x$$
$$\tfrac{1}{2}y + \tfrac{1}{2} = x$$

Since the ordered pair (x, y) is in f if and only if the reversed ordered pair (y, x) is in f^{-1}, this last equation defines f^{-1}:

$$x = f^{-1}(y) = \tfrac{1}{2}y + \tfrac{1}{2} \qquad (1)$$

Something interesting happens if we form the composition* of f and f^{-1} in either of the two possible orders

$$f^{-1}[f(x)] = f^{-1}[2x - 1] = \tfrac{1}{2}(2x - 1) + \tfrac{1}{2} = x - \tfrac{1}{2} + \tfrac{1}{2} = x$$

and

$$f[f^{-1}(y)] = f(\tfrac{1}{2}y + \tfrac{1}{2}) = 2(\tfrac{1}{2}y + \tfrac{1}{2}) - 1 = y + 1 - 1 = y$$

These compositions indicate that if f maps x into y, then f^{-1} maps y back into x and if f^{-1} maps y into x, then f maps x back into y. This is interpreted schematically in Figure 4.

FIGURE 4 Composition of f and f^{-1}.

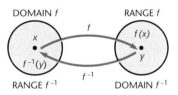

Finally, we note that we usually use x to represent the independent variable and y the dependent variable in an equation that defines a function. It is customary to do this for inverse functions also. Thus, interchanging the variables x and y in equation (1), we can state that the inverse of

$$y = f(x) = 2x - 1$$

is

$$y = f^{-1}(x) = \tfrac{1}{2}x + \tfrac{1}{2}$$

*When working with inverse functions, it is customary to write compositions as $f[g(x)]$ rather than as $f \circ g(x)$.

In general, we have the following result:

Theorem 5

Relationship between f and f^{-1}

If f^{-1} exists, then

1. $x = f^{-1}(y)$ if and only if $y = f(x)$.

2. $f^{-1}[f(x)] = x$ for all x in the domain of f.

3. $f[f^{-1}(y)] = y$ for all y in the domain of f^{-1} or, if x and y have been interchanged, $f[f^{-1}(x)] = x$ for all x in the domain of f^{-1}.

If f and g are one-to-one functions satisfying

$$f[g(x)] = x \qquad \text{for all } x \text{ in the domain of } g$$
$$g[f(x)] = x \qquad \text{for all } x \text{ in the domain of } f$$

then it can be shown that $g = f^{-1}$ and $f = g^{-1}$. Thus, the inverse function is the only function that satisfies both these compositions. We can use this fact to check that we have found the inverse correctly.

The procedure for finding the inverse of a function defined by an equation is given in the next box. This procedure can be applied whenever it is possible to solve $y = f(x)$ for x in terms of y.

Finding the Inverse of a Function f

Step 1. Find the domain of f and verify that f is one-to-one. If f is not one-to-one, then stop, since f^{-1} does not exist.

Step 2. Solve the equation $y = f(x)$ for x. The result is an equation of the form $x = f^{-1}(y)$.

Step 3. Interchange x and y in the equation found in step 2. This expresses f^{-1} as a function of x.

Step 4. Find the domain of f^{-1}. Remember, the domain of f^{-1} must be the same as the range of f.

Check your work by verifying that

$$f^{-1}[f(x)] = x \qquad \text{for all } x \text{ in the domain of } f$$

and

$$f[f^{-1}(x)] = x \qquad \text{for all } x \text{ in the domain of } f^{-1}$$

EXAMPLE 2   **Finding the Inverse of a Function**

Find f^{-1} for $f(x) = \sqrt{x - 1}$.

Solution   *Step 1. Find the domain of f and verify that f is one-to-one.* The domain of f is $[1, \infty)$. The graph of f in the margin shows that f is one-to-one, hence f^{-1} exists.

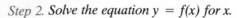

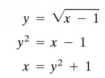

$f(x) = \sqrt{x-1},\ x \geq 1$

Step 2. Solve the equation $y = f(x)$ for x.

$$y = \sqrt{x - 1}$$
$$y^2 = x - 1$$
$$x = y^2 + 1$$

Thus,

$$x = f^{-1}(y) = y^2 + 1$$

Step 3. Interchange x and y.

$$y = f^{-1}(x) = x^2 + 1$$

Step 4. Find the domain of f^{-1}. The equation $f^{-1}(x) = x^2 + 1$ is defined for all values of x, but this does not tell us what the domain of f^{-1} is. Remember, the domain of f^{-1} must equal the range of f. From the graph of f, we see that the range of f is $[0, \infty)$. Thus, the domain of f^{-1} is also $[0, \infty)$. That is,

$$f^{-1}(x) = x^2 + 1 \qquad x \geq 0$$

Check   For x in $[1, \infty)$, the domain of f, we have

$$f^{-1}[f(x)] = f^{-1}(\sqrt{x - 1})$$
$$= (\sqrt{x - 1})^2 - 1$$
$$= x - 1 + 1$$
$$\overset{\checkmark}{=} x$$

For x in $[0, \infty)$, the domain of f^{-1}, we have

$$f[f^{-1}(x)] = f(x^2 + 1)$$
$$= \sqrt{(x^2 + 1) - 1}$$
$$= \sqrt{x^2}$$
$$= |x| \qquad\qquad \sqrt{x^2} = |x| \text{ for any real number } x.$$
$$\overset{\checkmark}{=} x \qquad\qquad |x| = x \text{ for } x \geq 0.$$

Problem 2   Find f^{-1} for $f(x) = \sqrt{x + 2}$.

There is an important relationship between the graph of any function and its inverse that is based on the following observation: In a rectangular coordinate system, the points (a, b) and (b, a) are symmetric with respect to the line $y = x$ [see Figure 5(a)]. Theorem 6 is an immediate consequence of this observation.

Theorem 6	**Symmetry Property for the Graphs of f and f^{-1}** The graphs of $y = f(x)$ and $y = f^{-1}(x)$ are symmetric with respect to the line $y = x$.

Knowledge of this symmetry property makes it easy to graph f^{-1} if the graph of f is known, and vice versa. Figures 5(b) and 5(c) illustrate this property for the two inverse functions we found earlier in this section.

FIGURE 5 Symmetry with respect to the line $y = x$.

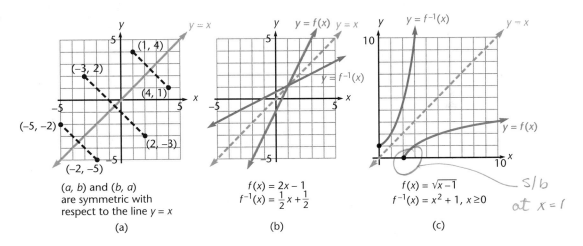

(a, b) and (b, a) are symmetric with respect to the line $y = x$

(a)

$f(x) = 2x - 1$
$f^{-1}(x) = \frac{1}{2}x + \frac{1}{2}$

(b)

$f(x) = \sqrt{x - 1}$
$f^{-1}(x) = x^2 + 1, x \geq 0$

(c)

s/b
at $x = 1$

If a function is not one-to-one, we usually can restrict the domain of the function to produce a new function that is one-to-one. Then we can find an inverse for the restricted function. Suppose we start with $f(x) = x^2 - 4$. Since f is not one-to-one, f^{-1} does not exist [Figure 6(a)]. But there are many ways the domain of f can be restricted to obtain a one-to-one function. Figures 6(b) and 6(c) illustrate two such restrictions.

FIGURE 6 Restricting the domain of a function.

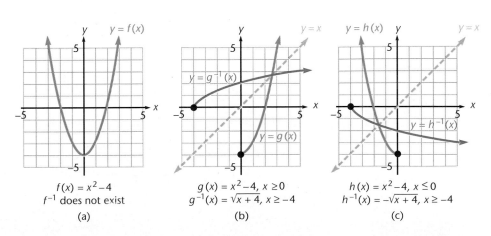

$f(x) = x^2 - 4$
f^{-1} does not exist

(a)

$g(x) = x^2 - 4, x \geq 0$
$g^{-1}(x) = \sqrt{x + 4}, x \geq -4$

(b)

$h(x) = x^2 - 4, x \leq 0$
$h^{-1}(x) = -\sqrt{x + 4}, x \geq -4$

(c)

Recall from Theorem 2 that increasing and decreasing functions are always one-to-one. This provides the basis for a convenient and popular method of restricting the domain of a function:

> If the domain of a function f is restricted to an interval on the x axis over which f is increasing (or decreasing), then the new function formed by this restriction is one-to-one and has an inverse.

We used this method to form the functions g and h in Figure 6.

EXAMPLE 3   Finding the Inverse of a Function

Find the inverse of $f(x) = 4x - x^2, x \leq 2$. Graph f, f^{-1}, and $y = x$ in the same coordinate system.

Solution   *Step 1. Find the domain of f and verify that f is one-to-one.* The graph of $y = 4x - x^2$ is the parabola shown in figure (a). Restricting the domain of f to $x \leq 2$ restricts the graph of f to the left side of this parabola [figure (b)]. Thus, f is a one-to-one function.

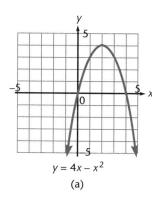

$y = 4x - x^2$

(a)

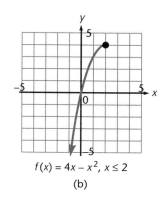

$f(x) = 4x - x^2, x \leq 2$

(b)

Step 2. Solve the equation y = f(x) for x.

$$y = 4x - x^2$$

$$x^2 - 4x = -y \qquad \text{Rearrange terms.}$$

$$x^2 - 4x + 4 = -y + 4 \qquad \text{Add 4 to complete the square on the left side.}$$

$$(x - 2)^2 = 4 - y$$

Taking the square root of both sides of this last equation, we obtain two possible solutions:

$$x - 2 = \pm\sqrt{4 - y}$$

The restricted domain of f tells us which solution to use. Since $x \leq 2$ implies $x - 2 \leq 0$, we must choose the negative square root. Thus,

$$x - 2 = -\sqrt{4 - y}$$

$$x = 2 - \sqrt{4 - y}$$

and we have found

$$x = f^{-1}(y) = 2 - \sqrt{4 - y}$$

Step 3. Interchange x and y.

$$y = f^{-1}(x) = 2 - \sqrt{4 - x}$$

Step 4. Find the domain of f^{-1}. The equation $f^{-1}(x) = 2 - \sqrt{4 - x}$ is defined for $x \leq 4$. From the graph in figure (b), the range of f also is $(-\infty, 4]$. Thus,

$$f^{-1}(x) = 2 - \sqrt{4 - x} \qquad x \leq 4$$

The check is left for the reader.

The graphs of f, f^{-1}, and $y = x$ are shown in figure (c). Sometimes it is difficult to visualize the reflection of the graph of f in the line $y = x$. Choosing some points on the graph of f and plotting their reflections first makes it easier to sketch the graph of f^{-1}.

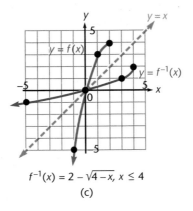

$f^{-1}(x) = 2 - \sqrt{4-x},\ x \leq 4$

(c)

Problem 3 Find the inverse of $f(x) = 4x - x^2, x \geq 2$. Graph f, f^{-1}, and $y = x$ in the same coordinate system.

Answers to Matched Problems
1. (A) Not one-to-one   (B) One-to-one
2. $f^{-1}(x) = x^2 - 2, x \geq 0$
3. $f^{-1}(x) = 2 + \sqrt{4 - x}, x \leq 4$

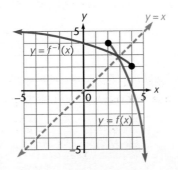

EXERCISE 3-8

A _____

Which of the functions in Problems 1–16 are one-to-one?

1. {(1, 2), (2, 1), (3, 4), (4, 3)}

2. {(−1, 0), (0, 1), (1, −1), (2, 1)}

3. {(5, 4), (4, 3), (3, 3), (2, 4)}

4. {(5, 4), (4, 3), (3, 2), (2, 1)}

5. Domain   Range

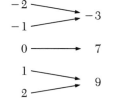

6. Domain   Range

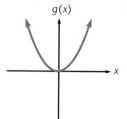

7. Domain   Range

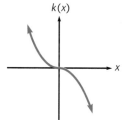

8. Domain   Range

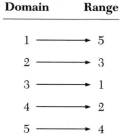

9. $f(x)$

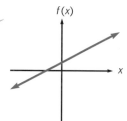

10. $g(x)$

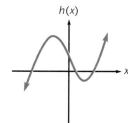

11. $h(x)$

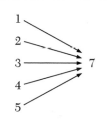

12. $k(x)$

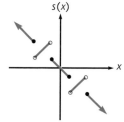

13. $m(x)$

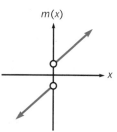

14. $n(x)$

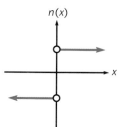

15. $r(x)$

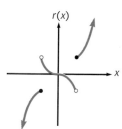

16. $s(x)$

B _____

Which of the functions in Problems 17–22 are one-to-one?

17. $F(x) = \frac{1}{2}x + 2$

18. $G(x) = -\frac{1}{3}x + 1$

19. $H(x) = 4 - x^2$

20. $K(x) = \sqrt{4 - x}$

21. $M(x) = \sqrt{x + 1}$

22. $N(x) = x^2 - 1$

In Problems 23–26, use the graph of the one-to-one function f to sketch the graph of f^{-1}. State the domain and range of f^{-1}.

23.

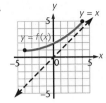

24.

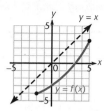

25.

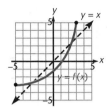

26.

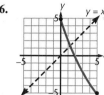

In Problems 27–32, verify that g is the inverse of the one-to-one function f by showing that $g[f(x)] = x$ and $f[g(x)] = x$. Sketch the graphs of f, g, and the line $y = x$ in the same coordinate system.

27. $f(x) = 3x + 6; g(x) = \frac{1}{3}x - 2$

28. $f(x) = -\frac{1}{2}x + 2; g(x) = -2x + 4$

29. $f(x) = 4 + x^2, x \geq 0; g(x) = \sqrt{x - 4}$

30. $f(x) = \sqrt{x + 2}; g(x) = x^2 - 2, x \geq 0$

31. $f(x) = -\sqrt{x - 2}; g(x) = x^2 + 2, x \leq 0$

32. $f(x) = 6 - x^2, x \leq 0; g(x) = -\sqrt{6 - x}$

The functions in Problems 33–52 are one-to-one. Find f^{-1}.

33. $f(x) = 3x$

34. $f(x) = \frac{1}{2}x$

35. $f(x) = 4x - 3$

36. $f(x) = -\frac{1}{3}x + \frac{5}{3}$

37. $f(x) = \frac{1}{10}x + \frac{3}{5}$

38. $f(x) = -2x - 7$

39. $f(x) = \dfrac{2}{x - 1}$

40. $f(x) = \dfrac{3}{x + 4}$

41. $f(x) = \dfrac{x}{x + 2}$

42. $f(x) = \dfrac{x - 3}{x}$

43. $f(x) = \dfrac{2x + 5}{3x - 4}$

44. $f(x) = \dfrac{5 - 3x}{7 - 4x}$

45. $f(x) = x^3 + 1$

46. $f(x) = x^5 - 2$

47. $f(x) = 4 - \sqrt[5]{x + 2}$

48. $f(x) = \sqrt[3]{x + 3} - 2$

49. $f(x) = \frac{1}{2}\sqrt{16 - x}$

50. $f(x) = -\frac{1}{3}\sqrt{36 - x}$

51. $f(x) = 3 - \sqrt{x - 2}$

52. $f(x) = 4 + \sqrt{5 - x}$

C

The functions in Problems 53–56 are one-to-one. Find f^{-1}.

53. $f(x) = (x - 1)^2 + 2, x \geq 1$

54. $f(x) = 3 - (x - 5)^2, x \leq 5$

55. $f(x) = x^2 + 2x - 2, x \leq -1$

56. $f(x) = x^2 + 8x + 7, x \geq 4$

The graph of each function in Problems 57–60 is one-quarter of the graph of the circle with radius 3 and center $(0, 0)$. Find f^{-1}, find the domain and range of f^{-1}, and sketch the graphs of f and f^{-1} in the same coordinate system.

57. $f(x) = -\sqrt{9 - x^2}, 0 \leq x \leq 3$

58. $f(x) = \sqrt{9 - x^2}, 0 \leq x \leq 3$

59. $f(x) = \sqrt{9 - x^2}, -3 \leq x \leq 0$

60. $f(x) = -\sqrt{9 - x^2}, -3 \leq x \leq 0$

The graph of each function in Problems 61–64 is one-quarter of the graph of the circle with radius 1 and center $(0, 1)$. Find f^{-1}, find the domain and range of f^{-1}, and sketch the graphs of f and f^{-1} in the same coordinate system.

61. $f(x) = 1 + \sqrt{1 - x^2},\ 0 \leq x \leq 1$

62. $f(x) = 1 - \sqrt{1 - x^2},\ 0 \leq x \leq 1$

63. $f(x) = 1 - \sqrt{1 - x^2},\ -1 \leq x \leq 0$

64. $f(x) = 1 + \sqrt{1 - x^2},\ -1 \leq x \leq 0$

65. Find $f^{-1}(x)$ for $f(x) = ax + b,\ a \neq 0$.

66. Find $f^{-1}(x)$ for $f(x) = \sqrt{a^2 - x^2},\ a > 0,\ 0 \leq x \leq a$.

67. Show that the line through the points (a, b) and (b, a), $a \neq b$, is perpendicular to the line $y = x$ (see the figure).

68. Show that the point $((a + b)/2, (a + b)/2)$ bisects the line segment from (a, b) to (b, a), $a \neq b$ (see the figure).

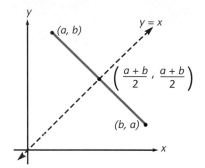

In Problems 69–72, the function f is not one-to-one. Find the inverses of the functions formed by restricting the domain of f as indicated.

69. $f(x) = (2 - x)^2$:
   (A) $x \leq 2$ (B) $x \geq 2$

70. $f(x) = (1 + x)^2$:
   (A) $x \leq -1$ (B) $x \geq -1$

71. $f(x) = \sqrt{4x - x^2}$:
   (A) $0 \leq x \leq 2$ (B) $2 \leq x \leq 4$

72. $f(x) = \sqrt{6x - x^2}$:
   (A) $0 \leq x \leq 3$ (B) $3 \leq x \leq 6$

Problems 73–80 require the use of a graphic calculator or a computer. Graph each function and use the graph to determine if the function is one-to-one.

73. $f(x) = \dfrac{x^2 + |x|}{x}$

74. $f(x) = \dfrac{x^2 - |x|}{x}$

75. $f(x) = \dfrac{x^3 + |x|}{x}$

76. $f(x) = \dfrac{|x|^3 + |x|}{x}$

77. $f(x) = \dfrac{x^2 - 4}{|x - 2|}$

78. $f(x) = \dfrac{1 - x^2}{|x + 1|}$

79. $f(x) = \dfrac{x^3 - 9x}{|x^2 - 9|}$

80. $f(x) = \dfrac{4x - x^3}{|x^2 - 4|}$

Chapter 3 Review

3-1   BASIC TOOLS; CIRCLES

A **Cartesian** or **rectangular coordinate system** is formed by the intersection of a horizontal real number line and a vertical real number line at their origins. These lines are called the **coordinate axes**. The **horizontal axis** is often referred to as the *x* **axis** and the **vertical axis** as the *y* **axis**. These axes divide the plane into four **quadrants**. Each point in the plane corresponds to its **coordinates**—an ordered pair (a, b) determined by passing horizontal and vertical lines through the point. The **abscissa** or *x* **coordinate** a is the coordinate of the intersection of the vertical line with the horizontal axis, and the **ordinate** or *y* **coordinate** b is the coordinate of the intersection of the horizontal line with the vertical axis. The point $(0, 0)$ is called the **origin**. The **solution set** of an equation in two variables is the set of all ordered pairs of real numbers that make the equation a true statement. The **graph of an equation in two variables** is the graph of its solution set.

A graph is **symmetric with respect to**:

1. **The y axis** if $(-a, b)$ is on the graph whenever (a, b) is on the graph.
2. **The x axis** if $(a, -b)$ is on the graph whenever (a, b) is on the graph.
3. **The origin** if $(-a, -b)$ is on the graph whenever (a, b) is on the graph.

Testing an Equation for Symmetry

Symmetry with respect to the:	Equation is equivalent when:
y axis	x is replaced with $-x$
x axis	y is replaced with $-y$
Origin	x and y are replaced with $-x$ and $-y$

The **distance between the two points** $P_1(x_1, y_1)$ and $P_2(x_2, y_2)$ is

$$d(P_1, P_2) = \sqrt{(x_2 - x_1)^2 + (y_2 - y_1)^2}$$

The **standard equations for a circle** are

$$(x - h)^2 + (y - k)^2 = r^2 \qquad \text{Radius: } r > 0,$$
$$\text{Center: } (h, k)$$
$$x^2 + y^2 = r^2 \qquad \text{Radius: } r > 0,$$
$$\text{Center: } (0, 0)$$

3-2 STRAIGHT LINES

The **standard form** for the equation of a line is $Ax + By = C$, where A, B and C are constants, A and B not both 0. The **y intercept** is the ordinate of the point where the graph crosses the y axis, and the **x intercept** is the abscissa of the point where the graph crosses the x axis. The **slope** of the line through the points (x_1, y_1) and (x_2, y_2) is

$$m = \frac{y_2 - y_1}{x_2 - x_1} \qquad \text{if } x_1 \neq x_2$$

The slope is not defined for a vertical line where $x_1 = x_2$. Two lines with slopes m_1 and m_2 are **parallel** if and only if $m_1 = m_2$ and **perpendicular** if and only if $m_1 m_2 = -1$.

Equations of a Line

Standard form	$Ax + By = C$	A and B not both 0
Slope–intercept form	$y = mx + b$	Slope: m; y intercept: b
Point–slope form	$y - y_1 = m(x - x_1)$	Slope: m; Point: (x_1, y_1)
Horizontal line	$y = b$	Slope: 0
Vertical line	$x = a$	Slope: Undefined

3-3   FUNCTIONS

A **function** is a **rule** that produces a correspondence between two sets, called the **domain** and **range**, such that to each element in the **domain** there corresponds one and only one element in the range. Equivalently, a **function** is a **set of ordered pairs** with the property that no two ordered pairs have the same first component and different second components. The **domain** is the set of all first components, and the **range** is the set of all second components. An **equation** in two variables **defines a function** if to each value of the **independent variable**, the placeholder for domain values, there corresponds exactly one value of the **dependent variable**, the placeholder for range values. A **vertical line** will intersect the graph of a function in at most one point. Unless otherwise specified, the **domain of a function defined by an equation** is assumed to be the set of all real number replacements for the independent variable that produce real values for the dependent variable. The symbol $f(x)$ represents the real number in the range of the function f corresponding to the domain value x. Equivalently, the ordered pair $(x, f(x))$ belongs to the function f.

3-4   GRAPHING FUNCTIONS

The **graph of a function** f is the graph of the equation $y = f(x)$. The abscissa of any point where the graph of a function f crosses the x axis is called an x **intercept** of f. The ordinate of a point where the graph crosses the y axis is called the y **intercept**.

Let I be an interval in the domain of a function f. Then:

1. f **is increasing** on I if $f(b) > f(a)$ whenever $b > a$ in I.

2. f **is decreasing** on I if $f(b) < f(a)$ whenever $b > a$ in I.

3. f **is constant** on I if $f(a) = f(b)$ for all a and b in I.

A function f is a **linear function** if $f(x) = mx + b, m \neq 0$, and a **quadratic function** if $f(x) = ax^2 + bx + c, a \neq 0$. The graph of a linear function is a straight line that is neither horizontal nor vertical.

Properties of $f(x) = ax^2 + bx + c, a \neq 0$ **and its graph:**

1. **Axis:** $x = -\dfrac{b}{2a}$ 2. **Vertex:** $\left(-\dfrac{b}{2a}, f\left(-\dfrac{b}{2a} \right) \right)$

3. $f\left(-\dfrac{b}{2a} \right) = \begin{cases} \textbf{Minimum value of } f(x) & \text{if } a > 0 \\ \textbf{Maximum value of } f(x) & \text{if } a < 0 \end{cases}$

4. **Graph is a parabola:**

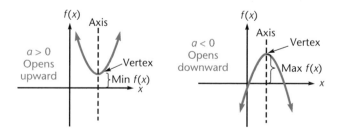

5. **Domain:** all real numbers; **range:** determine from graph.

A **piecewise-defined function** is a function whose definition involves more than one formula. The graph of a function is **continuous** if it has no holes or breaks and **discontinuous** at any point where it has a hole or break. The **greatest integer function** is defined by

$$f(x) = [\![x]\!] = n \qquad \text{where } n \text{ is an integer, } n \le x < n + 1$$

3-5   AIDS TO GRAPHING FUNCTIONS

Symmetry Tests:

$f(-x) = f(x)$	f is an **even function** whose graph is symmetric with respect to the vertical axis.
$f(-x) = -f(x)$	f is an **odd function** whose graph is symmetric with respect to the origin.

Vertical Translation:

$y = f(x) + k, k > 0$	Shift the graph of $y = f(x)$ upward k units.
$y = f(x) - k, k > 0$	Shift the graph of $y = f(x)$ downward k units.

Horizontal Translation:

$y = f(x - h), h > 0$	Shift the graph of $y = f(x)$ to the right h units.
$y = f(x + h), h > 0$	Shift the graph of $y = f(x)$ to the left h units.

Reflection:

$y = -f(x)$	Reflect the graph of $y = f(x)$ in the x axis.

Expansion and Contraction:

$y = Cf(x), C > 1$	Expand the graph of $y = f(x)$ by multiplying each ordinate value by C.
$y = Cf(x), 0 < C < 1$	Contract the graph of $y = f(x)$ by multiplying each ordinate value by C.

3-6   RATIONAL FUNCTIONS

An **nth degree polynomial function** is a function of the form

$$P(x) = a_n x^n + a_{n-1} x^{n-1} + \cdots + a_1 x + a_0 \qquad a_n \ne 0$$

where the coefficients a_i are real numbers and n is an integer, $n \ge 0$. A function of the form $f(x) = n(x)/d(x)$, where $n(x)$ and $d(x)$ are polynomials, is a **rational function**. The line $x = a$ is a **vertical asymptote** for the graph of $y = f(x)$ if $f(x) \to \infty$ or $f(x) \to -\infty$ as $x \to a^+$ or $x \to a^-$. If $d(a) = 0$ and $n(a) \ne 0$, then the line $x = a$ is a vertical asymptote. The line $y = b$ is a **horizontal asymptote** for the graph of $y = f(x)$ if $f(x) \to b$ as $x \to \infty$ or $x \to -\infty$. The line $y = mx + b$ is an **oblique asymptote** if $[f(x) - (mx + b)] \to 0$ as $x \to \infty$ or $x \to -\infty$.

Let

$$f(x) = \frac{a_m x^m + \cdots + a_1 x + a_0}{b_n x^n + \cdots + b_1 x + b_0} \qquad a_m, b_n \ne 0$$

1. If $m < n$, then the x axis is a horizontal asymptote.

2. If $m = n$, then the line $y = a_m/b_n$ is a horizontal asymptote.

3. If $m > n$, then there are no horizontal asymptotes.

> **Graphing a Rational Function: $f(x) = n(x)/d(x)$**
>
> *Step 1. Intercepts.* Find the real solutions of the equation $n(x) = 0$ and use these solutions to plot any x intercepts of the graph of f. Evaluate $f(0)$, if it exists, and plot the y intercept.
>
> *Step 2. Vertical Asymptotes.* Find the real solutions of the equation $d(x) = 0$ and use these solutions to determine the domain of f, the points of discontinuity, and the vertical asymptotes. Sketch any vertical asymptotes as dashed lines.
>
> *Step 3. Sign Chart.* Construct a sign chart for f and use it to determine the behavior of the graph near each vertical asymptote.
>
> *Step 4. Horizontal Asymptotes.* Determine whether there is a horizontal asymptote and if so, sketch it as a dashed line.
>
> *Step 5. Symmetry.* Determine symmetry with respect to the vertical axis and the origin.
>
> *Step 6. Complete the Sketch.* Complete the sketch of the graph by plotting additional points and joining these points with a smooth continuous curve over each interval in the domain of f. (Do not cross any points of discontinuity.)

3-7   OPERATIONS ON FUNCTIONS; COMPOSITION

The **sum**, **difference**, **product**, and **quotient** of the functions f and g are defined by

$$(f + g)(x) = f(x) + g(x) \qquad (f - g)(x) = f(x) - g(x)$$

$$(fg)(x) = f(x)g(x) \qquad \left(\frac{f}{g}\right)(x) = \frac{f(x)}{g(x)} \qquad g(x) \neq 0$$

The **domain** of each function is the intersection of the domains of f and g, with the exception that values of x where $g(x) = 0$ must be excluded from the domain of f/g.

The **composition** of functions f and g is defined by $(f \circ g)(x) = f[g(x)]$. The **domain** of $f \circ g$ is the set of all real numbers x in the domain of g where $g(x)$ is in the domain of f. The domain of $f \circ g$ is always a subset of the domain of g.

3-8   INVERSE FUNCTIONS

A function is **one-to-one** if no two ordered pairs in the function have the same second component and different first components. A **horizontal line** will intersect the graph of a one-to-one function in at most one point. A function that is increasing (decreasing) throughout its domain is one-to-one. The **inverse** of the one-to-one function f is the function f^{-1} formed by reversing all the ordered pairs in f. If f is not one-to-one, then f^{-1} **does not exist**.

Assuming that f^{-1} exists, then:

1. f^{-1} is one-to-one.

2. Domain of f^{-1} = Range of f.

3. Range of f^{-1} = Domain of f.

4. $x = f^{-1}(y)$ if and only if $y = f(x)$.

5. $f^{-1}[f(x)] = x$ for all x in the domain of f.

6. $f[f^{-1}(x)] = x$ for all x in the domain of f^{-1}.

7. To find f^{-1}, solve the equation $y = f(x)$ for x and then interchange x and y.

8. The graphs of $y = f(x)$ and $y = f^{-1}(x)$ are symmetric with respect to the line $y = x$.

Chapter 3 Review Exercise

Work through all the problems in this chapter review and check answers in the back of the book. Answers to all review problems are there, and following each answer is a number in italics indicating the section in which that type of problem is discussed. Where weaknesses show up, review appropriate sections in the text.

A

1. Given the points $A(-2, 3)$ and $B(4, 0)$, find:
   (A) Distance between A and B
   (B) Slope of the line through A and B
   (C) Slope of a line perpendicular to the line through A and B

2. Write the equation of a circle with radius $\sqrt{7}$ and center:
   (A) $(0, 0)$ (B) $(3, -2)$

3. Find the center and radius of the circle given by

$$(x + 3)^2 + (y - 2)^2 = 5$$

4. Graph $3x + 2y = 9$ and indicate its slope.

5. Write an equation of a line with x intercept 6 and y intercept 4. Write the final answer in the standard form $Ax + By = C$, where A, B, and C are integers.

6. Write the slope–intercept form of the equation of the line with slope $-\frac{2}{3}$ and y intercept 2.

7. Write the equations of the vertical and horizontal lines passing through the point $(-3, 4)$. What is the slope of each?

8. Which of the following equations define functions?
   (A) $y = x$ (B) $y^2 = x$
   (C) $y^3 = x$ (D) $|y| = x$

Problems 9–18 refer to the functions f, g, k, and m given by:

$$f(x) = 3x + 5 \qquad g(x) = 4 - x^2$$
$$k(x) = 5 \qquad m(x) = 2|x| - 1$$

Find the indicated quantities or expressions.

9. $f(2) + g(-2) + k(0)$

10. $\dfrac{m(-2) + 1}{g(2) + 4}$

11. $\dfrac{f(2 + h) - f(2)}{h}$

12. $\dfrac{g(a + h) - g(a)}{h}$

13. $(f + g)(x)$

14. $(f - g)(x)$

15. $(fg)(x)$

16. $\left(\dfrac{f}{g}\right)(x)$

17. $(f \circ g)(x)$

18. $(g \circ f)(x)$

19. Find the maximum or minimum value of $f(x) = x^2 - 6x + 11$ without graphing. What are the coordinates of the vertex of the graph?

20. How are the graphs of the following related to the graph of $y = x^2$?
   (A) $y = -x^2$ (B) $y = x^2 - 3$
   (C) $y = (x + 3)^2$

B

21. Find the domain and x intercept(s) for:
   (A) $f(x) = \dfrac{2x - 3}{x + 4}$ (B) $g(x) = \dfrac{3x}{x^2 - x - 6}$

22. Find the horizontal and vertical asymptotes for the functions in Problem 21.

Problems 23–27 refer to the function q given by the following graph. (Assume the graph continues as indicated beyond the part shown.)

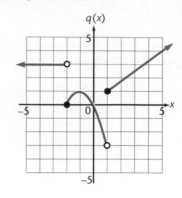

23. Find the domain and range of q.

24. Find the intervals over which q is increasing.

25. Find the intervals over which q is decreasing.

26. Find the intervals over which q is constant.

27. Identify any points of discontinuity.

28. (A) Find an equation of the line through $P(-4, 3)$ and $Q(0, -3)$. Write the final answer in the standard form $Ax + By = C$, where A, B, and C are integers with $A > 0$.
    (B) Find $d(P, Q)$.

29. Write equations of the lines
    (A) Parallel to   (B) Perpendicular to
    the line $6x + 3y = 5$ and passing through the point $(-2, 1)$. Write the final answers in the slope–intercept form $y = mx + b$.

30. Discuss the graph of $4x^2 + 9y^2 = 36$ relative to symmetry with respect to the x axis, y axis and origin.

31. Find the domain of $g(x) = 1/\sqrt{3 - x}$.

32. Graph $f(x) = x^2 - 6x + 5$. Show the axis of symmetry and vertex, and find the range, intercepts, and maximum or minimum value of $f(x)$.

33. Graph $f(x) = 1/(x + 2)$. Indicate any vertical or horizontal asymptotes with dashed lines.

34. Given $f(x) = \sqrt{x} - 8$ and $g(x) = |x|$:
    (A) Find $f \circ g$ and $g \circ f$.
    (B) Find the domains of $f \circ g$ and $g \circ f$.

35. Which of the following functions are one-to-one?
    (A) $f(x) = x^3$
    (B) $g(x) = (x - 2)^2$
    (C) $h(x) = 2x - 3$
    (D) $F(x) = (x + 3)^2, x \geq -3$

36. Given $f(x) = 3x - 7$:
    (A) Find $f^{-1}(x)$.   (B) Find $f^{-1}(5)$.
    (C) Find $f^{-1}[f(x)]$.
    (D) Is f increasing, decreasing, or constant on $(-\infty, \infty)$?

37. Graph, finding the domain, range, and any points of discontinuity:
    $$f(x) = \begin{cases} 2 - x & \text{if } -1 \leq x < 0 \\ x^2 & \text{if } 0 \leq x \leq 1 \end{cases}$$

38. Graph:
    (A) $y = |x| - 2$ (B) $y = |x + 1|$
    (C) $y = \frac{1}{2}|x|$

39. Let: $f(x) = \dfrac{x - 1}{2x + 2}$
    (A) Find the domain and the intercepts for f.
    (B) Find the vertical and horizontal asymptotes for f.
    (C) Sketch a graph of f. Draw vertical and horizontal asymptotes with dashed lines.

40. Given $f(x) = \sqrt{x - 1}$:
    (A) Find $f^{-1}(x)$.
    (B) Find the domain and range of f and f^{-1}.
    (C) Graph f, f^{-1}, and $y = x$ on the same coordinate system.

41. Find the equation of a circle that passes through the point $(-1, 4)$ with center at $(3, 0)$.

42. Find the center and radius of the circle given by $x^2 + y^2 + 4x - 6y = 3$.

43. Determine symmetry with respect to the x axis, y axis, and origin, and graph $xy = 4$.

44. If the slope of a line is negative, is the linear function represented by the graph of the line increasing, decreasing, or constant on $(-\infty, \infty)$?

45. Given $f(x) = x^2 - 1, x \geq 0$:
    (A) Find the domain and range of f and f^{-1}.
    (B) Find $f^{-1}(x)$.   (C) Find $f^{-1}(3)$.
    (D) Find $f^{-1}[f(4)]$.   (E) Find $f^{-1}[f(x)]$.

C

46. How is the graph of $f(x) = -(x - 2)^2 - 1$ related to the graph of $g(x) = x^2$?

47. Graph: $f(x) = -|x + 1| - 1$

48. Find the domain of $f(x) = \sqrt{25 - x^2}$.

49. Given $f(x) = x^2$ and $g(x) = \sqrt{1 - x}$, find each function and its domain.
(A) fg (B) f/g (C) $f \circ g$ (D) $g \circ f$

50. For the one-to-one function f given by

$$f(x) = \frac{x + 2}{x - 3}$$

(A) Find $f^{-1}(x)$.   (B) Find $f^{-1}(3)$.
(C) Find $f^{-1}[f(x)]$.

51. Find a piecewise definition of $f(x) = |x + 1| - |x - 1|$ that does not involve the absolute value function. Find the domain and range of f.

52. Find the equation of the set of points equidistant from $(3, 3)$ and $(6, 0)$. What is the name of the geometric figure formed by this set?

53. Prove that two nonvertical lines are parallel if and only if their slopes are the same.

54. Graph

$$f(x) = \frac{x^2 + 2x + 3}{x + 1}$$

Indicate any vertical, horizontal, or oblique asymptotes with dashed lines.

55. Graph:
(A) $f(x) = [\![\,|x|\,]\!]$ (B) $g(x) = |[\![x]\!]|$

APPLICATIONS

56. Linear Depreciation. A computer system was purchased by a small company for $12,000 and is assumed to have a depreciated value of $2,000 after 8 years. If the value is depreciated linearly from $12,000 to $2,000:
(A) Find the linear equation that relates value V (in dollars) to time t (in years).
(B) What would be the depreciated value of the system after 5 years?

57. Business—Pricing. A sporting goods store sells tennis shorts that cost $30 for $48 and sunglasses that cost $20 for $32.
(A) If the markup policy of the store for items that cost over $10 is assumed to be linear and is reflected in the pricing of these two items, write an equation that expresses retail price R as a function of cost C.
(B) What should be the retail price of a pair of skis that cost $105?

★ **58. Income.** A salesperson receives a base salary of $200 per week and a commission of 10% on all sales over $3,000 during the week. If x represents the salesperson's weekly sales, express the total weekly earnings $E(x)$ as a function of x. Find $E(2,000)$ and $E(5,000)$.

★ **59. Market Research.** If x units of a product are produced each week and sold for a price of $\$p$ per unit, then the weekly demand, revenue, and cost equations are, respectively,

$$x = 500 - 10p$$
$$R(x) = 50x - \tfrac{1}{10}x^2$$
$$C(x) = 20x + 4,000$$

Express the weekly profit as a function of the price p.

★ **60. Architecture.** A circular arc forms the top of an entryway with 6-foot vertical sides 8 feet apart. If the top of the arc is 2 feet above the ends, what is the radius of the arc?

61. Construction. A farmer has 120 feet of fencing to be used in the construction of two identical rectangular pens sharing a common side (see the figure).

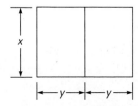

(A) Express the total area $A(x)$ enclosed by both pens as a function of the width x.

(B) From physical considerations, what is the domain of the function A?

(C) Find the dimensions of the pens that will make the total enclosed area maximum.

62. Computer Science. In computer programming, it is often necessary to check numbers for certain properties (even, odd, perfect square, etc.). The greatest integer function provides a convenient method for determining some of these properties. For example, the following function can be used to determine whether a number is the square of an integer:

$$f(x) = x - (\llbracket \sqrt{x} \rrbracket)^2$$

(A) Find $f(1)$.   (B) Find $f(2)$.   (C) Find $f(3)$.

(D) Find $f(4)$.   (E) Find $f(5)$.

(F) Find $f(n^2)$, where n is a positive integer.

Polynomial Functions: Graphs and Zeros

4

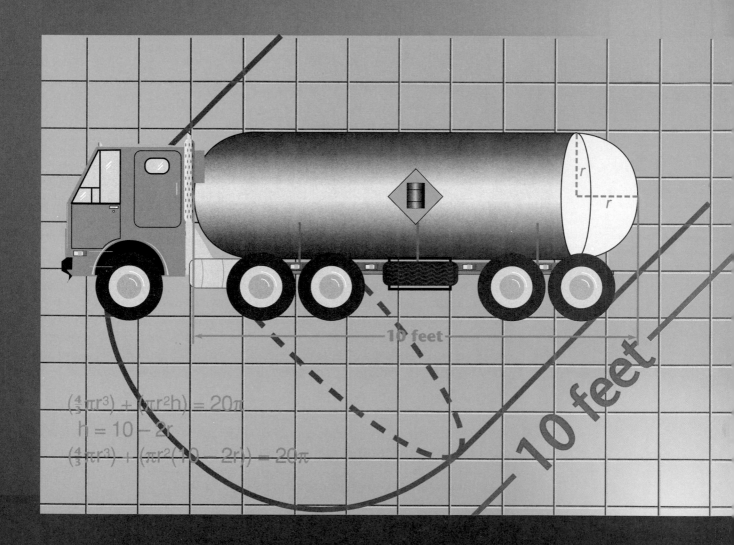

$$\left(\tfrac{4}{3}\pi r^3\right) + \left(\pi r^2 h\right) = 20\pi$$

$$h = 10 - 2r$$

$$\left(\tfrac{4}{3}\pi r^3\right) + \left(\pi r^2(10 - 2r)\right) = 20\pi$$

We know how to solve linear and quadratic equations with real coefficients for all real and imaginary solutions. For example, for the linear equation

$$ax + b = 0 \qquad a \neq 0$$

the solution is

$$x = -\frac{b}{a}$$

and for the quadratic equation

$$ax^2 + bx + c = 0 \qquad a \neq 0$$

the solutions are

$$x = \frac{-b \pm \sqrt{b^2 - 4ac}}{2a}$$

Recall that linear and quadratic equations are also called first- and second-degree polynomial equations, respectively. Thus, we know how to solve any first- or second-degree polynomial equation. What about third- and higher-degree polynomial equations? For example, how do we solve equations such as the following?

$$2x^3 - 3x^2 + x - 5 = 0 \qquad x^7 - 6x^4 + 3x - 1 = 0$$

It turns out that there are direct, though complicated, methods for finding all solutions for any third- or fourth-degree polynomial equation. However, the Frenchman Evariste Galois (1811–1832) proved at the age of 20 that for polynomial equations of degree greater than 4 there is no finite step-by-step process that will always yield all solutions.* This does not mean that we give up looking for solutions to higher-degree polynomials.

In this chapter we find that solutions always exist for all polynomial equations of degree greater than 0. We develop a step-by-step method for finding all rational solutions of polynomial equations with rational coefficients and for approximating all real solutions of polynomial equations with real coefficients. In the process, we also find some irrational and imaginary solutions of higher-degree polynomial equations.

---

*Galois' contribution, using the new concept of "group," was of the highest mathematical significance and originality. However, his contemporaries hardly read his papers, dismissing them as "almost unintelligible." At the age of 21, involved in political agitation, Galois met an untimely death in a duel. A short but fascinating account of Galois' tragic life can be found in E. T. Bell's *Men of Mathematics* (New York: Simon & Schuster, 1937), pp. 362–377.

We first switch our emphasis from polynomial equations to polynomial functions. This approach will enable us to uncover some important properties of polynomial functions that lead directly to solutions of certain polynomial equations. The following definitions provide a useful link between polynomial functions and polynomial equations.

Recall that the function

$$P(x) = a_n x^n + a_{n-1} x^{n-1} + \cdots + a_1 x + a_0 \qquad a_n \neq 0$$

is called an ***nth-degree polynomial function***. In our previous work with these functions, we only considered real coefficients and real values for the independent variable so that both the domain and range were subsets of the real numbers. Now we need to consider a more general situation.

> In this chapter the coefficients of a polynomial function are complex numbers and the domain of a polynomial function is the set of complex numbers.

However, when we discuss certain properties of polynomials, such as their graphs, we will once again restrict the coefficients and the independent variable to be real numbers.

The number r is said to be a **zero of the function P**, or a **zero of the polynomial $P(x)$**, or a **solution or root of the equation $P(x) = 0$**, if

$$P(r) = 0$$

A zero of a polynomial may or may not be the number 0. A zero of a polynomial is any number that makes the value of the polynomial 0. If the coefficients of a polynomial $P(x)$ are real numbers, then a real zero is simply an x intercept for the graph of $y = P(x)$. Consider the polynomial

$$P(x) = x^2 - 4x + 3$$

The graph of P is shown in Figure 1.

FIGURE 1 Zeros, roots, and x intercepts.

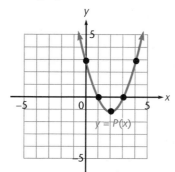

The x intercepts 1 and 3 are zeros of $P(x) = x^2 - 4x + 3$, since $P(1) = 0$ and $P(3) = 0$. The x intercepts 1 and 3 are also solutions or roots for the equation $x^2 - 4x + 3 = 0$.

In general:

Zeros and Roots

If the coefficients of a polynomial $P(x)$ are real, then the x intercepts of the graph of $y = P(x)$ are real **zeros** of P and $P(x)$ and real **solutions**, or **roots**, for the equation $P(x) = 0$.

In addition to increasing substantially your ability to solve polynomial equations, the techniques discussed in this chapter yield several very useful bonuses:

1. A fairly efficient process for graphing higher-degree polynomial functions

2. A process for factoring many third- and higher-degree polynomials

3. A process for solving many third- and higher-degree polynomial inequalities as well as many rational inequalities of a more general nature (a consequence of the second bonus and material discussed in Section 2-8)

We start the development with a digression to introduce *synthetic division*, a very practical tool that will be used throughout this chapter.

SECTION 4-1   Synthetic Division

- Algebraic Long Division
- Synthetic Division

After a brief review of algebraic long division, we will show how the division process can be streamlined substantially using synthetic division if the divisor is of the form $x - r$. The reasons for our interest in synthetic division will become clear in the sections that follow.

- **Algebraic Long Division**

We can find quotients of polynomials by a long-division process similar to that used in arithmetic. An example will illustrate the process.

EXAMPLE 1   Algebraic Long Division

Divide $5 + 4x^3 - 3x$ by $2x - 3$.

Solution

$$
\begin{array}{r}
2x^2 + 3x + 3 \\
2x - 3 \overline{\smash)4x^3 + 0x^2 - 3x + 5} \\
\underline{4x^3 - 6x^2} \\
6x^2 - 3x \\
\underline{6x^2 - 9x} \\
6x + 5 \\
\underline{6x - 9} \\
14 = R
\end{array}
$$

Remainder

Arrange the dividend and the divisor in descending powers of the variable. Insert, with 0 coefficients, any missing terms of degree less than 3.
Divide the first term of the divisor into the first term of the dividend.
Multiply the divisor by $2x^2$, line up like terms, subtract as in arithmetic, and bring down $-3x$. Repeat the process until the degree of the remainder is less than that of the divisor.

Thus,

$$
\frac{4x^3 - 3x + 5}{2x - 3} = 2x^2 + 3x + 3 + \frac{14}{2x - 3}
$$

Check

$$
(2x - 3)\left[(2x^2 + 3x + 3) + \frac{14}{2x - 3}\right] = (2x - 3)(2x^2 + 3x + 3) + 14
$$

$$
= 4x^3 - 3x + 5
$$

Problem 1   Divide $6x^2 - 30 + 9x^3$ by $3x - 4$.

• **Synthetic Division**   Being able to divide a polynomial $P(x)$ by a linear polynomial of the form $x - r$ quickly and accurately will be of great help in the search for zeros of higher-degree polynomial functions. This kind of division can be carried out more efficiently by a method called **synthetic division**. The method is most easily understood through an example. Let's start by dividing $P(x) = 2x^4 + 3x^3 - x - 5$ by $x + 2$, using ordinary long division. The critical parts of the process are indicated in color.

$$
\begin{array}{r}
2x^3 - 1x^2 + 2x - 5 \qquad \text{Quotient} \\
x + 2 \overline{)2x^4 + 3x^3 + 0x^2 - 1x - 5} \quad \text{Dividend} \\
\underline{2x^4 + 4x^3} \\
-1x^3 + 0x^2 \\
\underline{-1x^3 - 2x^2} \\
2x^2 - 1x \\
\underline{2x^2 + 4x} \\
-5x - 5 \\
\underline{-5x - 10} \\
5 \qquad \text{Remainder}
\end{array}
$$

The numerals printed in color, which represent the essential part of the division process, are arranged more conveniently as follows:

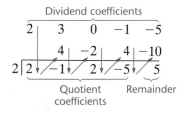

Mechanically, we see that the second and third rows of numerals are generated as follows. The first coefficient, 2, of the dividend is brought down and multiplied by 2 from the divisor; and the product, 4, is placed under the second dividend coefficient, 3, and subtracted. The difference, -1, is again multiplied by the 2 from the divisor; and the product is placed under the third coefficient from the dividend and subtracted. This process is repeated until the remainder is reached. The process can be made a little faster, and less prone to sign errors, by changing $+2$ from the divisor to -2 and adding instead of subtracting. Thus:

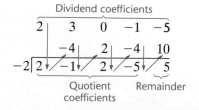

Key Steps in the Synthetic Division Process

To divide the polynomial $P(x)$ by $x - r$:

Step 1. Arrange the coefficients of $P(x)$ in order of descending powers of x. Write 0 as the coefficient for each missing power.

Step 2. After writing the divisor in the form $x - r$, use r to generate the second and third rows of numbers as follows. Bring down the first coefficient of the dividend and multiply it by r; then add the product to the second coefficient of the dividend. Multiply this sum by r, and add the product to the third coefficient of the dividend. Repeat the process until a product is added to the constant term of $P(x)$.

Step 3. The last number to the right in the third row of numbers is the remainder. The other numbers in the third row are the coefficients of the quotient, which is of degree 1 less than $P(x)$.

EXAMPLE 2   **Synthetic Division**

Use synthetic division to find the quotient and remainder resulting from dividing $P(x) = 4x^5 - 30x^3 - 50x - 2$ by $x + 3$. Write the answer in the form $Q(x) + R/(x - r)$, where R is a constant.

Solution   Since $x + 3 = x - (-3)$, we have $r = -3$, and

$$
\begin{array}{r|rrrrrr}
 & 4 & 0 & -30 & 0 & -50 & -2 \\
 &   & -12 & 36 & -18 & 54 & -12 \\
\hline
-3 & 4 & -12 & 6 & -18 & 4 & -14
\end{array}
$$

The quotient is $4x^4 - 12x^3 + 6x^2 - 18x + 4$ with a remainder of -14. Thus,

$$\frac{P(x)}{x + 3} = 4x^4 - 12x^3 + 6x^2 - 18x + 4 + \frac{-14}{x + 3}$$

Problem 2   Repeat Example 2 with $P(x) = 3x^4 - 11x^3 - 18x + 8$ and divisor $x - 4$.

We now show a step-by-step method for performing synthetic division using a calculator. After a few trials you should find the process almost automatic and be able to work problems involving decimal or large coefficients as easily as those involving small integer coefficients.

EXAMPLE 3   **Synthetic Division Using a Calculator**

Use synthetic division and a calculator to find the quotient and remainder resulting from dividing $P(x) = 2x^3 - 5x^2 + x - 3$ by $x - 0.3$. Assume all quantities are exact.

Solution We first show the completed synthetic division table and then show the calculator steps used to generate the table. You may be able to generate the table using a calculator immediately without referring to the calculator steps—the spelled-out steps make the process look far more complicated than it is!

Since $x - r = x - 0.3$, we have $r = 0.3$, and

$$
\begin{array}{r|rrrr}
 & 2 & -5 & 1 & -3 \\
 & & 0.6 & -1.32 & -0.096 \\
\hline
0.3 & 2 & -4.4 & -0.32 & -3.096
\end{array}
$$

Calculator Steps
Store $r = 0.3$ in a memory cell:

$\boxed{0.3}\ \boxed{\text{STO}}\ \boxed{\text{M}}$ See the manual for your calculator for the keystrokes used to store and recall numbers.

Overbars indicate repeating cycles

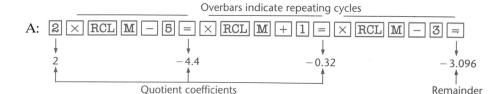

Overbars indicate repeating cycles

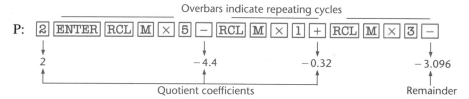

Thus, the answer is $2x^2 - 4.4x - 0.32$, $R = -3.096$.

Problem 3   Repeat Example 3 using $x + 0.3$ as a divisor.

Answers to Matched Problems

1. $3x^2 + 6x + 8 + \dfrac{2}{3x - 4}$

2. $\dfrac{P(x)}{x - 4} = 3x^3 + x^2 + 4x - 2 + \dfrac{0}{x - 4} = 3x^3 + x^2 + 4x - 2$

3. $2x^2 - 5.6x + 2.68$, $R = -3.804$

EXERCISE 4-1

A

In Problems 1–8, divide, using algebraic long division. Write the quotient, and indicate the remainder.

1. $(4m^2 - 1) \div (2m - 1)$

2. $(y^2 - 9) \div (y + 3)$

3. $(6 - 6x + 8x^2) \div (2x + 1)$

4. $(11x - 2 + 12x^2) \div (3x + 2)$

5. $(x^3 - 1) \div (x - 1)$

6. $(a^3 + 27) \div (a + 3)$

7. $(3y - y^2 + 2y^3 - 1) \div (y + 2)$

8. $(3 + x^3 - x) \div (x - 3)$

In Problems 9–14, use synthetic division to write the quotient $P(x) \div (x - r)$ in the form $P(x)/(x - r) = Q(x) + R/(x - r)$, where R is a constant.

9. $(x^2 + 3x - 7) \div (x - 2)$

10. $(x^2 + 3x - 3) \div (x - 3)$

11. $(4x^2 + 10x - 9) \div (x + 3)$

12. $(2x^2 + 7x - 5) \div (x + 4)$

13. $(2x^3 - 3x + 1) \div (x - 2)$

14. $(x^3 + 2x^2 - 3x - 4) \div (x + 2)$

B _____

In Problems 15–30, divide, using synthetic division. Write the quotient, and indicate the remainder. As coefficients get more involved, a calculator should prove helpful. Do not round off—all quantities are exact.

15. $(3x^4 - x - 4) \div (x + 1)$

16. $(5x^4 - 2x^2 - 3) \div (x - 1)$

17. $(x^5 + 1) \div (x + 1)$

18. $(x^4 - 16) \div (x - 2)$

19. $(3x^4 + 2x^3 - 4x - 1) \div (x + 3)$

20. $(x^4 - 3x^3 - 5x^2 + 6x - 3) \div (x - 4)$

21. $(2x^6 - 13x^5 + 75x^3 + 2x^2 - 50) \div (x - 5)$

22. $(4x^6 + 20x^5 - 24x^4 - 3x^2 - 13x + 30) \div (x + 6)$

23. $(4x^4 + 2x^3 - 6x^2 - 5x + 1) \div (x + \frac{1}{2})$

24. $(2x^3 - 5x^2 + 6x + 3) \div (x - \frac{1}{2})$

25. $(4x^3 + 4x^2 - 7x - 6) \div (x + \frac{3}{2})$

26. $(3x^3 - x^2 + x + 2) \div (x + \frac{2}{3})$

27. $(3x^4 - 2x^3 + 2x^2 - 3x + 1) \div (x - 0.4)$

28. $(4x^4 - 3x^3 + 5x^2 + 7x - 6) \div (x - 0.7)$

29. $(3x^5 + 2x^4 + 5x^3 - 7x - 3) \div (x + 0.8)$

30. $(7x^5 - x^4 + 3x^3 - 2x^2 - 5) \div (x + 0.9)$

In Problems 31–34, divide, using synthetic division and a calculator. Carry all numbers to the full capacity of your calculator as you proceed through the synthetic division process. However, write down the coefficients of the quotient and the remainder to two decimal places as you progress.

31. $(2.14x^3 - 5.23x^2 - 8.71x + 6.85) \div (x - 3.37)$

32. $(6.03x^3 - 35.67x^2 + 8.98x - 12.81) \div (x - 5.72)$

33. $(0.96x^4 + 4.09x^2 + 9.44x - 1.87) \div (x + 1.37)$

34. $(6.45x^4 - 1.07x^3 + 8.67x - 3.03) \div (x + 0.88)$

C _____

In Problems 35 and 36, divide, using algebraic long division. Write the quotient, and indicate the remainder.

35. $(16x - 5x^3 - 8 + 6x^4 - 8x^2) \div (2x - 4 + 3x^2)$

36. $(8x^2 - 7 - 13x + 24x^4) \div (3x + 5 + 6x^2)$

In Problems 37 and 38, divide, using synthetic division. Do not use a calculator.

37. $(x^3 - 3x^2 + x - 3) \div (x - i)$

38. $(x^3 - 2x^2 + x - 2) \div (x + i)$

39. (A) Divide $P(x) = a_2x^2 + a_1x + a_0$ by $x - r$, using both synthetic division and the long-division process, and compare the coefficients of the quotient and the remainder produced by each method.

(B) Expand the expression representing the remainder. What do you observe?

40. Repeat Problem 39 for

$$P(x) = a_3x^3 + a_2x^2 + a_1x + a_0$$

SECTION 4-2   Remainder and Factor Theorems

- Division Algorithm
- Remainder Theorem
- Graphing Polynomial Functions
- Factor Theorem

In this section we introduce two basic theorems that are very useful and easily proved.

• Division Algorithm   If we divide $P(x) = 2x^4 - 5x^3 - 4x^2 + 13$ by $x - 3$, we obtain

$$\frac{2x^4 - 5x^3 - 4x^2 + 13}{x - 3} = 2x^3 + x^2 - x - 3 + \frac{4}{x - 3} \qquad x \neq 3$$

If we multiply both sides of this equation by $x - 3$, then we get

$$2x^4 - 5x^3 - 4x^2 + 13 = (x - 3)(2x^3 + x^2 - x - 3) + 4$$

This last equation is an identity in that the left side is equal to the right side for *all* replacements of x by real or imaginary numbers, including $x = 3$. This example suggests the important **division algorithm**, which we state as Theorem 1 without proof.

Theorem 1   Division Algorithm

For each polynomial $P(x)$ of degree greater than 0 and each number r, there exists a unique polynomial $Q(x)$ of degree 1 less than $P(x)$ and a unique number R such that

$$P(x) = (x - r)Q(x) + R$$

The polynomial $Q(x)$ is called the **quotient**, $x - r$ is the **divisor**, and R is the **remainder**. Note that R may be 0.

• Remainder Theorem   We now use the division algorithm in Theorem 1 to prove the *remainder theorem*. The equation in Theorem 1,

$$P(x) = (x - r)Q(x) + R$$

is an identity; that is, it is true for all real or imaginary replacements for x. In particular, if we let $x = r$, then we observe a very interesting and useful relationship:

$$
\begin{aligned}
P(r) &= (r - r)Q(r) + R \\
&= 0 \cdot Q(r) + R \\
&= 0 + R \\
&= R
\end{aligned}
$$

In words, the value of a polynomial $P(x)$ at $x = r$ is the same as the remainder R obtained when we divide $P(x)$ by $x - r$. We have proved the well-known remainder theorem:

Theorem 2	**Remainder Theorem**
	If R is the remainder after dividing the polynomial $P(x)$ by $x - r$, then
	$$P(r) = R$$

EXAMPLE 1   Two Methods for Evaluating Polynomials

If $P(x) = 4x^4 + 10x^3 + 19x + 5$, find $P(-3)$ by:

(A) Using the remainder theorem and synthetic division
(B) Evaluating $P(-3)$ directly

Solutions   (A) Use synthetic division to divide $P(x)$ by $x - (-3)$.

$$
\begin{array}{r|rrrrr}
 & 4 & 10 & 0 & 19 & 5 \\
 & & -12 & 6 & -18 & -3 \\
\hline
-3 & 4 & -2 & 6 & 1 & 2 = R = P(-3)
\end{array}
$$

(B) $P(-3) = 4(-3)^4 + 10(-3)^3 + 19(-3) + 5$
        $= 2$

Problem 1   Repeat Example 1 for $P(x) = 3x^4 - 16x^2 - 3x + 7$ and $x = -2$.

● Graphing Polynomial Functions

Earlier, we graphed polynomial functions of degree 0, 1, and 2, and we found that the graphs are continuous with no holes or gaps. Table 1 gives a summary of these graphs.

In general, it can be shown using techniques of calculus that:

A polynomial function with real coefficients is continuous everywhere.

TABLE 1   Graphs of Polynomial Functions of Degree 0, 1, or 2

Degree	Polynomial function	Graph	
0	$P(x) = a$	Horizontal line	
1	$P(x) = ax + b$	Line with slope a	
2	$P(x) = ax^2 + bx + c$	Parabola	

That is, the graph of a polynomial function has no holes or gaps. Also, it can be shown that the graph contains no sharp corners.

Graphing polynomial functions of degree greater than 2 requires additional work. Fortunately, synthetic division and the remainder theorem provide effective tools for point-by-point graphing. We usually start by using synthetic division, and the fact that $P(r) = R$, to evaluate a polynomial function at integer values from its domain. Then, for any parts of the graph that are not clear, we evaluate the polynomial function for numbers between the selected integers. These calculations result in a table of ordered pairs of numbers. The total quantity of ordered pairs we compute depends on how accurate we want the final graph to be. Generally, the more ordered pairs, the greater the accuracy. Finally, we plot the ordered pairs of numbers and join the resulting points with a smooth curve. The following example illustrates the process.

EXAMPLE 2   Graphing a Polynomial

Graph $P(x) = x^3 + 3x^2 - x - 3$, $-4 \leq x \leq 2$. Find points by using synthetic division and the remainder theorem.

Solution   We evaluate $P(x)$ from $x = -4$ to $x = 2$, for selected values of x. The process is speeded by forming a synthetic division table. To simplify the form of the table, we dispense with writing the product of r with each coefficient in the quotient and perform the calculations mentally or with a calculator. A calculator becomes increasingly useful as the coefficients become more numerous or complicated. The table also provides other important information, as will be seen in subsequent sections.

Next, we plot the points found in the table and connect them with a smooth continuous curve. As we draw this curve, we observe that the graph changes direction twice, first between $x = -3$ and $x = -1$, and then between $x = -1$ and $x = 1$. Precise determination of the points where a graph changes direction requires calculus techniques. Lacking this precise information, we simply change direction at $x = -2$ and $x = 0$.

	1	3	-1	-3	
-4	1	-1	3	-15	$= P(-4)$
-3.5	1	-0.5	0.75	-5.625	$= P(-3.5)$
-3	1	0	-1	0	$= P(-3)$
-2	1	1	-3	3	$= P(-2)$
-1	1	2	-3	0	$= P(-1)$
0	1	3	-1	-3	$= P(0)$
1	1	4	3	0	$= P(1)$
1.5	1	4.5	5.75	5.625	$= P(1.5)$
2	1	5	9	15	$= P(2)$

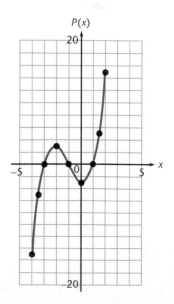

Problem 2   Graph $P(x) = x^3 - 4x^2 - 4x + 16$, $-3 \leq x \leq 5$. Find points by using synthetic division and the remainder theorem.

● **Factor Theorem**   The division algorithm in Theorem 1,

$$P(x) = (x - r)Q(x) + R$$

may, because of the remainder theorem (Theorem 2), be written in a form where R is replaced by $P(r)$:

$$P(x) = (x - r)Q(x) + P(r)$$

It is easy to see that $x - r$ is a factor of $P(x)$ if and only if $P(r) = 0$; that is, if and only if r is a zero of the polynomial $P(x)$. This result is known as the **factor theorem**:

Theorem 3   **Factor Theorem**

If r is a zero of the polynomial $P(x)$, then $x - r$ is a factor of $P(x)$. Conversely, if $x - r$ is a factor of $P(x)$, then r is a zero of $P(x)$.

If we can find a zero of a polynomial, then we can find a linear factor of the polynomial. Conversely, if we can find a linear factor of a polynomial, we can find a zero of the polynomial.

EXAMPLE 3   **Using the Factor Theorem**

(A) Use the factor theorem to show that $x + 1$ is a factor of $P(x) = x^{25} + 1$.
(B) What are the zeros of $P(x) = 3(x - 5)(x + 2)(x - 3)$?

Solutions   (A) Since $x + 1 = x - (-1)$ we have $r = -1$, and

$$P(r) = P(-1) = (-1)^{25} + 1 = -1 + 1 = 0$$

Hence, -1 is a zero of $P(x) = x^{25} + 1$. Thus, by the factor theorem, $x - (-1) = x + 1$ is a factor of $x^{25} + 1$.
(B) Since $(x - 5)$, $(x + 2)$, and $(x - 3)$ are all factors of $P(x)$, 5, -2, and 3 are zeros of $P(x)$.

Problem 3   (A) Use the factor theorem to show that $x - 1$ is a factor of $P(x) = x^{54} - 1$.
(B) What are the zeros of $P(x) = 2(x + 3)(x + 7)(x - 8)(x + 1)$?

Answers to Matched Problems

1. $P(-2) = -3$ for both parts, as it should.

2.

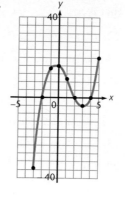

x	$P(x)$
-3	-35
-2	0
-1	15
0	16
1	9
2	0
3	-5
4	0
5	21

3. (A) $r = 1$ and $P(1) = 1^{54} - 1 = 1 - 1 = 0$; therefore, $x - r = x - 1$ is a factor of $P(x) = x^{54} - 1$.   (B) $-3, -7, 8, -1$.

EXERCISE 4-2

A

Use synthetic division and the remainder theorem in Problems 1–6.

1. Find $P(-2)$, given $P(x) = 3x^2 - x - 10$.

2. Find $P(-3)$, given $P(x) = 4x^2 + 10x - 8$.

3. Find $P(2)$, given $P(x) = 2x^3 - 5x^2 + 7x - 7$.

4. Find $P(5)$, given $P(x) = 2x^3 - 12x^2 - x + 30$.

5. Find $P(-4)$, given $P(x) = x^4 - 10x^2 + 25x - 2$.

6. Find $P(-7)$, given $P(x) = x^4 + 5x^3 - 13x^2 - 30$.

Find the zeros for the polynomials in Problems 7–10 using the factor theorem.

7. $P(x) = (x - 3)(x + 5)$

8. $P(x) = (x + 2)(x - 7)$

9. $P(x) = 2(x + \frac{1}{2})(x - 8)(x + 2)$

10. $P(x) = 3(x - \frac{2}{3})(x - 5)(x + 7)$

In Problems 11–14, determine whether the second polynomial is a factor of the first polynomial without dividing or using synthetic division. [Hint: Evaluate directly and use the factor theorem.]

11. $x^{18} - 1; x - 1$

12. $x^{18} - 1; x + 1$

13. $3x^3 - 7x^2 - 8x + 2; x + 1$

14. $3x^4 - 2x^3 + 5x - 6; x - 1$

B

Use synthetic division and the remainder theorem in Problems 15–18.

15. Find $P(\frac{1}{2})$, given $P(x) = 4x^3 - 8x^2 + 5x - 4$.

16. Find $P(\frac{1}{3})$, given $P(x) = 6x^3 + 4x^2 - 5x - 4$.

17. Find $P(0.3)$ for $P(x) = x^3 - 2x + 1$.

18. Find $P(0.7)$ for $P(x) = 2x^3 + 3x^2 - 5x + 2$.

In Problems 19–26, graph each polynomial function using synthetic division and the remainder theorem.

19. $P(x) = x^3 - 5x^2 + 2x + 8, -2 \le x \le 5$

20. $P(x) = x^3 + 2x^2 - 5x - 6, -4 \le x \le 3$

21. $P(x) = x^3 + 4x^2 - x - 4, -5 \le x \le 2$

22. $P(x) = x^3 - 2x^2 - 5x + 6, -3 \le x \le 4$

23. $P(x) = -x^3 + 2x^2 - 3, -2 \le x \le 3$

24. $P(x) = -x^3 - x + 4, -2 \le x \le 2$

25. $P(x) = -x^3 + 3x^2 - 3x + 2, -1 \le x \le 3$

26. $P(x) = -x^3 + x^2 + 4x + 6, -3 \le x \le 4$

In Problems 27–30, find three solutions for each equation.

27. $(x + 4)(x + 8)(x - 1) = 0$

28. $(x - 2)(x + 5)(x - 3) = 0$

29. $7(x - \frac{1}{8})(x + \frac{3}{5})(x + 4) = 0$

30. $4(x + \frac{3}{4})(x - 5)(x - \frac{2}{3}) = 0$

In Problems 31–34, use the quadratic formula and the factor theorem to factor each polynomial.

31. $P(x) = x^2 - 3x + 1$

32. $P(x) = x^2 - 4x - 2$

33. $P(x) = x^2 - 6x + 10$

34. $P(x) = x^2 - 4x + 5$

In Problems 35–38, determine whether the second polynomial is a factor of the first polynomial without dividing or using synthetic division.

35. $x^n - a^n; x - a$

36. $x^n - a^n, n$ even; $x + a$

37. $4x^7 - 2x^6 + x^2 + 2x + 5; x - 1$

38. $2x^5 - 5x^2 - x + 4; x + 1$

C

Graph each polynomial function in Problems 39–46 using synthetic division and the remainder theorem.

39. $P(x) = x^4 - 2x^3 - 2x^2 + 8x - 8$

40. $P(x) = x^4 - 2x^2 + 16x - 15$

41. $P(x) = x^4 + 4x^3 - x^2 - 20x - 20$

42. $P(x) = x^4 - 4x^2 - 4x - 1$

43. $P(x) = -x^4 + 2x^3 + x + 20$

44. $P(x) = -x^4 + 3x^3 - x + 16$

45. $P(x) = x^5 - 5x^4 - 10x^3 + 50x^2 + 9x - 45$

46. $P(x) = x^5 - 4x^4 - 5x^3 + 20x^2 + 4x - 16$

47. Polynomials also can be evaluated conveniently using a "nested factoring" scheme. For example, the polynomial $P(x) = 2x^4 - 3x^3 + 2x^2 - 5x + 7$ can be written in a "nested factored" form as follows:

$$
\begin{aligned}
P(x) &= 2x^4 - 3x^3 + 2x^2 - 5x + 7 \\
&= (2x - 3)x^3 + 2x^2 - 5x + 7 \\
&= [(2x - 3)x + 2]x^2 - 5x + 7 \\
&= \{[(2x - 3)x + 2]x - 5\}x + 7
\end{aligned}
$$

Use the "nested factored" form to find $P(-2)$ and $P(1.7)$. [*Hint:* To evaluate $P(-2)$, store -2 in your calculator's memory and proceed from left to right recalling -2 as needed.]

48. Find $P(-2)$ and $P(1.3)$ for $P(x) = 3x^4 + x^3 - 10x^2 + 5x - 2$ using the "nested factoring" scheme presented in Problem 47.

49. (A) Write $P(x) = a_2x^2 + a_1x + a_0$ in the "nested factored" form $P(x) = (a_2x + a_1)x + a_0$, and find $P(r)$ using the latter form.
    (B) Find $P(r)$ using synthetic division and the remainder theorem, and compare with part (A).

50. Repeat Problem 49 for $P(x) = a_3x^3 + a_2x^2 + a_1x + a_0$.

Problems 51–54 require the use of a graphic calculator or a computer. These problems illustrate some of the possible shapes for the graphs of polynomials of various degrees. Graph all parts of each problem in the same viewing rectangle for $-10 \le x \le 10$ and $-10 \le y \le 10$.

51. (A) $P(x) = (x + 5)(x^2 + 10x + 26)$
    (B) $P(x) = x^3$
    (C) $P(x) = (x - 3)(x - 5)(x - 7)$

52. (A) $P(x) = (x + 4)(x + 6)(x + 7)(x + 8)$
    (B) $P(x) = (x + 1)(x - 2)(x^2 + 3)$
    (C) $P(x) = (x - 5)^3(x - 8)$

53. (A) $P(x) = (x + 4)(x + 6)^4$
    (B) $P(x) = x^5 - 3x^3$
    (C) $P(x) = (x - 3)(x - 4)(x - 5)(x - 6)(x - 7)$

54. (A) $P(x) = (x + 9)(x + 8.5)(x + 8)(x + 6.5)$
    $\times (x + 6)(x + 5)$
    (B) $P(x) = (x^4 - 1)(x^2 - 4)$
    (C) $P(x) = (x - 6)^5(x - 8)$

SECTION 4-3 Fundamental Theorem of Algebra

- Fundamental Theorem of Algebra
- *n* Zeros Theorem
- Imaginary Zeros
- Remarks

In our search for zeros of polynomial functions it is useful to know at the outset how many zeros to expect for a given function. The following two theorems tell us exactly how many zeros exist for a polynomial function of a given degree. Even though the theorems don't tell us how to find the zeros, it is still very helpful to know that what we are looking for exists. These theorems were first proved in 1797 by Carl Friedrich Gauss, one of the greatest mathematicians of all time, at the age of 20.

• Fundamental Theorem of Algebra

Theorem 1, often referred to as the **fundamental theorem of algebra**, requires verification that is beyond the scope of this book, so we state it without proof.

Theorem 1	**Fundamental Theorem of Algebra**
	Every polynomial $P(x)$ of degree $n > 0$ has at least one zero.

• *n* Zeros Theorem

If $P(x) = a_nx^n + a_{n-1}x^{n-1} + \cdots + a_1x + a_0$ is a polynomial of degree $n > 0$ with complex coefficients, then, according to Theorem 1, it has at least one zero, say r_1. According to the factor theorem, $x - r_1$ is a factor of $P(x)$. Thus,

$$P(x) = (x - r_1)Q_1(x)$$

where $Q_1(x)$ is a polynomial of degree $n - 1$. If $n - 1 = 0$, then $Q_1(x) = a_n$. If $n - 1 > 0$, then, by Theorem 1, $Q_1(x)$ has at least one zero, say r_2. And

$$Q_1(x) = (x - r_2)Q_2(x)$$

where $Q_2(x)$ is a polynomial of degree $n - 2$. Thus,

$$P(x) = (x - r_1)(x - r_2)Q_2(x)$$

If $n - 2 = 0$, then $Q_2(x) = a_n$. If $n - 2 > 0$, then $Q_2(x)$ has at least one zero, say r_3. And

$$Q_2(x) = (x - r_3)Q_3(x)$$

where $Q_3(x)$ is a polynomial of degree $n - 3$.

We continue in this way until $Q_k(x)$ is of degree zero—that is, until $k = n$. At this point, $Q_n(x) = a_n$, and we have

$$P(x) = (x - r_1)(x - r_2) \cdot \cdots \cdot (x - r_n)a_n$$

Thus, $r_1, r_2, \ldots, r_n$ are n zeros, not necessarily distinct, of $P(x)$. Is it possible for $P(x)$ to have more than these n zeros? Let's assume that r is a number different from the zeros above. Then

$$P(r) = a_n(r - r_1)(r - r_2) \cdot \cdots \cdot (r - r_n) \neq 0$$

since r is not equal to any of the zeros. Hence, r is not a zero, and we conclude that $r_1, r_2, \ldots, r_n$ are the only zeros of $P(x)$. We have just sketched a proof of Theorem 2.

Theorem 2 **_n_ Zeros Theorem**

Every polynomial $P(x)$ of degree $n > 0$ can be expressed as the product of n linear factors. Hence, $P(x)$ has exactly n zeros—not necessarily distinct.

If $P(x)$ is represented as the product of linear factors and $x - r$ occurs m times, then r is called a **zero of multiplicity _m_**. For example, if

$$P(x) = 4(x - 5)^3(x + 1)^2(x - i)(x + i)$$

then this seventh-degree polynomial has seven zeros, not all distinct. That is, 5 is a zero of multiplicity 3, or a triple zero; -1 is a zero of multiplicity 2, or a double zero; and i and $-i$ are zeros of multiplicity 1, or simple zeros. Thus, this seventh-degree polynomial has exactly seven zeros if we count 5 and -1 with their respective multiplicities.

EXAMPLE 1   **Factoring a Polynomial**

If -2 is a double zero of $P(x) = x^4 - 7x^2 + 4x + 20$, write $P(x)$ as a product of first-degree factors.

Solution Since -2 is a double zero of $P(x)$, we can write

$$P(x) = (x + 2)^2 Q(x)$$
$$= (x^2 + 4x + 4)Q(x)$$

and find $Q(x)$ by dividing $P(x)$ by $x^2 + 4x + 4$. Carrying out the algebraic long division, we obtain

$$Q(x) = x^2 - 4x + 5$$

The zeros of $Q(x)$ are found, using the quadratic formula, to be $2 - i$ and $2 + i$. Thus, $P(x)$ written as a product of linear factors is

$$P(x) = (x + 2)^2[x - (2 - i)][x - (2 + i)]$$

[*Note:* Any time $Q(x)$ is a quadratic polynomial, its zeros can be found using the quadratic formula.]

Problem 1 If 3 is a double zero of $P(x) = x^4 - 12x^3 + 55x^2 - 114x + 90$, write $P(x)$ as a product of first-degree factors.

$$(x-3)(x-3)\, Q(x)$$
$$(x^2 - 6x + 9)(Q(x))$$

• **Imaginary Zeros** Something interesting happens if we restrict the coefficients of a polynomial to real numbers. Let's use the quadratic formula to find the zeros of the polynomial

$$P(x) = x^2 - 6x + 13$$

To find the zeros of $P(x)$, we solve $P(x) = 0$:

$$x^2 - 6x + 13 = 0$$

$$x = \frac{6 \pm \sqrt{36 - 52}}{2}$$

$$= \frac{6 \pm \sqrt{-16}}{2} = \frac{6 \pm 4i}{2} = 3 \pm 2i$$

The zeros of $P(x)$ are $3 - 2i$ and $3 + 2i$, conjugate imaginary numbers (see Section 2-5). Also observe that the imaginary zeros in Example 1 are the conjugate imaginary numbers $2 - i$ and $2 + i$.

This is generalized in the following theorem:

Theorem 3 **Imaginary Zeros Theorem**

Imaginary zeros of polynomials with real coefficients, if they exist, occur in conjugate pairs.

As a consequence of Theorems 2 and 3, we also know (think this through) the following:

Theorem 4   **Real Zeros and Odd-Degree Polynomials**

A polynomial of odd degree with real coefficients always has at least one real zero.

EXAMPLE 2   The Nature of the Zeros of a Polynomial

Let $P(x)$ be a third-degree polynomial with real coefficients. One of the following statements is false; indicate which one.

(A) $P(x)$ has at least one real zero.   (B) $P(x)$ has three zeros.
(C) $P(x)$ can have two real zeros and one imaginary zero.

Solution   Statement (C) is false, since imaginary zeros of polynomials with real coefficients *must* occur in conjugate pairs. If $P(x)$ has two real zeros, then we know that the third zero also must be real.

Problem 2   Let $P(x)$ be a polynomial of fourth degree with real coefficients. One of the following statements is false; indicate which one.

(A) $P(x)$ has four zeros.   (B) $P(x)$ has at least two real zeros.
(C) If we know $P(x)$ has three real zeros, then the fourth zero must be real.

EXAMPLE 3   Possible Combinations of Real and Imaginary Zeros

What are the possible combinations of real and imaginary zeros for the polynomial below?

$$P(x) = 3x^5 - 2x^4 + x^2 - \sqrt{2}x - 5$$

Solution   Since the polynomial has real coefficients and is of odd degree, there must be at least one real zero and imaginary zeros must occur in conjugate pairs. Knowing that the polynomial must have five zeros, the possible combinations of real and imaginary zeros are summarized in the table:

Real	Imaginary
1	4
3	2
5	0

Problem 3 What are the possible combinations of real and imaginary zeros for the polynomial below?

$$P(x) = 7x^4 - \sqrt{5}x^3 - 9x^2 + 1$$

• Remarks The fundamental theorem of algebra (Theorem 1) tells us that in the set of complex numbers not only does $x^2 + 1 = 0$ have a solution, but also that every polynomial equation of degree greater than zero, with real or imaginary coefficients, has a solution.

This important result does not come free. In extending the real numbers to a number system that provides solutions for all polynomial equations, we have to give up something—namely, an ordering of the number system. The complex numbers cannot be ordered. That is, in general, we cannot say that one complex number is less than or greater than another.

Answers to Matched Problems
1. $P(x) = (x - 3)^2[x - (3 - i)][x - (3 + i)]$
2. (B) is false. According to the theorems in this section, the possible combinations of real and imaginary zeros for $P(x)$ are as follows: (1) four imaginary (2) two real and two imaginary, (3) four real. So $P(x)$ may not have any real zeros.
3.

Real	Imaginary
0	4
2	2
4	0

EXERCISE 4-3

A

Write the zeros of each polynomial in Problems 1–4, and indicate the multiplicity of each if over 1. What is the degree of each polynomial?

1. $P(x) = (x + 8)^3(x - 6)^2$

2. $P(x) = (x - 5)(x + 7)^2$

3. $P(x) = 3(x + 4)^3(x - 3)^2(x + 1)$

4. $P(x) = 5(x - 2)^3(x + 3)^2(x - 1)$

In Problems 5–10, find a polynomial P(x) of lowest degree, with leading coefficient 1, that has the indicated set of zeros. Leave the answer in a factored form. Indicate the degree of the polynomial.

5. 3 (multiplicity 2) and -4

6. -2 (multiplicity 3) and 1 (multiplicity 2)

7. -7 (multiplicity 3), $-3 + \sqrt{2}$, $-3 - \sqrt{2}$

8. $\frac{1}{3}$ (multiplicity 2), $5 + \sqrt{7}$, $5 - \sqrt{7}$

9. $(2 - 3i)$, $(2 + 3i)$, -4 (multiplicity 2)

10. $i\sqrt{3}$ (multiplicity 2), $-i\sqrt{3}$ (multiplicity 2), and 4 (multiplicity 3)

B

Given the polynomials in Problems 11–18, what are the possible combinations of real and imaginary zeros?

11. $P(x) = 2x^3 - 3x^2 + x - 5$

12. $P(x) = 2x^4 - 2x^3 + x - 8$

13. $P(x) = 3x^6 - 5x^5 + 3x^2 - 4$

14. $P(x) = x^5 - 2x^3 + 5x^2 - 6$

15. $P(x) = x^5 - 2x^4 + \sqrt{5}x^2 - 7$

16. $P(x) = 7x^3 + 8x^2 - \sqrt{7}$

17. $P(x) = 5x^4 - 2x^2 + x - 8$

18. $P(x) = x^6 + 5x^4 - 3x^3 + x - 9$

In Problems 19–24, write P(x) as a product of first-degree factors.

19. $P(x) = x^3 + 9x^2 + 24x + 16$; -1 is a zero

20. $P(x) = x^3 - 4x^2 - 3x + 18$; 3 is a double zero

21. $P(x) = x^4 - 1$; 1 and -1 are zeros

22. $P(x) = x^4 + 2x^2 + 1$; i is a double zero

23. $P(x) = 2x^3 - 17x^2 + 90x - 41$; $\frac{1}{2}$ is a zero

24. $P(x) = 3x^3 - 10x^2 + 31x + 26$; $-\frac{2}{3}$ is a zero

Given the equations in Problems 25–32, what are the possible combinations of real and imaginary solutions?

25. $x^4 + 7x^3 - 2x^2 + 4 = 0$

26. $7x^6 - x^5 + 2x^2 - 3x + 1 = 0$

27. $4x^3 - \sqrt{7}x^2 + \sqrt{2}x - 1 = 0$

28. $8x^5 + \sqrt{7}x^4 - x^2 + 2x - \sqrt{10} = 0$

29. $5x^6 - x^5 - \sqrt{2}x^3 + 7x - 12 = 0$

30. $12x^4 - 7x^3 + 9x - \sqrt{11} = 0$

31. $1.2x^5 - 2.1x^3 + 3.8x^2 - 4.5 = 0$

32. $7.7x^3 + 1.8x^2 - 0.9x - 6.3 = 0$

In Problems 33–38, multiply.

33. $[x - (4 - 5i)][x - (4 + 5i)]$

34. $[x - (2 - 3i)][x - (2 + 3i)]$

35. $[x - (3 + 4i)][x - (3 - 4i)]$

36. $[x - (5 + 2i)][x - (5 - 2i)]$

37. $[x - (a + bi)][x - (a - bi)]$

38. $(x - bi)(x + bi)$

C

In Problems 39–44, find all other zeros of P(x), given the indicated zero.

39. $P(x) = x^3 - 5x^2 + 4x + 10$; $3 - i$ is one zero

40. $P(x) = x^3 + x^2 - 4x + 6$; $1 + i$ is one zero

41. $P(x) = x^3 - 3x^2 + 25x - 75$; $-5i$ is one zero

42. $P(x) = x^3 + 2x^2 + 16x + 32$; $4i$ is one zero

43. $P(x) = x^4 - 4x^3 + 3x^2 + 8x - 10$; $2 + i$ is one zero

44. $P(x) = x^4 - 2x^3 + 7x^2 - 18x - 18$; $-3i$ is one zero

45. The solutions to the equation $x^3 - 1 = 0$ are all the cube roots of 1.
(A) How many cube roots of 1 are there?
(B) 1 is obviously a cube root of 1; find all others.

46. The solutions to the equation $x^3 - 8 = 0$ are all the cube roots of 8.
(A) How many cube roots of 8 are there?
(B) 2 is obviously a cube root of 8; find all others.

47. If P is a polynomial function with real coefficients of degree n, with n odd, then what is the maximum number of times the graph of $y = P(x)$ can cross the x axis? What is the minimum number of times?

48. Answer the questions in Problem 47 for n even.

49. Given $P(x) = x^2 + 2ix - 5$ with $2 - i$ a zero, show that $2 + i$ is not a zero of $P(x)$. Does this contradict Theorem 3? Explain.

50. If $P(x)$ and $Q(x)$ are two polynomials of degree n, and if $P(x) = Q(x)$ for more than n values of x, then how are $P(x)$ and $Q(x)$ related?

SECTION 4-4 **Isolating Real Zeros**

- Descartes' Rule of Signs
- Bounding Real Zeros
- Sign Changes in $P(x)$

For the remainder of this chapter we focus on the problem of finding zeros of polynomials with real coefficients. Three theorems will be of particular help to us. Theorem 1, along with earlier theorems, gives information about the possible combinations of positive, negative, and imaginary zeros. Theorem 2 tells us how to determine a finite interval that contains all the real zeros. And Theorem 3 helps us isolate particular real zeros within this interval.

- **Descartes' Rule of Signs**

When the terms of a polynomial with real coefficients are arranged in order of descending powers, we say that a **variation in sign** occurs if two successive terms have opposite signs. Missing terms (terms with 0 coefficients) are ignored. For a given polynomial $P(x)$, we are interested in the total number of variations in sign in both $P(x)$ and $P(-x)$.

EXAMPLE 1 Variations in Sign

If $P(x) = 3x^4 - 2x^3 + 3x - 5$, how many variations in sign are in $P(x)$ and in $P(-x)$?

Solution

$$P(x) = 3x^4 \overset{1}{\frown} 2x^3 \overset{1}{\frown} 3x \overset{1}{\frown} 5 \qquad \text{Three variations in sign}$$

$$P(-x) = 3x^4 + 2x^3 \overset{1}{\frown} 3x - 5 \qquad \text{One variation in sign}$$

Problem 1 If $P(x) = 2x^5 - x^4 - x^3 + x + 5$, how many variations in sign are in $P(x)$ and in $P(-x)$?

The number of variations in sign for $P(x)$ and for $P(-x)$ gives us useful information about the number of real zeros of a polynomial with real coefficients. In 1636, René Descartes, a French philosopher-mathematician, gave the first proof of a simplified version of a theorem that now bears his name. We state Theorem 1 without proof, since a proof is beyond the scope of this book.

Theorem 1   **Descartes' Rule of Signs**

Given a polynomial $P(x)$ with real coefficients:

1. **Positive Zeros.** The number of positive zeros of $P(x)$ is never greater than the number of variations in sign in $P(x)$ and, if less, then always by an even number.

2. **Negative Zeros.** The number of negative zeros of $P(x)$ is never greater than the number of variations in sign in $P(-x)$ and, if less, then always by an even number.

It is important to understand that when we refer to positive zeros and negative zeros we are referring to real zeros. There are no positive or negative imaginary numbers.

EXAMPLE 2   **Possible Combinations of Zeros**

Construct a table showing the possible combinations of positive, negative, and imaginary zeros of:

(A) $P(x) = 3x^4 - 2x^3 + 3x - 5$ (B) $Q(x) = 2x^6 + x^4 - x + 3$

Solutions   (A) Apply Descartes' rule of signs:

$$P(x) = 3x^4 - 2x^3 + 3x - 5 \qquad \text{Three variations in sign}$$
$$P(-x) = 3x^4 + 2x^3 - 3x - 5 \qquad \text{One variation in sign}$$

Because $P(x)$ has three variations in sign, we conclude from Descartes' rule of signs that $P(x)$ has either three positive zeros or one positive zero. Because $P(-x)$ has one variation in sign, we conclude from Descartes' rule of signs that $P(x)$ has exactly one negative zero. Note that since 0 does not differ from 1 by an even integer, 0 negative zeros is not a possibility. Using the fact that imaginary zeros occur in conjugate pairs and $P(x)$ has a total of four zeros, the table in the margin summarizes the possible combinations of zeros, where $+$ = positive, $-$ = negative, and I = imaginary.

+	−	I
3	1	0
1	1	2

(B) Apply Descartes' rule of signs:

$$Q(x) = 2x^6 + x^4 - x + 3 \qquad \text{Two variations in sign}$$
$$Q(-x) = 2x^6 + x^4 + x + 3 \qquad \text{No variation in sign}$$

Possible combinations of zeros are given in the table in the margin.

+	−	I
2	0	4
0	0	6

Problem 2   Construct a table showing the possible combinations of positive, negative, and imaginary zeros of:

(A) $P(x) = 4x^5 + 2x^4 - x^3 + x - 5$ (B) $Q(x) = x^3 + 3x^2 + 5$

● Bounding Real Zeros

Any number that is greater than or equal to the largest zero of a polynomial is called an **upper bound of the zeros** of the polynomial. Similarly, any number that is less than or equal to the smallest zero of a polynomial is called a **lower bound of the zeros** of the polynomial. Knowing the upper and lower bounds of the real zeros of a polynomial helps us in our search to locate the real zeros of the polynomial. Theorem 2, based on the synthetic division process, enables us to determine upper and lower bounds of all real zeros of a polynomial with real coefficients.

Theorem 2

Upper and Lower Bounds of Real Zeros

Given an nth-degree polynomial $P(x)$ with real coefficients, $n > 0$, $a_n > 0$, and $P(x)$ divided by $x - r$ using synthetic division:

1. **Upper Bound.** If $r > 0$ and all numbers in the quotient row of the synthetic division, including the remainder, are nonnegative, then r is an upper bound of the real zeros of $P(x)$.

2. **Lower Bound.** If $r < 0$ and all numbers in the quotient row of the synthetic division, including the remainder, alternate in sign, then r is a lower bound of the real zeros of $P(x)$.

[*Note:* In the lower-bound test, if 0 appears in one or more places in the quotient row, including the remainder, the sign in front of it can be considered either positive or negative, but not both. For example, the numbers 1, 0, 1 can be considered to alternate in sign, while 1, 0, -1 cannot.]

We sketch a proof of part 1 of Theorem 2. The proof of part 2 is similar, only a little more difficult.

Proof

If all the numbers in the quotient row of the synthetic division are nonnegative after dividing $P(x)$ by $x - r$, then

$$P(x) = (x - r)Q(x) + R$$

where the coefficients of $Q(x)$ are nonnegative and R is nonnegative. If $x > r > 0$, then $x - r > 0$ and $Q(x) > 0$; hence,

$$P(x) = (x - r)Q(x) + R > 0$$

Thus, $P(x)$ cannot be 0 for any x greater than r, and r is an upper bound for the real zeros of $P(x)$.

EXAMPLE 3   Finding Upper and Lower Bounds for Real Zeros

Find the smallest positive integer and the largest negative integer that, by Theorem 2, are upper and lower bounds, respectively, for the real zeros of

$$P(x) = x^3 - 3x^2 - 18x + 4$$

Solution An easy way to locate these upper and lower bounds, particularly if the coefficients of $P(x)$ are not too large, is to test $r = 1, 2, 3, \ldots$ until the quotient row turns nonnegative; then test $r = -1, -2, -3, \ldots$ until the quotient row alternates in sign. The resulting table provides additional benefits, as we will see later.

		1	−3	−18	4
	1	1	−2	−20	−16
	2	1	−1	−20	−36
	3	1	0	−18	−50
	4	1	1	−14	−52
	5	1	2	−8	−36
UB	6	1	3	0	4
	−1	1	−4	−14	18
	−2	1	−5	−8	20
	−3	1	−6	0	4
LB	−4	1	−7	10	−36

$\leftarrow$ { This quotient row is nonnegative; hence, 6 is an upper bound (UB).

$\leftarrow$ { This quotient row alternates in sign; hence, −4 is a lower bound (LB).

Because of Theorem 2, we conclude that all real zeros of $P(x) = x^3 - 3x^2 - 18x + 4$ must lie between −4 and 6. Equivalently, all real solutions of the equation $x^3 - 3x^2 - 18x + 4 = 0$ also must lie between −4 and 6.

Problem 3 Repeat Example 3 for $P(x) = x^3 - 4x^2 - 5x + 8$.

● **Sign Changes in $P(x)$** Observing sign changes in a polynomial $P(x)$ with real coefficients as x is replaced with different real numbers leads to the further isolation of real zeros of $P(x)$. Recall that a polynomial function P with real coefficients is continuous everywhere; that is, the graph of $y = P(x)$ has no holes or breaks. This property of polynomial functions is the basis of Theorem 3.

Theorem 3 **Location Theorem**

If $P(x)$ is a polynomial with real coefficients, and if $P(a)$ and $P(b)$ are of opposite sign, then there is at least one real zero between a and b.

Notice in Figure 1(a) that $P(-3) < 0, P(4) > 0$, and there are three real zeros between −3 and 4. Also, in Figure 1(c), $P(-3) < 0, P(4) > 0$, and there is one real zero between −3 and 4. Since the graph of a polynomial function P with real coefficients is continuous, if $P(a)$ and $P(b)$ are of opposite sign, the graph of $y = P(x)$ must cross the x axis at least once for x between a and b.

The converse of Theorem 3 is false; that is, if $P(x)$ has at least one real zero, then $P(x)$ may or may not change sign at a zero. Compare Figure 1(b) and Figure

FIGURE 1 The x intercepts of the graphs of $y = P(x)$ are the real zeros of $P(x)$.

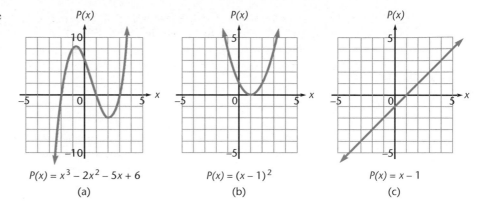

$P(x) = x^3 - 2x^2 - 5x + 6$
(a)

$P(x) = (x - 1)^2$
(b)

$P(x) = x - 1$
(c)

1(c). Both functions have zeros at $x = 1$, but the first is never negative, whereas the second is negative for $x < 1$ and positive for $x > 1$.

EXAMPLE 4 **Locating Real Zeros**

Show that there is at least one real zero of

$$P(x) = x^4 - 2x^3 - 6x^2 + 6x + 9$$

between 1 and 2.

Solution Show that $P(1)$ and $P(2)$ have opposite signs.

$$
\begin{array}{c|ccccc}
 & 1 & -2 & -6 & 6 & 9 \\
\hline
1 & 1 & -1 & -7 & -1 & 8 = P(1) \\
2 & 1 & 0 & -6 & -6 & -3 = P(2)
\end{array}
$$

Since $P(1)$ and $P(2)$ have opposite signs, there is at least one real zero between 1 and 2 (Theorem 3).

Problem 4 Show that there is at least one real zero of

$$P(x) = 2x^4 - 3x^3 - 3x - 4$$

between 2 and 3.

Answers to Matched Problems
1. Two in $P(x)$ and three in $P(-x)$

2. (A)

+	−	I
3	2	0
3	0	2
1	2	2
1	0	4

(B)

+	−	I
0	1	2

3. Lower bound: -2; Upper bound: 5
4. $P(2) = -2$ and $P(3) = 68$, and the conclusion follows from Theorem 3.

EXERCISE 4-4

A

In Problems 1–6, construct a table showing the possible combinations of positive, negative, and imaginary zeros using Descartes' rule of signs and earlier theorems.

1. $P(x) = 2x^2 + x - 4$

2. $Q(x) = 3x^2 - x - 5$

3. $M(x) = 7x^2 + 2x + 4$

4. $N(x) = -3x^2 - 2x - 1$

5. $Q(x) = 2x^3 - 4x^2 + x - 3$

6. $P(x) = x^3 + 7x^2 - x + 2$

Find the smallest positive integer and the largest negative integer that, by Theorem 2, are upper and lower bounds, respectively, for the real zeros of each of the polynomials given in Problems 7–12.

7. $P(x) = x^2 - 2x + 3$

8. $Q(x) = x^2 - 3x - 2$

9. $M(x) = x^3 - 3x + 5$

10. $R(x) = x^3 - 2x^2 + 3$

11. $M(x) = x^4 - x^2 + 3x + 2$

12. $N(x) = x^4 - 2x^3 + 4x - 3$

Show, using Theorem 3, that for each polynomial in Problems 13–18 there is at least one real zero between the given values of a and b.

13. $P(x) = x^2 - 3x - 2; a = 3, b = 4$

14. $Q(x) = x^2 - 3x - 2; a = -1, b = 0$

15. $P(x) = x^3 - 3x + 5; a = -3, b = -2$

16. $P(x) = x^3 - 2x^2 - 4; a = 2, b = 3$

17. $Q(x) = x^3 - 3x^2 - 3x + 9; a = 1, b = 2$

18. $G(x) = x^3 - 3x^2 - 3x + 9; a = -2, b = -1$

B

For each polynomial P(x) in Problems 19–30:
(A) Construct a table showing the possible combinations of positive, negative, and imaginary zeros using Descartes' rule of signs and earlier theorems.
(B) Find the smallest positive integer and largest negative integer that, according to Theorem 2, are, respectively, upper and lower bounds of the zeros of P(x).
(C) Discuss the location of real zeros within the lower and upper bound interval by applying Theorem 3 to integer values of x within this interval.

19. $P(x) = x^3 - x^2 - 6x + 6$

20. $P(x) = x^3 - 3x^2 - 2x + 6$

21. $P(x) = x^3 - 2x - 6$

22. $P(x) = x^3 - 3x^2 - 5$

23. $P(x) = x^4 + 4x^3 - 2x^2 - 12x - 3$

24. $P(x) = x^4 - 4x^3 + 8x - 4$

25. $P(x) = x^5 - 3x^3 + 2x - 5$

26. $P(x) = 2x^5 - 5x^4 - 2x + 5$

27. $P(x) = x^5 + 2x^4 + 3x^2 - 2x - 8$

28. $P(x) = x^5 - 2x^4 - x^3 + 3x - 2$

29. $P(x) = x^6 - x^5 - 2x^3 + 3x - 5$

30. $P(x) = x^6 - x^4 - 6x^2 + 5$

C

31. Prove that $P(x) = x^4 + 3x^2 - x - 5$ has two imaginary and two real zeros, without finding the zeros.

32. Prove that $P(x) = x^3 + 3x^2 + 5$ has one negative zero and two imaginary zeros, without finding the zeros.

33. Prove that the graph of $P(x) = x^5 + 3x^3 + x$ crosses the x axis only once without graphing $y = P(x)$.

34. Prove that the graph of $P(x) = x^4 + 3x^2 + 7$ does not cross the x axis at all. Do not graph $y = P(x)$.

SECTION 4-5 Rational Zero Theorem

- Rational Zero Theorem
- Strategy for Finding Rational Zeros

To find *all* the rational zeros of any polynomial with rational coefficients requires one more theorem in addition to the material we have studied so far in this chapter—the rational zero theorem presented in this section. We will provide guidelines and strategies that simplify the process of finding rational zeros of a polynomial with rational coefficients, but ingenuity and skill gained from problem solving will substantially reduce the time required in the process.

Before beginning our discussion, we note that a polynomial with rational coefficients can always be written as a constant times a polynomial with integer coefficients. For example,

$$P(x) = \tfrac{1}{2}x^3 - \tfrac{2}{3}x^2 + \tfrac{7}{4}x + 5$$
$$= \tfrac{1}{12}(6x^3 - 8x^2 + 21x + 60)$$

Thus, it is sufficient to confine our attention to polynomials with integer coefficients.

- **Rational Zero Theorem**

We introduce the rational zero theorem by examining the following quadratic polynomial whose zeros can be found easily by factoring:

$$P(x) = 6x^2 - 13x - 5 = (2x - 5)(3x + 1)$$

$$\text{Zeros of } P(x): \frac{5}{2} \quad \text{and} \quad -\frac{1}{3} = \frac{-1}{3}$$

Notice that the numerators, 5 and -1, of the zeros are both integer factors of -5, the constant term in $P(x)$. The denominators 2 and 3 of the zeros are both integer factors of 6, the coefficient of the highest-degree term in $P(x)$. These observations are generalized in Theorem 1.

Theorem 1

Rational Zero Theorem

If the rational number b/c, in lowest terms, is a zero of the polynomial

$$P(x) = a_n x^n + a_{n-1}x^{n-1} + \cdots + a_1 x + a_0 \qquad a_n \neq 0$$

with integer coefficients, then b must be an integer factor of a_0 and c must be an integer factor of a_n.

$$P(x) = a_n x^n + a_{n-1}x^{n-1} + \cdots + a_1 x + a_0$$

$$\frac{b}{c}$$

c must be a factor of a_n

b must be a factor of a_0

The proof of Theorem 1 is not difficult, and is instructive, so we sketch it here.

Proof   Since b/c is a zero of $P(x)$,

$$a_n\left(\frac{b}{c}\right)^n + a_{n-1}\left(\frac{b}{c}\right)^{n-1} + \cdots + a_1\left(\frac{b}{c}\right) + a_0 = 0 \qquad (1)$$

If we multiply both sides of equation (1) by c^n, we obtain

$$a_n b^n + a_{n-1} b^{n-1} c + \cdots + a_1 b c^{n-1} + a_0 c^n = 0 \qquad (2)$$

which can be written in the form

$$a_n b^n = c(-a_{n-1} b^{n-1} - \cdots - a_0 c^{n-1}) \qquad (3)$$

Since both sides of equation (3) are integers, c must be a factor of $a_n b^n$. And since the rational number b/c is given to be in lowest terms, b and c can have no common factors other than ± 1. That is, b and c are **relatively prime**. This implies that b^n and c also are relatively prime. Hence, c must be a factor of a_n.

Now, if we solve equation (2) for $a_0 c^n$ and factor b out of the right side, we have

$$a_0 c^n = b(-a_n b^{n-1} - \cdots - a_1 c^{n-1})$$

We see that b is a factor of $a_0 c^n$ and, hence, a factor of a_0, since b and c are relatively prime.

[*Remark:* Theorem 1 does not say that a polynomial $P(x)$ with integer coefficients has rational zeros. It simply states that if $P(x)$ does have a rational zero, then the numerator of the zero must be an integer factor of a_0 and the denominator of the zero must be an integer factor of a_n. In short, the theorem enables us to list a set of rational numbers that must include all rational zeros if any exist.]

EXAMPLE 1   **Listing All Possible Rational Zeros**

List all possible rational zeros for $P(x) = 2x^4 - 3x^3 + x - 9$.

Solution   If b/c in lowest terms is a rational zero of $P(x)$, then b must be a factor of -9 and c must be a factor of 2.

Possible values of b are the integer factors of -9: ± 1, ± 3, ± 9 (4)

Possible values of c are the integer factors of 2: ± 1, ± 2 (5)

Writing all possible fractions b/c where b is from (4) and c is from (5), we have

Possible rational zeros for $P(x)$: ± 1, ± 3, ± 9, $\pm \frac{1}{2}$, $\pm \frac{3}{2}$, $\pm \frac{9}{2}$

[*Note:* $\pm 1/\pm 1 = \pm 1$; $\pm 3/\pm 1 = \pm 3$; etc.]   (6)

Thus, if $P(x)$ has rational zeros, they must be in list (6).

Problem 1   List all possible rational zeros for $P(x) = x^3 + 2x^2 - 5x - 6$.

● **Strategy for Finding Rational Zeros**

If a polynomial is of first or second degree, then we can always find all its zeros using methods discussed in Chapter 2. With the tools developed in this chapter, we are now ready to state a strategy that efficiently leads to all existing rational zeros of polynomials of degree greater than 2 with integer coefficients. Of course, we could just test each of the possible rational zeros that result from the rational zero theorem. However, we can make the process more efficient by using many of the other properties of polynomials discussed earlier in this chapter. In addition, some of these properties and procedures can help locate irrational zeros, as we will see in the next section.

The following strategy for finding rational zeros is only a point of departure. As you gain experience through problem solving, you will most likely modify parts of the strategy to fit your own style.

Strategy for Finding Rational Zeros

Assume $P(x)$ is a polynomial with integer coefficients and is of degree greater than 2.

Step 1. List the possible rational zeros of $P(x)$ using the rational zero theorem (Theorem 1).

Step 2. Make a table of possible combinations of positive, negative, and imaginary zeros using Descartes' rule of signs and earlier theorems. Use the results to reduce the size of the list from step 1, if applicable.

Step 3. Construct a synthetic division table, keeping the results of steps 1 and 2 in mind. Generally, start at $r = 0$ and move either way through integer values, usually starting with the positive integers, until a zero is found, $P(x)$ changes sign, or lower and upper bounds are found.
(A) If a rational zero r is found at any time, stop, write

$$P(x) = (x - r)Q(x)$$

and immediately proceed to find the rational zeros for $Q(x)$, the **reduced polynomial** relative to $P(x)$. If $Q(x)$ is of degree greater than 2, return to step 1 using $Q(x)$ in place of $P(x)$. If $Q(x)$ is quadratic, find all its zeros using standard methods for solving quadratic equations.
(B) If $P(x)$ changes sign, try the possible rational zeros from the list in step 1 that are between the two integers producing the sign change. If a rational zero is found, proceed to step 3(A).
(C) If lower or upper bounds are found, modify the possible rational zeros list by discarding any numbers below the lower bound or above the upper bound.

> *Step 4.* Test any untested numbers from the list of remaining possible rational zeros. If a rational zero is found, proceed to step 3(A). If no rational number from the list is a zero, then conclude that $P(x)$ [or $Q(x)$] has no rational zeros. The remaining zeros must be irrational or imaginary.

Let's use this strategy in several concrete examples.

EXAMPLE 2   Finding Rational Zeros

Find all rational zeros for $P(x) = 2x^3 - x^2 - 8x + 4$.

Solution   *Step 1. List the possible rational zeros:*

$$\pm 1, \ \pm 2, \ \pm 4, \ \pm \tfrac{1}{2}$$

+	−	I
2	1	0
0	1	2

Step 2. List possible combinations of zeros:

$$P(x) = 2x^3 - x^2 - 8x + 4 \qquad \text{Two variations in sign}$$
$$P(-x) = -2x^3 - x^2 + 8x + 4 \qquad \text{One variation in sign}$$

Step 3. Construct a synthetic division table. We start with $r = 0$ and move either way through integer values, usually starting with positive integers, until a zero is found, $P(x)$ changes sign, or lower and upper bounds are reached.

$$
\begin{array}{r|rrrr}
 & 2 & -1 & -8 & 4 \\
\hline
0 & 2 & -1 & -8 & 4 \\
1 & 2 & 1 & -7 & -3 \quad \text{\footnotesize $P(x)$ changes sign. Go to step 3(B) in the strategy.} \\
\tfrac{1}{2} & 2 & 0 & -8 & 0 \quad \text{\footnotesize $\tfrac{1}{2}$ is a zero} \\
\end{array}
$$

We have found a zero. Stop! Referring to step 3(A) in the strategy and using the last line of the synthetic division table, we write

$$P(x) = (x - r)Q(x) = (x - \tfrac{1}{2})(2x^2 - 8)$$

The reduced polynomial $Q(x) = 2x^2 - 8$ is quadratic, so all the remaining zeros can be found:

$$2x^2 - 8 = 0$$
$$x^2 - 4 = 0$$
$$(x + 2)(x - 2) = 0$$
$$x + 2 = 0 \qquad \text{or} \qquad x - 2 = 0$$
$$x = -2 \qquad\qquad\qquad x = 2$$

Thus, the rational zeros of $P(x)$ are ± 2 and $\tfrac{1}{2}$.

Problem 2 Find all rational zeros for $P(x) = 3x^3 + 10x^2 + x - 6$.

EXAMPLE 3 **Finding All Zeros**

Find all zeros (rational, irrational, and imaginary) for $P(x) = 2x^3 - 5x^2 - 8x + 6$.

Solution **Step 1.** *List the possible rational zeros:*

$$\pm 1,\ \pm 2,\ \pm 3,\ \pm 6,\ \pm \tfrac{1}{2},\ \pm \tfrac{3}{2}$$

Step 2. *List possible combinations of zeros:*

$$P(x) = 2x^3 - 5x^2 - 8x + 6 \qquad \text{Two variations in sign}$$
$$P(-x) = -2x^3 - 5x^2 + 8x + 6 \qquad \text{One variation in sign}$$

+	−	I
2	1	0
0	1	2

Step 3. *Construct a synthetic division table:*

	2	−5	−8	6	
0	2	−5	−8	6	
1	2	−3	−11	−5	$P(x)$ changes sign. Go to step 3(B) in the strategy.
$\tfrac{1}{2}$	2	−4	−10	1	$\tfrac{1}{2}$ is not a zero, so there must be an irrational zero between 0 and 1 (actually, between $\tfrac{1}{2}$ and 1).
2	2	−1	−10	−14	
3	2	1	−5	−9	
UB 4	2	3	4	22	There is an irrational zero between 3 and 4. Also, 4 is an upper bound (UB).
−1	2	−7	−1	7	
LB −2	2	−9	10	−14	$P(x)$ changes sign. Go to step 3(B) in the strategy. Also, −2 is a lower bound (LB).
$-\tfrac{3}{2}$	2	−8	4	0	$-\tfrac{3}{2}$ is a rational zero.

We now know that $P(x)$ has three real zeros—one negative and two positive. We have found that the negative zero is the rational number $-\tfrac{3}{2}$, and the two positive zeros are irrational. We now proceed to step 3(A) in the strategy to write

$$P(x) = (x - r)Q(x) = (x + \tfrac{3}{2})(2x^2 - 8x + 4)$$

Because the reduced polynomial $Q(x) = 2x^2 - 8x + 4$ is quadratic, all the remaining zeros can be found:

$$2x^2 - 8x + 4 = 0$$
$$x^2 - 4x + 2 = 0$$
$$x = \frac{4 \pm \sqrt{16 - 4(1)(2)}}{2}$$
$$= \frac{4 \pm 2\sqrt{2}}{2} = 2 \pm \sqrt{2}$$

The zeros of $P(x)$ are $-\frac{3}{2}$ and $2 \pm \sqrt{2}$ or, to two decimal places:

$$-1.5, \ 0.59, \ 3.41$$ Notice that 0.59 is between $\frac{1}{2}$ and 1, and 3.41 is between 3 and 4, as predicted in the synthetic division table.

Problem 3   Find all zeros (rational, irrational, and imaginary) for $P(x) = 2x^3 - 7x^2 + 6x + 5$.

EXAMPLE 4   Finding Rational Zeros

Find all rational zeros for $P(x) = x^4 - 7x^3 + 17x^2 - 17x + 6$.

Solution   *Step 1. List the possible rational zeros:*

$$\pm 1, \ \pm 2, \ \pm 3, \ \pm 6$$

+	−	I
4	0	0
2	0	2
0	0	4

Step 2. List possible combinations of zeros:

$$P(x) = x^4 - 7x^3 + 17x^2 - 17x + 6$$ Four variations in sign
$$P(-x) = x^4 + 7x^3 + 17x^2 + 17x + 6$$ No variation in sign

Since $P(-x)$ has no variation in sign, there are no negative real zeros. The list of possible rational zeros in step 1 can be cut in half! Our new list of candidates is 1, 2, 3, 6. Because we are interested only in finding rational zeros, we proceed directly to test these candidates in step 3.

Step 3. Construct a synthetic division table:

$$
\begin{array}{r|rrrrr}
 & 1 & -7 & 17 & -17 & 6 \\
1 & 1 & -6 & 11 & -6 & 0 \\
\end{array}
$$ 1 is a zero

Write

$$P(x) = (x - r)Q(x) = (x - 1)(x^3 - 6x^2 + 11x - 6)$$

and return to step 1 in the strategy for the reduced polynomial

$$Q(x) = x^3 - 6x^2 + 11x - 6$$

Step 1. List the possible rational zeros of $Q(x)$:

$$1, 2, 3, 6$$ The negatives of these were eliminated in step 2 above, since any zero of $Q(x)$ must be a zero of $P(x)$.

+	−	I
3	0	0
1	0	2

Step 2. List possible combinations of zeros:

$$Q(x) = x^3 - 6x^2 + 11x - 6$$ Three variations in sign

Since $P(x)$ has no negative zeros, neither does $Q(x)$.

Step 3. Construct a synthetic division table:

$$1 \quad\begin{array}{|cccc} 1 & -6 & 11 & -6 \\ 1 & -5 & 6 & 0 \end{array}$$ 1 is a zero

Write

$$Q(x) = (x - r)Q_1(x) = (x - 1)(x^2 - 5x + 6)$$

and find the zeros of the reduced quadratic polynomial $Q_1(x)$ by solving $Q_1(x) = 0$:

$$x^2 - 5x + 6 = 0$$
$$(x - 2)(x - 3) = 0$$
$$x = 2, 3$$

The zeros of the original polynomial $P(x)$ are 1 (multiplicity 2), 2, and 3.

Problem 4   Find all rational zeros for $P(x) = x^4 + 8x^3 + 23x^2 + 28x + 12$.

With a little practice and ingenuity or educated guessing, you will be able to reduce the number of steps and effort required to find rational zeros, if they exist. To develop this efficiency, you must work problems yourself. The more you work, the easier and faster the process will become.

Answers to Matched Problems
1. $\pm 1, \pm 2, \pm 3, \pm 6$ **2.** $-3, -1, \frac{2}{3}$ **3.** $-\frac{1}{2}, 2 \pm i$ **4.** $-3, -2$ (multiplicity 2), -1

EXERCISE 4-5

A

For each polynomial in Problems 1–8:
(A) List all possible rational zeros (Theorem 1).
(B) Find all rational zeros. If there are no rational zeros, say so.

1. $P(x) = x^3 - 2x^2 - 5x + 6$

2. $P(x) = x^3 + 3x^2 - 6x - 8$

3. $P(x) = 3x^3 - 11x^2 + 8x + 4$

4. $P(x) = 2x^3 + x^2 - 4x - 3$

5. $P(x) = 12x^3 - 16x^2 - 5x + 3$

6. $P(x) = 2x^3 - 9x^2 + 14x - 5$

7. $P(x) = 3x^3 + 7x^2 - 10x - 4$

8. $P(x) = 2x^3 - 5x^2 - 2x + 15$

B

For each polynomial in Problems 9–14:
(A) List all possible rational zeros (Theorem 1).
(B) Find all rational zeros. If there are no rational zeros, say so.

9. $P(x) = x^3 - 3x^2 + 6$

10. $P(x) = x^3 - 3x + 1$

11. $P(x) = x^4 - 2x^3 - 2x^2 + 8x - 8$

12. $P(x) = 2x^4 + 5x^3 - 7x^2 - 6x + 4$

13. $P(x) = 3x^4 - 8x^3 - 6x^2 + 17x + 6$

14. $P(x) = 12x^4 - 8x^3 - 37x^2 + 7x + 6$

In Problems 15–22, find all roots (rational, irrational, and imaginary) for each polynomial equation.

15. $2x^3 - 5x^2 + 1 = 0$

16. $2x^3 - 10x^2 + 12x - 4 = 0$

17. $x^4 + 4x^3 - x^2 - 20x - 20 = 0$

18. $x^4 - 4x^2 - 4x - 1 = 0$

19. $x^4 - 2x^3 - 5x^2 + 8x + 4 = 0$

20. $x^4 - 2x^2 - 16x - 15 = 0$

21. $2x^5 - 3x^4 - 2x + 3 = 0$

22. $2x^5 + x^4 - 6x^3 - 3x^2 - 8x - 4 = 0$

In Problems 23–30, find all zeros (rational, irrational, and imaginary) for each polynomial.

23. $P(x) = x^3 - 19x + 30$

24. $P(x) = x^3 - 7x^2 + 36$

25. $P(x) = x^4 - \frac{21}{10}x^3 + \frac{2}{5}x$

26. $P(x) = x^4 + \frac{7}{6}x^3 - \frac{7}{3}x^2 - \frac{5}{2}x$

27. $P(x) = x^4 - 5x^3 + \frac{15}{2}x^2 - 2x - 2$

28. $P(x) = x^4 - \frac{13}{4}x^2 - \frac{5}{2}x - \frac{1}{4}$

29. $P(x) = 3x^5 - 5x^4 - 8x^3 + 16x^2 + 21x + 5$

30. $P(x) = 2x^5 - 3x^4 - 6x^3 + 23x^2 - 26x + 10$

In Problems 31–36, write each polynomial as a product of linear factors.

31. $P(x) = 6x^3 + 13x^2 - 4$

32. $P(x) = 6x^3 - 17x^2 - 4x + 3$

33. $P(x) = x^3 + 2x^2 - 9x - 4$

34. $P(x) = x^3 - 8x^2 + 17x - 4$

35. $P(x) = 4x^4 - 4x^3 - 9x^2 + x + 2$

36. $P(x) = 2x^4 + 3x^3 - 4x^2 - 3x + 2$

In Problems 37–42, solve each inequality (see Section 2-7).

37. $x^2 \le 4x - 1$

38. $x^2 > 2x + 1$

39. $x^3 + 3 \le 3x^2 + x$

40. $9x + 9 \le x^3 + x^2$

41. $2x^3 + 6 \ge 13x - x^2$

42. $5x^3 - 3x^2 < 10x - 6$

C

In Problems 43–46, solve each inequality (see Section 2-7).

43. $\dfrac{4}{2x^3 + 5x^2 - 2x - 5} \ge 0$

44. $\dfrac{7}{2x^3 - x^2 - 8x + 4} \le 0$

45. $\dfrac{x^2 - 3x - 10}{x^3 - 4x^2 + x + 6} \le 0$

46. $\dfrac{x^2 + 4x - 21}{x^3 + 7x^2 + 7x - 15} \ge 0$

Prove that each of the real numbers in Problems 47–50 is not rational by writing an appropriate polynomial and making use of Theorem 1.

47. $\sqrt{6}$

48. $\sqrt{12}$

49. $\sqrt[3]{5}$

50. $\sqrt[5]{8}$

Problems 51–56 require the use of a graphic calculator or a computer. Graph the polynomial and use the graph to help locate the real zeros. Then find all zeros (rational, irrational, and imaginary) exactly.

51. $P(x) = 3x^3 - 37x^2 + 84x - 24$

52. $P(x) = 2x^3 - 9x^2 - 2x + 30$

53. $P(x) = 4x^4 + 4x^3 + 49x^2 + 64x - 240$

54. $P(x) = 6x^4 + 35x^3 + 2x^2 - 233x - 360$

55. $P(x) = 4x^4 - 44x^3 + 145x^2 - 192x + 90$

56. $P(x) = x^5 - 6x^4 + 6x^3 + 28x^2 - 72x + 48$

APPLICATIONS *Find all rational solutions exactly, and find irrational solutions to two decimal places.*

57. Storage. A rectangular storage unit has dimensions 1 by 2 by 3 feet. If each dimension is increased by the same amount, how much should this amount be to create a new storage unit with volume ten times the old?

58. Construction. A rectangular box has dimensions 1 by 1 by 2 feet. If each dimension is increased by the same amount, how much should this amount be to create a new box with volume six times the old?

★**59. Packaging.** An open box is to be made from a rectangular piece of cardboard that measures 8 by 5 inches, by cutting out squares of the same size from each corner and bending up the sides (see the figure). If the volume of the box is to be 14 cubic inches, how large a square should be cut from each corner? [*Hint:* Determine the domain of x from physical considerations before starting.]

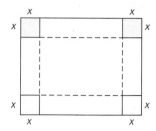

★**60. Fabrication.** An open metal chemical tank is to be made from a rectangular piece of stainless steel that measures 10 by 8 feet, by cutting out squares of the same size from each corner and bending up the sides (see the figure for Problem 59). If the volume of the tank is to be 48 cubic feet, how large a square should be cut from each corner?

SECTION 4-6 Approximating Real Zeros

- Finding All Real Zeros of a Polynomial
- Computer Approximations of Zeros
- Graphical Approximations of Zeros

If a polynomial has rational coefficients, then the strategy discussed in the last section can be used to find all rational zeros and to help locate irrational zeros. But how do we find the irrational zeros? For example, applying the strategy to $P(x) = x^3 - 4x^2 + 2x + 2$ would show that $P(x)$ has irrational zeros in $(-1, 0)$, $(1, 2)$, and $(3, 4)$, but it would not tell us how to find these zeros. Furthermore, what do we do if the coefficients of the polynomial are not rational? The rational zero theorem cannot be applied to the polynomial $Q(x) = x^4 + \sqrt{2}x^3 - 5x^2 - 3\sqrt{2}x - 2\sqrt{2} + 4$, yet it is easily verified that -1 and 2 are zeros of $Q(x)$. There are no simple answers to these questions. As we mentioned at the beginning of the chapter, there is no single step-by-step method that will find exactly all zeros of all polynomials. However, there are a number of methods that can be used to *approximate* the real zeros of a polynomial with real coefficients. In this section we introduce the *bisection method*. Once a zero has been isolated by the location theorem, the bisection method can be used to approximate the zero to any decimal accuracy desired.

Practically speaking, however, we will use this method only to obtain one- or two-decimal-place approximations. Because of the tedium involved in determining these approximations on an ordinary scientific calculator, more accurate results are usually obtained by using a computer. We also will discuss briefly the use of both computers and graphic calculators in approximating zeros.

• **Finding All Real Zeros of a Polynomial**

We outline a general strategy for finding the real zeros of a polynomial with real coefficients.

Strategy for Finding All Real Zeros of a Polynomial $P(x)$

Let $P(x)$ be a polynomial with real coefficients.

Step 1. Find any rational zeros by the methods of Section 4-5, if they apply.

Step 2. Write the final reduced polynomial $Q(x)$ resulting from step 1.

Step 3. Use Descartes' rule of signs to determine the possible number of positive and negative real zeros left after step 1.

Step 4. Isolate real zeros further. Form a synthetic division table using $Q(x)$ to:
  (A) Locate lower and upper bounds for any remaining zeros.
  (B) Isolate remaining zeros, if possible, between successive integers by observing sign changes in $Q(x)$.

Step 5. Approximate the remaining zeros isolated in step 4(B) by the bisection method described below.

EXAMPLE 1   Finding All the Zeros of a Polynomial

Find the rational zeros exactly and the irrational zeros to two decimal places for

$$P(x) = 2x^4 + x^3 + 4x^2 - 6x - 4$$

Solution   *Step 1.* Find rational zeros, if any. Using methods of the preceding section, we find $-\frac{1}{2}$ to be the only rational zero:

$$
\begin{array}{r|rrrrr}
 & 2 & 1 & 4 & -6 & -4 \\
-\frac{1}{2} & 2 & 0 & 4 & -8 & 0 \\
\end{array}
$$

We write

$$P(x) = (x + \tfrac{1}{2})(2x^3 + 4x - 8)$$
$$\quad\;\; = 2(x + \tfrac{1}{2})(x^3 + 2x - 4) \qquad \text{Factor 2 out of the second factor.}$$

Step 2. Write the final reduced polynomial from step 1.

$$Q(x) = x^3 + 2x - 4$$

+	−	I
1	0	2

Step 3. List the possible combinations of zeros.

$$Q(x) = x^3 + 2x - 4 \qquad \text{One variation in sign}$$
$$Q(-x) = -x^3 - 2x - 4 \qquad \text{No variation in sign}$$

Since Q has no rational zeros, the one real zero must be irrational.

Step 4. Isolate real zeros further. We form a synthetic division table to try to isolate the remaining zero between two integers by observing sign changes in $Q(x)$. We start with $x = 0$ because the remaining zero is positive and 0 is a lower bound.

$$
\begin{array}{c|cccc}
 & 1 & 0 & 2 & -4 \\
\hline
0 & 1 & 0 & 2 & -4 \\
1 & 1 & 1 & 3 & -1 \\
2 & 1 & 2 & 6 & 8 \quad \text{2 is an upper bound}
\end{array}
$$

Real zero $\left\{ \begin{matrix} 1 \\ 2 \end{matrix} \right.$

From the table we see that there must be a zero between 1 and 2.

Step 5. Approximate the remaining zero. Let r be the zero of $Q(x)$. We will use the *bisection method* to approximate r to two decimal places. For a first approximation, we use m, the midpoint of the interval containing r:

$$m = \frac{1 + 2}{2} = 1.5 \qquad \text{Midpoint of } [1, 2]$$

How accurate is this approximation? Since the distance between r and 1.5 can be no greater than half the length of the interval $[1, 2]$ (see the figure in the margin),

$$|r - 1.5| \le \tfrac{1}{2}(2 - 1) = 0.5$$

Thus, the maximum possible error in this approximation is 0.5. We will dispense with the absolute value inequality and write this error estimate simply as

$$r \approx 1.5 \pm 0.5$$

To improve on this approximation, we use the sign of $Q(1.5)$ to determine whether r is to the left or the right of 1.5. Omitting the details, we find that $Q(1.5) = 2.375 > 0$. Comparing this with the signs of $Q(1)$ and $Q(2)$ (see the figure in the margin), we see that r must be in the interval $[1, 1.5]$. The midpoint of this new interval will provide a better approximation to r:

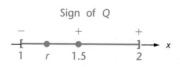
Sign of Q

$$m = \frac{1 + 1.5}{2} = 1.25 \qquad \text{Midpoint of } [1, 1.5]$$

$$r \approx 1.25 \pm 0.25 \qquad \text{Half the length of } [1, 1.5]\text{: } \frac{1}{2}(1.5 - 1) = 0.25$$

Notice that the maximum possible error in the approximation has been cut in half, from 0.5 to 0.25. It is helpful to organize the results we have obtained up to this point in a table:

Interval $[a, b]$	Midpoint m	Error $0.5(b - a)$	Sign of Q		
			$Q(a)$	$Q(m)$	$Q(b)$
$[1, 2]$	1.5	0.5	−	+	+
$[1, 1.5]$	1.25	0.25			

To complete the second line in the table, we need to determine the sign of Q at $a = 1, m = 1.25$, and $b = 1.5$. The signs at 1 and 1.5 can be determined from the preceding line. To find the sign at 1.25, we evaluate $Q(1.25)$ and find that $Q(1.25) = 0.453\ 125 > 0$. The completed line is shown below:

Interval $[a, b]$	Midpoint m	Error $0.5(b - a)$	Sign of Q		
			$Q(a)$	$Q(m)$	$Q(b)$
$[1, 2]$	1.5	0.5	−	+	+
$[1, 1.5]$	1.25	0.25	−	+	+

Examining the signs of Q, we see that r must lie in the interval $[1, 1.25]$. Adding this interval, its midpoint, and the maximum possible error to the table, we have:

Interval $[a, b]$	Midpoint m	Error $0.5(b - a)$	Sign of Q		
			$Q(a)$	$Q(m)$	$Q(b)$
$[1, 2]$	1.5	0.5	−	+	+
$[1, 1.5]$	1.25	0.25	−	+	+
$[1, 1.25]$	1.125	0.125			

From the last line of this table, we conclude that $r = 1.125 \pm 0.125$. To obtain an approximation that is accurate when rounded to two decimal places, we continue this process until the error is less than 0.001 and then round to two decimal places. The results are shown in Table 1.

From the last line in Table 1, we conclude that

$$r \approx 1.17871 \pm 0.00098$$

Since $0.00098 < 0.001$, the irrational zero to two decimal places is 1.18. Thus, the real zeros for the original polynomial

$$P(x) = 2x^4 + x^3 + 4x^2 - 6x - 4$$

are -0.5 and 1.18.

TABLE 1   **Approximating the Real Zero of $Q(x) = x^3 + 2x - 4$**

Interval $[a, b]$	Midpoint m	Error $0.5(b - a)$	Sign of Q $Q(a)$	$Q(m)$	$Q(b)$
$[1, 2]$	1.5	0.5	−	+	+
$[1, 1.5]$	1.25	0.25	−	+	+
$[1, 1.25]$	1.125	0.125	−	−	+
$[1.125, 1.25]$	1.1875	0.0625	−	+	+
$[1.125, 1.1875]$	1.15625	0.03125	−	−	+
$[1.15625, 1.1875]$	1.17188	0.01562	−	−	+
$[1.17188, 1.1875]$	1.17969	0.00781	−	+	+
$[1.17188, 1.17969]$	1.17578	0.00391	−	−	+
$[1.17578, 1.17969]$	1.17773	0.00195	−	−	+
$[1.17773, 1.17969]$	1.17871	0.00098			

Problem 1   Find all real zeros for $P(x) = x^3 + 2x - 7$. Construct a table like Table 1 to approximate irrational roots to one decimal place.

The bisection method illustrated in Example 1 is summarized in the next box.

The Bisection Method

Let $P(x)$ be a polynomial with real coefficients. If $P(x)$ has opposite signs at the endpoints of the interval $[a, b]$, then a real zero r that lies in this interval can be approximated as follows:

Step 1. Find $m = \frac{1}{2}(a + b)$, the midpoint of the interval $[a, b]$.

Step 2. Find $\frac{1}{2}(b - a)$, the maximum possible error if m is used to approximate r. The approximation determined at this step is

$$r \approx m \pm \tfrac{1}{2}(b - a)$$

Step 3. To improve this approximation, find $P(m)$ and consider the following three cases:
   (A) If $P(m)$ and $P(a)$ have opposite sign, let $b = m$ and repeat steps 1 and 2.
   (B) If $P(m)$ and $P(b)$ have opposite sign, let $a = m$ and repeat steps 1 and 2.
   (C) If $P(m) = 0$, stop, because $r = m$.

Step 4. To approximate the zero to d decimal places, we continue this process until the maximum possible error is less than $10^{-(d+1)}$ and then round the approximate value to d decimal places.

It is useful to note that the bisection method is a very general procedure that can be used to approximate both rational and irrational zeros and that can be applied to both polynomial and nonpolynomial functions. The only requirements are that the function be continuous on $[a, b]$ and of opposite sign at a and b. We included case (C) in step 3 of the bisection method to accommodate these more general applications. If $P(x)$ is a polynomial with integer coefficients and no rational zeros and if a and b are rational numbers, then case (C) will never occur.

● Computer Approximations of Zeros

Since each step in the bisection method halves the maximum possible error, this method can be used to produce a decimal approximation of a real zero to any accuracy desired. However, performing all the calculations on an ordinary calculator soon becomes tedious. Most real-world applications of the bisection method are performed on a programmable calculator or a computer. Table 2 shows a typical computer program written in the BASIC language that uses the bisection method to approximate zeros. The table also shows the output generated when we use this program to approximate the irrational zero of the polynomial $Q(x)$ in Example 1 to four decimal places. The last line of output in Table 2 indicates that $r \approx 1.1795$ to four decimal places.

● Graphical Approximations of Zeros

Graphic calculators employ a more intuitive approach to approximating zeros. Not only can these calculators graph functions, but with a few keystrokes, they will expand any portion of a graph. Thus, a real zero can be approximated by repeatedly expanding a graph near an intercept and displaying the coordinates of the point on the calculator's screen that is closest to the actual intercept. Figure 1 illustrates this approach for the polynomial $Q(x)$ from Example 1.

The coordinates of the point at the location of the cursor are shown at the bottom of each graph. From the graph in Figure 1(c), we would conclude that $r \approx$

TABLE 2   Computer Approximations of the Real Zero of $Q(x) = x^3 + 2x - 4$

Program

```
100 DEF FNF(X) = X^3 + 2*X - 4
110 INPUT "A = ";A
120 INPUT "B = ";B
130 IF FNF(A)*FNF(B) < 0 THEN GOTO 160
140 PRINT "F(A)*F(B) MUST BE NEGATIVE"
150 GOTO 310
160 INPUT "DESIRED MAXIMUM ERROR";E
170 IF E>0 THEN GOTO 200
180 PRINT "ERROR MUST BE POSITIVE"
190 GOTO 310
200 PRINT "A","B","M","E"
210 LET M = (A+B)/2
220 LET E1 = (B-A)/2
230 PRINT A,B,M,E1
240 IF E1<=E THEN GOTO 310
250 IF FNF(M) = 0 THEN GOTO 310
260 IF FNF(A)*FNF(M) < 0 THEN GOTO 290
270 LET A = M
280 GOTO 210
290 LET B = M
300 GOTO 210
310 END
```

Output

```
A = ? 1
B = ? 2
DESIRED MAXIMUM ERROR? .00001
```

A	B	M	E
1	2	1.5	.5
1	1.5	1.25	.25
1	1.25	1.125	.125
1.125	1.25	1.1875	.0625
1.125	1.1875	1.15625	.03125
1.15625	1.1875	1.171875	.015625
1.171875	1.1875	1.179688	.0078125
1.171875	1.179688	1.175781	3.90625E-03
1.175781	1.179688	1.177734	1.953125E-03
1.177734	1.179688	1.178711	9.765625E-04
1.178711	1.179688	1.179199	4.882813E-04
1.179199	1.179688	1.179443	2.441406E-04
1.179443	1.179688	1.179565	1.220703E-04
1.179443	1.179565	1.179504	6.103516E-05
1.179504	1.179565	1.179535	3.051758E-05
1.179504	1.179535	1.17952	1.525879E-05
1.179504	1.17952	1.179512	7.629395E-06

FIGURE 1 Graphical approximations of the real zero of $Q(x) = x^3 + 2x - 4$.

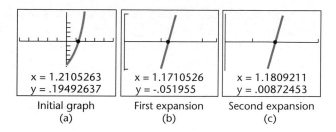

x = 1.2105263	x = 1.1710526	x = 1.1809211
y = .19492637	y = -.051955	y = .00872453
Initial graph	First expansion	Second expansion
(a)	(b)	(c)

1.180 921 1. We can see that each successive approximation is an improvement over the preceding one because the y coordinate at the cursor point becomes closer to 0. However, the graph gives no information concerning the accuracy of these approximations. Generally, one proceeds by comparing successive approximations and stopping when agreement is obtained in a couple of decimal places beyond the desired accuracy. One of the principal advantages of the bisection method is the precise control over the accuracy of the approximation.

Answers to Matched Problems

1. The only real zero is 1.6, to one decimal place.

Interval [a, b]	Midpoint m	Error 0.5(b − a)	P(a)	Sign of P P(m)	P(b)
[1, 2]	1.5	0.5	−	−	+
[1.5, 2]	1.75	0.25	−	+	+
[1.5, 1.75]	1.625	0.125	−	+	+
[1.5, 1.625]	1.562	0.062	−	−	+
[1.562, 1.625]	1.594	0.031	−	+	+
[1.562, 1.594]	1.578	0.016	−	+	+
[1.562, 1.578]	1.57	0.008			

EXERCISE 4-6

A

Show that each polynomial has a zero in the indicated interval and approximate the zero to one decimal place.

1. $P(x) = x^3 + x - 4$; [1, 2]

2. $P(x) = x^3 + x - 1$; [0, 1]

3. $P(x) = x^3 + 0.5x - 1$; [0, 1]

4. $P(x) = x^3 + 0.7x - 5$; [1, 2]

5. $P(x) = x^3 + \sqrt{2}x - 15$; [2, 3]

6. $P(x) = x^3 + \sqrt{3}x - 20$; [2, 3]

B

In Problems 7–12, find all rational zeros exactly and all irrational zeros correct to one decimal place.

7. $P(x) = x^3 + x + 1$

8. $P(x) = x^3 + 2x + 5$

9. $P(x) = x^4 - 2x^3 + x^2 - 5x + 6$

10. $P(x) = x^4 - x^3 + 10x^2 - 28x + 18$

11. $P(x) = 2x^5 - 5x^4 - 7x^3 + 4x^2 + 21x + 9$

12. $P(x) = 2x^4 + 3x^3 + 6x^2 - x - 15$

In Problems 13–16, approximate all real solutions to two decimal places.

13. $x^3 + \sqrt{5}x + 4 = 0$

14. $x^3 + \sqrt{6}x - 15 = 0$

15. $x^4 - 6x + 4 = 0$

16. $x^4 + 4x + 2 = 0$

C

In Problems 17 and 18, show that the equation has two real solutions, both irrational and both between 1 and 2. Then approximate each solution to two decimal places.

17. $x^4 - 12x + 12 = 0$

18. $2x^4 - 4x^3 + 3 = 0$

 Problems 19–26 require the use of a graphic calculator or computer. Approximate all the real zeros of each polynomial to four decimal places.

19. $P(x) = x^3 - 9x^2 + 24x - 17$

20. $P(x) = x^3 - 3x^2 - 9x + 5$

21. $P(x) = 2x^4 - 4x^3 + 1$

22. $P(x) = x^4 + 4x^3 + 10$

23. $P(x) = x^4 - 2\sqrt{5}x^3 + x^2 + 4\sqrt{5}x - 5$

24. $P(x) = x^4 - 4\sqrt{2}x^3 - x^2 + 18\sqrt{2}x + 14$

25. $P(x) = 8x^3 - 21x^2 + 15x - 3$

26. $P(x) = 2x^3 - 8x^2 + 6x - 1$

APPLICATIONS   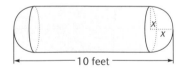 *In Problems 27–30, find all rational zeros exactly and approximate all irrational zeros to two decimal places.*

★**27. Geometry.** Find all points on the graph of $y = x^2$ that are 1 unit away from the point (1, 2). [*Hint:* Use the distance-between-two-points formula from Section 3-1.]

★**28. Geometry.** Find all points on the graph of $y = x^2$ that are 1 unit away from the point (2, 1).

★**29. Construction.** A propane gas tank is in the shape of a right circular cylinder with a hemisphere at each end (see the figure). If the overall length of the tank is 10 feet and the volume is 20π cubic feet, find the common radius of the hemispheres and the cylinder.

★**30. Shipping.** A shipping box is reinforced with steel bands in all three directions (see the figure). A total of 20.5 feet of steel tape is to be used, with 6 inches of waste because of a 2-inch overlap in each direction. If the box has a square base and a volume of 2 cubic feet, find its dimensions.

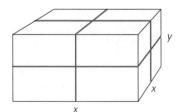

SECTION **4-7**   **Partial Fractions**

- Basic Theorems
- Partial Fraction Decomposition

You have now had considerable experience combining two or more rational expressions into a single rational expression. For example, problems such as

$$\frac{2}{x + 5} + \frac{3}{x - 4} = \frac{2(x - 4) + 3(x + 5)}{(x + 5)(x - 4)} = \frac{5x + 7}{(x + 5)(x - 4)}$$

should seem routine. Frequently in more advanced courses, particularly in calculus, it is advantageous to be able to reverse this process—that is, to be able to express a rational expression as the sum of two or more simpler rational expressions called **partial fractions**. As is often the case with reverse processes, the process of decomposing a rational expression into partial fractions is more difficult than combining rational expressions. Basic to the process is the factoring of polynomials, so the topics discussed earlier in this chapter can be put to effective use.

We confine our attention to rational expressions of the form $P(x)/D(x)$, where $P(x)$ and $D(x)$ are polynomials with real coefficients. In addition, we assume that the degree of $P(x)$ is less than the degree of $D(x)$. If the degree of $P(x)$ is greater than or equal to that of $D(x)$, we have only to divide $P(x)$ by $D(x)$ to obtain

$$\frac{P(x)}{D(x)} = Q(x) + \frac{R(x)}{D(x)}$$

where the degree of $R(x)$ is less than that of $D(x)$. For example,

$$\frac{x^4 - 3x^3 + 2x^2 - 5x + 1}{x^2 - 2x + 1} = x^2 - x - 1 + \frac{-6x + 2}{x^2 - 2x + 1}$$

If the degree of $P(x)$ is less than that of $D(x)$, then $P(x)/D(x)$ is called a **proper fraction**.

● **Basic Theorems**   Our task now is to establish a systematic way to decompose a proper fraction into the sum of two or more partial fractions. The following three theorems take care of the problem completely. Theorems 1 and 3 are stated without proof.

Theorem 1   **Equal Polynomials**

Two polynomials are equal to each other if and only if the coefficients of terms of like degree are equal.

For example, if

Equate the constant terms.

$$\underbrace{(A + 2B)}x + B = 5x - 3$$

Equate the coefficients of x.

then

$B = -3$ Substitute $B = -3$ into the second equation to solve for A.

$A + 2B = 5$

$A + 2(-3) = 5$

$A = 11$

Theorem 2	**Linear and Quadratic Factor Theorem**

For a polynomial with real coefficients, there always exists a complete factoring involving only linear and/or quadratic factors with real coefficients where the linear and quadratic factors are prime relative to the real numbers.

That Theorem 2 is true can be seen as follows: From earlier theorems in this chapter, we know that an nth-degree polynomial $P(x)$ has n zeros and n linear factors. The real zeros of $P(x)$ correspond to linear factors of the form $(x - r)$, where r is a real number. Since $P(x)$ has real coefficients, the imaginary zeros occur in conjugate pairs. Thus, the imaginary zeros correspond to pairs of factors of the form $[x - (a + bi)]$ and $[x - (a - bi)]$, where a and b are real numbers. Multiplying these two imaginary factors, we have

$$[x - (a + bi)][x - (a - bi)] = x^2 - 2ax + a^2 + b^2$$

This quadratic polynomial with real coefficients is a factor of $P(x)$. Thus, $P(x)$ can be factored into a product of linear factors and quadratic factors, all with real coefficients.

● **Partial Fraction Decomposition**

We are now ready to state Theorem 3, which forms the basis for partial fraction decomposition.

Theorem 3	**Partial Fraction Decomposition**

Any proper fraction $P(x)/D(x)$ reduced to lowest terms can be decomposed into the sum of partial fractions as follows:

1. If $D(x)$ has a nonrepeating linear factor of the form $ax + b$, then the partial fraction decomposition of $P(x)/D(x)$ contains a term of the form

$$\frac{A}{ax + b} \qquad A \text{ a constant}$$

2. If $D(x)$ has a k-repeating linear factor of the form $(ax + b)^k$, then the partial fraction decomposition of $P(x)/D(x)$ contains k terms of the form

$$\frac{A_1}{ax + b} + \frac{A_2}{(ax + b)^2} + \cdots + \frac{A_k}{(ax + b)^k} \qquad A_1, A_2, \ldots, A_k \text{ constants}$$

3. If $D(x)$ has a nonrepeating quadratic factor of the form $ax^2 + bx + c$, which is prime relative to the real numbers, then the partial fraction decomposition of $P(x)/D(x)$ contains a term of the form

$$\frac{Ax + B}{ax^2 + bx + c} \qquad A, B \text{ constants}$$

> **4.** If $D(x)$ has a k-repeating quadratic factor of the form $(ax^2 + bx + c)^k$, where $ax^2 + bx + c$ is prime relative to the real numbers, then the partial fraction decomposition of $P(x)/D(x)$ contains k terms of the form
>
> $$\frac{A_1 x + B_1}{ax^2 + bx + c} + \frac{A_2 x + B_2}{(ax^2 + bx + c)^2} + \cdots + \frac{A_k x + B_k}{(ax^2 + bx + c)^k}$$
>
> $$A_1, \ldots, A_k, \quad B_1, \ldots, B_k \text{ constants}$$

Let's see how the theorem is used to obtain partial fraction decompositions in several examples.

EXAMPLE 1   **Nonrepeating Linear Factors**

Decompose into partial fractions: $\dfrac{5x + 7}{x^2 + 2x - 3}$

Solution   We first try to factor the denominator. If it can't be factored in the real numbers, then we can't go any further. In this example, the denominator factors, so we apply part 1 from Theorem 3:

$$\frac{5x + 7}{(x - 1)(x + 3)} = \frac{A}{x - 1} + \frac{B}{x + 3} \tag{1}$$

To find the constants A and B, we combine the fractions on the right side of equation (1) to obtain

$$\frac{5x + 7}{(x - 1)(x + 3)} = \frac{A(x + 3) + B(x - 1)}{(x - 1)(x + 3)}$$

Since these fractions have the same denominator, their numerators must be equal. Thus,

$$5x + 7 = A(x + 3) + B(x - 1) \tag{2}$$

We could multiply the right side and find A and B by using Theorem 1, but in this case it is easier to take advantage of the fact that equation (2) is an identity—that is, it must hold for all values of x. In particular, we note that if we let $x = 1$, then the second term of the right side drops out and we can solve for A:

$$5 \cdot 1 + 7 = A(1 + 3) + B(1 - 1)$$

$$12 = 4A$$

$$A = 3$$

Similarly, if we let $x = -3$, the first term drops out and we find

$$-8 = -4B$$

$$B = 2$$

Hence,

$$\frac{5x + 7}{x^2 + 2x - 3} = \frac{3}{x - 1} + \frac{2}{x + 3}$$

as can easily be checked by adding the two fractions on the right.

Problem 1   Decompose into partial fractions: $\dfrac{7x + 6}{x^2 + x - 6}$

EXAMPLE 2   **Repeating Linear Factors**

Decompose into partial fractions: $\dfrac{6x^2 - 14x - 27}{(x + 2)(x - 3)^2}$

Solution   Using parts 1 and 2 from Theorem 3, we write

$$\begin{aligned}
\frac{6x^2 - 14x - 27}{(x + 2)(x - 3)^2} &= \frac{A}{x + 2} + \frac{B}{x - 3} + \frac{C}{(x - 3)^2} \\
&= \frac{A(x - 3)^2 + B(x + 2)(x - 3) + C(x + 2)}{(x + 2)(x - 3)^2}
\end{aligned}$$

Thus, for all x,

$$6x^2 - 14x - 27 = A(x - 3)^2 + B(x + 2)(x - 3) + C(x + 2)$$

If $x = 3$, then   If $x = -2$, then

$$\begin{aligned} -15 &= 5C & 25 &= 25A \\ C &= -3 & A &= 1 \end{aligned}$$

There are no other values of x that will cause terms on the right to drop out. Since any value of x can be substituted to produce an equation relating A, B, and C, we let $x = 0$ and obtain

$$\begin{aligned}
-27 &= 9A - 6B + 2C \quad \text{Substitute } A = 1 \text{ and } C = -3. \\
-27 &= 9 - 6B - 6 \\
B &= 5
\end{aligned}$$

Thus,

$$\frac{6x^2 - 14x - 27}{(x + 2)(x - 3)^2} = \frac{1}{x + 2} + \frac{5}{x - 3} - \frac{3}{(x - 3)^2}$$

Problem 2   Decompose into partial fractions: $\dfrac{x^2 + 11x + 15}{(x - 1)(x + 2)^2}$

EXAMPLE 3   **Nonrepeating Linear and Quadratic Factors**

Decompose into partial fractions: $\dfrac{5x^2 - 8x + 5}{(x - 2)(x^2 - x + 1)}$

Solution   First, we see that the quadratic in the denominator can't be factored further in the real numbers. Then, we use parts 1 and 3 from Theorem 3 to write

$$\frac{5x^2 - 8x + 5}{(x - 2)(x^2 - x + 1)} = \frac{A}{x - 2} + \frac{Bx + C}{x^2 - x + 1}$$

$$= \frac{A(x^2 - x + 1) + (Bx + C)(x - 2)}{(x - 2)(x^2 - x + 1)}$$

Thus, for all x,

$$5x^2 - 8x + 5 = A(x^2 - x + 1) + (Bx + C)(x - 2)$$

If $x = 2$, then

$$9 = 3A$$
$$A = 3$$

If $x = 0$, then, using $A = 3$, we have

$$5 = 3 - 2C$$
$$C = -1$$

If $x = 1$, then, using $A = 3$ and $C = -1$, we have

$$2 = 3 + (B - 1)(-1)$$
$$B = 2$$

Hence,

$$\frac{5x^2 - 8x + 5}{(x - 2)(x^2 - x + 1)} = \frac{3}{x - 2} + \frac{2x - 1}{x^2 - x + 1}$$

Problem 3   Decompose into partial fractions: $\dfrac{7x^2 - 11x + 6}{(x - 1)(2x^2 - 3x + 2)}$

EXAMPLE 4   **Repeating Quadratic Factors**

Decompose into partial fractions: $\dfrac{x^3 - 4x^2 + 9x - 5}{(x^2 - 2x + 3)^2}$

Solution   Since $x^2 - 2x + 3$ can't be factored further in the real numbers, we proceed to use part 4 from Theorem 3 to write

$$\frac{x^3 - 4x^2 + 9x - 5}{(x^2 - 2x + 3)^2} = \frac{Ax + B}{x^2 - 2x + 3} + \frac{Cx + D}{(x^2 - 2x + 3)^2}$$

$$= \frac{(Ax + B)(x^2 - 2x + 3) + Cx + D}{(x^2 - 2x + 3)^2}$$

Thus, for all x,

$$x^3 - 4x^2 + 9x - 5 = (Ax + B)(x^2 - 2x + 3) + Cx + D$$

Since the substitution of carefully chosen values of x doesn't lead to the immediate determination of A, B, C, or D, we multiply and rearrange the right side to obtain

$$x^3 - 4x^2 + 9x - 5 = Ax^3 + (B - 2A)x^2 + (3A - 2B + C)x + (3B + D)$$

Now we use Theorem 1 to equate coefficients of terms of like degree:

$$A = 1$$
$$B - 2A = -4$$
$$3A - 2B + C = 9$$
$$3B + D = -5$$

$$\underset{Ax^3}{\overset{1x^3}{\uparrow}} + \underset{(B - 2A)x^2}{\overset{-4x^2}{\uparrow}} + \underset{(3A - 2B + C)x}{\overset{+9x}{\uparrow}} + \underset{(3B + D)}{\overset{-5}{\uparrow}}$$

From these equations we easily find that $A = 1$, $B = -2$, $C = 2$, and $D = 1$. Now we can write

$$\frac{x^3 - 4x^2 + 9x - 5}{(x^2 - 2x + 3)^2} = \frac{x - 2}{x^2 - 2x + 3} + \frac{2x + 1}{(x^2 - 2x + 3)^2}$$

Problem 4   Decompose into partial fractions: $\dfrac{3x^3 - 6x^2 + 7x - 2}{(x^2 - 2x + 2)^2}$

Answers to Matched Problems

1. $\dfrac{4}{x - 2} + \dfrac{3}{x + 3}$ 2. $\dfrac{3}{x - 1} - \dfrac{2}{x + 2} + \dfrac{1}{(x + 2)^2}$ 3. $\dfrac{2}{x - 1} + \dfrac{3x - 2}{2x^2 - 3x + 2}$

4. $\dfrac{3x}{x^2 - 2x + 2} + \dfrac{x - 2}{(x^2 - 2x + 2)^2}$

EXERCISE 4-7

A

In Problems 1–10, find constants A, B, C, and D so that the right side is equal to the left.

1. $\dfrac{7x - 14}{(x - 4)(x + 3)} = \dfrac{A}{x - 4} + \dfrac{B}{x + 3}$

2. $\dfrac{9x + 21}{(x + 5)(x - 3)} = \dfrac{A}{x + 5} + \dfrac{B}{x - 3}$

3. $\dfrac{17x - 1}{(2x - 3)(3x - 1)} = \dfrac{A}{2x - 3} + \dfrac{B}{3x - 1}$

4. $\dfrac{x - 11}{(3x + 2)(2x - 1)} = \dfrac{A}{3x + 2} + \dfrac{B}{2x - 1}$

5. $\dfrac{3x^2 + 7x + 1}{x(x + 1)^2} = \dfrac{A}{x} + \dfrac{B}{x + 1} + \dfrac{C}{(x + 1)^2}$

6. $\dfrac{x^2 - 6x + 11}{(x + 1)(x - 2)^2} = \dfrac{A}{x + 1} + \dfrac{B}{x - 2} + \dfrac{C}{(x - 2)^2}$

7. $\dfrac{3x^2 + x}{(x - 2)(x^2 + 3)} = \dfrac{A}{x - 2} + \dfrac{Bx + C}{x^2 + 3}$

8. $\dfrac{5x^2 - 9x + 19}{(x - 4)(x^2 + 5)} = \dfrac{A}{x - 4} + \dfrac{Bx + C}{x^2 + 5}$

9. $\dfrac{2x^2 + 4x - 1}{(x^2 + x + 1)^2} = \dfrac{Ax + B}{x^2 + x + 1} + \dfrac{Cx + D}{(x^2 + x + 1)^2}$

10. $\dfrac{3x^3 - 3x^2 + 10x - 4}{(x^2 - x + 3)^2} =$

$$\dfrac{Ax + B}{x^2 - x + 3} + \dfrac{Cx + D}{(x^2 - x + 3)^2}$$

B _____

In Problems 11–22, decompose into partial fractions.

11. $\dfrac{-x + 22}{x^2 - 2x - 8}$

12. $\dfrac{-x - 21}{x^2 + 2x - 15}$

13. $\dfrac{3x - 13}{6x^2 - x - 12}$

14. $\dfrac{11x - 11}{6x^2 + 7x - 3}$

15. $\dfrac{x^2 - 12x + 18}{x^3 - 6x^2 + 9x}$

16. $\dfrac{5x^2 - 36x + 48}{x(x - 4)^2}$

17. $\dfrac{5x^2 + 3x + 6}{x^3 + 2x^2 + 3x}$

18. $\dfrac{6x^2 - 15x + 16}{x^3 - 3x^2 + 4x}$

19. $\dfrac{2x^3 + 7x + 5}{x^4 + 4x^2 + 4}$

20. $\dfrac{-5x^2 + 7x - 18}{x^4 + 6x^2 + 9}$

21. $\dfrac{x^3 - 7x^2 + 17x - 17}{x^2 - 5x + 6}$

22. $\dfrac{x^3 + x^2 - 13x + 11}{x^2 + 2x - 15}$

C _____

In Problems 23–30, decompose into partial fractions.

23. $\dfrac{4x^2 + 5x - 9}{x^3 - 6x - 9}$

24. $\dfrac{4x^2 - 8x + 1}{x^3 - x + 6}$

25. $\dfrac{x^2 + 16x + 18}{x^3 + 2x^2 - 15x - 36}$

26. $\dfrac{5x^2 - 18x + 1}{x^3 - x^2 - 8x + 12}$

27. $\dfrac{-x^2 + x - 7}{x^4 - 5x^3 + 9x^2 - 8x + 4}$

28. $\dfrac{-2x^3 + 12x^2 - 20x - 10}{x^4 - 7x^3 + 17x^2 - 21x + 18}$

29. $\dfrac{4x^5 + 12x^4 - x^3 + 7x^2 - 4x + 2}{4x^4 + 4x^3 - 5x^2 + 5x - 2}$

30. $\dfrac{6x^5 - 13x^4 + x^3 - 8x^2 + 2x}{6x^4 - 7x^3 + x^2 + x - 1}$

_____

Chapter 4 Review

In this chapter, unless indicated otherwise, the coefficients of the **nth-degree polynomial function** $P(x) = a_n x^n + a_{n-1} x^{n-1} + \cdots + a_1 x + a_0$ are complex numbers and the domain is the set of complex numbers. The number r is said to be a **zero of the function P**, or a **zero of the polynomial $P(x)$**, or a **solution or root of the equation $P(x) = 0$**, if $P(r) = 0$. If the coefficients of $P(x)$ are real numbers, then the x intercepts of the graph of $y = P(x)$ are real **zeros** of P and $P(x)$ and real **solutions** or **roots** for the equation $P(x) = 0$.

4-1   SYNTHETIC DIVISION

Synthetic division is an efficient method for dividing polynomials by linear terms of the form $x - r$ that is well-suited to calculator use.

4-2   REMAINDER AND FACTOR THEOREMS

Let $P(x)$ be a polynomial of degree greater than 0 and let r be a real number. Then we have the following important theorems:

Division Algorithm. $P(x) = (x - r)Q(x) + R$, where $x - r$ is the **divisor**, $Q(x)$, a unique polynomial of degree 1 less than $P(x)$, is the **quotient**, and R, a unique real number, is the **remainder**.

Remainder Theorem. $P(r) = R$.

Factor Theorem. The number r is a zero of $P(x)$ if and only if $(x - r)$ is a factor of $P(x)$.

A polynomial function with real coefficients is continuous everywhere, and its graph has no holes or gaps.

4-3   FUNDAMENTAL THEOREM OF ALGEBRA

If $P(x)$ is a polynomial of degree $n > 0$, then we have the following important theorems:

Fundamental Theorem of Algebra. $P(x)$ has at least one zero.

***n* Zeros Theorem.** $P(x)$ can be expressed as a product of n linear factors and has n zeros, not necessarily distinct.

Imaginary Zeros Theorem. If $P(x)$ has real coefficients, then imaginary zeros of $P(x)$, if they exist, must occur in conjugate pairs.

Real Zeros and Odd-Degree Polynomials. If $P(x)$ has real coefficients and is of odd degree, then $P(x)$ always has at least one real zero.

If $P(x)$ is represented as the product of linear factors and $x - r$ occurs m times, then r is called a **zero of multiplicity *m***.

4-4   ISOLATING REAL ZEROS

Let $P(x)$ be a polynomial with real coefficients. If the terms of $P(x)$ are arranged in order of descending powers, a **variation in sign** occurs if two successive terms have opposite signs. Missing terms are ignored. We have the following important theorems:

Descartes' Rule of Signs. The number of **positive zeros** of $P(x)$ is never greater than the number of variations in sign in $P(x)$ and, if less, then always by an even number. The number of **negative zeros** of $P(x)$ is never greater than the number of variations in sign in $P(-x)$ and, if less, then always by an even number.

Upper and Lower Bounds of Real Zeros. If $a_n > 0$ and $P(x)$ is divided by $x - r$ using synthetic division:

1. If $r > 0$ and all numbers in the quotient row of the synthetic division, including the remainder, are nonnegative, then r is greater than or equal to the largest zero of $P(x)$ and is called an **upper bound** of the real zeros of $P(x)$.

2. If $r < 0$ and all numbers in the quotient row of the synthetic division, including the remainder, alternate in sign, then r is less than or equal to the smallest zero of $P(x)$ and is called a **lower bound** of the real zeros of $P(x)$.

Location Theorem. If $P(a)$ and $P(b)$ are of opposite sign, then there is at least one real zero between a and b.

4-5   RATIONAL ZERO THEOREM

Rational Zero Theorem. If the rational number b/c, in lowest terms, is a zero of the polynomial

$$P(x) = a_n x^n + a_{n-1} x^{n-1} + \cdots + a_1 x + a_0 \qquad a_n \neq 0$$

with integer coefficients, then b must be an integer factor of a_0 and c must be an integer factor of a_n.

Strategy for Finding Rational Zeros

Assume $P(x)$ is a polynomial with integer coefficients and is of degree greater than 2.

Step 1. List the possible rational zeros of $P(x)$ using the rational zero theorem.

Step 2. Make a table of possible combinations of positive, negative, and imaginary zeros using Descartes' rule of signs and earlier theorems. Use the results to reduce the size of the list from step 1, if applicable.

Step 3. Construct a synthetic division table, keeping the results of steps 1 and 2 in mind. Generally, start at $r = 0$ and move either way through integer values, usually starting with the positive integers, until a zero is found, $P(x)$ changes sign, or lower and upper bounds are found.

(A) If a rational zero r is found at any time, stop, write

$$P(x) = (x - r)Q(x)$$

and immediately proceed to find the rational zeros for $Q(x)$, the **reduced polynomial** relative to $P(x)$. If $Q(x)$ is of degree greater than 2, return to step 1 using $Q(x)$ in place of $P(x)$. If $Q(x)$ is quadratic, find all its zeros using standard methods for solving quadratic equations.

(B) If $P(x)$ changes sign, try the possible rational zeros from the list in step 1 that are between the two integers producing the sign change. If a rational zero is found, proceed to step 3(A).

(C) If lower or upper bounds are found, modify the possible rational zeros list by discarding any numbers below the lower bound or above the upper bound.

Step 4. Test any untested numbers from the list of remaining possible rational zeros. If a rational zero is found, proceed to step 3(A). If no rational number from the list is a zero, then conclude that $P(x)$ [or $Q(x)$] has no rational zeros. The remaining zeros must be irrational or imaginary.

4-6   APPROXIMATING REAL ZEROS

Strategy for Finding All Real Zeros of a Polynomial $P(x)$. Let $P(x)$ be a polynomial with real coefficients.

Step 1. Find any rational roots by the methods of Section 4-5, if they apply.

Step 2. Write the final reduced polynomial $Q(x)$ resulting from step 1.

Step 3. Use Descartes' rule of signs to determine the possible number of positive and negative real zeros left after step 1.

Step 4. Isolate real zeros further. Form a synthetic division table using $Q(x)$ to:
   (A) Locate lower and upper bounds for any remaining zeros.
   (B) Isolate remaining zeros, if possible, between successive integers by observing sign changes in $Q(x)$.

Step 5. Approximate the remaining zeros isolated in step 4(B) by the bisection method.

The Bisection Method. Let $P(x)$ be a polynomial with real coefficients. If $P(x)$ has opposite signs at the endpoints of the interval $[a, b]$, then a real zero r that lies in this interval can be approximated as follows:

Step 1. Find $m = \frac{1}{2}(a + b)$, the midpoint of the interval $[a, b]$.

Step 2. Find $\frac{1}{2}(b - a)$, the maximum possible error if m is used to approximate r. The approximation determined at this step is

$$r \approx m \pm \tfrac{1}{2}(b - a)$$

Step 3. To improve this approximation, find $P(m)$ and consider the following three cases:
   (A) If $P(m)$ and $P(a)$ have opposite sign, let $b = m$ and repeat steps 1 and 2.
   (B) If $P(m)$ and $P(b)$ have opposite sign, let $a = m$ and repeat steps 1 and 2.
   (C) If $P(m) = 0$, stop, because $r = m$.

Step 4. To approximate the zero to d decimal places, we continue this process until the maximum possible error is less than $10^{-(d+1)}$ and then round the approximate value to d decimal places.

4-7 PARTIAL FRACTIONS

A rational function $P(x)/D(x)$ often can be decomposed into a sum of simpler rational functions called **partial fractions**. If the degree of $P(x)$ is less than the degree of $D(x)$, then $P(x)/D(x)$ is called a **proper fraction**. We have the following important theorems:

Equal Polynomials. Two polynomials are equal to each other if and only if the coefficients of terms of like degree are equal.

Linear and Quadratic Factor Theorem. A polynomial with real coefficients can be factored into a product of linear and/or quadratic factors with real coefficients where the linear and quadratic factors are prime relative to the real numbers.

Partial Fraction Decomposition. Any proper fraction $P(x)/D(x)$ reduced to lowest terms can be decomposed into the sum of partial fractions as follows:

1. If $D(x)$ has a nonrepeating linear factor of the form $ax + b$, then the partial fraction decomposition of $P(x)/D(x)$ contains a term of the form

$$\frac{A}{ax + b} \qquad A \text{ a constant}$$

2. If $D(x)$ has a k-repeating linear factor of the form $(ax + b)^k$, then the partial fraction decomposition of $P(x)/D(x)$ contains k terms of the form

$$\frac{A_1}{ax + b} + \frac{A_2}{(ax + b)^2} + \cdots + \frac{A_k}{(ax + b)^k} \qquad A_1, A_2, \ldots, A_k \text{ constants}$$

3. If $D(x)$ has a nonrepeating quadratic factor of the form $ax^2 + bx + c$, which is prime relative to the real numbers, then the partial fraction decomposition of $P(x)/D(x)$ contains a term of the form

$$\frac{Ax + B}{ax^2 + bx + c} \qquad A, B \text{ constants}$$

4. If $D(x)$ has a k-repeating quadratic factor of the form $(ax^2 + bx + c)^k$, where $ax^2 + bx + c$ is prime relative to the real numbers, then the partial fraction decomposition of $P(x)/D(x)$ contains k terms of the form

$$\frac{A_1x + B_1}{ax^2 + bx + c} + \frac{A_2x + B_2}{(ax^2 + bx + c)^2} + \cdots + \frac{A_kx + B_k}{(ax^2 + bx + c)^k}$$

$$A_1, \ldots, A_k, \quad B_1, \ldots, B_k \text{ constants}$$

Chapter 4 Review Exercise

Work through all the problems in this chapter review, and check answers in the back of the book. Answers to all review problems are there, and following each answer is a number in italics indicating the section in which that type of problem is discussed. Where weaknesses show up, review appropriate sections in the text.

A

1. Use synthetic division to divide $P(x) = 2x^3 + 3x^2 - 1$ by $D(x) = x + 2$, and write the answer in the form $P(x) = D(x)Q(x) + R$.

2. If $P(x) = x^5 - 4x^4 + 9x^2 - 8$, find $P(3)$ using the remainder theorem and synthetic division.

3. What are the zeros of $P(x) = 3(x - 2)(x + 4)(x + 1)$?

$3(x^3 + 3x^2 - 6x - 8)$ $(x - 2)(x^2 + 5x + 4)$
$x^3 + 5x^2 + 4x - 2x^2 - 10x - 8$

4. If $P(x) = x^2 - 2x + 2$ and $P(1 + i) = 0$, find another zero of $P(x)$.

5. Make a table of possible combinations of positive, negative, and imaginary zeros for the following polynomials using Descartes' rule of signs and other appropriate theorems:
   (A) $P(x) = x^3 - x^2 - x + 3$
   (B) $P(x) = x^5 + x^3 + 4$

6. According to the upper and lower bound theorem, which of the following are upper or lower bounds of the zeros of $P(x) = x^3 - 4x^2 + 2$?

$$-2, -1, 3, 4$$

7. How do you know that $P(x) = 2x^3 - 3x^2 + x - 5$ has at least one real zero between 1 and 2?

8. Write the possible rational zeros for $P(x) = x^3 - 4x^2 + x + 6$.

9. Find all rational zeros for $P(x) = x^3 - 4x^2 + x + 6$.

10. Show that $P(x) = x^4 - x^2 - 2$ has a real zero between 1 and 2, and approximate this zero to one decimal place.

11. Decompose into partial fractions: $\dfrac{7x - 11}{(x - 3)(x + 2)}$

B

12. If $P(x) = 8x^4 - 14x^3 - 13x^2 - 4x + 7$, find $Q(x)$ and R such that $P(x) = (x - \frac{1}{4})Q(x) + R$. What is $P(\frac{1}{4})$?

13. If $P(x) = 4x^3 - 8x^2 - 3x - 3$, find $P(-\frac{1}{2})$ using the remainder theorem and synthetic division.

14. Use the quadratic formula and the factor theorem to factor $P(x) = x^2 - 2x - 1$.

15. Is $x + 1$ a factor of $P(x) = 9x^{26} - 11x^{17} + 8x^{11} - 5x^4 - 7$? Explain, without dividing or using synthetic division.

16. For $P(x) = 2x^4 - 3x^3 - 14x^2 + 2x + 4$:
  (A) Construct a table showing the possible combinations of positive, negative, and imaginary zeros using Descartes' rule of signs and earlier theorems.
  (B) Find the smallest and largest integers that are upper and lower bounds, respectively, of the zeros of $P(x)$ according to Theorem 2 in Section 4-4 (the upper and lower bound theorem).
  (C) Discuss the location of real zeros within the lower and upper bound interval by applying Theorem 3 in Section 4-4 (the location theorem) to integer values of x within this interval.

17. Determine all rational zeros of $P(x) = 2x^3 - 3x^2 - 18x - 8$.

18. Factor the polynomial in Problem 17 into linear factors.

19. Find all rational zeros of $P(x) = x^3 - 3x^2 + 5$.

20. Find all zeros (rational, irrational, and imaginary) for $P(x) = 2x^4 - x^3 + 2x - 1$.

21. Factor the polynomial in Problem 20 into linear factors.

22. Solve $2x^3 + 3x^2 \le 11x + 6$. Write the answer in inequality and interval notation.

23. Find all real solutions of $x^3 + 3x - 6 = 0$ to one decimal place.

24. Decompose into partial fractions: $\dfrac{-x^2 + 3x + 4}{x(x - 2)^2}$

25. Decompose into partial fractions: $\dfrac{8x^2 - 10x + 9}{2x^3 - 3x^2 + 3x}$

C

26. Use synthetic division to divide $P(x) = x^3 + 3x + 2$ by $[x - (1 + i)]$, and write the answer in the form $P(x) = D(x)Q(x) + R$.

27. Find a polynomial of lowest degree with leading coefficient 1 that has zeros $-\frac{1}{2}$ (multiplicity 2), -3, and 1 (multiplicity 3). (Leave the answer in factored form.) What is the degree of the polynomial?

28. Repeat Problem 27 for a polynomial $P(x)$ with zeros -5, $2 - 3i$, and $2 + 3i$.

29. Find all zeros (rational, irrational, and imaginary) exactly for $P(x) = 2x^5 - 5x^4 - 8x^3 + 21x^2 - 4$.

30. Factor the polynomial in Problem 29 into linear factors.

31. Solve

$$\frac{4x^2 + 4x - 3}{2x^3 + 3x^2 - 11x - 6} \ge 0$$

Write the answer in both inequality and interval notation.

32. Find rational roots exactly and irrational roots to two decimal places for $P(x) = x^5 - 9x^3 + 4x + 12$.

33. Decompose into partial fractions:

$$\frac{5x^2 + 2x + 9}{x^4 - 3x^3 + x^2 - 3x}$$

APPLICATIONS   *Find rational solutions exactly and irrational solutions to one decimal place.*

34. Architecture. An entryway is formed by placing a rectangular door inside an arch in the shape of the parabola with graph $y = 16 - x^2$, x and y in feet (see the figure). If the area of the door is 48 square feet, find the dimensions of the door.

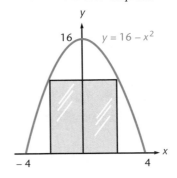

35. Construction. A grain silo is formed by attaching a hemisphere to the top of a right circular cylinder (see the figure). If the cylinder is 18 feet high and the volume of the silo is 486π cubic feet, find the common radius of the cylinder and the hemisphere.

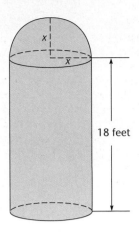

Exponential and Logarithmic Functions

5

$$y = \frac{e^{mx} + e^{-mx}}{2m}$$

Most of the functions we have considered so far have been polynomial and rational functions, with a few others involving roots or powers of polynomial or rational functions. The general class of functions defined by means of the algebraic operations of addition, subtraction, multiplication, division, and the taking of powers and roots on variables and constants are called *algebraic functions*.

In this chapter we define and investigate the properties of two new and important types of functions called *exponential functions* and *logarithmic functions*. These functions are not algebraic, but are members of another class of functions called *transcendental functions*. The exponential functions and logarithmic functions are used in describing and solving a wide variety of real-world problems, including growth of populations of people, animals, and bacteria; radioactive decay; growth of money at compound interest; absorption of light as it passes through air, water, or glass; and magnitudes of sounds and earthquakes. We consider applications in these areas plus many more in the sections that follow.

SECTION 5-1 Exponential Functions

- Exponential Functions
- Basic Exponential Graphs
- Additional Exponential Properties
- Applications

In this section we define exponential functions, look at some of their important properties—including graphs—and consider several significant applications.

● **Exponential Functions**

Let's start by noting that the functions f and g given by

$$f(x) = 2^x \qquad \text{and} \qquad g(x) = x^2$$

are not the same function. Whether a variable appears as an exponent with a constant base or as a base with a constant exponent makes a big difference. The function g is a quadratic function, which we have already discussed. The function f is a new type of function called an *exponential function*.

DEFINITION 1   **Exponential Function**

The equation

$$f(x) = b^x \qquad b > 0, b \neq 1$$

defines an **exponential function** for each different constant b, called the **base**. The independent variable x may assume any real value.

Thus, the **domain of f** is the set of all real numbers, and it can be shown that the **range of f** is the set of all positive real numbers. We require the base b to be positive to avoid imaginary numbers such as $(-2)^{1/2}$.

● **Basic Exponential Graphs**

Many students, if asked to graph an exponential function such as $g(x) = 2^x$, would not hesitate at all. They would likely make up a table by assigning integers to x, plot the resulting points, and then join these points with a smooth curve, as shown in Figure 1.

FIGURE 1 $g(x) = 2^x$.

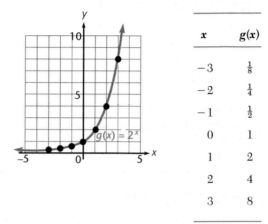

x	$g(x)$
-3	$\frac{1}{8}$
-2	$\frac{1}{4}$
-1	$\frac{1}{2}$
0	1
1	2
2	4
3	8

The only catch is that we have not defined 2^x for all real numbers x. We know what 2^5, 2^{-3}, $2^{2/3}$, $2^{-3/5}$, $2^{1.4}$, and $2^{-3.15}$ mean because we have defined 2^p for any rational number p, but what does

$$2^{\sqrt{2}}$$

mean? The question is not easy to answer at this time. In fact, a precise definition of $2^{\sqrt{2}}$ must wait for more advanced courses, where we can show that

$$b^x$$

names a real number for b a positive real number and x any real number and that the graph of $g(x) = 2^x$ is as indicated in Figure 1. We also can show that for x irrational, b^x can be approximated as closely as we like by using rational number approximations for x. Since $\sqrt{2} = 1.414213\ldots$, for example, the sequence

$$2^{1.4}, \ 2^{1.41}, \ 2^{1.414}, \ldots$$

approximates $2^{\sqrt{2}}$, and as we use more decimal places, the approximation improves.

It is useful to compare the graphs of $y = 2^x$ and $y = (\frac{1}{2})^x = 2^{-x}$ by plotting both on the same coordinate system, as shown in Figure 2(a) on the next page. The graph of

$$f(x) = b^x \qquad b > 1 \qquad \text{[Figure 2(b)]}$$

looks very much like the graph of $y = 2^x$, and the graph of

$$f(x) = b^x \qquad 0 < b < 1 \qquad \text{[Figure 2(b)]}$$

FIGURE 2 Basic exponential
graphs.

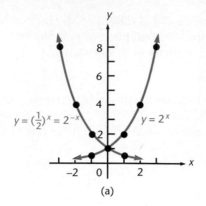

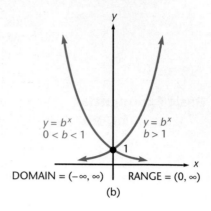

(a)

(b)

looks very much like the graph of $y = (\frac{1}{2})^x$. Note in both cases that the x axis is a *horizontal asymptote* for the graph.

The graphs in Figure 2 suggest the following important general properties of exponential functions, which we state without proof:

Basic Properties of the Graph of $f(x) = b^x$, $b > 0$, $b \neq 1$

1. All graphs pass through the point $(0, 1)$. $b^0 = 1$ for any permissible base b.

2. All graphs are continuous, with no holes or jumps.

3. The x axis is a horizontal asymptote.

4. If $b > 1$, then b^x increases as x increases.

5. If $0 < b < 1$, then b^x decreases as x increases.

6. The function f is one-to-one.

Property 6 implies that an exponential function has an inverse, called a *logarithmic function*, which we will discuss in Section 5-3.

The use of the $\boxed{y^x}$ key on a scientific calculator makes the creation of an accurate table of values to aid in graphing exponential functions almost routine. Example 1 illustrates the process.

EXAMPLE 1   **Graphing Exponential Functions**

Use integer values of x from -3 to 3 to construct a table of values for $y = \frac{1}{2}(4^x)$, and then graph this function.

Solution   Use a scientific calculator to create the table of values shown below. Then plot the points, and join these points with a smooth curve.

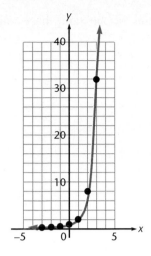

x	y
-3	0.01
-2	0.03
-1	0.13
0	0.50
1	2.00
2	8.00
3	32.00

A: $\boxed{4}\ \boxed{y^x}\ \boxed{3}\ \boxed{+/-}\ \boxed{=}\ \boxed{\div}\ \boxed{2}\ \boxed{=}$

P: $\boxed{4}\ \boxed{\text{ENTER}}\ \boxed{3}\ \boxed{+/-}\ \boxed{y^x}\ \boxed{2}\ \boxed{\div}$

Problem 1 Repeat Example 1 for $y\ \boxed{= \frac{1}{2}\left(\frac{1}{4}\right)^x}\ = \frac{1}{2}(4^{-x})$.

● **Additional Exponential Properties**

Exponential functions, which include irrational exponents, obey the familiar laws of exponents we discussed earlier for rational exponents. We summarize these exponent laws here and add two other important and useful properties.

Exponential Function Properties

For a and b positive, $a \neq 1$, $b \neq 1$, and x and y real:

1. Exponent laws:

$$a^x a^y = a^{x+y} \qquad (a^x)^y = a^{xy} \qquad (ab)^x = a^x b^x$$

$$\left(\frac{a}{b}\right)^x = \frac{a^x}{b^x} \qquad \frac{a^x}{a^y} = a^{x-y} \qquad \frac{2^{5x}}{2^{7x}}\ \boxed{= 2^{5x-7x}}\ = 2^{-2x}$$

2. $a^x = a^y$ if and only if $x = y$.   If $6^{4x} = 6^{2x+4}$, then $4x = 2x + 4$, and $x = 2$.

3. For $x \neq 0$, then $a^x = b^x$ if and only if $a = b$.   If $a^4 = 3^4$, then $a = 3$.

EXAMPLE 2   **Using Exponential Function Properties**

Solve $4^{x-3} = 8$ for x.

Solution   Express both sides in terms of the same base, and use property 2 to equate exponents.

$$4^{x-3} = 8$$

$$(2^2)^{x-3} = 2^3 \quad \text{Express 4 and 8 as powers of 2.}$$

$$2^{2x-6} = 2^3 \quad (a^x)^y = a^{xy}$$

$$2x - 6 = 3 \quad \text{Property 2}$$

$$2x = 9$$

$$x = \tfrac{9}{2}$$

Check

$$4^{(9/2)-3} = 4^{3/2} = (\sqrt{4})^3 = 2^3 \overset{\checkmark}{=} 8$$

Problem 2   Solve $27^{x+1} = 9$ for x.

• **Applications**   We now consider three applications that utilize exponential functions in their analysis: population growth, radioactive decay, and compound interest. Population growth and compound interest are examples of exponential growth, while radioactive decay is an example of negative exponential growth.

   Our first example involves the growth of populations, such as people, animals, insects, and bacteria. Populations tend to grow exponentially and at different rates. A convenient and easily understood measure of growth rate is the **doubling time**—that is, the time it takes for a population to double. Over short periods of time the **doubling time growth model** is often used to model population growth:

$$P = P_0 2^{t/d}$$

where $P = $ Population at time t
$\quad P_0 = $ Population at time $t = 0$
$\quad\quad d = $ Doubling time

Note that when $t = d$,

$$P = P_0 2^{d/d} = P_0 2$$

and the population is double the original, as it should be. We use this model to solve a population growth problem in Example 3.

EXAMPLE 3   **Population Growth**

Mexico has a population of around 100 million people, and it is estimated that the population will double in 21 years. If population growth continues at the same rate, what will be the population:

(A) 15 years from now?   (B) 30 years from now?

Calculate answers to 3 significant digits.

Solutions   We use the doubling time growth model:

$$P = P_0 2^{t/d}$$

Substituting $P_0 = 100$ and $d = 21$, we obtain

$$P = 100(2^{t/21})$$ See graph in margin.

P (millions)

Years

(A) Find P when $t = 15$ years:

$$P = 100(2^{15/21})$$
$$\approx 164 \text{ million people}$$ Use a scientific calculator.

(B) Find P when $t = 30$ years:

$$P = 100(2^{30/21})$$
$$\approx 269 \text{ million people}$$ Use a scientific calculator.

Problem 3   The bacterium *Escherichia coli* (*E. coli*) is found naturally in the intestines of many mammals. In a particular laboratory experiment, the doubling time for *E. coli* is found to be 25 minutes. If the experiment starts with a population of 1,000 *E. coli* and there is no change in the doubling time, how many bacteria will be present:

(A) In 10 minutes?   (B) In 5 hours?

Write answers to 3 significant digits.

Our second application involves radioactive decay, which is often referred to as negative growth. Radioactive materials are used extensively in medical diagnosis and therapy, as power sources in satellites, and as power sources in many countries. If we start with an amount A_0 of a particular radioactive isotope, the amount declines exponentially in time. The rate of decay varies from isotope to isotope. A convenient and easily understood measure of the rate of decay is the **half-life** of the isotope—that is, the time it takes for half of a particular material to decay. In this section we use the following **half-life decay model**:

$$A = A_0\left(\tfrac{1}{2}\right)^{t/h}$$
$$= A_0 2^{-t/h}$$

where A = Amount at time t
A_0 = Amount at time $t = 0$
h = Half-life

Note that when $t = h$,

$$A = A_0 2^{-h/h} = A_0 2^{-1} = \frac{A_0}{2}$$

and the amount of isotope is half the original amount, as it should be.

EXAMPLE 4   Radioactive Decay

The radioactive isotope gallium-67 (^{67}Ga), used in the diagnosis of malignant tumors, has a biological half-life of 46.5 hours. If we start with 100 milligrams of the isotope, how many milligrams will be left after:

(A) 24 hours?   (B) 1 week?

Compute answers to 3 significant digits.

Solutions   We use the half-life decay model:

$$A = A_0(\tfrac{1}{2})^{t/h} = A_0 2^{-t/h}$$

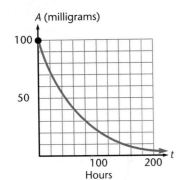

A (milligrams)

100

50

100   200
Hours

Using $A_0 = 100$ and $h = 46.5$, we obtain

$$A = 100(2^{-t/46.5})   \text{See the graph in the margin.}$$

(A) Find A when $t = 24$ hours:

$$A = 100(2^{-24/46.5})$$
$$= 69.9 \text{ milligrams}   \text{Use a scientific calculator.}$$

(B) Find A when $t = 168$ hours (1 week = 168 hours):

$$A = 100(2^{-168/46.5})$$
$$= 8.17 \text{ milligrams}   \text{Use a scientific calculator.}$$

Problem 4   Radioactive gold-198 (^{198}Au), used in imaging the structure of the liver, has a half-life of 2.67 days. If we start with 50 milligrams of the isotope, how many milligrams will be left after:

(A) $\frac{1}{2}$ day?   (B) 1 week?

Compute answers to 3 significant digits.

Our third application deals with the growth of money at compound interest. This topic is important to most people and is fundamental to many topics in the mathematics of finance.

The fee paid to use another's money is called **interest**. It is usually computed as a percent, called the **interest rate**, of the principal over a given period of time. If, at the end of a payment period, the interest due is reinvested at the same rate, then the interest earned as well as the principal will earn interest during the next payment period. Interest paid on interest reinvested is called **compound interest**.

Suppose you deposit $1,000 in a savings and loan that pays 8% compounded semiannually. How much will the savings and loan owe you at the end of 2 years?

Compounded semiannually means that interest is paid to your account at the end of each 6-month period, and the interest will in turn earn interest. The **interest rate per period** is the annual rate, $8\% = 0.08$, divided by the number of compounding periods per year, 2. If we let A_1, A_2, A_3, and A_4 represent the new amounts due at the end of the first, second, third, and fourth periods, respectively, then

$$A_1 = \$1{,}000 + \$1{,}000\left(\frac{0.08}{2}\right)$$

$$= \$1{,}000(1 + 0.04) \qquad\qquad P\left(1 + \frac{r}{n}\right)$$

$$A_2 = A_1(1 + 0.04)$$

$$= [\$1{,}000(1 + 0.04)](1 + 0.04)$$

$$= \$1{,}000(1 + 0.04)^2 \qquad\qquad P\left(1 + \frac{r}{n}\right)^2$$

$$A_3 = A_2(1 + 0.04)$$

$$= [\$1{,}000(1 + 0.04)^2](1 + 0.04)$$

$$= \$1{,}000(1 + 0.04)^3 \qquad\qquad P\left(1 + \frac{r}{n}\right)^3$$

$$A_4 = A_3(1 + 0.04)$$

$$= [\$1{,}000(1 + 0.04)^3](1 + 0.04)$$

$$= \$1{,}000(1 + 0.04)^4 \qquad\qquad P\left(1 + \frac{r}{n}\right)^4$$

What do you think the savings and loan will owe you at the end of 6 years? If you guessed

$$A = \$1{,}000(1 + 0.04)^{12}$$

you have observed a pattern that is generalized in the following compound interest formula:

Compound Interest

If a **principal** P is invested at an annual **rate** r compounded n times a year, then the **amount** A in the account at the end of t years is given by

$$A = P\left(1 + \frac{r}{n}\right)^{nt}$$

The annual rate r is expressed as a decimal.

Since the principal P represents the initial amount in the account and A represents the amount in the account t years later, we also call P the **present value** of the account and A the **future value** of the account.

___

EXAMPLE 5 Compound Interest

If you deposit \$5,000 in an account paying 9% compounded daily, how much will you have in the account in 5 years? Compute the answer to the nearest cent.

Solution   We use the compound interest formula as follows:

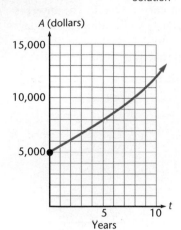

A (dollars)

$$A = P\left(1 + \frac{r}{n}\right)^{nt}$$

$$= 5{,}000\left(1 + \frac{0.09}{365}\right)^{(365)(5)}$$

$$= \$7{,}841.12 \qquad\qquad \text{Use a scientific calculator.}$$

The graph of

$$A = 5{,}000\left(1 + \frac{0.09}{365}\right)^{365t}$$

is shown in the margin.

Problem 5   If \$1,000 is invested in an account paying 10% compounded monthly, how much will be in the account at the end of 10 years? Compute the answer to the nearest cent.

Answers to Matched Problems

1. $y = \frac{1}{2}(4^{-x})$

x	y
-3	32.00
-2	8.00
-1	2.00
0	0.50
1	0.13
2	0.03
3	0.01

2. $x = -\frac{1}{3}$
3. (A) 1,320   (B) $4{,}100{,}000 = 4.10 \times 10^6$
4. (A) 43.9 mg   (B) 8.12 mg
5. \$2,707.04

EXERCISE 5-1

A

In Problems 1–10, construct a table of values for integer values of x over the indicated interval and then graph the function.

1. $y = 3^x;\ [-3, 3]$

2. $y = 5^x;\ [-2, 2]$

3. $y = \left(\frac{1}{3}\right)^x = 3^{-x};\ [-3, 3]$

4. $y = \left(\frac{1}{5}\right)^x = 5^{-x};\ [-2, 2]$

5. $g(x) = -3^{-x};\ [-3, 3]$

6. $f(x) = -5^x;\ [-2, 2]$

7. $h(x) = 5(3^x);\ [-3, 3]$

8. $f(x) = 4(5^x);\ [-2, 2]$

9. $y = 3^{x+3} - 5;\ [-6, 0]$

10. $y = 5^{x+2} + 4;\ [-4, 0]$

Simplify Problems 11–16.

11. $10^{3x-1}10^{4-x}$ **12.** $(4^{3x})^{2y}$ **13.** $\dfrac{3^x}{3^{1-x}}$ **14.** $\dfrac{5^{x-3}}{5^{x-4}}$ **15.** $\left(\dfrac{4^x}{5^y}\right)^{3z}$ **16.** $(2^x3^y)^z$

B _____

In Problems 17–28, solve for x.

17. $5^{3x} = 5^{4x-2}$ **18.** $10^{2-3x} = 10^{5x-6}$ **19.** $7^{x^2} = 7^{2x+3}$ **20.** $4^{5x-x^2} = 4^{-6}$

21. $(1-x)^5 = (2x-1)^5$ **22.** $5^3 = (x+2)^3$ **23.** $2^x = 4^{x+1}$ **24.** $9^{x-1} = 3^x$

25. $25^{x+1} = 125^{2x}$ **26.** $100^{x-1} = 1{,}000^{2x}$ **27.** $9^{x^2} = 3^{3x-1}$ **28.** $4^{x^2} = 2^{x+3}$

Graph each function in Problems 29–38.

29. $G(t) = 3^{t/100}$ **30.** $f(t) = 2^{t/10}$ **31.** $y = 11(3^{-x/2})$ **32.** $y = 7(2^{-2x})$

33. $g(x) = 2^{-|x|}$ **34.** $f(x) = 2^{|x|}$ **35.** $y = 1{,}000(1.08)^x$ **36.** $y = 100(1.03)^x$

37. $y = 2^{-x^2}$ **38.** $y = 3^{-x^2}$

Calculate Problems 39–44 using a scientific calculator. Compute answers to 4 significant digits.

39. $5^{\sqrt{3}}$ **40.** $3^{-\sqrt{2}}$ **41.** $\pi^{\sqrt{2}}$ **42.** $\pi^{-\sqrt{3}}$ **43.** $\dfrac{2^\pi + 2^{-\pi}}{2}$ **44.** $\dfrac{3^\pi - 3^{-\pi}}{2}$

C _____

Simplify Problems 45–48.

45. $(6^x + 6^{-x})(6^x - 6^{-x})$ **46.** $(3^x - 3^{-x})(3^x + 3^{-x})$

47. $(6^x + 6^{-x})^2 - (6^x - 6^{-x})^2$ **48.** $(3^x - 3^{-x})^2 + (3^x + 3^{-x})^2$

Graph each function in Problems 49–52.

49. $m(x) = x(3^{-x})$ **50.** $h(x) = x(2^x)$ **51.** $f(x) = \dfrac{2^x + 2^{-x}}{2}$ **52.** $g(x) = \dfrac{3^x + 3^{-x}}{2}$

_____

APPLICATIONS

53. Gaming. A person bets on red and black on a roulette wheel using a *Martingale strategy*. That is, a \$2 bet is placed on red, and the bet is doubled each time until a win occurs. The process is then repeated. If black occurs n times in a row, then $L = 2^n$ dollars is lost on the nth bet. Graph this function for $1 \le n \le 10$. Even though the function is defined only for positive integers, points on this type of graph are usually joined with a smooth curve as a visual aid.

54. Bacterial Growth. If bacteria in a certain culture double every $\frac{1}{2}$ hour, write an equation that gives the number of bacteria N in the culture after t hours, assuming the culture has 100 bacteria at the start. Graph the equation for $0 \le t \le 5$.

55. Population Growth. Because of its short life span and frequent breeding, the fruit fly *Drosophila* is used in some genetic studies. Raymond Pearl of Johns Hopkins University, for example, studied 300 successive generations of descendants of a single pair of *Drosophila* flies. In a laboratory situation with ample food supply and space, the doubling time for a particular population is 2.4 days. If we start with 5 male and 5 female flies, how many flies should we expect to have in:
(A) 1 week?   (B) 2 weeks?

56. Population Growth. If Kenya has a population of about 30,000,000 people and a doubling time of 19 years and if the growth continues at the same rate, find the population in:
(A) 10 years   (B) 30 years
Compute answers to 2 significant digits.

57. Insecticides. The use of the insecticide DDT is no longer allowed in many countries because of its long-term adverse effects. If a farmer uses 25 pounds of active DDT, assuming its half-life is 12 years, how much will still be active after:
(A) 5 years?   (B) 20 years?
Compute answers to 2 significant digits.

58. Radioactive Tracers. The radioactive isotope technetium-99m (^{99m}Tc) is used in imaging the brain. The isotope has a half-life of 6 hours. If 12 milligrams are used, how much will be present after:
(A) 3 hours?   (B) 24 hours?
Compute answers to 3 significant digits.

59. Finance. Suppose $4,000 is invested at 11% compounded weekly. How much money will be in the account in:
(A) $\frac{1}{2}$ year?   (B) 10 years?
Compute answers to the nearest cent.

60. Finance. Suppose $2,500 is invested at 7% compounded quarterly. How much money will be in the account in:
(A) $\frac{3}{4}$ year?   (B) 15 years?
Compute answers to the nearest cent.

★ **61. Finance.** A couple just had a new child. How much should they invest now at 8.25% compounded daily in order to have $40,000 for the child's education 17 years from now? Compute the answer to the nearest dollar.

★ **62. Finance.** A person wishes to have $15,000 cash for a new car 5 years from now. How much should be placed in an account now if the account pays 9.75% compounded weekly? Compute the answer to the nearest dollar.

SECTION 5-2   The Exponential Function with Base e

- Base e Exponential Function
- Growth and Decay Applications Revisited
- Continuous Compound Interest
- A Comparison of Exponential Growth Phenomena

Until now the number π has probably been the most important irrational number you have encountered. In this section we will introduce another irrational number, e, that is just as important in mathematics and its applications.

• Base e Exponential Function

The following expression is important to the study of calculus and, as we will see later in this section, also is closely related to the compound interest formula discussed in the preceding section:

$$\left(1 + \frac{1}{m}\right)^m$$

TABLE 1

m	$\left(1 + \dfrac{1}{m}\right)^m$
1	2
10	2.59374 . . .
100	2.70481 . . .
1,000	2.71692 . . .
10,000	2.71814 . . .
100,000	2.71827 . . .
1,000,000	2.71828 . . .
$\vdots$	$\vdots$

What happens to the value of the expression as m increases without bound? Think about this for a moment before proceeding. Maybe you guessed that the value approaches 1 using the following reasoning: As m gets large, $1 + (1/m)$ approaches 1, because $1/m$ approaches 0, and 1 raised to any power is 1. Let's see if this reasoning is correct by actually calculating the value of the expression for larger and larger values of m. Table 1 summarizes the results.

Interestingly, the value of $[1 + (1/m)]^m$ is never close to 1 but seems to be approaching a number close to 2.7183. In a calculus course we can show that as m increases without bound, the value of $[1 + (1/m)]^m$ approaches an irrational number that we call e. Just as irrational numbers such as π and $\sqrt{2}$ have nonending, nonrepeating decimal representations (see Section 1-1), *e* also has a nonending, nonrepeating decimal representation. To 12 decimal places,

$$e = 2.718\ 281\ 828\ 459$$

Exactly who discovered *e* is still being debated. It is named after the great Swiss mathematician Leonhard Euler (1707–1783), who computed *e* to 23 decimal places using $[1 + (1/m)]^m$.

The constant *e* turns out to be an ideal base for an exponential function because in calculus and higher mathematics many operations take on their simplest form using this base. This is why you will see *e* used extensively in expressions and formulas that model real-world phenomena.

DEFINITION 1   **Exponential Function with Base *e***

For *x* a real number, the equation

$$f(x) = e^x$$

defines the **exponential function with base *e*.**

The exponential function with base *e* is used so frequently that it is often referred to as *the* exponential function. Because of its importance, all scientific calculators have an $\boxed{e^x}$ key or its equivalent—consult your user's manual. The graphs of $y = e^x$ and $y = e^{-x}$ are shown in Figure 1.

FIGURE 1 Exponential functions with base *e*.

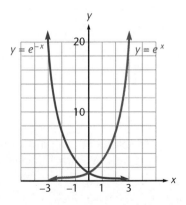

EXAMPLE 1   **Graphing Exponential Functions**

Graph $y = 4 - e^{x/2}$.

Solution   Use a scientific calculator to construct a table of values for integer values of *x*. Then plot the points and join these points with a smooth curve.

x	y
-4	3.86
-3	3.78
-2	3.63
-1	3.39
0	3
1	2.35
2	1.28
3	-0.48
4	-3.39

Notice that as x approaches $-\infty$, the values of $e^{x/2}$ approach 0 and the line $y = 4$ is a horizontal asymptote for the graph.

Problem 1 Graph $y = 2e^{x/2} - 5$.

● Growth and Decay Applications Revisited

Most exponential growth and decay problems are modeled using base e exponential functions. We present two applications here and many more in Exercise 5-2.

EXAMPLE 2 Medicine—Bacteria Growth

Cholera, an intestinal disease, is caused by a cholera bacterium that multiplies exponentially by cell division as given approximately by

$$N = N_0 e^{1.386t}$$

where N is the number of bacteria present after t hours and N_0 is the number of bacteria present at $t = 0$. If we start with 1 bacterium, how many bacteria will be present in:

(A) 5 hours?   (B) 12 hours?

Compute the answers to 3 significant digits.

Solutions   (A) Use $N_0 = 1$ and $t = 5$:

$$N = N_0 e^{1.386t}$$
$$= e^{1.386(5)}$$
$$= 1{,}020$$

A: $\boxed{1.386}\ \boxed{\times}\ \boxed{5}\ \boxed{=}\ \boxed{e^x}$
P: $\boxed{1.386}\ \boxed{\text{ENTER}}\ \boxed{5}\ \boxed{\times}\ \boxed{e^x}$

(B) Use $N_0 = 1$ and $t = 12$:

$$N = N_0 e^{1.386t}$$
$$= e^{1.386(12)}$$
$$= 16{,}700{,}000$$

Problem 2   Graph the exponential growth model for cholera bacteria over the indicated interval:

$$N = e^{1.386t} \qquad 0 \le t \le 5$$

EXAMPLE 3   **Carbon-14 Dating**

Cosmic-ray bombardment of the atmosphere produces neutrons, which in turn react with nitrogen to produce radioactive carbon-14. Radioactive carbon-14 enters all living tissues through carbon dioxide, which is first absorbed by plants. As long as a plant or animal is alive, carbon-14 is maintained in the living organism at a constant level. Once the organism dies, however, carbon-14 decays according to the equation

$$A = A_0 e^{-0.000124t}$$

where A is the amount present after t years and A_0 is the amount present at time $t = 0$. If 1,000 milligrams of carbon-14 are present at the start, how many milligrams will be present in:

(A) 10,000 years?   (B) 50,000 years?

Compute answers to 3 significant digits.

Solutions   Substituting $A_0 = 1{,}000$ in the decay equation, we have

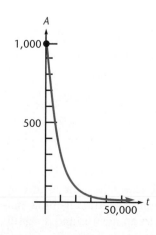

$$A = 1{,}000 e^{-0.000124t} \quad \text{See the graph in the margin.}$$

(A) Solve for A when $t = 10{,}000$:

$$A = 1{,}000 e^{-0.000124(10{,}000)}$$
$$= 289 \text{ milligrams} \quad \text{Use a scientific calculator.}$$

(B) Solve for A when $t = 50{,}000$:

$$A = 1{,}000 e^{-0.000124(50{,}000)}$$
$$= 2.03 \text{ milligrams} \quad \text{Use a scientific calculator.}$$

More will be said about carbon-14 dating in Exercise 5-5, where we will be interested in solving for t given information about A and A_0.

Problem 3 Referring to Example 3, how many milligrams of carbon-14 would have to be present at the beginning in order to have 10 milligrams present after 20,000 years? Compute the answers to 4 significant digits.

• Continuous Compound Interest The constant e occurs naturally in the study of compound interest. Returning to the compound interest formula discussed in Section 5-1,

$$A = P\left(1 + \frac{r}{n}\right)^{nt} \quad \text{Compound interest}$$

recall that P is the principal invested at an annual rate r compounded n times a year and A is the amount in the account after t years. Suppose P, r, and t are held fixed and n is increased without bound. Will the amount A increase without bound or will it tend to some limiting value? Let's perform a calculator experiment before we attack the general problem. If $P = \$100$, $r = 0.08$, and $t = 2$ years, then

$$A = 100\left(1 + \frac{0.08}{n}\right)^{2n}$$

The amount A is computed for several values of n in Table 2. Notice that the largest gain appears in going from annually to semiannually. Then, the gains slow down as n increases. In fact, it appears that A might be tending to something close to \$117.35 as n gets larger and larger.

TABLE 2 Effect of Compounding Frequency

Compounding frequency	n	$A = 100\left(1 + \dfrac{0.08}{n}\right)^{2n}$
Annually	1	\$116.6400
Semiannually	2	116.9859
Quarterly	4	117.1659
Weekly	52	117.3367
Daily	365	117.3490
Hourly	8,760	117.3501

We now return to the general problem to see if we can determine what happens to $A = P[1 + (r/n)]^{nt}$ as n increases without bound. A little algebraic manipulation of the compound interest formula will lead to an answer and a significant result in the mathematics of finance.

$$A = P\left(1 + \frac{r}{n}\right)^{nt}$$

$$= P\left(1 + \frac{1}{n/r}\right)^{(n/r)rt} \qquad \text{Change algebraically.}$$

$$= P\left[\left(1 + \frac{1}{m}\right)^{m}\right]^{rt} \qquad \text{Let } m = n/r.$$

The expression within the square brackets should look familiar. Recall from the first part of this section that

$$\left(1 + \frac{1}{m}\right)^{m} \quad \text{approaches } e \qquad \text{as} \qquad m \text{ approaches } \infty$$

Since r is fixed, $m = n/r$ approaches ∞ as n approaches ∞. Thus,

$$P\left(1 + \frac{r}{n}\right)^{nt} \quad \text{approaches } Pe^{rt} \qquad \text{as} \qquad n \text{ approaches } \infty$$

and we have arrived at the **continuous compound interest formula**, a very important and widely used formula in business, banking, and economics.

DEFINITION 2   **Continuous Compound Interest Formula**

If a principal P is invested at an annual rate r compounded continuously, then the amount A in the account at the end of t years is given by

$$A = Pe^{rt}$$

The annual rate r is expressed as a decimal.

EXAMPLE 4   Continuous Compound Interest

If \$100 is invested at an annual rate of 8% compounded continuously, what amount, to the nearest cent, will be in the account after 2 years?

Solution   Use the continuous compound interest formula to find A when $P = \$100$, $r = 0.08$, and $t = 2$:

$$A = Pe^{rt}$$
$$= \$100e^{(0.08)(2)} \qquad \text{8\% is equivalent to } r = 0.08.$$
$$= \$117.35$$

Compare this result with the values calculated in Table 2.

Problem 4 What amount will an account have after 5 years if \$100 is invested at an annual rate of 12% compounded annually? Quarterly? Continuously? Compute answers to the nearest cent.

● A Comparison of Exponential Growth Phenomena

The equations and graphs given in Table 3 compare several widely used growth models. These are divided basically into two groups: unlimited growth and limited growth. Following each equation and graph is a short, incomplete list of areas in which the models are used. We have only touched on a subject that has been extensively developed and which you are likely to study in greater depth in the future.

TABLE 3   Exponential Growth and Decay

Description	Equation	Graph	Uses
Unlimited growth	$y = ce^{kt}$ $c, k > 0$		Short-term population growth (people, bacteria, etc.); growth of money at continuous compound interest
Exponential decay	$y = ce^{-kt}$ $c, k > 0$		Radioactive decay; light absorption in water, glass, etc.; atmospheric pressure; electric circuits
Limited growth	$y = c(1 - e^{-kt})$ $c, k > 0$		Learning skills; sales fads; company growth; electric circuits
Logistic growth	$y = \dfrac{M}{1 + ce^{-kt}}$ $c, k, M > 0$		Long-term population growth; epidemics; sales of new products; company growth

Answers to Matched Problems

1.

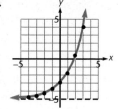

x	y
-4	-4.73
-3	-4.55
-2	-4.26
-1	-3.79
0	-3
1	-1.7
2	0.44
3	3.96

2.

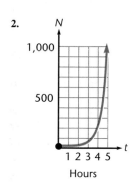

Hours

3. 119.4 mg

4. Annually: \$176.23; quarterly: \$180.61; continuously: \$182.21

EXERCISE 5-2

A

In Problems 1–6, construct a table of values for integer values of x over the indicated interval and then graph the function.

1. $y = -e^x;\ [-3, 3]$

2. $y = -e^{-x};\ [-3, 3]$

3. $y = 10e^{0.2x};\ [-5, 5]$

4. $y = 100e^{0.1x};\ [-5, 5]$

5. $f(t) = 100e^{-0.1t};\ [-5, 5]$

6. $g(t) = 10e^{-0.2t};\ [-5, 5]$

Simplify Problems 7–12.

7. $e^{2x}e^{-3x}$

8. $(e^{-x})^4$

9. $(e^x)^3$

10. $e^{-4x}e^{6x}$

11. $\dfrac{e^{5x}}{e^{2x+1}}$

12. $\dfrac{e^{4-3x}}{e^{2-5x}}$

B

Graph Problems 13–20.

13. $y = 2 + e^{x-2}$

14. $y = -3 + e^{1+x}$

15. $y = e^{-|x|}$

16. $y = e^{|x|}$

17. $M(x) = e^{x/2} + e^{-x/2}$

18. $C(x) = \dfrac{e^x + e^{-x}}{2}$

19. $N = \dfrac{200}{1 + 3e^{-t}}$

20. $N = \dfrac{100}{1 + e^{-t}}$

∫ Simplify Problems 21–26.

21. $\dfrac{-2x^3e^{-2x} - 3x^2e^{-2x}}{x^6}$

22. $\dfrac{5x^4e^{5x} - 4x^3e^{5x}}{x^8}$

23. $(e^x + e^{-x})^2 + (e^x - e^{-x})^2$

24. $e^x(e^{-x} + 1) - e^{-x}(e^x + 1)$

25. $\dfrac{e^{-x}(e^x - e^{-x}) + e^{-x}(e^x + e^{-x})}{e^{-2x}}$

26. $\dfrac{e^x(e^x + e^{-x}) - (e^x - e^{-x})e^x}{e^{2x}}$

In Problems 27–30, solve each equation. [Remember: $e^{-x} \neq 0$.]

27. $2xe^{-x} = 0$

28. $(x - 3)e^x = 0$

29. $x^2e^x - 5xe^x = 0$

30. $3xe^{-x} + x^2e^{-x} = 0$

C _____

*One of the most important functions in statistics is the **normal probability density function***

$$f(x) = \frac{1}{\sigma\sqrt{\pi}}\, e^{-(x-\mu)^2/2\sigma^2}$$

*where μ is the **mean** and σ is the **standard deviation**. The graph of this function is the "bell-shaped" curve that instructors refer to when they say that they are grading on a curve.*

Graph the related functions given in Problems 31 and 32.

31. $f(x) = e^{-x^2}$

32. $g(x) = \dfrac{1}{\sqrt{\pi}}\, e^{-x^2/2}$

 33. Given $f(s) = (1 + s)^{1/s}$, $s \neq 0$:
   (A) Complete the tables at the right to four decimal places.
   (B) What does $(1 + s)^{1/s}$ seem to tend to as s approaches 0?

34. Refer to Problem 33. Graph $f(s) = (1 + s)^{1/s}$ for $[-0.5, 0) \cup (0, 0.5]$.

s	$f(s)$	s	$f(s)$
-0.5	4.0000	0.5	2.2500
-0.2	3.0518	0.2	2.4883
-0.1		0.1	
-0.01		0.01	
-0.001		0.001	
-0.0001		0.0001	

 Problems 35–38 require the use of a graphic calculator or a computer.

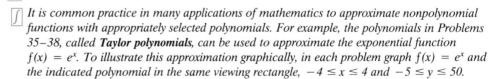 *It is common practice in many applications of mathematics to approximate nonpolynomial functions with appropriately selected polynomials. For example, the polynomials in Problems 35–38, called **Taylor polynomials**, can be used to approximate the exponential function $f(x) = e^x$. To illustrate this approximation graphically, in each problem graph $f(x) = e^x$ and the indicated polynomial in the same viewing rectangle, $-4 \leq x \leq 4$ and $-5 \leq y \leq 50$.*

35. $P_1(x) = 1 + x + \frac{1}{2}x^2$

36. $P_2(x) = 1 + x + \frac{1}{2}x^2 + \frac{1}{6}x^3$

37. $P_3(x) = 1 + x + \frac{1}{2}x^2 + \frac{1}{6}x^3 + \frac{1}{24}x^4$

38. $P_4(x) = 1 + x + \frac{1}{2}x^2 + \frac{1}{6}x^3 + \frac{1}{24}x^4 + \frac{1}{120}x^5$

_____

APPLICATIONS

39. Population Growth. If the world population is about 6 billion people now and if the population continues to grow at 1.7% compounded continuously, what will the population be in 10 years? Compute the answer to 2 significant digits.

40. Population Growth. If the population in Mexico is around 100 million people now and if the population continues to grow at 2.3% compounded continuously, what will the population be in 8 years? Compute the answer to 2 significant digits.

41. Space Science. Radioactive isotopes, as well as solar cells, are used to supply power to space vehicles. The isotopes gradually lose power because of radioactive decay. On a particular space vehicle the nuclear energy source has a power output of P watts after t days of use as given by

$$P = 75e^{-0.0035t}$$

Graph this function for $0 \leq t \leq 100$.

42. Earth Science. The atmospheric pressure P, in pounds per square inch, decreases exponentially with altitude h, in miles above sea level, as given by

$$P = 14.7e^{-0.21h}$$

Graph this function for $0 \leq h \leq 10$.

43. Marine Biology. Marine life is dependent upon the microscopic plant life that exists in the *photic zone*, a zone that goes to a depth where about 1% of the surface light still remains. Light intensity I relative to depth d, in feet, for one of the clearest bodies of water in the world, the Sargasso Sea in the West Indies, can be approximated by

$$I = I_0 e^{-0.00942d}$$

where I_0 is the intensity of light at the surface. What percentage of the surface light will reach a depth of:
(A) 50 feet?   (B) 100 feet?

44. Marine Biology. Refer to Problem 43. In some waters with a great deal of sediment, the photic zone may go down only 15 to 20 feet. In some murky harbors, the intensity of light d feet below the surface is given approximately by

$$I = I_0 e^{-0.23d}$$

What percentage of the surface light will reach a depth of:
(A) 10 feet?   (B) 20 feet?

45. Money Growth. If you invest $5,250 in an account paying 11.38% compounded continuously, how much money will be in the account at the end of:
(A) 6.25 years?   (B) 17 years?

46. Money Growth. If you invest $7,500 in an account paying 8.35% compounded continuously, how much money will be in the account at the end of:
(A) 5.5 years?   (B) 12 years?

47. Money Growth. In 1987, *Barron's*, a national business and financial weekly, published the following "Top Savings Deposit Yields" for $2\frac{1}{2}$-year certificate of deposit accounts:

Gill Saving	8.30% (CC)
Richardson Savings and Loan	8.40% (CQ)
USA Savings	8.25% (CD)

where CC represents compounded continuously, CQ compounded quarterly, and CD compounded daily. Compute the value of $1,000 invested in each account at the end of $2\frac{1}{2}$ years.

48. Money Growth. Refer to Problem 47. In another issue of *Barron's* in 1987, 1-year certificate of deposit accounts included:

| Alamo Savings | 8.25% (CQ) |
| Lamar Savings | 8.05% (CC) |

Compute the value of $10,000 invested in each account at the end of 1 year.

★49. Present Value. A promissory note will pay $30,000 at maturity 10 years from now. How much should you be willing to pay for the note now if money is worth 9% compounded continuously?

★50. Present Value. A promissory note will pay $50,000 at maturity $5\frac{1}{2}$ years from now. How much should you be willing to pay for the note now if money is worth 10% compounded continuously?

51. AIDS Epidemic. At the end of 1989 there were approximately 100,000 diagnosed cases of AIDS in the United States. As of this writing, the disease is estimated to be spreading among the general population at the rate of 9% compounded continuously. Assuming this rate doesn't change, how many cases of AIDS, to 2 significant digits, should be expected by the end of:
(A) 1996?   (B) 2000?

52. AIDS Epidemic. As of this writing, AIDS among intravenous drug users in the United States is spreading at the rate of about 21% compounded continuously. Assuming this rate doesn't change and there were 24,000 cases of AIDS among intravenous drug users at the end of 1989, how many cases should we expect, to 2 significant digits, by the end of:
(A) 1996?   (B) 2000?

★53. Learning Curve. People assigned to assemble circuit boards for a computer manufacturing company undergo on-the-job training. From past experience it was found that the learning curve for the average employee is given by

$$N = 40(1 - e^{-0.12t})$$

where N is the number of boards assembled per day after t days of training. Graph this function for $0 \leq t \leq 30$. What is the maximum number of boards an average employee can be expected to produce in 1 day?

★**54. Advertising.** A company is trying to expose a new product to as many people as possible through television advertising in a large metropolitan area with 2 million possible viewers. A model for the number of people N, in millions, who are aware of the product after t days of advertising was found to be

$$N = 2(1 - e^{-0.037t})$$

Graph this function for $0 \le t \le 50$. What value does N tend to as t increases without bound?

55. Newton's Law of Cooling. This law states that the rate at which an object cools is proportional to the difference in temperature between the object and its surrounding medium. The temperature T of the object t hours later is given by

$$T = T_m + (T_0 - T_m)e^{-kt}$$

where T_m is the temperature of the surrounding medium and T_0 is the temperature of the object at $t = 0$. Suppose a bottle of wine at a room temperature of 72°F is placed in the refrigerator to cool before a dinner party. If the temperature in the refrigerator is kept at 40°F and $k = 0.4$, find the temperature of the wine, to the nearest degree, after 3 hours. (In Exercise 5-5 we will find out how to determine k.)

56. Newton's Law of Cooling. Refer to Problem 55. What is the temperature, to the nearest degree, of the wine after 5 hours in the refrigerator?

★**57. Photography.** An electronic flash unit for a camera is activated when a capacitor is discharged through a filament of wire. After the flash is triggered, and the capacitor is discharged, the circuit (see the figure) is connected and the battery pack generates a current to recharge the capacitor. The time it takes for the capacitor to recharge is called the *recycle time*. For a particular flash unit using a 12-volt battery pack, the charge q, in coulombs, on the capacitor t seconds after recharging has started is given by

$$q = 0.0009(1 - e^{-0.2t})$$

Graph this function for $0 \le t \le 10$. Estimate the maximum charge on the capacitor.

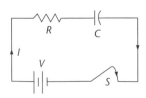

★**58. Medicine.** An electronic heart pacemaker utilizes the same type of circuit as the flash unit in Problem 57, but it is designed so that the capacitor discharges 72 times a minute. For a particular pacemaker, the charge on the capacitor t seconds after it starts recharging is given by

$$q = 0.000\ 008(1 - e^{-2t})$$

Graph this function for $0 \le t \le \frac{5}{6}$. Estimate the maximum charge on the capacitor.

★**59. Wildlife Management.** A herd of 20 white-tailed deer is introduced to a coastal island where there had been no deer before. Their population is predicted to increase according to the logistic curve

$$N = \frac{100}{1 + 4e^{-0.14t}}$$

where N is the number of deer expected in the herd after t years. Graph the equation for $0 \le t \le 30$. Estimate the herd size that the island can support.

★**60. Training.** A trainee is hired by a computer manufacturing company to learn to test a particular model of a personal computer after it comes off the assembly line. The learning curve for an average trainee is given by

$$N = \frac{200}{4 + 21e^{-0.1t}}$$

where N is the number of computers tested per day after t days on the job. Graph the equation for $0 \le t \le 50$. What is the maximum number of computers that an average tester can be expected to test per day after training?

★**61. Catenary.** A free-hanging power line between two supporting towers looks something like a parabola, but it is actually a curve called a **catenary**. A catenary has an equation of the form

$$y = \frac{e^{mx} + e^{-mx}}{2m}$$

where m is a constant. Graph the equation for $m = 0.25$.

★**62. Catenary.** Graph the equation in Problem 61 for $m = 0.4$.

SECTION 5-3 Logarithmic Functions

- Definition of Logarithmic Function
- From Logarithmic Form to Exponential Form, and Vice Versa
- Properties of Logarithmic Functions

• Definition of Logarithmic Function

We now define a new class of functions, called **logarithmic functions**, as inverses of exponential functions. Since exponential functions are one-to-one, their inverses exist. Here you will see why we placed special emphasis on the general concept of inverse functions in Section 3-8. If you know quite a bit about a function, then, based on a knowledge of inverses in general, you will automatically know quite a bit about its inverse. For example, the graph of f^{-1} is the graph of f reflected across the line $y = x$, and the domain and range of f^{-1} are, respectively, the range and domain of f.

If we start with the exponential function,

$$f: \quad y = 2^x$$

and interchange the variables x and y, we obtain the inverse of f:

$$f^{-1}: \quad x = 2^y$$

The graphs of f, f^{-1}, and the line $y = x$ are shown in Figure 1. This new function is given the name **logarithmic function with base 2**. Since we cannot solve the equation $x = 2^y$ for y using the algebra properties discussed so far, we introduce a new symbol to represent this inverse function:

$$y = \log_2 x \qquad \text{Read "log to the base 2 of } x.\text{"}$$

Thus,

$$y = \log_2 x \qquad \text{is equivalent to} \qquad x = 2^y$$

that is, $\log_2 x$ is the exponent to which 2 must be raised to obtain x. Symbolically, $x = 2^y = 2^{\log_2 x}$.

FIGURE 1 Logarithmic function with base 2.

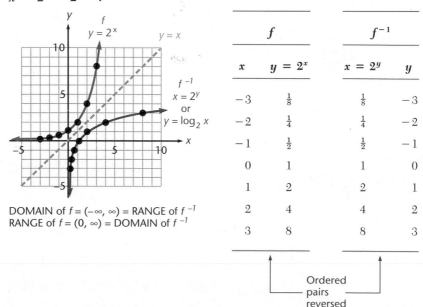

DOMAIN of $f = (-\infty, \infty)$ = RANGE of f^{-1}
RANGE of $f = (0, \infty)$ = DOMAIN of f^{-1}

\multicolumn{2}{c}{f}		\multicolumn{2}{c}{f^{-1}}
x	$y = 2^x$	$x = 2^y$	y
-3	$\frac{1}{8}$	$\frac{1}{8}$	-3
-2	$\frac{1}{4}$	$\frac{1}{4}$	-2
-1	$\frac{1}{2}$	$\frac{1}{2}$	-1
0	1	1	0
1	2	2	1
2	4	4	2
3	8	8	3

Ordered pairs reversed

In general, we define the **logarithmic function with base b** to be the inverse of the exponential function with base b ($b > 0$, $b \neq 1$).

DEFINITION 1

Definition of Logarithmic Function

For $b > 0$ and $b \neq 1$,

Logarithmic form		Exponential form
$y = \log_b x$	is equivalent to	$x = b^y$

The log to the base b of x is the exponent to which b must be raised to obtain x.

$$y = \log_{10} x \quad \text{is equivalent to} \quad x = 10^y$$
$$y = \log_e x \quad \text{is equivalent to} \quad x = e^y$$

Remember: A logarithm is an exponent.

It is very important to remember that $y = \log_b x$ and $x = b^y$ define the same function, and as such can be used interchangeably.

Since the domain of an exponential function includes all real numbers and its range is the set of positive real numbers, the **domain** of a logarithmic function is the set of all positive real numbers and its **range** is the set of all real numbers. Thus, $\log_{10} 3$ is defined, but $\log_{10} 0$ and $\log_{10}(-5)$ are not defined. That is, 3 is a logarithmic domain value, but 0 and -5 are not. Typical logarithmic curves are shown in Figure 2.

FIGURE 2 Typical logarithmic graphs.

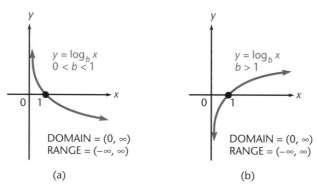

(a)

(b)

● From Logarithmic Form to Exponential Form, and Vice Versa

We now look into the matter of converting logarithmic forms to equivalent exponential forms, and vice versa.

EXAMPLE 1   **Logarithmic–Exponential Conversions**

Change each logarithmic form to an equivalent exponential form:

(A) $\log_2 8 = 3$ (B) $\log_{25} 5 = \frac{1}{2}$ (C) $\log_2 \left(\frac{1}{4}\right) = -2$

Solutions   (A) $\log_2 8 = 3$ is equivalent to $8 = 2^3$
                (B) $\log_{25} 5 = \frac{1}{2}$ is equivalent to $5 = 25^{1/2}$
                (C) $\log_2 \left(\frac{1}{4}\right) = -2$ is equivalent to $\frac{1}{4} = 2^{-2}$

Problem 1   Change each logarithmic form to an equivalent exponential form:

(A) $\log_3 27 = 3$ (B) $\log_{36} 6 = \frac{1}{2}$ (C) $\log_3 \left(\frac{1}{9}\right) = -2$

EXAMPLE 2   **Logarithmic–Exponential Conversions**

Change each exponential form to an equivalent logarithmic form:

(A) $49 = 7^2$ (B) $3 = \sqrt{9}$ (C) $\frac{1}{5} = 5^{-1}$

Solutions   (A) $49 = 7^2$ is equivalent to $\log_7 49 = 2$
                (B) $3 = \sqrt{9}$ is equivalent to $\log_9 3 = \frac{1}{2}$
                (C) $\frac{1}{5} = 5^{-1}$ is equivalent to $\log_5 \left(\frac{1}{5}\right) = -1$

Problem 2   Change each exponential form to an equivalent logarithmic form:

(A) $64 = 4^3$ (B) $2 = \sqrt[3]{8}$ (C) $\frac{1}{16} = 4^{-2}$

To gain a little deeper understanding of logarithmic functions and their relationship to the exponential functions, we consider a few problems where we want to find x, b, or y in $y = \log_b x$, given the other two values. All values were chosen so that the problems can be solved without tables or a calculator.

EXAMPLE 3   **Solutions of the Equation $y = \log_b x$**

Find x, b, or y as indicated:

(A) Find y: $y = \log_4 8$ (B) Find x: $\log_3 x = -2$
(C) Find b: $\log_b 1,000 = 3$

Solutions   (A) Write $y = \log_4 8$ in equivalent exponential form:

$$8 = 4^y$$

$$2^3 = 2^{2y} \quad \text{Write each number to the same base 2.}$$

$$2y = 3 \quad \text{Recall that } b^m = b^n \text{ if and only if } m = n.$$

$$y = \tfrac{3}{2}$$

Thus, $\frac{3}{2} = \log_4 8$.

(B) Write $\log_3 x = -2$ in equivalent exponential form:

$$x = 3^{-2}$$

$$= \frac{1}{3^2} = \frac{1}{9}$$

Thus, $\log_3 \left(\tfrac{1}{9}\right) = -2$.

(C) Write $\log_b 1,000 = 3$ in equivalent exponential form:

$$1,000 = b^3$$

$$10^3 = b^3 \quad \text{Write 1,000 as a third power.}$$

$$b = 10$$

Thus, $\log_{10} 1,000 = 3$.

Problem 3   Find x, b, or y as indicated.

(A) Find y: $y = \log_9 27$ (B) Find x: $\log_2 x = -3$
(C) Find b: $\log_b 100 = 2$

● **Properties of Logarithmic Functions**

Logarithmic functions have many powerful and useful properties. We list eight basic properties in Theorem 1.

Theorem 1

Properties of Logarithmic Functions

If b, M, and N are positive real numbers, $b \neq 1$, and p and x are real numbers, then:

1. $\log_b 1 = 0$ **5.** $\log_b MN = \log_b M + \log_b N$

2. $\log_b b = 1$ **6.** $\log_b \dfrac{M}{N} = \log_b M - \log_b N$

3. $\log_b b^x = x$ **7.** $\log_b M^p = p \log_b M$

4. $b^{\log_b x} = x, x > 0$ **8.** $\log_b M = \log_b N$ if and only if $M = N$

The first two properties in Theorem 1 follow directly from the definition of a logarithmic function:

$$\log_b 1 = 0 \quad \text{since} \quad b^0 = 1$$

$$\log_b b = 1 \quad \text{since} \quad b^1 = b$$

The third and fourth properties look more complicated than they are. They follow directly from the fact that exponential and logarithmic functions are inverses of each other. Recall from Section 3-8 that if f is one-to-one, then f^{-1} is a one-to-one function satisfying

$$f^{-1}[f(x)] = x \quad \text{and} \quad f[f^{-1}(x)] = x$$

Applying these general properties to $f(x) = b^x$ and $f^{-1}(x) = \log_b x$, we see that

$$f^{-1}[f(x)] = x \qquad f[f^{-1}(x)] = x$$
$$\log_b [f(x)] = x \qquad b^{f^{-1}(x)} = x$$
$$\log_b b^x = x \qquad b^{\log_b x} = x$$

Properties 5 to 7 enable us to convert multiplication into addition, division into subtraction, and power and root problems into multiplication. The proofs of these properties are based on properties of exponents. A sketch of a proof of the fifth property follows: To bring exponents into the proof, we let

$$u = \log_b M \quad \text{and} \quad v = \log_b N$$

and convert these to the equivalent exponential forms

$$M = b^u \quad \text{and} \quad N = b^v$$

Now, see if you can provide the reasons for each of the following steps:

$$\log_b MN = \log_b b^u b^v = \log_b b^{u+v} = u + v = \log_b M + \log_b N$$

The other properties are established in a similar manner (see Problems 109 and 110 in Exercise 5-3.)

Finally, the eighth property follows from the fact that logarithmic functions are one-to-one.

We now illustrate the use of these properties in several examples.

EXAMPLE 4   **Using Logarithmic Properties**

Simplify using the properties in Theorem 1:

(A) $\log_e 1$ (B) $\log_{10} 10$ (C) $\log_e e^{2x+1}$
(D) $\log_{10} 0.01$ (E) $10^{\log_{10} 7}$ (F) $e^{\log_e x^2}$

Solutions   (A) $\log_e 1 = 0$ (B) $\log_{10} 10 = 1$
(C) $\log_e e^{2x+1} = 2x + 1$ (D) $\log_{10} 0.01 = \log_{10} 10^{-2} = -2$
(E) $10^{\log_{10} 7} = 7$ (F) $e^{\log_e x^2} = x^2$

Problem 4   Simplify using the properties in Theorem 1:

(A) $\log_{10} 10^{-5}$ (B) $\log_5 25$ (C) $\log_{10} 1$
(D) $\log_e e^{m+n}$ (E) $10^{\log_{10} 4}$ (F) $e^{\log_e (x^4 + 1)}$

EXAMPLE 5   **Using Logarithmic Properties**

Write in terms of simpler logarithmic forms:

(A) $\log_b 3x$ (B) $\log_b \dfrac{x}{5}$ (C) $\log_b x^7$

(D) $\log_b \dfrac{mn}{pq}$ (E) $\log_b (mn)^{2/3}$ (F) $\log_b \dfrac{x^8}{y^{1/5}}$

Solutions   (A) $\log_b 3x = \log_b 3 + \log_b x$ $\log_b MN = \log_b M + \log_b N$

(B) $\log_b \dfrac{x}{5} = \log_b x - \log_b 5$ $\log_b \dfrac{M}{N} = \log_b M - \log_b N$

(C) $\log_b x^7 = 7 \log_b x$ $\log_b M^p = p \log_b M$

(D) $\log_b \dfrac{mn}{pq} = \log_b mn - \log_b pq$ $\log_b \dfrac{M}{N} = \log_b M - \log_b N$

$\qquad\qquad = \log_b m + \log_b n - (\log_b p + \log_b q)$ $\log_b MN = \log_b M + \log_b N$

$\qquad\qquad = \log_b m + \log_b n - \log_b p - \log_b q$

(E) $\log_b (mn)^{2/3} = \tfrac{2}{3} \log_b mn$ $\log_b M^p = p \log_b M$

$\qquad\qquad\quad = \tfrac{2}{3}(\log_b m + \log_b n)$ $\log_b MN = \log_b M + \log_b N$

(F) $\log_b \dfrac{x^8}{y^{1/5}} = \log_b x^8 - \log_b y^{1/5}$ $\log_b \dfrac{M}{N} = \log_b M - \log_b N$

$\qquad\qquad = 8 \log_b x - \tfrac{1}{5} \log_b y$ $\log_b M^p = p \log_b M$

Problem 5   Write in terms of simpler logarithmic forms, as in Example 5.

(A) $\log_b \dfrac{r}{uv}$ (B) $\log_b \left(\dfrac{m}{n}\right)^{3/5}$ (C) $\log_b \dfrac{u^{1/3}}{v^5}$

EXAMPLE 6   **Using Logarithmic Properties**

If $\log_e 3 = 1.10$ and $\log_e 7 = 1.95$, find:

(A) $\log_e \left(\tfrac{7}{3}\right)$ (B) $\log_e \sqrt[3]{21}$

Solutions   (A) $\log_e \left(\tfrac{7}{3}\right) = \log_e 7 - \log_e 3 = 1.95 - 1.10 = 0.85$

(B) $\log_e \sqrt[3]{21} = \log_e (21)^{1/3} = \tfrac{1}{3} \log_e (3 \cdot 7) = \tfrac{1}{3}(\log_e 3 + \log_e 7)$

$\qquad\qquad\qquad = \tfrac{1}{3}(1.10 + 1.95) = 1.02$

Problem 6   If $\log_e 5 = 1.609$ and $\log_e 8 = 2.079$, find:

(A) $\log_e \dfrac{5^{10}}{8}$ (B) $\log_e \sqrt[4]{\tfrac{8}{5}}$

The following example and problem, though somewhat artificial, will give you additional practice in using the properties in Theorem 1.

EXAMPLE 7   **Using Logarithmic Properties**

Find x so that $\log_b x = \frac{2}{3} \log_b 27 + 2 \log_b 2 - \log_b 3$ without using a calculator or table.

Solution   First we use properties from Theorem 1 to express the right side as the logarithm of a single number.

$$
\begin{aligned}
\log_b x &= \tfrac{2}{3} \log_b 27 + 2 \log_b 2 - \log_b 3 \\
&= \log_b 27^{2/3} + \log_b 2^2 - \log_b 3 \\
&= \log_b 9 + \log_b 4 - \log_b 3 \quad 27^{2/3} = 9; \ 2^2 = 4 \\
&= \log_b \frac{9 \cdot 4}{3} = \log_b 12 \qquad \text{Properties 5 and 6 of Theorem 1}
\end{aligned}
$$

Thus,

$$\log_b x = \log_b 12$$

Now we use property 8 of Theorem 1 to find x:

$$x = 12$$

Problem 7   Find x so that $\log_b x = \frac{2}{3} \log_b 8 + \frac{1}{2} \log_b 9 - \log_b 6$ without using a calculator or table.

CAUTION   We conclude this section by noting two common errors:

1. $\dfrac{\log_b M}{\log_b N} \neq \log_b M - \log_b N$ $\log_b M - \log_b N = \log_b \dfrac{M}{N}$;

$\dfrac{\log_b M}{\log_b N}$ cannot be simplified.

2. $\log_b (M + N) \neq \log_b M + \log_b N$ $\log_b M + \log_b N = \log_b MN$;
$\log_b (M + N)$ cannot be simplified.

Answers to Matched Problems
1. (A) $27 = 3^3$ (B) $6 = 36^{1/2}$ (C) $\frac{1}{9} = 3^{-2}$
2. (A) $\log_4 64 = 3$ (B) $\log_8 2 = \frac{1}{3}$ (C) $\log_4 (\frac{1}{16}) = -2$
3. (A) $y = \frac{3}{2}$ (B) $x = \frac{1}{8}$ (C) $b = 10$
4. (A) -5 (B) 2 (C) 0 (D) $m + n$ (E) 4 (F) $x^4 + 1$
5. (A) $\log_b r - \log_b u - \log_b v$ (B) $\frac{3}{5}(\log_b m - \log_b n)$ (C) $\frac{1}{3}\log_b u - 5\log_b v$
6. (A) 14.01 (to 4 significant digits) (B) 0.1175 (to 4 significant digits)
7. $x = 2$

EXERCISE 5-3

A

Rewrite Problems 1–8 in equivalent exponential form.

1. $\log_3 81 = 4$
2. $\log_5 125 = 3$
3. $\log_{10} 0.001 = -3$
4. $\log_{10} 1,000 = 3$
5. $\log_{81} 3 = \frac{1}{4}$
6. $\log_4 2 = \frac{1}{2}$
7. $\log_{1/2} 16 = -4$
8. $\log_{1/3} 27 = -3$

Rewrite Problems 9–16 in equivalent logarithmic form.

9. $0.0001 = 10^{-4}$
10. $10,000 = 10^4$
11. $8 = 4^{3/2}$
12. $9 = 27^{2/3}$
13. $\frac{1}{2} = 32^{-1/5}$
14. $\frac{1}{8} = 2^{-3}$
15. $7 = \sqrt{49}$
16. $4 = \sqrt[3]{64}$

In Problems 17–30, simplify each expression using Theorem 1.

17. $\log_{16} 1$
18. $\log_{25} 1$
19. $\log_{0.5} 0.5$
20. $\log_7 7$
21. $\log_e e^4$
22. $\log_{10} 10^5$
23. $\log_{10} 0.01$
24. $\log_{10} 100$
25. $\log_5 \sqrt[3]{5}$
26. $\log_2 \sqrt{8}$
27. $e^{\log_e \sqrt{x}}$
28. $e^{\log_e(x-1)}$
29. $e^{2\log_e x}$
30. $10^{-3\log_{10} u}$

B

Find x, y, or b, as indicated in Problems 31–44.

31. $\log_2 x = 2$
32. $\log_3 x = 3$
33. $\log_4 16 = y$
34. $\log_8 64 = y$
35. $\log_b 16 = 2$
36. $\log_b 10^{-3} = -3$
37. $\log_b 1 = 0$
38. $\log_b b = 1$
39. $\log_4 x = \frac{1}{2}$
40. $\log_8 x = \frac{1}{3}$
41. $\log_{1/3} 9 = y$
42. $\log_{49} (\frac{1}{7}) = y$
43. $\log_b 1,000 = \frac{3}{2}$
44. $\log_b 4 = \frac{2}{3}$

Write Problems 45–58 in terms of simpler logarithmic forms (see Example 5).

45. $\log_b u^2 v^7$
46. $\log_b u^{1/2} v^{1/3}$
47. $\log_b \dfrac{m^{2/3}}{n^{1/2}}$
48. $\log_b \dfrac{u^3}{v^5}$

49. $\log_b \dfrac{u}{vw}$
50. $\log_b \dfrac{uv}{w}$
51. $\log_b \dfrac{1}{a^2}$
52. $\log_b \dfrac{1}{M^5}$

53. $\log_b \sqrt[3]{x^2 - y^2}$
54. $\log_b \sqrt{u^2 + 1}$
55. $\log_b \dfrac{\sqrt[3]{N}}{p^2 q^3}$
56. $\log_b \dfrac{m^5 n^3}{\sqrt{p}}$

57. $\log_b \sqrt[4]{\dfrac{x^2 y^3}{\sqrt{z}}}$
58. $\log_b \sqrt[5]{\left(\dfrac{x}{y^4 z^9}\right)^3}$

In Problems 59–68, write each expression in terms of a single logarithm with a coefficient of 1. Example: $\log_b u^2 - \log_b v = \log_b (u^2/v)$.

59. $2 \log_b x - \log_b y$

60. $\log_b m = \frac{1}{2} \log_b n$

61. $\log_b w - \log_b x - \log_b y$

62. $\log_b w + \log_b x - \log_b y$

63. $3 \log_b x + 2 \log_b y - \frac{1}{4} \log_b z$

64. $\frac{1}{3} \log_b w - 3 \log_b x - 5 \log_b y$

65. $5(\frac{1}{2} \log_b u - 2 \log_b v)$

66. $7(4 \log_b m + \frac{1}{3} \log_b n)$

67. $\frac{1}{5}(2 \log_b x + 3 \log_b y)$

68. $\frac{1}{3}(4 \log_b x - 2 \log_b y)$

$\int$ *In Problems 69–76, write each expression in terms of logarithms of first-degree polynomials. Example:*

$$\log_b \frac{(2x + 1)^3}{(3x - 5)^4} = 3 \log_b (2x + 1) - 4 \log_b (3x - 5)$$

69. $\log_b [(x + 3)^5 (2x - 7)^2]$

70. $\log_b [(5x - 4)^3 (3x + 2)^4]$

71. $\log_b \dfrac{(x + 10)^7}{(1 + 10x)^2}$

72. $\log_b \dfrac{(x - 3)^5}{(5 + x)^3}$

73. $\log_b \dfrac{x^2}{\sqrt{x + 1}}$

74. $\log_b \dfrac{\sqrt{x - 1}}{x^3}$

75. $\log_b (x^4 + x^3 - 20x^2)$

76. $\log_b (x^5 + 5x^4 - 14x^3)$

In Problems 77–86, solve for x without using a calculator or table.

77. $\log_2 (x + 5) = 2 \log_2 3$

78. $\log_{10} (5 - x) = 3 \log_{10} 2$

79. $2 \log_5 x = \log_5 (x^2 - 6x + 2)$

80. $\log_{10} (x^2 - 2x - 2) = 2 \log_{10} (x - 2)$

81. $\log_e (x + 8) - \log_e x = 3 \log_e 2$

82. $\log_7 4x - \log_7 (x + 1) = \frac{1}{2} \log_7 4$

83. $2 \log_3 x = \log_3 2 + \log_3 (4 - x)$

84. $\log_4 x + \log_4 (x + 2) = \frac{1}{2} \log_4 9$

85. $3 \log_b 2 + \frac{1}{2} \log_b 25 - \log_b 20 = \log_b x$

86. $\frac{3}{2} \log_b 4 - \frac{2}{3} \log_b 8 + 2 \log_b 2 = \log_b x$

If $\log_b 2 = 0.69$, $\log_b 3 = 1.10$, *and* $\log_b 5 = 1.61$, *find the value of each expression in Problems 87–96.*

87. $\log_b 30$

88. $\log_b 12$

89. $\log_b \frac{2}{5}$

90. $\log_b \frac{5}{3}$

91. $\log_b 27$

92. $\log_b 16$

93. $\log_b \sqrt[3]{2}$

94. $\log_b \sqrt{3}$

95. $\log_b \sqrt{0.9}$

96. $\log_b \sqrt[3]{1.5}$

C _____

Graph Problems 97–100.

97. $y = \log_2 (x - 2)$

98. $y = \log_2 (x + 3)$

99. $y = \log_2 x - 2$

100. $y = \log_2 x + 3$

101. (A) For $f = \{(x, y) | y = (\frac{1}{2})^x = 2^{-x}\}$, graph f, f^{-1}, and $y = x$ on the same coordinate system.
(B) Indicate the domain and range of f and f^{-1}.
(C) What other name can you use for the inverse of f?

102. (A) For $f = \{(x, y) | y = (\frac{1}{3})^x = 3^{-x}\}$, graph f, f^{-1}, and $y = x$ on the same coordinate system.
(B) Indicate the domain and range of f and f^{-1}.
(C) What other name can you use for the inverse of f?

Find the inverse of each function in Problems 103–106.

103. $f(x) = 5^{3x - 1} + 4$

104. $g(x) = 3^{2x - 3} - 2$

105. $g(x) = 3 \log_e (5x - 2)$

106. $f(x) = 2 + \log_e (5x - 3)$

107. Write $\log_e x - \log_e 100 = -0.08t$ in an exponential form that is free of logarithms.

108. Write $\log_e x - \log_e C + kt = 0$ in an exponential form that is free of logarithms.

109. Prove that $\log_b (M/N) = \log_b M - \log_b N$ under the hypotheses of Theorem 1.

110. Prove that $\log_b M^p = p \log_b M$ under the hypotheses of Theorem 1.

SECTION 5-4 Common and Natural Logarithms

- Common and Natural Logarithms—Definition and Evaluation
- Applications

John Napier (1550–1617) is credited with the invention of logarithms, which evolved out of an interest in reducing the computational strain in research in astronomy. This new computational tool was immediately accepted by the scientific world. Now, with the availability of inexpensive calculators, logarithms have lost most of their importance as a computational device. However, the logarithmic concept has been greatly generalized since its conception, and logarithmic functions are used widely in both theoretical and applied sciences. For example, even with a very good scientific calculator, we still need logarithmic functions to solve the following simple-looking exponential equation from population growth studies and the mathematics of finance:

$$2 = 1.08^x$$

Of all possible logarithmic bases, the base e and the base 10 are used almost exclusively. Before we can use logarithms in certain practical problems, we need to be able to approximate the logarithm of any positive number to either base 10 or base e. And conversely, if we are given the logarithm of a number to base 10 or base e, we need to be able to approximate the number. Historically, tables such as Table I and Table II at the back of the book were used for this purpose, but now with inexpensive scientific calculators readily available, most people will use a calculator, since it is faster and can find far more values than any table can possibly include.

• Common and Natural Logarithms— Definition and Evaluation

Common logarithms, also called **Briggsian logarithms**, are logarithms with base 10. **Natural logarithms**, also called **Napierian logarithms**, are logarithms with base e. Most scientific calculators have a function key labeled "log" and a function key labeled "ln." The former represents a common logarithm and the latter a natural logarithm. In fact, "log" and "ln" are both used extensively in mathematical literature, and whenever you see either used in this book without a base indicated, they should be interpreted as follows:

Logarithmic Notation

$$\log x = \log_{10} x \qquad \textbf{Common Logarithm}$$

$$\ln x = \log_e x \qquad \textbf{Natural Logarithm}$$

Finding the common or natural logarithm using a scientific calculator is very easy: You simply enter a number from the domain of the function and push $\boxed{\log}$ or $\boxed{\ln}$

EXAMPLE 1   **Calculator Evaluation of Logarithms**

Use a scientific calculator to evaluate each to six decimal places:

(A) log 3,184   (B) ln 0.000 349   (C) log (-3.24)

Solutions

Enter	Press	Display
(A) 3184	$\boxed{\log}$	3.502973
(B) 0.000349	$\boxed{\ln}$	-7.960439
(C) -3.24	$\boxed{\log}$	Error

Why is an error indicated in part (C)? Because -3.24 is not in the domain of the log function. [*Note:* The manner in which error messages are displayed varies from one brand of calculator to another.]

Problem 1   Use a scientific calculator to evaluate each to six decimal places:

(A) log 0.013 529   (B) ln 28.693 28   (C) ln (-0.438)

When working with common and natural logarithms, we follow the common practice of using the equal sign "=" where it might be more appropriate to use the approximately equal sign "≈." No harm is done as long as we keep in mind that in a statement such as log 3.184 = 3.503, the number on the right is only assumed accurate to three decimal places and is not exact.

EXAMPLE 2   **Calculator Evaluation of Logarithms**

Use a scientific calculator to evaluate each to three decimal places:

(A) $n = \dfrac{\log 2}{\log 1.1}$ (B) $n = \dfrac{\ln 3}{\ln 1.08}$

Solutions

(A) First, note that $(\log 2)/(\log 1.1) \neq \log 2 - \log 1.1$ (see Theorem 1, Section 5-3).

$$n = \frac{\log 2}{\log 1.1} = 7.273$$

A: $\boxed{2}\;\boxed{\log}\;\boxed{\div}\;\boxed{1.1}\;\boxed{\log}\;\boxed{=}$
P: $\boxed{2}\;\boxed{\log}\;\boxed{1.1}\;\boxed{\log}\;\boxed{\div}$

$$(B)\; n = \frac{\ln 3}{\ln 1.08} = 14.275$$

A: $\boxed{3}\;\boxed{\ln}\;\boxed{\div}\;\boxed{1.08}\;\boxed{\ln}\;\boxed{=}$
P: $\boxed{3}\;\boxed{\ln}\;\boxed{1.08}\;\boxed{\ln}\;\boxed{\div}$

Problem 2

(A) $n = \dfrac{\ln 2}{\ln 1.1}$ (B) $n = \dfrac{\log 3}{\log 1.08}$

We now turn to the second problem: Given the logarithm of a number, find the number. To solve this problem, we make direct use of the logarithmic–exponential relationships discussed in Section 5-3.

Logarithmic–Exponential Relationships

$\log x = y$ is equivalent to $x = 10^y$

$\ln x = y$ is equivalent to $x = e^y$

EXAMPLE 3   Solving $\log_b x = y$ for x

Find x to 3 significant digits, given the indicated logarithms.

(A) $\log x = -9.315$ (B) $\ln x = 2.386$

Solutions   (A) $\log x = -9.315$

$\qquad x = 10^{-9.315}$ Change to equivalent exponential form.

$\qquad\quad = 4.84 \times 10^{-10}$ $\boxed{9.315}$ $\boxed{+/-}$ $\boxed{10^x}$

Notice that the answer is displayed in scientific notation in the calculator.

(B) $\ln x = 2.386$

$\qquad x = e^{2.386}$ Change to equivalent exponential form.

$\qquad\quad = 10.9$ $\boxed{2.386}$ $\boxed{e^x}$

Problem 3   Find x to 4 significant digits, given the indicated logarithms.

(A) $\ln x = -5.062$ (B) $\log x = 12.0821$

● **Applications**   We now consider three applications that are solved using common and natural logarithms. The first application concerns sound intensity; the second, earthquake intensity; and the third, rocket flight theory.

Sound Intensity

TABLE 1
Typical Sound Intensities

Sound intensity, W/m^2	Sound source
1.0×10^{-12}	Threshold of hearing
5.2×10^{-10}	Whisper
3.2×10^{-6}	Normal conversation
8.5×10^{-4}	Heavy traffic
3.2×10^{-3}	Jackhammer
1.0×10^{0}	Threshold of pain
8.3×10^{2}	Jet plane with afterburner

The human ear is able to hear sound over an incredible range of intensities. The loudest sound a healthy person can hear without damage to the eardrum has an intensity 1 trillion (1,000,000,000,000) times that of the softest sound a person can hear. Working directly with numbers over such a wide range is very cumbersome. Since the logarithm, with base greater than 1, of a number increases much more slowly than the number itself, logarithms are often used to create more convenient compressed scales. The decibel scale for sound intensity is an example of such a scale. The **decibel**, named after the inventor of the telephone, Alexander Graham Bell (1847–1922), is defined as follows:

$$D = 10 \log \frac{I}{I_0} \quad \text{Decibel scale} \qquad (1)$$

where D is the **decibel level** of the sound, I is the **intensity** of the sound measured in watts per square meter, (W/m^2), and I_0 is the intensity of the least audible sound that an average healthy young person can hear. The latter is standardized to be $I_0 = 10^{-12}$ watt per square meter. Table 1 lists some typical sound intensities from familiar sources.

EXAMPLE 4 Sound Intensity

Find the number of decibels from a whisper with sound intensity 5.20×10^{-10} watt per square meter. Compute the answer to two decimal places.

Solution   We use the decibel formula (1):

$$D = 10 \log \frac{I}{I_0}$$
$$= 10 \log \frac{5.2 \times 10^{-10}}{10^{-12}}$$
$$= 10 \log (5.2 \times 10^2)$$
$$= 10(\log 5.2 + \log 10^2)$$
$$= 10(0.716 + 2) \qquad \log 5.2 = 0.716$$
$$= 27.16 \text{ decibels}$$

Problem 4 Find the number of decibels from a jackhammer with sound intensity 3.2×10^{-3} watt per square meter. Compute the answer to two decimal places.

Earthquake Intensity

The energy released by the largest earthquake recorded, measured in joules, is about 100 billion (100,000,000,000) times the energy released by a small earthquake that is barely felt. Over the past 150 years several people from various countries have devised different types of measures of earthquake magnitudes so that their severity could be easily compared. In 1935 the California seismologist

Charles Richter devised a logarithmic scale that bears his name and is still widely used in the United States. The **magnitude** M on the **Richter scale*** is given as follows:

$$M = \frac{2}{3} \log \frac{E}{E_0} \quad \text{Richter scale} \tag{2}$$

where E is the energy released by the earthquake, measured in joules, and E_0 is the energy released by a very small reference earthquake which has been standardized to be

$$E_0 = 10^{4.40} \text{ joules}$$

The destructive power of earthquakes relative to magnitudes on the Richter scale is indicated in Table 2.

TABLE 2
The Richter Scale

Magnitude on Richter scale	Destructive power
$M < 4.5$	Small
$4.5 < M < 5.5$	Moderate
$5.5 < M < 6.5$	Large
$6.5 < M < 7.5$	Major
$7.5 < M$	Greatest

EXAMPLE 5   Earthquake Intensity

The 1906 San Francisco earthquake released approximately 5.96×10^{16} joules of energy. What was its magnitude on the Richter scale? Compute the answer to two decimal places.

Solution   We use the magnitude formula (2):

$$
\begin{aligned}
M &= \frac{2}{3} \log \frac{E}{E_0} \\[2mm]
&= \frac{2}{3} \log \frac{5.96 \times 10^{16}}{10^{4.40}} \\[2mm]
&= \frac{2}{3} \log (5.96 \times 10^{11.6}) \\[2mm]
&= \frac{2}{3} (\log 5.96 + \log 10^{11.6}) \\[2mm]
&= \frac{2}{3} (0.775 + 11.6) \\[2mm]
&= 8.25
\end{aligned}
$$

Problem 5   The 1985 earthquake in central Chile released approximately 1.26×10^{16} joules of energy. What was its magnitude on the Richter scale? Compute the answer to two decimal places.

---

*Originally, Richter defined the magnitude of an earthquake in terms of logarithms of the maximum seismic wave amplitude, in thousandths of a millimeter, measured on a standard seismograph. Formula (2) gives essentially the same magnitude that Richter obtained for a given earthquake but in terms of logarithms of the energy released by the earthquake.

EXAMPLE 6   **Earthquake Intensity**

If the energy release of one earthquake is 1,000 times that of another, how much larger is the Richter scale reading of the larger than the smaller?

Solution   Let

$$M_1 = \frac{2}{3} \log \frac{E_1}{E_0} \quad \text{and} \quad M_2 = \frac{2}{3} \log \frac{E_2}{E_0}$$

be the Richter equations for the smaller and larger earthquakes, respectively. Substituting $E_2 = 1,000E_1$ into the second equation, we obtain

$$M_2 = \frac{2}{3} \log \frac{1,000E_1}{E_0}$$

$$= \frac{2}{3} \left(\log 10^3 + \log \frac{E_1}{E_0} \right)$$

$$= \frac{2}{3} (3) + \frac{2}{3} \log \frac{E_1}{E_0}$$

$$= 2 + M_1$$

Thus, an earthquake with 1,000 times the energy of another has a Richter scale reading of 2 more than the other.

Problem 6   If the energy release of one earthquake is 10,000 times that of another, how much larger is the Richter scale reading of the larger than the small?

Rocket Flight Theory   The theory of rocket flight uses advanced mathematics and physics to show that the velocity v of a rocket at burnout (depletion of fuel supply) is given by

$$v = c \ln \frac{W_t}{W_b} \quad \text{Rocket equation} \tag{3}$$

where c is the exhaust velocity of the rocket engine, W_t is the takeoff weight (fuel, structure, and payload), and W_b is the burnout weight (structure and payload).

Because of the Earth's atmospheric resistance, a launch vehicle velocity of at least 9.0 kilometers per second is required in order to achieve the minimum altitude needed for a stable orbit. It is clear that to increase velocity v, either the weight ratio W_t/W_b must be increased or the exhaust velocity c must be increased. The weight ratio can be increased by the use of solid fuels, and the exhaust velocity can be increased by improving the fuels, solid or liquid.

EXAMPLE 7   **Rocket Flight Theory**

A typical single-stage, solid-fuel rocket may have a weight ratio $W_t/W_b = 18.7$ and an exhaust velocity $c = 2.38$ kilometers per second. Would this rocket reach a launch velocity of 9.0 kilometers per second?

Solution   We use the rocket equation (3):

$$v = c \ln \frac{W_t}{W_b}$$
$$= 2.38 \ln 18.7$$
$$= 6.97 \text{ kilometers per second}$$

The velocity of the launch vehicle is far short of the 9.0 kilometers per second required to achieve orbit. This is why multiple-stage launchers are used—the dead weight from a preceding stage can be jettisoned into the ocean when the next stage takes over.

Problem 7   A launch vehicle using liquid fuel, such as a mixture of liquid hydrogen and liquid oxygen, can produce an exhaust velocity of $c = 4.7$ kilometers per second. However, the weight ratio W_t/W_b must be low—around 5.5 for some vehicles—because of the increased structural weight to accommodate the liquid fuel. How much more or less than the 9.0 kilometers per second required to reach orbit will be achieved by this vehicle?

Answers to Matched Problems
1. (A) $-1.868\ 734$ (B) $3.356\ 663$ (C) Not possible   2. (A) 7.27   (B) 14.27
3. (A) $x = 0.006\ 333$ (B) $x = 1.21 \times 10^{12}$ 4. 95.05 decibels   5. 7.80   6. 2.67
7. 1 km/s less

EXERCISE 5-4

A

Use a calculator to evaluate Problems 1–8 to four decimal places.

1. $\log 82{,}734$ 2. $\log 843{,}250$ 3. $\log 0.001\ 439$ 4. $\log 0.035\ 604$

5. $\ln 43.046$ 6. $\ln 2{,}843{,}100$ 7. $\ln 0.081\ 043$ 8. $\ln 0.000\ 032\ 4$

In Problems 9–16, use a calculator to evaluate x to 4 significant digits, given:

9. $\log x = 5.3027$ 10. $\log x = 1.9168$ 11. $\log x = -3.1773$ 12. $\log x = -2.0411$

13. $\ln x = 3.8655$ 14. $\ln x = 5.0884$ 15. $\ln x = -0.3916$ 16. $\ln x = -4.1083$

B

Evaluate Problems 17–24 to three decimal places using a calculator.

17. $n = \dfrac{\log 2}{\log 1.15}$ 18. $n = \dfrac{\log 2}{\log 1.12}$ 19. $n = \dfrac{\ln 3}{\ln 1.15}$ 20. $n = \dfrac{\ln 4}{\ln 1.2}$

21. $x = \dfrac{\ln 0.5}{-0.21}$ 22. $x = \dfrac{\ln 0.1}{-0.0025}$ 23. $t = \dfrac{\ln 150}{\ln 3}$ 24. $t = \dfrac{\log 200}{\log 2}$

In Problems 25–32, evaluate x to 5 significant digits using a calculator.

25. $x = \log (5.3147 \times 10^{12})$

26. $x = \log (2.0991 \times 10^{17})$

27. $x = \ln (6.7917 \times 10^{-12})$

28. $x = \ln (4.0304 \times 10^{-8})$

29. $\log x = 32.068\ 523$

30. $\log x = -12.731\ 64$

31. $\ln x = -14.667\ 13$

32. $\ln x = 18.891\ 143$

Graph each function in Problems 33–40 using a calculator.

33. $y = \ln x$

34. $y = -\ln x$

35. $y = |\ln x|$

36. $y = \ln |x|$

37. $y = 2 \ln (x + 2)$

38. $y = 2 \ln x + 2$

39. $y = 4 \ln x - 3$

40. $y = 4 \ln (x - 3)$

C

41. Find the fallacy:

$$1 < 3$$
$$\tfrac{1}{27} < \tfrac{3}{27} \qquad \text{Divide both sides by 27.}$$
$$\tfrac{1}{27} < \tfrac{1}{9}$$
$$(\tfrac{1}{3})^3 < (\tfrac{1}{3})^2$$
$$\log (\tfrac{1}{3})^3 < \log (\tfrac{1}{3})^2$$
$$3 \log \tfrac{1}{3} < 2 \log \tfrac{1}{3}$$
$$3 < 2 \qquad \text{Divide both sides by } \log \tfrac{1}{3}.$$

42. Find the fallacy:

$$3 > 2$$
$$3 \log \tfrac{1}{2} > 2 \log \tfrac{1}{2} \qquad \text{Multiply both sides by } \log \tfrac{1}{2}.$$
$$\log (\tfrac{1}{2})^3 > \log (\tfrac{1}{2})^2$$
$$(\tfrac{1}{2})^3 > (\tfrac{1}{2})^2$$
$$\tfrac{1}{8} > \tfrac{1}{4}$$
$$1 > 2 \qquad \text{Multiply both sides by 8.}$$

 Problems 43–46 require the use of a graphic calculator or a computer.

The polynomials in Problems 43–46, called **Taylor polynomials**, *can be used to approximate the function g(x) = ln (1 + x). To illustrate this approximation graphically, in each problem, graph g(x) = ln (1 + x) and the indicated polynomial in the same viewing rectangle,* $-1 \leq x \leq 3$ *and* $-2 \leq y \leq 2$.

43. $P_1(x) = x - \tfrac{1}{2}x^2$

44. $P_2(x) = x - \tfrac{1}{2}x^2 + \tfrac{1}{3}x^3$

45. $P_3(x) = x - \tfrac{1}{2}x^2 + \tfrac{1}{3}x^3 - \tfrac{1}{4}x^4$

46. $P_4(x) = x - \tfrac{1}{2}x^2 + \tfrac{1}{3}x^3 - \tfrac{1}{4}x^4 + \tfrac{1}{5}x^5$

APPLICATIONS

47. **Sound.** What is the decibel level of:
(A) The threshold of hearing, 1.0×10^{-12} watt per square meter?
(B) The threshold of pain, 1.0 watt per square meter?
Compute answers to 2 significant digits.

48. **Sound.** What is the decibel level of:
(A) A normal conversation, 3.2×10^{-6} watt per square meter?
(B) A jet plane with an afterburner, 8.3×10^2 watts per square meter?
Compute answers to 2 significant digits.

49. **Sound.** If the intensity of a sound from one source is 1,000 times that of another, how much more is the decibel level of the louder sound than the quieter one?

50. **Sound.** If the intensity of a sound from one source is 10,000 times that of another, how much more is the decibel level of the louder sound than the quieter one?

51. **Earthquakes.** The largest recorded earthquake to date was in Colombia in 1906, with an energy release of 1.99×10^{17} joules. What was its magnitude on the Richter scale? Compute the answer to one decimal place.

52. Earthquakes. Anchorage, Alaska, had a major earthquake in 1964 that released 7.08×10^{16} joules of energy. What was its magnitude on the Richter scale? Compute the answer to one decimal place.

★★**53. Earthquakes.** The 1933 Long Beach, California, earthquake had a Richter scale reading of 6.3 and the 1964 Anchorage, Alaska, earthquake had a Richter scale reading of 8.3. How many times more powerful was the Anchorage earthquake than the Long Beach earthquake?

★★**54. Earthquakes.** Generally, an earthquake requires a magnitude of over 5.6 on the Richter scale to inflict serious damage. How many times more powerful than this was the great 1906 Colombia earthquake, which registered a magnitude of 8.6 on the Richter scale?

55. Space Vehicles. A new solid-fuel rocket has a weight ratio $W_t/W_b = 19.8$ and an exhaust velocity $c = 2.57$ kilometers per second. What is its velocity at burnout? Compute the answer to two decimal places.

56. Space Vehicles. A liquid-fuel rocket has a weight ratio $W_t/W_b = 6.2$ and an exhaust velocity $c = 5.2$ kilometers per second. What is its velocity at burnout? Compute the answer to two decimal places.

57. Chemistry. The hydrogen ion concentration of a substance is related to its acidity and basicity. Because hydrogen ion concentrations vary over a very wide range, logarithms are used to create a compressed **pH scale**, which is defined as follows:

$$pH = -\log [H^+]$$

where $[H^+]$ is the hydrogen ion concentration, in moles per liter. Pure water has a pH of 7, which means it is neutral. Substances with a pH less than 7 are acid, and those with a pH greater than 7 are basic. Compute the pH of each substance listed, given the indicated hydrogen ion concentration.
(A) Seawater, 4.63×10^{-9}
(B) Vinegar, 9.32×10^{-4}
Also, indicate whether it is acid or basic. Compute answers to one decimal place.

58. Chemistry. Refer to Problem 57. Compute the pH of each substance below, given the indicated hydrogen ion concentration. Also, indicate whether it is acid or basic. Compute answers to one decimal place.
(A) Milk, 2.83×10^{-7}
(B) Garden mulch, 3.78×10^{-6}

★**59. Ecology.** Refer to Problem 57. Many lakes in Canada and the United States will no longer sustain some forms of wildlife because of the increase in acidity of the water from acid rain and snow caused by sulfur dioxide emissions from industry. If the pH of a sample of rainwater is 5.2, what is its hydrogen ion concentration in moles per liter? Compute the answer to one decimal place.

★**60. Ecology.** Refer to Problem 57. If normal rainwater has a pH of 5.7, what is its hydrogen ion concentration in moles per liter? Compute the answer to one decimal place.

SECTION 5-5 Exponential and Logarithmic Equations

- Exponential Equations
- Logarithmic Equations
- Change of Base

Equations involving exponential and logarithmic functions, such as

$$2^{3x-2} = 5 \quad \text{and} \quad \log (x + 3) + \log x = 1$$

are called **exponential** and **logarithmic equations**, respectively. Logarithmic properties play a central role in their solution.

● **Exponential Equations** The following examples illustrate the use of logarithmic properties in solving exponential equations.

EXAMPLE 1   **Solving an Exponential Equation**

Solve $2^{3x-2} = 5$ for x to four decimal places.

Solution   How can we get x out of the exponent? Use logs! Since the logarithm function is one-to-one, if two positive quantities are equal, their logs are equal. See Theorem 1 in Section 5-3.

$$2^{3x-2} = 5$$

$$\log 2^{3x-2} = \log 5 \qquad \text{Take the common or natural log of both sides.}$$

$$(3x - 2) \log 2 = \log 5 \qquad \text{Use } \log_b N^p = p \log_b N \text{ to get } 3x - 2 \text{ out of the exponent position.}$$

$$3x - 2 = \frac{\log 5}{\log 2}$$

$$x = \frac{1}{3}\left(2 + \frac{\log 5}{\log 2}\right) \qquad \text{Remember: } \frac{\log 5}{\log 2} \neq \log 5 - \log 2.$$

A: $\boxed{3}\,\boxed{1/x}\,\boxed{\times}\,\boxed{(}\,\boxed{(}\,\boxed{2}\,\boxed{+}\,\boxed{(}\,\boxed{(}\,\boxed{5}\,\boxed{\log}\,\boxed{\div}\,\boxed{2}\,\boxed{\log}\,\boxed{)}\,\boxed{)}\,\boxed{=}$

P: $\boxed{5}\,\boxed{\log}\,\boxed{2}\,\boxed{\log}\,\boxed{\div}\,\boxed{2}\,\boxed{+}\,\boxed{3}\,\boxed{\div}$

$$x = 1.4406 \qquad \text{To four decimal places}$$

Problem 1   Solve $35^{1-2x} = 7$ for x to four decimal places.

EXAMPLE 2   **Compound Interest**

A certain amount of money P (principal) is invested at an annual rate r compounded annually. The amount of money A in the account after t years, assuming no withdrawals, is given by

$$A = P\left(1 + \frac{r}{n}\right)^{nt} = P(1 + r)^t \qquad n = 1 \text{ for annual compounding}$$

How many years to the nearest year will it take the money to double if it is invested at 6% compounded annually?

Solution   To find the doubling time, we replace A in $A = P(1.06)^t$ with $2P$ and solve for t.

$$2P = P(1.06)^t$$

$$2 = 1.06^t \qquad \text{Divide both sides by } P.$$

$$\log 2 = \log 1.06^t \qquad \text{Take the common or natural log of both sides.}$$

$$= t \log 1.06 \qquad \text{Note how log properties are used to get } t \text{ out of the exponent position.}$$

$$t = \frac{\log 2}{\log 1.06}$$

$$= 12 \text{ years} \qquad \text{To the nearest year}$$

Problem 2   Repeat Example 2, changing the interest rate to 9% compounded annually.

EXAMPLE 3   **Atmospheric Pressure**

The atmospheric pressure P, in pounds per square inch, at x miles above sea level is given approximately by

$$P = 14.7e^{-0.21x}$$

At what height will the atmospheric pressure be half the sea-level pressure? Compute the answer to 2 significant digits.

Solution   Sea-level pressure is the pressure at $x = 0$. Thus,

$$P = 14.7e^0 = 14.7$$

One-half of sea-level pressure is $\frac{14.7}{2} = 7.35$. Now our problem is to find x so that $P = 7.35$; that is, we solve $7.35 = 14.7e^{-0.21x}$ for x:

$$7.35 = 14.7e^{-0.21x}$$

$$0.5 = e^{-0.21x} \qquad \text{Divide both sides by 14.7 to simplify.}$$

$$\ln 0.5 = \ln e^{-0.21x} \qquad \text{Since the base is } e \text{, take the natural log of both sides.}$$

$$= -0.21x \qquad \ln e = 1$$

$$x = \frac{\ln 0.5}{-0.21}$$

$$= 3.3 \text{ miles} \qquad \text{To 2 significant digits}$$

Problem 3   Using the formula in Example 3, find the altitude in miles so that the atmospheric pressure will be one-eighth that at sea level. Compute the answer to 2 significant digits.

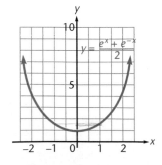

FIGURE 1 Catenary.

The graph of

$$y = \frac{e^x + e^{-x}}{2} \qquad (1)$$

is a curve called a **catenary** (Figure 1). A uniform cable suspended between two fixed points is a physical example of such a curve.

EXAMPLE 4   Solving an Exponential Equation

∫ Given equation (1), find x for $y = 2.5$. Compute the answer to four decimal places.

Solution

$$y = \frac{e^x + e^{-x}}{2}$$

$$2.5 = \frac{e^x + e^{-x}}{2}$$

$$5 = e^x + e^{-x}$$

$$5e^x = e^{2x} + 1 \qquad \text{Multiply both sides by } e^x.$$

$$e^{2x} - 5e^x + 1 = 0 \qquad \text{This is a quadratic in } e^x.$$

Let $u = e^x$; then

$$u^2 - 5u + 1 = 0$$

$$u = \frac{5 \pm \sqrt{25 - 4(1)(1)}}{2}$$

$$= \frac{5 \pm \sqrt{21}}{2}$$

$$e^x = \frac{5 \pm \sqrt{21}}{2} \qquad \text{Replace } u \text{ with } e^x \text{ and solve for } x.$$

$$\ln e^x = \ln \frac{5 \pm \sqrt{21}}{2} \qquad \begin{array}{l}\text{Take the natural log of both sides}\\ \text{(both values on the right are positive).}\end{array}$$

$$x = \ln \frac{5 \pm \sqrt{21}}{2} \qquad \log_b b^x = x.$$

$$= -1.5668, \ 1.5668$$

Problem 4   Given $y = (e^x - e^{-x})/2$, find x for $y = 1.5$. Compute the answer to three decimal places.

● Logarithmic Equations

We now illustrate the solution of several types of logarithmic equations.

EXAMPLE 5   Solving a Logarithmic Equation

Solve $\log (x + 3) + \log x = 1$, and check.

Solution   First use properties of logarithms to express the left side as a single logarithm, then convert to exponential form and solve for x.

$$\log (x + 3) + \log x = 1$$

$$\log [x(x + 3)] = 1 \qquad \text{Combine left side using } \log M + \log N = \log MN.$$

$$x(x + 3) = 10^1 \qquad \text{Change to equivalent exponential form.}$$

$$x^2 + 3x - 10 = 0 \qquad \text{Write in } ax^2 + bx + c = 0 \text{ form and solve.}$$

$$(x + 5)(x - 2) = 0$$

$$x = -5, 2$$

Check $x = -5$: $\log (-5 + 3) + \log (-5)$ is not defined because the domain of the log function is $(0, \infty)$.

$x = 2$: $\log (2 + 3) + \log 2 = \log 5 + \log 2$

$$= \log (5 \cdot 2) = \log 10 \overset{\checkmark}{=} 1$$

Thus, the only solution to the original equation is $x = 2$. Remember, answers should be checked in the original equation to see whether any should be discarded.

Problem 5   Solve $\log (x - 15) = 2 - \log x$, and check.

EXAMPLE 6   **Solving a Logarithmic Equation**

Solve $(\ln x)^2 = \ln x^2$.

Solution   There are no logarithmic properties for simplifying $(\ln x)^2$. However, we can simplify $\ln x^2$, obtaining an equation involving $\ln x$ and $(\ln x)^2$.

$$(\ln x)^2 = \ln x^2$$

$$= 2 \ln x \qquad \text{This is a quadratic equation in } \ln x.$$
$$\text{Move all nonzero terms to the left and factor.}$$

$$(\ln x)^2 - 2 \ln x = 0$$

$$(\ln x)(\ln x - 2) = 0$$

$$\ln x = 0 \qquad \text{or} \qquad \ln x - 2 = 0$$

$$x = e^0 \qquad\qquad \ln x = 2$$

$$= 1 \qquad\qquad\quad x = e^2$$

Checking that both $x = 1$ and $x = e^2$ are solutions to the original equation is left to you.

Problem 6   Solve $\log x^2 = (\log x)^2$.

CAUTION

Note that

$$(\log_b x)^2 \neq \log_b x^2 \qquad \begin{array}{l} (\log_b x)^2 = (\log_b x)(\log_b x) \\ \log_b x^2 = 2 \log_b x \end{array}$$

EXAMPLE 7   **Earthquake Intensity**

Recall from Section 5-4 that the magnitude of an earthquake on the Richter scale is given by

$$M = \frac{2}{3} \log \frac{E}{E_0}$$

Solve for E in terms of the other symbols.

Solution

$$M = \frac{2}{3} \log \frac{E}{E_0}$$

$$\log \frac{E}{E_0} = \frac{3M}{2} \qquad \text{Multiply both sides by } \tfrac{3}{2}.$$

$$\frac{E}{E_0} = 10^{3M/2} \qquad \text{Change to exponential form.}$$

$$E = E_0 10^{3M/2}$$

Problem 7   Solve the rocket equation from Section 5-4 for W_b in terms of the other symbols:

$$v = c \ln \frac{W_t}{W_b}$$

● **Change of Base**   How would you find the logarithm of a positive number to a base other than 10 or e? For example, how would you find $\log_3 5.2$? In Example 8 we evaluate this logarithm using a direct process. Then we develop a change-of-base formula to find such logarithms in general. You may find it easier to remember the process than the formula.

EXAMPLE 8   **Evaluating a Base 3 Logarithm**

Evaluate $\log_3 5.2$ to four decimal places.

Solution   Let $y = \log_3 5.2$ and proceed as follows:

$$\log_3 5.2 = y$$

$$5.2 = 3^y \qquad \text{Change to exponential form.}$$

$$\ln 5.2 = \ln 3^y \qquad \text{Take the natural log (or common log) of each side.}$$

$$= y \ln 3 \qquad \log_b M^p = p \log_b M$$

$$y = \frac{\ln 5.2}{\ln 3} \qquad \text{Solve for } y.$$

Replace y with $\log_3 5.2$ from the first step, and use a calculator to evaluate the right side:

$$\log_3 5.2 = \frac{\ln 5.2}{\ln 3} = 1.5007$$

Problem 8   Evaluate $\log_{0.5} 0.0372$ to four decimal places.

To develop a change-of-base formula for arbitrary positive bases, with neither base equal to 1, we proceed as above. Let $y = \log_b N$, where N and b are positive and $b \neq 1$. Then

$$\log_b N = y$$

$$N = b^y \qquad \text{Write in exponential form.}$$

$$\log_a N = \log_a b^y \qquad \text{Take the log of each side to another positive base } a, a \neq 1.$$

$$= y \log_a b \qquad \log_b M^p = p \log_b M$$

$$y = \frac{\log_a N}{\log_a b} \qquad \text{Solve for } y.$$

Replacing y with $\log_b N$ from the first step, we obtain the **change-of-base formula**:

$$\log_b N = \frac{\log_a N}{\log_a b}$$

In words, this formula states that the logarithm of a number to a given base is the logarithm of that number to a new base divided by the logarithm of the old base to the new base. In practice, we usually choose either e or 10 for the new base so that a calculator can be used to evaluate the necessary logarithms (see Example 8).

Answers to Matched Problems
1. $x = 0.2263$ 2. More than double in 9 years, but not quite double in 8 years
3. 9.9 miles   4. $x = 1.195$ 5. $x = 20$ 6. $x = 1,100$ 7. $W_b = W_t e^{-v/c}$
8. 4.7486

$x = .96$

EXERCISE 5-5

A _____

Solve Problems 1–12 to 3 significant digits.

1. $10^{-x} = 0.0347$ **2.** $10^x = 14.3$ **3.** $10^{3x+1} = 92$ **4.** $10^{5x-2} = 348$

5. $e^x = 3.65$ **6.** $e^{-x} = 0.0142$ **7.** $e^{2x-1} = 405$ **8.** $e^{3x+5} = 23.8$

9. $5^x = 18$ **10.** $3^x = 4$ **11.** $2^{-x} = 0.238$ **12.** $3^{-x} = 0.074$

Solve Problems 13–18 exactly.

13. $\log 5 + \log x = 2$ **14.** $\log x - \log 8 = 1$ **15.** $\log x + \log (x - 3) = 1$

16. $\log (x - 9) + \log 100x = 3$ **17.** $\log (x + 1) - \log (x - 1) = 1$ **18.** $\log (2x + 1) = 1 + \log (x - 2)$

B _____

Solve Problems 19–26 to 3 significant digits.

19. $2 = 1.05^x$ **20.** $3 = 1.06^x$ **21.** $e^{-1.4x} = 13$ **22.** $e^{0.32x} = 632$

23. $123 = 500e^{-0.12x}$ **24.** $438 = 200e^{0.25x}$ **25.** $e^{-x^2} = 0.23$ **26.** $e^{x^2} = 125$

Solve Problems 27–38 exactly.

27. $\log x - \log 5 = \log 2 - \log (x - 3)$ **28.** $\log (6x + 5) - \log 3 = \log 2 - \log x$

29. $\ln x = \ln (2x - 1) - \ln (x - 2)$ **30.** $\ln (x + 1) = \ln (3x + 1) - \ln x$

31. $\log (2x + 1) = 1 - \log (x - 1)$ **32.** $1 - \log (x - 2) = \log (3x + 1)$

33. $(\ln x)^3 = \ln x^4$ **34.** $(\log x)^3 = \log x^4$

35. $\ln (\ln x) = 1$ **36.** $\log (\log x) = 1$

37. $x^{\log x} = 100x$ **38.** $3^{\log x} = 3x$

Evaluate Problems 39–44 to four decimal places.

39. $\log_5 372$ **40.** $\log_4 23$ **41.** $\log_8 0.0352$

42. $\log_2 0.005\ 439$ **43.** $\log_3 0.1483$ **44.** $\log_{12} 435.62$

C _____

Solve Problems 45–52 for the indicated variable in terms of the remaining symbols. Use the natural log for solving exponential equations.

45. $A = Pe^{rt}$ for r (finance)   **46.** $A = P\left(1 + \dfrac{r}{n}\right)^{nt}$ for t (finance)

47. $D = 10 \log \dfrac{I}{I_0}$ for I (sound)   **48.** $t = \dfrac{-1}{k} (\ln A - \ln A_0)$ for A (decay)

49. $M = 6 - 2.5 \log \dfrac{I}{I_0}$ for I (astronomy)   **50.** $L = 8.8 + 5.1 \log D$ for D (astronomy)

51. $I = \dfrac{E}{R}\left(1 - e^{-Rt/L}\right)$ for t (circuitry)   **52.** $S = R\left[\dfrac{(1 + i)^n - 1}{i}\right]$ for n (annuity)

∫ *The following combinations of exponential functions define four of six* **hyperbolic functions,** *an important class of functions in calculus and higher mathematics. Solve Problems 53–56 for x in terms of y. The results are used to define* **inverse hyperbolic functions,** *another important class of functions in calculus and higher mathematics.*

53. $y = \dfrac{e^x + e^{-x}}{2}$ **54.** $y = \dfrac{e^x - e^{-x}}{2}$ **55.** $y = \dfrac{e^x - e^{-x}}{e^x + e^{-x}}$ **56.** $y = \dfrac{e^x + e^{-x}}{e^x - e^{-x}}$

Problems 57–72 require the use of a graphic calculator or a computer.

In Problems 57–60, use a graphic calculator or a computer to graph each function. [*Hint: Use the change-of-base formula first.*]

57. $y = 3 + \log_2(2 - x)$ **58.** $y = \log_3(4 + x) - 5$ **59.** $y = \log_3 x - \log_2 x$ **60.** $y = \log_3 x + \log_2 x$

In Problems 61–72, use a graphic calculator or a computer to approximate to two decimal places any solutions of the equation in the interval $0 \le x \le 1$. None of these equations can be solved exactly using any step-by-step algebraic process.

61. $2^{-x} - 2x = 0$ **62.** $3^{-x} - 3x = 0$ **63.** $x3^x - 1 = 0$ **64.** $x2^x - 1 = 0$

65. $e^{-x} - x = 0$ **66.** $xe^{2x} - 1 = 0$ **67.** $xe^x - 2 = 0$ **68.** $e^{-x} - 2x = 0$

69. $\ln x + 2x = 0$ **70.** $\ln x + x^2 = 0$ **71.** $\ln x + e^x = 0$ **72.** $\ln x + x = 0$

APPLICATIONS

73. Compound Interest. How many years, to the nearest year, will it take a sum of money to double if it is invested at 15% compounded annually? Use the formula $A = P[1 + (r/n)]^{nt}$.

74. Compound Interest. How many years, to the nearest year, will it take money to quadruple if it is invested at 20% compounded annually? Use the formula $A = P[1 + (r/n)]^{nt}$.

75. Compound Interest. At what annual rate compounded continuously will $1,000 have to be invested to amount to $2,500 in 10 years? Use the formula $A = Pe^{rt}$. Compute the answer to 3 significant digits.

76. Compound Interest. How many years will it take $5,000 to amount to $8,000 if it is invested at an annual rate of 9% compounded continuously? Use the formula $A = Pe^{rt}$. Compute the answer to 3 significant digits.

★★ **77. Astronomy.** The brightness of stars is expressed in terms of magnitudes on a numerical scale that increases as the brightness decreases. The magnitude m is given by the formula

$$m = 6 - 2.5 \log \frac{L}{L_0}$$

where L is the light flux of the star and L_0 is the light flux of the dimmest stars visible to the naked eye.
(A) What is the magnitude of the dimmest stars visible to the naked eye?

(B) How many times brighter is a star of magnitude 1 than a star of magnitude 6?

78. Astronomy. An optical instrument is required to observe stars beyond the sixth magnitude, the limit of ordinary vision. However, even optical instruments have their limitations. The limiting magnitude L of any optical telescope with lens diameter D, in inches, is given by

$$L = 8.8 + 5.1 \log D$$

(A) Find the limiting magnitude for a homemade 6-inch reflecting telescope.
(B) Find the diameter of a lens that would have a limiting magnitude of 20.6.
Compute answers to 3 significant digits.

79. World Population. A mathematical model for world population growth over short periods of time is given by

$$P = P_0 e^{rt}$$

where P is the population after t years, P_0 is the population at $t = 0$, and r is the annual growth rate compounded continuously. How many years, to the nearest year, will it take the world population to double if it grows at 2% compounded continuously?

★ **80. World Population.** Refer to Problem 79. Starting with a world population of 4 billion people and assuming an annual growth rate of 2% compounded continuously, how many years, to the nearest year, will it be before there is only 1 square yard of land per person? Earth contains approximately 1.7×10^{14} square yards of land.

★ **81. Archaeology—Carbon-14 Dating.** As long as a plant or animal is alive, carbon-14 is maintained in a constant amount in its tissues. Once dead, however, the plant or animal ceases taking in carbon, and carbon-14 diminishes by radioactive decay according to the equation

$$A = A_0 e^{-0.000124t}$$

where A is the amount after t years and A_0 is the amount when $t = 0$. Estimate the age of a skull uncovered in an archaeological site if 10% of the original amount of carbon-14 is still present. Compute the answer to 3 significant digits.

★ **82. Archaeology—Carbon-14 Dating.** Refer to Problem 81. What is the half-life of carbon-14? That is, how long will it take for half of a sample of carbon-14 to decay? Compute the answer to 3 significant digits.

★ **83. Photography.** An electronic flash unit for a camera is activated when a capacitor is discharged through a filament of wire. After the flash is triggered and the capacitor is discharged, the circuit (see the figure) is connected and the battery pack generates a current to recharge the capacitor. The time it takes for the capacitor to recharge is called the *recycle time*. For a particular flash unit using a 12-volt battery pack, the charge q, in coulombs, on the capacitor t seconds after recharging has started is given by

$$q = 0.0009(1 - e^{-0.2t})$$

How many seconds will it take the capacitor to reach a charge of 0.0007 coulomb? Compute the answer to 3 significant digits.

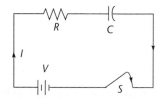

★ **84. Advertising.** A company is trying to expose a new product to as many people as possible through television advertising in a large metropolitan area with 2 million possible viewers. A model for the number of people N, in millions, who are aware of the product after t days of advertising was found to be

$$N = 2(1 - e^{-0.037t})$$

How many days, to the nearest day, will the advertising campaign have to last so that 80% of the possible viewers will be aware of the product?

★★ **85. Newton's Law of Cooling.** This law states that the rate at which an object cools is proportional to the difference in temperature between the object and its surrounding medium. The temperature T of the object t hours later is given by

$$T = T_m + (T_0 - T_m)e^{-kt}$$

where T_m is the temperature of the surrounding medium and T_0 is the temperature of the object at $t = 0$. Suppose a bottle of wine at a room temperature of 72°F is placed in a refrigerator at 40°F to cool before a dinner party. After an hour the temperature of the wine is found to be 61.5°F. Find the constant k, to two decimal places, and the time, to one decimal place, it will take the wine to cool from 72 to 50°F.

★ **86. Marine Biology.** Marine life is dependent upon the microscopic plant life that exists in the *photic zone*, a zone that goes to a depth where about 1% of the surface light still remains. Light intensity is reduced according to the exponential function

$$I = I_0 e^{-kd}$$

where I is the intensity d feet below the surface and I_0 is the intensity at the surface. The constant k is called the *coefficient of extinction*. At Crystal Lake in Wisconsin it was found that half the surface light remained at a depth of 14.3 feet. Find k, and find the depth of the photic zone. Compute answers to 3 significant digits.

★ **87. Wildlife Management.** A herd of 20 white-tailed deer is introduced to a coastal island where there had been no deer before. Their population is predicted to increase according to the logistic curve

$$N = \frac{100}{1 + 4e^{-0.14t}}$$

where N is the number of deer expected in the herd after t years. In how many years, to the nearest year, will the herd number 50?

★ **88. Training.** A trainee is hired by a computer manufacturing company to learn to test a particular model of a personal computer after it comes off the assembly line. The learning curve for an average trainee is given by

$$N = \frac{200}{4 + 21e^{-0.1t}}$$

where N is the number of computers tested per day after t days on the job. How many days, to the nearest day, will it take an average trainee to achieve 40 tested computers per day?

Chapter 5 Review

5-1 EXPONENTIAL FUNCTIONS

The equation $f(x) = b^x$, $b > 0$, $b \neq 1$, defines an **exponential function** with **base** b. The domain of f is $(-\infty, \infty)$ and the range is $(0, \infty)$. The **graph** of an exponential function is a continuous curve that always passes through the point $(0, 1)$ and has the x axis as a horizontal asymptote. If $b > 1$, then b^x increases as x increases, and if $0 < b < 1$, then b^x decreases as x increases. The function f is one-to-one and has an inverse. We have the following **exponential function properties**:

1. $a^x a^y = a^{x+y}$ $(a^x)^y = a^{xy}$ $(ab)^x = a^x b^x$

   $\left(\dfrac{a}{b}\right)^x = \dfrac{a^x}{b^x}$ $\dfrac{a^x}{a^y} = a^{x-y}$

2. $a^x = a^y$ if and only if $x = y$.

3. For $x \neq 0$, then $a^x = b^x$ if and only if $a = b$.

Exponential functions are used to describe various types of **growth**.

1. **Population growth** can be modeled by using the **doubling time growth model** $P = P_0 2^{t/d}$, where P is population at time t, P_0 is the population at time $t = 0$, and d is the **doubling time**—the time it takes for the population to double.

2. **Radioactive decay** can be modeled by using the **half-life decay model** $A = A_0\left(\frac{1}{2}\right)^{t/h} = A_0 2^{-t/h}$, where A is the amount at time t, A_0 is the amount at time $t = 0$, and h is the **half-life**—the time it takes for half the material to decay.

3. The growth of money in an account paying **compound interest** is described by $A = P(1 + r/n)^{nt}$, where P is the **principal**, r is the annual **rate**, n is the number of compounding periods in one year, and A is the **amount** in the account after t years. We also call P the **present value** and A the **future value** of the account.

5-2 THE EXPONENTIAL FUNCTION WITH BASE e

As m approaches ∞, the expression $(1 + 1/m)^m$ approaches the irrational number $e \approx 2.718\ 281\ 828\ 459$. The function $f(x) = e^x$ is called the **exponential function with base e**. Exponential functions with base e are used to model a variety of different types of exponential growth and decay, including growth of money in accounts that pay **continuous compound interest**. If a principal P is invested at an annual rate r compounded continuously, then the amount A in the account after t years is given by $A = Pe^{rt}$.

5-3 LOGARITHMIC FUNCTIONS

The **logarithmic function with base b** is defined to be the inverse of the exponential function with base b and is denoted by $y = \log_b x$. Thus, $y = \log_b x$ if and only if $x = b^y$, $b > 0$, $b \neq 1$. The domain of a logarithmic function is $(0, \infty)$ and the range is $(-\infty, \infty)$. The graph of a logarithmic function is a continuous curve that always passes through the point $(1, 0)$ and has the y axis as a vertical asymptote. We have the following **properties of logarithmic functions**:

1. $\log_b 1 = 0$

2. $\log_b b = 1$

3. $\log_b b^x = x$

4. $b^{\log_b x} = x, x > 0$

5. $\log_b MN = \log_b M + \log_b N$

6. $\log_b \dfrac{M}{N} = \log_b M - \log_b N$

7. $\log_b M^p = p \log_b M$

8. $\log_b M = \log_b N$ if and only if $M = N$

5-4   COMMON AND NATURAL LOGARITHMS

Logarithms to the base 10 are called **common logarithms** and are denoted by $\log x$. Logarithms to the base e are called **natural logarithms** and are denoted by $\ln x$. Thus, $\log x = y$ is equivalent to $x = 10^y$, and $\ln x = y$ is equivalent to $x = e^y$.

The following applications involve logarithms:

1. The **decibel** is defined by $D = 10 \log (I/I_0)$, where D is the **decibel level** of the sound, I is the **intensity** of the sound, and $I_0 = 10^{-12}$ watt per square meter is a standardized sound level.

2. The **magnitude** M of an earthquake on the **Richter scale** is given by $M = \frac{2}{3} \log (E/E_0)$, where E is the energy released by the earthquake and $E_0 = 10^{4.40}$ joules is a standardized energy level.

3. The **velocity** v of a rocket at burnout is given by the **rocket equation** $v = c \ln (W_t/W_b)$, where c is the exhaust velocity, W_t is the takeoff weight, and W_b is the burnout weight.

5-5   EXPONENTIAL AND LOGARITHMIC EQUATIONS

Various techniques for solving **exponential equations**, such as $2^{3x-2} = 5$, and **logarithmic equations**, such as $\log (x + 3) + \log x = 1$, are illustrated by examples. The **change-of-base formula**, $\log_b N = (\log_a N)/(\log_a b)$, relates logarithms to two different bases and can be used, along with a calculator, to evaluate logarithms to bases other than e or 10.

Chapter 5 Review Exercise

Work through all the problems in this chapter review and check answers in the back of the book. Answers to all review problems are there, and following each answer is a number in italics indicating the section in which that type of problem is discussed. Where weaknesses show up, review appropriate sections in the text.

A

1. Write in logarithmic form using base 10: $m = 10^n$

2. Write in logarithmic form using base e: $x = e^y$

Write Problems 3 and 4 in exponential form.

3. $\log x = y$

4. $\ln y = x$

Simplify Problems 5 and 6.

5. $\dfrac{7^{x+2}}{7^{2-x}}$

6. $\left(\dfrac{e^x}{e^{-x}}\right)^x$

Solve Problems 7–9 for x exactly. Do not use a calculator or table.

7. $\log_2 x = 3$

8. $\log_x 25 = 2$

9. $\log_3 27 = x$

Solve Problems 10–13 for x to 3 significant digits.

10. $10^x = 17.5$

11. $e^x = 143{,}000$

12. $\ln x = -0.015\,73$

13. $\log x = 2.013$

B

Solve Problems 14–24 for x exactly. Do not use a calculator or table.

14. $\ln (2x - 1) = \ln (x + 3)$

15. $\log (x^2 - 3) = 2 \log (x - 1)$

16. $e^{x^2-3} = e^{2x}$

17. $4^{x-1} = 2^{1-x}$

18. $2x^2 e^{-x} = 18 e^{-x}$

19. $\log_{1/4} 16 = x$

20. $\log_x 9 = -2$

21. $\log_{16} x = \frac{3}{2}$

22. $\log_x e^5 = 5$

23. $10^{\log_{10} x} = 33$

24. $\ln x = 0$

Evaluate Problems 25–28 to 4 significant digits using a calculator.

25. $\ln \pi$

26. $\log (-e)$

27. $\pi^{\ln 2}$

28. $\dfrac{e^{\pi} + e^{-\pi}}{2}$

Solve Problems 29–38 for x to 3 significant digits.

29. $x = 2(10^{1.32})$

30. $x = \log_5 23$

31. $\ln x = -3.218$

32. $x = \log (2.156 \times 10^{-7})$

33. $x = \dfrac{\ln 4}{\ln 2.31}$

34. $25 = 5(2^x)$

35. $4{,}000 = 2{,}500(e^{0.12x})$

36. $0.01 = e^{-0.05x}$

37. $5^{2x-3} = 7.08$

38. $\dfrac{e^x - e^{-x}}{2} = 1$

Solve Problems 39–44 for x exactly. Do not use a calculator or table.

39. $\log 3x^2 - \log 9x = 2$

40. $\log x - \log 3 = \log 4 - \log (x + 4)$

41. $\ln (x + 3) - \ln x = 2 \ln 2$

42. $\ln (2x + 1) - \ln (x - 1) = \ln x$

43. $(\log x)^3 = \log x^9$

44. $\ln (\log x) = 1$

Simplify Problems 45 and 46.

45. $(e^x + 1)(e^{-x} - 1) - e^x(e^{-x} - 1)$

46. $(e^x + e^{-x})(e^x - e^{-x}) - (e^x - e^{-x})^2$

Graph each function in Problems 47–50.

47. $y = 2^{x-1}$

48. $f(t) = 10e^{-0.08t}$

49. $y = \ln (x + 1)$

50. $N = \dfrac{100}{1 + 3e^{-t}}$

C

Solve Problems 51–54 for the indicated variable in terms of the remaining symbols. Use natural logs for exponential equations.

51. $D = 10 \log \dfrac{I}{I_0}$ for I (sound intensity)

52. $y = \dfrac{1}{\sqrt{2\pi}} e^{-x^2/2}$ for x (probability)

53. $x = -\dfrac{1}{k} \ln \dfrac{I}{I_0}$ for I (x-ray intensity)

54. $r = P\left[\dfrac{i}{1 - (1 + i)^{-n}}\right]$ for n (finance)

Find the inverse of each function in Problems 55 and 56.

55. $f(x) = 2 \ln (x - 1)$

56. $f(x) = \dfrac{e^x - e^{-x}}{2}$

57. Write $\ln y = -5t + \ln c$ in an exponential form free of logarithms; then solve for y in terms of the remaining symbols.

59. Explain why 1 cannot be used as a logarithmic base.

58. For $f = \{(x, y)|y = \log_2 x\}$, graph f and f^{-1} on the same coordinate system. What are the domains and ranges for f and f^{-1}?

60. Prove that $\log_b (M/N) = \log_b M - \log_b N$.

APPLICATIONS

61. Population Growth. Many countries have a population growth rate of 3% (or more) per year. At this rate, how many years will it take a population to double? Use the annual compounding growth model $P = P_0(1 + r)^t$. Compute the answer to 3 significant digits.

62. Population Growth. Repeat Problem 61 using the continuous compounding growth model $P = P_0 e^{rt}$.

63. Carbon-14 Dating. How many years will it take for carbon-14 to diminish to 1% of the original amount after the death of a plant or animal? Use the formula $A = A_0 e^{-0.000124t}$. Compute the answer to 3 significant digits.

★**64. Medicine.** One leukemic cell injected into a healthy mouse will divide into two cells in about $\frac{1}{2}$ day. At the end of the day these two cells will divide into four. This doubling continues until 1 billion cells are formed; then the animal dies with leukemic cells in every part of the body.
   (A) Write an equation that will give the number N of leukemic cells at the end of t days.
   (B) When, to the nearest day, will the mouse die?

65. Money Growth. Assume $1 had been invested at an annual rate of 3% compounded continuously at the birth of Christ. What would be the value of the account in the year 2000? Compute the answer to 2 significant digits.

66. Present Value. Solving $A = Pe^{rt}$ for P, we obtain $P = Ae^{-rt}$, which is the **present value** of the amount A due in t years if money is invested at a rate r compounded continuously.
   (A) Graph $P = 1,000(e^{-0.08t})$, $0 \le t \le 30$.
   (B) What does it appear that P tends to as t tends to infinity? [*Conclusion:* The longer the time until the amount A is due, the smaller its present value, as we would expect.]

67. Earthquakes. The 1971 San Fernando, California, earthquake released 1.99×10^{14} joules of energy. Compute its magnitude on the Richter scale using the formula $M = \frac{2}{3} \log (E/E_0)$, where $E_0 = 10^{4.40}$ joules. Compute the answer to one decimal place.

68. Earthquakes. Refer to Problem 67. If the 1906 San Francisco earthquake had a magnitude of 8.3 on the Richter scale, how much energy was released? Compute the answer to 3 significant digits.

★**69. Sound.** If the intensity of a sound from one source is 100,000 times that of another, how much more is the decibel level of the louder sound than the softer one? Use the formula $D = 10 \log (I/I_0)$.

★★**70. Marine Biology.** The intensity of light entering water is reduced according to the exponential function

$$I = I_0 e^{-kd}$$

where I is the intensity d feet below the surface, I_0 is the intensity at the surface, and k is the coefficient of extinction. Measurements in the Sargasso Sea in the West Indies have indicated that half the surface light reaches a depth of 73.6 feet. Find k, and find the depth at which 1% of the surface light remains. Compute answers to 3 significant digits.

★**71. Wildlife Management.** A lake formed by a newly constructed dam is stocked with 1,000 fish. Their population is expected to increase according to the logistic curve

$$N = \frac{30}{1 + 29e^{-1.35t}}$$

where N is the number of fish, in thousands, expected after t years. The lake will be open to fishing when the number of fish reaches 20,000. How many years, to the nearest year, will this take?

Cumulative Review Exercise
Chapters 3–5

Work through all the problems in this cumulative review and check answers in the back of the book. Answers to all review problems are there, and following each answer is a number in italics indicating the section in which that type of problem is discussed. Where weaknesses show up, review appropriate sections in the text.

A _____

1. Given the points $A(3, 2)$ and $B(5, 6)$, find:
   (A) Distance between A and B
   (B) Slope of the line through A and B
   (C) Slope of a line perpendicular to the line through A and B

2. Find the equation of the circle with radius $\sqrt{2}$ and center:
   (A) $(0, 0)$ (B) $(-3, 1)$

3. Graph $2x - 3y = 6$ and indicate its slope and intercepts.

4. For $f(x) = x^2 - 2x + 5$ and $g(x) = 3x - 2$, find:
   (A) $f(-2) + g(3)$
   (B) $(f + g)(x)$
   (C) $(f \circ g)(x)$
   (D) $\dfrac{f(a + h) - f(a)}{h}$

5. How are the graphs of the following related to the graph of $y = |x|$?
   (A) $y = 2|x|$
   (B) $y = |x - 2|$
   (C) $y = |x| - 2$

6. For $P(x) = 3x^3 + 5x^2 - 18x - 3$ and $D(x) = x + 3$, use synthetic division to divide $P(x)$ by $D(x)$, and write the answer in the form $P(x) = D(x)Q(x) + R$.

7. Let $P(x) = 2(x + 2)(x - 3)(x - 5)$. What are the zeros of $P(x)$?

8. Let $P(x) = 4x^3 - 5x^2 - 3x - 1$. How do you know that $P(x)$ has at least one real zero between 1 and 2?

9. Let $P(x) = x^3 + x^2 - 10x + 8$. Find all rational zeros for $P(x)$.

10. Decompose into partial fractions:
$$\frac{5x - 4}{(x - 2)(x + 1)}$$

11. Solve for x:
    (A) $y = 10^x$ (B) $y = \ln x$

12. Simplify:
    (A) $(2e^x)^3$ (B) $\dfrac{e^{3x}}{e^{-2x}}$

13. Solve for x exactly. Do not use a calculator or a table.
    (A) $\log_3 x = 2$
    (B) $\log_3 81 = x$
    (C) $\log_x 4 = -2$

14. Solve for x to 3 significant digits.
    (A) $10^x = 2.35$
    (B) $e^x = 87{,}500$
    (C) $\log x = -1.25$
    (D) $\ln x = 2.75$

B _____

15. Find each of the following for the function f given by the graph shown below.
    (A) The domain of f
    (B) The range of f
    (C) $f(-3) + f(-2) + f(2)$
    (D) The intervals over which f is increasing
    (E) The x coordinates of any points of discontinuity

16. Write equations of the lines
    (A) Parallel to
    (B) Perpendicular to
    the line $3x + 2y = 12$ and passing through the point $(-6, 1)$. Write the final answers in the slope–intercept form $y = mx + b$.

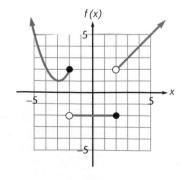

17. Find the domain of $g(x) = \sqrt{x + 4}$.

18. Graph $f(x) = x^2 - 2x - 8$. Show the axis of symmetry and vertex, and find the range, intercepts, and maximum or minimum value of $f(x)$.

19. Given $f(x) = 1/(x - 2)$ and $g(x) = (x + 3)/x$, find $f \circ g$. What is the domain of $f \circ g$?

20. Find $f^{-1}(x)$ for $f(x) = 2x + 5$.

21. Graph, finding the domain, range, and any points of discontinuity:

$$f(x) = \begin{cases} x - 1 & \text{if } x < 0 \\ x^2 + 1 & \text{if } x \geq 0 \end{cases}$$

22. Graph:
(A) $y = 2\sqrt{x} + 1$
(B) $y = -\sqrt{x + 1}$

23. Let $f(x) = \dfrac{2x + 8}{x + 2}$.
(A) Find the domain and the intercepts for f.
(B) Find the vertical and horizontal asymptotes for f.
(C) Sketch the graph of f. Draw vertical and horizontal asymptotes with dashed lines.

24. Let $f(x) = \sqrt{x + 4}$.
(A) Find $f^{-1}(x)$.
(B) Find the domain and range of f and f^{-1}.
(C) Graph f, f^{-1}, and $y = x$ on the same coordinate system.

25. Find the center and radius of the circle given by $x^2 - 6x + y^2 + 2y = 0$. Graph the circle and show the center and the radius.

26. Discuss symmetry with respect to the x axis, y axis, and the origin for the equation

$$xy + |xy| = 5$$

27. If $P(x) = 2x^3 - 5x^2 + 3x + 2$, find $P(\frac{1}{2})$ using the remainder theorem and synthetic division.

28. Which of the following is a factor of

$$P(x) = x^{25} - x^{20} + x^{15} + x^{10} - x^5 + 1$$

(A) $x - 1$ (B) $x + 1$

29. Let $P(x) = 2x^4 - 9x^3 + 10x^2 + x - 4$.
(A) Construct a table showing possible combinations of positive, negative, and imaginary zeros.
(B) Find the smallest and largest integers that are upper and lower bounds, respectively, of the zeros of $P(x)$.
(C) Locate real zeros between successive integers.

30. Find all zeros (rational, irrational, and imaginary) for $P(x) = 4x^3 - 20x^2 + 29x - 15$.

31. Find all zeros (rational, irrational, and imaginary) for $P(x) = x^4 + 5x^3 + x^2 - 15x - 12$, and factor $P(x)$ into linear factors.

32. Solve $x^3 + 36 \leq 7x^2$. Write answers in inequality and interval notation.

33. Approximate all real solutions to one decimal place:

$$x^3 + 4x - 20 = 0$$

34. Decompose into partial fractions:

$$\frac{3x^2 - x + 1}{x(x + 1)^2}$$

35. Decompose into partial fractions:

$$\frac{x^2 + x - 2}{x^3 - x^2 + x}$$

Solve Problems 36–45 for x exactly. Do not use a calculator or a table.

36. $2^{x^2} = 4^{x+4}$

37. $2x^2 e^{-x} + xe^{-x} = e^{-x}$

38. $e^{\ln x} = 2.5$

39. $\log_x 10^4 = 4$

40. $\log_9 x = -\frac{3}{2}$

41. $\ln (x + 4) - \ln (x - 4) = 2 \ln 3$

42. $\ln (2x^2 + 2) = 2 \ln (2x - 4)$

43. $\log x + \log (x + 15) = 2$

44. $\log (\ln x) = -1$

45. $4 (\ln x)^2 = \ln x^2$

Solve Problems 46–50 for x to 3 significant digits.

46. $x = \log_3 41$

47. $\ln x = 1.45$

48. $4(2^x) = 20$

49. $10e^{-0.5x} = 1.6$

50. $\dfrac{e^x - e^{-x}}{e^x + e^{-x}} = \dfrac{1}{2}$

Graph each function in Problems 51–55.

51. $y = 3^{1-x}$

52. $f(x) = \ln(2 - x)$

53. $A(t) = 100e^{-0.3t}$

54. $y = -2\sqrt{x + 1} + 3$

55. $y = -2e^{-x} + 3$

C

56. Let $f(x) = |x + 2| + |x - 2|$. Find a piecewise definition of $f(x)$ that does not involve the absolute value function. Graph f and find the domain and range.

57. Given $f(x) = x^2$ and $g(x) = \sqrt{4 - x^2}$, find:
(A) Domain of g
(B) f/g and its domain
(C) $f \circ g$ and its domain

58. Let $f(x) = x^2 - 2x - 3, x \geq 1$.
(A) Find $f^{-1}(x)$.
(B) Find the domain and range of f^{-1}.
(C) Graph f, f^{-1}, and $y = x$ on the same coordinate system.

59. Graph f and indicate any horizontal, vertical, or oblique asymptotes with dashed lines:

$$f(x) = \frac{x^2 + 4x + 8}{x + 2}$$

60. Write a piecewise definition for $f(x) = 2x - [\![2x]\!]$, and sketch the graph of f. Include sufficient intervals to clearly illustrate both the definition and the graph. Find the domain, range, and any points of discontinuity.

61. Find a polynomial of lowest degree with leading coefficient 1 that has zeros -1 (multiplicity 2), 0 (multiplicity 3), $3 + 5i$, and $3 - 5i$. Leave the answer in factored form. What is the degree of the polynomial?

62. Find all zeros (rational, irrational, and imaginary) for

$$P(x) = x^5 - 4x^4 + 3x^3 + 10x^2 - 10x - 12$$

and factor $P(x)$ into linear factors.

63. Find rational roots exactly and irrational roots to two decimal places for

$$P(x) = x^5 + 4x^4 + 6x^3 + 7x^2 + 4x - 4$$

64. Decompose into partial fractions:

$$\frac{x^2 - 4x + 11}{x^4 - x^3 + x^2 - 3x + 2}$$

65. Let $f(x) = 3 \ln(x - 2)$.
(A) Find $f^{-1}(x)$.
(B) Find the domain and range of f and f^{-1}.
(C) Graph f, f^{-1}, and $y = x$ on the same coordinate system.

66. Use natural logarithms to solve for n:

$$A = P \frac{(1 + i)^n - 1}{i}$$

67. Solve $\ln y = 5x + \ln A$ for y. Express the answer in a form that is free of logarithms.

68. Solve for x:

$$y = \frac{e^x - 2e^{-x}}{2}$$

APPLICATIONS

69. Price and Demand. The weekly demand for mouthwash in a chain of drugstores is 1,160 bottles at a price of $3.79 each. If the price is lowered to $3.59, the weekly demand increases to 1,340 bottles. Assuming that the relationship between the weekly demand x and the price per bottle p is linear, express x as a function of p. How many bottles would the store sell each week if the price were lowered to $3.29?

70. Business—Pricing. A telephone company begins a new pricing plan that charges customers for local calls as follows: The first 60 calls each month are 6 cents each, the next 90 are 5 cents each, the next 150 are 4 cents each, and any additional calls are 3 cents each. If C is the cost, in dollars, of placing x calls per month, write a piecewise definition of C as a function of x and graph.

71. Construction. A homeowner has 80 feet of chain-link fencing to be used to construct a dog pen adjacent to a house (see the figure).
(A) Express the area $A(x)$ enclosed by the pen as a function of the width x.
(B) From physical considerations, what is the domain of the function A?
(C) Graph A and determine the dimensions of the pen that will make the area maximum.

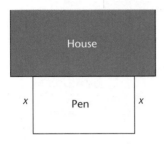

72. Computer Science. Let $f(x) = x - 2[\![x/2]\!]$. This function can be used to determine if an integer is odd or even.
(A) Find $f(1)$, $f(2)$, $f(3)$, $f(4)$.
(B) Find $f(n)$ for any integer n. [*Hint:* Consider two cases, $n = 2k$ and $n = 2k + 1$, k an integer.]

73. Shipping. A mailing service provides customers with rectangular shipping containers. The length plus the girth of one of these containers is 10 feet (see the figure in the next column). If the end of the container is square and the volume is 8 cubic feet, find the dimensions. Find rational solutions exactly and irrational solutions to one decimal place.

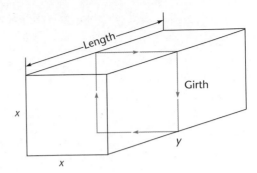

74. Geometry. The diagonal of a rectangle is 2 feet longer than one of the sides, and the area of the rectangle is 6 square feet. Find the dimensions of the rectangle. Find rational solutions exactly and irrational solutions to one decimal place.

75. Population Growth. If Zaire has a population of about 40 million people and a doubling time of 22 years, find the population in:
(A) 5 years   (B) 30 years
Compute answers to 3 significant digits.

76. Compound Interest. How long will it take money invested in an account earning 7% compounded annually to double? Use the annual compounding growth model $P = P_0(1 + r)^t$, and compute the answer to 3 significant digits.

77. Compound Interest. Repeat Problem 76 using the continuous compound interest model $P = P_0 e^{rt}$.

78. Earthquakes. If the 1906 and 1989 San Francisco earthquakes registered 8.3 and 7.1, respectively, on the Richter scale, how many times more powerful was the 1906 earthquake than the 1989 earthquake? Use the formula $M = \frac{2}{3} \log (E/E_0)$, where $E_0 = 10^{4.40}$ joules, and compute the answer to one decimal place.

79. Sound. If the decibel level at a rock concert is 88, find the intensity of the sound at the concert. Use the formula $D = 10 \log (I/I_0)$, where $I_0 = 10^{-12}$ watt per square meter, and compute the answer to 2 significant digits.

Systems of Equations and Inequalities

6

$$C = 7x + 9y$$
$$10x + 5y \geq 350$$
$$8x + 24y \geq 840$$
$$9x + 6y \leq 630$$
$$x, y \geq 0$$

In this chapter we move from the standard methods of solving two linear equations with two variables to a method that can be used to solve linear systems with any number of variables and equations. This method is well-suited to computer use in the solution of larger systems. We also discuss systems involving nonlinear equations and systems of linear inequalities. Finally, we introduce a relatively new and powerful mathematical tool called *linear programming*. Many applications in mathematics involve systems of equations or inequalities. The mathematical techniques discussed in this chapter are applied to a variety of interesting and significant applications.

SECTION 6-1   Systems of Linear Equations in Two Variables; Augmented Matrices

- Solving Systems in Two Variables
- Augmented Matrices
- Solving Linear Systems Using Augmented Matrix Methods

In this section we review the standard methods for solving two linear equations with two variables. The *elimination-by-addition method* is then transformed, through the use of augmented matrices, into a solution process that is well-suited for computer use in the solutions of linear systems involving large numbers of equations and variables.

To keep the introduction to augmented matrices in this section as simple as possible, we restrict the discussion to two equations with two variables. In the next section, Section 6-2, the solution process introduced here in a simple setting is generalized and applied to larger-scale linear systems.

- **Solving Systems in Two Variables**

Any equation that can be written in the form

$$ax + by = c$$

where x and y are variables and a, b, and c are real constants* with a and b not both 0, is called a **linear equation in two variables**. The numbers a and b are called the **coefficients** of x and y, respectively, and the number c is called the **constant term** of the equation.

We are interested in systems of two linear equations with two variables. To establish basic concepts, consider the following simple example. At a computer fair, student tickets cost \$2 and general admission tickets cost \$3. If a total of 7 tickets is purchased for \$18, how many of each type were purchased?

Let

$$x = \text{Number of student tickets}$$

$$y = \text{Number of general admission tickets}$$

*Unless noted otherwise, we assume that the constants in an equation are real numbers and that the replacement set for each of the variables in the equation is the set of real numbers.

Then

$$x + y = 7 \qquad \text{Total number of tickets purchased}$$

$$2x + 3y = 18 \qquad \text{Total purchase cost}$$

We now have a system of two linear equations in two variables. Variables in this type of problem are often referred to as **unknowns**. To solve this system, we find all ordered pairs of real numbers that satisfy both equations.

In general, we are interested in solving linear systems of the type

$$ax + by = h \qquad \text{System of two linear equations in two variables}$$

$$cx + dy = k$$

where x and y are variables and a, b, c, d, h, and k are real constants. A pair of numbers $x = x_0$ and $y = y_0$, also written as an ordered pair (x_0, y_0), is a **solution** of this system if each equation is satisfied by the pair. The set of all such ordered pairs of numbers is called the **solution set** for the system. To **solve** a system is to find its solution set. We will consider three methods of solving such systems: *graphing, substitution,* and *elimination by addition.* Each method has certain advantages, depending on the situation. First, we use the graphing method to solve the ticket problem.

Solution by Graphing

To solve the ticket problem by graphing, we graph both equations in the same rectangular coordinate system. Then the coordinates of any points that the graphs have in common must be solutions to the system, since they must satisfy both equations.

EXAMPLE 1   Solving a System by Graphing

Solve the ticket problem by graphing: $x + y = 7$
$2x + 3y = 18$

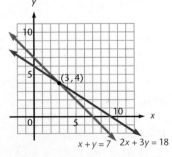

Solution

From the graph we see that:

$x = 3$ Student tickets

$y = 4$ General admission tickets

Check $\begin{aligned} x + y &= 7 \\ 3 + 4 &\overset{?}{=} 7 \\ 7 &\overset{\checkmark}{=} 7 \end{aligned}$ $\begin{aligned} 2x + 3y &= 18 \\ 2(3) + 3(4) &\overset{?}{=} 18 \\ 18 &\overset{\checkmark}{=} 18 \end{aligned}$

Problem 1   Solve by graphing and check: $x - y = 3$
$x + 2y = -3$

It is clear that the preceding example has exactly one solution, since the lines have exactly one point of intersection. In general, lines in a rectangular coordinate system are related to each other in one of the three ways illustrated in the next example.

EXAMPLE 2 **Solving Three Important Types of Systems by Graphing**

Solve each of the following systems by graphing:

(A) $2x - 3y = 2$ (B) $4x + 6y = 12$ (C) $2x - 3y = -6$
     $x + 2y = 8$ $2x + 3y = -6$ $-x + \frac{3}{2}y = 3$

Solutions   (A)

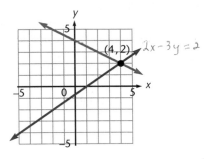

(B)

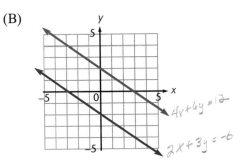

Lines intersect at one point only.
Exactly one solution: $x = 4, y = 2$

Lines are parallel (each has slope $-\frac{2}{3}$).
No solution.

(C)

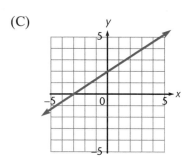

Lines coincide.
Infinitely many solutions.

Problem 2 Solve each of the following systems by graphing:

(A) $2x + 3y - 12$ (B) $x - 3y = -3$ (C) $2x - 3y = 12$
     $x - 3y = -3$ $-2x + 6y = 12$ $-x + \frac{3}{2}y = -6$

We now define some terms that can be used to describe the different types of solutions to systems of equations illustrated in Example 2. A system of linear equations is **consistent** if it has one or more solutions and **inconsistent** if no solutions exist. Furthermore, a consistent system is said to be **independent** if it has exactly one solution—often referred to as the **unique solution**—and **dependent** if it has more than one solution. Referring to the three systems in Example 2, the system in part (A) is a consistent and independent system with the unique solution $x = 4$ and $y = 2$. The system in part (B) is inconsistent. And the system in part (C) is consistent and dependent.

By geometrically interpreting a system of two linear equations in two variables, we gain useful information about what to expect in the way of solutions to the system. In general, any two lines in a rectangular coordinate plane must intersect in exactly one point, be parallel, or coincide (have identical graphs). Thus, the systems in Example 2 illustrate the only three possible types of solutions for systems of two linear equations in two variables. These ideas are summarized in Theorem 1.

Theorem 1

Possible Solutions to a Linear System

The linear system

$$ax + by = h$$
$$cx + dy = k$$

must have:

1. Exactly one solution   Consistent and independent
   or
2. No solution   Inconsistent
   or
3. Infinitely many solutions   Consistent and dependent

There are no other possibilities.

Graphs frequently reveal relationships in problems that might otherwise be hidden. Generally, however, *graphic methods give us only approximations of solutions.* The methods of substitution and elimination by addition to be considered next will yield exact solutions, assuming solutions exist.

Solution by Substitution

We solve the system in the next example by a three-step process called the **substitution method**.

EXAMPLE 3   Solving a System by Substitution

Solve by substitution: $2x - 3y = 7$
                          $3x - y = 7$

Solution   *Step 1.* Solve either equation for one variable in terms of the other. In this problem, we can avoid fractions if we solve the second equation for y in terms of x.

$$3x - y = 7 \qquad \text{Solve for } y \text{ in terms of } x.$$

$$-y = -3x + 7$$

$$y = 3x - 7$$

Step 2. Substitute the expression obtained in step 1 into the other equation in the system and solve the resulting linear equation in one variable.

$$2x - 3y = 7$$

$$2x - 3(3x - 7) = 7 \qquad \text{Substitute } y = 3x - 7 \text{ and solve for } x.$$

$$2x - 9x + 21 = 7$$

$$-7x = -14$$

$$x = 2$$

Step 3. Substitute the value of the variable determined in step 2 into the expression found in step 1 to determine the value of the other variable.

$$y = 3x - 7$$

$$= 3(2) - 7 \qquad \text{Substitute } x = 2.$$

$$y = -1$$

Thus, $(2, -1)$ is the unique solution to the original system.

Check

$$
\begin{array}{ll}
2x - 3y = 7 & 3x - y = 7 \\
2(2) - 3(-1) \overset{?}{=} 7 & 3(2) - (-1) \overset{?}{=} 7 \\
7 \overset{\checkmark}{=} 7 & 7 \overset{\checkmark}{=} 7
\end{array}
$$

Problem 3   Solve by substitution: $3x - 4y = 18$
 $2x + y = 1$

Solution Using Elimination by Addition

Now we turn to **elimination by addition**. This is probably the most important method of solution, since it is readily generalized to higher-order systems. The method involves the replacement of systems of equations with simpler *equivalent systems*, by performing appropriate operations, until we obtain a system with an obvious solution. **Equivalent systems** of equations are, as you would expect, systems that have exactly the same solution set. Theorem 2 lists operations that produce equivalent systems.

Theorem 2	**Elementary Equation Operations Producing Equivalent Systems**
	A system of linear equations is transformed into an equivalent system if:
	1. Two equations are interchanged.
	2. An equation is multiplied by a nonzero constant.
	3. A constant multiple of another equation is added to a given equation.

Any one of the three operations in Theorem 2 can be used to produce an equivalent system, but operations 2 and 3 will be of most use to us now. Operation 1 becomes more important later in the section. The use of Theorem 2 is best illustrated by examples.

EXAMPLE 4 Solving a System Using Elimination by Addition

Solve using elimination by addition:
$$3x - 2y = 8$$
$$2x + 5y = -1$$

Solution   We use Theorem 2 to eliminate one of the variables and thus obtain a system with an obvious solution.

$$3x - 2y = 8$$
$$2x + 5y = -1$$

If we multiply the top equation by 5, the bottom by 2, and then add, we can eliminate y.

$$
\begin{array}{rcr}
15x - 10y &=& 40 \\
4x + 10y &=& -2 \\
\hline
19x &=& 38 \\
x &=& 2
\end{array}
$$

The equation $x = 2$ paired with either of the two original equations produces an equivalent system. Thus, we can substitute $x = 2$ back into either of the two original equations to solve for y. We choose the second equation.

$$2(2) + 5y = -1$$
$$5y = -5$$
$$y = -1$$

Solution: $x = 2, y = -1$ or $(2, -1)$.

Check

$$
\begin{array}{cc}
3x - 2y = 8 & 2x + 5y = -1 \\
3(2) - 2(-1) \overset{?}{=} 8 & 2(2) + 5(-1) \overset{?}{=} -1 \\
8 \overset{\checkmark}{=} 8 & -1 \overset{\checkmark}{=} -1
\end{array}
$$

Problem 4 Solve using elimination by addition:
$$6x + 3y = 3$$
$$5x + 4y = 7$$

Let's see what happens in the elimination process when a system either has no solution or has infinitely many solutions. Consider the following system:

$$2x + 6y = -3$$
$$x + 3y = 2$$

Multiplying the second equation by -2 and adding, we obtain

$$\begin{array}{r} 2x + 6y = -3 \\ -2x - 6y = -4 \\ \hline 0 = -7 \end{array}$$

We have obtained a contradiction. An assumption that the original system has solutions must be false, otherwise, we have proved that $0 = -7$! Thus, the system has no solution. The graphs of the equations are parallel and the system is inconsistent.

Now consider the system

$$x - \tfrac{1}{2}y = 4$$
$$-2x + y = -8$$

If we multiply the top equation by 2 and add the result to the bottom equation, we get

$$\begin{array}{r} 2x - y = 8 \\ -2x + y = -8 \\ \hline 0 = 0 \end{array}$$

Obtaining $0 = 0$ by addition implies that the two original equations are equivalent. That is, their graphs coincide and the system is dependent. If we let $x = t$, where t is any real number, and solve either equation for y, we obtain $y = 2t - 8$. Thus,

$$(t, 2t - 8) \qquad t \text{ a real number}$$

describes the solution set for the system. Some particular solutions to this system are

$$\begin{array}{cccc} t = -1 & t = 2 & t = 5 & t = 9.4 \\ (-1, -10) & (2, -4) & (5, 2) & (9.4, 10.8) \end{array}$$

Many real-world problems are readily solved by applying two equation–two variable methods. Example 5 provides an illustration of an application that leads to such a system.

EXAMPLE 5   **Airspeed**

A private airplane makes the 2,400-mile trip from Washington, D.C., to San Francisco in 7.5 hours and makes the return trip in 6 hours. Assuming that the plane travels at a constant airspeed and that the wind blows at a constant rate from San Francisco to Washington, D.C., find the plane's airspeed and the wind rate.

Solution   Let x represent the airspeed of the plane and let y represent the rate at which the wind is blowing, both in miles per hour. The ground speed of the plane is determined by combining these two rates; that is,

$x - y$ = Ground speed flying from Washington, D.C., to San Francisco (headwind)

$x + y$ = Ground speed flying from San Francisco to Washington, D.C. (tailwind)

The distance traveled, the ground speed, and the duration of the trip are related by the familiar formula

$$\text{Distance} = \text{Rate} \times \text{Time}$$

Applying this formula to each leg of the trip leads to the following system of equations:

$$2,400 = 7.5(x - y) \quad \text{From Washington, D.C., to San Francisco}$$

$$2,400 = 6(x + y) \quad \text{From San Francisco to Washington, D.C.}$$

After simplification, we have

$$x - y = 320$$

$$x + y = 400$$

Solve using elimination by addition:

$$x - y = 320$$
$$\underline{x + y = 400}$$
$$2x \quad\quad = 720$$
$$x = 360 \text{ miles per hour} \quad \text{Airspeed}$$

$$360 + y = 400$$
$$y = 40 \text{ miles per hour} \quad \text{Wind rate}$$

Check

$$2,400 = 7.5(x - y) \quad\quad\quad 2,400 = 6(x + y)$$
$$2,400 \overset{?}{=} 7.5(360 - 40) \quad\quad 2,400 \overset{?}{=} 6(360 + 40)$$
$$2,400 \overset{\checkmark}{=} 2,400 \quad\quad\quad\quad\quad 2,400 \overset{\checkmark}{=} 2,400$$

Problem 5   Repeat Example 5 if the trip from Washington, D.C., to San Francisco takes 6 hours and the return trip takes 5 hours.

• **Augmented Matrices**

In solving systems of equations using elimination by addition the coefficients of the variables and constant terms played a central role. The process can be made more efficient for generalization and computer work by the introduction of a mathematical form called a *matrix*. A **matrix** is a rectangular array of numbers written within brackets. Some examples are

$$\begin{bmatrix} 3 & 5 \\ 0 & -2 \end{bmatrix} \qquad \begin{bmatrix} 2 \\ -3 \\ 0 \end{bmatrix} \qquad \begin{bmatrix} 1 & -1 & 0 & 5 \end{bmatrix}$$

$$\begin{bmatrix} -1 & 2 & -5 & 0 \\ 0 & 3 & 2 & 1 \end{bmatrix} \qquad \begin{bmatrix} 1 & 0 & 0 \\ 0 & 1 & 0 \\ 0 & 0 & 1 \end{bmatrix}$$

Each number in a matrix is called an **element** of the matrix.

Associated with the system

$$\begin{aligned} 3x - 2y &= 6 \\ 4x + y &= -7 \end{aligned} \qquad (1)$$

is the *augmented matrix*

$$\begin{bmatrix} 3 & -2 & 6 \\ 4 & 1 & -7 \end{bmatrix}$$

which contains the essential parts of the system—namely, the coefficients of the variables and the constant terms. The vertical bar is included only to separate the coefficients of the variables from the constant terms.

For ease of generalization to the larger systems in Section 6-2, we are now going to change the notation for variables in system (1) to a subscript form. For larger systems we could soon run out of letters, but we can't run out of subscripts. Thus, in place of x and y, we use x_1 and x_2, and system (1) is then written as

$$\begin{aligned} 3x_1 - 2x_2 &= 6 \\ 4x_1 + x_2 &= -7 \end{aligned}$$

In general, associated with each linear system of the form

$$\begin{aligned} a_1 x_1 + b_1 x_2 &= k_1 \\ a_2 x_1 + b_2 x_2 &= k_2 \end{aligned} \qquad (2)$$

where x_1 and x_2 are variables, is the **augmented matrix** of the system:

Our objective is to learn how to manipulate augmented matrices in such a way that a solution to system (2) will result, if a solution exists. The manipulative process is a direct outgrowth of the elimination-by-addition process discussed above.

Recall that two linear systems are said to be **equivalent** if they have exactly the same solution set. How did we transform linear systems into equivalent linear systems? We used Theorem 2, which we restate here for convenient reference.

Theorem 2   **Elementary Equation Operations Producing Equivalent Systems**

A system of linear equations is transformed into an equivalent system if:

1. Two equations are interchanged.

2. An equation is multiplied by a nonzero constant.

3. A constant multiple of another equation is added to a given equation.

Paralleling the earlier discussion, we say that two augmented matrices are **row-equivalent**, denoted by the symbol ~ between the two matrices, if they are augmented matrices of equivalent systems of equations. Stop and think about this for a moment. How do we transform augmented matrices into row-equivalent matrices? We use Theorem 3, which is a direct consequence of Theorem 2.

Theorem 3   **Elementary Row Operations Producing Row-Equivalent Matrices**

An augmented matrix is transformed into a row-equivalent matrix if any of the following **row operations** is performed:

1. Two rows are interchanged.

2. A row is multiplied by a nonzero constant.

3. A constant multiple of another row is added to a given row.

The following symbols are used to describe these row operations:

1. $R_i \leftrightarrow R_j$ means "interchange row i and row j."

2. $kR_i \rightarrow R_i$ means "multiply row i by the constant k."

3. $R_i + kR_j \rightarrow R_i$ means "multiply row j by the constant k and add to R_i."

● **Solving Linear Systems Using Augmented Matrix Methods**

The use of Theorem 3 in solving systems in the form of (2) is best illustrated by examples.

EXAMPLE 6   Solving a System Using Augmented Matrix Methods

Solve, using augmented matrix methods:

$$3x_1 + 4x_2 = 1$$
$$x_1 - 2x_2 = 7 \tag{3}$$

Solution   We start by writing the augmented matrix corresponding to system (3):

$$\begin{bmatrix} 3 & 4 & | & 1 \\ 1 & -2 & | & 7 \end{bmatrix} \tag{4}$$

Our objective is to use row operations from Theorem 3 to try to transform matrix (4) into the form

$$\begin{bmatrix} 1 & 0 & | & m \\ 0 & 1 & | & n \end{bmatrix} \tag{5}$$

where m and n are real numbers. The solution to system (3) will then be obvious, since matrix (5) will be the augmented matrix of the following system:

$$x_1 = m \qquad x_1 + 0x_2 = m$$
$$x_2 = n \qquad 0x_1 + x_2 = n$$

We now proceed to use row operations to transform (4) into form (5).

Step 1. To get a 1 in the upper left corner, we interchange rows 1 and 2—Theorem 3, part 1:

$$\begin{bmatrix} 3 & 4 & | & 1 \\ 1 & -2 & | & 7 \end{bmatrix} \overset{R_1 \leftrightarrow R_2}{\sim} \begin{bmatrix} 1 & -2 & | & 7 \\ 3 & 4 & | & 1 \end{bmatrix} \quad \text{Now you see why we wanted Theorem 2, part 1.}$$

Step 2. To get a 0 in the lower left corner, we multiply R_1 by -3 and add to R_2—Theorem 3, part 3. This changes R_2 but not R_1. Some people find it useful to write $(-3)R_1$ outside the matrix to help reduce errors in arithmetic:

$$\begin{bmatrix} 1 & -2 & | & 7 \\ 3 & 4 & | & 1 \end{bmatrix} \overset{R_2 + (-3)R_1 \rightarrow R_2}{\sim} \begin{bmatrix} 1 & -2 & | & 7 \\ 0 & 10 & | & -20 \end{bmatrix} \quad ,$$

$$-3 \qquad 6 \quad -21$$

Step 3. To get a 1 in the second row, second column, we multiply R_2 by $\frac{1}{10}$—Theorem 3, part 2:

$$\begin{bmatrix} 1 & -2 & | & 7 \\ 0 & 10 & | & -20 \end{bmatrix} \overset{\frac{1}{10}R_2 \rightarrow R_2}{\sim} \begin{bmatrix} 1 & -2 & | & 7 \\ 0 & 1 & | & -2 \end{bmatrix}$$

Step 4. To get a 0 in the first row, second column, we multiply R_2 by 2 and add the result to R_1—Theorem 3, part 3. This changes R_1 but not R_2.

$$\begin{matrix} 0 & 2 & -4 \end{matrix}$$
$$\begin{bmatrix} 1 & -2 & | & 7 \\ 0 & 1 & | & -2 \end{bmatrix} \quad R_1 + 2R_2 \rightarrow R_1 \quad \begin{bmatrix} 1 & 0 & | & 3 \\ 0 & 1 & | & -2 \end{bmatrix}$$

We have accomplished our objective! The last matrix is the augmented matrix for the system

$$\begin{aligned} x_1 &= 3 \\ x_2 &= -2 \end{aligned} \tag{6}$$

Since system (6) is equivalent to the original system (3), we have solved system (3). That is, $x_1 = 3$ and $x_2 = -2$.

Check

$$\begin{array}{ll} 3x_1 + 4x_2 = 1 & x_1 - 2x_2 = 7 \\ 3(3) + 4(-2) \stackrel{?}{=} 1 & 3 - 2(-2) \stackrel{?}{=} 7 \\ 1 \stackrel{\checkmark}{=} 1 & 7 \stackrel{\checkmark}{=} 7 \end{array}$$

The above process is written more compactly as follows:

Step 1: Need a 1 here
$$\begin{bmatrix} 3 & 4 & | & 1 \\ 1 & -2 & | & 7 \end{bmatrix} \quad R_1 \leftrightarrow R_2$$

Step 2: Need a 0 here
$$\sim \begin{bmatrix} 1 & -2 & | & 7 \\ 3 & 4 & | & 1 \end{bmatrix} \quad R_2 + (-3)R_1 \rightarrow R_2$$
$$\begin{matrix} -3 & 6 & -21 \end{matrix}$$

Step 3: Need a 1 here
$$\sim \begin{bmatrix} 1 & -2 & | & 7 \\ 0 & 10 & | & -20 \end{bmatrix} \quad \tfrac{1}{10}R_2 \rightarrow R_2$$

$$\begin{matrix} 0 & 2 & -4 \end{matrix}$$
Step 4: Need a 0 here
$$\sim \begin{bmatrix} 1 & -2 & | & 7 \\ 0 & 1 & | & -2 \end{bmatrix} \quad R_1 + 2R_2 \rightarrow R_1$$

$$\sim \begin{bmatrix} 1 & 0 & | & 3 \\ 0 & 1 & | & -2 \end{bmatrix}$$

Therefore, $x_1 = 3$ and $x_2 = -2$.

Problem 6   Solve, using augmented matrix methods: $\begin{aligned} 2x_1 - x_2 &= -7 \\ x_1 + 2x_2 &= 4 \end{aligned}$

EXAMPLE 7   Solving a System Using Augmented Matrix Methods

Solve, using augmented matrix methods: $\begin{aligned} 2x_1 - 3x_2 &= 7 \\ 3x_1 + 4x_2 &= 2 \end{aligned}$

Solution

Step 1:
Need a 1 here
$\begin{bmatrix} 2 & -3 & | & 7 \\ 3 & 4 & | & 2 \end{bmatrix}$ $\frac{1}{2}R_1 \to R_1$

Step 2:
Need a 0 here
$\sim \begin{bmatrix} 1 & -\frac{3}{2} & | & \frac{7}{2} \\ 3 & 4 & | & 2 \end{bmatrix}$ $R_2 + (-3)R_1 \to R_2$

$\begin{matrix} -3 & \frac{9}{2} & -\frac{21}{2} \end{matrix}$

Step 3:
Need a 1 here
$\sim \begin{bmatrix} 1 & -\frac{3}{2} & | & \frac{7}{2} \\ 0 & \frac{17}{2} & | & -\frac{17}{2} \end{bmatrix}$ $\frac{2}{17}R_2 \to R_2$

$\begin{matrix} 0 & \frac{3}{2} & -\frac{3}{2} \end{matrix}$

Step 4:
Need a 0 here
$\sim \begin{bmatrix} 1 & -\frac{3}{2} & | & \frac{7}{2} \\ 0 & 1 & | & -1 \end{bmatrix}$ $R_1 + \frac{3}{2}R_2 \to R_1$

$\sim \begin{bmatrix} 1 & 0 & | & 2 \\ 0 & 1 & | & -1 \end{bmatrix}$

Thus, $x_1 = 2$ and $x_2 = -1$. You should check this solution in the original system.

Problem 7 Solve, using augmented matrix methods: $\begin{aligned} 5x_1 - 2x_2 &= 12 \\ 2x_1 + 3x_2 &= 1 \end{aligned}$

EXAMPLE 8 Solving a System Using Augmented Matrix Methods

Solve, using augmented matrix methods: $\begin{aligned} 2x_1 - x_2 &= 4 \\ -6x_1 + 3x_2 &= -12 \end{aligned}$ (7)

Solution

$\begin{bmatrix} 2 & -1 & | & 4 \\ -6 & 3 & | & -12 \end{bmatrix}$ $\frac{1}{2}R_1 \to R_1$ (This produces a 1 in the upper left corner.)
$\frac{1}{3}R_2 \to R_2$ (This simplifies R_2.)

$\sim \begin{bmatrix} 1 & -\frac{1}{2} & | & 2 \\ -2 & 1 & | & -4 \end{bmatrix}$ $R_2 + 2R_1 \to R_2$ (This produces a 0 in the lower left corner.)

$\begin{matrix} 2 & -1 & 4 \end{matrix}$

$\sim \begin{bmatrix} 1 & -\frac{1}{2} & | & 2 \\ 0 & 0 & | & 0 \end{bmatrix}$

The last matrix corresponds to the system

$$x_1 - \tfrac{1}{2}x_2 = 2 \qquad x_1 - \tfrac{1}{2}x_2 = 2$$
$$0 = 0 \qquad 0x_1 + 0x_2 = 0$$

Thus, $x_1 = \tfrac{1}{2}x_2 + 2$. Hence, for any real number t, if $x_2 = t$, then $x_1 = \tfrac{1}{2}t + 2$. That is, the solution set is described by

$$(\tfrac{1}{2}t + 2, t) \qquad t \text{ a real number} \tag{8}$$

For example, if $t = 6$, then $(5, 6)$ is a particular solution; if $t = -2$, then $(1, -2)$ is another particular solution; and so on. Geometrically, the graphs of the two original equations coincide and there are infinitely many solutions.

In general, if we end up with a row of 0's in an augmented matrix for a two equation–two variable system, the system is dependent and there are infinitely many solutions.

Check The following is a check that (8) provides a solution for system (7) for any real number t:

$$2x_1 - x_2 = 4 \qquad\qquad -6x_1 + 3x_2 = -12$$
$$2(\tfrac{1}{2}t + 2) - t \overset{?}{=} 4 \qquad -6(\tfrac{1}{2}t + 2) + 3t \overset{?}{=} -12$$
$$t + 4 - t \overset{?}{=} 4 \qquad\qquad -3t - 12 + 3t \overset{?}{=} -12$$
$$4 \overset{\checkmark}{=} 4 \qquad\qquad\qquad -12 \overset{\checkmark}{=} -12$$

Problem 8 Solve, using augmented matrix methods:
$$-2x_1 + 6x_2 = 6$$
$$3x_1 - 9x_2 = -9$$

EXAMPLE 9 **Solving a System Using Augmented Matrix Methods**

Solve, using augmented matrix methods:
$$2x_1 + 6x_2 = -3$$
$$x_1 + 3x_2 = 2$$

Solution

$$\begin{bmatrix} 2 & 6 & | & -3 \\ 1 & 3 & | & 2 \end{bmatrix} \quad R_1 \leftrightarrow R_2$$

$$\sim \begin{bmatrix} 1 & 3 & | & 2 \\ 2 & 6 & | & -3 \end{bmatrix} \quad R_2 + (-2)R_1 \to R_2$$
$$\qquad\quad -2 \quad -6 \quad -4$$

$$\sim \begin{bmatrix} 1 & 3 & | & 2 \\ 0 & 0 & | & -7 \end{bmatrix} \quad R_2 \text{ implies the contradiction: } 0 = -7$$

This is the augmented matrix of the system

$$x_1 + 3x_2 = 2 \qquad x_1 + 3x_2 = 2$$
$$0 = -7 \qquad 0x_1 + 0x_2 = -7$$

The second equation is not satisfied by any ordered pair of real numbers. Hence, the original system is inconsistent and has no solution. Otherwise, we have proved that $0 = -7$!

Thus, if we obtain all 0's to the left of the vertical bar and a nonzero number to the right of the bar in a row of an augmented matrix, then the system is inconsistent and there are no solutions.

Problem 9 Solve, using augmented matrix methods:

$$2x_1 - x_2 = 3$$
$$4x_1 - 2x_2 = -1$$

Summary

For m, n, p real numbers: $p \neq 0$:

Form 1: A Unique Solution (Consistent and Independent)

$$\begin{bmatrix} 1 & 0 & | & m \\ 0 & 1 & | & n \end{bmatrix}$$

Form 2: Infinitely Many Solutions (Consistent and Dependent)

$$\begin{bmatrix} 1 & m & | & n \\ 0 & 0 & | & 0 \end{bmatrix}$$

Form 3: No Solution (Inconsistent)

$$\begin{bmatrix} 1 & m & | & n \\ 0 & 0 & | & p \end{bmatrix}$$

The process of solving systems of equations described in this section is referred to as *Gauss–Jordan elimination*. We will use this method to solve larger-scale systems in the next section, including systems where the number of equations and the number of variables are not the same.

Answers to Matched Problems

1.

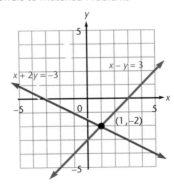

$x = 1, y = -2$

Check:

$$x - y = 3$$
$$1 - (-2) \overset{?}{=} 3$$
$$3 \overset{\checkmark}{=} 3$$

$$x + 2y = -3$$
$$1 + 2(-2) \overset{?}{=} -3$$
$$-3 \overset{\checkmark}{=} -3$$

2. (A) $(3, 2)$, or $x = 3$ and $y = 2$ (B) No solution   (C) Infinite number of solutions

3. $(2, -3)$, or $x = 2$ and $y = -3$ **4.** $(-1, 3)$, or $x = -1$ and $y = 3$

5. Airspeed = 440 mph, wind rate = 40 mph

6. $x_1 = -2, x_2 = 3$ **7.** $x_1 = 2, x_2 = -1$

8. The system is dependent. For t any real number, $x_2 = t, x_1 = 3t - 3$ is a solution.

9. Inconsistent—no solution

EXERCISE 6-1

A

Solve Problems 1–6 by graphing.

1. $x + y = 7$
$x - y = 3$

2. $x - y = 2$
$x + y = 4$

3. $3x - 2y = 12$
$7x + 2y = 8$

4. $3x - y = 2$
$x + 2y = 10$

5. $3u + 5v = 15$
$6u + 10v = -30$

6. $m + 2n = 4$
$2m + 4n = -8$

Solve Problems 7–12 by substitution.

7. $x - y = 4$
$x + 3y = 12$

8. $2x - y = 3$
$x + 2y = 14$

9. $3x - y = 7$
$2x + 3y = 1$

10. $2x + y = 6$
$x - y = -3$

11. $y = 0.08x$
$y = 100 + 0.04x$

12. $y = 0.07x$
$y = 80 + 0.05x$

Solve Problems 13–20 using elimination by addition.

13. $2x + 3y = 1$
$3x - y = 7$

14. $2m - n = 10$
$m - 2n = -4$

15. $4x + 3y = 26$
$3x - 11y = -7$

16. $9x - 3y = 24$
$11x + 2y = 1$

17. $7m + 12n = -1$
$5m - 3n = 7$

18. $3x + 8y = 4$
$15x + 10y = -10$

19. $4x - 3y = 15$
$3x + 4y = 5$

20. $5x + 2y = 1$
$2x - 3y = -11$

Perform each of the row operations indicated in Problems 21–32 on the following matrix:

$$\begin{bmatrix} 1 & -3 & | & 2 \\ 4 & -6 & | & -8 \end{bmatrix}$$

21. $R_1 \leftrightarrow R_2$

22. $\frac{1}{2}R_2 \to R_2$

23. $-4R_1 \to R_1$

24. $-2R_1 \to R_1$

25. $2R_2 \to R_2$

26. $-1R_2 \to R_2$

27. $R_2 + (-4)R_1 \to R_2$

28. $R_1 + (-\frac{1}{2})R_2 \to R_1$

29. $R_2 + (-2)R_1 \to R_2$

30. $R_2 + (-3)R_1 \to R_2$

31. $R_2 + (-1)R_1 \to R_2$

32. $R_2 + 1R_1 \to R_2$

B

Solve Problems 33–48 using augmented matrix methods:

33. $x_1 + x_2 = 5$
$x_1 - x_2 = 1$

34. $x_1 - x_2 = 2$
$x_1 + x_2 = 6$

35. $x_1 - 2x_2 = 1$
$2x_1 - x_2 = 5$

36. $x_1 + 3x_2 = 1$
$3x_1 - 2x_2 = 14$

37. $x_1 - 4x_2 = -2$
$-2x_1 + x_2 = -3$

38. $x_1 - 3x_2 = -5$
$-3x_1 - x_2 = 5$

39. $3x_1 - x_2 = 2$
$x_1 + 2x_2 = 10$

40. $2x_1 + x_2 = 0$
$x_1 - 2x_2 = -5$

41. $x_1 + 2x_2 = 4$
$2x_1 + 4x_2 = -8$

42. $2x_1 - 3x_2 = -2$
$-4x_1 + 6x_2 = 7$

43. $2x_1 + x_2 = 6$
$x_1 - x_2 = -3$

44. $3x_1 - x_2 = -5$
$x_1 + 3x_2 = 5$

45. $3x_1 - 6x_2 = -9$
$-2x_1 + 4x_2 = 6$

46. $2x_1 - 4x_2 = -2$
$-3x_1 + 6x_2 = 3$

47. $4x_1 - 2x_2 = 2$
$-6x_1 + 3x_2 = -3$

48. $-6x_1 + 2x_2 = 4$
$3x_1 - x_2 = -2$

C _____

Solve Problems 49–54 using augmented matrix methods:

49. $3x_1 - x_2 = 7$
$2x_1 + 3x_2 = 1$

50. $2x_1 - 3x_2 = -8$
$5x_1 + 3x_2 = 1$

51. $3x_1 + 2x_2 = 4$
$2x_1 - x_2 = 5$

52. $4x_1 + 3x_2 = 26$
$3x_1 - 11x_2 = -7$

53. $0.2x_1 - 0.5x_2 = 0.07$
$0.8x_1 - 0.3x_2 = 0.79$

54. $0.3x_1 - 0.6x_2 = 0.18$
$0.5x_1 - 0.2x_2 = 0.54$

APPLICATIONS

55. Puzzle. A friend of yours came out of the post office having spent $18.75 on 29¢ and 23¢ stamps. If she bought 75 stamps in all, how many of each type did she buy?

56. Puzzle. A parking meter contains only nickels and dimes worth $6.05. If there are 89 coins in all, how many of each type are there?

57. Airspeed. It takes a private airplane 8.75 hours to make the 2,100-mile flight from Atlanta to Los Angeles and 5 hours to make the return trip. Assuming that the plane travels at a constant airspeed and that the wind blows at a constant rate from Los Angeles to Atlanta, find the airspeed of the plane and the wind rate.

58. Airspeed. Repeat Problem 57 if the trip from Atlanta to Los Angeles takes 5 hours and the return trip takes 4.2 hours.

59. Chemistry. A chemist has two solutions of hydrochloric acid in stock: a 50% solution and an 80% solution. How much of each should she mix to obtain 100 milliliters of a 68% solution?

60. Chemistry. Repeat Problem 59 assuming the 50% stock solution is replaced with a 60% stock solution.

★**61. Business.** A jeweler has two bars of gold alloy in stock, one 12 carat and the other 18 carat (24-carat gold is pure gold, 12-carat gold is $\frac{12}{24}$ pure, 18-carat gold is $\frac{18}{24}$ pure, and so on). How many grams of each alloy must be mixed to obtain 10 grams of 14-carat gold?

★**62. Business.** Repeat Problem 61 assuming the jeweler has only 10-carat and pure gold in stock.

★**63. Physics.** An object dropped off the top of a tall building falls vertically with constant acceleration. If s is the distance of the object above the ground (in feet) t seconds after its release, then s and t are related by an equation of the form

$$s = a + bt^2$$

where a and b are constants. Suppose the object is 180 feet above the ground 1 second after its release and 132 feet above the ground 2 seconds after its release.
(A) Find the constants a and b.
(B) How high is the building?
(C) How long does the object fall?

★**64. Physics.** Repeat Problem 63 if the object is 240 feet above the ground after 1 second and 192 feet above the ground after 2 seconds.

★**65. Earth Science.** An earthquake emits a primary wave and a secondary wave. Near the surface of the Earth the primary wave travels at about 5 miles per second and the secondary wave at about 3 miles per second. From the time lag between the two waves arriving at a given receiving station, it is possible to estimate the distance to the quake. (The *epicenter* can be located by obtaining distance bearings at three or more stations.) Suppose a station measured a time difference of 16 seconds between the arrival of the two waves. How long did each wave travel, and how far was the earthquake from the station?

★**66. Earth Science.** A ship using sound-sensing devices above and below water recorded a surface explosion 6 seconds sooner by its underwater device than its above-water device. Sound travels in air at about 1,100 feet per second and in seawater at about 5,000 feet per second.
(A) How long did it take each sound wave to reach the ship?
(B) How far was the explosion from the ship?

SECTION 6-2 Gauss–Jordan Elimination

- Reduced Matrices
- Solving Systems by Gauss–Jordan Elimination
- Application

Now that you have had some experience with row operations on simple augmented matrices, we will consider systems involving more than two variables. In addition, we will not require that a system have the same number of equations as variables. It turns out that the results for two-variable, two-equation linear systems, stated in Theorem 1, Section 6-1, actually hold for linear systems of any size. That is, it can be shown that

Any linear system must have exactly one solution, no solution, or an infinite number of solutions,

regardless of the number of equations or the number of variables in the system. The terms *unique solution, consistent, inconsistent, dependent,* and *independent* are used to describe these solutions, just as they are for systems with two variables.

• Reduced Matrices

Again, our objective is to start with the augmented matrix of a linear system and transform it using row operations from Theorem 3 in Section 6-1 into a simple form where the solution can be read by inspection. The simple form obtained is called the *reduced form*, which we define below.

DEFINITION 1

Reduced Matrix

A matrix is in **reduced form** if:

1. Each row consisting entirely of 0's is below any row having at least one nonzero element.

2. The leftmost nonzero element in each row is 1.

3. The column containing the leftmost 1 of a given row has 0's above and below the 1.

4. The leftmost 1 in any row is to the right of the leftmost 1 in the preceding row.

EXAMPLE 1 Reduced Matrices

The following matrices are in reduced form. Check each one carefully to convince yourself that the conditions in the definition are met.

$$\begin{bmatrix} 1 & 0 & | & 2 \\ 0 & 1 & | & -3 \end{bmatrix} \qquad \begin{bmatrix} 1 & 0 & 0 & | & 2 \\ 0 & 1 & 0 & | & -1 \\ 0 & 0 & 1 & | & 3 \end{bmatrix} \qquad \begin{bmatrix} 1 & 0 & | & 3 \\ 0 & 1 & | & -1 \\ 0 & 0 & | & 0 \end{bmatrix}$$

$$\begin{bmatrix} 1 & 4 & 0 & 0 & | & -3 \\ 0 & 0 & 1 & 0 & | & 2 \\ 0 & 0 & 0 & 1 & | & 6 \end{bmatrix} \qquad \begin{bmatrix} 1 & 0 & 4 & | & 0 \\ 0 & 1 & 3 & | & 0 \\ 0 & 0 & 0 & | & 1 \end{bmatrix}$$

Problem 1   The matrices below are not in reduced form. Indicate which condition in the definition is violated for each matrix.

(A) $\begin{bmatrix} 1 & 0 & | & 2 \\ 0 & 3 & | & -6 \end{bmatrix}$ (B) $\begin{bmatrix} 1 & 5 & 4 & | & 3 \\ 0 & 1 & 2 & | & -1 \\ 0 & 0 & 0 & | & 0 \end{bmatrix}$

(C) $\begin{bmatrix} 0 & 1 & 0 & | & -3 \\ 1 & 0 & 0 & | & 0 \\ 0 & 0 & 1 & | & 2 \end{bmatrix}$ (D) $\begin{bmatrix} 1 & 2 & 0 & | & 3 \\ 0 & 0 & 0 & | & 0 \\ 0 & 0 & 1 & | & 4 \end{bmatrix}$

Example 2 illustrates methods of writing solutions to reduced systems. A **reduced system** is a linear system that corresponds to a reduced augmented matrix.

EXAMPLE 2   **Writing Solutions to Reduced Systems**

Write the linear system corresponding to each reduced augmented matrix and solve.

(A) $\begin{bmatrix} 1 & 0 & 0 & | & 2 \\ 0 & 1 & 0 & | & -1 \\ 0 & 0 & 1 & | & 3 \end{bmatrix}$ (B) $\begin{bmatrix} 1 & 0 & 4 & | & 0 \\ 0 & 1 & 3 & | & 0 \\ 0 & 0 & 0 & | & 1 \end{bmatrix}$

(C) $\begin{bmatrix} 1 & 0 & 2 & | & -3 \\ 0 & 1 & -1 & | & 8 \\ 0 & 0 & 0 & | & 0 \end{bmatrix}$ (D) $\begin{bmatrix} 1 & 4 & 0 & 0 & 3 & | & -2 \\ 0 & 0 & 1 & 0 & -2 & | & 0 \\ 0 & 0 & 0 & 1 & 2 & | & 4 \end{bmatrix}$

Solutions   **(A)**
$$\begin{aligned} x_1 &= 2 \\ x_2 &= -1 \\ x_3 &= 3 \end{aligned}$$

The solution is obvious: $x_1 = 2, x_2 = -1, x_3 = 3$.

(B)
$$\begin{aligned} x_1 + 4x_3 &= 0 \\ x_2 + 3x_3 &= 0 \\ 0x_1 + 0x_2 + 0x_3 &= 1 \end{aligned}$$

The last equation implies $0 = 1$, which is a contradiction. Hence, the system is inconsistent and has no solution.

(C) The system corresponding to the reduced augmented matrix is

$$
\begin{aligned}
x_1 \quad + 2x_3 &= -3 \\
x_2 - \ x_3 &= \ \ 8
\end{aligned}
$$

Discard the equation corresponding to the third all-zero row in the matrix, since it is satisfied by all values of x_1, x_2, and x_3.

Note that the leftmost variable in each equation appears in one and only one equation. We solve for the leftmost variables x_1 and x_2 in terms of the remaining variable x_3:

$$
\begin{aligned}
x_1 &= -2x_3 - 3 \\
x_2 &= x_3 + 8
\end{aligned}
$$

If we let $x_3 = t$, then for any real number t

$$
\begin{aligned}
x_1 &= -2t - 3 \\
x_2 &= t + 8 \\
x_3 &= t
\end{aligned}
$$

is a solution. Some particular solutions are

$$
\begin{array}{ccc}
t = 0 & t = -2 & t = 3.5 \\
(-3, 8, 0) & (1, 6, -2) & (-10, 11.5, 3.5)
\end{array}
$$

In general, **when a reduced system has more variables than equations and contains no contradictions, the system is dependent and has infinitely many solutions**. It is not difficult to describe the solution set to this type of system. In a reduced system the **leftmost variables** are the variables that correspond to the leftmost 1's in the corresponding reduced augmented matrix. The definition of reduced form for an augmented matrix ensures that each leftmost variable in the corresponding reduced system appears in one and only one equation of the system. Thus, for a reduced system with more variables than equations and no contradictions, it is usually easy to solve for each leftmost variable in terms of the remaining variables and to write a general solution to the system, as we did above. Part (D) illustrates a slightly more involved case.

(D)
$$
\begin{aligned}
x_1 + 4x_2 \quad + 3x_5 &= -2 \\
x_3 \quad - 2x_5 &= \ \ \ 0 \\
x_4 + 2x_5 &= \ \ \ 4
\end{aligned}
$$

Solve for the leftmost variables x_1, x_3, and x_4 in terms of the remaining variables x_2 and x_5:

$$
\begin{aligned}
x_1 &= -4x_2 - 3x_5 - 2 \\
x_3 &= 2x_5 \\
x_4 &= -2x_5 + 4
\end{aligned}
$$

If we let $x_2 = s$ and $x_5 = t$, then for any real numbers s and t,

$$x_1 = -4s - 3t - 2$$
$$x_2 = s$$
$$x_3 = 2t$$
$$x_4 = -2t + 4$$
$$x_5 = t$$

is a solution. The system is dependent and has infinitely many solutions.

Problem 2   Write the linear system corresponding to each reduced augmented matrix and solve.

(A) $\begin{bmatrix} 1 & 0 & 0 & | & -5 \\ 0 & 1 & 0 & | & 3 \\ 0 & 0 & 1 & | & 6 \end{bmatrix}$ (B) $\begin{bmatrix} 1 & 2 & -3 & | & 0 \\ 0 & 0 & 0 & | & 1 \\ 0 & 0 & 0 & | & 0 \end{bmatrix}$

(C) $\begin{bmatrix} 1 & 0 & -2 & | & 4 \\ 0 & 1 & 3 & | & -2 \\ 0 & 0 & 0 & | & 0 \end{bmatrix}$ (D) $\begin{bmatrix} 1 & 0 & 3 & 2 & | & 5 \\ 0 & 1 & -2 & -1 & | & 3 \\ 0 & 0 & 0 & 0 & | & 0 \end{bmatrix}$

● **Solving Systems by Gauss–Jordan Elimination**

We are now ready to outline a step-by-step procedure for solving systems of linear equations called the **Gauss–Jordan elimination*** method. The method provides us with a systematic way of transforming augmented matrices into a reduced form from which we can write the solution to the original system by inspection, if a solution exists. The method also reveals when a solution fails to exist. (see Example 5).

EXAMPLE 3   **Solving a System Using Gauss–Jordan Elimination**

Solve by Gauss–Jordan elimination:
$$2x_1 - 2x_2 + x_3 = 3$$
$$3x_1 + x_2 - x_3 = 7$$
$$x_1 - 3x_2 + 2x_3 = 0$$

Solution   Write the augmented matrix and follow the steps indicated at the right.

$\begin{bmatrix} 2 & -2 & 1 & | & 3 \\ 3 & 1 & -1 & | & 7 \\ 1 & -3 & 2 & | & 0 \end{bmatrix}$ $R_1 \leftrightarrow R_3$

Need a 1 here

Step 1. Choose leftmost nonzero column and get a 1 at the top.

---

*Named after the German mathematician Carl Friedrich Gauss (1777–1855) and the French mathematician Camille Jordan (1838–1922). Gauss is considered by many to be one of the three greatest mathematicians of all time (along with Archimedes and Newton). He used a method of solving systems of equations in his research that was later generalized by Jordan to the method we present here.

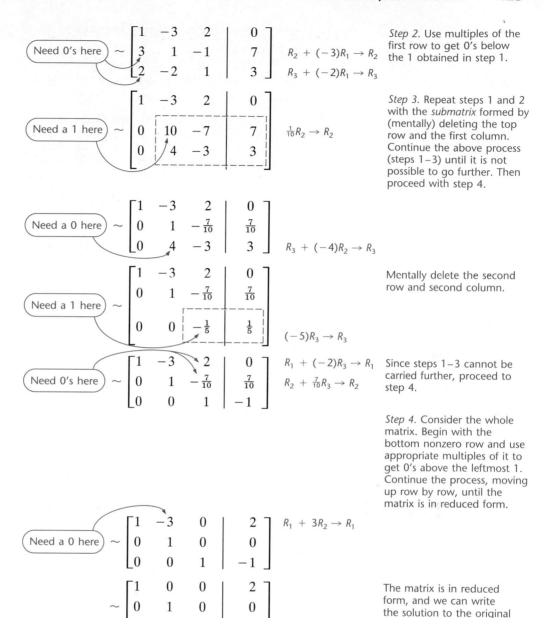

Step 2. Use multiples of the first row to get 0's below the 1 obtained in step 1.

Step 3. Repeat steps 1 and 2 with the *submatrix* formed by (mentally) deleting the top row and the first column. Continue the above process (steps 1–3) until it is not possible to go further. Then proceed with step 4.

Mentally delete the second row and second column.

Since steps 1–3 cannot be carried further, proceed to step 4.

Step 4. Consider the whole matrix. Begin with the bottom nonzero row and use appropriate multiples of it to get 0's above the leftmost 1. Continue the process, moving up row by row, until the matrix is in reduced form.

The matrix is in reduced form, and we can write the solution to the original system by inspection.

Solution: $x_1 = 2, x_2 = 0, x_3 = -1$. It is left to you to check this solution.

Steps 1–4 outlined in the solution of Example 3 are referred to as **Gauss–Jordan elimination**. The steps are summarized on the next page for easy reference.

Gauss–Jordan Elimination

Step 1. Choose the leftmost nonzero column and use appropriate row operations to get a 1 at the top.

Step 2. Use multiples of the first row to get 0's in all places below the 1 obtained in step 1.

Step 3. Delete (mentally) the top row and first column of the matrix. Repeat steps 1 and 2 with the **submatrix** (the matrix that remains after deleting the top row and first column). Continue this process (steps 1–3) until it is not possible to go further.

Step 4. Now consider the whole matrix. Begin with the bottom nonzero row and use appropriate multiples of it to get 0's above the leftmost 1. Continue this process, moving up row by row, until the matrix is finally in reduced form.

[*Note:* If at any point in the process we obtain a row with all 0's to the left of the vertical line and a nonzero number n to the right, we can stop, since we have a contradiction: $0 = n, n \neq 0$. We can then conclude that the system has no solution.]

Problem 3 Solve by Gauss–Jordan elimination:
$$\begin{aligned} 3x_1 + x_2 - 2x_3 &= 2 \\ x_1 - 2x_2 + x_3 &= 3 \\ 2x_1 - x_2 - 3x_3 &= 3 \end{aligned}$$

EXAMPLE 4 Solving a System Using Gauss–Jordan Elimination

Solve by Gauss–Jordan elimination:
$$\begin{aligned} 2x_1 - x_2 + 4x_3 &= -2 \\ 3x_1 + 2x_2 - x_3 &= 1 \end{aligned}$$

Solution

Need a 1 here
$$\begin{bmatrix} 2 & -1 & 4 & | & -2 \\ 3 & 2 & -1 & | & 1 \end{bmatrix} \quad \tfrac{1}{2}R_1 \to R_1$$

Need a 0 here
$$\sim \begin{bmatrix} 1 & -\tfrac{1}{2} & 2 & | & -1 \\ 3 & 2 & -1 & | & 1 \end{bmatrix} \quad R_2 + (-3)R_1 \to R_2$$

Need a 1 here
$$\sim \begin{bmatrix} 1 & -\tfrac{1}{2} & 2 & | & -1 \\ 0 & \tfrac{7}{2} & -7 & | & 4 \end{bmatrix} \quad \tfrac{2}{7}R_2 \to R_2$$

Need a 0 here
$$\sim \begin{bmatrix} 1 & -\tfrac{1}{2} & 2 & | & -1 \\ 0 & 1 & -2 & | & \tfrac{8}{7} \end{bmatrix} \quad R_1 + \tfrac{1}{2}R_2 \to R_1$$

$$\sim \begin{bmatrix} 1 & 0 & 1 & | & -\tfrac{3}{7} \\ 0 & 1 & -2 & | & \tfrac{8}{7} \end{bmatrix}$$

The matrix is now in reduced form. Write the corresponding system and the solution.

$$\begin{aligned} x_1 \quad + \quad x_3 &= -\tfrac{3}{7} \\ x_2 - 2x_3 &= \tfrac{8}{7} \end{aligned}$$

Solve for the leftmost variables x_1 and x_2 in terms of the remaining variable x_3:

$$x_1 = -x_3 - \tfrac{3}{7}$$
$$x_2 = 2x_3 + \tfrac{8}{7}$$

If $x_3 = t$, then for t any real number,

$$x_1 = -t - \tfrac{3}{7}$$
$$x_2 = 2t + \tfrac{8}{7}$$
$$x_3 = t$$

is a solution. You should check this solution in the original system.

Problem 4   Solve by Gauss–Jordan elimination:
$$\begin{aligned} 3x_1 + 6x_2 - 3x_3 &= 2 \\ 2x_1 - x_2 + 2x_3 &= -1 \end{aligned}$$

EXAMPLE 5   **Solving a System Using Gauss–Jordan Elimination**

Solve by Gauss–Jordan elimination:
$$\begin{aligned} 2x_1 - x_2 &= -4 \\ x_1 + 2x_2 &= 3 \\ 3x_1 - x_2 &= -1 \end{aligned}$$

Solution

$$\begin{bmatrix} 2 & -1 & | & -4 \\ 1 & 2 & | & 3 \\ 3 & -1 & | & -1 \end{bmatrix} \quad R_1 \leftrightarrow R_2$$

$$\sim \begin{bmatrix} 1 & 2 & | & 3 \\ 2 & -1 & | & -4 \\ 3 & -1 & | & -1 \end{bmatrix} \quad \begin{aligned} R_2 + (-2)R_1 &\rightarrow R_2 \\ R_3 + (-3)R_1 &\rightarrow R_3 \end{aligned}$$

$$\sim \begin{bmatrix} 1 & 2 & | & 3 \\ 0 & -5 & | & -10 \\ 0 & -7 & | & -10 \end{bmatrix} \quad -\tfrac{1}{5}R_2 \rightarrow R_2$$

$$\sim \begin{bmatrix} 1 & 2 & | & 3 \\ 0 & 1 & | & 2 \\ 0 & -7 & | & -10 \end{bmatrix} \quad R_3 + 7R_2 \rightarrow R_3$$

$$\sim \begin{bmatrix} 1 & 2 & | & 3 \\ 0 & 1 & | & 2 \\ 0 & 0 & | & 4 \end{bmatrix} \quad \begin{array}{l}\text{We stop the Gauss–Jordan elimination} \\ \text{even though the matrix is not in reduced form,} \\ \text{since the last row produces a contradiction.}\end{array}$$

The last row implies $0 = 4$, which is a contradiction; therefore, the system has no solution.

Problem 5   Solve by Gauss–Jordan elimination:
$$
\begin{aligned}
3x_1 + x_2 &= 5 \\
2x_1 + 3x_2 &= 1 \\
x_1 - x_2 &= 3
\end{aligned}
$$

• **Application**   We now consider an application that involves a dependent system of equations.

EXAMPLE 6   **Purchasing Decision**

A university computer center wants to purchase 8 hard disk drives with a combined storage capacity of 300 megabytes of data (a megabyte of data is approximately 1 million characters). The three types of disk drives available for their computer system have the capacity of storing 20, 40, and 80 megabytes of data, respectively. How many disk drives of each type should be purchased?

Solution   Let

$$x_1 = \text{Number of 20-megabyte disk drives}$$
$$x_2 = \text{Number of 40-megabyte disk drives}$$
$$x_3 = \text{Number of 80-megabyte disk drives}$$

Then

$$x_1 + x_2 + x_3 = 8 \quad \text{Total number of disk drives}$$
$$20x_1 + 40x_2 + 80x_3 = 300 \quad \text{Total storage capacity}$$

Now we can form the augmented matrix of the system and solve by using Gauss–Jordan elimination:

$$
\begin{bmatrix}
1 & 1 & 1 & \Big| & 8 \\
20 & 40 & 80 & \Big| & 300
\end{bmatrix}
\quad \tfrac{1}{20}R_2 \rightarrow R_2 \ (\text{simplify } R_2)
$$

$$
\sim
\begin{bmatrix}
1 & 1 & 1 & \Big| & 8 \\
1 & 2 & 4 & \Big| & 15
\end{bmatrix}
\quad R_2 + (-1)R_1 \rightarrow R_2
$$

$$
\sim
\begin{bmatrix}
1 & 1 & 1 & \Big| & 8 \\
0 & 1 & 3 & \Big| & 7
\end{bmatrix}
\quad R_1 + (-1)R_2 \rightarrow R_1
$$

$$
\sim
\begin{bmatrix}
1 & 0 & -2 & \Big| & 1 \\
0 & 1 & 3 & \Big| & 7
\end{bmatrix}
\quad \text{Matrix is in reduced form.}
$$

$$
\begin{aligned}
x_1 \quad - 2x_3 &= 1 \qquad \text{or} \qquad x_1 = 2x_3 + 1 \\
x_2 + 3x_3 &= 7 \qquad\qquad\quad\ x_2 = -3x_3 + 7
\end{aligned}
$$

Let $x_3 = t$. Then for t any real number,

$$x_1 = 2t + 1$$

$$x_2 = -3t + 7$$

$$x_3 = t$$

is a solution—or is it? Since we can't purchase a negative or fractional number of disk drives, x_1, x_2, and x_3 must be nonnegative whole numbers. Thus, in the third equation ($x_3 = t$), t must also be a nonnegative whole number. And in the second equation ($x_2 = -3t + 7$), t is further restricted to the values 0, 1, and 2 to avoid negative values of x_2. Thus, there are only three possible configurations that meet the computer center's specifications of 8 disk drives with a total storage capacity of 300 megabytes, as shown in the table:

t	20-megabyte drives, x_1	40-megabyte drives, x_2	80-megabyte drives, x_3
0	1	7	0
1	3	4	1
2	5	1	2

The final choice would probably be influenced by other factors. For example, the computer center might want to minimize the total purchase cost of the 8 disk drives.

Problem 6　Repeat Example 6 if the computer center wants to purchase 10 disk drives with a total storage capacity of 360 megabytes.

Answers to Matched Problems
1. (A) Condition 2 is violated: The 3 in the second row should be a 1.
   (B) Condition 3 is violated: In the second column, the 5 should be a 0.
   (C) Condition 4 is violated: The leftmost 1 in the second row is not to the right of the leftmost 1 in the first row.
   (D) Condition 1 is violated: The all-zero second row should be at the bottom.
2. (A)　　　$x_1 = -5$　　　　　(B)　　$x_1 + 2x_2 - 3x_3 = 0$

　　　　　　　$x_2 = 3$　　　　　　　　　$0x_1 + 0x_2 + 0x_3 = 1$

　　　　　　　$x_3 = 6$　　　　　　　　　$0x_1 + 0x_2 + 0x_3 = 0$

　　Solution: $x_1 = -5, x_2 = 3, x_3 = 6$　　　Inconsistent; no solution

　(C)　　　$x_1 - 2x_3 = 4$　　　　　(D)　　$x_1 + 3x_3 + 2x_4 = 5$

　　　　　　　$x_2 + 3x_3 = -2$　　　　　　　$x_2 - 2x_3 - x_4 = 3$

　　Dependent: Let $x_3 = t$.　　　　　　Dependent: Let $x_3 = s$ and $x_4 = t$.

　　Then for any real t,　　　　　　　Then for any real s and t,

　　　　　$x_1 = 2t + 4$　　　　　　　　$x_1 = -3s - 2t + 5$

　　　　　$x_2 = -3t - 2$　　　　　　　$x_2 = 2s + t + 3$

　　　　　$x_3 = t$　　　　　　　　　　$x_3 = s$

　　is a solution.　　　　　　　　　$x_4 = t$

　　　　　　　　　　　　　　　　is a solution.

3. $x_1 = 1, x_2 = -1, x_3 = 0$ **4.** $x_1 = -\frac{3}{5}t - \frac{4}{15}, x_2 = \frac{4}{5}t + \frac{7}{15}, x_3 = t, t$ any real number
5. $x_1 = 2, x_2 = -1$
6. $x_1 = 2t + 2$ (20-megabyte drives), $x_2 = -3t + 8$ (40-megabyte drives), $x_3 = t$ (80-megabyte drives), where $t = 0, 1,$ or 2

EXERCISE 6-2

A

In Problems 1–8, indicate whether each matrix is in reduced form.

1. $\begin{bmatrix} 1 & 0 & 2 \\ 0 & 1 & -1 \end{bmatrix}$ yes

2. $\begin{bmatrix} 0 & 1 & 2 \\ 1 & 0 & -1 \end{bmatrix}$ no

3. $\begin{bmatrix} 1 & 0 & 2 & 3 \\ 0 & 0 & 0 & 0 \\ 0 & 1 & -1 & 4 \end{bmatrix}$ no

4. $\begin{bmatrix} 1 & 0 & 0 & -2 \\ 0 & 1 & 0 & 0 \\ 0 & 0 & 1 & 1 \end{bmatrix}$ yes

5. $\begin{bmatrix} 0 & 1 & 0 & 2 \\ 0 & 0 & 3 & -1 \\ 0 & 0 & 0 & 0 \end{bmatrix}$ NO

6. $\begin{bmatrix} 1 & 3 & 0 & 0 \\ 0 & 0 & 1 & 0 \\ 0 & 0 & 0 & 1 \end{bmatrix}$ yes

7. $\begin{bmatrix} 1 & 2 & 0 & 3 & 2 \\ 0 & 0 & 1 & -1 & 0 \end{bmatrix}$ yes

8. $\begin{bmatrix} 0 & 1 & 2 & 1 \\ 1 & 0 & -3 & 2 \end{bmatrix}$ no

In Problems 9–16, write the linear system corresponding to each reduced augmented matrix and solve.

9. $\begin{bmatrix} 1 & 0 & 0 & -2 \\ 0 & 1 & 0 & 3 \\ 0 & 0 & 1 & 0 \end{bmatrix}$

10. $\begin{bmatrix} 1 & 0 & 0 & 0 & -2 \\ 0 & 1 & 0 & 0 & 0 \\ 0 & 0 & 1 & 0 & 1 \\ 0 & 0 & 0 & 1 & 3 \end{bmatrix}$

11. $\begin{bmatrix} 1 & 0 & -2 & 3 \\ 0 & 1 & 1 & -5 \\ 0 & 0 & 0 & 0 \end{bmatrix}$

12. $\begin{bmatrix} 1 & -2 & 0 & -3 \\ 0 & 0 & 1 & 5 \\ 0 & 0 & 0 & 0 \end{bmatrix}$

13. $\begin{bmatrix} 1 & 0 & 0 \\ 0 & 1 & 0 \\ 0 & 0 & 1 \end{bmatrix}$

14. $\begin{bmatrix} 1 & 0 & 5 \\ 0 & 1 & -3 \\ 0 & 0 & 0 \end{bmatrix}$

15. $\begin{bmatrix} 1 & -2 & 0 & -3 & -5 \\ 0 & 0 & 1 & 3 & 2 \end{bmatrix}$

16. $\begin{bmatrix} 1 & 0 & -2 & 3 & 4 \\ 0 & 1 & -1 & 2 & -1 \end{bmatrix}$

B

Use row operations to change each matrix in Problems 17–22 to reduced form.

17. $\begin{bmatrix} 1 & 2 & -1 \\ 0 & 1 & 3 \end{bmatrix}$

18. $\begin{bmatrix} 1 & 3 & 1 \\ 0 & 2 & -4 \end{bmatrix}$

19. $\begin{bmatrix} 1 & 0 & -3 & 1 \\ 0 & 1 & 2 & 0 \\ 0 & 0 & 3 & -6 \end{bmatrix}$

20. $\begin{bmatrix} 1 & 0 & 4 & | & 0 \\ 0 & 1 & -3 & | & -1 \\ 0 & 0 & -2 & | & 2 \end{bmatrix}$

21. $\begin{bmatrix} 1 & 2 & -2 & | & -1 \\ 0 & 3 & -6 & | & 1 \\ 0 & -1 & 2 & | & -\frac{1}{3} \end{bmatrix}$

22. $\begin{bmatrix} 0 & -2 & 8 & | & 1 \\ 2 & -2 & 6 & | & -4 \\ 0 & -1 & 4 & | & \frac{1}{2} \end{bmatrix}$

Solve Problems 23–40 using Gauss–Jordan elimination.

23. $2x_1 + 4x_2 - 10x_3 = -2$
$3x_1 + 9x_2 - 21x_3 = 0$
$x_1 + 5x_2 - 12x_3 = 1$

24. $3x_1 + 5x_2 - x_3 = -7$
$x_1 + x_2 + x_3 = -1$
$2x_1 + 11x_3 = 7$

25. $3x_1 + 8x_2 - x_3 = -18$
$2x_1 + x_2 + 5x_3 = 8$
$2x_1 + 4x_2 + 2x_3 = -4$

26. $2x_1 + 7x_2 + 15x_3 = -12$
$4x_1 + 7x_2 + 13x_3 = -10$
$3x_1 + 6x_2 + 12x_3 = -9$

27. $2x_1 - x_2 - 3x_3 = 8$
$x_1 - 2x_2 = 7$

28. $2x_1 + 4x_2 - 6x_3 = 10$
$3x_1 + 3x_2 - 3x_3 = 6$

29. $2x_1 + 3x_2 - x_3 = 1$
$x_1 - 2x_2 + 2x_3 = -2$

30. $x_1 - 3x_2 + 2x_3 = -1$
$3x_1 + 2x_2 - x_3 = 2$

31. $2x_1 + 2x_2 = 2$
$x_1 + 2x_2 = 3$
$-3x_2 = -6$

32. $2x_1 - x_2 = 0$
$3x_1 + 2x_2 = 7$
$x_1 - x_2 = -1$

33. $2x_1 - x_2 = 0$
$3x_1 + 2x_2 = 7$
$x_1 - x_2 = -2$

34. $x_1 - 3x_2 = 5$
$2x_1 + x_2 = 3$
$x_1 - 2x_2 = 5$

35. $3x_1 - 4x_2 - x_3 = 1$
$2x_1 - 3x_2 + x_3 = 1$
$x_1 - 2x_2 + 3x_3 = 2$

36. $3x_1 + 7x_2 - x_3 = 11$
$x_1 + 2x_2 - x_3 = 3$
$2x_1 + 4x_2 - 2x_3 = 10$

37. $-2x_1 + x_2 + 3x_3 = -7$
$x_1 - 4x_2 + 2x_3 = 0$
$x_1 - 3x_2 + x_3 = 1$

38. $3x_1 + 5x_2 + 4x_3 = -7$
$-x_1 - 5x_2 + 2x_3 = 9$
$x_1 - x_2 + 4x_3 = 3$

39. $2x_1 - 2x_2 - 4x_3 = -2$
$-3x_1 + 3x_2 + 6x_3 = 3$

40. $2x_1 + 8x_2 - 6x_3 = 4$
$-3x_1 - 12x_2 + 9x_3 = -6$

C

Solve Problems 41–46 using Gauss–Jordan elimination.

41. $2x_1 - 3x_2 + 3x_3 = -15$
$3x_1 + 2x_2 - 5x_3 = 19$
$5x_1 - 4x_2 - 2x_3 = -2$

42. $3x_1 - 2x_2 - 4x_3 = -8$
$4x_1 + 3x_2 - 5x_3 = -5$
$6x_1 - 5x_2 + 2x_3 = -17$

43. $5x_1 - 3x_2 + 2x_3 = 13$
$2x_1 + 4x_2 - 3x_3 = -9$
$4x_1 - 2x_2 + 5x_3 = 13$

44. $4x_1 - 2x_2 + 3x_3 = 0$
$3x_1 - 5x_2 - 2x_3 = -12$
$2x_1 + 4x_2 - 3x_3 = -4$

45. $x_1 + 2x_2 - 4x_3 - x_4 = 7$
$2x_1 + 5x_2 - 9x_3 - 4x_4 = 16$
$x_1 + 5x_2 - 7x_3 - 7x_4 = 13$

46. $2x_1 + 4x_2 + 5x_3 + 4x_4 = 8$
$x_1 + 2x_2 + 2x_3 + x_4 = 3$

APPLICATIONS   *Solve Problems 47–68 using Gauss–Jordan elimination.*

★**47. Puzzle.** A friend of yours came out of the post office after spending $14.00 on 15¢, 20¢, and 35¢ stamps. If she bought 45 stamps in all, how many of each type did she buy?

★**48. Puzzle.** A parking meter contains only nickels, dimes, and quarters with a total value of $6.80. If there are 32 coins in all, how many of each type are there?

★★**49. Chemistry.** A chemist can purchase a 10% saline solution in 500 cubic centimeter containers, a 20% saline solution in 500 cubic centimeter containers, and a 50% saline solution in 1,000 cubic centimeter containers. He needs 12,000 cubic centimeters of 30% saline solution. How many containers of each type of solution should he purchase in order to form this solution?

★★**50. Chemistry.** Repeat Problem 49 if the 50% saline solution is available only in 1,500 cubic centimeter containers.

51. Geometry. Find a, b, and c so that the graph of the parabola with equation $y = a + bx + cx^2$ passes through the points $(-2, 3)$, $(-1, 2)$, and $(1, 6)$.

52. Geometry. Find a, b, and c so that the graph of the parabola with equation $y = a + bx + cx^2$ passes through the points $(1, 3)$, $(2, 2)$, and $(3, 5)$.

53. Geometry. Find a, b, and c so that the graph of the circle with equation $x^2 + y^2 + ax + by + c = 0$ passes through the points $(6, 2)$, $(4, 6)$, and $(-3, -1)$.

54. Geometry. Find a, b, and c so that the graph of the circle with equation $x^2 + y^2 + ax + by + c = 0$ passes through the points $(-4, 1)$, $(-1, 2)$, and $(3, -6)$.

55. Production Scheduling. A small manufacturing plant makes three types of inflatable boats: one-person, two-person, and four-person models. Each boat requires the services of three departments, as listed in the table. The cutting, assembly, and packaging departments have available a maximum of 380, 330, and 120 labor-hours per week, respectively. How many boats of each type must be produced each week for the plant to operate at full capacity?

	One-person boat	Two-person boat	Four-person boat
Cutting department	0.5 h	1.0 h	1.5 h
Assembly department	0.6 h	0.9 h	1.2 h
Packaging department	0.2 h	0.3 h	0.5 h

56. Production Scheduling. Repeat Problem 55 assuming the cutting, assembly, and packaging departments have available a maximum of 350, 330, and 115 labor-hours per week, respectively.

★**57. Production Scheduling.** Rework Problem 55 assuming the packaging department is no longer used.

★**58. Production Scheduling.** Rework Problem 56 assuming the packaging department is no longer used.

★**59. Production Scheduling.** Rework Problem 55 assuming the four-person boat is no longer produced.

★**60. Production Scheduling.** Rework Problem 56 assuming the four-person boat is no longer produced.

61. Nutrition. A dietitian in a hospital is to arrange a special diet using three basic foods. The diet is to include exactly 340 units of calcium, 180 units of iron, and 220 units of vitamin A. The number of units per ounce of each special ingredient for each of the foods is indicated in the table. How many ounces of each food must be used to meet the diet requirements?

	Units per ounce		
	Food A	Food B	Food C
Calcium	30	10	20
Iron	10	10	20
Vitamin A	10	30	20

62. Nutrition. Repeat Problem 61 if the diet is to include exactly 400 units of calcium, 160 units of iron, and 240 units of vitamin A.

★**63. Nutrition.** Solve Problem 61 with the assumption that food C is no longer available.

★**64. Nutrition.** Solve Problem 62 with the assumption that food C is no longer available.

★**65. Nutrition.** Solve Problem 61 assuming the vitamin A requirement is deleted.

★**66. Nutrition.** Solve Problem 62 assuming the vitamin A requirement is deleted.

67. Sociology. Two sociologists have grant money to study school busing in a particular city. They wish to conduct an opinion survey using 600 telephone contacts and 400 house contacts. Survey company A has personnel to do 30 telephone and 10 house contacts per hour; survey company B can handle 20 telephone and 20 house contacts per hour. How many hours should be scheduled for each firm to produce exactly the number of contacts needed?

68. Sociology. Repeat Problem 67 if 650 telephone contacts and 350 house contacts are needed.

SECTION 6-3 Systems Involving Second-Degree Equations

- Solution by Substitution
- Other Solution Methods

If a system of equations contains any equations that are not linear, then the system is called a **nonlinear system**. In this section we investigate nonlinear systems involving second-degree terms such as

$$x^2 + y^2 = 5 \qquad x^2 - 2y^2 = 2 \qquad x^2 + 3xy + y^2 = 20$$

$$3x + y = 1 \qquad xy = 2 \qquad xy - y^2 = 0$$

It can be shown that such systems have at most four solutions, some of which may be imaginary. Since we are interested in finding both real and imaginary solutions to the systems we consider, we now assume that the replacement set for each variable is the set of complex numbers, rather than the set of real numbers.

● Solution by Substitution

The substitution method used to solve linear systems of two equations in two variables is also an effective method for solving nonlinear systems. This process is best illustrated by examples.

EXAMPLE 1 Solving a Nonlinear System by Substitution

Solve the system: $x^2 + y^2 = 5$
$\qquad\qquad\qquad\quad 3x + y = 1$

Solution Solve the second equation for y in terms of x; then substitute for y in the first equation to obtain an equation that involves x alone.

$$3x + y = 1$$

$$y = 1 - 3x \qquad \text{Substitute this expression for } y \text{ in the first equation.}$$

$$x^2 + y^2 = 5$$

$$x^2 + (1 - 3x)^2 = 5$$

$$10x^2 - 6x - 4 = 0 \qquad \text{Simplify and write in standard quadratic form.}$$

$$5x^2 - 3x - 2 = 0 \qquad \text{Divide through by 2 to simplify further.}$$

$$(x - 1)(5x + 2) = 0$$

$$x = 1, \ -\tfrac{2}{5}$$

If we substitute these values back into the equation $y = 1 - 3x$, we obtain two solutions to the system:

$$x = 1 \qquad\qquad\qquad x = -\tfrac{2}{5}$$

$$y = 1 - 3(1) = -2 \qquad y = 1 - 3(-\tfrac{2}{5}) = \tfrac{11}{5}$$

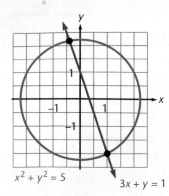

$x^2 + y^2 = 5$

$3x + y = 1$

A check, which you should provide, verifies that $(1, -2)$ and $(-\frac{2}{5}, \frac{11}{5})$ are both solutions to the system. These solutions are illustrated in the figure. However, if we substitute the values of x back into the equation $x^2 + y^2 = 5$, we obtain

$$x = 1 \qquad\qquad x = -\tfrac{2}{5}$$
$$1^2 + y^2 = 5 \qquad (-\tfrac{2}{5})^2 + y^2 = 5$$
$$y^2 = 4 \qquad\qquad y^2 = \tfrac{121}{25}$$
$$y = \pm 2 \qquad\qquad y = \pm \tfrac{11}{5}$$

It appears that we have found two additional solutions, $(1, 2)$ and $(-\frac{2}{5}, -\frac{11}{5})$. But neither of these solutions satisfies the equation $3x + y = 1$, which you should verify. So, neither is a solution of the original system. We have produced two **extraneous roots**, apparent solutions that do not actually satisfy both equations in the system. This is a common occurrence when solving nonlinear systems.

> It is always very important to check the solutions of any nonlinear system to ensure that extraneous roots have not been introduced.

Problem 1   Solve the system: $x^2 + y^2 = 10$
$$2x + y = 1$$

EXAMPLE 2   Solving a Nonlinear System by Substitution

Solve: $x^2 - 2y^2 = 2$
$$xy = 2$$

Solution   Solve the second equation for y, substitute in the first equation, and proceed as before.

$$xy = 2$$
$$y = \frac{2}{x}$$
$$x^2 - 2\left(\frac{2}{x}\right)^2 - 2$$
$$x^2 - \frac{8}{x^2} = 2$$
$$x^4 - 2x^2 - 8 = 0 \qquad \text{Multiply both sides by } x^2 \text{ and simplify.}$$
$$u^2 - 2u - 8 = 0 \qquad \text{Substitute } u = x^2 \text{ to transform to quadratic form and solve.}$$
$$(u - 4)(u + 2) = 0$$
$$u = 4, -2$$

Thus,

$$x^2 = 4 \qquad \text{or} \qquad x^2 = -2$$
$$x = \pm 2 \qquad\qquad x = \pm\sqrt{-2} = \pm i\sqrt{2}$$

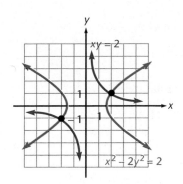

For $x = 2$, $y = \dfrac{2}{2} = 1$.   For $x = i\sqrt{2}$, $y = \dfrac{2}{i\sqrt{2}} = -i\sqrt{2}$.

For $x = -2$, $y = \dfrac{2}{-2} = -1$.   For $x = -i\sqrt{2}$, $y = \dfrac{2}{-i\sqrt{2}} = i\sqrt{2}$.

Thus, the four solutions to this system are $(2, 1)$, $(-2, -1)$, $(i\sqrt{2}, -i\sqrt{2})$, and $(-i\sqrt{2}, i\sqrt{2})$. Notice that two of the solutions involve imaginary numbers. These imaginary solutions cannot be illustrated graphically (see the figure); however, they do satisfy both equations in the system (verify this).

Problem 2   Solve: $3x^2 - y^2 = 6$
$$xy = 3$$

EXAMPLE 3   Design

An engineer is to design a rectangular computer screen with a 19-inch diagonal and a 175 square inch area. Find the dimensions of the screen to the nearest tenth of an inch.

Solution   Sketch a figure letting x be the width and y the height:

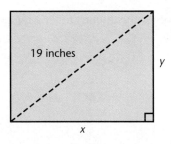

We obtain the following system using the Pythagorean theorem and the area-of-a-rectangle formula:

$$x^2 + y^2 = 19^2$$
$$xy = 175$$

This system is solved using the procedures outlined in Example 2. However, in this case, we are only interested in real solutions. We start by solving the second equation for y in terms of x and substituting the result into the first equation.

$$y = \frac{175}{x}$$

$$x^2 + \frac{175^2}{x^2} = 19^2$$

$$x^4 + 30{,}625 = 361x^2 \qquad \text{Multiply both sides by } x^2 \text{ and simplify}$$

$$x^4 - 361x^2 + 30{,}625 = 0 \qquad \text{Quadratic in } x^2$$

Solve the last equation for x^2 using the quadratic formula, then solve for x:

$$x = \sqrt{\frac{361 \pm \sqrt{361^2 - 4(1)(30{,}625)}}{2}}$$

$$= 15.0 \text{ inches or } 11.7 \text{ inches}$$

Substitute each choice of x into $y = 175/x$ to find the corresponding y values:

For $x = 15.0$ inches,   For $x = 11.7$ inches,

$$y = \frac{175}{15} = 11.7 \text{ inches} \qquad\qquad y = \frac{175}{11.7} = 15.0 \text{ inches}$$

Assuming the screen is wider than it is high, the dimensions are 15.0 by 11.7 inches.

Problem 3   An engineer is to design a rectangular television screen with a 21-inch diagonal and a 209 square inch area. Find the dimensions of the screen to the nearest tenth of an inch.

● **Other Solution Methods**   We now look at some other techniques for solving nonlinear systems of equations.

EXAMPLE 4   Solving a Nonlinear System by Elimination

Solve: $x^2 - y^2 = 5$
        $x^2 + 2y^2 = 17$

Solution   This type of system can be solved using elimination by addition. Multiply the second equation by -1 and add:

$$\begin{aligned} x^2 - y^2 &= 5 \\ -x^2 - 2y^2 &= -17 \\ \hline -3y^2 &= -12 \\ y^2 &= 4 \\ y &= \pm 2 \end{aligned}$$

Now substitute $y = 2$ and $y = -2$ back into either original equation to find x.

For $y = 2$,   For $y = -2$,

$$x^2 - (2)^2 = 5 \qquad\qquad x^2 - (-2)^2 = 5$$
$$x = \pm 3 \qquad\qquad\qquad x = \pm 3$$

Thus, $(3, -2)$, $(3, 2)$, $(-3, -2)$, and $(-3, 2)$, are the four solutions to the system. The check of the solutions is left to you.

Problem 4 Solve: $2x^2 - 3y^2 = 5$
$3x^2 + 4y^2 = 16$

EXAMPLE 5 **Solving a Nonlinear System Using Factoring and Substitution**

Solve: $x^2 + 3xy + y^2 = 20$
$xy - y^2 = 0$

Solution Factor the left side of the equation that has a 0 constant term:

$$xy - y^2 = 0$$
$$y(x - y) = 0$$
$$y = 0 \qquad \text{or} \qquad y = x$$

Thus, the original system is equivalent to the two systems:

$$y = 0 \quad \text{or} \qquad\qquad y = x$$
$$x^2 + 3xy + y^2 = 20 \qquad\qquad x^2 + 3xy + y^2 = 20$$

These systems are solved by substitution.

First System
$$y = 0$$
$$x^2 + 3xy + y^2 = 20$$

Substitute $y = 0$ in the second equation, and solve for x.

$$x^2 + 3x(0) + (0)^2 = 20$$
$$x^2 = 20$$
$$x = \pm\sqrt{20} = \pm 2\sqrt{5}$$

Second System
$$y = x$$
$$x^2 + 3xy + y^2 = 20$$

Substitute $y = x$ in the second equation and solve for x.

$$x^2 + 3xx + x^2 = 20$$
$$5x^2 = 20$$
$$x^2 = 4$$
$$x = \pm 2$$

Substitute these values back into $y = x$ to find y.
For $x = 2, y = 2$.   For $x = -2, y = -2$.
    The solutions for the original system are $(2\sqrt{5}, 0)$, $(-2\sqrt{5}, 0)$, $(2, 2)$, and $(-2, -2)$. The check of the solutions is left to you.

Problem 5   Solve: $x^2 - xy + y^2 = 9$
                      $2x^2 - xy = 0$

    Example 5 is somewhat specialized. However, it suggests a procedure that is effective for some problems.

Answers to Matched Problems
1. $(-1, 3), (\frac{9}{5}, -\frac{13}{5})$ **2.** $(\sqrt{3}, \sqrt{3}), (-\sqrt{3}, -\sqrt{3}), (i, -3i), (-i, 3i)$ **3.** 17.1 by 12.2 in
4. $(2, 1), (2, -1), (-2, 1), (-2, -1)$ **5.** $(0, 3), (0, -3), (\sqrt{3}, 2\sqrt{3}), (-\sqrt{3}, -2\sqrt{3})$

EXERCISE 6-3

A

Solve each system in Problems 1–12.

1. $x^2 + y^2 = 169$
    $x = -12$

2. $x^2 + y^2 = 25$
    $y = -4$

3. $8x^2 - y^2 = 16$
    $y = 2x$

4. $y^2 = 2x$
    $x = y - \frac{1}{2}$

5. $3x^2 - 2y^2 = 25$
    $x + y = 0$

6. $x^2 + 4y^2 = 32$
    $x + 2y = 0$

7. $y^2 = x$
    $x - 2y = 2$

8. $x^2 = 2y$
    $3x = y + 2$

9. $2x^2 + y^2 = 24$
    $x^2 - y^2 = -12$

10. $x^2 - y^2 = 3$
    $x^2 + y^2 = 5$

11. $x^2 + y^2 = 10$
    $16x^2 + y^2 = 25$

12. $x^2 - 2y^2 = 1$
    $x^2 + 4y^2 = 25$

B

Solve each system in Problems 13–24.

13. $xy - 4 = 0$
    $x - y = 2$

14. $xy - 6 = 0$
    $x - y = 4$

15. $x^2 + 2y^2 = 6$
    $xy = 2$

16. $2x^2 + y^2 = 18$
    $xy = 4$

17. $2x^2 + 3y^2 = -4$
    $4x^2 + 2y^2 = 8$

18. $2x^2 - 3y^2 = 10$
    $x^2 + 4y^2 = -17$

19. $x^2 - y^2 = 2$
    $y^2 = x$

20. $x^2 + y^2 = 20$
    $x^2 = y$

21. $x^2 + y^2 = 9$
    $x^2 = 9 - 2y$

22. $x^2 + y^2 = 16$
    $y^2 = 4 - x$

23. $x^2 - y^2 = 3$
    $xy = 2$

24. $y^2 = 5x^2 + 1$
    $xy = 2$

∫ *An important type of calculus problem is to find the area between the graphs of two functions. To solve some of these problems it is necessary to find the coordinates of the points of intersections of the two graphs. In Problems 25–32, find the coordinates of the points of intersections of the two given equations.*

25. $y = 5 - x^2, y = 2 - 2x$

26. $y = 5x - x^2, y = x + 3$

27. $y = x^2 - x, y = 2x$

28. $y = x^2 + 2x, y = 3x$

29. $y = x^2 - 6x + 9, y = 5 - x$

30. $y = x^2 + 2x + 3, y = 2x + 4$

31. $y = 8 + 4x - x^2, y = x^2 - 2x$

32. $y = x^2 - 4x - 10, y = 14 - 2x - x^2$

C _____

Solve each system in Problems 33–40.

33. $2x + 5y + 7xy = 8$
$xy - 3 = 0$

34. $2x + 3y + xy = 16$
$xy - 5 = 0$

35. $x^2 - 2xy + y^2 = 1$
$x - 2y = 2$

36. $x^2 + xy - y^2 = -5$
$y - x = 3$

37. $2x^2 - xy + y^2 = 8$
$x^2 - y^2 = 0$

38. $x^2 + 2xy + y^2 = 36$
$x^2 - xy = 0$

39. $x^2 + xy - 3y^2 = 3$
$x^2 + 4xy + 3y^2 = 0$

40. $x^2 - 2xy + 2y^2 = 16$
$x^2 - y^2 = 0$

⊞ *In Problems 41–46, use a graphic calculator or a computer to estimate the coordinates of the points of intersection of the graphs of the given equations to two decimal places.*

41. $y = -x^2 + 6x - 5, y = 0.5x + 1$

42. $y = x^2 - 8x + 12, y = -0.5x + 5$

43. $y = x^2 - 8x + 12, y = -x^2 + 6x - 5$

44. $y = x^2 - 7x + 11, y = -x^2 + 5x - 4$

45. $y = e^x, y = x + 2$

46. $y = \ln x, y = x - 2$

_____

APPLICATIONS   🌐

47. Numbers. Find two numbers such that their sum is 3 and their product is 1.

48. Numbers. Find two numbers such that their difference is 1 and their product is 1. (Let x be the larger number and y the smaller number.)

49. Geometry. Find the lengths of the legs of a right triangle with an area of 30 square inches if its hypotenuse is 13 inches long.

50. Geometry. Find the dimensions of a rectangle with an area of 32 square meters if its perimeter is 36 meters long.

51. Design. An engineer is designing a small portable television set. According to the design specifications, the set must have a rectangular screen with a 7.5-inch diagonal and an area of 27 square inches. Find the dimensions of the screen.

52. Design. An artist is designing a logo for a business in the shape of a circle with an inscribed rectangle. The diameter of the circle is 6.5 inches, and the area of the rectangle is 15 square inches. Find the dimensions of the rectangle.

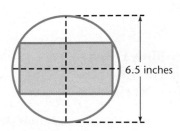

6.5 inches

★ **53. Construction.** A rectangular swimming pool with a deck 5 feet wide is enclosed by a fence as shown in the figure. The surface area of the pool is 572 square feet, and the total area enclosed by the fence (including the pool and the deck) is 1,152 square feet. Find the dimensions of the pool.

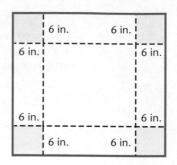

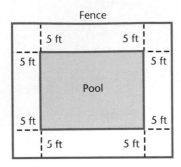

★★ **55. Transportation.** Two boats leave Bournemouth, England, at the same time and follow the same route on the 75-mile trip across the English Channel to Cherbourg, France. The average speed of boat *A* is 5 miles per hour greater than the average speed of boat *B*. Consequently, boat *A* arrives at Cherbourg 30 minutes before boat *B*. Find the average speed of each boat.

★ **54. Construction.** An open-topped rectangular box is formed by cutting a 6-inch square from each corner of a rectangular piece of cardboard and bending up the ends and sides. The area of the cardboard before the corners are removed is 768 square inches, and the volume of the box is 1,440 cubic inches. Find the dimensions of the original piece of cardboard.

★★ **56. Transportation.** Bus *A* leaves Milwaukee at noon and travels west on Interstate 94. Bus *B* leaves Milwaukee 30 minutes later, travels the same route, and overtakes bus *A* at a point 210 miles west of Milwaukee. If the average speed of bus *B* is 10 miles per hour greater than the average speed of bus *A*, at what time did bus *B* overtake bus *A*?

SECTION $6\text{-}4$ **Systems of Linear Inequalities in Two Variables**

- Graphing Linear Inequalities in Two Variables
- Solving Systems of Linear Inequalities Graphically
- Application

Many applications of mathematics involve systems of inequalities rather than systems of equations. A graph is often the most convenient way to represent the solutions of a system of inequalities in two variables. In this section, we discuss techniques for graphing both a single linear inequality in two variables and a system of linear inequalities in two variables.

● **Graphing Linear Inequalities in Two Variables**

We know how to graph first-degree equations such as

$$y = 2x - 3 \quad \text{and} \quad 2x - 3y = 5$$

but how do we graph first-degree inequalities such as

$$y \le 2x - 3 \quad \text{and} \quad 2x - 3y > 5$$

Actually, graphing these inequalities is almost as easy as graphing the equations. The following discussion leads to a simple solution of the problem.

A vertical line divides a plane into **left** and **right half-planes**. A nonvertical line divides a plane into **upper** and **lower half-planes** (see Figure 1).

FIGURE 1 Upper and lower half-planes.

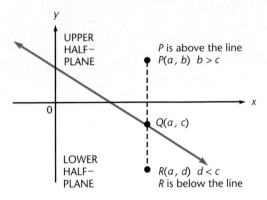

Now let's compare the graphs of the following:

$$y < 2x - 3 \qquad y = 2x - 3 \qquad y > 2x - 3$$

We start by graphing $y = 2x - 3$. For a fixed x, equality holds if a point is on the line. For the same x, if a point is below the line, then $y < 2x - 3$, and if a point is above the line, then $y > 2x - 3$. See Figure 2. Since the same results are obtained for each point on the x axis, we conclude that the graph of $y > 2x - 3$ is the upper half-plane determined by the line $y = 2x - 3$, and the graph of $y < 2x - 3$ is the lower half-plane determined by the same line.

FIGURE 2 The graphs of $y > 2x - 3$, $y = 2x - 3$, and $y < 2x - 3$.

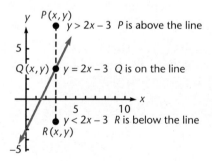

In graphing $y > 2x - 3$, we show the line $y = 2x - 3$ as a broken line, indicating that it is not part of the graph, and then shade in the upper half-plane to complete the graph. In graphing $y \geq 2x - 3$, we show the line $y = 2x - 3$ as a solid line, indicating that it is part of the graph, and then shade in the upper half-plane to complete the graph. Figure 3 illustrates four typical cases.

FIGURE 3 Four typical cases.

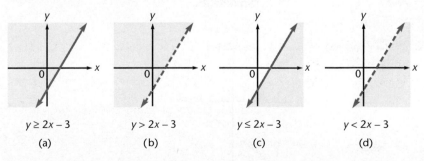

Theorem 1   **Graphs of Linear Inequalities in Two Variables**

Let A, B, and C be real numbers with A and B not both zero. The graph of a linear inequality

$$Ax + By < C \quad \text{or} \quad Ax + By > C$$

with $B \neq 0$, is either the upper half-plane or the lower half-plane (but not both) determined by the line $Ax + By = C$. If $B = 0$, then the graph of

$$Ax < C \quad \text{or} \quad Ax > C$$

is either the left half-plane or the right half-plane (but not both) determined by the line $Ax = C$.

As a consequence of Theorem 1, we state a simple and fast mechanical procedure for graphing linear inequalities.

Procedure for Graphing Linear Inequalities in Two Variables

Step 1. Graph $Ax + By = C$ as a broken line if equality is not included in the original statement or as a solid line if equality is included.

Step 2. Choose a test point anywhere in the plane not on the line and substitute the coordinates into the inequality. The origin $(0, 0)$ often requires the least computation.

Step 3. The graph of the original inequality includes the half-plane containing the test point if the inequality is satisfied by that point, or the half-plane not containing that point if the inequality is not satisfied by that point.

EXAMPLE 1   Graphing a Linear Inequality

Graph: $3x - 4y \leq 12$

Solution   *Step 1.* Graph $3x - 4y = 12$ as a solid line, since equality is included in the original statement.

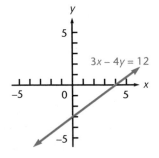

Step 2. Pick a convenient test point above or below the line. The origin $(0, 0)$ requires the least computation. Substituting $(0, 0)$ into the inequality

$$3x - 4y \leq 12$$

$$3(0) - 4(0) = 0 \leq 12$$

produces a true statement; therefore, $(0, 0)$ is in the solution set.

Step 3. The line $3x - 4y = 12$ and the half-plane containing the origin form the graph of $3x - 4y \leq 12$.

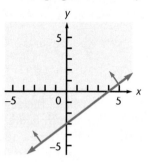

Problem 1   Graph: $2x + 3y < 6$

EXAMPLE 2   **Graphing a Linear Inequality**

Graph: (A) $y > -3$ (B) $2x \leq 5$

Solutions   (A)

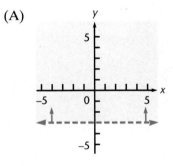

(B)

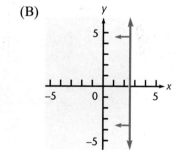

Problem 2   Graph: (A) $y \leq 2$ (B) $3x > -8$

● Solving Systems of Linear Inequalities Graphically

We now consider systems of linear inequalities such as

$$x + y \geq 6 \qquad \text{and} \qquad 2x + y \leq 22$$
$$2x - y \geq 0 \qquad\qquad\qquad x + y \leq 13$$
$$\qquad\qquad\qquad\qquad\qquad 2x + 5y \leq 50$$
$$\qquad\qquad\qquad\qquad\qquad x \geq 0$$
$$\qquad\qquad\qquad\qquad\qquad y \geq 0$$

We wish to **solve** such systems **graphically**—that is, to find the graph of all ordered pairs of real numbers (x, y) that simultaneously satisfy all the inequalities in the system. The graph is called the **solution region** for the system. To find the solution region, we graph each inequality in the system and then take the intersection of all the graphs. To simplify the discussion that follows, *we will consider only systems of linear inequalities where equality is included in each statement in the system.*

EXAMPLE 3   **Solving a System of Linear Inequalities Graphically**

Solve the following system of linear inequalities graphically:

$$x + y \geq 6$$
$$2x - y \geq 0$$

Solution   First, graph the line $x + y = 6$ and shade the region that satisfies the inequality $x + y \geq 6$. This region is shaded in blue in figure (a). Next, graph the line $2x - y = 0$ and shade the region that satisfies the inequality $2x - y \geq 0$. This region is shaded in red in figure (a). The solution region for the system of inequalities is the intersection of these two regions. This is the region shaded in both red and blue in figure (a), which is redrawn in figure (b) with only the solution region shaded for clarity. The coordinates of any point in the shaded region of figure (b) specify a solution to the system. For example, the points $(2, 4)$, $(6, 3)$, and $(7.43, 8.56)$ are three of infinitely many solutions, as can be easily checked. The intersection point $(2, 4)$ can be obtained by solving the equations $x + y = 6$ and $2x - y = 0$ simultaneously.

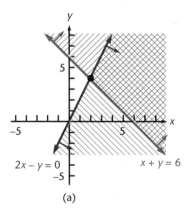

(a)

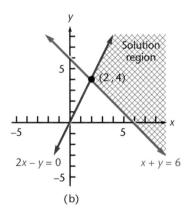

(b)

Problem 3   Solve the following system of linear inequalities graphically:

$$3x + y \leq 21$$
$$x - 2y \leq 0$$

The points of intersection of the lines that form the boundary of a solution region play a fundamental role in the solution of linear programming problems, which are discussed in the next section.

DEFINITION 1	**Corner Point**
	A **corner point** of a solution region is a point in the solution region that is the intersection of two boundary lines.

The point (2, 4) is the only corner point of the solution region in Example 3; see figure (b).

EXAMPLE 4   **Solving a System of Linear Inequalities Graphically**

Solve the following system of linear inequalities graphically, and find the corner points.

$$2x + y \leq 22$$
$$x + y \leq 13$$
$$2x + 5y \leq 50$$
$$x \geq 0$$
$$y \geq 0$$

Solution   The inequalities $x \geq 0$ and $y \geq 0$, called **nonnegative restrictions**, occur frequently in applications involving systems of inequalities since x and y often represent quantities that can't be negative—number of units produced, number of hours worked, etc. The solution region lies in the first quadrant, and we can restrict our attention to that portion of the plane. First, we graph the lines

$$2x + y = 22$$ Find the x and y intercepts of each line;
     then sketch the line through these points.
$$x + y = 13$$
$$2x + 5y = 50$$

Next, choosing $(0, 0)$ as a test point, we see that the graph of each of the first three inequalities in the system consists of its corresponding line and the half-plane lying below the line, as indicated by the arrows in the figure. Thus, the solution region of the system consists of the points in the first quadrant that simultaneously lie on or below all three of these lines—see the figure.

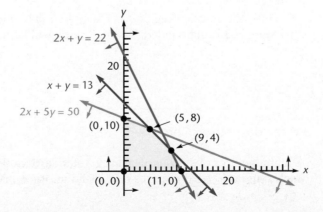

The corner points $(0, 0)$, $(0, 10)$, and $(11, 0)$ can be determined from the graph. The other two corner points are determined as follows:

Solve the system

$$2x + 5y = 50$$
$$x + y = 13$$

to obtain $(5, 8)$.

Solve the system

$$2x + y = 22$$
$$x + y = 13$$

to obtain $(9, 4)$.

Note that the lines $2x + 5y = 50$ and $2x + y = 22$ also intersect, but the intersection point is not part of the solution region, and hence, is not a corner point.

Problem 4   Solve the following system of linear inequalities graphically, and find the corner points:

$$5x + y \geq 20$$
$$x + y \geq 12$$
$$x + 3y \geq 18$$
$$x \geq 0$$
$$y \geq 0$$

If we compare the solution regions of Examples 3 and 4, we see that there is a fundamental difference between these two regions. We can draw a circle around the solution region in Example 4. However, it is impossible to include all the points in the solution region in Example 3 in any circle, no matter how large we draw it. This leads to the following definition.

DEFINITION 2   **Bounded and Unbounded Solution Regions**

A solution region of a system of linear inequalities is **bounded** if it can be enclosed within a circle. If it cannot be enclosed within a circle, then it is **unbounded**.

Thus, the solution region for Example 4 is bounded and the solution region for Example 3 is unbounded. This definition will be important in the next section.

● **Application**

EXAMPLE 5   Production Scheduling

A manufacturer of surfboards makes a standard model and a competition model. Each standard board requires 6 labor-hours for fabricating and 1 labor-hour for finishing. Each competition board requires 8 labor-hours for fabricating and 3 labor-hours for finishing. The maximum labor-hours available per week in the fabricating and finishing departments are 120 and 30, respectively. What combinations of boards can be produced each week so as not to exceed the number of labor-hours available in each department per week?

Solution   To clarify relationships, we summarize the information in the following table:

	Standard model (labor-hours per board)	Competition model (labor-hours per board)	Maximum labor-hours available per week
Fabricating	6	8	120
Finishing	1	3	30

Let

$$x = \text{Number of standard boards produced per week}$$

$$y = \text{Number of competition boards produced per week}$$

These variables are restricted as follows:

Fabricating department restriction:

$$\begin{pmatrix} \text{Weekly fabricating time} \\ \text{for } x \text{ standard boards} \end{pmatrix} + \begin{pmatrix} \text{Weekly fabricating time} \\ \text{for } y \text{ competition boards} \end{pmatrix} \leq \begin{pmatrix} \text{Maximum labor-hours} \\ \text{available per week} \end{pmatrix}$$
$$6x + 8y \leq 120$$

Finishing department restriction:

$$\begin{pmatrix} \text{Weekly finishing time} \\ \text{for } x \text{ standard boards} \end{pmatrix} + \begin{pmatrix} \text{Weekly finishing time} \\ \text{for } y \text{ competition boards} \end{pmatrix} \leq \begin{pmatrix} \text{Maximum labor-hours} \\ \text{available per week} \end{pmatrix}$$
$$1x + 3y \leq 30$$

Since it is not possible to manufacture a negative number of boards, x and y also must satisfy the nonnegative restrictions

$$x \geq 0$$

$$y \geq 0$$

Thus, x and y must satisfy the following system of linear inequalities:

$$6x + 8y \leq 120 \quad \text{Fabricating department restriction}$$

$$x + 3y \leq 30 \quad \text{Finishing department restriction}$$

$$x \geq 0 \quad \text{Nonnegative restriction}$$

$$y \geq 0 \quad \text{Nonnegative restriction}$$

Graphing this system of linear inequalities, we obtain the set of **feasible solutions**, or the **feasible region**, as shown in the figure. For problems of this type and for the linear programming problems we consider in the next section, solution regions are often referred to as feasible regions. Any point within the shaded area, including the boundary lines, represents a possible production schedule. Any point outside the shaded area represents an impossible schedule. For example, it would be possible to produce 12 standard boards and 5 competition boards per week, but it would not be possible to produce 12 standard boards and 7 competition boards per week (see the figure).

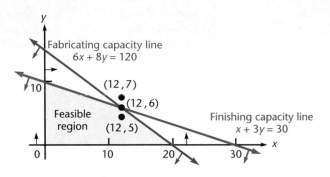

Problem 5   Repeat Example 5 using 5 hours for fabricating a standard board and a maximum of 27 labor-hours for the finishing department.

Answers to Matched Problems

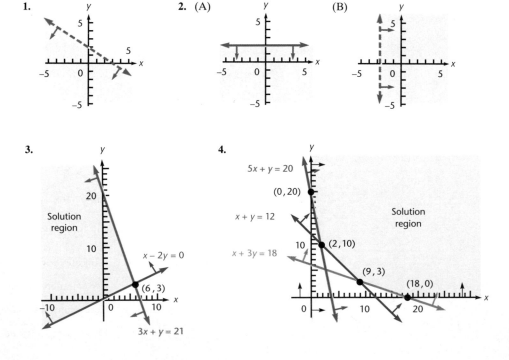

5.

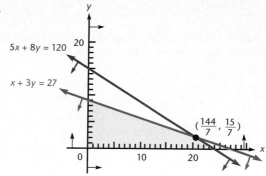

EXERCISE 6-4

A _____

Graph each inequality in Problems 1–10.

1. $2x - 3y < 6$ **2.** $3x + 4y < 12$ **3.** $3x + 2y \geq 18$ **4.** $3y - 2x \geq 24$ **5.** $y \leq \frac{2}{3}x + 5$

6. $y \geq \frac{1}{3}x - 2$ **7.** $y < 8$ **8.** $x > -5$ **9.** $-3 \leq y < 2$ **10.** $-1 < x \leq 3$

In Problems 11–14, match the solution region of each system of linear inequalities with one of the four regions shown in the figure below.

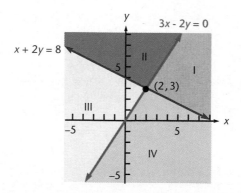

11. $x + 2y \leq 8$
$3x - 2y \geq 0$

12. $x + 2y \geq 8$
$3x - 2y \leq 0$

13. $x + 2y \geq 8$
$3x - 2y \geq 0$

14. $x + 2y \leq 8$
$3x - 2y \leq 0$

In Problems 15–20, solve each system of linear inequalities geographically.

15. $x \geq 5$
$y \leq 6$

16. $x \leq 4$
$y \geq 2$

17. $3x + y \geq 6$
$x \leq 4$

18. $3x + 4y \leq 12$
$y \geq -3$

19. $x - 2y \leq 12$
$2x + y \geq 4$

20. $2x + 5y \leq 20$
$x - 5y \leq -5$

B

In Problems 21–24, match the solution region of each system of linear inequalities with one of the four regions shown in the figure below. Identify the corner points of each solution region.

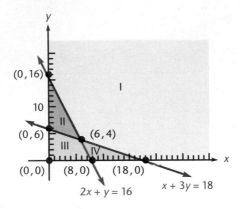

21. $x + 3y \leq 18$
$2x + y \geq 16$
$x \geq 0$
$y \geq 0$

22. $x + 3y \leq 18$
$2x + y \leq 16$
$x \geq 0$
$y \geq 0$

23. $x + 3y \geq 18$
$2x + y \geq 16$
$x \geq 0$
$y \geq 0$

24. $x + 3y \geq 18$
$2x + y \leq 16$
$x \geq 0$
$y \geq 0$

In Problems 25–36, solve the systems graphically, and indicate whether each solution region is bounded or unbounded. Find the coordinates of each corner point.

25. $2x + 3y \leq 6$
$x \geq 0$
$y \geq 0$

26. $4x + 3y \leq 12$
$x \geq 0$
$y \geq 0$

27. $4x + 5y \geq 20$
$x \geq 0$
$y \geq 0$

28. $5x + 6y \geq 30$
$x \geq 0$
$y \geq 0$

29. $2x + y \leq 8$
$x + 3y \leq 12$
$x \geq 0$
$y \geq 0$

30. $x + 2y \leq 10$
$3x + y \leq 15$
$x \geq 0$
$y \geq 0$

31. $4x + 3y \geq 24$
$2x + 3y \geq 18$
$x \geq 0$
$y \geq 0$

32. $x + 2y \geq 8$
$2x + y \geq 10$
$x \geq 0$
$y \geq 0$

33. $2x + y \leq 12$
$x + y \leq 7$
$x + 2y \leq 10$
$x \geq 0$
$y \geq 0$

34. $3x + y \leq 21$
$x + y \leq 9$
$x + 3y \leq 21$
$x \geq 0$
$y \geq 0$

35. $x + 2y \geq 16$
$x + y \geq 12$
$2x + y \geq 14$
$x \geq 0$
$y \geq 0$

36. $3x + y \geq 30$
$x + y \geq 16$
$x + 3y \geq 24$
$x \geq 0$
$y \geq 0$

C

In Problems 37–42, solve the systems graphically, and indicate whether each solution region is bounded or unbounded. Find the coordinates of each corner point.

37. $x + y \leq 11$
$5x + y \geq 15$
$x + 2y \geq 12$

38. $4x + y \leq 32$
$x + 3y \leq 30$
$5x + 4y \geq 51$

39. $3x + 2y \geq 24$
$3x + y \leq 15$
$x \geq 4$

40. $3x + 4y \leq 48$
$x + 2y \geq 24$
$y \leq 9$

41. $x + y \leq 10$
$3x + 5y \geq 15$
$3x - 2y \leq 15$
$-5x + 2y \leq 6$

42. $3x - y \geq 1$
$-x + 5y \geq 9$
$x + y \leq 9$
$y \leq 5$

APPLICATIONS

43. Manufacturing—Resource Allocation. A manufacturing company makes two types of water skis: a trick ski and a slalom ski. The trick ski requires 6 labor-hours for fabricating and 1 labor-hour for finishing. The slalom ski requires 4 labor-hours for fabricating and 1 labor-hour for finishing. The maximum labor-hours available per day for fabricating and finishing are 108 and 24, respectively. If x is the number of trick skis and y is the number of slalom skis produced per day, write a system of inequalities that indicates appropriate restraints on x and y. Find the set of feasible solutions graphically for the number of each type of ski that can be produced.

44. Psychology. In an experiment on conditioning, a psychologist uses two types of Skinner (conditioning) boxes with mice and rats. Each mouse spends 10 minutes per day in box A and 20 minutes per day in box B. Each rat spends 20 minutes per day in box A and 10 minutes per day in box B. The total maximum time available per day is 800 minutes for box A and 640 minutes for box B. We are interested in the various numbers of mice and rats that can be used in the experiment under the conditions stated. If x is the number of mice used and y is the number of rats used, write a system of linear inequalities that indicates appropriate restrictions on x and y. Find the set of feasible solutions graphically.

45. Sociology. A city council voted to conduct a study on inner-city community problems. A nearby university was contacted to provide sociologists and research assistants. Each sociologist will spend 10 hours per week collecting data in the field and 30 hours per week analyzing data in the research center. Each research assistant will spend 30 hours per week in the field and 10 hours per week in the research center. The minimum weekly labor-hour requirements are 280 hours in the field and 360 hours in the research center. If x is the number of sociologists hired for the study and y is the number of research assistants hired for the study, write a system of linear inequalities that indicates appropriate restrictions on x and y. Find the set of feasible solutions graphically.

46. Nutrition. A dietitian in a hospital is to arrange a special diet using two foods. Each ounce of food M contains 30 units of calcium, 10 units of iron, and 10 units of vitamin A. Each ounce of food N contains 10 units of calcium, 10 units of iron, and 30 units of vitamin A. The minimum requirements in the diet are 360 units of calcium, 160 units of iron, and 240 units of vitamin A. If x is the number of ounces of food M used and y is the number of ounces of food N used, write a system of linear inequalities that reflects the conditions indicated. Find the set of feasible solutions graphically for the amount of each kind of food that can be used.

SECTION 6-5 Linear Programming

* A Linear Programming Problem
* Linear Programming—A General Description
* Application

Several problems in Section 6-4 are related to the general type of problems called *linear programming problems*. Linear programming is a mathematical process that has been developed to help management in decision making, and it has become one of the most widely used and best-known tools of management science and industrial engineering. We will use an intuitive graphical approach based on the techniques discussed in Section 6-4 to illustrate this process for problems involving two variables.

The American mathematician George B. Dantzig (1914–   ) formulated the first linear programming problem in 1947 and introduced a solution technique, called the *simplex method*, that does not rely on graphing and is readily adaptable to computer solutions. Today, it is quite common to use a computer to solve applied linear programming problems involving thousands of variables and hundreds of inequalities.

• A Linear Programming Problem

We begin our discussion with an example that will lead to a general procedure for solving linear programming problems in two variables.

EXAMPLE 1   **Production Scheduling**

A manufacturer of fiberglass camper tops for pickup trucks makes a compact model and a regular model. Each compact top requires 5 hours from the fabricating department and 2 hours from the finishing department. Each regular top requires 4 hours from the fabricating department and 3 hours from the finishing department. The maximum labor-hours available per week in the fabricating department and the finishing department are 200 and 108, respectively. If the company makes a profit of $40 on each compact top and $50 on each regular top, how many tops of each type should be manufactured each week to maximize the total weekly profit, assuming all tops can be sold?

Solution   This is an example of a linear programming problem. To see relationships more clearly, we summarize the manufacturing requirements, objectives, and restrictions in the table:

	Compact model (labor hours per top)	Regular model (labor-hours per top)	Maximum labor-hours available per week
Fabricating	5	4	200
Finishing	2	3	108
Profit per top	$40	$50	

We now proceed to formulate a *mathematical model* for the problem and then to solve it using graphical methods.

Objective Function   The *objective* of management is to *decide* how many of each camper top model should be produced each week in order to *maximize* profit. Let

x = Number of compact tops produced per week   ⎱
y = Number of regular tops produced per week   ⎰   Decision variables

The following function gives the total profit P for x compact tops and y regular tops manufactured each week:

$$P = 40x + 50y \quad \text{Objective function}$$

Mathematically, management needs to decide on values for the **decision variables** (x and y) that achieve its objective, that is, maximizing the **objective function** (profit) $P = 40x + 50y$. It appears that the profit can be made as large as we like by manufacturing more and more tops—or can it?

Constraints   Any manufacturing company, no matter how large or small, has manufacturing limits imposed by available resources, plant capacity, demand, and so forth. These limits are referred to as **problem constraints**.

Fabricating department constraint:

$$\begin{pmatrix} \text{Weekly fabricating} \\ \text{time for } x \\ \text{compact tops} \end{pmatrix} + \begin{pmatrix} \text{Weekly fabricating} \\ \text{time for } y \\ \text{regular tops} \end{pmatrix} \leq \begin{pmatrix} \text{Maximum labor-hours} \\ \text{available per week} \end{pmatrix}$$

$$5x \quad + \quad 4y \quad \leq \quad 200$$

Finishing department constraint:

$$\begin{pmatrix} \text{Weekly finishing} \\ \text{time for } x \\ \text{compact tops} \end{pmatrix} + \begin{pmatrix} \text{Weekly finishing} \\ \text{time for } y \\ \text{regular tops} \end{pmatrix} \leq \begin{pmatrix} \text{Maximum labor-hours} \\ \text{available per week} \end{pmatrix}$$

$$2x \quad + \quad 3y \quad \leq \quad 108$$

Nonnegative constraints: It is not possible to manufacture a negative number of tops; thus, we have the **nonnegative constraints**

$$x \geq 0$$
$$y \geq 0$$

which we usually write in the form

$$x, y \geq 0$$

Mathematical Model   We now have a **mathematical model** for the problem under consideration:

$$\begin{aligned} \text{Maximize} \quad & P = 40x + 50y && \text{Objective function} \\ \text{Subject to} \quad & \left. \begin{matrix} 5x + 4y \leq 200 \\ 2x + 3y \leq 108 \end{matrix} \right\} && \text{Problem constraints} \\ & x, y \geq 0 && \text{Nonnegative constraints} \end{aligned}$$

Graphic Solution   **Solving** the system of linear inequality constraints **graphically**, as in Section 6-4, we obtain the feasible region for production schedules, as shown in the figure.

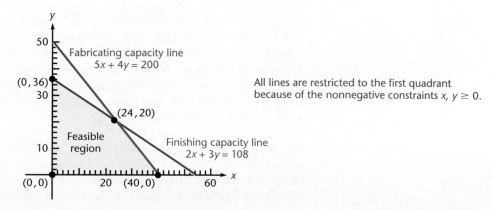

All lines are restricted to the first quadrant because of the nonnegative constraints $x, y \geq 0$.

By choosing a production schedule (x, y) from the feasible region, a profit can be determined using the objective function $P = 40x + 50y$. For example, if $x = 24$ and $y = 10$, then the profit for the week is

$$P = 40(24) + 50(10) = \$1,460$$

Or if $x = 15$ and $y = 20$, then the profit for the week is

$$P = 40(15) + 50(20) = \$1,600$$

The question is, out of all possible production schedules (x, y) from the feasible region, which schedule(s) produces the maximum profit? Such a schedule, if it exists, is called an **optimal solution** to the problem because it produces the maximum value of the objective function and is in the feasible region. It is not practical to use point-by-point checking to find the optimal solution. Even if we consider only points with integer coordinates, there are over 800 such points in the feasible region for this problem. Instead, we use the theory that has been developed to solve linear programming problems. Using advanced techniques, it can be shown that:

If the feasible region is bounded, then one or more of the corner points of the feasible region is an optimal solution to the problem.

Corner point (x, y)	Objective function $P = 40x + 50y$	
$(0, 0)$	0	
$(0, 36)$	$1,800$	
$(24, 20)$	$1,960$	Maximum value of P
$(40, 0)$	$1,600$	

The maximum value of the objective function is unique; however, there can be more than one feasible production schedule that will produce this unique value. We will have more to say about this later in this section.

Since the feasible region for this problem is bounded, at least one of the corner points, $(0, 0)$, $(0, 36)$, $(24, 20)$, or $(40, 0)$, is an optimal solution. To find which one, we evaluate $P = 40x + 50y$ at each corner point and choose the corner point that produces the largest value of P. It is convenient to organize these calculations in a table, as shown in the margin.

Examining the values in the table, we see that the maximum value of P at a corner point is $P = 1,960$ at $x = 24$ and $y = 20$. Since the maximum value of P over the entire feasible region must always occur at a corner point, we conclude that the maximum profit is $\$1,960$ when 24 compact tops and 20 regular tops are produced each week.

Problem 1 We now convert the surfboard problem discussed in Section 6-4 into a linear programming problem. A manufacturer of surfboards makes a standard model and a competition model. Each standard board requires 6 labor-hours for fabricating and 1 labor-hour for finishing. Each competition board requires 8 labor-hours for fabricating and 3 labor-hours for finishing. The maximum labor-hours available per week in the fabricating and finishing departments are 120 and 30, respectively. If the company makes a profit of $40 on each standard board and $75 on each competition board, how many boards of each type should be manufactured each week to maximize the total weekly profit?

(A) Identify the decision variables.
(B) Write the objective function P.
(C) Write the problem constraints and the nonnegative constraints.
(D) Graph the feasible region, identify the corner points, and evaluate P at each corner point.

(E) How many boards of each type should be manufactured each week to maximize the profit? What is the maximum profit?

• Linear Programming—A General Description

The linear programming problems considered in Example 1 and Problem 1 were *maximization problems* where we wanted to maximize profits. The same technique can be used to solve *minimization problems* where, for example, we may want to minimize costs. Before considering additional examples, we state a few general definitions.

A **linear programming problem** is one that is concerned with finding the **optimal value** (maximum or minimum value) of a linear **objective function** of the form

$$z = ax + by$$

where the **decision variables** x and y are subject to **problem constraints** in the form of linear inequalities and the **nonnegative constraints** $x, y \geq 0$. The set of points satisfying both the problem constraints and the nonnegative constraints is called the **feasible region** for the problem. Any point in the feasible region that produces the optimal value of the objective function over the feasible region is called an **optimal solution**.

Theorem 1 is fundamental to the solving of linear programming problems.

Theorem 1

Fundamental Theorem of Linear Programming

Let S be the feasible region for a linear programming problem, and let $z = ax + by$ be the objective function. If S is bounded, then z has both a maximum and a minimum value on S and each of these occurs at a corner point of S. If S is unbounded, then a maximum or minimum value of z on S may not exist. However, if either does exist, then it must occur at a corner point of S.

We will not consider any problems with unbounded feasible regions in this brief introduction. If a feasible region is bounded, then Theorem 1 provides the basis for the following simple procedure for solving the associated linear programming problem:

Solution of Linear Programming Problems

Step 1. Form a mathematical model for the problem:
   (A) Introduce decision variables and write a linear objective function.
   (B) Write problem constraints in the form of linear inequalities.
   (C) Write nonnegative constraints.

Step 2. Graph the feasible region and find the corner points.

Step 3. Evaluate the objective function at each corner point to determine the optimal solution.

Before considering additional applications, we use this procedure to solve a linear programming problem where the model has already been determined.

EXAMPLE 2   **Solving a Linear Programming Problem**

Minimize and maximize $z = 5x + 15y$
Subject to $x + 3y \leq 60$
$\qquad\qquad x + y \geq 10$
$\qquad\qquad x - y \leq 0$
$\qquad\qquad x, y \geq 0$

Solution   This problem is a combination of two linear programming problems—a minimization problem and a maximization problem. Since the feasible region is the same for both problems, we can solve these problems together. To begin, we graph the feasible region S, as shown in the figure, and find the coordinates of each corner point.

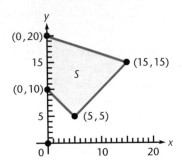

Next, we evaluate the objective function at each corner point, with the results given in the table:

Corner point (x, y)	Objective function $z = 5x + 15y$		
$(0, 10)$	150		
$(0, 20)$	300	Maximum value	Multiple optimal solutions
$(15, 15)$	300	Maximum value	
$(5, 5)$	100	Minimum value	

Examining the values in the table, we see that the minimum value of z on the feasible region S is 100 at $(5, 5)$. Thus, $(5, 5)$ is the optimal solution to the minimization problem. The maximum value of z on the feasible region S is 300, which occurs at $(0, 20)$ and at $(15, 15)$. Thus, the maximization problem has **multiple optimal solutions**. In general:

> If two corner points are both optimal solutions of the same type (both produce the same maximum value or both produce the same minimum value) to a linear programming problem, then any point on the line segment joining the two corner points is also an optimal solution of that type.

It can be shown that this is the only time that an optimal value occurs at more than one point.

Problem 2   Minimize and maximize $z = 10x + 5y$
Subject to $2x + y \geq 40$
$\qquad\qquad 3x + y \leq 150$
$\qquad\qquad 2x - y \geq 0$
$\qquad\qquad x, y \geq 0$

● **Application**   Now we consider another application where we must first find the mathematical model and then find its solution.

EXAMPLE 3   Agriculture

	Pounds per cubic yard	
	Mix A	**Mix B**
Nitrogen	10	5
Potash	8	24
Phosphoric acid	9	6

A farmer can use two types of plant food, mix A and mix B. The amounts (in pounds) of nitrogen, phosphoric acid, and potash in a cubic yard of each mix are given in the table. Tests performed on the soil in a large field indicate that the field needs at least 840 pounds of potash and at least 350 pounds of nitrogen. The tests also indicate that no more than 630 pounds of phosphoric acid should be added to the field. A cubic yard of mix A costs \$7, and a cubic yard of mix B costs \$9. How many cubic yards of each mix should the farmer add to the field in order to supply the necessary nutrients at minimal cost?

Solution   Let

x = Number of cubic yards of mix A added to the field ⎫
y = Number of cubic yards of mix B added to the field ⎭   Decision variables

We form the linear objective function

$$C = 7x + 9y$$

which gives the cost of adding x cubic yards of mix A and y cubic yards of mix B to the field. Using the data in the table and proceeding as in Example 1, we formulate the mathematical model for the problem:

Minimize	$C = 7x + 9y$	Objective function
Subject to	$10x + 5y \geq 350$	Nitrogen constraint
	$8x + 24y \geq 840$	Potash constraint
	$9x + 6y \leq 630$	Phosphoric acid constraint
	$x, y \geq 0$	Nonnegative constraints

Solving the system of constraint inequalities graphically, we obtain the feasible region S shown in the figure, and then we find the coordinates of each corner point.

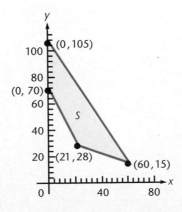

Corner point (x, y)	Objective function $C = 7x + 9y$	
(0, 105)	945	
(0, 70)	630	
(21, 28)	399	Minimum value of C
(60, 15)	555	

Next, we evaluate the objective function at each corner point, as shown in the table in the margin.

The optimal value is $C = 399$ at the corner point (21, 28). Thus, the farmer should add 21 cubic yards of mix A and 28 cubic yards of mix B at a cost of \$399. This will result in adding the following nutrients to the field:

Nitrogen: $10(21) + 5(28) = 350$ pounds

Potash: $8(21) + 24(28) = 840$ pounds

Phosphoric acid: $9(21) + 6(28) = 357$ pounds

All the nutritional requirements are satisfied.

Problem 3 Repeat Example 3 if the tests indicate that the field needs at least 400 pounds of nitrogen with all other conditions remaining the same.

Answers to Matched Problems

1. (A) x = Number of standard boards manufactured each week
   y = Number of competition boards manufactured each week
   (B) $P = 40x + 75y$ (C) $6x + 8y \leq 120$ Fabricating constraint
   $x + 3y \leq 30$ Finishing constraint
   $x, y \geq 0$ Nonnegative constraints

(D)

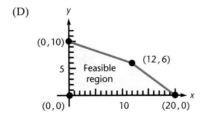

Corner point (x, y)	Objective function $P = 40x + 75y$
(0, 0)	0
(0, 10)	750
(12, 6)	930
(20, 0)	800

(E) 12 standard boards and 6 competition boards for a maximum profit of \$930

2. Max $z = 600$ at (30, 60); min $z = 200$ at (10, 20) and (20, 0) (multiple optimal solutions)
3. 27 cubic yards of mix A, 26 cubic yards of mix B; min $C = \$423$

EXERCISE 6-5

A _____

In Problems 1–4, find the maximum value of each objective function over the feasible region S shown in the figure below.

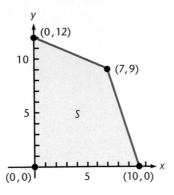

1. $z = x + y$ **2.** $z = 4x + y$ **3.** $z = 3x + 7y$ **4.** $z = 9x + 3y$

In Problems 5–8, find the minimum value of each objective function over the feasible region T shown in the figure below.

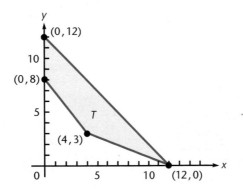

5. $z = 7x + 4y$ **6.** $z = 7x + 9y$ **7.** $z = 3x + 8y$ **8.** $z = 5x + 4y$

B _____

In Problems 9–20, solve the linear programming problems.

9. Maximize $z = 3x + 2y$
   Subject to $x + 2y \le 10$
   $3x + \ y \le 15$
   $x, y \ge 0$

10. Maximize $z = 4x + 5y$
   Subject to $2x + \ y \le 12$
   $x + 3y \le 21$
   $x, y \ge 0$

11. Minimize $z = 3x + 4y$
   Subject to $2x + \ y \ge 8$
   $x + 2y \le 10$
   $x, y \ge 0$

12. Minimize $z = 2x + y$
   Subject to $4x + 3y \ge 24$
   $4x + \ y \le 16$
   $x, y \ge 0$

13. Maximize $z = 3x + 4y$
Subject to
$$x + 2y \le 24$$
$$x + y \le 14$$
$$2x + y \le 24$$
$$x, y \ge 0$$

14. Maximize $z = 5x + 3y$
Subject to
$$3x + y \le 24$$
$$x + y \le 10$$
$$x + 3y \le 24$$
$$x, y \ge 0$$

15. Minimize $z = 5x + 6y$
Subject to
$$x + 4y \ge 20$$
$$4x + y \ge 20$$
$$x + y \le 20$$
$$x, y \ge 0$$

16. Minimize $z = x + 2y$
Subject to
$$2x + 3y \ge 30$$
$$3x + 2y \ge 30$$
$$x + y \le 15$$
$$x, y \ge 0$$

17. Minimize and maximize $z = 25x + 50y$
Subject to
$$x + 2y \le 120$$
$$x + y \ge 60$$
$$x - 2y \ge 0$$
$$x, y \ge 0$$

18. Minimize and maximize $z = 15x + 30y$
Subject to
$$x + 2y \ge 100$$
$$2x - y \le 0$$
$$2x + y \le 200$$
$$x, y \ge 0$$

19. Minimize and maximize $z = 25x + 15y$
Subject to
$$4x + 5y \ge 100$$
$$3x + 4y \le 240$$
$$x \le 60$$
$$y \le 45$$
$$x, y \ge 0$$

20. Minimize and maximize $z = 25x + 30y$
Subject to
$$2x + 3y \ge 120$$
$$3x + 2y \le 360$$
$$x \le 80$$
$$y \le 120$$
$$x, y \ge 0$$

C

21. The corner points for the feasible region determined by the problem constraints

$$2x + y \le 10$$
$$x + 3y \le 15$$
$$x, y \ge 0$$

are $O = (0, 0), A = (5, 0), B = (3, 4)$, and $C = (0, 5)$. If $z = ax + by$ and $a, b > 0$, determine conditions on a and b that ensure that the maximum value of z occurs:
(A) Only at A (B) Only at B
(C) Only at C (D) At both A and B
(E) At both B and C

22. The corner points for the feasible region determined by the problem constraints

$$x + y \ge 4$$
$$x + 2y \ge 6$$
$$2x + 3y \le 12$$
$$x, y \ge 0$$

are $A = (6, 0), B = (2, 2)$, and $C = (0, 4)$. If $z = ax + by$ and $a, b > 0$, determine conditions on a and b that ensure that the minimum value of z occurs:
(A) Only at A (B) Only at B
(C) Only at C (D) At both A and B
(E) At both B and C

APPLICATIONS

23. Resource Allocation. A manufacturing company makes two types of water skis, a trick ski and a slalom ski. The relevant manufacturing data are given in the table at the top of the next page. How many of each type of ski should be manufactured each day to realize a maximum profit? What is the maximum profit?

	Trick ski (labor-hours per ski)	Slalom ski (labor-hours per ski)	Maximum labor-hours available per day
Fabricating department	6	4	108
Finishing department	1	1	24
Profit per ski	$40	$30	

8 hours from the assembly department and 2 hours from the finishing department and contributes a profit of $90. Each chair requires 2 hours from the assembly department and 1 hour from the finishing department and contributes a profit of $25. The maximum labor-hours available each day in the assembly and finishing departments are 400 and 120, respectively.

(A) How many tables and how many chairs should be manufactured each day to maximize the daily profit? What is the maximum daily profit?

(B) Repeat part (A) if the marketing department of the company has decided that the number of chairs produced should be at least four times the number of tables produced.

24. Psychology. In an experiment on conditioning, a psychologist uses two types of Skinner boxes with mice and rats. The amount of time (in minutes) each mouse and each rat spends in each box per day is given in the table. What is the maximum total number of mice and rats that can be used in this experiment? How many mice and how many rats produce this maximum?

	Mice (minutes)	Rats (minutes)	Maximum time available per day (minutes)
Skinner box A	10	20	800
Skinner box B	20	10	640

25. Purchasing. A trucking firm wants to purchase a maximum of 15 new trucks that will provide at least 36 tons of additional shipping capacity. A model A truck holds 2 tons and costs $15,000. A model B truck holds 3 tons and costs $24,000. How many trucks of each model should the company purchase to provide the additional shipping capacity at minimal cost? What is the minimal cost?

26. Transportation. The officers of a high school senior class are planning to rent buses and vans for a class trip. Each bus can transport 40 students, requires 3 chaperones, and costs $1,200 to rent. Each van can transport 8 students, requires 1 chaperone, and costs $100 to rent. The officers want to be able to accommodate at least 400 students with no more than 36 chaperones. How many vehicles of each type should they rent in order to minimize the transportation costs? What are the minimal transportation costs?

★ **27. Resource Allocation.** A furniture company manufactures dining room tables and chairs. Each table requires

★ **28. Resource Allocation.** An electronics firm manufactures two types of personal computers, a desktop model and a portable model. The production of a desktop computer requires a capital expenditure of $400 and 40 hours of labor. The production of a portable computer requires a capital expenditure of $250 and 30 hours of labor. The firm has $20,000 capital and 2,160 labor-hours available for production of desktop and portable computers.

(A) What is the maximum number of computers the company is capable of producing?

(B) If each desktop computer contributes a profit of $320 and each portable computer contributes a profit of $220, how many of each type of computer should the firm produce in order to maximize profit? What is the maximum profit?

★★ **29. Purchasing.** A company is installing a large computer that can accommodate a maximum of 24 external disk drives. A disk drive that can store 40 megabytes of data costs $600 and one that can store 80 megabytes of data costs $1,500. The company needs at least 1,200 megabytes of storage.

(A) How many disk drives of each type should the company purchase in order to minimize the purchase cost? What is the minimum purchase cost?

(B) Repeat part (A) if the company's systems engineer decides that there should be at least twice as many 80-megabyte disk drives as 40-megabyte disk drives.

★★ **30. Sociology.** A city council voted to conduct a study on inner-city community problems. A nearby university was contacted to provide a maximum of 40 sociologists and research assistants. Allocation of time and cost per week are given in the table on the next page.

(A) How many sociologists and research assistants should be hired to meet the weekly labor-hour requirements and minimize the weekly cost? What is the weekly cost?

(B) Repeat part (A) if the council decides that they should not hire more sociologists than research assistants.

	Sociologist (labor-hours)	Research assistant (labor-hours)	Minimum labor-hours needed per week
Fieldwork	10	30	280
Research center	30	10	360
Cost per week	$500	$300	

31. Plant Nutrition. A fruit grower can use two types of fertilizer in her orange grove, brand A and brand B. The amounts (in pounds) of nitrogen, phosphoric acid, potash, and chlorine in a bag of each mix are given in the table. Tests indicate that the grove needs at least 480 pounds of phosphoric acid, at least 540 pounds of potash, and at most 620 pounds of chlorine.

(A) If the grower wants to maximize the amount of nitrogen added to the grove, how many bags of each mix should be added? How much nitrogen will be added?

(B) If the grower wants to minimize the amount of nitrogen added to the grove, how many bags of each mix should be added? How much nitrogen will be addcd?

	Pounds per bag	
	Brand A	Brand B
Nitrogen	6	7
Phosphoric acid	2	4
Potash	6	3
Chlorine	3	4

32. Diet. A dietitian in a hospital is to arrange a special diet composed of two foods, M and N. Each ounce of food M contains 16 units of calcium, 5 units of iron, 6 units of cholesterol, and 8 units of vitamin A. Each ounce of food N contains 4 units of calcium, 25 units of iron, 4 units of cholesterol, and 4 units of vitamin A. The diet requires at least 320 units of calcium, at least 575 units of iron, and at most 300 units of cholesterol.

(A) How many ounces of each food should be used to maximize the amount of vitamin A in the diet? What is the maximum amount of vitamin A?

(B) How many ounces of each food should be used to minimize the amount of vitamin A in the diet? What is the minimum amount of vitamin A?

Chapter 6 Review

6-1   SYSTEMS OF LINEAR EQUATIONS IN TWO VARIABLES; AUGMENTED MATRICES

A system of two linear equations with two variables is a system of the form

$$ax + by = h$$
$$cx + dy = k \tag{1}$$

where x and y are the two variables. The ordered pair of numbers (x_0, y_0) is a **solution** to system (1) if each equation is satisfied by the pair. The set of all such ordered pairs of numbers is called the **solution set** for the system. To **solve** a system is to find its solution set.

In general, a system of linear equations has exactly one solution, no solution, or infinitely many solutions. A system of linear equations is **consistent** if it has one or more solutions and **inconsistent** if no solutions exist. A consistent system is said to be **independent** if it has exactly one solution and **dependent** if it has more than one solution.

Three standard methods for solving system (1) were reviewed: **solution by graphing, solution by substitution**, and **solution using elimination by addition**.

Two systems of equations are **equivalent** if both have the same solution set. A system of linear equations is transformed into an equivalent system if:

1. Two equations are interchanged.

2. An equation is multiplied by a nonzero constant.

3. A constant multiple of another equation is added to a given equation.

These operations form the basis of solution using elimination by addition. The method of solution using elimination by addition can be transformed into a more efficient method for larger-scale systems by the introduction of an *augmented matrix*. A **matrix** is a rectangular array of numbers written within brackets. Each number in a matrix is called an **element** of the matrix.

For ease of generalization to larger systems, we change the notation for variables and constants in system (1) to a subscript form:

$$a_1 x_1 + b_1 x_2 = k_1$$
$$a_2 x_1 + b_2 x_2 = k_2 \tag{2}$$

Associated with each linear system of the form (2), where x_1 and x_2 are variables, is the **augmented matrix** of the system:

$$
\begin{array}{l}
\text{Column 1 } (C_1) \\
\text{Column 2 } (C_2) \\
\text{Column 3 } (C_3)
\end{array}
\begin{bmatrix} a_1 & b_1 & k_1 \\ a_2 & b_2 & k_2 \end{bmatrix}
\begin{array}{l}
\text{Row 1 } (R_1) \\
\text{Row 2 } (R_2)
\end{array} \tag{3}
$$

Two augmented matrices are **row-equivalent**, denoted by the symbol $\sim$ between the two matrices, if they are augmented matrices of equivalent systems of equations. An augmented matrix is transformed into a row-equivalent matrix if any of the following **row operations** is performed:

1. Two rows are interchanged.

2. A row is multiplied by a nonzero constant.

3. A constant multiple of another row is added to a given row.

The following symbols are used to describe these row operations:

1. $R_i \leftrightarrow R_j$ means "interchange row i with row j."

2. $kR_i \rightarrow R_i$ means "multiply row i by the constant k."

3. $R_i + kR_j \rightarrow R_i$ means "multiply row j by the constant k and add to R_i."

In solving system (2) using row operations, the objective is to transform the augmented matrix (3) into the form

$$\begin{bmatrix} 1 & 0 & m \\ 0 & 1 & n \end{bmatrix}$$

If this can be done, then (m, n) is the unique solution of system (2). If (3) is transformed into the form

$$\begin{bmatrix} 1 & m & n \\ 0 & 0 & 0 \end{bmatrix}$$

then system (2) has infinitely many solutions. If (3) is transformed into the form

$$\begin{bmatrix} 1 & m & n \\ 0 & 0 & p \end{bmatrix} \quad p \neq 0$$

then system (2) does not have a solution.

6-2   GAUSS–JORDAN ELIMINATION

In the last part of Section 6-1 we were actually using *Gauss–Jordan elimination* to solve a system of two equations with two variables. The method generalizes completely for systems with more than two variables, and the number of variables does not have to be the same as the number of equations.

As before, our objective is to start with the augmented matrix of a linear system and transform it using row operations into a simple form where the solution can be read by inspection. The simple form, called the **reduced form**, is achieved if:

1. Each row consisting entirely of 0's is below any row having at least one nonzero element.

2. The leftmost nonzero element in each row is 1.

3. The column containing the leftmost 1 of a given row has 0's above and below the 1.

4. The leftmost 1 in any row is to the right of the leftmost 1 in the preceding row.

A **reduced system** is a system of linear equations that corresponds to a reduced augmented matrix. When a reduced system has more variables than equations and contains no contradictions, the system is dependent and has infinitely many solutions.

The **Gauss–Jordan elimination** procedure for solving a system of linear equations is given in step-by-step form as follows:

Step 1. Choose the leftmost nonzero column and use appropriate row operations to get a 1 at the top.

Step 2. Use multiples of the first row to get 0's in all places below the 1 obtained in step 1.

Step 3. Mentally delete the top row and first column of the matrix. Repeat steps 1 and 2 with the **submatrix** (the matrix that remains after deleting the top row and first column). Continue this process (steps 1–3) until it is not possible to go further.

Step 4. Now consider the whole matrix. Begin with the bottom nonzero row and use appropriate multiples of it to get 0's above the leftmost 1. Continue this process, moving up row by row, until the matrix is finally in reduced form.

If at any point in the above process we obtain a row with all 0's to the left of the vertical line and a nonzero number n to the right, we can stop, since we have a contradiction: $0 = n, n \neq 0$. We can then conclude that the system has no solution. If this does not happen and we obtain an augmented matrix in reduced form without any contradictions, the solution can be read by inspection.

6-3   SYSTEMS INVOLVING SECOND-DEGREE EQUATIONS

If a system of equations contains any equations that are not linear, then the system is called a **nonlinear system**. In this section we investigated nonlinear systems involving second-degree terms such as

$$x^2 + y^2 = 5 \qquad x^2 - 2y^2 = 2 \qquad x^2 + 3xy + y^2 = 20$$

$$3x + y = 1 \qquad \qquad xy = 2 \qquad \qquad xy - y^2 = 0$$

It can be shown that such systems have at most four solutions, some of which may be imaginary.

Several methods were used to solve nonlinear systems of the indicated form: **solution by substitution, solution using elimination by addition**, and **solution using factoring and substitution**. It is always important to check the solutions of any nonlinear system to ensure that extraneous roots have not been introduced.

6-4   SYSTEMS OF LINEAR INEQUALITIES IN TWO VARIABLES

A graph is often the most convenient way to represent the solution of a linear inequality in two variables or of a system of linear inequalities in two variables.

A vertical line divides a plane into **left** and **right half-planes**. A nonvertical line divides a plane into **upper** and **lower half-planes**. Let A, B, and C be real numbers with A and B not both zero, then the **graph of the linear inequality**

$$Ax + By < C \quad \text{or} \quad Ax + By > C$$

with $B \neq 0$, is either the upper half-plane or the lower half-plane (but not both) determined by the line $Ax + By = C$. If $B = 0$, then the graph of

$$Ax < C \quad \text{or} \quad Ax > C$$

is either the left half-plane or the right half-plane (but not both) determined by the line $Ax = C$. Out of these results follow an easy **step-by-step procedure for graphing a linear inequality in two variables**:

Step 1. Graph $Ax + By = C$ as a broken line if equality is not included in the original statement or as a solid line if equality is included.

Step 2. Choose a test point anywhere in the plane not on the line and substitute the coordinates into the inequality. The origin $(0, 0)$ often requires the least computation.

Step 3. The graph of the original inequality includes the half-plane containing the test point if the inequality is satisfied by that point, or the half-plane not containing that point if the inequality is not satisfied by that point.

We now turn to systems of linear inequalities in two variables. The **solution to a system of linear inequalities in two variables** is the set of all ordered pairs of real numbers that simultaneously satisfy all the inequalities in the system. The graph is called the **solution region**. In many applications the solution region is also referred to as the **feasible region**. To **find the solution region**, we graph each inequality in the system and then take the intersection of all the graphs. A **corner point** of a solution region is a point in the solution region that is the intersection of two boundary lines. A solution region is **bounded** if it can be enclosed within a circle. If it cannot be enclosed within a circle, then it is **unbounded**.

6-5   LINEAR PROGRAMMING

Linear programming is a mathematical process that has been developed to help management in decision making, and it has become one of the most widely used and best-known tools of management science and industrial engineering.

A **linear programming problem** is one that is concerned with finding the **optimal value** (maximum or minimum value) of a linear **objective function** of the form $z = ax + by$, where the **decision variables** x and y are subject to **problem constraints** in the form of linear inequalities and **nonnegative constraints** $x, y \geq 0$. The set of points satisfying both the problem constraints and the nonnegative constraints is called the **feasible region** for the problem. Any point in the feasible region that produces the optimal value of the objective function over the feasible region is called an **optimal solution**. The **fundamental theorem** of linear programming is basic to the solving of linear programming problems: Let S be the feasible region for a linear programming problem, and let $z = ax + by$ be the objective function. If S is bounded, then z has both a maximum and a minimum value on S and each of these occurs at a corner point of S. If S is unbounded, then a maximum or minimum value of z on S may not exist. However, if either does exist, then it must occur at a corner point of S.

Problems with unbounded feasible regions are not considered in this brief introduction. The theorem leads to a simple **step-by-step solution to linear programming problems with a bounded feasible region**:

Step 1. Form a mathematical model for the problem:
  (A) Introduce decision variables and write a linear objective function.
  (B) Write problem constraints in the form of linear inequalities.
  (C) Write nonnegative constraints.

Step 2. Graph the feasible region and find the corner points.

Step 3. Evaluate the objective function at each corner point to determine the optimal solution.

If two corner points are both optimal solutions of the same type (both produce the same maximum value or both produce the same minimum value) to a linear programming problem, then any point on the line segment joining the two corner points is also an optimal solution of that type.

Chapter 6 Review Exercise

Work through all the problems in this chapter review and check answers in the back of the book. Answers to all review problems are there, and following each answer is a number in italics indicating the section in which that type of problem is discussed. Where weaknesses show up, review appropriate sections in the text.

A

Solve Problems 1–6 using substitution or elimination by addition.

1. $2x + y = 7$
$3x - 2y = 0$

2. $3x - 6y = 5$
$-2x + 4y = 1$

3. $4x - 3y = -8$
$-2x + \frac{3}{2}y = 4$

4. $y = x^2 - 5x - 3$
$y = -x + 2$

5. $x^2 + y^2 = 2$
$2x - y = 3$

6. $3x^2 - y^2 = -6$
$2x^2 + 3y^2 = 29$

Solve Problems 7–9 by graphing.

7. $3x - 2y = 8$
$x + 3y = -1$

8. $3x - 4y \geq 24$

9. $2x + y \leq 2$
$x + 2y \geq -2$

Perform each of the row operations indicated in Problems 10–12 on the following augmented matrix:

$$\begin{bmatrix} 1 & -4 & | & 5 \\ 3 & -6 & | & 12 \end{bmatrix}$$

10. $R_1 \leftrightarrow R_2$

11. $\frac{1}{3}R_2 \to R_2$

12. $R_2 + (-3)R_1 \to R_2$

In Problems 13–15, write the linear system corresponding to each reduced augmented matrix and solve.

13. $\begin{bmatrix} 1 & 0 & | & 4 \\ 0 & 1 & | & -7 \end{bmatrix}$

14. $\begin{bmatrix} 1 & -1 & | & 4 \\ 0 & 0 & | & 1 \end{bmatrix}$

15. $\begin{bmatrix} 1 & -1 & | & 4 \\ 0 & 0 & | & 0 \end{bmatrix}$

16. Find the maximum and minimum values of $z = 5x + 3y$ over the feasible region S shown in the figure.

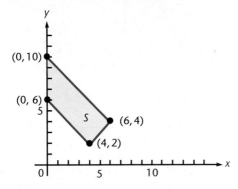

B _____

Solve Problems 17–22 using Gauss–Jordan elimination.

17. $3x_1 + 2x_2 = 3$
$\quad\; x_1 + 3x_2 = 8$

18. $x_1 + x_2 = 1$
$\quad x_1 - x_3 = -2$
$\quad x_2 + 2x_3 = 4$

19. $x_1 + 2x_2 + 3x_3 = 1$
$\quad 2x_1 + 3x_2 + 4x_3 = 3$
$\quad\; x_1 + 2x_2 + x_3 = 3$

20. $x_1 + 2x_2 - x_3 = 2$
$\quad 2x_1 + 3x_2 + x_3 = -3$
$\quad 3x_1 + 5x_2 = -1$

21. $x_1 - 2x_2 = 1$
$\quad 2x_1 - x_2 = 0$
$\quad\; x_1 - 3x_2 = -2$

22. $x_1 + 2x_2 - x_3 = 2$
$\quad 3x_1 - x_2 + 2x_3 = -3$

Solve Problems 23–25.

23. $x^2 - y^2 = 2$
$\quad\;\; y^2 = x$

24. $x^2 + 2xy + y^2 = 1$
$\quad\quad\quad\quad xy = -2$

25. $2x^2 + xy + y^2 = 8$
$\quad\; x^2 - y^2 = 0$

Solve the systems in Problems 26–28 graphically, and indicate whether each solution region is bounded or unbounded. Find the coordinates of each corner point.

26. $2x + y \le 8$
$\quad 2x + 3y \le 12$
$\quad\quad\quad x, y \ge 0$

27. $2x + y \ge 8$
$\quad\; x + 3y \ge 12$
$\quad\quad\; x, y \ge 0$

28. $x + y \le 20$
$\quad x + 4y \ge 20$
$\quad x - y \ge 0$

Solve the linear programming problems in Problems 29–31.

29. Maximize $\;z = 7x + 9y$
Subject to $\quad x + 2y \le 8$
$\quad\quad\quad\quad\; 2x + y \le 10$
$\quad\quad\quad\quad\quad\quad x, y \ge 0$

30. Minimize $\;z = 5x + 10y$
Subject to $\quad x + y \le 20$
$\quad\quad\quad\quad 3x + y \ge 15$
$\quad\quad\quad\quad\; x + 2y \ge 15$
$\quad\quad\quad\quad\quad\; x, y \ge 0$

31. Minimize and maximize
$z = 5x + 8y$
Subject to $\quad x + 2y \le 20$
$\quad\quad\quad\quad 3x + y \le 15$
$\quad\quad\quad\quad\; x + y \ge 7$
$\quad\quad\quad\quad\quad x, y \ge 0$

C

32. Solve using Gauss–Jordan elimination:

$$x_1 + x_2 + x_3 = 7{,}000$$
$$0.04x_1 + 0.05x_2 + 0.06x_3 = 360$$
$$0.04x_1 + 0.05x_2 - 0.06x_3 = 120$$

33. Solve:

$$x^2 - xy + y^2 = 4$$
$$x^2 + xy - 2y^2 = 0$$

34. Maximize $z = 30x + 20y$
    Subject to $1.2x + 0.6y \leq 960$
                $0.04x + 0.03y \leq 36$
                $0.2x + 0.3y \leq 270$
                     $x, y \geq 0$

APPLICATIONS

35. Business. A container holds 120 packages. Some of the packages weigh $\frac{1}{2}$ pound each, and the rest weigh $\frac{1}{3}$ pound each. If the total contents of the container weigh 48 pounds, how many are there of each type of package? Solve using two equation–two variable methods.

★ **36. Geometry.** Find the dimensions of a rectangle with an area of 48 square meters and a perimeter of 28 meters. Solve using two equation–two variable methods.

★ **37. Diet.** A laboratory assistant wishes to obtain a food mix that contains, among other things, 27 grams of protein, 5.4 grams of fat, and 19 grams of moisture. He has available mixes *A, B,* and *C* with the compositions listed in the table. How many grams of each mix should be used to get the desired diet mix? Set up a system of equations and solve using Gauss–Jordan elimination.

Mix	Protein (%)	Fat (%)	Moisture (%)
A	30	3	10
B	20	5	20
C	10	4	10

★★ **38. Puzzle.** A piggy bank contains 30 coins worth $1.90.
  (A) If the bank contains only nickels and dimes, how many coins of each type does it contain?
  (B) If the bank contains nickels, dimes, and quarters, how many coins of each type does it contain?

★ **39. Resource Allocation.** North Star Sail Loft manufactures regular and competition sails. Each regular sail takes 1 labor-hour to cut and 3 labor-hours to sew. Each competition sail takes 2 labor-hours to cut and 4 labor-hours to sew. The Loft makes a profit of $60 on each regular sail and $100 on each competition sail. If there are 140 labor-hours available in the cutting department and 360 labor-hours available in the sewing department, how many sails of each type should the company manufacture in order to maximize its profit? What is the maximum profit?

7

Matrices and Determinants

In this chapter we discuss matrices in more detail. In the first four sections we define and study some algebraic operations on matrices, including addition, multiplication, and inversion. The next three sections deal with the determinant of a matrix.

In the last chapter we used row operations and Gauss–Jordan elimination to solve systems of linear equations. Row operations play a prominent role in the development of several topics in this chapter. One consequence of our discussion will be the development of two additional methods for solving systems of linear equations: one method involves inverse matrices and the other determinants.

Matrices are both a very ancient and a very topical mathematical concept. References to matrices and systems of equations can be found in Chinese manuscripts dating back to around 200 B.C. Over the years, mathematicians and scientists have found many applications of matrices. More recently, the advent of personal and large-scale computers has increased the use of matrices in a wide variety of applications. In 1979 Dan Bricklin and Robert Frankston introduced VisiCalc, the first electronic spreadsheet program for personal computers. Simply put, a *spreadsheet* is a computer program that allows the user to enter and manipulate numbers, often using matrix notation and operations. Spreadsheets were initially used by businesses in areas such as budgeting, sales projections, and cost estimation. However, many other applications have begun to appear. For example, a scientist can use a spreadsheet to analyze the results of an experiment, or a teacher can use one to record and average grades. There are even spreadsheets that can be used to help compute an individual's income tax.

SECTION 7-1   **Matrix Addition; Multiplication by a Number**

- Dimension of a Matrix
- Matrix Addition and Subtraction
- Multiplication of a Matrix by a Number
- Application

In this section we introduce some basic matrix terminology and operations.

• Dimension of a Matrix

Recall that we defined a **matrix** as any rectangular array of real numbers enclosed within brackets, and that each number in the array is called an **element** of the matrix. The **size**, or **dimension**, **of a matrix** is important relative to operations on matrices. We define an $m \times n$ **matrix** (read "m by n matrix") to be one with m rows and n columns. It is important to note that the number of rows is always given first. If a matrix has the same number of rows and columns, it is called a **square matrix**. A matrix with only one column is called a **column matrix**, and one with only one row is called a **row matrix**. These definitions are illustrated by the following:

$$
\underset{3 \times 2}{\begin{bmatrix} -2 & 5 \\ 0 & -2 \\ 3 & 6 \end{bmatrix}}
\qquad
\underset{\substack{3 \times 3 \\ \text{Square matrix}}}{\begin{bmatrix} 0.5 & 0.2 & 1.0 \\ 0.0 & 0.3 & 0.5 \\ 0.7 & 0.0 & 0.2 \end{bmatrix}}
\qquad
\underset{\substack{4 \times 1 \\ \text{Column matrix}}}{\begin{bmatrix} 3 \\ -2 \\ 1 \\ 0 \end{bmatrix}}
\qquad
\underset{\substack{1 \times 4 \\ \text{Row matrix}}}{\begin{bmatrix} 2 & \tfrac{1}{2} & 0 & -\tfrac{2}{3} \end{bmatrix}}
$$

Two matrices are **equal** if they have the same dimension and their corresponding elements are equal. For example,

$$
\underset{2 \times 3}{\begin{bmatrix} a & b & c \\ d & e & f \end{bmatrix}}
=
\underset{2 \times 3}{\begin{bmatrix} u & v & w \\ x & y & z \end{bmatrix}}
\qquad \text{if and only if} \qquad
\begin{array}{ccc} a = u & b = v & c = w \\ d = x & e = y & f = z \end{array}
$$

• Matrix Addition and Subtraction

The **sum of two matrices** of the same dimension is a matrix with elements that are the sums of the corresponding elements of the two given matrices.

Addition is not defined for matrices with different dimensions.

EXAMPLE 1   Matrix Addition

(A) $\begin{bmatrix} a & b \\ c & d \end{bmatrix} + \begin{bmatrix} w & x \\ y & z \end{bmatrix} = \begin{bmatrix} (a + w) & (b + x) \\ (c + y) & (d + z) \end{bmatrix}$

(B) $\begin{bmatrix} 2 & -3 & 0 \\ 1 & 2 & -5 \end{bmatrix} + \begin{bmatrix} 3 & 1 & 2 \\ -3 & 2 & 5 \end{bmatrix} = \begin{bmatrix} 5 & -2 & 2 \\ -2 & 4 & 0 \end{bmatrix}$

Problem 1   Add: $\begin{bmatrix} 3 & 2 \\ -1 & -1 \\ 0 & 3 \end{bmatrix} + \begin{bmatrix} -2 & 3 \\ 1 & -1 \\ 2 & -2 \end{bmatrix}$

Because we add two matrices by adding their corresponding elements, it follows from the properties of real numbers that matrices of the same dimension are commutative and associative relative to addition. That is, if A, B, and C are matrices of the same dimension, then

$$A + B = B + A \qquad \text{Commutative}$$

$$(A + B) + C = A + (B + C) \qquad \text{Associative}$$

A matrix with elements that are all 0's is called a **zero matrix**. For example, the following are zero matrices of different dimensions:

$$[0 \quad 0 \quad 0] \qquad \begin{bmatrix} 0 & 0 \\ 0 & 0 \end{bmatrix} \qquad \begin{bmatrix} 0 \\ 0 \\ 0 \\ 0 \end{bmatrix} \qquad \begin{bmatrix} 0 & 0 & 0 & 0 \\ 0 & 0 & 0 & 0 \\ 0 & 0 & 0 & 0 \end{bmatrix}$$

[*Note:* "0" may be used to denote the zero matrix of any dimension.]

The **negative of a matrix** M, denoted by $-M$, is a matrix with elements that are the negatives of the elements in M. Thus, if

$$M = \begin{bmatrix} a & b \\ c & d \end{bmatrix}$$

then

$$-M = \begin{bmatrix} -a & -b \\ -c & -d \end{bmatrix}$$

Note that $M + (-M) = 0$ (a zero matrix).

If A and B are matrices of the same dimension, then we define **subtraction** as follows:

$$A - B = A + (-B)$$

Thus, to subtract matrix B from matrix A, we simply subtract corresponding elements.

EXAMPLE 2   **Matrix Subtraction**

$$\begin{bmatrix} 3 & -2 \\ 5 & 0 \end{bmatrix} - \begin{bmatrix} -2 & 2 \\ 3 & 4 \end{bmatrix} \quad \boxed{= \begin{bmatrix} 3 & -2 \\ 5 & 0 \end{bmatrix} + \begin{bmatrix} 2 & -2 \\ -3 & -4 \end{bmatrix}} = \begin{bmatrix} 5 & -4 \\ 2 & -4 \end{bmatrix}$$

Problem 2   Subtract: $[2 \quad -3 \quad 5] - [3 \quad -2 \quad 1]$

● **Multiplication of a Matrix by a Number**

Finally, the **product of a number** k **and a matrix** M, denoted by kM, is a matrix formed by multiplying each element of M by k. This definition is partly motivated by the fact that if M is a matrix, then we would like $M + M$ to equal $2M$.

EXAMPLE 3   Multiplication of a Matrix by a Number

$$-2 \begin{bmatrix} 3 & -1 & 0 \\ -2 & 1 & 3 \\ 0 & -1 & -2 \end{bmatrix} = \begin{bmatrix} -6 & 2 & 0 \\ 4 & -2 & -6 \\ 0 & 2 & 4 \end{bmatrix}$$

Problem 3   Find: $10 \begin{bmatrix} 1.3 \\ 0.2 \\ 3.5 \end{bmatrix}$

● **Application**   We now consider an application that uses various operations.

EXAMPLE 4   Sales and Commissions

Ms. Fong and Mr. Petris are salespeople for a new car agency that sells only two models. August was the last month for this year's models, and next year's models were introduced in September. Gross dollar sales for each month are given in the following matrices:

	AUGUST SALES	
	Compact	Luxury
Fong	$36,000	$72,000
Petris	$72,000	$0

$= A$

	SEPTEMBER SALES	
	Compact	Luxury
Fong	$144,000	$288,000
Petris	$180,000	$216,000

$= B$

For example, Ms. Fong had $36,000 in compact sales in August and Mr. Petris had $216,000 in luxury car sales in September.

(A) What were the combined dollar sales in August and September for each salesperson and each model?
(B) What was the increase in dollar sales from August to September?
(C) If both salespeople receive a 3% commission on gross dollar sales, compute the commission for each salesperson for each model sold in September.

Solutions   We use matrix addition for part (A), matrix subtraction for part (B), and multiplication of a matrix by a number for part (C).

(A) $A + B = \begin{bmatrix} \$180,000 & \$360,000 \\ \$252,000 & \$216,000 \end{bmatrix}$ Fong, Petris

	Compact	Luxury

(B) $B - A = \begin{bmatrix} \$108,000 & \$216,000 \\ \$108,000 & \$216,000 \end{bmatrix}$ Fong, Petris

$$(C) \quad 0.03B = \begin{bmatrix} (0.03)(\$144{,}000) & (0.03)(\$288{,}000) \\ (0.03)(\$180{,}000) & (0.03)(\$216{,}000) \end{bmatrix}$$

Compact   Luxury

$$= \begin{bmatrix} \$4{,}320 & \$8{,}640 \\ \$5{,}400 & \$6{,}480 \end{bmatrix} \begin{array}{l} \text{Fong} \\ \text{Petris} \end{array}$$

Problem 4   Repeat Example 4 with

$$A = \begin{bmatrix} \$72{,}000 & \$72{,}000 \\ \$36{,}000 & \$72{,}000 \end{bmatrix} \quad \text{and} \quad B = \begin{bmatrix} \$180{,}000 & \$216{,}000 \\ \$144{,}000 & \$216{,}000 \end{bmatrix}$$

Example 4 involved an agency with only two salespeople and two models. A more realistic problem might involve 20 salespeople and 15 models. Problems of this size are often solved with the aid of an electronic spreadsheet program on a personal computer. Figure 1 illustrates a computer solution for Example 4.

FIGURE 1 Spreadsheet.

	A	B	C	D	E	F	G	H	I
1		August Sales			September Sales				
2		Compact	Luxury		Compact	Luxury			
3	Wong	$36,000	$72,000		$144,000	$288,000			
4	Petris	$72,000	$0		$180,000	$216,000			
5									
6		Combined Sales			Sales Increase			September Commissions	
7	Wong	$180,000	$360,000		$108,000	$216,000		$4,320	$8,640
8	Petris	$252,000	$216,000		$108,000	$216,000		$5,400	$6,480

Answers to Matched Problems

1. $\begin{bmatrix} 1 & 5 \\ 0 & -2 \\ 2 & 1 \end{bmatrix}$ **2.** $\begin{bmatrix} -1 & -1 & 4 \end{bmatrix}$ **3.** $\begin{bmatrix} 13 \\ 2 \\ 35 \end{bmatrix}$

4. (A) $\begin{bmatrix} \$252{,}000 & \$288{,}000 \\ \$180{,}000 & \$288{,}000 \end{bmatrix}$ (B) $\begin{bmatrix} \$108{,}000 & \$144{,}000 \\ \$108{,}000 & \$144{,}000 \end{bmatrix}$ (C) $\begin{bmatrix} \$5{,}400 & \$6{,}480 \\ \$4{,}320 & \$6{,}480 \end{bmatrix}$

EXERCISE 7-1

A

Problems 1–20 refer to the following matrices:

$$A = \begin{bmatrix} 2 & -1 \\ 3 & 0 \end{bmatrix} \quad B = \begin{bmatrix} -3 & 1 \\ 2 & -3 \end{bmatrix} \quad C = \begin{bmatrix} 2 \\ -3 \\ 0 \end{bmatrix}$$

$$D = \begin{bmatrix} 1 \\ 3 \\ 5 \end{bmatrix} \quad E = \begin{bmatrix} -4 & 1 & 0 & -2 \end{bmatrix} \quad F = \begin{bmatrix} 2 & -3 \\ -2 & 0 \\ 1 & 2 \\ 3 & 5 \end{bmatrix}$$

1. What are the dimensions of B? Of E?

2. What are the dimensions of F? Of D?

3. What element is in the third row and second column of matrix F?

4. What element is in the second row and first column of matrix F?

5. Write a zero matrix of the same dimension as B.

6. Write a zero matrix of the same dimension as E.

7. Which of the above matrices are column matrices?

8. Which of the above matrices are row matrices?

9. Which of the above matrices are square matrices?

10. How many additional columns would F have to have to be a square matrix?

11. Find $A + B$.   12. Find $C + D$.

13. Find $E + F$.   14. Find $B + C$.

15. Write the negative of matrix C.

16. Write the negative of matrix B.

17. Find $D - C$.   18. Find $A - A$.

19. Find $5B$.   20. Find $-2E$.

B _____

In Problems 21–34, perform the indicated operations, if possible.

21. $\begin{bmatrix} 4 & 0 & -2 & 1 \\ 3 & -2 & -3 & 5 \\ 0 & 2 & -1 & 4 \end{bmatrix} + \begin{bmatrix} -3 & -1 & 6 & 0 \\ -5 & -4 & 7 & -3 \\ 2 & 5 & 1 & -6 \end{bmatrix}$

22. $\begin{bmatrix} 5 & -3 & 7 \\ 5 & 2 & -4 \\ -7 & 4 & 1 \\ 2 & 3 & -5 \end{bmatrix} + \begin{bmatrix} -8 & -1 & -2 \\ 0 & -1 & 4 \\ 9 & 2 & -6 \\ 2 & -7 & 0 \end{bmatrix}$

23. $\begin{bmatrix} 1 & -2 & 4 \\ -3 & 4 & -6 \end{bmatrix} - \begin{bmatrix} -1 & 3 & 4 \\ 2 & 1 & -7 \end{bmatrix}$

24. $\begin{bmatrix} 0 & 1 & -1 \\ 4 & -5 & -3 \end{bmatrix} - \begin{bmatrix} -3 & -1 & 1 \\ 8 & -5 & 0 \end{bmatrix}$

25. $\begin{bmatrix} 3 & -1 \\ 0 & 2 \\ -1 & 4 \end{bmatrix} - \begin{bmatrix} -2 & 1 \\ 1 & 2 \\ -1 & 3 \end{bmatrix}$

26. $\begin{bmatrix} -3 \\ 1 \\ 5 \\ 0 \end{bmatrix} - \begin{bmatrix} 7 \\ -2 \\ 6 \\ -1 \end{bmatrix}$

27. $1,000 \begin{bmatrix} 0.25 & 0.36 \\ 0.04 & 0.35 \end{bmatrix}$

28. $100 \begin{bmatrix} 0.32 & 0.05 & 0.17 \\ 0.22 & 0.03 & 0.21 \end{bmatrix}$

29. $2 \begin{bmatrix} 1 & 2 \\ -1 & 5 \end{bmatrix} + 3 \begin{bmatrix} 4 & 1 & -2 \\ 3 & 2 & -1 \end{bmatrix}$

30. $4 \begin{bmatrix} 2 & -1 \\ 1 & 2 \\ 0 & 1 \end{bmatrix} + 5 \begin{bmatrix} 1 \\ -1 \\ 2 \end{bmatrix}$

31. $2 \begin{bmatrix} 3 & -1 \\ 0 & 1 \end{bmatrix} - 3 \begin{bmatrix} -1 & 1 \\ 2 & -1 \end{bmatrix}$

32. $3 \begin{bmatrix} 2 & 0 \\ -1 & -1 \end{bmatrix} - 2 \begin{bmatrix} 1 & 2 \\ -4 & -1 \end{bmatrix}$

33. $0.05 \begin{bmatrix} 400 & 600 \\ 200 & 700 \end{bmatrix} + 0.04 \begin{bmatrix} 350 & 200 \\ 450 & 800 \end{bmatrix}$

34. $0.06 \begin{bmatrix} 3,000 & 4,500 \\ 4,800 & 2,700 \end{bmatrix} + 0.08 \begin{bmatrix} 2,000 & 4,000 \\ 2,500 & 5,500 \end{bmatrix}$

35. Find $a, b, c,$ and d so that

$$\begin{bmatrix} a & b \\ c & d \end{bmatrix} + \begin{bmatrix} 2 & -3 \\ 0 & 1 \end{bmatrix} = \begin{bmatrix} 1 & -2 \\ 3 & -4 \end{bmatrix}$$

36. Find $w, x, y,$ and z so that

$$\begin{bmatrix} 4 & -2 \\ -3 & 0 \end{bmatrix} + \begin{bmatrix} w & x \\ y & z \end{bmatrix} = \begin{bmatrix} 2 & -3 \\ 0 & 5 \end{bmatrix}$$

37. Find x and y so that

$$\begin{bmatrix} 3x & 5 \\ -1 & 4x \end{bmatrix} + \begin{bmatrix} 2y & -3 \\ -6 & -y \end{bmatrix} = \begin{bmatrix} 7 & 2 \\ -7 & 2 \end{bmatrix}$$

38. Find x and y so that

$$\begin{bmatrix} 4 & 2x \\ -4x & -3 \end{bmatrix} + \begin{bmatrix} -5 & -3y \\ 5y & 3 \end{bmatrix} = \begin{bmatrix} -1 & 1 \\ 1 & 0 \end{bmatrix}$$

C

In Problems 39–42, is it possible to find 2×2 matrices A, B, and C and a real number k such that the indicated equation fails to hold? If so, state an example.

39. $A + B = B + A$

40. $(A + B) + C = A + (B + C)$

41. $k(A + B) = kA + kB$

42. $k(A - B) = kA - kB$

APPLICATIONS

43. Cost Analysis. A company with two different plants manufactures guitars and banjos. Its production costs for each instrument are given in the following matrices:

$$
\begin{array}{cc}
& \begin{array}{c}\text{PLANT } X \\ \text{Guitar} \quad \text{Banjo}\end{array} \\
\begin{array}{c}\text{Materials} \\ \text{Labor}\end{array} & \begin{bmatrix} \$30 & \$25 \\ \$60 & \$80 \end{bmatrix} = A
\end{array}
\qquad
\begin{array}{cc}
\begin{array}{c}\text{PLANT } Y \\ \text{Guitar} \quad \text{Banjo}\end{array} \\
\begin{bmatrix} \$36 & \$27 \\ \$54 & \$74 \end{bmatrix} = B
\end{array}
$$

Find $\frac{1}{2}(A + B)$, the average cost of production for the two plants.

44. Cost Analysis. If both labor and materials at plant X in Problem 43 are increased 20%, find $\frac{1}{2}(1.2A + B)$, the new average cost of production for the two plants.

45. Psychology. Two psychologists independently carried out studies on the relationship between height and aggressive behavior in women over 18 years of age. The results of the studies are summarized in the following matrices:

$$
\begin{array}{c}
\text{PROFESSOR ALDQUIST} \\
\begin{array}{cccc}
& \text{Under 5 ft} & \text{5–5}\frac{1}{2}\text{ ft} & \text{Over 5}\frac{1}{2}\text{ ft}
\end{array} \\
\begin{array}{c}\text{Passive} \\ \text{Aggressive}\end{array}
\begin{bmatrix} 70 & 122 & 20 \\ 30 & 118 & 80 \end{bmatrix} = A
\end{array}
$$

$$
\begin{array}{c}
\text{PROFESSOR KELLEY} \\
\begin{array}{cccc}
& \text{Under 5 ft} & \text{5–5}\frac{1}{2}\text{ ft} & \text{Over 5}\frac{1}{2}\text{ ft}
\end{array} \\
\begin{array}{c}\text{Passive} \\ \text{Aggressive}\end{array}
\begin{bmatrix} 65 & 160 & 30 \\ 25 & 140 & 75 \end{bmatrix} = B
\end{array}
$$

The two psychologists decided to combine their results and publish a joint paper. Write the matrix $A + B$ illustrating their combined results. There are 935 women in the combined study. Use a calculator to compute the decimal fraction of the total sample in each category of the combined study. [*Hint:* Compute $\frac{1}{935}(A + B)$.]

46. Heredity. Gregor Mendel (1822–1884), a Bavarian monk and botanist, made discoveries that revolutionized the science of heredity. In one experiment he crossed dihybrid yellow round peas (yellow and round are dominant characteristics; the peas also contained green and wrinkled as recessive genes) and obtained 560 peas of the types indicated in the matrix:

$$
\begin{array}{cc}
& \begin{array}{cc}\text{Round} & \text{Wrinkled}\end{array} \\
\begin{array}{c}\text{Yellow} \\ \text{Green}\end{array} & \begin{bmatrix} 319 & 101 \\ 108 & 32 \end{bmatrix} = M
\end{array}
$$

Suppose he carried out a second experiment of the same type and obtained 640 peas of the types indicated in this matrix:

$$
\begin{array}{cc}
& \begin{array}{cc}\text{Round} & \text{Wrinkled}\end{array} \\
\begin{array}{c}\text{Yellow} \\ \text{Green}\end{array} & \begin{bmatrix} 370 & 124 \\ 110 & 36 \end{bmatrix} = N
\end{array}
$$

If the results of the two experiments are combined, write the resulting matrix $M + N$. Compute the decimal fraction of the total number of peas (1,200) in each category of the combined results. [*Hint:* Compute $\frac{1}{1,200}(M + N)$.]

SECTION 7-2   Matrix Multiplication

- Dot Product
- Matrix Product
- Multiplication Properties
- Application

In this section we introduce two types of matrix multiplication that may at first seem rather strange. In spite of this apparent strangeness, these operations are well-founded in the general theory of matrices and are extremely useful in practical problems.

Historically, matrix multiplication was introduced by the English mathematician Arthur Cayley (1821–1895) in studies of systems of linear equations and linear transformations. In Section 7-4 you will see that matrix multiplication is central to the process of expressing systems of linear equations as matrix equations and to the process of solving matrix equations.

● **Dot Product**   We start by defining the *dot product* of two special matrices.

DEFINITION 1   **Dot Product**

The **dot product** of a $1 \times n$ row matrix and an $n \times 1$ column matrix is a real number given by

$$[a_1 \quad a_2 \quad \cdots \quad a_n] \cdot \begin{bmatrix} b_1 \\ b_2 \\ \vdots \\ b_n \end{bmatrix} = a_1b_1 + a_2b_2 + \cdots + a_nb_n \quad \text{A real number}$$

The dot between the two matrices is important. If the dot is omitted, the multiplication is of another type, which we consider later.

EXAMPLE 1   Dot Product

$$[2 \quad -3 \quad 0] \cdot \begin{bmatrix} -5 \\ 2 \\ -2 \end{bmatrix} = (2)(-5) + (-3)(2) + (0)(-2)$$
$$= -10 - 6 + 0 = -16$$

Problem 1
$$[-1 \quad 0 \quad 3 \quad 2] \cdot \begin{bmatrix} 2 \\ 3 \\ 4 \\ -1 \end{bmatrix} = \,?$$

EXAMPLE 2   **Production Scheduling**

A factory produces a slalom water ski that requires 4 labor-hours in the fabricating department and 1 labor-hour in the finishing department. Fabricating personnel receive $10 per hour, and finishing personnel receive $8 per hour. Total labor cost per ski is given by the dot product

$$[4 \quad 1] \cdot \begin{bmatrix} 10 \\ 8 \end{bmatrix} = (4)(10) + (1)(8) = 40 + 8 = \$48 \text{ per ski}$$

Problem 2 If the factory in Example 2 also produces a trick water ski that requires 6 labor-hours in the fabricating department and 1.5 labor-hours in the finishing department, write a dot product between appropriate row and column matrices that gives the total labor cost for this ski. Compute the cost.

It is important to remember that the dot product of a row matrix and a column matrix is a real number and not a matrix.

● Matrix Product We now use the dot product to define a *matrix product* for certain matrices.

DEFINITION 2   **Matrix Product**

The **product of two matrices** A and B is defined only on the assumption that the number of columns in A is equal to the number of rows in B. If A is an $m \times p$ matrix and B is a $p \times n$ matrix, then the matrix product of A and B, denoted by AB, with *no* dot, is an $m \times n$ matrix whose element in the ith row and jth column is the dot product of the ith row matrix of A and the jth column matrix of B.

It is important to check dimensions before starting the multiplication process. If matrix A has dimension $a \times b$ and matrix B has dimension $c \times d$, then if $b = c$, the product AB exists and has dimension $a \times d$. This is shown schematically in Figure 1.

FIGURE 1 Matrix product dimensions.

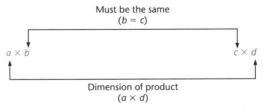

Must be the same
$(b = c)$

$a \times b$ $c \times d$

Dimension of product
$(a \times d)$

The definition is not as complicated as it might first seem. An example should help clarify the process. For

$$A = \begin{bmatrix} 2 & 3 & -1 \\ -2 & 1 & 2 \end{bmatrix} \quad \text{and} \quad B = \begin{bmatrix} 1 & 3 \\ 2 & 0 \\ -1 & 2 \end{bmatrix}$$

A is 2×3, B is 3×2, and so AB is 2×2. The four dot products used to produce the four elements in AB (usually calculated mentally for small whole numbers) are shown in the large matrix in the dashed box at the top of the next page:

$$
\begin{bmatrix} \overset{2 \times 3}{2} & 3 & -1 \\ -2 & 1 & 2 \end{bmatrix}
\begin{bmatrix} \overset{3 \times 2}{1} & 3 \\ 2 & 0 \\ -1 & 2 \end{bmatrix}
=
\begin{bmatrix}
[2 \quad 3 \quad -1] \cdot \begin{bmatrix} 1 \\ 2 \\ -1 \end{bmatrix} & [2 \quad 3 \quad -1] \cdot \begin{bmatrix} 3 \\ 0 \\ 2 \end{bmatrix} \\
[-2 \quad 1 \quad 2] \cdot \begin{bmatrix} 1 \\ 2 \\ -1 \end{bmatrix} & [-2 \quad 1 \quad 2] \cdot \begin{bmatrix} 3 \\ 0 \\ 2 \end{bmatrix}
\end{bmatrix}
= \begin{bmatrix} \overset{2 \times 2}{9} & 4 \\ -2 & -2 \end{bmatrix}
$$

EXAMPLE 3 Matrix Product

(A) $\begin{bmatrix} \overset{3 \times 2}{2} & 1 \\ 1 & 0 \\ -1 & 2 \end{bmatrix} \begin{bmatrix} \overset{2 \times 4}{1} & -1 & 0 & 1 \\ 2 & 1 & 2 & 0 \end{bmatrix} = \begin{bmatrix} \overset{3 \times 4}{4} & -1 & 2 & 2 \\ 1 & -1 & 0 & 1 \\ 3 & 3 & 4 & -1 \end{bmatrix}$

(B) $\begin{bmatrix} \overset{2 \times 4}{1} & -1 & 0 & 1 \\ 2 & 1 & 2 & 0 \end{bmatrix} \begin{bmatrix} \overset{3 \times 2}{2} & 1 \\ 1 & 0 \\ -1 & 2 \end{bmatrix}$

Product is not defined

(C) $\begin{bmatrix} 2 & 6 \\ -1 & -3 \end{bmatrix}\begin{bmatrix} 1 & 2 \\ 3 & 6 \end{bmatrix} = \begin{bmatrix} 20 & 40 \\ -10 & -20 \end{bmatrix}$

(D) $\begin{bmatrix} 1 & 2 \\ 3 & 6 \end{bmatrix}\begin{bmatrix} 2 & 6 \\ -1 & -3 \end{bmatrix} = \begin{bmatrix} 0 & 0 \\ 0 & 0 \end{bmatrix}$

(E) $[2 \quad -3 \quad 0]\begin{bmatrix} -5 \\ 2 \\ -2 \end{bmatrix} = [-16]$

(F) $\begin{bmatrix} -5 \\ 2 \\ -2 \end{bmatrix}[2 \quad -3 \quad 0] = \begin{bmatrix} -10 & 15 & 0 \\ 4 & -6 & 0 \\ -4 & 6 & 0 \end{bmatrix}$

Problem 3 Find each product, if it is defined:

(A) $\begin{bmatrix} -1 & 0 & 3 & -2 \\ 1 & 2 & 2 & 0 \end{bmatrix}\begin{bmatrix} -1 & 1 \\ 2 & 3 \\ 1 & 0 \end{bmatrix}$

(B) $\begin{bmatrix} -1 & 1 \\ 2 & 3 \\ 1 & 0 \end{bmatrix}\begin{bmatrix} -1 & 0 & 3 & -2 \\ 1 & 2 & 2 & 0 \end{bmatrix}$

(C) $\begin{bmatrix} 1 & 2 \\ -1 & -2 \end{bmatrix}\begin{bmatrix} -2 & 4 \\ 1 & -2 \end{bmatrix}$

(D) $\begin{bmatrix} -2 & 4 \\ 1 & -2 \end{bmatrix}\begin{bmatrix} 1 & 2 \\ -1 & -2 \end{bmatrix}$

(E) $[3 \quad -2 \quad 1]\begin{bmatrix} 4 \\ 2 \\ 3 \end{bmatrix}$

(F) $\begin{bmatrix} 4 \\ 2 \\ 3 \end{bmatrix}[3 \quad -2 \quad 1]$

Notice that the matrix product of a $1 \times n$ row matrix and an $n \times 1$ column matrix is a 1×1 matrix. Compare the dot product in Example 1 with the matrix product in Example 3E. This is a technical distinction, and it is common to see 1×1 matrices written as real numbers, but we will not follow this practice in this course.

● **Multiplication Properties**

In the arithmetic of real numbers the order in which we multiply doesn't matter; for example, $5 \times 7 = 7 \times 5$. In matrix multiplication, order does make a difference. That is, AB does not always equal BA, even if both multiplications are defined [see Examples 3C and 3D]. Thus:

Matrix multiplication is not commutative.

Also, AB may be 0 with neither A nor B equal to 0, as in Example 3D. Thus:

The zero property does not hold for matrix multiplication.

See Section 1-1 for a discussion of the zero property for real numbers.

Matrix multiplication does have general properties, many of which are similar to the properties of real numbers. Some of these are listed in Theorem 1. In Section 7-4 we will use these properties to solve matrix equations. Notice that since matrix multiplication is not commutative, there are two distributive laws and two multiplication properties.

Theorem 1

Properties of Matrix Multiplication

Assuming all products and sums are defined for the indicated matrices A, B, and C, then for k a real number:

1. $A(BC) = (AB)C$ **Associative Property**
2. $A(B + C) = AB + AC$ **Left Distributive Property**
3. $(B + C)A = BA + CA$ **Right Distributive Property**
4. If $A = B$, then $CA = CB$ **Left Multiplication Property**
5. If $A = B$, then $AC = BC$ **Right Multiplication Property**
6. $k(AB) = (kA)B = A(kB)$

● **Application**

The next example illustrates the use of the dot and matrix product in a business application.

EXAMPLE 4   Production Scheduling

Let's combine the time requirements discussed in Example 2 and Problem 2 into one matrix:

$$
\begin{array}{cc}
 & \text{Fabricating} \quad \text{Finishing} \\
 & \text{department} \quad \text{department}
\end{array}
$$

$$
\begin{array}{c}
\text{Trick ski} \\
\text{Slalom ski}
\end{array}
\begin{bmatrix}
6\ \text{h} & 1.5\ \text{h} \\
4\ \text{h} & 1\ \text{h}
\end{bmatrix} = A
$$

Now suppose the company has two manufacturing plants, X and Y, in different parts of the country and that their hourly rates for each department are given in the following matrix:

$$
\begin{array}{cc}
& \text{Plant } X \quad \text{Plant } Y \\
\begin{array}{r} \text{Fabricating department} \\ \text{Finishing department} \end{array} &
\begin{bmatrix} \$10 & \$12 \\ \$8 & \$10 \end{bmatrix} = B
\end{array}
$$

To find the total labor costs for each ski at each factory, we multiply A and B:

$$
AB = \overset{2 \times 2}{\begin{bmatrix} 6 & 1.5 \\ 4 & 1 \end{bmatrix}} \overset{2 \times 2}{\begin{bmatrix} 10 & 12 \\ 8 & 10 \end{bmatrix}} = \overset{X \qquad Y}{\begin{bmatrix} \$72 & \$87 \\ \$48 & \$58 \end{bmatrix}} \begin{array}{l} \text{Trick ski} \\ \text{Slalom ski} \end{array}
$$

Notice that the dot product of the first row matrix of A and the first column matrix of B gives us the labor costs, $72, for a trick ski manufactured at plant X. The dot product of the second row matrix of A and the second column matrix of B gives us the labor costs, $58, for manufacturing a slalom ski at plant Y. And so on.

Example 4 is, of course, oversimplified. Companies that manufacture many different items in many different plants deal with matrices that have very large numbers of rows and columns.

Problem 4   Repeat Example 4 with

$$
A = \begin{bmatrix} 7\,h & 2\,h \\ 5\,h & 1.5\,h \end{bmatrix} \quad \text{and} \quad B = \begin{bmatrix} \$11 & \$10 \\ \$8 & \$6 \end{bmatrix}
$$

Answers to Matched Problems

1. 8   **2.** $[6 \ \ 1.5] \cdot \begin{bmatrix} 10 \\ 8 \end{bmatrix} = \$72$

3. (A) Not defined   (B) $\begin{bmatrix} 2 & 2 & -1 & 2 \\ 1 & 6 & 12 & -4 \\ -1 & 0 & 3 & -2 \end{bmatrix}$ (C) $\begin{bmatrix} 0 & 0 \\ 0 & 0 \end{bmatrix}$ (D) $\begin{bmatrix} -6 & -12 \\ 3 & 6 \end{bmatrix}$

(E) $[11]$ (F) $\begin{bmatrix} 12 & -8 & 4 \\ 6 & -4 & 2 \\ 9 & -6 & 3 \end{bmatrix}$

4. $\begin{array}{cc} X & Y \end{array}$
$\begin{bmatrix} \$93 & \$82 \\ \$67 & \$59 \end{bmatrix} \begin{array}{l} \text{Trick} \\ \text{Slalom} \end{array}$

EXERCISE 7-2

A _____

Find the dot products in Problems 1–4.

1. $[2 \quad 4] \cdot \begin{bmatrix} 3 \\ 1 \end{bmatrix}$

2. $[3 \quad 1] \cdot \begin{bmatrix} 2 \\ 4 \end{bmatrix}$

3. $[-3 \quad 2] \cdot \begin{bmatrix} -1 \\ -2 \end{bmatrix}$

4. $[3 \quad -2] \cdot \begin{bmatrix} -4 \\ -1 \end{bmatrix}$

Find the matrix products in Problems 5–16.

5. $[2 \quad 5]\begin{bmatrix} 1 & -1 \\ 2 & 3 \end{bmatrix}$

6. $[1 \quad 3]\begin{bmatrix} 2 & 3 \\ 1 & -4 \end{bmatrix}$

7. $\begin{bmatrix} 3 & 4 \\ -1 & -2 \end{bmatrix}\begin{bmatrix} -1 \\ 2 \end{bmatrix}$

8. $\begin{bmatrix} -1 & 1 \\ 2 & -3 \end{bmatrix}\begin{bmatrix} 4 \\ -2 \end{bmatrix}$

9. $\begin{bmatrix} 2 & -3 \\ 1 & 2 \end{bmatrix}\begin{bmatrix} 1 & -1 \\ 0 & -2 \end{bmatrix}$

10. $\begin{bmatrix} -3 & 2 \\ 4 & -1 \end{bmatrix}\begin{bmatrix} -2 & 5 \\ -1 & 3 \end{bmatrix}$

11. $\begin{bmatrix} 1 & -1 \\ 0 & -2 \end{bmatrix}\begin{bmatrix} 2 & -3 \\ 1 & 2 \end{bmatrix}$

12. $\begin{bmatrix} -2 & 5 \\ -1 & 3 \end{bmatrix}\begin{bmatrix} -3 & 2 \\ 4 & -1 \end{bmatrix}$

13. $[4 \quad -2]\begin{bmatrix} -5 \\ -3 \end{bmatrix}$

14. $[2 \quad -1]\begin{bmatrix} 3 \\ -4 \end{bmatrix}$

15. $\begin{bmatrix} -5 \\ -3 \end{bmatrix}[4 \quad -2]$

16. $\begin{bmatrix} 3 \\ -4 \end{bmatrix}[2 \quad -1]$

B _____

Find the dot products in Problems 17–20.

17. $[-1 \quad -2 \quad 2] \cdot \begin{bmatrix} 2 \\ -1 \\ 3 \end{bmatrix}$

18. $[-2 \quad 4 \quad 0] \cdot \begin{bmatrix} -1 \\ -3 \\ 2 \end{bmatrix}$

19. $[-1 \quad -3 \quad 0 \quad 5] \cdot \begin{bmatrix} 4 \\ -3 \\ -1 \\ 2 \end{bmatrix}$

20. $[-1 \quad 2 \quad 3 \quad -2] \cdot \begin{bmatrix} 3 \\ -2 \\ 0 \\ 4 \end{bmatrix}$

In Problems 21–34, find each matrix product, if it is defined.

21. $\begin{bmatrix} 2 & -1 & 1 \\ 1 & 3 & -2 \end{bmatrix}\begin{bmatrix} 1 & 3 \\ 0 & -1 \\ -2 & 2 \end{bmatrix}$

22. $\begin{bmatrix} -1 & -4 & 3 \\ 2 & 0 & 1 \end{bmatrix}\begin{bmatrix} 2 & -3 \\ 1 & 2 \\ 0 & -1 \end{bmatrix}$

23. $\begin{bmatrix} 1 & 3 \\ 0 & -1 \\ -2 & 2 \end{bmatrix}\begin{bmatrix} 2 & -1 & 1 \\ 1 & 3 & -2 \end{bmatrix}$

24. $\begin{bmatrix} 2 & -3 \\ 1 & 2 \\ 0 & -1 \end{bmatrix}\begin{bmatrix} -1 & -4 & 3 \\ 2 & 0 & 1 \end{bmatrix}$

25. $[3 \quad -2 \quad -4]\begin{bmatrix} 1 \\ 2 \\ -3 \end{bmatrix}$

26. $[1 \quad -2 \quad 2]\begin{bmatrix} 2 \\ -1 \\ 1 \end{bmatrix}$

27. $\begin{bmatrix} 1 \\ 2 \\ -3 \end{bmatrix}[3 \quad -2 \quad -4]$

28. $\begin{bmatrix} 2 \\ -1 \\ 1 \end{bmatrix}[1 \quad -2 \quad 2]$

29. $\begin{bmatrix} 1 & 2 \\ 2 & -1 \\ -3 & 1 \end{bmatrix}[3 \quad -2 \quad -4]$

30. $\begin{bmatrix} 1 \\ 2 \\ -3 \end{bmatrix}\begin{bmatrix} 3 & -2 & 4 \\ 1 & -2 & 2 \end{bmatrix}$

31. $\begin{bmatrix} 1 & 2 & -1 \\ 3 & -1 & 4 \\ 2 & -4 & 5 \end{bmatrix} \begin{bmatrix} 4 \\ 5 \\ 7 \end{bmatrix}$

32. $\begin{bmatrix} 2 & 1 & -1 \\ 1 & 3 & 1 \\ -4 & 2 & 5 \end{bmatrix} \begin{bmatrix} -1 \\ 2 \\ -3 \end{bmatrix}$

33. $\begin{bmatrix} 1 & -1 & 2 & 0 \\ 3 & -4 & -2 & 1 \\ 1 & 0 & -1 & 2 \end{bmatrix} \begin{bmatrix} 1 & -2 & 1 \\ 3 & 0 & -4 \\ -1 & 2 & -2 \\ 3 & -3 & 4 \end{bmatrix}$

34. $\begin{bmatrix} 1 & -2 & 1 \\ 3 & 0 & -4 \\ -1 & 2 & -2 \\ 3 & -3 & 4 \end{bmatrix} \begin{bmatrix} 1 & -1 & 2 & 0 \\ 3 & -4 & -2 & 1 \\ 1 & 0 & -1 & 2 \end{bmatrix}$

C

In Problems 35–38, find $A^2 = AA$.

35. $\begin{bmatrix} 6 & 9 \\ -4 & -6 \end{bmatrix}$

36. $\begin{bmatrix} -2 & 4 \\ -1 & 2 \end{bmatrix}$

37. $\begin{bmatrix} \frac{1}{3} & \frac{1}{3} \\ \frac{2}{3} & \frac{2}{3} \end{bmatrix}$

38. $\begin{bmatrix} \frac{3}{4} & \frac{3}{4} \\ \frac{1}{4} & \frac{1}{4} \end{bmatrix}$

In Problems 39–44, verify each statement using the following matrices:

$$A = \begin{bmatrix} 1 & 2 \\ 0 & 1 \end{bmatrix} \quad B = \begin{bmatrix} 1 & 1 \\ 2 & 3 \end{bmatrix} \quad C = \begin{bmatrix} -3 & 1 \\ -1 & 2 \end{bmatrix}$$

39. $AB \neq BA$

40. $(AB)C = A(BC)$

41. $A(B + C) = AB + AC$

42. $(B + C)A = BA + CA$

43. $A^2 - B^2 \neq (A - B)(A + B)$

44. $A^2 + 2AB + B^2 \neq (A + B)(A + B)$

APPLICATIONS

45. Labor Costs. A company with manufacturing plants located in different parts of the country has labor-hour and wage requirements for the manufacturing of three types of inflatable boats as given in the following two matrices:

LABOR-HOURS PER BOAT

	Cutting department	Assembly department	Packaging department	
$M = $	0.6 h	0.6 h	0.2 h	One-person boat
	1.0 h	0.9 h	0.3 h	Two-person boat
	1.5 h	1.2 h	0.4 h	Four-person boat

HOURLY WAGES

	Plant I	Plant II	
$N = $	\$8	\$9	Cutting department
	\$10	\$12	Assembly department
	\$5	\$6	Packaging department

(A) Find the labor costs for a one-person boat manufactured at plant I; that is, find the dot product

$$[0.6 \quad 0.6 \quad 0.2] \cdot \begin{bmatrix} 8 \\ 10 \\ 5 \end{bmatrix}$$

(B) Find the labor costs for a four-person boat manufactured at plant II. Set up a dot product as in part (A) and multiply.

(C) What is the dimension of MN?

(D) Find MN and interpret.

46. Inventory Value. A personal computer retail company sells five different computer models through three stores located in a large metropolitan area. The inventory of each model on hand in each store is summarized in matrix M. Wholesale (W) and retail (R) values of each model computer are summarized in matrix N.

Model

	A	B	C	D	E	
$M = $	4	2	3	7	1	Store 1
	2	3	5	0	6	Store 2
	10	4	3	4	3	Store 3

	W	R	
$N = $	\$700	\$840	A
	\$1,400	\$1,800	B
	\$1,800	\$2,400	C
	\$2,700	\$3,300	D
	\$3,500	\$4,900	E

(A) What is the retail value of the inventory at store 2?

(B) What is the wholesale value of the inventory at store 3?

(C) Compute *MN* and interpret.

47. Politics. In a local election, a group hired a public relations firm to promote its candidate in three ways: telephone, house calls, and letters. The cost per contact is given in matrix *M*:

$$M = \begin{bmatrix} \$0.40 \\ \$0.75 \\ \$0.25 \end{bmatrix} \begin{matrix} \text{Telephone} \\ \text{House call} \\ \text{Letter} \end{matrix}$$

Cost per contact

The number of contacts of each type made in two adjacent cities is given in matrix *N*:

$$N = \begin{bmatrix} 1{,}000 & 500 & 5{,}000 \\ 2{,}000 & 800 & 8{,}000 \end{bmatrix} \begin{matrix} \text{Berkeley} \\ \text{Oakland} \end{matrix}$$

Telephone   House call   Letter

(A) Find the total amount spent in Berkeley by computing the dot product

$$[1{,}000 \quad 500 \quad 5{,}000] \cdot \begin{bmatrix} \$0.40 \\ \$0.75 \\ \$0.25 \end{bmatrix}$$

(B) Find the total amount spent in Oakland by computing the dot product of appropriate matrices.

(C) Compute *NM* and interpret.

(D) Multiply *N* by the matrix [1   1] and interpret. (The multiplication only makes sense in one order.)

48. Nutrition. A nutritionist for a cereal company blends two cereals in different mixes. The amounts of protein, carbohydrate, and fat (in grams per ounce) in each cereal are given by matrix *M*. The amounts of each cereal used in the three mixes are given by matrix *N*.

$$M = \begin{bmatrix} 4\text{ g} & 2\text{ g} \\ 20\text{ g} & 16\text{ g} \\ 3\text{ g} & 1\text{ g} \end{bmatrix} \begin{matrix} \text{Protein} \\ \text{Carbohydrate} \\ \text{Fat} \end{matrix}$$

Cereal A   Cereal B

$$N = \begin{bmatrix} 15\text{ oz} & 10\text{ oz} & 5\text{ oz} \\ 5\text{ oz} & 10\text{ oz} & 15\text{ oz} \end{bmatrix} \begin{matrix} \text{Cereal A} \\ \text{Cereal B} \end{matrix}$$

Mix X   Mix Y   Mix Z

(A) Find the amount of protein in mix *X* by computing the dot product

$$[4 \quad 2] \cdot \begin{bmatrix} 15 \\ 5 \end{bmatrix}$$

(B) Find the amount of fat in mix *Z*. Set up a dot product as in part (A) and multiply.

(C) What is the dimension of *MN*?

(D) Find *MN* and interpret.

(E) Find $\frac{1}{20}MN$ and interpret.

SECTION $7\text{-}3$ **Inverse of a Square Matrix**

- Identity Matrix for Multiplication
- Inverse of a Square Matrix

In this section we introduce the identity matrix and the inverse of a square matrix. These matrix forms, along with matrix multiplication, are then used to solve some systems of equations written in matrix form in Section 7-4.

• Identity Matrix for Multiplication

We know that for any real number *a*

$$(1)a = a(1) = a$$

The number 1 is called the *identity* for real number multiplication. Does the set of all matrices of a given dimension have an identity element for multiplication? That is, if *M* is an arbitrary $m \times n$ matrix, does *M* have an identity element *I* such that

$IM = MI = M$? The answer, in general, is no. However, the set of all **square matrices of order n** (dimension $n \times n$) does have an identity, and it is given as follows: The **identity matrix for multiplication** for the set of all square matrices of order n is the square matrix of order n, denoted by I, with 1's along the **principal diagonal** (from upper left corner to lower right corner) and 0's elsewhere. For example,

$$\begin{bmatrix} 1 & 0 \\ 0 & 1 \end{bmatrix} \quad \text{and} \quad \begin{bmatrix} 1 & 0 & 0 \\ 0 & 1 & 0 \\ 0 & 0 & 1 \end{bmatrix}$$

are the identity matrices for square matrices of order 2 and 3, respectively.

EXAMPLE 1   **Identity Matrix Multiplication**

(A) $\begin{bmatrix} 1 & 0 & 0 \\ 0 & 1 & 0 \\ 0 & 0 & 1 \end{bmatrix}\begin{bmatrix} a & b & c \\ d & e & f \\ g & h & i \end{bmatrix} = \begin{bmatrix} a & b & c \\ d & e & f \\ g & h & i \end{bmatrix}$

(B) $\begin{bmatrix} a & b & c \\ d & e & f \\ g & h & i \end{bmatrix}\begin{bmatrix} 1 & 0 & 0 \\ 0 & 1 & 0 \\ 0 & 0 & 1 \end{bmatrix} = \begin{bmatrix} a & b & c \\ d & e & f \\ g & h & i \end{bmatrix}$

(C) $\begin{bmatrix} 1 & 0 \\ 0 & 1 \end{bmatrix}\begin{bmatrix} a & b & c \\ d & e & f \end{bmatrix} = \begin{bmatrix} a & b & c \\ d & e & f \end{bmatrix}$

(D) $\begin{bmatrix} a & b & c \\ d & e & f \end{bmatrix}\begin{bmatrix} 1 & 0 & 0 \\ 0 & 1 & 0 \\ 0 & 0 & 1 \end{bmatrix} = \begin{bmatrix} a & b & c \\ d & e & f \end{bmatrix}$

Problem 1   Multiply:

(A) $\begin{bmatrix} 1 & 0 \\ 0 & 1 \end{bmatrix}\begin{bmatrix} 3 & -5 \\ 4 & 6 \end{bmatrix} \quad \text{and} \quad \begin{bmatrix} 3 & -5 \\ 4 & 6 \end{bmatrix}\begin{bmatrix} 1 & 0 \\ 0 & 1 \end{bmatrix}$

(B) $\begin{bmatrix} 1 & 0 & 0 \\ 0 & 1 & 0 \\ 0 & 0 & 1 \end{bmatrix}\begin{bmatrix} 5 & -7 \\ 2 & 4 \\ 6 & -8 \end{bmatrix} \quad \text{and} \quad \begin{bmatrix} 5 & -7 \\ 2 & 4 \\ 6 & -8 \end{bmatrix}\begin{bmatrix} 1 & 0 \\ 0 & 1 \end{bmatrix}$

In general, we can show that if M is a square matrix of order n and I is the identity matrix of order n, then

$$IM = MI = M$$

If M is an $n \times m$ matrix that is not square ($n \neq m$), it is not possible to find a *single* identity matrix I satisfying both $IM = M$ and $MI = M$ [see Examples 1C and 1D]. For this reason we restrict our attention in this section to square matrices.

● **Inverse of a Square Matrix**

In the set of real numbers, we know that for each real number a, except 0, there exists a real number a^{-1} such that

$$a^{-1}a = 1$$

The number a^{-1} is called the *inverse* of the number a relative to multiplication, or the *multiplicative inverse* of a. For example, 2^{-1} is the multiplicative inverse of 2, since $2^{-1}(2) = 1$. We use this idea to define the *inverse of a square matrix*.

DEFINITION 1   **Inverse of a Square Matrix**

If M is a square matrix of order n and if there exists a matrix M^{-1} (read "M inverse") such that

$$M^{-1}M = MM^{-1} = I$$

then M^{-1} is called the **multiplicative inverse of M** or, more simply, the **inverse of M**.

The multiplicative inverse of a nonzero real number a also can be written as $1/a$. This notation is not used for matrix inverses.

Let's use Definition 1 to find M^{-1}, if it exists, for

$$M = \begin{bmatrix} 2 & 3 \\ 1 & 2 \end{bmatrix}$$

We are looking for

$$M^{-1} = \begin{bmatrix} a & c \\ b & d \end{bmatrix}$$

such that

$$MM^{-1} = M^{-1}M = I$$

Thus, we write

$$\overset{M}{\begin{bmatrix} 2 & 3 \\ 1 & 2 \end{bmatrix}} \overset{M^{-1}}{\begin{bmatrix} a & c \\ b & d \end{bmatrix}} = \overset{I}{\begin{bmatrix} 1 & 0 \\ 0 & 1 \end{bmatrix}}$$

and try to find a, b, c, and d so that the product of M and M^{-1} is the identity matrix I. Multiplying M and M^{-1} on the left side, we obtain

$$\begin{bmatrix} (2a + 3b) & (2c + 3d) \\ (a + 2b) & (c + 2d) \end{bmatrix} = \begin{bmatrix} 1 & 0 \\ 0 & 1 \end{bmatrix}$$

which is true only if

$$2a + 3b = 1 \qquad 2c + 3d = 0$$
$$a + 2b = 0 \qquad c + 2d = 1$$

Solving these two systems, we find that $a = 2, b = -1, c = -3,$ and $d = 2.$ Thus,

$$M^{-1} = \begin{bmatrix} 2 & -3 \\ -1 & 2 \end{bmatrix}$$

as is easily checked:

$$\overset{M}{\begin{bmatrix} 2 & 3 \\ 1 & 2 \end{bmatrix}} \overset{M^{-1}}{\begin{bmatrix} 2 & -3 \\ -1 & 2 \end{bmatrix}} = \overset{I}{\begin{bmatrix} 1 & 0 \\ 0 & 1 \end{bmatrix}} = \overset{M^{-1}}{\begin{bmatrix} 2 & -3 \\ -1 & 2 \end{bmatrix}} \overset{M}{\begin{bmatrix} 2 & 3 \\ 1 & 2 \end{bmatrix}}$$

Unlike nonzero real numbers, inverses do not always exist for nonzero square matrices. For example, if

$$N = \begin{bmatrix} 2 & 1 \\ 4 & 2 \end{bmatrix}$$

then, proceeding as before, we are led to the systems

$$2a + b = 1 \qquad 2c + d = 0$$
$$4a + 2b = 0 \qquad 4c + 2d = 1$$

These systems are both inconsistent and have no solution. Hence, N^{-1} does not exist.

Being able to find inverses, when they exist, leads to direct and simple solutions to many practical problems. In the next section, for example, we will show how inverses can be used to solve systems of linear equations.

The method outlined above for finding the inverse, if it exists, gets very involved for matrices of order larger than 2. Now that we know what we are looking for, we can use augmented matrices as in Section 6-2 to make the process more efficient. Details are illustrated in Example 2.

EXAMPLE 2   Finding an Inverse

Find the inverse, if it exists, of

$$M = \begin{bmatrix} 1 & -1 & 1 \\ 0 & 2 & -1 \\ 2 & 3 & 0 \end{bmatrix}$$

Solution   We start as before and write

$$\overset{M}{\begin{bmatrix} 1 & -1 & 1 \\ 0 & 2 & -1 \\ 2 & 3 & 0 \end{bmatrix}} \overset{M^{-1}}{\begin{bmatrix} a & d & g \\ b & e & h \\ c & f & i \end{bmatrix}} = \overset{I}{\begin{bmatrix} 1 & 0 & 0 \\ 0 & 1 & 0 \\ 0 & 0 & 1 \end{bmatrix}}$$

This is true only if

$$\begin{array}{rcl}
a - b + c = 1 & \qquad d - e + f = 0 & \qquad g - h + i = 0 \\
2b - c = 0 & \qquad 2e - f = 1 & \qquad 2h - i = 0 \\
2a + 3b \quad\;\; = 0 & \qquad 2d + 3e \quad\;\; = 0 & \qquad 2g + 3h \quad\;\; = 1
\end{array}$$

Now we write augmented matrices for each of the three systems:

$$\overset{\text{First}}{\left[\begin{array}{rrr|r} 1 & -1 & 1 & 1 \\ 0 & 2 & -1 & 0 \\ 2 & 3 & 0 & 0 \end{array}\right]} \quad \overset{\text{Second}}{\left[\begin{array}{rrr|r} 1 & -1 & 1 & 0 \\ 0 & 2 & -1 & 1 \\ 2 & 3 & 0 & 0 \end{array}\right]} \quad \overset{\text{Third}}{\left[\begin{array}{rrr|r} 1 & -1 & 1 & 0 \\ 0 & 2 & -1 & 0 \\ 2 & 3 & 0 & 1 \end{array}\right]}$$

Since each matrix to the left of the vertical bar is the same, exactly the same row operations can be used on each augmented matrix to transform it into a reduced form. We can speed up the process substantially by combining all three augmented matrices into the single augmented matrix form

$$\left[\begin{array}{rrr|rrr} 1 & -1 & 1 & 1 & 0 & 0 \\ 0 & 2 & -1 & 0 & 1 & 0 \\ 2 & 3 & 0 & 0 & 0 & 1 \end{array}\right] = [M \mid I] \qquad (1)$$

We now try to perform row operations on matrix (1) until we obtain a row-equivalent matrix that looks like matrix (2):

$$\left[\begin{array}{rrr|rrr} 1 & 0 & 0 & a & d & g \\ 0 & 1 & 0 & b & e & h \\ 0 & 0 & 1 & c & f & i \end{array}\right] = [I \mid B] \qquad (2)$$

If this can be done, then the new matrix to the right of the vertical bar is M^{-1}! Now let's try to transform (1) into a form like (2). We follow the same sequence of steps as in the solution of linear systems by Gauss–Jordan elimination (see Section 6-2):

$$
\begin{array}{c}
MI\\
\left[\begin{array}{rrr|rrr}
1 & -1 & 1 & 1 & 0 & 0\\
0 & 2 & -1 & 0 & 1 & 0\\
2 & 3 & 0 & 0 & 0 & 1
\end{array}\right] \quad R_3 + (-2)R_1 \to R_3
\end{array}
$$

$$
\sim \left[\begin{array}{rrr|rrr}
1 & -1 & 1 & 1 & 0 & 0\\
0 & 2 & -1 & 0 & 1 & 0\\
0 & 5 & -2 & -2 & 0 & 1
\end{array}\right] \quad \tfrac{1}{2}R_2 \to R_2
$$

$$
\sim \left[\begin{array}{rrr|rrr}
1 & -1 & 1 & 1 & 0 & 0\\
0 & 1 & -\tfrac{1}{2} & 0 & \tfrac{1}{2} & 0\\
0 & 5 & -2 & -2 & 0 & 1
\end{array}\right] \quad R_3 + (-5)R_2 \to R_3
$$

$$
\sim \left[\begin{array}{rrr|rrr}
1 & -1 & 1 & 1 & 0 & 0\\
0 & 1 & -\tfrac{1}{2} & 0 & \tfrac{1}{2} & 0\\
0 & 0 & \tfrac{1}{2} & -2 & -\tfrac{5}{2} & 1
\end{array}\right] \quad 2R_3 \to R_3
$$

$$
\sim \left[\begin{array}{rrr|rrr}
1 & -1 & 1 & 1 & 0 & 0\\
0 & 1 & -\tfrac{1}{2} & 0 & \tfrac{1}{2} & 0\\
0 & 0 & 1 & -4 & -5 & 2
\end{array}\right] \quad \begin{array}{l} R_1 + (-1)R_3 \to R_1\\ R_2 + \tfrac{1}{2}R_3 \to R_2 \end{array}
$$

$$
\sim \left[\begin{array}{rrr|rrr}
1 & -1 & 0 & 5 & 5 & -2\\
0 & 1 & 0 & -2 & -2 & 1\\
0 & 0 & 1 & -4 & -5 & 2
\end{array}\right] \quad R_1 + R_2 \to R_1
$$

$$
\begin{array}{c}
IB\\
\sim \left[\begin{array}{rrr|rrr}
1 & 0 & 0 & 3 & 3 & -1\\
0 & 1 & 0 & -2 & -2 & 1\\
0 & 0 & 1 & -4 & -5 & 2
\end{array}\right]
\end{array}
$$

Converting back to systems of equations equivalent to our three original systems (we won't have to do this step in practice), we have

$$
\begin{array}{lll}
a = 3 & d = 3 & g = -1\\
b = -2 & e = -2 & h = 1\\
c = -4 & f = -5 & i = 2
\end{array}
$$

And these are just the elements of M^{-1} that we are looking for! Hence,

$$
M^{-1} = \begin{bmatrix}
3 & 3 & -1\\
-2 & -2 & 1\\
-4 & -5 & 2
\end{bmatrix}
$$

Note that this is the matrix to the right of the vertical line in the last augmented matrix.

Check   Since the definition of matrix inverse requires that

$$
M^{-1}M = I \quad \text{and} \quad MM^{-1} = I \tag{3}
$$

it appears that we must compute both $M^{-1}M$ and MM^{-1} to check our work. However, it can be shown that if one of the equations in (3) is satisfied, then the other is also satisfied. Thus, for checking purposes it is sufficient to compute either $M^{-1}M$ or MM^{-1}—we don't need to do both.

$$M^{-1}M = \begin{bmatrix} 3 & 3 & -1 \\ -2 & -2 & 1 \\ -4 & -5 & 2 \end{bmatrix}\begin{bmatrix} 1 & -1 & 1 \\ 0 & 2 & -1 \\ 2 & 3 & 0 \end{bmatrix} = \begin{bmatrix} 1 & 0 & 0 \\ 0 & 1 & 0 \\ 0 & 0 & 1 \end{bmatrix} = I$$

Problem 2 Let: $M = \begin{bmatrix} 3 & -1 & 1 \\ -1 & 1 & 0 \\ 1 & 0 & 1 \end{bmatrix}$

(A) Form the augmented matrix $[M \mid I]$.
(B) Use row operations to transform $[M \mid I]$ into $[I \mid B]$.
(C) Verify by multiplication that $B = M^{-1}$.

The procedure used in Example 2 can be used to find the inverse of any square matrix, if the inverse exists, and will also indicate when the inverse does not exist. These ideas are summarized in Theorem 1.

Theorem 1   **Inverse of a Square Matrix M**

If $[M \mid I]$ is transformed by row operations into $[I \mid B]$, then the resulting matrix B is M^{-1}. If, however, we obtain all 0's in one or more rows to the left of the vertical line, then M^{-1} does not exist.

EXAMPLE 3   **Finding an Inverse**

Find M^{-1}, given: $M = \begin{bmatrix} 3 & -1 \\ -4 & 2 \end{bmatrix}$

Solution   Can you identify the row operations used in the transformations?

$$\begin{bmatrix} 3 & -1 & \bigm| & 1 & 0 \\ -4 & 2 & \bigm| & 0 & 1 \end{bmatrix} \sim \begin{bmatrix} 1 & -\frac{1}{3} & \bigm| & \frac{1}{3} & 0 \\ -4 & 2 & \bigm| & 0 & 1 \end{bmatrix}$$

$$\sim \begin{bmatrix} 1 & -\frac{1}{3} & \bigm| & \frac{1}{3} & 0 \\ 0 & \frac{2}{3} & \bigm| & \frac{4}{3} & 1 \end{bmatrix}$$

$$\sim \begin{bmatrix} 1 & -\frac{1}{3} & \bigm| & \frac{1}{3} & 0 \\ 0 & 1 & \bigm| & 2 & \frac{3}{2} \end{bmatrix}$$

$$\sim \begin{bmatrix} 1 & 0 & \bigm| & 1 & \frac{1}{2} \\ 0 & 1 & \bigm| & 2 & \frac{3}{2} \end{bmatrix}$$

Thus,

$$M^{-1} = \begin{bmatrix} 1 & \frac{1}{2} \\ 2 & \frac{3}{2} \end{bmatrix}$$ Check by showing that $M^{-1}M = I$.

Problem 3 Find M^{-1}, given: $M = \begin{bmatrix} 2 & -6 \\ 1 & -2 \end{bmatrix}$

EXAMPLE 4 **Finding an Inverse**

Find M^{-1}, if it exists, given: $M = \begin{bmatrix} 10 & -2 \\ -5 & 1 \end{bmatrix}$

Solution

$$\begin{bmatrix} 10 & -2 & | & 1 & 0 \\ -5 & 1 & | & 0 & 1 \end{bmatrix} \sim \begin{bmatrix} 1 & -\frac{1}{5} & | & \frac{1}{10} & 0 \\ -5 & 1 & | & 0 & 1 \end{bmatrix}$$

$$\sim \begin{bmatrix} 1 & -\frac{1}{5} & | & \frac{1}{10} & 0 \\ 0 & 0 & | & \frac{1}{2} & 1 \end{bmatrix}$$

We have all 0's in the second row to the left of the vertical line. Therefore, M^{-1} does not exist.

Problem 4 Find M^{-1}, if it exists, given: $M = \begin{bmatrix} 6 & -3 \\ -2 & 1 \end{bmatrix}$

Answers to Matched Problems

1. (A) $\begin{bmatrix} 3 & -5 \\ 4 & 6 \end{bmatrix}$ (B) $\begin{bmatrix} 5 & -7 \\ 2 & 4 \\ 6 & -8 \end{bmatrix}$

2. (A) $\begin{bmatrix} 3 & -1 & 1 & | & 1 & 0 & 0 \\ -1 & 1 & 0 & | & 0 & 1 & 0 \\ 1 & 0 & 1 & | & 0 & 0 & 1 \end{bmatrix}$ (B) $\begin{bmatrix} 1 & 0 & 0 & | & 1 & 1 & -1 \\ 0 & 1 & 0 & | & 1 & 2 & -1 \\ 0 & 0 & 1 & | & -1 & -1 & 2 \end{bmatrix}$

(C) $\begin{bmatrix} 1 & 1 & -1 \\ 1 & 2 & -1 \\ -1 & -1 & 2 \end{bmatrix}\begin{bmatrix} 3 & -1 & 1 \\ -1 & 1 & 0 \\ 1 & 0 & 1 \end{bmatrix} = \begin{bmatrix} 1 & 0 & 0 \\ 0 & 1 & 0 \\ 0 & 0 & 1 \end{bmatrix}$

3. $\begin{bmatrix} -1 & 3 \\ -\frac{1}{2} & 1 \end{bmatrix}$ 4. Does not exist

EXERCISE 7-3

A

Perform the indicated operations in Problems 1–8.

1. $\begin{bmatrix} 1 & 0 \\ 0 & 1 \end{bmatrix}\begin{bmatrix} 2 & -3 \\ 4 & 5 \end{bmatrix}$

2. $\begin{bmatrix} 1 & 0 \\ 0 & 1 \end{bmatrix}\begin{bmatrix} 4 & -3 \\ 0 & 2 \end{bmatrix}$

3. $\begin{bmatrix} 2 & -3 \\ 4 & 5 \end{bmatrix}\begin{bmatrix} 1 & 0 \\ 0 & 1 \end{bmatrix}$

4. $\begin{bmatrix} 4 & -3 \\ 0 & 2 \end{bmatrix}\begin{bmatrix} 1 & 0 \\ 0 & 1 \end{bmatrix}$

5. $\begin{bmatrix} 1 & 0 & 0 \\ 0 & 1 & 0 \\ 0 & 0 & 1 \end{bmatrix}\begin{bmatrix} -2 & 1 & 3 \\ 2 & 4 & -2 \\ 5 & 1 & 0 \end{bmatrix}$

6. $\begin{bmatrix} 1 & 0 & 0 \\ 0 & 1 & 0 \\ 0 & 0 & 1 \end{bmatrix}\begin{bmatrix} -3 & 0 & 2 \\ 1 & 1 & 5 \\ 2 & -1 & 7 \end{bmatrix}$

7. $\begin{bmatrix} -2 & 1 & 3 \\ 2 & 4 & -2 \\ 5 & 1 & 0 \end{bmatrix}\begin{bmatrix} 1 & 0 & 0 \\ 0 & 1 & 0 \\ 0 & 0 & 1 \end{bmatrix}$

8. $\begin{bmatrix} -3 & 0 & 2 \\ 1 & 1 & 5 \\ 2 & -1 & 7 \end{bmatrix}\begin{bmatrix} 1 & 0 & 0 \\ 0 & 1 & 0 \\ 0 & 0 & 1 \end{bmatrix}$

For Problems 9–14, show that the two matrices are inverses of each other by showing that their product is the identity matrix I.

9. $\begin{bmatrix} 3 & -4 \\ -2 & 3 \end{bmatrix}\begin{bmatrix} 3 & 4 \\ 2 & 3 \end{bmatrix}$

10. $\begin{bmatrix} 5 & -7 \\ -2 & 3 \end{bmatrix}\begin{bmatrix} 3 & 7 \\ 2 & 5 \end{bmatrix}$

11. $\begin{bmatrix} 3 & -4 \\ 4 & -5 \end{bmatrix}\begin{bmatrix} -5 & 4 \\ -4 & 3 \end{bmatrix}$

12. $\begin{bmatrix} 2 & -3 \\ 3 & -5 \end{bmatrix}\begin{bmatrix} 5 & -3 \\ 3 & -2 \end{bmatrix}$

13. $\begin{bmatrix} 1 & -1 & 1 \\ 0 & 2 & -1 \\ 2 & 3 & 0 \end{bmatrix}\begin{bmatrix} 3 & 3 & -1 \\ -2 & -2 & 1 \\ -4 & -5 & 2 \end{bmatrix}$

14. $\begin{bmatrix} 3 & 3 & -1 \\ -2 & -2 & 1 \\ -4 & -5 & 2 \end{bmatrix}\begin{bmatrix} 1 & -1 & 1 \\ 0 & 2 & -1 \\ 2 & 3 & 0 \end{bmatrix}$

B

Given M in Problems 15–24, find M^{-1}, and show that $M^{-1}M = I$.

15. $\begin{bmatrix} 0 & -1 \\ 1 & 4 \end{bmatrix}$

16. $\begin{bmatrix} -1 & 5 \\ 0 & -1 \end{bmatrix}$

17. $\begin{bmatrix} 1 & 2 \\ 1 & 3 \end{bmatrix}$

18. $\begin{bmatrix} 2 & 1 \\ 5 & 3 \end{bmatrix}$

19. $\begin{bmatrix} 1 & 3 \\ 2 & 7 \end{bmatrix}$

20. $\begin{bmatrix} 2 & 1 \\ 1 & 1 \end{bmatrix}$

21. $\begin{bmatrix} 1 & -3 & 0 \\ 0 & 3 & 1 \\ 2 & -1 & 2 \end{bmatrix}$

22. $\begin{bmatrix} 2 & 9 & 0 \\ 1 & 2 & 3 \\ 0 & -1 & 1 \end{bmatrix}$

23. $\begin{bmatrix} 1 & 1 & 0 \\ 0 & 3 & -1 \\ 1 & 0 & 1 \end{bmatrix}$

24. $\begin{bmatrix} 1 & 0 & -1 \\ 2 & -1 & 0 \\ 1 & 1 & 1 \end{bmatrix}$

Find the inverse of each matrix in Problems 25–28, if it exists.

25. $\begin{bmatrix} 3 & 9 \\ 2 & 6 \end{bmatrix}$

26. $\begin{bmatrix} 2 & -4 \\ -3 & 6 \end{bmatrix}$

27. $\begin{bmatrix} 3 & 5 \\ 2 & 4 \end{bmatrix}$

28. $\begin{bmatrix} -5 & 4 \\ 3 & -3 \end{bmatrix}$

C

Find the inverse of each matrix in Problems 29–34, if it exists.

29. $\begin{bmatrix} 2 & 2 & -1 \\ 0 & 3 & -1 \\ -1 & -2 & 1 \end{bmatrix}$

30. $\begin{bmatrix} 4 & 1 & -1 \\ 1 & 2 & -1 \\ -3 & -1 & 1 \end{bmatrix}$

31. $\begin{bmatrix} 2 & 1 & 1 \\ 1 & 1 & 0 \\ -1 & -1 & 0 \end{bmatrix}$

32. $\begin{bmatrix} 1 & -1 & 0 \\ 2 & -1 & 1 \\ 0 & 1 & 1 \end{bmatrix}$

33. $\begin{bmatrix} 1 & 5 & 10 \\ 0 & 1 & 4 \\ 1 & 6 & 15 \end{bmatrix}$

34. $\begin{bmatrix} 1 & -5 & -10 \\ 0 & 1 & 6 \\ 1 & -4 & -3 \end{bmatrix}$

35. Show that $(A^{-1})^{-1} = A$ for

$$A = \begin{bmatrix} 3 & 4 \\ 2 & 3 \end{bmatrix}$$

36. Show that $(AB)^{-1} = B^{-1}A^{-1}$ for

$$A = \begin{bmatrix} 3 & 4 \\ 2 & 3 \end{bmatrix} \quad \text{and} \quad B = \begin{bmatrix} 3 & 7 \\ 2 & 5 \end{bmatrix}$$

SECTION 7-4 Matrix Equations and Systems of Linear Equations

- Matrix Equations
- Matrix Equations and Systems of Linear Equations
- Application

The identity matrix and inverse matrix discussed in the last section can be put to immediate use in the solving of certain simple matrix equations. Being able to solve a matrix equation gives us another important method of solving a system of equations having the same number of variables as equations. If the system either has fewer variables than equations or more variables than equations, then we must return to the Gauss–Jordan method of elimination.

• Matrix Equations

Solving certain simple types of matrix equations follows very much the same procedures used in solving real number equations. But we have less freedom with matrix equations. For one thing, matrix multiplication is not commutative. In solving matrix equations, we will be guided by the results from Section 7-3 and from Theorem 1 in Section 7-2, which are restated for convenient reference.

From Section 7-3: If A is a square matrix and A^{-1} is its inverse, and if I is an identity matrix of the same dimension as A, then

$$IA = AI = A \quad \text{and} \quad A^{-1}A = AA^{-1} = I$$

From Section 7-2:

Theorem 1 **Properties of Matrix Multiplication**

Assuming all products and sums are defined for the indicated matrices A, B, and C, then for k a real number:

1. $A(BC) = (AB)C$ **Associative Property**
2. $A(B + C) = AB + AC$ **Left Distributive Property**
3. $(B + C)A = BA + CA$ **Right Distributive Property**
4. If $A = B$, then $CA = CB$ **Left Multiplication Property**
5. If $A = B$, then $AC = BC$ **Right Multiplication Property**
6. $k(AB) = (kA)B = A(kB)$

The process of solving certain types of simple matrix equations is best illustrated by an example.

EXAMPLE 1 **Solving a Matrix Equation**

Given an $n \times n$ matrix A and $n \times 1$ column matrices B and X, solve $AX = B$ for X. Assume all necessary inverses exist.

Solution We are interested in finding a column matrix X that satisfies the matrix equation $AX = B$. To solve this equation, we multiply both sides, on the left, by A^{-1}, assuming it exists, to isolate X on the left side.

$$AX = B$$
$$A^{-1}(AX) = A^{-1}B \qquad \text{Use the left multiplication property.}$$
$$(A^{-1}A)X = A^{-1}B \qquad \text{Associative property}$$
$$IX = A^{-1}B \qquad A^{-1}A = I$$
$$X = A^{-1}B \qquad IX = X$$

CAUTION Do not mix the left multiplication property and the right multiplication property. If $AX = B$, then

$$A^{-1}(AX) \neq BA^{-1}$$

Problem 1 Given an $n \times n$ matrix A and $n \times 1$ column matrices B, C, and X, solve $AX + C = B$ for X. Assume all necessary inverses exist.

● **Matrix Equations and Systems of Linear Equations**

We now show how independent systems of linear equations with the same number of variables as equations can be solved by first converting the system into a matrix equation of the form $AX = B$ and using $X = A^{-1}B$ as obtained in Example 1.

EXAMPLE 2   Solving a System of Linear Equations Using Inverses

Solve the system

$$\begin{aligned} x_1 - x_2 + x_3 &= k_1 \\ 2x_2 - x_3 &= k_2 \\ 2x_1 + 3x_2 \qquad &= k_3 \end{aligned} \tag{1}$$

for:

(A) $k_1 = 1, k_2 = 1, k_3 = 1$ (B) $k_1 = 3, k_2 = 1, k_3 = 4$
(C) $k_1 = -5, k_2 = 2, k_3 = -3$

Solutions   The inverse of the coefficient matrix

$$A = \begin{bmatrix} 1 & -1 & 1 \\ 0 & 2 & -1 \\ 2 & 3 & 0 \end{bmatrix}$$

can be used to solve all three parts very easily. To see how, we convert system (1) into the following **matrix equation**:

$$\overset{A}{\begin{bmatrix} 1 & -1 & 1 \\ 0 & 2 & -1 \\ 2 & 3 & 0 \end{bmatrix}} \overset{X}{\begin{bmatrix} x_1 \\ x_2 \\ x_3 \end{bmatrix}} = \overset{B}{\begin{bmatrix} k_1 \\ k_2 \\ k_3 \end{bmatrix}} \tag{2}$$

You should check that matrix equation (2) is equivalent to system (1) by finding the product on the left side and then equating corresponding elements on the left with those on the right. Now you can see another important reason for defining matrix multiplication as we did in Section 7-2.

We are interested in finding a column matrix X that satisfies the matrix equation $AX = B$. In Example 1 we found that if $AX = B$ and if A^{-1} exists, then

$$X = A^{-1}B$$

The inverse of A was found in Example 2 in Section 7-3 to be

$$A^{-1} = \begin{bmatrix} 3 & 3 & -1 \\ -2 & -2 & 1 \\ -4 & -5 & 2 \end{bmatrix}$$

Thus,

$$\overset{X}{\begin{bmatrix} x_1 \\ x_2 \\ x_3 \end{bmatrix}} = \overset{A^{-1}}{\begin{bmatrix} 3 & 3 & -1 \\ -2 & -2 & 1 \\ -4 & -5 & 2 \end{bmatrix}} \overset{B}{\begin{bmatrix} k_1 \\ k_2 \\ k_3 \end{bmatrix}}$$

To solve parts (A), (B), and (C), we simply replace k_1, k_2, and k_3 with the given values and multiply.

(A) $\begin{bmatrix} x_1 \\ x_2 \\ x_3 \end{bmatrix} = \begin{bmatrix} 3 & 3 & -1 \\ -2 & -2 & 1 \\ -4 & -5 & 2 \end{bmatrix} \begin{bmatrix} 1 \\ 1 \\ 1 \end{bmatrix} = \begin{bmatrix} 5 \\ -3 \\ -7 \end{bmatrix}$

Thus, $x_1 = 5$, $x_2 = -3$, and $x_3 = -7$.

(B) $\begin{bmatrix} x_1 \\ x_2 \\ x_3 \end{bmatrix} = \begin{bmatrix} 3 & 3 & -1 \\ -2 & -2 & 1 \\ -4 & -5 & 2 \end{bmatrix} \begin{bmatrix} 3 \\ 1 \\ 4 \end{bmatrix} = \begin{bmatrix} 8 \\ -4 \\ -9 \end{bmatrix}$

Thus, $x_1 = 8$, $x_2 = -4$, and $x_3 = -9$.

(C) $\begin{bmatrix} x_1 \\ x_2 \\ x_3 \end{bmatrix} = \begin{bmatrix} 3 & 3 & -1 \\ -2 & -2 & 1 \\ -4 & -5 & 2 \end{bmatrix} \begin{bmatrix} -5 \\ 2 \\ -3 \end{bmatrix} = \begin{bmatrix} -6 \\ 3 \\ 4 \end{bmatrix}$

Thus, $x_1 = -6$, $x_2 = 3$, and $x_3 = 4$.

Problem 2   Solve the system

$$3x_1 - x_2 + x_3 = k_1$$
$$-x_1 + x_2 \qquad = k_2$$
$$x_1 \qquad + x_3 = k_3$$

for:

(A) $k_1 = 1, k_2 = 3, k_3 = 2$ (B) $k_1 = 3, k_2 = -3, k_3 = 2$
(C) $k_1 = -5, k_2 = 1, k_3 = -4$

[*Note:* The inverse of the coefficient matrix was found in Problem 2 in Section 7-3.]

Computer programs are readily available for finding inverses of square matrices. A great advantage of using an inverse to solve a system of linear equations is that once the inverse is found, it can be used to solve any new system formed by changing the constant terms. However, as we indicated earlier, this method is not suited for a system where the number of equations and the number of variables is not the same, since the coefficient matrix for such a system is not square.

● **Application**   The following application illustrates the usefulness of the inverse method.

EXAMPLE 3 Investment Allocation

An investment adviser currently has two types of investments available for clients: an investment A that pays 10% per year and an investment B of higher risk that pays 20% per year. Clients may divide their investments between the two to achieve any total return desired between 10 and 20%. However, the higher the desired return, the higher the risk. How should each client listed in the table invest to achieve the indicated return?

	Client			
	1	**2**	**3**	**k**
Total investment	$20,000	$50,000	$10,000	k_1
Annual return desired	$2,400	$7,500	$1,300	k_2
	(12%)	(15%)	(13%)	

Solution   We first solve the problem for an arbitrary client k using inverses, and then apply the result to the three specific clients.

Let

$$x_1 = \text{Amount invested in } A$$

$$x_2 = \text{Amount invested in } B$$

Then

$$x_1 + \quad x_2 = k_1 \quad \text{Total invested}$$

$$0.1x_1 + 0.2x_2 = k_2 \quad \text{Total annual return}$$

Write as a matrix equation:

$$\overset{A}{\begin{bmatrix} 1 & 1 \\ 0.1 & 0.2 \end{bmatrix}} \overset{X}{\begin{bmatrix} x_1 \\ x_2 \end{bmatrix}} = \overset{B}{\begin{bmatrix} k_1 \\ k_2 \end{bmatrix}}$$

If A^{-1} exists, then

$$X = A^{-1}B$$

We now find A^{-1} by starting with $[A \mid I]$ and proceeding as discussed in Section 10-3:

$$\begin{bmatrix} 1 & 1 & \bigm| & 1 & 0 \\ 0.1 & 0.2 & \bigm| & 0 & 1 \end{bmatrix} \quad 10R_2 \to R_2$$

$$\sim \begin{bmatrix} 1 & 1 & \bigm| & 1 & 0 \\ 1 & 2 & \bigm| & 0 & 10 \end{bmatrix} \quad R_2 + (-1)R_1 \to R_2$$

$$\sim \begin{bmatrix} 1 & 1 & | & 1 & 0 \\ 0 & 1 & | & -1 & 10 \end{bmatrix} \quad R_1 + (-1)R_2 \rightarrow R_1$$

$$\sim \begin{bmatrix} 1 & 0 & | & 2 & -10 \\ 0 & 1 & | & -1 & 10 \end{bmatrix}$$

Thus,

$$A^{-1} = \begin{bmatrix} 2 & -10 \\ -1 & 10 \end{bmatrix} \quad \text{Check} \quad \overset{A^{-1}}{\begin{bmatrix} 2 & -10 \\ -1 & 10 \end{bmatrix}} \overset{A}{\begin{bmatrix} 1 & 1 \\ 0.1 & 0.2 \end{bmatrix}} = \overset{I}{\begin{bmatrix} 1 & 0 \\ 0 & 1 \end{bmatrix}}$$

and

$$\overset{X}{\begin{bmatrix} x_1 \\ x_2 \end{bmatrix}} = \overset{A^{-1}}{\begin{bmatrix} 2 & -10 \\ -1 & 10 \end{bmatrix}} \overset{B}{\begin{bmatrix} k_1 \\ k_2 \end{bmatrix}}$$

To solve each client's investment problem, we replace k_1 and k_2 with appropriate values from the table and multiply by A^{-1}:

Client 1
$$\begin{bmatrix} x_1 \\ x_2 \end{bmatrix} = \begin{bmatrix} 2 & -10 \\ -1 & 10 \end{bmatrix} \begin{bmatrix} 20{,}000 \\ 2{,}400 \end{bmatrix} = \begin{bmatrix} 16{,}000 \\ 4{,}000 \end{bmatrix}$$

Solution: $x_1 = \$16{,}000$ in A, $x_2 = \$4{,}000$ in B

Client 2
$$\begin{bmatrix} x_1 \\ x_2 \end{bmatrix} = \begin{bmatrix} 2 & -10 \\ -1 & 10 \end{bmatrix} \begin{bmatrix} 50{,}000 \\ 7{,}500 \end{bmatrix} = \begin{bmatrix} 25{,}000 \\ 25{,}000 \end{bmatrix}$$

Solution: $x_1 = \$25{,}000$ in A, $x_2 = \$25{,}000$ in B

Client 3
$$\begin{bmatrix} x_1 \\ x_2 \end{bmatrix} = \begin{bmatrix} 2 & -10 \\ -1 & 10 \end{bmatrix} \begin{bmatrix} 10{,}000 \\ 1{,}300 \end{bmatrix} = \begin{bmatrix} 7{,}000 \\ 3{,}000 \end{bmatrix}$$

Solution: $x_1 = \$7{,}000$ in A, $x_2 = \$3{,}000$ in B

Problem 3 Repeat Example 3 with investment A paying 8% and investment B paying 24%.

Answers to Matched Problems

1.

$$AX + C = B$$

$$(AX + C) - C = B - C$$
$$AX + (C - C) = B - C$$
$$AX + 0 = B - C$$

$$AX = B - C$$

$$A^{-1}(AX) = A^{-1}(B - C)$$
$$(A^{-1}A)X = A^{-1}(B - C)$$
$$IX = A^{-1}(B - C)$$

$$X = A^{-1}(B - C)$$

2. (A) $x_1 = 2, x_2 = 5, x_3 = 0$ (B) $x_1 = -2, x_2 = -5, x_3 = 4$ (C) $x_1 = 0, x_2 = 1, x_3 = -4$

3. $A^{-1} = \begin{bmatrix} 1.5 & -6.25 \\ -0.5 & 6.25 \end{bmatrix}$; Client 1: \$15,000 in A and \$5,000 in B; Client 2: \$28,125 in A and \$21,875 in B; Client 3: \$6,875 in A and \$3,125 in B

EXERCISE 7-4

A

Write Problems 1–4 as systems of linear equations without matrices.

1. $\begin{bmatrix} 2 & -1 \\ 1 & 3 \end{bmatrix} \begin{bmatrix} x_1 \\ x_2 \end{bmatrix} = \begin{bmatrix} 3 \\ -2 \end{bmatrix}$

2. $\begin{bmatrix} -3 & 1 \\ -1 & 2 \end{bmatrix} \begin{bmatrix} x_1 \\ x_2 \end{bmatrix} = \begin{bmatrix} -2 \\ 5 \end{bmatrix}$

3. $\begin{bmatrix} -2 & 0 & 1 \\ 1 & 2 & 1 \\ 0 & 1 & -1 \end{bmatrix} \begin{bmatrix} x_1 \\ x_2 \\ x_3 \end{bmatrix} = \begin{bmatrix} 3 \\ -4 \\ 2 \end{bmatrix}$

4. $\begin{bmatrix} 1 & -2 & 0 \\ -3 & 1 & -1 \\ 2 & 0 & 4 \end{bmatrix} \begin{bmatrix} x_1 \\ x_2 \\ x_3 \end{bmatrix} = \begin{bmatrix} 3 \\ -2 \\ 5 \end{bmatrix}$

Write each system in Problems 5–8 as a matrix equation of the form $AX = B$.

5. $4x_1 - 3x_2 = 2$
$\quad x_1 + 2x_2 = 1$

6. $\quad x_1 - 2x_2 = 7$
$-3x_1 + \quad x_2 = -3$

7. $\quad x_1 - 2x_2 + x_3 = -1$
$-x_1 + \quad x_2 \quad\quad = 2$
$2x_1 + 3x_2 + x_3 = -3$

8. $2x_1 \quad\quad + 3x_3 = 5$
$\quad x_1 - 2x_2 + \quad x_3 = -4$
$-x_1 + 3x_2 \quad\quad = 2$

In Problems 9–12, find x_1 and x_2.

9. $\begin{bmatrix} x_1 \\ x_2 \end{bmatrix} = \begin{bmatrix} 3 & -2 \\ 1 & 4 \end{bmatrix} \begin{bmatrix} -2 \\ 1 \end{bmatrix}$

10. $\begin{bmatrix} x_1 \\ x_2 \end{bmatrix} = \begin{bmatrix} -2 & 1 \\ -1 & 2 \end{bmatrix} \begin{bmatrix} 3 \\ -2 \end{bmatrix}$

11. $\begin{bmatrix} x_1 \\ x_2 \end{bmatrix} = \begin{bmatrix} -2 & 3 \\ 2 & -1 \end{bmatrix} \begin{bmatrix} 3 \\ 2 \end{bmatrix}$

12. $\begin{bmatrix} x_1 \\ x_2 \end{bmatrix} = \begin{bmatrix} 3 & -1 \\ 0 & 2 \end{bmatrix} \begin{bmatrix} -2 \\ 1 \end{bmatrix}$

B

Write each system in Problems 13–20 as a matrix equation and solve using inverses. [Note: The inverses were found in Problems 17–24 in Exercise 7-3.]

13. $x_1 + 2x_2 = k_1$
$x_1 + 3x_2 = k_2$
(A) $k_1 = 1, k_2 = 3$
(B) $k_1 = 3, k_2 = 5$
(C) $k_1 = -2, k_2 = 1$

14. $2x_1 + x_2 = k_1$
$5x_1 + 3x_2 = k_2$
(A) $k_1 = 2, k_2 = 13$
(B) $k_1 = -2, k_2 = 4$
(C) $k_1 = 1, k_2 = -3$

15. $x_1 + 3x_2 = k_1$
$2x_1 + 7x_2 = k_2$
(A) $k_1 = 2, k_2 = -1$
(B) $k_1 = 1, k_2 = 0$
(C) $k_1 = 3, k_2 = -1$

16. $2x_1 + x_2 = k_1$
$x_1 + x_2 = k_2$
(A) $k_1 = -1, k_2 = -2$
(B) $k_1 = 2, k_2 = 3$
(C) $k_1 = 2, k_2 = 0$

17. $x_1 - 3x_2 \qquad = k_1$
$3x_2 + x_3 = k_2$
$2x_1 - x_2 + 2x_3 = k_3$
(A) $k_1 = 1, k_2 = 0, k_3 = 2$
(B) $k_1 = -1, k_2 = 1, k_3 = 0$
(C) $k_1 = 2, k_2 = -2, k_3 = 1$

18. $2x_1 + 9x_2 \qquad = k_1$
$x_1 + 2x_2 + 3x_3 = k_2$
$-x_2 + x_3 = k_3$
(A) $k_1 = 0, k_2 = 2, k_3 = 1$
(B) $k_1 = -2, k_2 = 0, k_3 = 1$
(C) $k_1 = 3, k_2 = 1, k_3 = 0$

19. $x_1 + x_2 \qquad = k_1$
$3x_2 - x_3 = k_2$
$x_1 \qquad + x_3 = k_3$
(A) $k_1 = 2, k_2 = 0, k_3 = 4$
(B) $k_1 = 0, k_2 = 4, k_3 = -2$
(C) $k_1 = 4, k_2 = 2, k_3 = 0$

20. $x_1 \qquad -x_3 = k_1$
$2x_1 - x_2 \qquad = k_2$
$x_1 + x_2 + x_3 = k_3$
(A) $k_1 = 4, k_2 = 8, k_3 = 0$
(B) $k_1 = 4, k_2 = 0, k_3 = -4$
(C) $k_1 = 0, k_2 = 8, k_3 = -8$

C

For $n \times n$ matrices A and B and $n \times 1$ matrices C, D, and X, solve each matrix equation in Problems 21–26 for X. Assume all necessary inverses exist.

21. $AX - BX = C$

22. $AX + BX = C$

23. $AX + X = C$

24. $AX - X = C$

25. $AX - C = D - BX$

26. $AX + C = BX + D$

APPLICATIONS   *Solve using systems of equations and inverses.*

27. Resource Allocation. A concert hall has 10,000 seats. If tickets are $4 and $8, how many of each type of ticket should be sold (assuming all seats can be sold) to bring in each of the returns indicated in the table? Use decimals in computing the inverse.

	Concert		
	1	**2**	**3**
Tickets sold	10,000	10,000	10,000
Return required	$56,000	$60,000	$68,000

28. Production Scheduling. Labor and material costs for manufacturing two guitar models are given in the table at the top of the next column:

Guitar Model	Labor Cost	Material Cost
A	$30	$20
B	$40	$30

If a total of $3,000 a week is allowed for labor and material, how many of each model should be produced each week to exactly use each of the allocations of the $3,000 indicated in the following table? Use decimals in computing the inverse.

	Weekly Allocation		
	1	**2**	**3**
Labor	$1,800	$1,750	$1,720
Material	$1,200	$1,250	$1,280

★**29. Circuit Analysis.** A direct current electrical circuit consisting of conductors (wires), resistors, and batteries is diagrammed below:

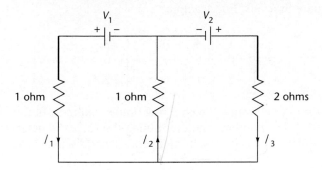

If $I_1, I_2,$ and I_3 are the currents (in amperes) in the three branches of the circuit and V_1 and V_2 are the voltages (in volts) of the two batteries, then *Kirchhoff's* * *laws* can be used to show that the currents satisfy the following system of equations:

$$I_1 - I_2 + I_3 = 0$$
$$I_1 + I_2 = V_1$$
$$I_2 + 2I_3 = V_2$$

Solve this system for:
(A) $V_1 = 10$ volts, $V_2 = 10$ volts
(B) $V_1 = 10$ volts, $V_2 = 15$ volts
(C) $V_1 = 15$ volts, $V_2 = 10$ volts

★**30. Circuit Analysis.** Repeat Problem 29 for the electrical circuit shown at the top of the next column.

$$I_1 - I_2 + I_3 = 0$$
$$I_1 + 2I_2 = V_1$$
$$2I_2 + 2I_3 = V_2$$

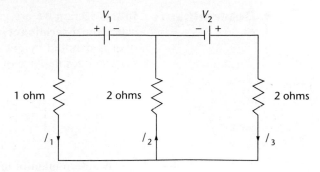

★★**31. Geometry.** The graph of $f(x) = ax^2 + bx + c$ passes through the points $(1, k_1), (2, k_2),$ and $(3, k_3)$. Determine $a, b,$ and c for:
(A) $k_1 = -2, k_2 = 1, k_3 = 6$
(B) $k_1 = 4, k_2 = 3, k_3 = -2$
(C) $k_1 = 8, k_2 = -5, k_3 = 4$

★★**32. Geometry.** Repeat Problem 31 if the graph passes through the points $(-1, k_1), (0, k_2),$ and $(1, k_3)$.

33. Diets. A biologist has available two commercial food mixes with the following percentages of protein and fat:

Mix	Protein (%)	Fat (%)
A	20	2
B	10	6

How many ounces of each mix should be used to prepare each of the diets listed in the following table?

	Diet		
	1	2	3
Protein	20 oz	10 oz	10 oz
Fat	6 oz	4 oz	6 oz

SECTION 7-5 Determinants

- Determinants
- Second-Order Determinants
- Third-Order Determinants
- Higher-Order Determinants

*Gustav Kirchhoff (1824–1887), a German physicist, was among the first to apply theoretical mathematics to physics. He is best-known for his development of certain properties of electrical circuits, which are now known as **Kirchhoff's laws**.

• **Determinants**   In this section we are going to associate with each square matrix a real number, called the **determinant** of the matrix. If A is a square matrix, then the determinant of A is denoted by **det A**, or simply by writing the array of elements in A using vertical lines in place of square brackets. For example,

$$\det \begin{bmatrix} 2 & -3 \\ 5 & 1 \end{bmatrix} = \begin{vmatrix} 2 & -3 \\ 5 & 1 \end{vmatrix} \qquad \det \begin{bmatrix} 1 & -2 & 3 \\ 0 & 5 & -7 \\ -2 & 1 & 6 \end{bmatrix} = \begin{vmatrix} 1 & -2 & 3 \\ 0 & 5 & -7 \\ -2 & 1 & 6 \end{vmatrix}$$

A determinant of **order** n is a determinant with n rows and n columns. In this section we concentrate most of our attention on determining the values of determinants of orders 2 and 3. But many of the results and procedures discussed can be generalized completely to determinants of order n.

• **Second-Order Determinants**   In general, we can symbolize a **second-order determinant** as follows:

$$\begin{vmatrix} a_{11} & a_{12} \\ a_{21} & a_{22} \end{vmatrix}$$

where we use a single letter with a **double subscript** to facilitate generalization to higher-order determinants. The first subscript number indicates the row in which the element appears, and the second subscript number indicates the column. Thus, a_{21} is the element in the second row and first column, and a_{12} is the element in the first row and second column. Each second-order determinant represents a real number given by Definition 1.

DEFINITION 1   **Value of a Second-Order Determinant**

$$\begin{vmatrix} a_{11} & a_{12} \\ a_{21} & a_{22} \end{vmatrix} = a_{11}a_{22} - a_{21}a_{12} \tag{1}$$

Formula (1) is easily remembered if you notice that the expression on the right is the product of the **principal diagonal**, from upper left to lower right, minus the product of the **secondary diagonal**, from lower left to upper right.

EXAMPLE 1   Evaluating a Second-Order Determinant

$$\begin{vmatrix} -1 & 2 \\ -3 & -4 \end{vmatrix} = (-1)(-4) - (-3)(2) = 4 - (-6) = 10$$

Problem 1   Find: $\begin{vmatrix} 3 & -5 \\ 4 & -2 \end{vmatrix}$

● **Third-Order Determinants**

A determinant of order 3 is a square array of nine elements and represents a real number given by Definition 2, which is a special case of the general definition of the value of an nth-order determinant. Note that each term in the expansion on the right of equation (2) contains exactly one element from each row and each column.

DEFINITION 2

Value of a Third-Order Determinant

$$\begin{vmatrix} a_{11} & a_{12} & a_{13} \\ a_{21} & a_{22} & a_{23} \\ a_{31} & a_{32} & a_{33} \end{vmatrix} = \begin{aligned} a_{11}a_{22}a_{33} - a_{11}a_{32}a_{23} + a_{21}a_{32}a_{13} - a_{21}a_{12}a_{33} \\ + a_{31}a_{12}a_{23} - a_{31}a_{22}a_{13} \end{aligned} \quad (2)$$

Don't panic! You don't need to memorize formula (2). After we introduce the ideas of *minor* and *cofactor* below, we will state a theorem that can be used to obtain the same result with much less trouble.

The **minor of an element** in a third-order determinant is a second-order determinant obtained by deleting the row and column that contains the element. For example, in the determinant in formula (2),

$$\text{Minor of } a_{23} = \begin{vmatrix} a_{11} & a_{12} \\ a_{31} & a_{32} \end{vmatrix}$$

Deletions are usually done mentally.

$$\text{Minor of } a_{32} = \begin{vmatrix} a_{11} & a_{13} \\ a_{21} & a_{23} \end{vmatrix}$$

A quantity closely associated with the minor of an element is the **cofactor of an element** a_{ij} (from the ith row and jth column), which is the product of the minor of a_{ij} and $(-1)^{i+j}$.

DEFINITION 3

Cofactor

$$\text{Cofactor of } a_{ij} = (-1)^{i+j}(\text{Minor of } a_{ij})$$

Thus, a cofactor of an element is nothing more than a signed minor. The sign is determined by raising -1 to a power that is the sum of the numbers indicating the row and column in which the element appears. Note that $(-1)^{i+j}$ is 1 if $i + j$ is even and -1 if $i + j$ is odd. Thus, if we are given the determinant

$$\begin{vmatrix} a_{11} & a_{12} & a_{13} \\ a_{21} & a_{22} & a_{23} \\ a_{31} & a_{32} & a_{33} \end{vmatrix}$$

then

$$\text{Cofactor of } a_{23} = (-1)^{2+3} \begin{vmatrix} a_{11} & a_{12} \\ a_{31} & a_{32} \end{vmatrix} = - \begin{vmatrix} a_{11} & a_{12} \\ a_{31} & a_{32} \end{vmatrix}$$

$$\text{Cofactor of } a_{11} = (-1)^{1+1} \begin{vmatrix} a_{22} & a_{23} \\ a_{32} & a_{33} \end{vmatrix} = \begin{vmatrix} a_{22} & a_{23} \\ a_{32} & a_{33} \end{vmatrix}$$

EXAMPLE 2   **Finding Cofactors**

Find the cofactors of -2 and 5 in the determinant

$$\begin{vmatrix} -2 & 0 & 3 \\ 1 & -6 & 5 \\ -1 & 2 & 0 \end{vmatrix}$$

Solution

$$\text{Cofactor of } -2 = (-1)^{1+1} \begin{vmatrix} -6 & 5 \\ 2 & 0 \end{vmatrix} = \begin{vmatrix} -6 & 5 \\ 2 & 0 \end{vmatrix}$$

$$= (-6)(0) - (2)(5) = -10$$

$$\text{Cofactor of } 5 = (-1)^{2+3} \begin{vmatrix} -2 & 0 \\ -1 & 2 \end{vmatrix} = - \begin{vmatrix} -2 & 0 \\ -1 & 2 \end{vmatrix}$$

$$= -[(-2)(2) - (-1)(0)] = 4$$

Problem 2   Find the cofactors of 2 and 3 in the determinant in Example 2.

[*Note:* The sign in front of the minor, $(-1)^{i+j}$, can be determined rather mechanically by using a checkerboard pattern of $+$ and $-$ signs over the determinant, starting with $+$ in the upper left-hand corner:

$$\begin{matrix} + & - & + \\ - & + & - \\ + & - & + \end{matrix}$$

Use either the checkerboard or the exponent method—whichever is easier for you—to determine the sign in front of the minor.]

Now we are ready for the key theorem of this section, Theorem 1. This theorem provides us with an efficient step-by-step procedure, called an algorithm, for evaluating third-order determinants.

Theorem 1   **Value of a Third-Order Determinant**

The value of a determinant of order 3 is the sum of three products obtained by multiplying each element of any one row (or each element of any one column) by its cofactor.

To prove this theorem we must show that the expansions indicated by the theorem for any row or any column (six cases) produce the expression on the right of formula (2). Proofs of special cases of this theorem are left to the C problems in Exercise 7-5.

EXAMPLE 3   **Evaluating a Third-Order Determinant**

Evaluate

$$\begin{vmatrix} 2 & -2 & 0 \\ -3 & 1 & 2 \\ 1 & -3 & -1 \end{vmatrix}$$

by expanding by:

(A) The first row   (B) The second column

Solutions   (A) $\begin{vmatrix} 2 & -2 & 0 \\ -3 & 1 & 2 \\ 1 & -3 & -1 \end{vmatrix} = a_{11}\left(\begin{matrix}\text{Cofactor}\\\text{of } a_{11}\end{matrix}\right) + a_{12}\left(\begin{matrix}\text{Cofactor}\\\text{of } a_{12}\end{matrix}\right) + a_{13}\left(\begin{matrix}\text{Cofactor}\\\text{of } a_{13}\end{matrix}\right)$

$= 2\left((-1)^{1+1}\begin{vmatrix} 1 & 2 \\ -3 & -1 \end{vmatrix}\right) + (-2)\left((-1)^{1+2}\begin{vmatrix} -3 & 2 \\ 1 & -1 \end{vmatrix}\right) + 0$

$= (2)(1)[(1)(-1) - (-3)(2)] + (-2)(-1)[(-3)(-1) - (1)(2)]$

$= (2)(5) + (2)(1) = 12$

(B) $\begin{vmatrix} 2 & -2 & 0 \\ -3 & 1 & 2 \\ 1 & -3 & -1 \end{vmatrix} = a_{12}\left(\begin{matrix}\text{Cofactor}\\\text{of } a_{12}\end{matrix}\right) + a_{22}\left(\begin{matrix}\text{Cofactor}\\\text{of } a_{22}\end{matrix}\right) + a_{32}\left(\begin{matrix}\text{Cofactor}\\\text{of } a_{32}\end{matrix}\right)$

$= (-2)\left((-1)^{1+2}\begin{vmatrix} -3 & 2 \\ 1 & -1 \end{vmatrix}\right) + (1)\left((-1)^{2+2}\begin{vmatrix} 2 & 0 \\ 1 & -1 \end{vmatrix}\right)$

$\quad + (-3)\left((-1)^{3+2}\begin{vmatrix} 2 & 0 \\ -3 & 2 \end{vmatrix}\right)$

$= (-2)(-1)[(-3)(-1) - (1)(2)] + (1)(1)[(2)(-1) - (1)(0)]$
$\quad + (-3)(-1)[(2)(2) - (-3)(0)]$

$= (2)(1) + (1)(-2) + (3)(4) = 12$

Problem 3   Evaluate

$$\begin{vmatrix} 2 & 1 & -1 \\ -2 & -3 & 0 \\ -1 & 2 & 1 \end{vmatrix}$$

by expanding by:

(A) The first row   (B) The third column

● **Higher-Order Determinants**   Theorem 1 and the definitions of minor and cofactor generalize completely for determinants of order higher than 3. These concepts are illustrated for a fourth-order determinant in the next example.

EXAMPLE 4   **Evaluating a Fourth-Order Determinant**

Given the fourth-order determinant

$$\begin{vmatrix} 0 & -1 & 0 & 2 \\ -5 & -6 & 0 & -3 \\ 4 & 5 & -2 & 6 \\ 0 & 3 & 0 & -4 \end{vmatrix}$$

(A) Find the minor in determinant form of the element 3.
(B) Find the cofactor in determinant form of the element -5.
(C) Find the value of the fourth-order determinant.

Solutions   (A) Minor of 3 $= \begin{vmatrix} 0 & 0 & 2 \\ -5 & 0 & -3 \\ 4 & -2 & 6 \end{vmatrix}$

(B) Cofactor of $-5 = (-1)^{2+1} \begin{vmatrix} -1 & 0 & 2 \\ 5 & -2 & 6 \\ 3 & 0 & -4 \end{vmatrix} = -\begin{vmatrix} -1 & 0 & 2 \\ 5 & -2 & 6 \\ 3 & 0 & -4 \end{vmatrix}$

(C) Generalizing Theorem 1, the value of this fourth-order determinant is the sum of four products obtained by multiplying each element of any one row (or each element of any one column) by its cofactor. The work involved in this evaluation is greatly reduced if we choose the row or column with the greatest number of 0's. Since column 3 has three 0's, we expand along this column:

$$\begin{vmatrix} 0 & -1 & 0 & 2 \\ -5 & -6 & 0 & -3 \\ 4 & 5 & -2 & 6 \\ 0 & 3 & 0 & -4 \end{vmatrix} = 0 + 0 + (-2)(-1)^{3+3} \begin{vmatrix} 0 & -1 & 2 \\ -5 & -6 & -3 \\ 0 & 3 & -4 \end{vmatrix} + 0$$

$$= (-2) \begin{vmatrix} 0 & -1 & 2 \\ -5 & -6 & -3 \\ 0 & 3 & -4 \end{vmatrix} \quad \text{Expand this determinant along the first column.}$$

$$= (-2)\left(0 + (-5)(-1)^{2+1} \begin{vmatrix} -1 & 2 \\ 3 & -4 \end{vmatrix} + 0 \right)$$

$$= (-2)(-5)(-1)(-2) = 20$$

Problem 4 Repeat Example 4 for the following fourth-order determinant:

$$\begin{vmatrix} 0 & 4 & -2 & 0 \\ -3 & 3 & -1 & 2 \\ 0 & 6 & 0 & 0 \\ 5 & -6 & -5 & -4 \end{vmatrix}$$

Remark: Where are determinants used? Many equations and formulas have particularly simple and compact representations in determinant form that are easily remembered. (See Problems 44–48 in Exercise 7-6). Also, in Section 7-7 we will see that the solutions to certain systems of equations can be expressed in terms of determinants. In addition, determinants are involved in theoretical work in advanced mathematics courses. For example, it can be shown that the inverse of a square matrix exists if and only if its determinant is not 0.

Answers to Matched Problems

1. 14   **2.** Cofactor of 2 = 13; cofactor of 3 = -4 **3.** (A) 3   (B) 3

4. (A) $\begin{vmatrix} 0 & -2 & 0 \\ 0 & 0 & 0 \\ 5 & -5 & -4 \end{vmatrix}$ (B) $-\begin{vmatrix} 0 & 4 & 0 \\ -3 & 3 & 2 \\ 0 & 6 & 0 \end{vmatrix}$ (C) -24

EXERCISE 7-5

A

Evaluate each second-order determinant in Problems 1–6.

1. $\begin{vmatrix} 2 & 2 \\ -3 & 1 \end{vmatrix}$

2. $\begin{vmatrix} 2 & 4 \\ 3 & -1 \end{vmatrix}$

3. $\begin{vmatrix} 6 & -2 \\ -1 & -3 \end{vmatrix}$

4. $\begin{vmatrix} 5 & -4 \\ -2 & 2 \end{vmatrix}$

5. $\begin{vmatrix} 1.8 & -1.6 \\ -1.9 & 1.2 \end{vmatrix}$

6. $\begin{vmatrix} 0.5 & -3.2 \\ 1.4 & -6.7 \end{vmatrix}$

Problem 7–14 pertain to the determinant below:

$$\begin{vmatrix} a_{11} & a_{12} & a_{13} \\ a_{21} & a_{22} & a_{23} \\ a_{31} & a_{32} & a_{33} \end{vmatrix}$$

Write the minor of each element given in Problems 7–10.

7. a_{11} **8.** a_{33} **9.** a_{23} **10.** a_{22}

Write the cofactor of each element given in Problems 11–14.

11. a_{11} **12.** a_{33} **13.** a_{23} **14.** a_{22}

Problems 15–22 pertain to the determinant below:

$$\begin{vmatrix} -2 & 3 & 0 \\ 5 & 1 & -2 \\ 7 & -4 & 8 \end{vmatrix}$$

Write the minor of each element given in Problems 15–18. Leave the answer in determinant form.

15. a_{11} **16.** a_{22} **17.** a_{32} **18.** a_{21}

Write the cofactor of each element given in Problems 19–22, and evaluate each.

19. a_{11} **20.** a_{22} **21.** a_{32} **22.** a_{21}

Evaluate Problems 23–28 using cofactors.

23. $\begin{vmatrix} 1 & 0 & 0 \\ -2 & 4 & 3 \\ 5 & -2 & 1 \end{vmatrix}$

24. $\begin{vmatrix} 2 & -3 & 5 \\ 0 & -3 & 1 \\ 0 & 6 & 2 \end{vmatrix}$

25. $\begin{vmatrix} 0 & 1 & 5 \\ 3 & -7 & 6 \\ 0 & -2 & -3 \end{vmatrix}$

26. $\begin{vmatrix} 4 & -2 & 0 \\ 9 & 5 & 4 \\ 1 & 2 & 0 \end{vmatrix}$

27. $\begin{vmatrix} -1 & 2 & -3 \\ -2 & 0 & -6 \\ 4 & -3 & 2 \end{vmatrix}$

28. $\begin{vmatrix} 0 & 2 & -1 \\ -6 & 3 & 1 \\ 7 & -9 & -2 \end{vmatrix}$

B _____

Given the determinant

$$\begin{vmatrix} a_{11} & a_{12} & a_{13} & a_{14} \\ a_{21} & a_{22} & a_{23} & a_{24} \\ a_{31} & a_{32} & a_{33} & a_{34} \\ a_{41} & a_{42} & a_{43} & a_{44} \end{vmatrix}$$

write the cofactor in determinant form of each element in Problems 29–32.

29. a_{11} **30.** a_{44} **31.** a_{43} **32.** a_{23}

Evaluate each determinant in Problems 33–42 using cofactors.

33. $\begin{vmatrix} 3 & -2 & -8 \\ -2 & 0 & -3 \\ 1 & 0 & -4 \end{vmatrix}$

34. $\begin{vmatrix} 4 & -4 & 6 \\ 2 & 8 & -3 \\ 0 & -5 & 0 \end{vmatrix}$

35. $\begin{vmatrix} 1 & 4 & 1 \\ 1 & 1 & -2 \\ 2 & 1 & -1 \end{vmatrix}$

36. $\begin{vmatrix} 3 & 2 & 1 \\ -1 & 5 & 1 \\ 2 & 3 & 1 \end{vmatrix}$

37. $\begin{vmatrix} 1 & 4 & 3 \\ 2 & 1 & 6 \\ 3 & -2 & 9 \end{vmatrix}$

38. $\begin{vmatrix} 4 & -6 & 3 \\ -1 & 4 & 1 \\ 5 & -6 & 3 \end{vmatrix}$

39. $\begin{vmatrix} 2 & 6 & 1 & 7 \\ 0 & 3 & 0 & 0 \\ 3 & 4 & 2 & 5 \\ 0 & 9 & 0 & 2 \end{vmatrix}$

40. $\begin{vmatrix} 0 & 1 & 0 & 1 \\ 2 & 4 & 7 & 6 \\ 0 & 3 & 0 & 1 \\ 0 & 6 & 2 & 5 \end{vmatrix}$

41. $\begin{vmatrix} -2 & 0 & 0 & 0 & 0 \\ 9 & -1 & 0 & 0 & 0 \\ 2 & 1 & 3 & 0 & 0 \\ -1 & 4 & 2 & 2 & 0 \\ 7 & -2 & 3 & 5 & 5 \end{vmatrix}$

42. $\begin{vmatrix} 2 & 0 & 0 & 0 & 0 \\ 0 & 3 & 0 & 0 & 0 \\ 0 & 0 & 2 & 0 & 0 \\ 0 & 0 & 0 & 1 & 0 \\ 0 & 0 & 0 & 0 & 4 \end{vmatrix}$

If A is a 3×3 matrix, det A can be evaluated by the following **diagonal expansion**. Form a 3×5 matrix by augmenting A on the right with its first two columns, and compute the diagonal products $p_1, p_2, \ldots, p_6$ indicated by the arrows:

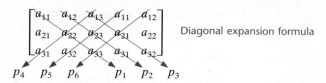

Diagonal expansion formula

The determinant of A is given by [compare with formula (2)]

$$\det A = p_1 + p_2 + p_3 - p_4 - p_5 - p_6$$
$$= a_{11}a_{22}a_{33} + a_{12}a_{23}a_{31} + a_{13}a_{21}a_{32} - a_{13}\,a_{22}a_{31} - a_{11}a_{23}a_{32} - a_{12}a_{21}a_{33}$$

[*Caution: The diagonal expansion procedure works only for 3×3 matrices. Do not apply it to matrices of any other size.*]

Use the diagonal expansion formula to evaluate the determinants in Problems 43 and 44.

43. $\begin{vmatrix} 2 & 6 & -1 \\ 5 & 3 & -7 \\ -4 & -2 & 1 \end{vmatrix}$

44. $\begin{vmatrix} 4 & 1 & -5 \\ 1 & 2 & -6 \\ -3 & -1 & 7 \end{vmatrix}$

C

If all the letters in Problems 45–48 represent real numbers, show that each statement is true.

45. $\begin{vmatrix} a & b \\ ka & kb \end{vmatrix} = 0$

46. $\begin{vmatrix} a & b \\ c & d \end{vmatrix} = -\begin{vmatrix} b & a \\ d & c \end{vmatrix}$

47. $\begin{vmatrix} a & b \\ c & d \end{vmatrix} = \begin{vmatrix} a & c \\ b & d \end{vmatrix}$

48. $\begin{vmatrix} ka & kb \\ c & d \end{vmatrix} = k\begin{vmatrix} a & b \\ c & d \end{vmatrix}$

49. Show that the expansion of the determinant

$$\begin{vmatrix} a_{11} & a_{12} & a_{13} \\ a_{21} & a_{22} & a_{23} \\ a_{31} & a_{32} & a_{33} \end{vmatrix}$$

by the first column is the same as its expansion by the third row.

50. Repeat Problem 49, using the second row and the third column.

51. If

$$A = \begin{bmatrix} 2 & 3 \\ 1 & -2 \end{bmatrix} \quad \text{and} \quad B = \begin{bmatrix} -1 & 3 \\ 2 & 1 \end{bmatrix}$$

show that $\det(AB) = \det A \cdot \det B$.

52. If

$$A = \begin{bmatrix} a & b \\ c & d \end{bmatrix} \quad \text{and} \quad B = \begin{bmatrix} w & x \\ y & z \end{bmatrix}$$

show that $\det(AB) = \det A \cdot \det B$.

If A is an $n \times n$ matrix and I is the $n \times n$ identity matrix, then the function $f(x) = |xI - A|$ is called the **characteristic polynomial** of A, and the zeros of $f(x)$ are called the **eigenvalues** of A. Characteristic polynomials and eigenvalues have many important applications that are discussed in more advanced treatments of matrices. In Problems 53–56, find the characteristic polynomial and the eigenvalues of each matrix.

53. $\begin{bmatrix} 5 & -4 \\ 2 & -1 \end{bmatrix}$

54. $\begin{bmatrix} 8 & -6 \\ 3 & -1 \end{bmatrix}$

55. $\begin{bmatrix} 4 & -4 & 0 \\ 2 & -2 & 0 \\ 4 & -8 & -4 \end{bmatrix}$

56. $\begin{bmatrix} -2 & 2 & 0 \\ -1 & 1 & 0 \\ -2 & 4 & 2 \end{bmatrix}$

SECTION 7-6   Properties of Determinants

- Discussion of Determinant Properties
- Summary of Determinant Properties

Determinants have a number of useful properties that can greatly reduce the labor in evaluating determinants of order 3 or greater. These properties and their use are the subject matter for this section.

• Discussion of Determinant Properties

We now state and discuss five general determinant properties in the form of theorems. Because the proofs for the general cases of these theorems are involved and notationally difficult, we will sketch only informal proofs for determinants of order 3. The theorems, however, apply to determinants of any order.

Theorem 1   **Multiplying a Row or Column by a Constant**

If each element of any row (or column) of a determinant is multiplied by a constant k, the new determinant is k times the original.

Partial Proof   Let C_{ij} be the cofactor of a_{ij}. Then expanding by the first row, we have

$$\begin{vmatrix} ka_{11} & ka_{12} & ka_{13} \\ a_{21} & a_{22} & a_{23} \\ a_{31} & a_{32} & a_{33} \end{vmatrix} = ka_{11}C_{11} + ka_{12}C_{12} + ka_{13}C_{13}$$

$$= k(a_{11}C_{11} + a_{12}C_{12} + a_{13}C_{13})$$

$$= k \begin{vmatrix} a_{11} & a_{12} & a_{13} \\ a_{21} & a_{22} & a_{23} \\ a_{31} & a_{32} & a_{33} \end{vmatrix}$$

Theorem 1 also states that a factor common to all elements of a row (or column) can be taken out as a factor of the determinant.

EXAMPLE 1   Taking Out a Common Factor of a Column

$$\begin{vmatrix} 6 & 1 & 3 \\ -2 & 7 & -2 \\ 4 & 5 & 0 \end{vmatrix} = 2 \begin{vmatrix} 3 & 1 & 3 \\ -1 & 7 & -2 \\ 2 & 5 & 0 \end{vmatrix}$$

where 2 is a common factor of the first column.

Problem 1   Take out factors common to any row or any column:

$$\begin{vmatrix} 3 & 2 & 1 \\ 6 & 3 & -9 \\ 1 & 0 & -5 \end{vmatrix}$$

Theorem 2   **Row or Column of Zeros**

If every element in a row (or column) is 0, the value of the determinant is 0.

Theorem 2 is an immediate consequence of Theorem 1, and its proof is left as an exercise. It is illustrated in the following example:

$$\begin{vmatrix} 3 & -2 & 5 \\ 0 & 0 & 0 \\ -1 & 4 & 9 \end{vmatrix} = 0$$

Theorem 3   **Interchanging Rows or Columns**

If two rows (or two columns) of a determinant are interchanged, the new determinant is the negative of the original.

A proof of Theorem 3 even for a determinant of order 3 is notationally involved. We suggest that you partially prove the theorem by direct expansion of the determinants before and after the interchange of two rows (or columns). The theorem is illustrated by the following example, where the second and third columns are interchanged:

$$\begin{vmatrix} 1 & 0 & 9 \\ -2 & 1 & 5 \\ 3 & 0 & 7 \end{vmatrix} = - \begin{vmatrix} 1 & 9 & 0 \\ -2 & 5 & 1 \\ 3 & 7 & 0 \end{vmatrix}$$

Theorem 4   **Equal Rows or Columns**

If the corresponding elements are equal in two rows (or columns), the value of the determinant is 0.

Proof   The general proof of Theorem 4 follows directly from Theorem 3. If we start with a determinant D that has two rows (or columns) equal and we interchange the equal rows (or columns), the new determinant will be the same as the original. But by Theorem 3,

$$D = -D$$

hence,

$$2D = 0$$

$$D = 0$$

Theorem 5   **Addition of Rows or Columns**

If a multiple of any row (or column) of a determinant is added to any other row (or column), the value of the determinant is not changed.

Partial Proof   If, in a general third-order determinant, we add a k multiple of the second column to the first and then expand by the first column, we obtain (where C_{ij} is the cofactor of a_{ij} in the original determinant)

$$\begin{vmatrix} a_{11} + ka_{12} & a_{12} & a_{13} \\ a_{21} + ka_{22} & a_{22} & a_{23} \\ a_{31} + ka_{32} & a_{32} & a_{33} \end{vmatrix} = (a_{11} + ka_{12})C_{11} + (a_{21} + ka_{22})C_{21} + (a_{31} + ka_{32})C_{31}$$

$$= (a_{11}C_{11} + a_{21}C_{21} + a_{31}C_{31}) + k(a_{12}C_{11} + a_{22}C_{21} + a_{32}C_{31})$$

$$= \begin{vmatrix} a_{11} & a_{12} & a_{13} \\ a_{21} & a_{22} & a_{23} \\ a_{31} & a_{32} & a_{33} \end{vmatrix} + k \begin{vmatrix} a_{12} & a_{12} & a_{13} \\ a_{22} & a_{22} & a_{23} \\ a_{32} & a_{32} & a_{33} \end{vmatrix} = \begin{vmatrix} a_{11} & a_{12} & a_{13} \\ a_{21} & a_{22} & a_{23} \\ a_{31} & a_{32} & a_{33} \end{vmatrix}$$

The determinant following k is 0 because the first and second columns are equal.

Note the similarity in the process described in Theorem 5 to that used to obtain row-equivalent matrices. We use this theorem to transform a determinant without 0 elements into one that contains a row or column with all elements 0 but one. The transformed determinant can then be easily expanded by this row (or column). An example best illustrates the process.

EXAMPLE 2   **Evaluating a Determinant**

Evaluate the determinant

$$\begin{vmatrix} 3 & -1 & 2 \\ -2 & 4 & -3 \\ 4 & -2 & 5 \end{vmatrix}$$

Solution   We use Theorem 5 to obtain two 0's in the first row, and then expand the determinant by this row. To start, we replace the third column with the sum of it and 2 times the second column to obtain a 0 in the a_{13} position:

$$\begin{vmatrix} 3 & -1 & 2 \\ -2 & 4 & -3 \\ 4 & -2 & 5 \end{vmatrix} = \begin{vmatrix} 3 & -1 & 0 \\ -2 & 4 & 5 \\ 4 & -2 & 1 \end{vmatrix} \qquad C_3 + 2C_2 \rightarrow C_3{}^*$$

Next, to obtain a 0 in the a_{11} position, we replace the first column with the sum of it and 3 times the second column:

$$\begin{vmatrix} 3 & -1 & 0 \\ -2 & 4 & 5 \\ 4 & -2 & 1 \end{vmatrix} = \begin{vmatrix} 0 & -1 & 0 \\ 10 & 4 & 5 \\ -2 & -2 & 1 \end{vmatrix} \qquad C_1 + 3C_2 \rightarrow C_1$$

Now it is an easy matter to expand this last determinant by the first row to obtain

$$0 + (-1)\left((-1)^{1+2} \begin{vmatrix} 10 & 5 \\ -2 & 1 \end{vmatrix} \right) + 0 = 20$$

Problem 2   Evaluate the following determinant by first using Theorem 5 to obtain 0's in the a_{11} and a_{31} positions, and then expand by the first column.

$$\begin{vmatrix} 3 & 10 & -5 \\ 1 & 6 & -3 \\ 2 & 3 & 4 \end{vmatrix}$$

● **Summary of Determinant Properties**

We now summarize the five determinant properties discussed above in table form for convenient reference. Even though these properties hold for determinants of any order, for simplicity, we illustrate each property in terms of second-order determinants.

$^*C_1$, C_2, and C_3 represent columns 1, 2, and 3, respectively.

TABLE 1   Summary of Determinant Properties

Property	Examples
1. If each element of any row (or column) of a determinant is multiplied by a constant k, the new determinant is k times the original.	$\begin{vmatrix} 2a & 2b \\ c & d \end{vmatrix} = 2\begin{vmatrix} a & b \\ c & d \end{vmatrix}$ $3\begin{vmatrix} a & b \\ c & d \end{vmatrix} = \begin{vmatrix} 3a & b \\ 3c & d \end{vmatrix}$
2. If every element in a row (or column) is 0, the value of the determinant is 0.	$\begin{vmatrix} a & b \\ 0 & 0 \end{vmatrix} = 0$ $\begin{vmatrix} 0 & b \\ 0 & d \end{vmatrix} = 0$
3. If two rows (or two columns) of a determinant are interchanged, the new determinant is the negative of the original.	$\begin{vmatrix} a & b \\ c & d \end{vmatrix} = -\begin{vmatrix} c & d \\ a & b \end{vmatrix}$ $\begin{vmatrix} a & b \\ c & d \end{vmatrix} = -\begin{vmatrix} b & a \\ d & c \end{vmatrix}$ *(Continued)*

Property	Examples
4. If the corresponding elements are equal in two rows (or columns), the value of the determinant is 0.	$\begin{vmatrix} a & b \\ a & b \end{vmatrix} = 0$ $\begin{vmatrix} a & a \\ c & c \end{vmatrix} = 0$
5. If a multiple of any row (or column) of a determinant is added to any other row (or column), the value of the determinant is not changed.	$\begin{vmatrix} a & b \\ c & d \end{vmatrix} = \begin{vmatrix} a & b \\ c + ka & d + kb \end{vmatrix}$ $\begin{vmatrix} a & b \\ c & d \end{vmatrix} = \begin{vmatrix} a + kb & b \\ c + kd & d \end{vmatrix}$

Answers to Matched Problems

1. $3\begin{vmatrix} 3 & 2 & 1 \\ 2 & 1 & -3 \\ 1 & 0 & -5 \end{vmatrix}$ **2.** 44

EXERCISE 7-6

A

For each statement in Problems 1–10, identify the theorem from this section that justifies it. Do not evaluate.

1. $\begin{vmatrix} 16 & 8 \\ 0 & -1 \end{vmatrix} = 8\begin{vmatrix} 2 & 1 \\ 0 & -1 \end{vmatrix}$

2. $\begin{vmatrix} 1 & -9 \\ 0 & -6 \end{vmatrix} = -3\begin{vmatrix} 1 & 3 \\ 0 & 2 \end{vmatrix}$

3. $-2\begin{vmatrix} 2 & 1 \\ -3 & 4 \end{vmatrix} = \begin{vmatrix} -4 & 1 \\ 6 & 4 \end{vmatrix}$

4. $4\begin{vmatrix} -1 & 3 \\ 2 & 1 \end{vmatrix} = \begin{vmatrix} -4 & 12 \\ 2 & 1 \end{vmatrix}$

5. $\begin{vmatrix} 3 & 0 \\ -2 & 0 \end{vmatrix} = 0$

6. $\begin{vmatrix} 5 & -7 \\ 0 & 0 \end{vmatrix} = 0$

7. $\begin{vmatrix} 5 & -1 \\ 8 & 0 \end{vmatrix} = -\begin{vmatrix} -1 & 5 \\ 0 & 8 \end{vmatrix}$

8. $\begin{vmatrix} 6 & 9 \\ 0 & 1 \end{vmatrix} = -\begin{vmatrix} 0 & 1 \\ 6 & 9 \end{vmatrix}$

9. $\begin{vmatrix} 4 & 3 \\ 1 & 2 \end{vmatrix} = \begin{vmatrix} 4-4 & 3-8 \\ 1 & 2 \end{vmatrix}$

10. $\begin{vmatrix} 3 & 2 \\ 5 & 1 \end{vmatrix} = \begin{vmatrix} 3+4 & 2 \\ 5+2 & 1 \end{vmatrix}$

In Problems 11–14, Theorem 5 was used to transform the determinant on the left to that on the right. Replace each letter x with an appropriate numeral to complete the transformation.

11. $\begin{vmatrix} -1 & 3 \\ 2 & -4 \end{vmatrix} = \begin{vmatrix} -1 & x \\ 2 & 2 \end{vmatrix}$

12. $\begin{vmatrix} -1 & 3 \\ 5 & -2 \end{vmatrix} = \begin{vmatrix} -1 & 3 \\ x & 13 \end{vmatrix}$

13. $\begin{vmatrix} -1 & 2 & 3 \\ 2 & 1 & 4 \\ 1 & 3 & 2 \end{vmatrix} = \begin{vmatrix} -1 & 2 & 0 \\ 2 & 1 & 10 \\ 1 & 3 & x \end{vmatrix}$

14. $\begin{vmatrix} -1 & 2 & 3 \\ 2 & 1 & 4 \\ 1 & 3 & 2 \end{vmatrix} = \begin{vmatrix} -1 & 0 & 3 \\ 2 & x & 4 \\ 1 & 5 & 2 \end{vmatrix}$

In Problems 15–18, transform each determinant into one that contains a row (or column) with all elements 0 but one, if possible. Then expand the transformed determinant by this row (or column).

15. $\begin{vmatrix} -1 & 0 & 3 \\ 2 & 5 & 4 \\ 1 & 5 & 2 \end{vmatrix}$
16. $\begin{vmatrix} -1 & 2 & 0 \\ 2 & 1 & 10 \\ 1 & 3 & 5 \end{vmatrix}$
17. $\begin{vmatrix} 3 & 5 & 0 \\ 1 & 1 & -2 \\ 2 & 1 & -1 \end{vmatrix}$
18. $\begin{vmatrix} 2 & 0 & 1 \\ -1 & -3 & 4 \\ 1 & 2 & 3 \end{vmatrix}$

B _____

For each statement in Problems 19–24, identify the theorem from this section that justifies it.

19. $-2\begin{vmatrix} 1 & 0 & 2 \\ 3 & -2 & 4 \\ 0 & 1 & 1 \end{vmatrix} = \begin{vmatrix} 1 & 0 & 2 \\ -6 & 4 & -8 \\ 0 & 1 & 1 \end{vmatrix}$

20. $\begin{vmatrix} 8 & 0 & 1 \\ 12 & -1 & 0 \\ 4 & 3 & 2 \end{vmatrix} = 4\begin{vmatrix} 2 & 0 & 1 \\ 3 & -1 & 0 \\ 1 & 3 & 2 \end{vmatrix}$

21. $\begin{vmatrix} 1 & 2 & 0 \\ -1 & 3 & 0 \\ 0 & 1 & 0 \end{vmatrix} = 0$

22. $\begin{vmatrix} -2 & 5 & 13 \\ 1 & 7 & 12 \\ 0 & 8 & 15 \end{vmatrix} = -\begin{vmatrix} 5 & -2 & 13 \\ 7 & 1 & 12 \\ 8 & 0 & 15 \end{vmatrix}$

23. $\begin{vmatrix} 4 & 2 & -1 \\ 2 & 0 & 2 \\ -3 & 5 & -2 \end{vmatrix} = \begin{vmatrix} 4-4 & 2 & -1 \\ 2+8 & 0 & 2 \\ -3-8 & 5 & -2 \end{vmatrix}$

24. $\begin{vmatrix} 7 & 7 & 1 \\ -3 & -3 & 11 \\ 2 & 2 & 0 \end{vmatrix} = 0$

In Problems 25–28, Theorem 5 was used to transform the determinant on the left to that on the right. Replace each letter x and y with an appropriate numeral to complete the transformation.

25. $\begin{vmatrix} 2 & 1 & -1 \\ 3 & 4 & 1 \\ 1 & 2 & -2 \end{vmatrix} = \begin{vmatrix} 0 & 0 & -1 \\ x & 5 & 1 \\ -3 & y & -2 \end{vmatrix}$

26. $\begin{vmatrix} 3 & -1 & 1 \\ -2 & 4 & 3 \\ 1 & 5 & 2 \end{vmatrix} = \begin{vmatrix} 0 & -1 & 0 \\ 10 & 4 & 7 \\ x & 5 & y \end{vmatrix}$

27. $\begin{vmatrix} 7 & 9 & 4 \\ 2 & 3 & 1 \\ 3 & 4 & -2 \end{vmatrix} = \begin{vmatrix} -1 & x & 0 \\ 2 & 3 & 1 \\ 7 & y & 0 \end{vmatrix}$

28. $\begin{vmatrix} 5 & 2 & 3 \\ 3 & 1 & 2 \\ -4 & -3 & 5 \end{vmatrix} = \begin{vmatrix} x & 0 & -1 \\ 3 & 1 & 2 \\ 5 & 0 & y \end{vmatrix}$

In Problems 29–36, transform each determinant into one that contains a row (or column) with all elements 0 but one, if possible. Then expand the transformed determinant by this row (or column).

29. $\begin{vmatrix} 1 & 5 & 3 \\ 4 & 2 & 1 \\ 3 & 1 & 2 \end{vmatrix}$
30. $\begin{vmatrix} -1 & 5 & 1 \\ 2 & 3 & 1 \\ 3 & 2 & 1 \end{vmatrix}$
31. $\begin{vmatrix} 5 & 2 & -3 \\ -2 & 4 & 4 \\ 1 & -1 & 3 \end{vmatrix}$
32. $\begin{vmatrix} 5 & 3 & -6 \\ -1 & 1 & 4 \\ 4 & 3 & -6 \end{vmatrix}$

33. $\begin{vmatrix} 3 & -4 & 1 \\ 6 & -1 & 2 \\ 9 & 2 & 3 \end{vmatrix}$
34. $\begin{vmatrix} 2 & 3 & -1 \\ 5 & 4 & 7 \\ -4 & -6 & 2 \end{vmatrix}$
35. $\begin{vmatrix} 0 & 1 & 0 & 1 \\ 1 & -2 & 4 & 3 \\ 2 & 1 & 5 & 4 \\ 1 & 2 & 1 & 2 \end{vmatrix}$
36. $\begin{vmatrix} 2 & 3 & 1 & -1 \\ 3 & 1 & 2 & 1 \\ 0 & 5 & 4 & 0 \\ -1 & 2 & 3 & 0 \end{vmatrix}$

C _____

Transform each determinant in Problems 37 and 38 into one that contains a row (or column) with all elements 0 but one, if possible. Then expand the transformed determinant by this row (or column).

37.
$$\begin{vmatrix} 3 & 2 & 3 & 1 \\ 3 & -2 & 8 & 5 \\ 2 & 1 & 3 & 1 \\ 4 & 5 & 4 & -3 \end{vmatrix}$$

38.
$$\begin{vmatrix} -1 & 4 & 2 & 1 \\ 5 & -1 & -3 & -1 \\ 2 & -1 & -2 & 3 \\ -3 & 3 & 3 & 3 \end{vmatrix}$$

Problems 39–42 are representative cases of theorems discussed in this section. Use cofactor expansions to verify each statement directly, without reference to the theorem it represents.

39.
$$\begin{vmatrix} a & b & a \\ d & e & d \\ g & h & g \end{vmatrix} = 0$$

40.
$$\begin{vmatrix} a & b & c \\ kd & ke & kf \\ g & h & i \end{vmatrix} = k \begin{vmatrix} a & b & c \\ d & e & f \\ g & h & i \end{vmatrix}$$

41.
$$\begin{vmatrix} a_1 & b_1 & c_1 \\ a_2 & b_2 & c_2 \\ a_3 & b_3 & c_3 \end{vmatrix} = - \begin{vmatrix} b_1 & a_1 & c_1 \\ b_2 & a_2 & c_2 \\ b_3 & a_3 & c_3 \end{vmatrix}$$

42.
$$\begin{vmatrix} a_1 & b_1 & c_1 \\ a_2 & b_2 & c_2 \\ a_3 & b_3 & c_3 \end{vmatrix} = \begin{vmatrix} a_1 + kc_1 & b_1 & c_1 \\ a_2 + kc_2 & b_2 & c_2 \\ a_3 + kc_3 & b_3 & c_3 \end{vmatrix}$$

43. Show, without expanding, that $(2, 5)$ and $(-3, 4)$ satisfy the equation

$$\begin{vmatrix} x & y & 1 \\ 2 & 5 & 1 \\ -3 & 4 & 1 \end{vmatrix} = 0$$

44. Show that

$$\begin{vmatrix} x & y & 1 \\ 2 & 3 & 1 \\ -1 & 2 & 1 \end{vmatrix} = 0$$

is the equation of a line that passes through $(2, 3)$ and $(-1, 2)$.

45. Show that

$$\begin{vmatrix} x & y & 1 \\ x_1 & y_1 & 1 \\ x_2 & y_2 & 1 \end{vmatrix} = 0$$

is the equation of a line that passes through (x_1, y_1) and (x_2, y_2).

46. In analytic geometry it is shown that the area of a triangle with vertices (x_1, y_1), (x_2, y_2), and (x_3, y_3) is the absolute value of

$$\frac{1}{2} \begin{vmatrix} x_1 & y_1 & 1 \\ x_2 & y_2 & 1 \\ x_3 & y_3 & 1 \end{vmatrix}$$

Use this result to find the area of a triangle with vertices $(-1, 4)$, $(4, 8)$, and $(1, 1)$.

47. What can we say about the three points (x_1, y_1), (x_2, y_2), and (x_3, y_3) if the following equation is true?

$$\begin{vmatrix} x_1 & y_1 & 1 \\ x_2 & y_2 & 1 \\ x_3 & y_3 & 1 \end{vmatrix} = 0$$

[*Hint:* See Problem 46.]

48. If the three points (x_1, y_1), (x_2, y_2), and (x_3, y_3) are all on the same line, what can we say about the value of the determinant below?

$$\begin{vmatrix} x_1 & y_1 & 1 \\ x_2 & y_2 & 1 \\ x_3 & y_3 & 1 \end{vmatrix}$$

SECTION 7-7 Cramer's Rule

- Two Equations–Two Variables
- Three Equations–Three Variables

Now let's see how determinants arise rather naturally in the process of solving systems of linear equations. We start by investigating two equations and two variables, and then extend our results to three equations and three variables.

● **Two Equations– Two Variables**

Instead of thinking of each system of linear equations in two variables as a different problem, let's see what happens when we attempt to solve the general system

$$a_{11}x + a_{12}y = k_1 \qquad (1A)$$

$$a_{21}x + a_{22}y = k_2 \qquad (1B)$$

once and for all, in terms of the unspecified real constants a_{11}, a_{12}, a_{21}, a_{22}, k_1, and k_2.

We proceed by multiplying equations (1A) and (1B) by suitable constants so that when the resulting equations are added, left side to left side and right side to right side, one of the variables drops out. Suppose we choose to eliminate y. What constant should we use to make the coefficients of y the same except for the signs? Multiply equation (1A) by a_{22} and (1B) by $-a_{12}$; then add:

$$
\begin{aligned}
a_{22}(1A): \qquad & a_{11}a_{22}x + a_{12}a_{22}y = k_1a_{22} \\
-a_{12}(1B): \qquad & \underline{-a_{21}a_{12}x - a_{12}a_{22}y = -k_2a_{12}} \\
& a_{11}a_{22}x - a_{21}a_{12}x + 0y = k_1a_{22} - k_2a_{12} \\
& (a_{11}a_{22} - a_{21}a_{12})x = k_1a_{22} - k_2a_{12} \\
& x = \frac{k_1a_{22} - k_2a_{12}}{a_{11}a_{22} - a_{21}a_{12}} \qquad a_{11}a_{22} - a_{21}a_{12} \neq 0
\end{aligned}
$$

What do the numerator and denominator remind you of? From your experience with determinants in the last two sections, you should recognize these expressions as

$$x = \frac{\begin{vmatrix} k_1 & a_{12} \\ k_2 & a_{22} \end{vmatrix}}{\begin{vmatrix} a_{11} & a_{12} \\ a_{21} & a_{22} \end{vmatrix}}$$

Similarly, starting with system (1A) and (1B) and eliminating x (this is left as an exercise), we obtain

$$y = \frac{\begin{vmatrix} a_{11} & k_1 \\ a_{21} & k_2 \end{vmatrix}}{\begin{vmatrix} a_{11} & a_{12} \\ a_{21} & a_{22} \end{vmatrix}}$$

These results are summarized in Theorem 1, **Cramer's Rule**, which is named after the Swiss mathematician G. Cramer (1704–1752).

Theorem 1   **Cramer's Rule for Two Equations and Two Variables**

Given the system

$$\begin{matrix} a_{11}x + a_{12}y = k_1 \\ a_{21}x + a_{22}y = k_2 \end{matrix} \quad \text{with} \quad D = \begin{vmatrix} a_{11} & a_{12} \\ a_{21} & a_{22} \end{vmatrix} \neq 0$$

then

$$x = \frac{\begin{vmatrix} k_1 & a_{12} \\ k_2 & a_{22} \end{vmatrix}}{D} \quad \text{and} \quad y = \frac{\begin{vmatrix} a_{11} & k_1 \\ a_{21} & k_2 \end{vmatrix}}{D}$$

The determinant D is called the **coefficient determinant**. If $D \neq 0$, then the system has exactly one solution, which is given by Cramer's rule. If, on the other hand, $D = 0$, then it can be shown that the system is either inconsistent and has no solutions or dependent and has an infinite number of solutions. We must use other methods, such as those discussed in Chapter 6, to determine the exact nature of the solutions when $D = 0$.

EXAMPLE 1   **Solving a System with Cramer's Rule**

Solve using Cramer's rule: $3x - 5y = 2$
$-4x + 3y = -1$

Solution
$$D = \begin{vmatrix} 3 & -5 \\ -4 & 3 \end{vmatrix} = -11$$

$$x = \frac{\begin{vmatrix} 2 & -5 \\ -1 & 3 \end{vmatrix}}{-11} = -\frac{1}{11} \qquad y = \frac{\begin{vmatrix} 3 & 2 \\ -4 & -1 \end{vmatrix}}{-11} = -\frac{5}{11}$$

Problem 1   Solve using Cramer's rule: $3x + 2y = -4$
$-4x + 3y = -10$

● **Three Equations– Three Variables**   Cramer's rule can be generalized completely for any size linear system that has the *same number of variables as equations*. However, it cannot be used to solve systems where the number of variables is not equal to the number of equations. In Theorem 2 we state without proof Cramer's rule for three equations and three variables.

Theorem 2   **Cramer's Rule for Three Equations and Three Variables**

Given the system

$$
\begin{aligned}
a_{11}x + a_{12}y + a_{13}z &= k_1 \\
a_{21}x + a_{22}y + a_{23}z &= k_2 \\
a_{31}x + a_{32}y + a_{33}z &= k_3
\end{aligned}
\quad \text{with} \quad
D = \begin{vmatrix} a_{11} & a_{12} & a_{13} \\ a_{21} & a_{22} & a_{23} \\ a_{31} & a_{32} & a_{33} \end{vmatrix} \neq 0
$$

then

$$
x = \frac{\begin{vmatrix} k_1 & a_{12} & a_{13} \\ k_2 & a_{22} & a_{23} \\ k_3 & a_{32} & a_{33} \end{vmatrix}}{D}
\qquad
y = \frac{\begin{vmatrix} a_{11} & k_1 & a_{13} \\ a_{21} & k_2 & a_{23} \\ a_{31} & k_3 & a_{33} \end{vmatrix}}{D}
\qquad
z = \frac{\begin{vmatrix} a_{11} & a_{12} & k_1 \\ a_{21} & a_{22} & k_2 \\ a_{31} & a_{32} & k_3 \end{vmatrix}}{D}
$$

You can easily remember these determinant formulas for x, y, and z if you observe the following:

1. Determinant D is formed from the coefficients of x, y, and z, keeping the same relative position in the determinant as found in the system of equations.

2. Determinant D appears in the denominators for x, y, and z.

3. The numerator for x can be obtained from D by replacing the coefficients of x (a_{11}, a_{21}, a_{31}) with the constants k_1, k_2, and k_3, respectively. Similar statements can be made for the numerators for y and z.

EXAMPLE 2   **Solving a System with Cramer's Rule**

Solve using Cramer's rule:
$$
\begin{aligned}
x + y \quad\;\; &= 2 \\
3y - z &= -4 \\
x \quad\;\; + z &= 3
\end{aligned}
$$

Solution

$$
D = \begin{vmatrix} 1 & 1 & 0 \\ 0 & 3 & -1 \\ 1 & 0 & 1 \end{vmatrix} = 2
$$

$$
x = \frac{\begin{vmatrix} 2 & 1 & 0 \\ -4 & 3 & -1 \\ 3 & 0 & 1 \end{vmatrix}}{2} = \frac{7}{2}
\qquad
y = \frac{\begin{vmatrix} 1 & 2 & 0 \\ 0 & -4 & -1 \\ 1 & 3 & 1 \end{vmatrix}}{2} = -\frac{3}{2}
$$

$$
z = \frac{\begin{vmatrix} 1 & 1 & 2 \\ 0 & 3 & -4 \\ 1 & 0 & 3 \end{vmatrix}}{2} = -\frac{1}{2}
$$

Problem 2 Solve using Cramer's rule:
$$\begin{aligned} 3x \quad\quad - z &= 5 \\ x - y + z &= 0 \\ x + y \quad\quad &= 1 \end{aligned}$$

In practice, Cramer's rule is rarely used to solve systems of order higher than 2 or 3 by hand, since more efficient methods are available utilizing computer methods. However, Cramer's rule is a valuable tool in more advanced theoretical and applied mathematics.

Answers to Matched Problems
1. $x = \frac{8}{17}, y = -\frac{46}{17}$
2. $x = \frac{6}{5}, y = -\frac{1}{5}, z = -\frac{7}{5}$

EXERCISE 7-7

A

Solve Problems 1–8 using Cramer's rule.

1. $\begin{aligned} x + 2y &= 1 \\ x + 3y &= -1 \end{aligned}$

2. $\begin{aligned} x + 2y &= 3 \\ x + 3y &= 5 \end{aligned}$

3. $\begin{aligned} 2x + y &= 1 \\ 5x + 3y &= 2 \end{aligned}$

4. $\begin{aligned} x + 3y &= 1 \\ 2x + 8y &= 0 \end{aligned}$

5. $\begin{aligned} 2x - y &= -3 \\ -x + 3y &= 3 \end{aligned}$

6. $\begin{aligned} -3x + 2y &= 1 \\ 2x - 3y &= -3 \end{aligned}$

7. $\begin{aligned} 4x - 3y &= 4 \\ 3x + 2y &= -2 \end{aligned}$

8. $\begin{aligned} 5x + 2y &= -1 \\ 2x - 3y &= 2 \end{aligned}$

B

Solve Problems 9–12 to 2 significant digits using Cramer's rule.

9. $\begin{aligned} 0.9925x - 0.9659y &= 0 \\ 0.1219x + 0.2588y &= 2{,}500 \end{aligned}$

10. $\begin{aligned} 0.9877x - 0.9744y &= 0 \\ 0.1564x + 0.2250y &= 1{,}900 \end{aligned}$

11. $\begin{aligned} 0.9954x - 0.9942y &= 0 \\ 0.0958x + 0.1080y &= 155 \end{aligned}$

12. $\begin{aligned} 0.9973x - 0.9957y &= 0 \\ 0.0732x + 0.0924y &= 112 \end{aligned}$

Solve Problems 13–20 using Cramer's rule:

13. $\begin{aligned} x + y \quad\quad &= 0 \\ 2y + z &= -5 \\ -x \quad\quad + z &= -3 \end{aligned}$

14. $\begin{aligned} x + y \quad\quad &= -4 \\ 2y + z &= 0 \\ -x \quad\quad + z &= 5 \end{aligned}$

15. $\begin{aligned} x + y \quad\quad &= 1 \\ 2y + z &= 0 \\ -y + z &= 1 \end{aligned}$

16. $\begin{aligned} x + 3y \quad\quad &= -3 \\ 2y + z &= 3 \\ -x \quad\quad + 3z &= 7 \end{aligned}$

17. $\begin{aligned} 3y + z &= -1 \\ x \quad\quad + 2z &= 3 \\ x - 3y \quad\quad &= -2 \end{aligned}$

18. $\begin{aligned} x \quad\quad - z &= 3 \\ 2x - y \quad\quad &= -3 \\ x + y + z &= 1 \end{aligned}$

19. $\begin{aligned} 2y - z &= -3 \\ x - y - z - 2 \\ x - y + 2z &= 4 \end{aligned}$

20. $\begin{aligned} 2x + y \quad\quad &= 2 \\ x - y + z &= -1 \\ x + y + z &= 2 \end{aligned}$

C

In Problems 21 and 22, use Cramer's rule to solve for x only.

21. $\begin{aligned} 2x - 3y + z &= -3 \\ -4x + 3y + 2z &= -11 \\ x - y - z &= 3 \end{aligned}$

22. $\begin{aligned} x + 4y - 3z &= 25 \\ 3x + y - z &= 2 \\ -4x + y + 2z &= 1 \end{aligned}$

In Problems 23 and 24, use Cramer's rule to solve for y only.

23. $\begin{aligned} 12x - 14y + 11z &= 5 \\ 15x + 7y - 9z &= -13 \\ 5x - 3y + 2z &= 0 \end{aligned}$

24. $\begin{aligned} 2x - y + 4z &= 15 \\ -x + y + 2z &= 5 \\ 3x + 4y - 2z &= 4 \end{aligned}$

In Problems 25 and 26, use Cramer's rule to solve for z only.

25. $\begin{aligned} 3x - 4y + 5z &= 18 \\ -9x + 8y + 7z &= -13 \\ 5x - 7y + 10z &= 33 \end{aligned}$

26. $\begin{aligned} 13x + 11y + 10z &= 2 \\ 10x + 8y + 7z &= 1 \\ 8x + 5y + 4z &= 4 \end{aligned}$

It is clear that $x = 0$, $y = 0$, $z = 0$ is a solution to each of the systems given in Problems 27 and 28. Use Cramer's rule to determine whether this solution is unique. [Hint: If $D \neq 0$, what can you conclude? If $D = 0$, what can you conclude?]

27. $\begin{aligned} x - 4y + 9z &= 0 \\ 4x - y + 6z &= 0 \\ x - y + 3z &= 0 \end{aligned}$

28. $\begin{aligned} 3x - y + 3z &= 0 \\ 5x + 5y - 9z &= 0 \\ -2x + y - 3z &= 0 \end{aligned}$

29. Prove Theorem 1 for y.

Chapter 7 Review

7-1   MATRIX ADDITION; MULTIPLICATION BY A NUMBER

A **matrix** is a rectangular array of real numbers enclosed within brackets. Each number in the array is called an **element** of the matrix. An $m \times n$ **matrix** (read "m by n matrix") is a matrix with m rows and n columns. If a matrix has the same number of rows and columns, it is called a **square matrix**. A matrix with only one column is called a **column matrix**, and a matrix with only one row is called a **row matrix**. Two matrices are **equal** if they have the same dimension and their corresponding elements are equal.

The **sum of two matrices** of the same dimension is a matrix with elements that are the sums of the corresponding elements of the two given matrices. Matrix addition is **commutative** and **associative**. A matrix with all zero elements is called the **zero matrix**. The **negative of a matrix M**, denoted by $-M$, is a matrix with elements that are the negatives of the elements in M. If A and B are matrices of the same dimension, then we define **subtraction** as follows: $A - B = A + (-B)$. The **product of a number k and a matrix M**, denoted by kM, is a matrix formed by multiplying each element of M by k.

7-2   MATRIX MULTIPLICATION

The **dot product** of a $1 \times n$ row matrix and an $n \times 1$ column matrix is a real number given by

$$[a_1 \quad a_2 \quad \cdots \quad a_n] \cdot \begin{bmatrix} b_1 \\ b_2 \\ \vdots \\ b_n \end{bmatrix} = a_1b_1 + a_2b_2 + \cdots + a_nb_n \quad \text{A real number}$$

The dot between the two matrices is important. If the dot is omitted, the multiplication is called a *matrix product*.

The **product of two matrices** A and B is defined only on the assumption that the number of columns in A is equal to the number of rows in M. If A is an $m \times p$ matrix and B is a $p \times n$ matrix, then the matrix product of A and B, denoted by AB, is an $m \times n$ matrix whose element in the ith row and jth column is the dot product of the ith row matrix of A and the jth column matrix of B.

Matrix multiplication is not commutative, and the **zero property does not hold for matrix multiplication**. That is, for matrices A and B, the matrix product AB can be zero without either A or B being the zero matrix. Matrix multiplication does have other **general properties**, some of which are similar to the properties of real numbers. Assuming all products and sums are defined for the indicated matrices A, B, and C, then for k a real number:

1. $A(BC) = (AB)C$ **Associative Property**
2. $A(B + C) = AB + AC$ **Left Distributive Property**
3. $(B + C)A = BA + CA$ **Right Distributive Property**
4. If $A = B$, then $CA = CB$ **Left Multiplication Property**
5. If $A = B$, then $AC = BC$ **Right Multiplication Property**
6. $k(AB) = (kA)B = A(kB)$

7-3   INVERSE OF A SQUARE MATRIX

The **identity matrix for multiplication** for the set of all square matrices of order n is the square matrix of order n, denoted by I, with 1's along the **principal diagonal** (from upper left corner to lower right corner) and 0's elsewhere. If M is a square matrix of order n and I is the identity matrix of order n, then

$$IM = MI = M$$

If M is a square matrix of order n and if there exists a matrix M^{-1} (read "M inverse") such that

$$M^{-1}M = MM^{-1} = I$$

then M^{-1} is called the **multiplicative inverse of M** or, more simply, the **inverse of M**. If the augmented matrix $[M \mid I]$ is transformed by row operations into $[I \mid B]$, then the resulting matrix B is M^{-1}. If, however, we obtain all 0's in one or more rows to the left of the vertical line, then M^{-1} does not exist.

7-4   MATRIX EQUATIONS AND SYSTEMS OF LINEAR EQUATIONS

A system of linear equations with the same number of variables as equations such as

$$a_{11}x_1 + a_{12}x_2 + a_{13}x_3 = k_1$$

$$a_{21}x_1 + a_{22}x_2 + a_{23}x_3 = k_2$$

$$a_{31}x_1 + a_{32}x_2 + a_{33}x_3 = k_3$$

can be written as the matrix equation

$$\overset{A}{\begin{bmatrix} a_{11} & a_{12} & a_{13} \\ a_{21} & a_{22} & a_{23} \\ a_{31} & a_{32} & a_{33} \end{bmatrix}} \overset{X}{\begin{bmatrix} x_1 \\ x_2 \\ x_3 \end{bmatrix}} = \overset{B}{\begin{bmatrix} k_1 \\ k_2 \\ k_3 \end{bmatrix}}$$

If the inverse of A exists, then the matrix equation has a unique solution given by

$$X = A^{-1}B$$

After multiplying B by A^{-1} from the left, it is easy to read the solution to the original system of equations.

7-5 DETERMINANTS

Associated with each square matrix A is a real number called the **determinant** of the matrix. The determinant of A is denoted by **det** A, or simply by writing the array of elements in A using vertical lines in place of square brackets. For example,

$$\det\begin{bmatrix} a_{11} & a_{12} \\ a_{21} & a_{22} \end{bmatrix} = \begin{vmatrix} a_{11} & a_{12} \\ a_{21} & a_{22} \end{vmatrix}$$

A determinant of **order** n is a determinant with n rows and n columns.

The **value of a second-order determinant** is the real number given by

$$\begin{vmatrix} a_{11} & a_{12} \\ a_{21} & a_{22} \end{vmatrix} = a_{11}a_{22} - a_{21}a_{12}$$

The **value of a third-order determinant** is the sum of three products obtained by multiplying each element of any one row (or each element of any one column) by its cofactor. The **cofactor of an element** a_{ij} (from the ith row and jth column) is the product of the minor of a_{ij} and $(-1)^{i+j}$. The **minor of an element** a_{ij} is the determinant remaining after deleting the ith row and jth column. A similar process can be used to evaluate determinants of order higher than 3.

7-6 PROPERTIES OF DETERMINANTS

The use of the following five determinant properties can greatly reduce the effort in evaluating determinants of order 3 or greater:

1. If each element of any row (or column) of a determinant is multiplied by a constant k, the new determinant is k times the original.

$$\begin{vmatrix} 2a & 2b \\ c & d \end{vmatrix} = 2\begin{vmatrix} a & b \\ c & d \end{vmatrix}$$

$$3\begin{vmatrix} a & b \\ c & d \end{vmatrix} = \begin{vmatrix} 3a & b \\ 3c & d \end{vmatrix}$$

2. If every element in a row (or column) is 0, the value of the determinant is 0.

$$\begin{vmatrix} a & b \\ 0 & 0 \end{vmatrix} = 0$$

$$\begin{vmatrix} 0 & b \\ 0 & d \end{vmatrix} = 0$$

3. If two rows (or two columns) of a determinant are interchanged, the new determinant is the negative of the original.

$$\begin{vmatrix} a & b \\ c & d \end{vmatrix} = -\begin{vmatrix} c & d \\ a & b \end{vmatrix}$$

$$\begin{vmatrix} a & b \\ c & d \end{vmatrix} = -\begin{vmatrix} b & a \\ d & c \end{vmatrix}$$

4. If the corresponding elements are equal in two rows (or columns), the value of the determinant is 0.

$$\begin{vmatrix} a & b \\ a & b \end{vmatrix} = 0$$

$$\begin{vmatrix} a & a \\ c & c \end{vmatrix} = 0$$

5. If a multiple of any row (or column) of a determinant is added to any other row (or column), the value of the determinant is not changed.

$$\begin{vmatrix} a & b \\ c & d \end{vmatrix} = \begin{vmatrix} a & b \\ c + ka & d + kb \end{vmatrix}$$

$$\begin{vmatrix} a & b \\ c & d \end{vmatrix} = \begin{vmatrix} a + kb & b \\ c + kd & d \end{vmatrix}$$

7-7   CRAMER'S RULE

Systems of equations having the same number of variables as equations can also be solved using determinants and Cramer's rule. **Cramer's rule for three equations and three variables** is as follows: Given the system

$$a_{11}x + a_{12}y + a_{13}z = k_1$$
$$a_{21}x + a_{22}y + a_{23}z = k_2 \quad \text{with} \quad D = \begin{vmatrix} a_{11} & a_{12} & a_{13} \\ a_{21} & a_{22} & a_{23} \\ a_{31} & a_{32} & a_{33} \end{vmatrix} \neq 0$$
$$a_{31}x + a_{32}y + a_{33}z = k_3$$

then

$$x = \frac{\begin{vmatrix} k_1 & a_{12} & a_{13} \\ k_2 & a_{22} & a_{23} \\ k_3 & a_{32} & a_{33} \end{vmatrix}}{D} \qquad y = \frac{\begin{vmatrix} a_{11} & k_1 & a_{13} \\ a_{21} & k_2 & a_{23} \\ a_{31} & k_3 & a_{33} \end{vmatrix}}{D} \qquad z = \frac{\begin{vmatrix} a_{11} & a_{12} & k_1 \\ a_{21} & a_{22} & k_2 \\ a_{31} & a_{32} & k_3 \end{vmatrix}}{D}$$

Cramer's rule can be generalized completely for any size linear system that has the same number of variables as equations. The formulas are easily remembered if you observe the following:

1. Determinant D is formed from the coefficients of x, y, and z, keeping the same relative position in the determinant as found in the system of equations.

2. Determinant D appears in the denominators for x, y, and z.

3. The numerator for x can be obtained from D by replacing the coefficients of x (a_{11}, a_{21}, a_{31}) with the constants k_1, k_2, and k_3, respectively. Similar statements can be made for the numerators for y and z.

Cramer's rule is rarely used to solve systems of order higher than 3 by hand, since more efficient methods are available. Cramer's rule, however, is a valuable tool in more advanced theoretical and applied mathematics.

Chapter 7 Review Exercise

Work through all the problems in this chapter review and check answers in the back of the book. Answers to all review problems are there, and following each answer is a number in italics indicating the section in which that type of problem is discussed. Where weaknesses show up, review appropriate sections in the text.

A

In Problems 1–9, perform the operations that are defined, given the following matrices:

$$A = \begin{bmatrix} 1 & 2 \\ 3 & 1 \end{bmatrix} \qquad B = \begin{bmatrix} 2 & 1 \\ 1 & 1 \end{bmatrix} \qquad C = \begin{bmatrix} 2 & 3 \end{bmatrix} \qquad D = \begin{bmatrix} 1 \\ 2 \end{bmatrix}$$

1. $A + B$ **2.** $B + D$ **3.** $A - 2B$ **4.** AB **5.** AC

6. AD **7.** DC **8.** $C \cdot D$ **9.** $C + D$

10. Find the inverse of

$$A = \begin{bmatrix} 3 & 2 \\ 4 & 3 \end{bmatrix}$$

Show that $A^{-1}A = I$.

11. Write the system

$$3x_1 + 2x_2 = k_1$$
$$4x_1 + 3x_2 = k_2$$

as a matrix equation, and solve using the inverse found in Problem 10 for:
(A) $k_1 = 3, k_2 = 5$ (B) $k_1 = 7, k_2 = 10$
(C) $k_1 = 4, k_2 = 2$

Evaluate the determinants in Problems 12 and 13.

12. $\begin{vmatrix} 2 & -3 \\ -5 & -1 \end{vmatrix}$

13. $\begin{vmatrix} 2 & 3 & -4 \\ 0 & 5 & 0 \\ 1 & -4 & -2 \end{vmatrix}$

14. Solve the system using Cramer's rule:

$$3x - 2y = 8$$
$$x + 3y = -1$$

B _____

In Problems 15–20, perform the operations that are defined, given the following matrices:

$$A = \begin{bmatrix} 2 & -2 \\ 1 & 0 \\ 3 & 2 \end{bmatrix} \quad B = \begin{bmatrix} -1 \\ 2 \\ 3 \end{bmatrix} \quad C = [2 \quad 1 \quad 3]$$

$$D = \begin{bmatrix} 3 & -2 & 1 \\ -1 & 1 & 2 \end{bmatrix} \quad E = \begin{bmatrix} 3 & -4 \\ -1 & 0 \end{bmatrix}$$

15. $A + D$ **16.** $E + DA$ **17.** $DA - 3E$ **18.** $C \cdot B$ **19.** CB **20.** $AD - BC$

21. Find the inverse of

$$A = \begin{bmatrix} 1 & 2 & 3 \\ 2 & 3 & 4 \\ 1 & 2 & 1 \end{bmatrix}$$

Show that $A^{-1}A = I$.

22. Write the system

$$x_1 + 2x_2 + 3x_3 = k_1$$
$$2x_1 + 3x_2 + 4x_3 = k_2$$
$$x_1 + 2x_2 + x_3 = k_3$$

as a matrix equation, and solve using the inverse found in Problem 21 for:
(A) $k_1 = 1, k_2 = 3, k_3 = 3$
(B) $k_1 = 0, k_2 = 0, k_3 = -2$
(C) $k_1 = -3, k_2 = -4, k_3 = 1$

Evaluate the determinants in Problems 23 and 24.

23. $\begin{vmatrix} -\frac{1}{4} & \frac{3}{2} \\ \frac{1}{2} & \frac{2}{3} \end{vmatrix}$

24. $\begin{vmatrix} 2 & -1 & 1 \\ -3 & 5 & 2 \\ 1 & -2 & 4 \end{vmatrix}$

25. Solve for y only using Cramer's rule:

$$x - 2y + z = -6$$
$$y - z = 4$$
$$2x + 2y + z = 2$$

(Find the numerator and denominator first; then reduce.)

C

26. For $n \times n$ matrices A and C and $n \times 1$ column matrices B and X, solve for X assuming all necessary inverses exist:

$$AX - B = CX.$$

27. Find the inverse of

$$A = \begin{bmatrix} 4 & 5 & 6 \\ 4 & 5 & -6 \\ 1 & 1 & 1 \end{bmatrix}$$

Show that $A^{-1}A = I$.

28. Clear the decimals in the system

$$0.04x_1 + 0.05x_2 + 0.06x_3 = 360$$
$$0.04x_1 + 0.05x_2 - 0.06x_3 = 120$$
$$x_1 + x_2 + x_3 = 7{,}000$$

by multiplying the first two equations by 100. Then write the resulting system as a matrix equation and solve using the inverse found in Problem 27.

29. $\begin{vmatrix} -1 & 4 & 1 & 1 \\ 5 & -1 & 2 & -1 \\ 2 & -1 & 0 & 3 \\ -3 & 3 & 0 & 3 \end{vmatrix} = \; ?$

30. Show that

$$\begin{vmatrix} u & v \\ w & x \end{vmatrix} = \begin{vmatrix} u + kv & v \\ w + kx & x \end{vmatrix}$$

APPLICATIONS

31. **Resource Allocation.** A Colorado mining company operates mines at Big Bend and Saw Pit. The Big Bend mine produces ore that is 5% nickel and 7% copper. The Saw Pit mine produces ore that is 3% nickel and 4% copper. How many tons of ore should be produced at each mine to obtain the amounts of nickel and copper listed in the table? Set up a matrix equation and solve using matrix inverses.

	Nickel	Copper
(A)	3.6 tons	5 tons
(B)	3 tons	4.1 tons
(C)	3.2 tons	4.4 tons

32. **Labor Costs.** A company with manufacturing plants in North and South Carolina has labor-hour and wage requirements for the manufacturing of computer desks and printer stands as given in matrices L and H:

LABOR-HOUR REQUIREMENTS

$$L = \begin{bmatrix} 1.7\ h & 2.4\ h & 0.8\ h \\ 0.9\ h & 1.8\ h & 0.6\ h \end{bmatrix} \begin{array}{l} \text{Desk} \\ \text{Stand} \end{array}$$

with columns Fabricating department, Assembly department, Packaging department.

HOURLY WAGES

$$H = \begin{bmatrix} \$11.50 & \$10.00 \\ \$9.50 & \$8.50 \\ \$5.00 & \$4.50 \end{bmatrix} \begin{array}{l} \text{Fabricating department} \\ \text{Assembly department} \\ \text{Packaging department} \end{array}$$

with columns North Carolina plant, South Carolina plant.

(A) What is the labor cost of producing one printer stand at the South Carolina plant? Set up a dot product and multiply.

(B) Find LH and interpret.

33. **Labor Costs.** The monthly production of computer desks and printer stands for the company in Problem 32 for the months of January and February are given in matrices J and F:

JANUARY PRODUCTION

$$J = \begin{bmatrix} 1,500 & 1,650 \\ 850 & 700 \end{bmatrix} \begin{array}{l} \text{Desks} \\ \text{Stands} \end{array}$$

with columns North Carolina plant, South Carolina plant.

FEBRUARY PRODUCTION

$$F = \begin{bmatrix} 1,700 & 1,810 \\ 930 & 740 \end{bmatrix} \begin{array}{l} \text{Desks} \\ \text{Stands} \end{array}$$

with columns North Carolina plant, South Carolina plant.

(A) Find the average monthly production for the months of January and February.

(B) Find the increase in production from January to February.

(C) Find $J \begin{bmatrix} 1 \\ 1 \end{bmatrix}$ and interpret.

Cumulative Review Exercise
Chapters 6 and 7

Work through all the problems in this cumulative review and check answers in the back of the book. Answers to all review problems are there, and following each answer is a number in italics indicating the section in which that type of problem is discussed. Where weaknesses show up, review appropriate sections in the text.

A

1. Solve using substitution or elimination by addition:

$$3x - 5y = 11$$
$$2x + 3y = 1$$

2. Solve by graphing: $\quad 2x - y = -4$
$$3x + y = -1$$

3. Solve by substitution or elimination by addition:

$$x^2 + y^2 = 2$$
$$2x - y = 1$$

4. Solve by graphing: $\quad 3x + 5y \leq 15$
$$x, y \geq 0$$

5. Find the maximum and minimum value of $z = 2x + 3y$ over the feasible region S:

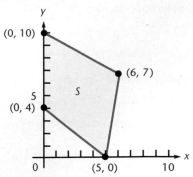

6. Perform the operations that are defined, given the following matrices:

$$M = \begin{bmatrix} 2 & 1 \\ 1 & -3 \end{bmatrix} \quad N = \begin{bmatrix} 1 & 2 \\ -1 & 3 \end{bmatrix}$$

$$P = [1 \quad 2] \quad Q = \begin{bmatrix} -1 \\ 2 \end{bmatrix}$$

(A) $M - 2N$ (B) $P + Q$ (C) $P \cdot Q$
(D) MN (E) PN (F) QM

7. Evaluate: $\begin{vmatrix} 0 & 2 & 0 \\ 1 & 3 & 2 \\ -1 & 4 & 3 \end{vmatrix}$

8. Write the linear system corresponding to each augmented matrix and solve:

(A) $\begin{bmatrix} 1 & 0 & | & 3 \\ 0 & 1 & | & -4 \end{bmatrix}$

(B) $\begin{bmatrix} 1 & -2 & | & 3 \\ 0 & 0 & | & 0 \end{bmatrix}$

(C) $\begin{bmatrix} 1 & -2 & | & 3 \\ 0 & 0 & | & 1 \end{bmatrix}$

9. Given the system: $\begin{aligned} x_1 + x_2 &= 3 \\ -x_1 + x_2 &= 5 \end{aligned}$

(A) Write the augmented matrix for the system.
(B) Transform the augmented matrix into reduced form.
(C) Write the solution to the system.

10. Given the system: $\begin{aligned} x_1 - 3x_2 &= k_1 \\ 2x_1 - 5x_2 &= k_2 \end{aligned}$

(A) Write the system as a matrix equation of the form $AX = B$.
(B) Find the inverse of the coefficient matrix A.
(C) Use A^{-1} to find the solution for $k_1 = -2$ and $k_2 = 1$.
(D) Use A^{-1} to find the solution for $k_1 = 1$ and $k_2 = -2$.

11. Given the system: $\begin{aligned} 2x - 3y &= 1 \\ 4x - 5y &= 2 \end{aligned}$

(A) Find the determinant of the coefficient matrix.
(B) Solve the system using Cramer's rule.

B _____

Solve Problems 12–14 using Gauss–Jordan elimination.

12. $\begin{aligned} x_1 + 2x_2 - x_3 &= 3 \\ x_2 + x_3 &= -2 \\ 2x_1 + 3x_2 + x_3 &= 0 \end{aligned}$

13. $\begin{aligned} x_1 + x_2 - x_3 &= 2 \\ 4x_2 + 6x_3 &= -1 \\ 6x_2 + 9x_3 &= 0 \end{aligned}$

14. $\begin{aligned} x_1 - 2x_2 + x_3 &= 1 \\ 3x_1 - 2x_2 - x_3 &= -5 \end{aligned}$

In Problems 15 and 16, solve each system.

15. $\begin{aligned} x^2 - 3xy + 3y^2 &= 1 \\ xy &= 1 \end{aligned}$

16. $\begin{aligned} x^2 - 3xy + y^2 &= -1 \\ x^2 - xy &= 0 \end{aligned}$

17. Given $M = [1 \quad 2 \quad -1]$ and $N = \begin{bmatrix} 1 \\ -1 \\ 2 \end{bmatrix}$. Find:

(A) $M \cdot N$ (B) MN (C) NM

18. Given

$$L = \begin{bmatrix} 2 & -1 & 0 \\ 1 & 2 & 1 \end{bmatrix} \quad M = \begin{bmatrix} 1 & 2 \\ -1 & 0 \\ 1 & 1 \end{bmatrix} \quad N = \begin{bmatrix} 2 & 1 \\ -1 & 0 \end{bmatrix}$$

Find, if defined: (A) $LM - 2N$ (B) $ML + N$

19. Solve graphically and indicate whether the solution region is bounded or unbounded. Find the coordinates of each corner point.

$$3x + 2y \geq 12$$
$$x + 2y \geq 8$$
$$x, y \geq 0$$

20. Solve the linear programming problem:

Maximize $z = 4x + 9y$

Subject to $x + 2y \leq 14$
$$2x + y \leq 16$$
$$x, y \geq 0$$

21. Given the system: $x_1 + 4x_2 + 2x_3 = k_1$
$$2x_1 + 6x_2 + 3x_3 = k_2$$
$$2x_1 + 5x_2 + 2x_3 = k_3$$

(A) Write the system as a matrix equation in the form $AX = B$.
(B) Find the inverse of the coefficient matrix A.
(C) Use A^{-1} to solve the system when $k_1 = -1$, $k_2 = 2$, and $k_3 = 1$.
(D) Use A^{-1} to solve the system when $k_1 = 2, k_2 = 0$, and $k_3 = -1$.

22. Given the system: $x_1 + 2x_2 - x_3 = 1$
$$2x_1 + 8x_2 + x_3 = -2$$
$$-x_1 + 3x_2 + 5x_3 = 2$$

(A) Evaluate the coefficient determinant D.
(B) Solve for z using Cramer's rule.

C

23. Which of the following augmented matrices are in reduced form?

$$L = \begin{bmatrix} 1 & 0 & 0 & | & 2 \\ 0 & 1 & 0 & | & 0 \\ 0 & 0 & 1 & | & -1 \end{bmatrix} \quad M = \begin{bmatrix} 1 & 0 & 3 & | & 3 \\ 0 & 1 & -2 & | & 2 \\ 0 & 0 & 0 & | & 0 \end{bmatrix}$$

$$N = \begin{bmatrix} 0 & 0 & | & 0 \\ 1 & 0 & | & 2 \\ 0 & 1 & | & -3 \end{bmatrix} \quad P = \begin{bmatrix} 1 & 2 & 0 & 2 & | & -2 \\ 0 & 0 & 1 & 3 & | & 1 \end{bmatrix}$$

24. Show that

$$k \begin{vmatrix} a & b \\ c & d \end{vmatrix} = \begin{vmatrix} ka & b \\ kc & d \end{vmatrix}$$

25. Show that

$$\begin{vmatrix} a & b \\ c & d \end{vmatrix} = \begin{vmatrix} a & b \\ c + ka & d + kb \end{vmatrix}$$

26. If $M = \begin{vmatrix} a & b \\ c & d \end{vmatrix}$ and $\det M \neq 0$, show that

$$M^{-1} = \frac{1}{\det M} \begin{bmatrix} d & -b \\ -c & a \end{bmatrix}$$

APPLICATIONS

27. Finance. An investor has $12,000 to invest. If part is invested at 8% and the rest in a higher-risk investment at 14%, how much should be invested at each rate to produce the same yield as if all had been invested at 10%?

28. Diet. In an experiment involving mice, a zoologist needs a food mix that contains, among other things, 23 grams of protein, 6.2 grams of fat, and 16 grams of moisture. She has on hand mixes of the following compositions: Mix A contains 20% protein, 2% fat, and 15% moisture;

mix B contains 10% protein, 6% fat, and 10% moisture; and mix C contains 15% protein, 5% fat, and 5% moisture. How many grams of each mix should be used to get the desired diet mix?

29. **Geometry.** Find the dimensions of a rectangle with perimeter 24 meters and area 32 square meters.

30. **Manufacturing.** A manufacturer makes two types of day packs, a standard model and a deluxe model. Each standard model requires 0.5 labor-hour from the fabricating department and 0.3 labor-hour from the sewing department. Each deluxe model requires 0.5 labor-hour from the fabricating department and 0.6 labor-hour from the sewing department. The maximum number of labor-hours available per week in the fabricating department and the sewing department are 300 and 240, respectively. The company makes a profit of $8 on each standard model and $12 on each deluxe model. How many of each type of day pack should be produced per week to maximize the weekly profit, assuming all the packs that are produced can be sold? What is the maximum profit?

Sequences and Series

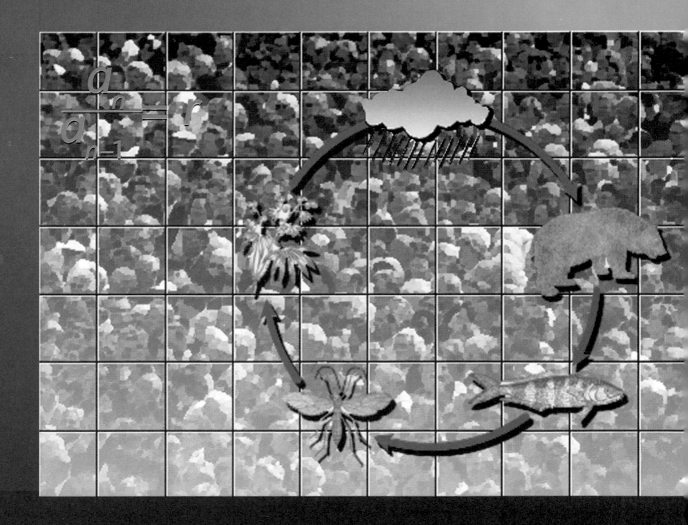

$$\frac{a_n}{a_{n-1}} = r$$

If someone asked you to list all natural numbers that are perfect squares, you might begin by writing

$$1, \quad 4, \quad 9, \quad 16, \quad 25, \quad 36$$

But you would soon realize that it is impossible to actually list all the perfect squares, since there are an infinite number of them. However, this collection of numbers can be represented in several different ways. One common method is to write

$$1, 4, 9, \ldots, n^2, \ldots \qquad n \in N$$

where N is the set of natural numbers. A list of numbers such as this is generally called a *sequence*. Sequences and related topics form the subject matter of this chapter. One of the related topics involves a method of proof we have referred to several times earlier in this book—*mathematical induction*. This method enables us to prove conjectures involving infinite sets of successive integers.

SECTION 8-1   Sequences and Series

- Sequences
- Series

In this section we introduce special notation and formulas for representing and generating sequences and sums of sequences.

● Sequences   Consider the function f given by

$$f(n) = 2n - 1 \tag{1}$$

where the domain of f is the set of natural numbers N. Note that

$$f(1) = 1, f(2) = 3, f(3) = 5, \ldots$$

The function f is an example of a sequence. A **sequence** is a function with the domain a set of successive integers. However, we hardly ever see a sequence represented in the form of equation (1). A special notation for sequences has evolved, which we describe here.

To start, the range value $f(n)$ is usually symbolized more compactly with a symbol such as a_n. Thus, in place of equation (1) we write

$$a_n = 2n - 1$$

The domain is understood to be the set of natural numbers N unless stated to the contrary or the context indicates otherwise. The elements in the range are called **terms of the sequence**: a_1 is the first term, a_2 the second term, and a_n the nth term, or the **general term**:

$$a_1 = 2(1) - 1 = 1 \quad \text{First term}$$

$$a_2 = 2(2) - 1 = 3 \quad \text{Second term}$$

$$a_3 = 2(3) - 1 = 5 \quad \text{Third term}$$
$$\vdots \qquad\qquad \vdots$$

When the terms in a sequence are written in their natural order with respect to domain values

$$a_1, a_2, a_3, \ldots, a_n, \ldots$$

or

$$1, 3, 5, \ldots, 2n - 1, \ldots$$

this ordered list of elements is often informally referred to as a sequence. A sequence is also represented in the abbreviated form $\{a_n\}$, where a symbol for the nth term is placed between braces. For example, we can refer to the sequence

$$1, 3, 5, \ldots, 2n - 1, \ldots$$

as the sequence $\{2n - 1\}$.

If the domain of a function is a finite set of successive integers, then the sequence is called a **finite sequence**. If the domain is an infinite set of successive integers, then the sequence is called an **infinite sequence**. The sequence $\{2n - 1\}$ above is an example of an infinite sequence.

Some sequences are specified by a **recursion formula**—that is, a formula that defines each term in terms of one or more preceding terms. The sequence we have chosen to illustrate a recursion formula is a very famous sequence in the history of mathematics called the **Fibonacci sequence**. It is named after the most celebrated mathematician of the thirteenth century, Leonardo Fibonacci from Italy (1180?–1250?).

EXAMPLE 1   Fibonacci Sequence

List the first six terms of the sequence specified by

$$a_1 = 1$$

$$a_2 = 1$$

$$a_n = a_{n-1} + a_{n-2} \qquad n \geq 3$$

Solution

$$a_1 \;\; = 1$$

$$a_2 \;\; = 1$$

$$a_3 \;\; = a_2 + a_1 = 1 + 1 \;\; = 2$$

$$a_4 \;\; = a_3 + a_2 = 2 + 1 \;\; = 3$$

$$a_5 \;\; = a_4 + a_3 = 3 + 2 \;\; = 5$$

$$a_6 \;\; = a_5 + a_4 = 5 + 3 \;\; = 8$$

The formula $a_n = a_{n-1} + a_{n-2}$ is a recursion formula that can be used to generate the terms of a sequence in terms of preceding terms. Of course, starting terms a_1 and a_2 must be provided in order to use the formula. Recursion formulas are particularly suitable for use with calculators and computers (see Problems 53 and 54 in Exercise 8-1).

Problem 1   List the first five terms of the sequence specified by

$$a_1 = 4$$

$$a_n = \tfrac{1}{2}a_{n-1} \qquad n \ge 2$$

Now we consider the reverse problem. That is, can a sequence be defined just by listing the first three or four terms of the sequence? And can we then use these initial terms to find a formula for the nth term? In general, without other information, the answer to the first question is no. Many different sequences may start off with the same terms. For example, each of the following sequences starts off with the same three terms:

$$1, 3, 9, \ldots, 3^{n-1}, \ldots$$

$$1, 3, 9, \ldots, 1 + 2(n - 1)^2, \ldots$$

$$1, 3, 9, \ldots, 8n + \frac{12}{n} - 19, \ldots$$

However, these are certainly different sequences. You should verify that these sequences agree for the first three terms and differ in the fourth term by evaluating the general term for each sequence at $n = 1, 2, 3, 4$. Thus, simply listing the first three terms, or any other finite number of terms, does not specify a particular sequence. In fact, it can be shown that given any list of m numbers, there are an infinite number of sequences whose first m terms agree with these given numbers. What about the second question? That is, given a few terms, can we find the general formula for at least one sequence whose first few terms agree with the given terms? The answer to this question is a qualified yes. If we can observe a simple pattern in the given terms, then we may be able to construct a general term that will produce the pattern. The next example illustrates this approach.

EXAMPLE 2   Finding the General Term of a Sequence

Find the general term of a sequence whose first four terms are:

(A) $5, 6, 7, 8, \ldots$ (B) $2, -4, 8, -16, \ldots$

Solutions   (A) Since these terms are consecutive integers, one solution is $a_n = n, n \geq 5$. If we want the domain of the sequence to be all natural numbers, then another solution is $b_n = n + 4$.

(B) Each of these terms can be written as the product of a power of 2 and a power of -1:

$$2 = (-1)^0 2^1$$
$$-4 = (-1)^1 2^2$$
$$8 = (-1)^2 2^3$$
$$-16 = (-1)^3 2^4$$

If we choose the domain to be all natural numbers, then a solution is

$$a_n = (-1)^{n-1} 2^n$$

Problem 2   Find the general term of a sequence whose first four terms are:

(A) $2, 4, 6, 8, \ldots$ (B) $1, -\frac{1}{2}, \frac{1}{4}, -\frac{1}{8}, \ldots$

In general, there is usually more than one way of representing the nth term of a given sequence. This was seen in the solution of Example 2A. However, unless stated to the contrary, we assume the domain of the sequence is the set of natural numbers N.

● Series   The sum of the terms of a sequence is called a **series**. If the sequence is finite, the corresponding series is a **finite series**. If the sequence is infinite, the corresponding series is an **infinite series**. For example,

$$1, 2, 4, 8, 16 \qquad \text{Finite sequence}$$
$$1 + 2 + 4 + 8 + 16 \qquad \text{Finite series}$$

We restrict our discussion to finite series in this section.

Series are often represented in a compact form called **summation notation** using the symbol $\sum$, which is a stylized version of the Greek letter sigma. Consider the following examples:

$$\sum_{k=1}^{4} a_k = a_1 + a_2 + a_3 + a_4$$

$$\sum_{k=3}^{7} b_k = b_3 + b_4 + b_5 + b_6 + b_7$$

$$\sum_{k=0}^{n} c_k = c_0 + c_1 + c_2 + \cdots + c_n \qquad \text{Domain is the set of integers greater than or equal to 0.}$$

The terms on the right are obtained from the expression on the left by successively replacing the **summing index** k with integers, starting with the first number indicated below $\sum$ and ending with the number that appears above $\sum$. Thus, for example, if we are given the sequence

$$\frac{1}{2}, \frac{1}{4}, \frac{1}{8}, \ldots, \frac{1}{2^n}$$

the corresponding series is

$$\frac{1}{2} + \frac{1}{4} + \frac{1}{8} + \cdots + \frac{1}{2^n}$$

or, more compactly,

$$\sum_{k=1}^{n} \frac{1}{2^k}$$

EXAMPLE 3   **Writing the Terms of a Series**

Write without summation notation: $\displaystyle\sum_{k=1}^{5} \frac{k-1}{k}$

Solution

$$\sum_{k=1}^{5} \frac{k-1}{k} = \frac{1-1}{1} + \frac{2-1}{2} + \frac{3-1}{3} + \frac{4-1}{4} + \frac{5-1}{5}$$

$$= 0 + \frac{1}{2} + \frac{2}{3} + \frac{3}{4} + \frac{4}{5}$$

Problem 3   Write without summation notation: $\displaystyle\sum_{k=0}^{5} \frac{(-1)^k}{2k+1}$

If the terms of a series are alternately positive and negative, we call the series an **alternating series**. Example 4 deals with the representation of such a series.

EXAMPLE 4   Writing a Series in Summation Notation

Write the following series using summation notation:

$$1 - \frac{1}{2} + \frac{1}{3} - \frac{1}{4} + \frac{1}{5} - \frac{1}{6}$$

(A) Start the summing index at $k = 1$.
(B) Start the summing index at $k = 0$.

Solutions   (A) $(-1)^{k-1}$ provides the alternation of sign, and $1/k$ provides the other part of each term. Thus, we can write

$$\sum_{k=1}^{6} \frac{(-1)^{k-1}}{k}$$

as can be easily checked.

(B) $(-1)^k$ provides the alternation of sign, and $1/(k + 1)$ provides the other part of each term. Thus, we write

$$\sum_{k=0}^{5} \frac{(-1)^k}{k + 1}$$

as can be checked.

Problem 4   Write the following series using summation notation:

$$1 - \frac{2}{3} + \frac{4}{9} - \frac{8}{27} + \frac{16}{81}$$

(A) Start with $k = 1$.   (B) Start with $k = 0$.

Answers to Matched Problems
1. $4, 2, 1, \frac{1}{2}, \frac{1}{4}$ 2. (A) $a_n = 2n$ (B) $a_n = (-1)^{n-1}(\frac{1}{2})^{n-1}$ 3. $1 - \frac{1}{3} + \frac{1}{5} - \frac{1}{7} + \frac{1}{9} - \frac{1}{11}$
4. (A) $\sum_{k=1}^{5} (-1)^{k-1}\left(\frac{2}{3}\right)^{k-1}$ (B) $\sum_{k=0}^{4} (-1)^k\left(\frac{2}{3}\right)^k$

EXERCISE 8-1

A

Write the first four terms for each sequence in Problems 1–6.

1. $a_n = n - 2$

2. $a_n = n + 3$

3. $a_n = \frac{n - 1}{n + 1}$

4. $a_n = \left(1 + \frac{1}{n}\right)^n$

5. $a_n = (-2)^{n+1}$

6. $a_n = \frac{(-1)^{n+1}}{n^2}$

7. Write the eighth term in the sequence in Problem 1.

8. Write the tenth term in the sequence in Problem 2.

9. Write the one-hundredth term in the sequence in Problem 3.

10. Write the two-hundredth term in the sequence in Problem 4.

In Problems 11–16, write each series in expanded form without summation notation.

11. $\displaystyle\sum_{k=1}^{5} k$

12. $\displaystyle\sum_{k=1}^{4} k^2$

13. $\displaystyle\sum_{k=1}^{3} \frac{1}{10^k}$

14. $\displaystyle\sum_{k=1}^{5} \left(\frac{1}{3}\right)^k$

15. $\displaystyle\sum_{k=1}^{4} (-1)^k$

16. $\displaystyle\sum_{k=1}^{6} (-1)^{k+1}k$

B

Write the first five terms of each sequence in Problems 17–26.

17. $a_n = (-1)^{n+1}n^2$

18. $a_n = (-1)^{n+1}\left(\dfrac{1}{2^n}\right)$

19. $a_n = \dfrac{1}{3}\left(1 - \dfrac{1}{10^n}\right)$

20. $a_n = n[1 - (-1)^n]$

21. $a_n = (-\tfrac{1}{2})^{n-1}$

22. $a_n = (-\tfrac{3}{2})^{n-1}$

23. $a_1 = 7; a_n = a_{n-1} - 4, n \geq 2$

24. $a_1 = a_2 = 1; a_n = a_{n-1} + a_{n-2}, n \geq 3$

25. $a_1 = 4; a_n = \tfrac{1}{4}a_{n-1}, n \geq 2$

26. $a_1 = 2; a_n = 2a_{n-1}, n \geq 2$

In Problems 27–38, find the general term of a sequence whose first four terms are given.

27. $4, 5, 6, 7, \ldots$

28. $-2, -1, 0, 1, \ldots$

29. $3, 6, 9, 12, \ldots$

30. $-2, -4, -6, -8, \ldots$

31. $\tfrac{1}{2}, \tfrac{2}{3}, \tfrac{3}{4}, \tfrac{4}{5}, \ldots$

32. $\tfrac{1}{2}, \tfrac{3}{4}, \tfrac{5}{6}, \tfrac{7}{8}, \ldots$

33. $1, -1, 1, -1, \ldots$

34. $1, -2, 3, -4, \ldots$

35. $-2, 4, -8, 16, \ldots$

36. $1, -3, 5, -7, \ldots$

37. $x, \dfrac{x^2}{2}, \dfrac{x^3}{3}, \dfrac{x^4}{4}, \ldots$

38. $x, -x^3, x^5, -x^7, \ldots$

In Problems 39–44, write each series in expanded form without summation notation.

39. $\displaystyle\sum_{k=1}^{4} \frac{(-2)^{k+1}}{k}$

40. $\displaystyle\sum_{k=1}^{5} (-1)^{k+1}(2k - 1)^2$

41. $\displaystyle\sum_{k=1}^{3} \frac{1}{k}x^{k+1}$

42. $\displaystyle\sum_{k=1}^{5} x^{k-1}$

43. $\displaystyle\sum_{k=1}^{5} \frac{(-1)^{k+1}}{k}x^k$

44. $\displaystyle\sum_{k=0}^{4} \frac{(-1)^k x^{2k+1}}{2k + 1}$

In Problems 45–52, write each series using summation notation with the summing index k starting at $k = 1$.

45. $1^2 + 2^2 + 3^2 + 4^2$

46. $2 + 3 + 4 + 5 + 6$

47. $\dfrac{1}{2} + \dfrac{1}{2^2} + \dfrac{1}{2^3} + \dfrac{1}{2^4} + \dfrac{1}{2^5}$

48. $1 - \dfrac{1}{2} + \dfrac{1}{3} - \dfrac{1}{4}$

49. $1 + \dfrac{1}{2^2} + \dfrac{1}{3^2} + \cdots + \dfrac{1}{n^2}$

50. $2 + \dfrac{3}{2} + \dfrac{4}{3} + \cdots + \dfrac{n + 1}{n}$

51. $1 - 4 + 9 - \cdots + (-1)^{n+1}n^2$

52. $\dfrac{1}{2} - \dfrac{1}{4} + \dfrac{1}{8} - \cdots + \dfrac{(-1)^{n+1}}{2^n}$

C _____

The sequence

$$a_n = \frac{a_{n-1}^2 + M}{2a_{n-1}} \qquad n \geq 2, M \text{ a positive real number}$$

can be used to find $\sqrt{M}$ to any decimal-place accuracy desired. To start the sequence, choose a_1 arbitrarily from the positive real numbers. Problems 53 and 54 are related to this sequence.

53. (A) Find the first four terms of the sequence

$$a_1 = 3 \qquad a_n = \frac{a_{n-1}^2 + 2}{2a_{n-1}} \qquad n \geq 2$$

   (B) Compare the terms with $\sqrt{2}$ from a calculator.
   (C) Repeat parts (A) and (B) letting a_1 be any other positive number, say 1.

54. (A) Find the first four terms of the sequence

$$a_1 = 2 \qquad a_n = \frac{a_{n-1}^2 + 5}{2a_{n-1}} \qquad n \geq 2$$

   (B) Find $\sqrt{5}$ with a calculator, and compare with the results of part (A).
   (C) Repeat parts (A) and (B) letting a_1 be any other positive number, say 3.

∫ *In calculus, it can be shown that*

$$e^x = \sum_{k=0}^{\infty} \frac{x^k}{k!} \approx 1 + \frac{x}{1!} + \frac{x^2}{2!} + \frac{x^3}{3!} + \cdots + \frac{x^n}{n!}$$

where the larger n is, the better the approximation. Problem 55 and 56 refer to this series. Note that n!, read "n factorial", is defined by $0! = 1$ and $n! = 1 \cdot 2 \cdot 3 \cdots \cdot n$ for $n \in N$.

55. Approximate $e^{0.2}$ using the first five terms of the series. Compare this approximation with your calculator evaluation of $e^{0.2}$.

56. Approximate $e^{-0.5}$ using the first five terms of the series. Compare this approximation with your calculator evaluation of $e^{-0.5}$.

57. Show that: $\displaystyle\sum_{k=1}^{n} ca_k = c \sum_{k=1}^{n} a_k$

58. Show that: $\displaystyle\sum_{k=1}^{n} (a_k + b_k) = \sum_{k=1}^{n} a_k + \sum_{k=1}^{n} b_k$

_____

SECTION 8-2 Mathematical Induction

- Introduction
- Mathematical Induction
- Additional Examples of Mathematical Induction
- Three Famous Problems

• Introduction

In common usage, the word "induction" means the generalization from particular cases or facts. The ability to formulate general hypotheses from a limited number of facts is a distinguishing characteristic of a creative mathematician. The creative process does not stop here, however. These hypotheses must then be proved or disproved. In mathematics, a special method of proof called **mathematical induction** ranks among the most important basic tools in a mathematician's toolbox. In this section we will use this method to prove a variety of mathematical statements, some new and some that up to now we have just assumed to be true.

We illustrate the process of formulating hypotheses by an example. Suppose we are interested in the sum of the first n consecutive odd integers, where n is a positive integer. We begin by writing the sums for the first few values of n to see if we can observe a pattern:

$$1 = 1 \qquad n = 1$$
$$1 + 3 = 4 \qquad n = 2$$
$$1 + 3 + 5 = 9 \qquad n = 3$$
$$1 + 3 + 5 + 7 = 16 \qquad n = 4$$
$$1 + 3 + 5 + 7 + 9 = 25 \qquad n = 5$$

Is there any pattern to the sums 1, 4, 9, 16, and 25? You no doubt observed that each is a perfect square and, in fact, each is the square of the number of terms in the sum. Thus, the following conjecture seems reasonable:

Conjecture P_n: For each positive integer n,

$$1 + 3 + 5 + \cdots + (2n - 1) = n^2$$

That is, the sum of the first n odd integers is n^2 for each positive integer n.

So far we have used ordinary induction to generalize the pattern observed in the first few cases listed above. But at this point conjecture P_n is simply that—a conjecture. How do we prove that P_n is a true statement? Continuing to list specific cases will never provide a general proof—not in your lifetime or all your descendants' lifetimes! Mathematical induction is the tool we will use to establish the validity of conjecture P_n.

Before we discuss this method of proof, let's consider another conjecture:

Conjecture Q_n: For each positive integer n, the number $n^2 - n + 41$ is a prime number.

It is important to recognize that a conjecture can be proved false if it fails for only one case. A single case or example for which a conjecture fails is called a **counterexample**. We check the conjecture for a few particular cases in Table 1. From the table, it certainly appears that conjecture Q_n has a good chance of being true. You may want to check a few more cases. If you persist, you will find that conjecture Q_n is true for n up to 41. What happens at $n = 41$?

$$41^2 - 41 + 41 = 41^2$$

which is not prime. Thus, since $n = 41$ provides a counterexample, conjecture Q_n is false. Here we see the danger of generalizing without proof from a few special cases. This example was discovered by Euler (1707–1783).

TABLE 1

n	$n^2 - n + 41$	Prime?
1	41	Yes
2	43	Yes
3	47	Yes
4	53	Yes
5	61	Yes

● **Mathematical Induction**

We begin by stating the *principle of mathematical induction*, which forms the basis for all our work in this section.

Theorem 1

Principle of Mathematical Induction

Let P_n be a statement associated with each positive integer n, and suppose the following conditions are satisfied:

1. P_1 is true.

2. For any positive integer k, if P_k is true, then P_{k+1} is also true.

Then the statement P_n is true for all positive integers n.

Theorem 1 must be read very carefully. At first glance, it seems to say that if we assume a statement is true, then it is true. But that is not the case at all. If the two conditions in Theorem 1 are satisfied, then we can reason as follows:

P_1 is true.	Condition 1
P_2 is true, because P_1 is true.	Condition 2
P_3 is true, because P_2 is true.	Condition 2
P_4 is true, because P_3 is true.	Condition 2
$\vdots$	$\vdots$

Since this chain of implications never ends, we will eventually reach P_n for any positive integer n.

To help visualize this process, picture a row of dominoes that goes on forever (see Figure 1) and interpret the conditions in Theorem 1 as follows: Condition 1 says that the first domino can be pushed over. Condition 2 says that if the kth domino falls, then so does the $(k + 1)$st domino. Together, these two conditions imply that all the dominoes must fall.

Now, to illustrate the process of proof by mathematical induction, we return to the conjecture P_n discussed earlier, which we restate below:

$$P_n: \quad 1 + 3 + 5 + \cdots + (2n - 1) = n^2 \qquad n \text{ any positive integer}$$

We already know that P_1 is a true statement. In fact, we demonstrated that P_1 through P_5 are all true by direct calculation. Thus, condition 1 in Theorem 1 is satisfied. To show that condition 2 is satisfied, we assume that P_k is a true statement:

$$P_k: \quad 1 + 3 + 5 + \cdots + (2k - 1) = k^2$$

Condition 1: The first domino can be pushed over.

(a)

Condition 2: If the kth domino falls, then so does the $(k + 1)$ st.

(b)

Conclusion: All the dominoes will fall.

(c)

FIGURE 1 Interpreting mathematical induction.

Now we must show that this assumption implies that P_{k+1} is also a true statement:

$$P_{k+1}: 1 + 3 + 5 + \cdots + (2k - 1) + (2k + 1) = (k + 1)^2$$

Since we have assumed that P_k is true, we can perform operations on this equation. We note that the left side of P_{k+1} is the left side of P_k plus $(2k + 1)$. So we start by adding $(2k + 1)$ to both sides of P_k:

$$1 + 3 + 5 + \cdots + (2k - 1) = k^2 \qquad P_k$$

$$1 + 3 + 5 + \cdots + (2k - 1) + (2k + 1) = k^2 + (2k + 1) \qquad \text{Add } 2k + 1 \text{ to both sides.}$$

Factoring the right side of this equation, we have

$$1 + 3 + 5 + \cdots + (2k - 1) + (2k + 1) = (k + 1)^2 \quad {}_{P_{k+1}}$$

But this last equation is P_{k+1}. Thus, we have started with P_k, the statement we assumed true, and performed valid operations to produce P_{k+1}, the statement we want to be true. In other words, we have shown that if P_k is true, then P_{k+1} is also true. Since we have shown that both conditions in Theorem 1 are satisfied, we conclude that P_n is true for all positive integers n.

• **Additional Examples of Mathematical Induction**

Now we will consider some additional examples of proof by induction. The first is another summation formula. Mathematical induction is the primary tool for proving that formulas of this type are true.

EXAMPLE 1   Proving a Summation Formula

Prove that for all positive integers n

$$\frac{1}{2} + \frac{1}{4} + \frac{1}{8} + \cdots + \frac{1}{2^n} = \frac{2^n - 1}{2^n}$$

Proof   State the conjecture:

$$P_n: \quad \frac{1}{2} + \frac{1}{4} + \frac{1}{8} + \cdots + \frac{1}{2^n} = \frac{2^n - 1}{2^n}$$

Part 1   Show that P_1 is true.

$$P_1: \quad \frac{1}{2} = \frac{2^1 - 1}{2^1}$$

$$= \frac{1}{2}$$

Thus, P_1 is true.

Part 2 Show that if P_k is true, then P_{k+1} is true. It is a good practice to always write out both P_k and P_{k+1} at the beginning of any induction proof to see what is assumed and what must be proved:

$$P_k: \quad \frac{1}{2} + \frac{1}{4} + \frac{1}{8} + \cdots + \frac{1}{2^k} = \frac{2^k - 1}{2^k} \qquad \text{We assume } P_k \text{ is true.}$$

$$P_{k+1}: \quad \frac{1}{2} + \frac{1}{4} + \frac{1}{8} + \cdots + \frac{1}{2^k} + \frac{1}{2^{k+1}} = \frac{2^{k+1} - 1}{2^{k+1}} \qquad \text{We must show that } P_{k+1} \text{ follows from } P_k.$$

We start with the true statement P_k, add $1/2^{k+1}$ to both sides, and simplify the right side:

$$\frac{1}{2} + \frac{1}{4} + \frac{1}{8} + \cdots + \frac{1}{2^k} = \frac{2^k - 1}{2^k} \qquad P_k$$

$$\frac{1}{2} + \frac{1}{4} + \frac{1}{8} + \cdots + \frac{1}{2^k} + \frac{1}{2^{k+1}} = \frac{2^k - 1}{2^k} + \frac{1}{2^{k+1}}$$

$$= \frac{2^k - 1}{2^k} \cdot \frac{2}{2} + \frac{1}{2^{k+1}}$$

$$= \frac{2^{k+1} - 2 + 1}{2^{k+1}}$$

$$= \frac{2^{k+1} - 1}{2^{k+1}}$$

Thus,

$$\frac{1}{2} + \frac{1}{4} + \frac{1}{8} + \cdots + \frac{1}{2^k} + \frac{1}{2^{k+1}} = \frac{2^{k+1} - 1}{2^{k+1}} \qquad P_{k+1}$$

and we have shown that if P_k is true, then P_{k+1} is true.

Conclusion Both conditions in Theorem 1 are satisfied. Thus, P_n is true for all positive integers n.

Problem 1 Prove that for all positive integers n

$$1 + 2 + 3 + \cdots + n = \frac{n(n + 1)}{2}$$

The next example provides a proof of a law of exponents that previously we had to assume was true. First we redefine a^n for n a positive integer, using a recursion formula:

DEFINITION 1	**Recursive Definition of a^n**

For n a positive integer

$$a^1 = a$$
$$a^{n+1} = a^n a \qquad n > 1$$

EXAMPLE 2 **Proving a Law of Exponents**

Prove that $(xy)^n = x^n y^n$ for all positive integers n.

Proof   State the conjecture:

$$P_n: \quad (xy)^n = x^n y^n$$

Part 1   Show that P_1 is true.

$$(xy)^1 = xy \qquad \text{Definition 1}$$
$$= x^1 y^1 \qquad \text{Definition 1}$$

Thus, P_1 is true.

Part 2   Show that if P_k is true, then P_{k+1} is true.

$$P_k: \quad (xy)^k = x^k y^k \qquad \text{Assume } P_k \text{ is true.}$$
$$P_{k+1}: \quad (xy)^{k+1} = x^{k+1} y^{k+1} \qquad \text{Show that } P_{k+1} \text{ follows from } P_k.$$

Here we start with the left side of P_{k+1} and use P_k to find the right side of P_{k+1}:

$$(xy)^{k+1} = (xy)^k (xy)^1 \qquad \text{Definition 1}$$
$$= x^k y^k xy \qquad \text{Use } P_k: \ (xy)^k = x^k y^k$$
$$= (x^k x)(y^k y) \qquad \text{Property of real numbers}$$
$$= x^{k+1} y^{k+1} \qquad \text{Definition 1}$$

Thus, $(xy)^{k+1} = x^{k+1} y^{k+1}$, and we have shown that if P_k is true, then P_{k+1} is true.

Conclusion   Both conditions in Theorem 1 are satisfied. Thus, P_n is true for all positive integers n.

Problem 2   Prove that $(x/y)^n = x^n / y^n$ for all positive integers n.

Our last example deals with factors of integers. Before we start, recall that an integer p is *divisible* by an integer q if $p = qr$ for some integer r.

EXAMPLE 3   **Proving a Divisibility Property**

Prove that $4^{2n} - 1$ is divisible by 5 for all positive integers n.

Proof   Use the definition of divisibility to state the conjecture as follows:

$$P_n: \quad 4^{2n} - 1 = 5r \qquad \text{for some integer } r$$

Part 1   Show that P_1 is true.

$$P_1: \quad 4^2 - 1 = 15 = 5 \cdot 3$$

Thus, P_1 is true.

Part 2   Show that if P_k is true, then P_{k+1} is true.

$$P_k: \quad 4^{2k} - 1 = 5r \qquad \text{for some integer } r \qquad \text{Assume } P_k \text{ is true.}$$

$$P_{k+1}: \quad 4^{2(k+1)} - 1 = 5s \qquad \text{for some integer } s \qquad \text{Show that } P_{k+1} \text{ must follow.}$$

As before, we start with the true statement P_k:

$$4^{2k} - 1 = 5r \qquad\qquad P_k$$

$$4^2(4^{2k} - 1) = 4^2(5r) \qquad\qquad \text{Multiply both sides by } 4^2.$$

$$4^{2k+2} - 16 = 80\,r \qquad\qquad \text{Simplify.}$$

$$4^{2(k+1)} - 1 = 80r + 15 \qquad\qquad \text{Add 15 to both sides.}$$

$$= 5(16r + 3) \qquad\qquad \text{Factor out 5.}$$

Thus,

$$4^{2(k+1)} - 1 = 5s \qquad\qquad P_{k+1}$$

where $s = 16r + 3$ is an integer, and we have shown that if P_k is true, then P_{k+1} is true.

Conclusion   Both conditions in Theorem 1 are satisfied. Thus, P_n is true for all positive integers n.

Problem 3   Prove that $8^n - 1$ is divisible by 7 for all positive integers n.

In some cases, a conjecture may be true only for $n \geq m$, where m is a positive integer, rather than for all $n \geq 0$. For example, see Problems 45 and 46 in Exercise

8-2. The principle of mathematical induction can be extended to cover cases like this as follows:

Theorem 2 **Extended Principle of Mathematical Induction**

Let m be a positive integer, let P_n be a statement associated with each integer $n \geq m$, and suppose the following conditions are satisfied:

1. P_m is true.

2. For any integer $k \geq m$, if P_k is true, then P_{k+1} is also true.

Then the statement P_n is true for all integers $n \geq m$.

● **Three Famous Problems**

We conclude this section by stating three famous problems.

1. Each positive integer can be expressed as the sum of four or fewer squares of positive integers.

2. **Fermat's Last Theorem, 1637:** For $n > 2$, $x^n + y^n = z^n$ does not have solutions in the natural numbers.

3. **Goldbach's Problem, 1742:** Every positive even integer greater than 2 is the sum of two prime numbers.

The first problem was considered by the early Greeks and finally proved in 1772 by Lagrange. No one has been able to prove or disprove the second and third problems, either by mathematical induction or by other methods.

Answers to Matched Problems

1. Sketch of proof. State the conjecture: P_n: $\quad 1 + 2 + 3 + \cdots + n = \dfrac{n(n + 1)}{2}$

Part 1. $1 = \dfrac{1(1 + 1)}{2}$. P_1 is true.

Part 2. Show that if P_k is true, then P_{k+1} is true.

$$1 + 2 + 3 + \cdots + k = \frac{k(k + 1)}{2} \qquad \qquad P_k$$

$$1 + 2 + 3 + \cdots + k + (k + 1) = \frac{k(k + 1)}{2} + (k + 1)$$

$$= \frac{(k + 1)(k + 2)}{2} \qquad P_{k+1}$$

Conclusion: P_n is true.

2. Sketch of proof. State the conjecture: P_n:　$\left(\dfrac{x}{y}\right)^n = \dfrac{x^n}{y^n}$

Part 1. $\left(\dfrac{x}{y}\right)^1 = \dfrac{x}{y} = \dfrac{x^1}{y^1}$. P_1 is true.

Part 2. Show that if P_k is true, then P_{k+1} is true.

$$\left(\frac{x}{y}\right)^{k+1} = \left(\frac{x}{y}\right)^k\left(\frac{x}{y}\right) = \frac{x^k}{y^k}\left(\frac{x}{y}\right) = \frac{x^k x}{y^k y} = \frac{x^{k+1}}{y^{k+1}}$$

Conclusion: P_n is true.

3. Sketch of proof. State the conjecture: P_n:　$8^n - 1 = 7r$　　for some integer r

Part 1. $8^1 - 1 = 7 = 7 \cdot 1$. P_1 is true.

Part 2. Show that if P_k is true, then P_{k+1} is true.

$$8^k - 1 = 7r \qquad\qquad\qquad P_k$$

$$8(8^k - 1) = 8(7r)$$

$$8^{k+1} - 1 = 56r + 7 = 7(8r + 1) = 7s \quad P_{k+1}$$

Conclusion: P_n is true.

EXERCISE 8-2

A

In Problems 1–4, find the first positive integer n that causes the statement to fail.

1. $(3 + 5)^n = 3^n + 5^n$　　　**2.** $n < 10$　　　　　**3.** $n^2 = 3n - 2$　　　　**4.** $n^3 + 11n = 6n^2 + 6$

Verify each statement P_n in Problems 5–10 for $n = 1, 2,$ and 3.

5. P_n: $2 + 6 + 10 + \cdots + (4n - 2) = 2n^2$　　　　**6.** P_n: $4 + 8 + 12 + \cdots + 4n = 2n(n + 1)$

7. P_n: $a^5 a^n = a^{5+n}$　　　　　　　　　　　**8.** P_n: $(a^5)^n = a^{5n}$

9. P_n: $9^n - 1$ is divisible by 4　　　　　　**10.** P_n: $4^n - 1$ is divisible by 3

Write P_k and P_{k+1} for P_n as indicated in Problems 11–16.

11. P_n in Problem 5　　　　**12.** P_n in Problem 6　　　　**13.** P_n in Problem 7

14. P_n in Problem 8　　　　**15.** P_n in Problem 9　　　　**16.** P_n in Problem 10

In Problems 17–22, use mathematical induction to prove that each P_n holds for all positive integers n.

17. P_n in Problem 5　　　　**18.** P_n in Problem 6　　　　**19.** P_n in Problem 7

20. P_n in Problem 8　　　　**21.** P_n in Problem 9　　　　**22.** P_n in Problem 10

B

In Problems 23–38, use mathematical induction to prove each proposition for all positive integers n, unless restricted otherwise.

23. $2 + 2^2 + 2^3 + \cdots + 2^n = 2^{n+1} - 2$

24. $\dfrac{1}{2} + \dfrac{1}{4} + \dfrac{1}{8} + \cdots + \dfrac{1}{2^n} = 1 - \left(\dfrac{1}{2}\right)^n$

25. $1^2 + 3^2 + 5^2 + \cdots + (2n - 1)^2 = \frac{1}{3}(4n^3 - n)$

26. $1 + 8 + 16 + \cdots + 8(n - 1) = (2n - 1)^2; n > 1$

27. $1^2 + 2^2 + 3^2 + \cdots + n^2 = \dfrac{n(n + 1)(2n + 1)}{6}$

28. $1 \cdot 2 + 2 \cdot 3 + 3 \cdot 4 + \cdots$
$$+ n(n + 1) = \dfrac{n(n + 1)(n + 2)}{3}$$

29. $\dfrac{a^n}{a^3} = a^{n-3}; n > 3$

30. $\dfrac{a^5}{a^n} = \dfrac{1}{a^{n-5}}; n > 5$

31. $a^m a^n = a^{m+n}; m, n \in N$
[*Hint:* Choose m as an arbitrary element of N, and then use induction on n.]

32. $(a^n)^m = a^{mn}; m, n \in N$

33. $x^n - 1$ is divisible by $x - 1; x \neq 1$
[*Hint:* Divisible means that $x^n - 1 = (x - 1)Q(x)$ for some polynomial $Q(x)$.]

34. $x^n - y^n$ is divisible by $x - y; x \neq y$

35. $x^{2n} - 1$ is divisible by $x - 1; x \neq 1$

36. $x^{2n} - 1$ is divisible by $x + 1; x \neq -1$

37. $1^3 + 2^3 + 3^3 + \cdots + n^3 = (1 + 2 + 3 + \cdots + n)^2$
[*Hint:* See Problem 1 following Example 1.]

38. $\dfrac{1}{1 \cdot 2 \cdot 3} + \dfrac{1}{2 \cdot 3 \cdot 4} + \dfrac{1}{3 \cdot 4 \cdot 5} + \cdots$
$$+ \dfrac{1}{n(n+ 1)(n + 2)} = \dfrac{n(n + 3)}{4(n + 1)(n + 2)}$$

C

In Problems 39–42, suggest a formula for each expression, and prove your hypothesis using mathematical induction, $n \in N$.

39. $2 + 4 + 6 + \cdots + 2n$

40. $\dfrac{1}{1 \cdot 2} + \dfrac{1}{2 \cdot 3} + \dfrac{1}{3 \cdot 4} + \cdots + \dfrac{1}{n(n + 1)}$

41. The number of lines determined by n points in a plane, no three of which are collinear

42. The number of diagonals in a polygon with n sides

Prove Problems 43–46 true for all integers n as specified.

43. $a > 1 \Rightarrow a^n > 1; n \in N$

44. $0 < a < 1 \Rightarrow 0 < a^n < 1; n \in N$

45. $n^2 > 2n; n \geq 3$

46. $2^n > n^2; n \geq 5$

47. Prove or disprove the generalization of the following two facts:

$$3^2 + 4^2 = 5^2$$
$$3^3 + 4^3 + 5^3 = 6^3$$

48. Prove or disprove: $n^2 + 21n + 1$ is a prime number for all natural numbers n.

If $\{a_n\}$ and $\{b_n\}$ are two sequences, we write $\{a_n\} = \{b_n\}$ if and only if $a_n = b_n$, $n \in N$. In Problems 49–52, use mathematical induction to show that $\{a_n\} = \{b_n\}$.

49. $a_1 = 1, a_n = a_{n-1} + 2; b_n = 2n - 1$

50. $a_1 = 2, a_n = a_{n-1} + 2; b_n = 2n$

51. $a_1 = 2, a_n = 2^2 a_{n-1}; b_n = 2^{2n-1}$

52. $a_1 = 2, a_n = 3a_{n-1}; b_n = 2 \cdot 3^{n-1}$

SECTION 8-3 Arithmetic Sequences and Series

- Arithmetic Sequences
- nth-Term Formula
- Finite Arithmetic Series

For most sequences it is difficult to sum an arbitrary number of terms of the sequence without adding term by term. A particular type of sequence, however, called an *arithmetic sequence*, has certain properties that lead to convenient and very useful summing formulas for the corresponding *arithmetic series*.

• Arithmetic Sequences

Consider the sequence

$$5, 9, 13, 17, \ldots, 5 + 4(n - 1), \ldots$$

where each term after the first is obtained from the preceding one by adding 4 to it. This is an example of an arithmetic sequence.

DEFINITION 1

Arithmetic Sequence

A sequence

$$a_1, a_2, a_3, \ldots, a_n, \ldots$$

is called an **arithmetic sequence**, or **arithmetic progression**, if there exists a constant d, called the **common difference**, such that

$$a_n - a_{n-1} = d$$

That is,

$$a_n = a_{n-1} + d \qquad \text{for every } n > 1$$

EXAMPLE 1 **Recognizing Arithmetic Sequences**

Which of the following can be the first four terms of an arithmetic sequence, and what is its common difference?

(A) $1, 2, 3, 5, \ldots$ (B) $3, 5, 7, 9, \ldots$ (C) $10, 8.5, 7, 5.5, \ldots$

Solution   The terms in (B), with $d = 2$, and the terms in (C), with $d = -1.5$.

Problem 1   Repeat Example 1 with:

(A) $-4, -1, 2, 5, \ldots$ (B) $2, 4, 8, 16, \ldots$ (C) $\frac{9}{5}, \frac{7}{5}, 1, \frac{3}{5}, \ldots$

● *n*th-Term Formula   Arithmetic sequences have several convenient properties. For example, we can derive formulas for the *n*th term and the sum of any number of consecutive terms. To obtain an *n*th-term formula, we note that if $\{a_n\}$ is an arithmetic sequence, then

$$a_2 = a_1 + d$$
$$a_3 = a_2 + d = a_1 + 2d$$
$$a_4 = a_3 + d = a_1 + 3d$$

This suggests the following theorem:

Theorem 1   **The *n*th Term of an Arithmetic Sequence**

$$a_n = a_1 + (n - 1)d \qquad \text{for every } n > 1$$

We have arrived at this formula for the *n*th term of an arithmetic sequence by ordinary induction. Its proof requires mathematical induction, which we leave as an exercise (see Problem 29 in Exercise 8-3).

EXAMPLE 2   Finding a Term in an Arithmetic Sequence

If the first and tenth terms of an arithmetic sequence are 3 and 30, respectively, find the fiftieth term of the sequence.

Solution   First use $a_1 = 3$ and $a_{10} = 30$ to find d: Now find a_{50}:

$$a_n = a_1 + (n - 1)d$$
$$a_{10} = a_1 + (10 - 1)d$$
$$30 = 3 + 9d$$
$$d = 3$$

$$a_{50} = a_1 + (50 - 1)3$$
$$= 3 + 49 \cdot 3$$
$$= 150$$

Problem 2   If the first and fifteenth terms of an arithmetic sequence are -5 and 23, respectively, find the seventy-third term of the sequence.

Finite Arithmetic Series

The sum of the terms of an arithmetic sequence is called an **arithmetic series**. We will derive two simple and very useful formulas for finding the sum of an arithmetic series. Using ordinary induction does not suggest a general formula—try it and see for yourself. Instead, we proceed as follows: Let

$$S_n = a_1 + (a_1 + d) + \cdots + [a_1 + (n - 2)d] + [a_1 + (n - 1)d]$$

which is the sum of the first n terms of an arithmetic sequence. Reversing the order of the sum, we obtain

$$S_n = [a_1 + (n - 1)d] + [a_1 + (n - 2)d] + \cdots + (a_1 + d) + a_1$$

Adding the left sides of these two equations and corresponding elements of the right sides, we see that

$$2S_n = [2a_1 + (n - 1)d] + [2a_1 + (n - 1)d] + \cdots + [2a_1 + (n - 1)d]$$
$$= n[2a_1 + (n - 1)d]$$

This can be restated as in Theorem 2:

Theorem 2

Sum of an Arithmetic Series—First Form

$$S_n = \frac{n}{2}[2a_1 + (n - 1)d]$$

By replacing $a_1 + (n - 1)d$ with a_n, we obtain a second useful formula for the sum:

Theorem 3

Sum of an Arithmetic Series—Second Form

$$S_n = \frac{n}{2}(a_1 + a_n)$$

The proof of the first sum formula by mathematical induction is left as an exercise (see Problem 30 in Exercise 8-3).

EXAMPLE 3 Finding the Sum of an Arithmetic Series

Find the sum of the first 26 terms of an arithmetic series if the first term is -7 and $d = 3$.

Solution   Let $n = 26$, $a_1 = -7$, $d = 3$, and use Theorem 2.

$$S_n = \frac{n}{2}[2a_1 + (n - 1)d]$$

$$S_{26} = \frac{26}{2}[2(-7) + (26 - 1)3]$$

$$= 793$$

Problem 3   Find the sum of the first 52 terms of an arithmetic series if the first term is 23 and $d = -2$.

EXAMPLE 4   **Finding the Sum of an Arithmetic Series**

Find the sum of all the odd numbers between 51 and 99, inclusive.

Solution   First, use $a_1 = 51$, $a_n = 99$, and   Now use Theorem 3 to find S_{25}:
Theorem 1 to find n:

$$a_n = a_1 + (n - 1)d$$

$$99 = 51 + (n - 1)2$$

$$n = 25$$

$$S_n = \frac{n}{2}(a_1 + a_n)$$

$$S_{25} = \frac{25}{2}(51 + 99)$$

$$= 1{,}875$$

Problem 4   Find the sum of all the even numbers between -22 and 52, inclusive.

EXAMPLE 5   **Prize Money**

A 16-team bowling league has $8,000 to be awarded as prize money. If the last-place team is awarded $275 in prize money and the award increases by the same amount for each successive finishing place, how much will the first-place team receive?

Solution   If a_1 is the award for the first-place team, a_2 is the award for the second-place team, and so on, then the prize money awards form an arithmetic sequence with $n = 16$, $a_{16} = 275$, and $S_{16} = 8{,}000$. Use Theorem 3 to find a_1.

$$S_n = \frac{n}{2}(a_1 + a_n)$$

$$8{,}000 = \frac{16}{2}(a_1 + 275)$$

$$a_1 = 725$$

Thus, the first-place team receives $725.

Problem 5 Refer to Example 5. How much prize money is awarded to the second-place team?

Answers to Matched Problems
1. The terms in (A), with $d = 3$, and the terms in (C), with $d = -\frac{2}{5}$ **2.** 139   **3.** $-1,456$
4. 570   **5.** \$695

EXERCISE 8-3

A

1. Determine which of the following can be the first three terms of an arithmetic sequence. Find d and the next two terms for those that are.
(A) $2, 4, 8, \ldots$ (B) $7, 6.5, 6, \ldots$
(C) $-11, -16, -21, \ldots$ (D) $\frac{1}{2}, \frac{1}{6}, \frac{1}{18}, \ldots$

2. Repeat Problem 1 for:
(A) $5, -1, -7, \ldots$ (B) $12, 4, \frac{4}{3}, \ldots$
(C) $\frac{1}{2}, \frac{2}{3}, \frac{3}{4}, \ldots$ (D) $16, 48, 80, \ldots$

Let $a_1, a_2, a_3, \ldots, a_n, \ldots$ be an arithmetic sequence. In Problems 3–10, find the indicated quantities.

3. $a_1 = -5, d = 4; a_2 = ?, a_3 = ?, a_4 = ?$

4. $a_1 = -18, d = 3; a_2 = ?, a_3 = ?, a_4 = ?$

5. $a_1 = -3, d = 5; a_{15} = ?, S_{11} = ?$

6. $a_1 = 3, d = 4; a_{22} = ?, S_{21} = ?$

7. $a_1 = 1, a_2 = 5; S_{21} = ?$

8. $a_1 = 5, a_2 = 11; S_{11} = ?$

9. $a_1 = 7, a_2 = 5; a_{15} = ?$

10. $a_1 = -3, d = -4; a_{10} = ?$

B

Let $a_1, a_2, a_3, \ldots, a_n, \ldots$ be an arithmetic sequence. In Problems 11–18, find the indicated quantities.

11. $a_1 = 3, a_{20} = 117; d = ?, a_{101} = ?$

12. $a_1 = 7, a_8 = 28; d = ?, a_{25} = ?$

13. $a_1 = -12, a_{40} = 22; S_{40} = ?$

14. $a_1 = 24, a_{24} = -28; S_{24} = ?$

15. $a_1 = \frac{1}{3}, a_2 = \frac{1}{2}; a_{11} = ?, S_{11} = ?$

16. $a_1 = \frac{1}{6}, a_2 = \frac{1}{4}; a_{19} = ?, S_{19} = ?$

17. $a_3 = 13, a_{10} = 55; a_1 = ?$

18. $a_9 = -12, a_{13} = 3; a_1 = ?$

19. $S_{51} = \displaystyle\sum_{k=1}^{51} (3k + 3) = ?$

20. $S_{40} = \displaystyle\sum_{k=1}^{40} (2k - 3) = ?$

21. Find $g(1) + g(2) + g(3) + \cdots + g(51)$ if $g(t) = 5 - t$.

22. Find $f(1) + f(2) + f(3) + \cdots + f(20)$ if $f(x) = 2x - 5$.

23. Find the sum of all the even integers between 21 and 135.

24. Find the sum of all the odd integers between 100 and 500.

25. Show that the sum of the first n odd natural numbers is n^2, using appropriate formulas from this section.

26. Show that the sum of the first n even natural numbers is $n + n^2$, using appropriate formulas from this section.

27. For a given sequence in which $a_1 = -3$ and $a_n = a_{n-1} + 3, n > 1$, find a_n in terms of n.

28. For the sequence in Problem 27, find $S_n = \displaystyle\sum_{k=1}^{n} a_k$ in terms of n.

C

29. Prove, using mathematical induction, that if $\{a_n\}$ is an arithmetic sequence, then

$$a_n = a_1 + (n - 1)d \quad \text{for every } n > 1$$

30. Prove, using mathematical induction, that if $\{a_n\}$ is an arithmetic sequence, then

$$S_n = \frac{n}{2}[2a_1 + (n - 1)d]$$

31. Show that $(x^2 + xy + y^2)$, $(z^2 + xz + x^2)$, and $(y^2 + yz + z^2)$ are consecutive terms of an arithmetic progression if $x, y,$ and z form an arithmetic progression. (From USSR Mathematical Olympiads, 1955–1956, Grade 9.)

32. Take 121 terms of each arithmetic progression 2, 7, 12, . . . and 2, 5, 8, ... How many numbers will there be in common? (From USSR Mathematical Olympiads, 1955–1956, Grade 9.)

33. Given the system of equations

$$ax + by = c$$

$$dx + ey = f$$

where a, b, c, d, e, f is any arithmetic progression with a nonzero constant difference, show that the system has a unique solution.

34. The sum of the first and fourth terms of an arithmetic sequence is 2, and the sum of their squares is 20. Find the sum of the first eight terms of the sequence.

APPLICATIONS

35. Business. In investigating different job opportunities, you find that firm A will start you at $25,000 per year and guarantee you a raise of $1,200 each year while firm B will start you at $28,000 per year but will guarantee you a raise of only $800 each year. Over a period of 15 years, how much would you receive from each firm?

36. Business. In Problem 35, what would be your annual salary at each firm for the tenth year?

37. Finance. Eleven years ago an investment earned $7,000 for the year. Last year the investment earned $14,000. If the earnings from the investment have increased the same amount each year, what is the yearly increase and how much income has accrued from the investment over the past 11 years?

38. Air Temperature. As dry air moves upward, it expands. In so doing, it cools at the rate of about 5°F for each 1,000-foot rise. This is known as the **adiabatic process**.
(A) Temperatures at altitudes that are multiples of 1,000 feet form what kind of a sequence?

(B) If the ground temperature is 80°F, write a formula for the temperature T_n in terms of n, if n is in thousands of feet.

★**39. Physics.** An object falling from rest in a vacuum near the surface of the Earth falls 16 feet during the first second, 48 feet during the second second, 80 feet during the third second, and so on.
(A) How far will the object fall during the eleventh second?
(B) How far will the object fall in 11 seconds?
(C) How far will the object fall in t seconds?

★**40. Physics.** In Problem 39, how far will the object fall during:
(A) The twentieth second?   (B) The tth second?

★★**41. Geometry.** We know that the sum of the interior angles of a triangle is 180°. Show that the sums of the interior angles of polygons with 3, 4, 5, 6, . . . sides form an arithmetic sequence. Find the sum of the interior angles for a 21-sided polygon.

SECTION 8-4 Geometric Sequences and Series

- Geometric Sequences
- nth-Term Formula
- Finite Geometric Series
- Infinite Geometric Series

As we indicated in the last section, for most sequences it is difficult to sum an arbitrary number of terms of the sequence without adding term by term. For a special type of sequence—an arithmetic sequence—we were able to find convenient summation formulas. We can do the same for another type of sequence, called a *geometric sequence*. Geometric sequences are of considerable importance in mathematics and are involved in the solution of many applied problems.

• **Geometric Sequences**

Consider the sequence

$$2, -4, 8, -16, \ldots, (-1)^{n+1}2^n, \ldots$$

where each term after the first is obtained from the preceding one by multiplying it by -2. This is an example of a geometric sequence.

DEFINITION 1

Geometric Sequence

A sequence

$$a_1, a_2, a_3, \ldots, a_n, \ldots$$

is called a **geometric sequence**, or **geometric progression**, if there exists a nonzero constant r, called the **common ratio**, such that

$$\frac{a_n}{a_{n-1}} = r$$

That is,

$$a_n = r a_{n-1} \quad \text{for every } n > 1$$

EXAMPLE 1   Recognizing a Geometric Sequence

Which of the following can be the first four terms of a geometric sequence, and what is its common ratio?

(A) $2, 6, 8, 10, \ldots$ (B) $-1, 3, -9, 27, \ldots$ (C) $8, 4, 2, 1, \ldots$

Solution   The terms in (B), with $r = -3$, and the terms in (C), with $r = \frac{1}{2}$.

Problem 1   Repeat Example 1 with:

(A) $\frac{1}{4}, \frac{1}{2}, 1, 2, \ldots$ (B) $\frac{1}{2}, \frac{1}{4}, \frac{1}{16}, \frac{1}{256}, \ldots$ (C) $2, -2, 2, -2, \ldots$

- ## *n*th-Term Formula

Just as with arithmetic sequences, geometric sequences have several convenient properties. It is easy to derive formulas for the *n*th term and the sum of the first *n* terms. To obtain an *n*th-term formula, we note that if $\{a_n\}$ is a geometric sequence, then

$$a_2 = ra_1$$
$$a_3 = ra_2 = r^2a_1$$
$$a_4 = ra_3 = r^3a_1$$

This suggests the following theorem:

Theorem 1   **nth Term of a Geometric Sequence**

$$a_n = a_1 r^{n-1} \qquad \text{for every } n > 1$$

We have arrived at this formula for the *n*th term of a geometric sequence using ordinary induction. Its proof requires mathematical induction, which we leave as an exercise (see Problem 35, Exercise 8-4).

EXAMPLE 2   **Finding a Term in a Geometric Sequence**

Find the seventh term of the geometric sequence $1, \frac{1}{2}, \frac{1}{4}, \ldots.$

Solution   Let $a_1 = 1, r = \frac{1}{2}, n = 7$, and use Theorem 1.

$$a_n = a_1 r^{n-1}$$
$$a_7 = 1\left(\tfrac{1}{2}\right)^{7-1} = \tfrac{1}{64}$$

Problem 2   Find the eighth term of the geometric sequence $\frac{1}{64}, -\frac{1}{32}, \frac{1}{16}, \ldots.$

EXAMPLE 3   **Finding the Common Ratio for a Geometric Sequence**

If the first and tenth terms of a geometric sequence are 1 and 2, respectively, find the common ratio r to two decimal places.

Solution   Let $n = 10, a_1 = 1, a_{10} = 2$, and use Theorem 1.

$$a_n = a_1 r^{n-1}$$
$$2 = 1r^{10-1}$$
$$r = 2^{1/9} = 1.08 \qquad \text{Calculation by calculator using } \boxed{y^x} \text{ key.}$$

Problem 3 If the first and eighth terms of a geometric sequence are 2 and 16, respectively, find the common ratio r to three decimal places.

● **Finite Geometric Series** The sum of the terms of a geometric sequence is called a **geometric series**. As was the case with an arithmetic series, we can derive two simple and very useful formulas for finding the **sum of a geometric series**. Let

$$S_n = a_1 + a_1 r + a_1 r^2 + a_1 r^3 + \cdots + a_1 r^{n-2} + a_1 r^{n-1}$$

which is the sum of the first n terms of a geometric sequence. Multiply both sides of this equation by r to obtain

$$r S_n = a_1 r + a_1 r^2 + a_1 r^3 + \cdots + a_1 r^{n-1} + a_1 r^n$$

Now subtract the left side of the second equation from the left side of the first, and the right side of the second equation from the right side of the first to obtain

$$S_n - r S_n = a_1 - a_1 r^n$$
$$S_n (1 - r) = a_1 - a_1 r^n$$

Thus, solving for S_n, we obtain the following formula for the sum of a geometric series:

Theorem 2	**Sum of a Geometric Series—First Form**
	$$S_n = \frac{a_1 - a_1 r^n}{1 - r} \qquad r \neq 1$$

Since $a_n = a_1 r^{n-1}$, or $r a_n = a_1 r^n$, the sum formula also can be written in the following form:

Theorem 3	**Sum of a Geometric Series—Second Form**
	$$S_n = \frac{a_1 - r a_n}{1 - r} \qquad r \neq 1$$

The proof of the first sum formula by mathematical induction is left as an exercise (see Problem 36, Exercise 8-4).
If $r = 1$, then

$$S_n = a_1 + a_1(1) + a_1(1^2) + \cdots + a_1(1^{n-1}) = n a_1$$

EXAMPLE 4   Finding the Sum of a Geometric Series

Find the sum of the first 20 terms of a geometric series if the first term is 1 and $r = 2$.

Solution   Let $n = 20$, $a_1 = 1$, $r = 2$, and use Theorem 3.

$$S_n = \frac{a_1 - a_1 r^n}{1 - r}$$

$$= \frac{1 - 1 \cdot 2^{20}}{1 - 2} = 1{,}048{,}575 \qquad \text{Calculation using a calculator}$$

Problem 4   Find the sum, to two decimal places, of the first 14 terms of a geometric series if the first term is $\frac{1}{64}$ and $r = -2$.

● **Infinite Geometric Series**   Consider a geometric series with $a_1 = 5$ and $r = \frac{1}{2}$. What happens to the sum S_n as n increases? To answer this question, we first write the sum formula in the more convenient form

$$S_n = \frac{a_1 - a_1 r^n}{1 - r} = \frac{a_1}{1 - r} - \frac{a_1 r^n}{1 - r} \tag{1}$$

For $a_1 = 5$ and $r = \frac{1}{2}$,

$$S_n = 10 - 10\left(\frac{1}{2}\right)^n$$

Thus,

$$S_2 = 10 - 10\left(\frac{1}{4}\right)$$

$$S_4 = 10 - 10\left(\frac{1}{16}\right)$$

$$S_{10} = 10 - 10\left(\frac{1}{1{,}024}\right)$$

$$S_{20} = 10 - 10\left(\frac{1}{1{,}048{,}576}\right)$$

It appears that $\left(\frac{1}{2}\right)^n$ becomes smaller and smaller as n increases and that the sum gets closer and closer to 10.

In general, it is possible to show that, if $|r| < 1$, then r^n will get closer and closer to 0 as n increases. Symbolically, $r^n \to 0$ as $n \to \infty$. Thus, the term

$$\frac{a_1 r^n}{1 - r}$$

in equation (1) will tend to 0 as n increases, and S_n will tend to

$$\frac{a_1}{1 - r}$$

In other words, if $|r| < 1$, then S_n can be made as close to

$$\frac{a_1}{1 - r}$$

as we wish by taking n sufficiently large. Thus, we define the **sum of an infinite geometric series** by the following formula:

DEFINITION 2   **Sum of an Infinite Geometric Series**

$$S_\infty = \frac{a_1}{1 - r} \qquad |r| < 1$$

If $|r| \geq 1$, an infinite geometric series has no sum.

EXAMPLE 5   **Expressing a Repeating Decimal as a Fraction**

Represent the repeating decimal $0.454\ 545 \cdots = 0.\overline{45}$ as the quotient of two integers. Recall that a repeating decimal names a rational number and that any rational number can be represented as the quotient of two integers.

Solution $0.\overline{45} = 0.45 + 0.0045 + 0.000\ 045 + \cdots$

The right side of the equation is an infinite geometric series with $a_1 = 0.45$ and $r = 0.01$. Thus,

$$S_\infty = \frac{a_1}{1 - r} = \frac{0.45}{1 - 0.01} = \frac{0.45}{0.99} = \frac{5}{11}$$

Hence, $0.\overline{45}$ and $\frac{5}{11}$ name the same rational number. Check the result by dividing 5 by 11.

Problem 5   Repeat Example 5 for $0.818\ 181 \cdots = 0.\overline{81}$.

EXAMPLE 6   **Economy Stimulation**

A state government uses proceeds from a lottery to provide a tax rebate for property owners. Suppose an individual receives a $500 rebate and spends 80% of this, and each of the recipients of the money spent by this individual also spends 80%

of what he or she receives, and this process continues without end. According to the **multiplier doctrine** in economics, the effect of the original $500 tax rebate on the economy is multiplied many times. What is the total amount spent if the process continues as indicated?

Solution The individual receives $500 and spends $0.8(500) = \$400$. The recipients of this $400 spend $0.8(400) = \$320$, the recipients of this $320 spend $0.8(320) = \$256$, and so on. Thus, the total spending generated by the $500 rebate is

$$400 + 320 + 256 + \cdots = 400 + 0.8(400) + (0.8)^2(400) + \cdots$$

which we recognize as an infinite geometric series with $a_1 = 400$ and $r = 0.8$. Thus, the total amount spent is

$$S_\infty = \frac{a_1}{1 - r} = \frac{400}{1 - 0.8} = \frac{400}{0.2} = \$2,000$$

Problem 6 Repeat Example 6 if the tax rebate is $1,000 and the percentage spent by all recipients is 90%.

Answers to Matched Problems
1. The terms in (A), with $r = 2$, and the terms in (C), with $r = -1$ 2. -2 3. $r = 1.346$
4. -85.33 5. $\frac{9}{11}$ 6. $9,000

EXERCISE 8-4

A

1. Determine which of the following can be the first three terms of a geometric sequence. Find r and the next two terms for those that are.
   (A) $2, -4, 8, \ldots$
   (B) $7, 6.5, 6, \ldots$
   (C) $-11, -16, -21, \ldots$
   (D) $\frac{1}{2}, \frac{1}{6}, \frac{1}{18}, \ldots$

2. Repeat Problem 1 for:
   (A) $5, -1, -7, \ldots$
   (B) $12, 4, \frac{4}{3}, \ldots$
   (C) $\frac{1}{2}, \frac{2}{3}, \frac{3}{4}, \ldots$
   (D) $16, 48, 80, \ldots$

Let $a_1, a_2, a_3, \ldots, a_n, \ldots$ be a geometric sequence. Find each of the indicated quantities in Problems 3–8.

3. $a_1 = -6, r = -\frac{1}{2}; a_2 = ?, a_3 = ?, a_4 = ?$

4. $a_1 = 12, r = \frac{2}{3}; a_2 = ?, a_3 = ?, a_4 = ?$

5. $a_1 = 81, r = \frac{1}{3}; a_{10} = ?$

6. $a_1 = 64, r = \frac{1}{2}; a_{13} = ?$

7. $a_1 = 3, a_7 = 2,187, r = 3; S_7 = ?$

8. $a_1 = 1, a_7 = 729, r = -3; S_7 = ?$

B

Let $a_1, a_2, a_3, \ldots, a_n, \ldots$ be a geometric sequence. Find each of the indicated quantities in Problems 9–14.

9. $a_1 = 100, a_6 = 1; r = ?$

10. $a_1 = 10, a_{10} = 30; r = ?$

11. $a_1 = 5, r = -2; S_{10} = ?$

12. $a_1 = 3, r = 2; S_{10} = ?$

13. $a_1 = 9, a_4 = \frac{8}{3}; a_2 = ?, a_3 = ?$

14. $a_1 = 12, a_4 = -\frac{4}{9}; a_2 = ?, a_3 = ?$

15. $S_7 = \sum_{k=1}^{7} (-3)^{k-1} = ?$

16. $S_7 = \sum_{k=1}^{7} 3^k = ?$

17. Find $g(1) + g(2) + \cdots + g(10)$ if $g(x) = (\frac{1}{2})^x$.

18. Find $f(1) + f(2) + \cdots + f(10)$ if $f(x) = 2^x$.

19. Find a positive number x so that $-2 + x - 6$ is a three-term geometric series.

20. Find a positive number x so that $6 + x + 8$ is a three-term geometric series.

In Problems 21–26, find the sum of each infinite geometric series that has a sum.

21. $3 + 1 + \frac{1}{3} + \cdots$

22. $16 + 4 + 1 + \cdots$

23. $2 + 4 + 8 + \cdots$

24. $4 + 6 + 9 + \cdots$

25. $2 - \frac{1}{2} + \frac{1}{8} - \cdots$

26. $21 - 3 + \frac{3}{7} - \cdots$

In Problems 27–32, represent each repeating decimal as the quotient of two integers.

27. $0.\overline{7} = 0.7777 \cdots$

28. $0.\overline{5} = 0.5555 \cdots$

29. $0.\overline{54} = 0.545\ 454 \cdots$

30. $0.\overline{27} = 0.272\ 727 \cdots$

31. $3.\overline{216} = 3.216\ 216\ 216 \cdots$

32. $5.\overline{63} = 5.636\ 363 \cdots$

C

33. If in a given sequence, $a_1 = -2$ and $a_n = -3a_{n-1}$, $n > 1$, find a_n in terms of n.

34. For the sequence in Problem 33, find $S_n = \sum_{k=1}^{n} a_k$ in terms of n.

35. Prove, using mathematical induction, that if $\{a_n\}$ is a geometric sequence, then

$$a_n = a_1 r^{n-1} \qquad n \in N$$

36. Prove, using mathematical induction, that if $\{a_n\}$ is a geometric sequence, then

$$S_n = \frac{a_1 - a_1 r^n}{1 - r} \qquad n \in N, r \neq 1$$

APPLICATIONS

37. Economics. The government, through a subsidy program, distributes $1,000,000. If we assume that each individual or agency spends 0.8 of what is received, and 0.8 of this is spent, and so on, how much total increase in spending results from this government action?

38. Economics. Due to reduced taxes, an individual has an extra $600 in spendable income. If we assume that the individual spends 70% of this on consumer goods, that the producers of these goods in turn spend 70% of what they receive on consumer goods, and that this process continues indefinitely, what is the total amount spent on consumer goods?

★39. Business. If P is invested at $r\%$ compound annually, the amount A present after n years forms a geometric progression with a common ratio $1 + r$. Write a formula

for the amount present after n years. How long will it take a sum of money P to double if invested at 6% interest compounded annually?

★40. Population Growth. If a population of A_0 people grows at the constant rate of $r\%$ per year, the population after t years forms a geometric progression with a common ratio $1 + r$. Write a formula for the total population after t years. If the world's population is increasing at the rate of 2% per year, how long will it take to double?

41. Engineering. A rotating flywheel coming to rest rotates 300 revolutions the first minute (see figure on next page). If in each subsequent minute it rotates two-thirds as many times as in the preceding minute, how many revolutions will the wheel make before coming to rest?

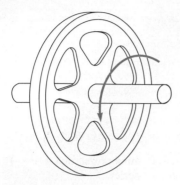

42. Physics. The first swing of a bob on a pendulum is 10 inches. If on each subsequent swing it travels 0.9 as far as on the preceding swing, how far will the bob travel before coming to rest?

43. Food Chain. A plant is eaten by an insect, an insect by a trout, a trout by a salmon, a salmon by a bear, and the bear is eaten by you. If only 20% of the energy is transformed from one stage to the next, how many calories must be supplied by plant food to provide you with 2,000 calories from the bear meat?

★ **44. Genealogy.** If there are 30 years in a generation, how many direct ancestors did each of us have 600 years ago? By *direct* ancestors we mean parents, grandparents, great-grandparents, and so on.

★ **45. Bacteria Growth.** A single cholera bacterium divides every $\frac{1}{2}$ hour to produce two complete cholera bacteria. If we start with a colony of A_0 bacteria, how many bacteria will we have in t hours, assuming adequate food supply?

★ **46. Cell Division.** One leukemic cell injected into a healthy mouse will divide into two cells in about $\frac{1}{2}$ day. At the end of the day these two cells will divide again, with the doubling process continuing each $\frac{1}{2}$ day until there are 1 billion cells, at which time the mouse dies. On which day after the experiment is started does this happen?

★★ **47. Astronomy.** Ever since the time of the Greek astronomer Hipparchus, second century B.C., the brightness of stars has been measured in terms of magnitude. The brightest stars, excluding the sun, are classed as magnitude 1, and the dimmest visible to the eye are classed as magnitude 6. In 1856, the English astronomer N. R. Pogson showed that first-magnitude stars are 100 times brighter than sixth-magnitude stars. If the ratio of brightness between consecutive magnitudes is constant, find this ratio. [*Hint:* If b_n is the brightness of an nth-magnitude star, find r for the geometric progression b_1, b_2, b_3, . . . , given $b_1 = 100b_6$.]

★ **48. Music.** The notes on a piano, as measured in cycles per second, form a geometric progression.
(A) If A is 400 cycles per second and A′, 12 notes higher, is 800 cycles per second, find the constant ratio r.
(B) Find the cycles per second for C, three notes higher than A.

49. Geometry. If the midpoints of the sides of an equilateral triangle are joined by straight lines, the new figure will be an equilateral triangle with a perimeter equal to half the original. If we start with an equilateral triangle with perimeter 1 and form a sequence of "nested" equilateral triangles proceeding as described, what will be the total perimeter of all the triangles that can be formed in this way?

50. Photography. The shutter speeds and f-stops on a camera are given as follows:

Shutter speeds: $1, \frac{1}{2}, \frac{1}{4}, \frac{1}{8}, \frac{1}{15}, \frac{1}{30}, \frac{1}{60}, \frac{1}{125}, \frac{1}{250}, \frac{1}{500}$

f-stops: 1.4, 2, 2.8, 4, 5.6, 8, 11, 16, 22

These are very close to being geometric progressions. Estimate their common ratios.

51. Puzzle. If you place 1¢ on the first square of a chessboard, 2¢ on the second square, 4¢ on the third, and so on, continuing to double the amount until all 64 squares are covered, how much money will be on the sixty-fourth square? How much money will there be on the whole board?

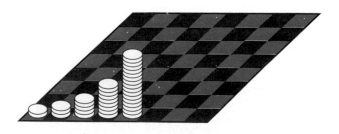

★52. **Puzzle.** If a sheet of very thin paper 0.001 inch thick is torn in half, and each half is again torn in half, and this process is repeated for a total of 32 times, how high will the stack of paper be if the pieces are placed one on top of the other? Give the answer to the nearest mile.

★53. **Atmospheric Pressure.** If atmospheric pressure decreases roughly by a factor of 10 for each 10-mile increase in altitude up to 60 miles, and if the pressure is 15 pounds per square inch at sea level, what will the pressure be 40 miles up?

54. **Zeno's Paradox.** Visualize a hypothetical 440-yard oval racetrack that has tapes stretched across the track at the halfway point and at each point that marks the halfway point of each remaining distance thereafter. A runner running around the track has to break the first tape before the second, the second before the third, and so on. From this point of view it appears that he will never finish the race. This famous paradox is attributed to the Greek philosopher Zeno, 495–435 B.C. If we assume the runner runs at 440 yards per minute, the times between tape breakings form an infinite geometric progression. What is the sum of this progression?

SECTION 8-5 Binomial Formula

- Factorial
- Binomial Formula

The binomial form

$$(a + b)^n$$

where n is a natural number, appears more frequently than you might expect. The coefficients in the expansion play an important role in probability studies. The *binomial formula*, which we derive below, enables us to expand $(a + b)^n$ directly for n any natural number. Since the formula involves *factorials*, we digress for a moment to introduce this important concept.

● **Factorial**   For n a natural number, **n factorial**—denoted by $n!$—is the product of the first n natural numbers. **Zero factorial** is defined to be 1.

DEFINITION 1   **n Factorial**

For n a natural number

$$n! = n(n - 1) \cdots \cdot 2 \cdot 1$$
$$1! = 1$$
$$0! = 1$$

It is also useful to note that:

Theorem 1   **Recursion Formula for n Factorial**

$$n! = n \cdot (n - 1)!$$

EXAMPLE 1   Evaluating Factorials

(A) $4! = 4 \cdot 3! = 4 \cdot 3 \cdot 2! = 4 \cdot 3 \cdot 2 \cdot 1! = 4 \cdot 3 \cdot 2 \cdot 1 = 24$
(B) $5! = 5 \cdot 4 \cdot 3 \cdot 2 \cdot 1 = 120$
(C) $\dfrac{7!}{6!} = \dfrac{7 \cdot 6!}{6!} = 7$
(D) $\dfrac{8!}{5!} = \dfrac{8 \cdot 7 \cdot 6 \cdot 5!}{5!} = 336$

Problem 1   Find: (A) $6!$ (B) $\dfrac{6!}{5!}$ (C) $\dfrac{9!}{6!}$

CAUTION   When reducing fractions involving factorials, don't confuse the single integer n with the symbol $n!$, which represents the product of n consecutive integers.

$$\frac{6!}{3!} \neq 2! \qquad \frac{6!}{3!} = \frac{6 \cdot 5 \cdot 4 \cdot 3!}{3!} = 6 \cdot 5 \cdot 4 = 120$$

Factorials are used in the definition of the important symbol $\dbinom{n}{r}$. This symbol is frequently used in probability studies. It is called the **combinatorial symbol** and is defined for nonnegative r and n, as follows:

DEFINITION 2   **Combinatorial Symbol**

For nonnegative integers r and n, $0 \le r \le n$.

$$\binom{n}{r} = \frac{n!}{r!(n-r)!}$$
$$= \frac{n(n-1)(n-2) \cdot \cdots \cdot (n-r+1)}{r(r-1) \cdot \cdots \cdot 2 \cdot 1}$$

The combinatorial symbol $\dbinom{n}{r}$ also can be denoted by $C_{n,r}$, $_nC_r$, or $C(n, k)$ and read as "n choose r." Your calculator may have a key labeled $\boxed{C_{n,r}}$ or $\boxed{_nC_r}$ which can be used to evaluate the combinatorial symbol.

EXAMPLE 2 Evaluating the Combinatorial Symbol

(A) $\dbinom{8}{3} = \dfrac{8!}{3!(8-3)!} = \dfrac{8!}{3!5!} = \dfrac{8 \cdot 7 \cdot 6 \cdot 5!}{3 \cdot 2 \cdot 1 \cdot 5!} = 56$

(B) $\dbinom{7}{0} = \dfrac{7!}{0!(7-0)!} = \dfrac{7!}{7!} = 1$ Remember, $0! = 1$.

Problem 2 Find: (A) $\dbinom{9}{2}$ (B) $\dbinom{5}{5}$

● **Binomial Formula**

We are now ready to try to discover a formula for the expansion of $(a + b)^n$ using ordinary induction; that is, we will look at a few special cases and postulate a general formula from them. We will then try to prove that the formula holds for all natural numbers, using mathematical induction. To start, we calculate directly the first five natural number powers of $(a + b)^n$, arranging the terms in decreasing powers of a:

$$(a + b)^1 = a + b$$
$$(a + b)^2 = a^2 + 2ab + b^2$$
$$(a + b)^3 = a^3 + 3a^2b + 3ab^2 + b^3$$
$$(a + b)^4 = a^4 + 4a^3b + 6a^2b^2 + 4ab^3 + b^4$$
$$(a + b)^5 = a^5 + 5a^4b + 10a^3b^2 + 10a^2b^3 + 5ab^4 + b^5$$

Observations

1. The expansion of $(a + b)^n$ has $n + 1$ terms.

2. The power of a decreases by 1 for each term as we move from left to right.

3. The power of b increases by 1 for each term as we move from left to right.

4. In each term, the sum of the powers of a and b always adds up to n.

5. Starting with a given term, we can get the coefficient of the next term by multiplying the coefficient of the given term by the exponent of a and dividing by the number that represents the position of the term in the series of terms. For example, in the expansion of $(a + b)^4$, the coefficient of the third term is found from the second term by multiplying 4 and 3 and then dividing by 2. Thus, the coefficient of the third term is $(4 \cdot 3)/2 = 6$.

We now postulate the properties for the general case:

$$(a + b)^n = a^n + \frac{n}{1} a^{n-1}b + \frac{n(n-1)}{1 \cdot 2} a^{n-2}b^2$$

$$+ \frac{n(n-1)(n-2)}{1 \cdot 2 \cdot 3} a^{n-3}b^3 + \cdots + b^n$$

$$= \frac{n!}{0!(n-0)!} a^n + \frac{n!}{1!(n-1)!} a^{n-1}b + \frac{n!}{2!(n-2)!} a^{n-2}b^2$$

$$+ \frac{n!}{3!(n-3)!} a^{n-3}b^3 + \cdots + \frac{n!}{n!(n-n)!} b^n$$

$$= \binom{n}{0}a_n + \binom{n}{1}a^{n-1}b + \binom{n}{2}a^{n-2}b^2 + \binom{n}{3} a^{n-3}b^3 + \cdots + \binom{n}{n}b^n$$

Thus, we have arrived at the **binomial formula** using ordinary induction:

Theorem 2   **Binomial Formula**

For n a positive integer

$$(a + b)^n = \sum_{k=0}^{n} \binom{n}{k}a^{n-k}b^k$$

We now proceed to prove that the binomial formula holds for all natural numbers n using mathematical induction.

Proof   State the conjecture.

$$P_n: \quad (a + b)^n = \sum_{j=0}^{n} \binom{n}{j}a^{n-j}b^j$$

Part 1   Show that P_1 is true.

$$\sum_{j=0}^{1} \binom{1}{j}a^{1-j}b^j = \binom{1}{0}a + \binom{1}{1}b = a + b = (a + b)^1$$

Thus, P_1 is true.

Part 2   Show that if P_k is true, then P_{k+1} is true.

$$P_k: \quad (a + b)^k = \sum_{j=0}^{k} \binom{k}{j}a^{k-j}b^j \qquad \text{Assume } P_k \text{ is true.}$$

$$P_{k+1}: \quad (a + b)^{k+1} = \sum_{j=0}^{k+1} \binom{k + 1}{j}a^{k+1-j}b^j \qquad \text{Show } P_{k+1} \text{ is true.}$$

We begin by multiplying both sides of P_k by $(a + b)$:

$$(a + b)^k(a + b) = \left[\sum_{j=0}^{k}\binom{k}{j}a^{k-j}b^j\right](a + b)$$

The left side of this equation is the left side of P_{k+1}. Now we multiply out the right side of the equation and try to obtain the right side of P_{k+1}:

$$(a + b)^{k+1} = \left[\binom{k}{0}a^k + \binom{k}{1}a^{k-1}b + \binom{k}{2}a^{k-2}b^2 + \cdots + \binom{k}{k}b^k\right](a + b)$$

$$= \left[\binom{k}{0}a^{k+1} + \binom{k}{1}a^k b + \binom{k}{2}a^{k-1}b^2 + \cdots + \binom{k}{k}ab^k\right]$$

$$+ \left[\binom{k}{0}a^k b + \binom{k}{1}a^{k-1}b^2 + \cdots + \binom{k}{k-1}ab^k + \binom{k}{k}b^{k+1}\right]$$

$$= \binom{k}{0}a^{k+1} + \left[\binom{k}{0} + \binom{k}{1}\right]a^k b + \left[\binom{k}{1} + \binom{k}{2}\right]a^{k-1}b^2 + \cdots$$

$$+ \left[\binom{k}{k-1} + \binom{k}{k}\right]ab^k + \binom{k}{k}b^{k+1}$$

We now use the following facts (the proofs are left as exercises; see Problems 49–51, Exercise 8-5)

$$\binom{k}{r-1} + \binom{k}{r} = \binom{k+1}{r} \qquad \binom{k}{0} = \binom{k+1}{0} \qquad \binom{k}{k} = \binom{k+1}{k+1}$$

to rewrite the right side as

$$\binom{k+1}{0}a^{k+1} + \binom{k+1}{1}a^k b + \binom{k+1}{2}a^{k-1}b^2 + \cdots$$

$$+ \binom{k+1}{k}ab^k + \binom{k+1}{k+1}b^{k+1} = \sum_{j=0}^{k+1}\binom{k+1}{j}a^{k+1-j}b^j$$

Since the right side of the last equation is the right side of P_{k+1}, we have shown that P_{k+1} follows from P_k.

Conclusion P_n is true. That is, the binomial formula holds for all positive integers n.

EXAMPLE 3   **Using the Binomial Formula**

Use the binomial formula to expand $(x + y)^6$.

Solution

$$(x + y)^6 = \sum_{k=0}^{6} \binom{6}{k} x^{6-k} y^k$$

$$= \binom{6}{0} x^6 + \binom{6}{1} x^5 y + \binom{6}{2} x^4 y^2 + \binom{6}{3} x^3 y^3$$

$$+ \binom{6}{4} x^2 y^4 + \binom{6}{5} x y^5 + \binom{6}{6} y^6$$

$$= x^6 + 6x^5 y + 15x^4 y^2 + 20x^3 y^3 + 15x^2 y^4 + 6xy^5 + y^6$$

Problem 3   Use the binomial formula to expand $(x + 1)^5$.

EXAMPLE 4   **Using the Binomial Formula**

Use the binomial formula to expand $(3p - 2q)^4$.

Solution

$$(3p - 2q)^4 = [(3p) + (-2q)]^4 \qquad a = 3p,\ b = -2q$$

$$= \sum_{k=0}^{4} \binom{4}{k} (3p)^{4-k} (-2q)^k$$

$$= \binom{4}{0} (3p)^4 + \binom{4}{1} (3p)^3 (-2q) + \binom{4}{2} (3p)^2 (-2q)^2$$

$$+ \binom{4}{3} (3p)(-2q)^3 + \binom{4}{4} (-2q)^4$$

$$= 81p^4 - 216p^3 q + 216p^2 q^2 - 96pq^3 + 16q^4$$

Problem 4   Use the binomial formula to expand $(2m - 5n)^3$.

EXAMPLE 5   **Using the Binomial Formula**

Use the binomial formula to find the fourth and sixteenth terms in the expansion of $(x - 2)^{20}$.

Solution   In the expansion of $(a + b)^n$, the exponent of b in the rth term is $r - 1$ and the exponent of a is $n - (r - 1)$. Thus,

Fourth term:

$$\binom{20}{3} x^{17} (-2)^3$$

$$= \frac{20 \cdot 19 \cdot 18}{3 \cdot 2 \cdot 1} x^{17} (-8)$$

$$= -9{,}120x^{17}$$

Sixteenth term:

$$\binom{20}{15} x^5 (-2)^{15}$$

$$= \frac{20 \cdot 19 \cdot 18 \cdot 17 \cdot 16}{5 \cdot 4 \cdot 3 \cdot 2 \cdot 1} x^5 (-32{,}768)$$

$$= 508{,}035{,}072x^5$$

Problem 5   Use the binomial formula to find the fifth and twelfth terms in the expansion of $(u - 1)^{18}$.

EXERCISE 8-5

A

Evaluate each expression in Problems 1–12.

1. 6!

2. 4!

3. $\dfrac{20!}{19!}$

4. $\dfrac{5!}{4!}$

5. $\dfrac{10!}{7!}$

6. $\dfrac{9!}{6!}$

7. $\dfrac{6!}{4!2!}$

8. $\dfrac{5!}{2!3!}$

9. $\dfrac{9!}{0!(9 - 0)!}$

10. $\dfrac{8!}{8!(8 - 8)!}$

11. $\dfrac{8!}{2!(8 - 2)!}$

12. $\dfrac{7!}{3!(7 - 3)!}$

Write each expression in Problems 13–16 as the quotient of two factorials.

13. 9

14. 12

15. $6 \cdot 7 \cdot 8$

16. $9 \cdot 10 \cdot 11 \cdot 12$

B

Evaluate each expression in Problems 17–24.

17. $\dbinom{9}{5}$

18. $\dbinom{5}{2}$

19. $\dbinom{6}{5}$

20. $\dbinom{7}{1}$

21. $\dbinom{9}{9}$

22. $\dbinom{5}{0}$

23. $\dbinom{17}{13}$

24. $\dbinom{20}{16}$

Expand Problems 25–36 using the binomial formula.

25. $(m + n)^3$

26. $(x + 2)^3$

27. $(2x - 3y)^3$

28. $(3u + 2v)^3$

29. $(x - 2)^4$

30. $(x - y)^4$

31. $(m + 3n)^4$

32. $(3p - q)^4$

33. $(2x - y)^5$

34. $(2x - 1)^5$

35. $(m + 2n)^6$

36. $(2x - y)^6$

In Problems 37–44, find the indicated term in each expansion.

37. $(u + v)^{15}$; seventh term

38. $(a + b)^{12}$; fifth term

39. $(2m + n)^{12}$; eleventh term

40. $(x + 2y)^{20}$; third term

41. $[(w/2) - 2]^{12}$; seventh term

42. $(x - 3)^{10}$; fourth term

43. $(3x - 2y)^8$; sixth term

44. $(2p - 3q)^7$; fourth term

C
_____

45. Evaluate $(1.01)^{10}$ to four decimal places, using the binomial formula. [*Hint:* Let $1.01 = 1 + 0.01$.]

46. Evaluate $(0.99)^6$ to four decimal places, using the binomial formula.

47. Show that: $\dbinom{n}{r} = \dbinom{n}{n-r}$

48. Show that: $\dbinom{n}{0} = \dbinom{n}{n}$

49. Show that: $\dbinom{k}{r-1} + \dbinom{k}{r} = \dbinom{k+1}{r}$

50. Show that: $\dbinom{k}{0} = \dbinom{k+1}{0}$

51. Show that: $\dbinom{k}{k} = \dbinom{k+1}{k+1}$

52. Show that $\dbinom{n}{r}$ is given by the recursion formula

$$\dbinom{n}{r} = \frac{n-r+1}{r}\dbinom{n}{r-1}$$

where $\dbinom{n}{0} = 1$.

53. Write $2^n = (1 + 1)^n$ and expand, using the binomial formula to obtain

$$2^n = \dbinom{n}{0} + \dbinom{n}{1} + \dbinom{n}{2} + \cdots + \dbinom{n}{n}$$

54. Can you guess what the next two rows in **Pascal's triangle**, shown below, are? Compare the numbers in the triangle with the binomial coefficients obtained with the binomial formula.

$$
\begin{array}{ccccccccc}
 & & & & 1 & & & & \\
 & & & 1 & & 1 & & & \\
 & & 1 & & 2 & & 1 & & \\
 & 1 & & 3 & & 3 & & 1 & \\
1 & & 4 & & 6 & & 4 & & 1 \\
\end{array}
$$

_____

Chapter 8 Review

8-1 SEQUENCES AND SERIES

A **sequence** is a function with the domain a set of successive integers. The symbol a_n, called the **nth term**, or **general term**, represents the range value associated with the domain value n. Unless specified otherwise, the domain is understood to be the set of natural numbers. A **finite sequence** has a finite domain, and an **infinite sequence** has an infinite domain. A **recursion formula** defines each term of a sequence in terms of one or more of the preceding terms. For example, the **Fibonacci sequence** is defined by $a_n = a_{n-1} + a_{n-2}$ for $n \geq 3$, where $a_1 = a_2 = 1$. The sum of the terms in a sequence is called a **series**. A finite sequence produces a **finite series**, and an infinite sequence produces an **infinite series**. Series can be represented using summation notation:

$$\sum_{k=m}^{n} a_k = a_m + a_{m+1} + \cdots + a_n$$

where k is called the **summing index**. If the terms in the series are alternately positive and negative, the series is called an **alternating series**.

8-2 MATHEMATICAL INDUCTION

A wide variety of statements can be proven using the **principle of mathematical induction**:

Let P_n be a statement associated with each positive integer n and suppose the following conditions are satisfied:

1. P_1 is true.

2. For any positive integer k, if P_k is true, then P_{k+1} is also true.

Then the statement P_n is true for all positive integers n.

To use mathematical induction to prove statements involving laws of exponents, it is convenient to state a **recursive definition of a^n**:

$$a^1 = a \quad \text{and} \quad a^{n+1} = a^n a \quad \text{for any integer } n > 1$$

To deal with conjectures that may be true only for $n \geq m$, where m is a positive integer, we use the **extended principle of mathematical induction**: Let m be a positive integer, let P_n be a statement associated with each integer $n \geq m$, and suppose the following conditions are satisfied:

1. P_m is true.

2. For any integer $k \geq m$, if P_k is true, then P_{k+1} is also true.

Then the statement P_n is true for all integers $n \geq m$.

8-3 ARITHMETIC SEQUENCES AND SERIES

A sequence is called an **arithmetic sequence**, or **arithmetic progression**, if there exists a constant d, called the **common difference**, such that

$$a_n - a_{n-1} = d \quad \text{or} \quad a_n = a_{n-1} + d \quad \text{for every } n > 1$$

The following formulas are useful when working with arithmetic sequences:

$$a_n = a_1 + (n - 1)d \qquad \textbf{\textit{n}th-Term Formula}$$

$$S_n = \frac{n}{2}[2a_1 + (n - 1)d] \qquad \textbf{Sum Formula—First Form}$$

$$S_n = \frac{n}{2}(a_1 + a_n) \qquad \textbf{Sum Formula—Second Form}$$

8-4 GEOMETRIC SEQUENCES AND SERIES

A sequence is called a **geometric sequence**, or a **geometric progression**, if there exists a nonzero constant r, called the **common ratio**, such that

$$\frac{a_n}{a_{n-1}} = r \quad \text{or} \quad a_n = ra_{n-1} \quad \text{for every } n > 1$$

The following formulas are useful when working with geometric sequences:

$$a_n = a_1 r^{n-1} \qquad \textbf{\textit{n}th-Term Formula}$$

$$S_n = \frac{a_1 - a_1 r^n}{1 - r} \quad r \neq 1 \qquad \textbf{Sum Formula—First Form}$$

$$S_n = \frac{a_1 - ra_n}{1 - r} \quad r \neq 1 \qquad \textbf{Sum Formula—Second Form}$$

$$S_\infty = \frac{a_1}{1 - r} \quad |r| < 1 \qquad \textbf{Sum of an Infinite Geometric Series}$$

8-5   BINOMIAL FORMULA

For n a natural number, n **factorial**—denoted $n!$—is defined by

$$n! = n(n - 1) \cdot \cdots \cdot 2 \cdot 1 \qquad 1! = 1 \qquad 0! = 1$$

Also, n factorial is given by the **recursion formula**

$$n! = n \cdot (n - 1)!$$

For nonnegative integers r and n, $0 \le r \le n$, the **combinatorial symbol** $\binom{n}{r}$ is defined by

$$\binom{n}{r} = \frac{n!}{r!(n - r)!} = \frac{n(n - 1) \cdot \cdots \cdot (n - r + 1)}{r(r - 1) \cdot \cdots \cdot 2 \cdot 1}$$

For n a positive integer, the **binomial formula** is

$$(a + b)^n = \sum_{k=0}^{n} \binom{n}{k} a^{n-k}b^k$$

Chapter 8 Review Exercise

Work through all the problems in this chapter review and check answers in the back of the book. Answers to all review problems are there, and following each answer is a number in italics indicating the section in which that type of problem is discussed. Where weaknesses show up, review appropriate sections in the text.

A

1. Determine whether each of the following can be the first three terms of a geometric sequence, an arithmetic sequence, or neither.
   (A) $16, -8, 4, \ldots$ (B) $5, 7, 9, \ldots$
   (C) $-8, -5, -2, \ldots$ (D) $2, 3, 5, \ldots$
   (E) $-1, 2, -4, \ldots$

In Problems 2–5:
(A) Write the first four terms of each sequence.
(B) Find a_{10}.   *(C) Find S_{10}.*

2. $a_n = 2n + 3$

3. $a_n = 32(\tfrac{1}{2})^n$

4. $a_1 = -8; a_n = a_{n-1} + 3, n \ge 2$

5. $a_1 = -1; a_n = (-2)a_{n-1}, n \ge 2$

6. Find S_∞ in Problem 3.

Evaluate Problems 7–9.

7. $6!$

8. $\dfrac{22!}{19!}$

9. $\dfrac{7!}{2!(7 - 2)!}$

Verify Problems 10–12 for $n = 1, 2,$ and 3.

10. P_n: $5 + 7 + 9 + \cdots + (2n + 3) = n^2 + 4n$

11. P_n: $2 + 4 + 8 + \cdots + 2^n = 2^{n+1} - 2$

12. P_n: $49^n - 1$ is divisible by 6

In Problems 13–15, write P_k and P_{k+1}.

13. For P_n in Problem 10

14. For P_n in Problem 11

15. For P_n in Problem 12

B _____

Write Problems 16 and 17 without summation notation, and find the sum.

16. $S_{10} = \sum_{k=1}^{10} (2k - 8)$

17. $S_7 = \sum_{k=1}^{7} \dfrac{16}{2^k}$

18. $S_\infty = 27 - 18 + 12 + \cdots = ?$

19. Write

$$S_n = \frac{1}{3} - \frac{1}{9} + \frac{1}{27} + \cdots + \frac{(-1)^{n+1}}{3^n}$$

using summation notation, and find S_∞.

20. In an arithmetic sequence, $a_1 = 13$ and $a_7 = 31$. Find the common difference d and the fifth term a_5.

21. Use the formula for the sum of an infinite geometric series to write $0.727\ 272 \cdots = 0.\overline{72}$ as the quotient of two integers.

Evaluate Problems 22–24.

22. $\dfrac{20!}{18!(20 - 18)!}$

23. $\dbinom{16}{12}$

24. $\dbinom{11}{11}$

25. Expand $(x - y)^5$ using the binomial formula.

26. Find the tenth term in the expansion of $(2x - y)^{12}$.

Establish each statement in Problems 27–29 for all natural numbers, using mathematical induction.

27. P_n in Problem 10

28. P_n in Problem 11

29. P_n in Problem 12

C _____

30. A free-falling body travels $g/2$ feet in the first second, $3g/2$ feet during the next second, $5g/2$ feet the next, and so on. Find the distance fallen during the twenty-fifth second and the total distance fallen from the start to the end of the twenty-fifth second.

31. Expand $(x + i)^6$, where i is the imaginary unit, using the binomial formula.

Prove that each statement in Problems 32–36 holds for all positive integers, using mathematical induction.

32. $\sum_{k=1}^{n} k^3 = \left(\sum_{k=1}^{n} k \right)^2$

33. $x^{2n} - y^{2n}$ is divisible by $x - y, x \neq y$

34. $\dfrac{a^n}{a^m} = a^{n-m}; n > m, n, m$ positive integers

35. $\{a_n\} = \{b_n\}$, where $a_n = a_{n-1} + 2, a_1 = -3, b_n = -5 + 2n$

36. $(1!)1 + (2!)2 + (3!)3 + \cdots + (n!)n = (n + 1)! - 1$
(From USSR Mathematical Olympiads, 1955–1956, Grade 10.)

Additional Topics in Analytic Geometry

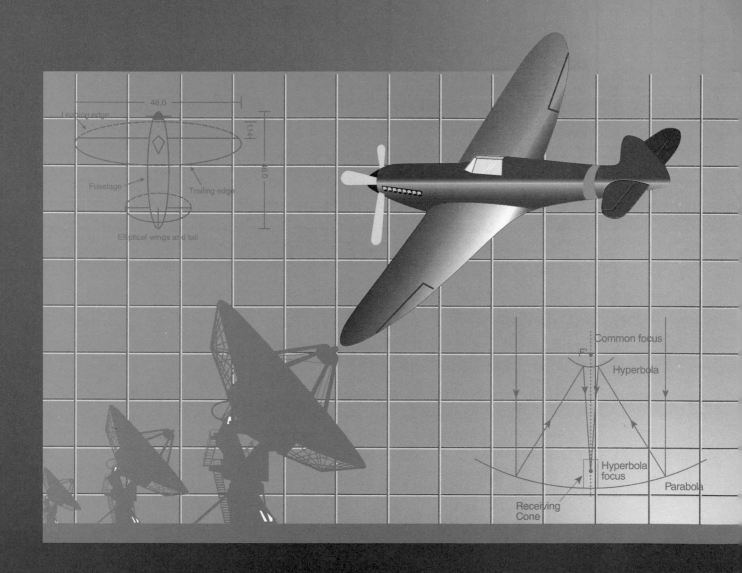

Analytic geometry, a union of geometry and algebra, enables us to analyze certain geometric concepts algebraically and to interpret certain algebraic relationships geometrically. Our two main concerns center around graphing algebraic equations and finding equations of useful geometric figures. We have discussed a number of topics in analytic geometry, such as straight lines and circles, in earlier chapters. In this chapter we discuss additional analytic geometry topics: conic sections and translation of axes.

René Descartes (1596–1650), the French philosopher–mathematician, is generally recognized as the founder of analytic geometry.

SECTION 9-1   Conic Sections; Parabola

- Conic Sections
- Definition of a Parabola
- Drawing a Parabola
- Standard Equations and Their Graphs
- Applications

In this section we introduce the general concept of a conic section and then discuss the particular conic section called a *parabola*. In the next two sections we will discuss two other conic sections called *ellipses* and *hyperbolas*.

● **Conic Sections**   In Section 3-2 we found that the graph of a first-degree equation in two variables,

$$Ax + By = C \tag{1}$$

where A and B are not both 0, is a straight line, and every straight line in a rectangular coordinate system has an equation of this form. What kind of graph will a second-degree equation in two variables,

$$Ax^2 + Bxy + Cy^2 + Dx + Ey + F = 0 \tag{2}$$

where A, B, and C are not all 0, yield for different sets of values of the coefficients? The graphs of equation (2) for various choices of the coefficients are plane curves obtainable by intersecting a cone* with a plane, as shown in Figure 1. These curves are called **conic sections**.

*Starting with a fixed line L and a fixed point V on L, the surface formed by all straight lines through V making a constant angle θ with L is called a **right circular cone**. The fixed line L is called the axis of the cone, and V is its **vertex**. The two parts of the cone separated by the vertex are called **nappes**.

FIGURE 1 Conic sections.

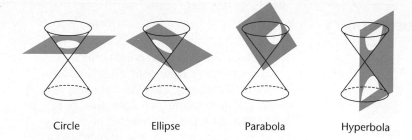

Circle   Ellipse   Parabola   Hyperbola

If a plane cuts clear through one nappe, then the intersection curve is called a **circle** if the plane is perpendicular to the axis and an **ellipse** if the plane is not perpendicular to the axis. If a plane cuts only one nappe, but does not cut clear through, then the intersection curve is called a **parabola**. Finally, if a plane cuts through both nappes, but not through the vertex, the resulting intersection curve is called a **hyperbola**. A plane passing through the vertex of the cone produces a **degenerate conic**—a point, a line, or a pair of lines.

Conic sections are very useful and are readily observed in your immediate surroundings: wheels (circle), the path of water from a garden hose (parabola), some serving platters (ellipses), and the shadow on a wall from a light surrounded by a cylindrical or conical lamp shade (hyperbola) are some examples (see Figure 2). We will discuss many applications of conics throughout the remainder of this chapter.

FIGURE 2 Examples of conics.

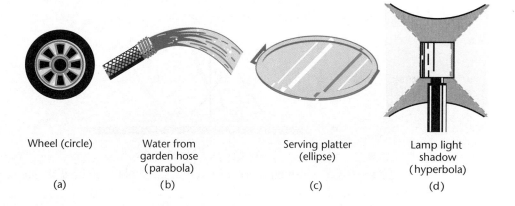

Wheel (circle)   Water from   Serving platter   Lamp light
                      garden hose   (ellipse)   shadow
                      (parabola)   (hyperbola)

(a)   (b)   (c)   (d)

A definition of a conic section that does not depend on the coordinates of points in any coordinate system is called a **coordinate-free definition**. In Section 3-1 we gave a coordinate-free definition of a circle and developed its standard equation in a rectangular coordinate system. In this and the next two sections we will give coordinate-free definitions of a parabola, ellipse, and hyperbola, and we will develop standard equations for each of these conics in a rectangular coordinate system.

• Definition of a Parabola

The following definition of a parabola does not depend on the coordinates of points in any coordinate system:

DEFINITION 1   **Parabola**

A **parabola** is the set of all points in a plane equidistant from a fixed point F and a fixed line L in the plane. The fixed point F is called the **focus**, and the fixed line L is called the **directrix**. A line through the focus perpendicular to the directrix is called the **axis**, and the point on the axis halfway between the directrix and focus is called the **vertex**.

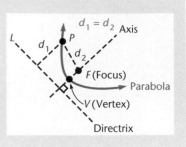

• **Drawing a Parabola**

Using the definition, we can draw a parabola with fairly simple equipment—a straightedge, a right-angle drawing triangle, a piece of string, a thumbtack, and a pencil. Referring to Figure 3, tape the straightedge along the line AB and place the thumbtack above the line AB. Place one leg of the triangle along the straightedge as indicated, then take a piece of string the same length as the other leg, tie one end to the thumbtack and fasten the other end with tape at C on the triangle. Now press the string to the edge of the triangle, and keeping the string taut, slide the triangle along the straightedge. Since DE will always equal DF, the resulting curve will be part of a parabola with directrix AB lying along the straightedge and focus F at the thumbtack.

FIGURE 3 Drawing a parabola.

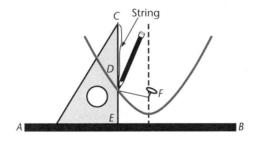

• **Standard Equations and Their Graphs**

Using the definition of a parabola and the distance-between-two-points formula

$$d = \sqrt{(x_2 - x_1)^2 + (y_2 - y_1)^2} \tag{3}$$

FIGURE 4 Parabola with center at the origin and axis the x axis.

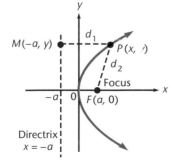

$a > 0$, focus on positive x axis

(a)

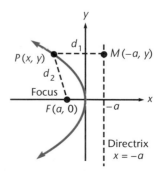

$a < 0$, focus on negative x axis

(b)

we can derive simple standard equations for a parabola located in a rectangular coordinate system with its vertex at the origin and its axis along a coordinate axis. We start with the axis of the parabola along the x axis and the focus at $F(a, 0)$. We locate the parabola in a coordinate system as in Figure 4 and label key lines and points. This is an important step in finding an equation of a geometric figure in a coordinate system. Note that the parabola opens to the right if $a > 0$ and to the left if $a < 0$. The vertex is at the origin, the directrix is $x = -a$, and the coordinates of M are $(-a, y)$.

The point $P(x, y)$ is a point on the parabola if and only if

$$d_1 = d_2$$

$$d(P, M) = d(P, F)$$

$$\sqrt{(x + a)^2 + (y - y)^2} = \sqrt{(x - a)^2 + (y - 0)^2} \quad \text{Use equation (3).}$$

$$(x + a)^2 = (x - a)^2 + y^2 \quad \text{Square both sides.}$$

$$x^2 + 2ax + a^2 = x^2 - 2ax + a^2 + y^2 \quad \text{Simplify.}$$

$$y^2 = 4ax \qquad (4)$$

Equation (4) is the standard equation of a parabola with vertex at the origin, axis the x axis, and focus at $(a, 0)$.

Now we locate the vertex at the origin and focus on the y axis at $(0, a)$. Looking at Figure 5, we note that the parabola opens upward if $a > 0$ and downward if $a < 0$. The directrix is $y = -a$, and the coordinates of N are $(x, -a)$. The point $P(x, y)$ is a point on the parabola if and only if

$$d_1 = d_2$$

$$d(P, N) = d(P, F)$$

$$\sqrt{(x - x)^2 + (y + a)^2} = \sqrt{(x - 0)^2 + (y - a)^2} \quad \text{Use equation (3).}$$

$$(y + a)^2 = x^2 + (y - a)^2 \quad \text{Square both sides.}$$

$$y^2 + 2ay + a^2 = x^2 + y^2 - 2ay + a^2 \quad \text{Simplify.}$$

$$x^2 = 4ay \qquad (5)$$

Equation (5) is the standard equation of a parabola with vertex at the origin, axis the y axis, and focus at $(0, a)$.

FIGURE 5 Parabola with center at the origin and axis the y axis.

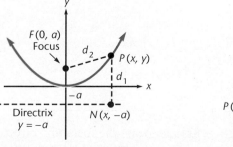

$a > 0$, focus on positive y axis

(a)

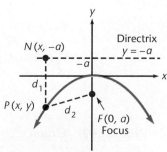

$a < 0$, focus on negative y axis

(b)

We summarize these results for easy reference in Theorem 1:

Theorem 1   **Standard Equations of a Parabola with Vertex at (0, 0)**

1. $y^2 = 4ax$
   Vertex: $(0, 0)$
   Focus: $(a, 0)$
   Directrix: $x = -a$
   Symmetric with respect
   to the x axis.
   Axis the x axis

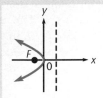

$a < 0$ (opens left)

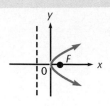

$a > 0$ (opens right)

2. $x^2 = 4ay$
   Vertex: $(0, 0)$
   Focus: $(0, a)$
   Directrix: $y = -a$
   Symmetric with respect
   to the y axis.
   Axis the y axis

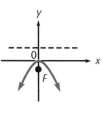

$a < 0$ (opens down)

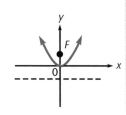

$a > 0$ (opens up)

EXAMPLE 1   **Graphing $x^2 = 4ay$**

Graph $x^2 = -16y$, and locate the focus and directrix.

Solution   To graph $x^2 = -16y$, it is convenient to assign y values that make the right side a perfect square, and solve for x. Note that y must be 0 or negative for x to be real. Since the coefficient of y is negative, a must be negative, and the parabola opens down.

x	0	± 4	± 8
y	0	-1	-4

Focus: $x^2 = -16y \overset{a}{=} 4(-4)y$
          $F(0, a) = F(0, -4)$
Directrix: $y = -a$
          $= -(-4) = 4$

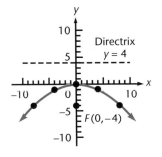

Problem 1   Graph $y^2 = -8x$, and locate the focus and directrix.

CAUTION

A common error in making a quick sketch of $y^2 = 4ax$ or $x^2 = 4ay$ is to sketch the first with the y axis as its axis and the second with the x axis as its axis. The graph of $y^2 = 4ax$ is symmetric with respect to the x axis, and the graph of $x^2 = 4ay$ is symmetric with respect to the y axis, as a quick symmetry check will reveal.

EXAMPLE 2   Finding the Equation of a Parabola

(A) Find the equation of a parabola having the origin as its vertex, the y axis as its axis, and $(-10, -5)$ on its graph.

(B) Find the coordinates of its focus and the equation of its directrix.

Solutions   (A) The parabola is opening down and has an equation of the form $x^2 = 4ay$. Since $(-10, -5)$ is on the graph, we have

$$x^2 = 4ay$$
$$(-10)^2 = 4a(-5)$$
$$100 = -20a$$
$$a = -5$$

Thus, the equation of the parabola is

$$x^2 = 4(-5)y$$
$$= -20y$$

(B) Focus: $x^2 = -20y \overset{a}{\;= 4(-5)y}$
$$F(0, a) = F(0, -5)$$
Directrix: $y = -a$
$$= -(-5)$$
$$= 5$$

Problem 2   (A) Find the equation of a parabola having the origin as its vertex, the x axis as its axis, and $(4, -8)$ on its graph.

(B) Find the coordinates of its focus and the equation of its directrix.

● **Applications**   Parabolic forms are frequently encountered in the physical world. Suspension bridges, arch bridges, microphones, symphony shells, satellite antennas, radio and optical telescopes, radar equipment, solar furnaces, and searchlights are only a few of many items that utilize parabolic forms in their design.

Figure 6(a) on the next page illustrates a parabolic reflector used in all reflecting telescopes—from 3- to 6-inch home type to the 200-inch research instrument

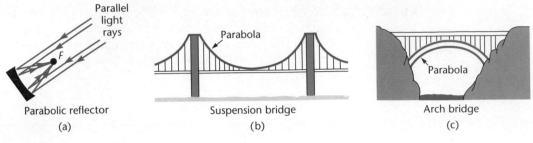

FIGURE 6 Uses of parabolic forms.

on Mount Palomar in California. Parallel light rays from distant celestial bodies are reflected to the focus off a parabolic mirror. If the light source is the sun, then the parallel rays are focused at F and we have a solar furnace. Temperatures of over 6,000°C have been achieved by such furnaces. If we locate a light source at F, then the rays in Figure 6(a) reverse, and we have a spotlight or a searchlight. Automobile headlights can use parabolic reflectors with special lenses over the light to diffuse the rays into useful patterns.

Figure 6(b) shows a suspension bridge, such as the Golden Gate Bridge in San Francisco. The suspension cable is a parabola. It is interesting to note that a free-hanging cable, such as a telephone line, does not form a parabola. It forms another curve called a *catenary*.

Figure 6(c) shows a concrete arch bridge. If all the loads on the arch are to be compression loads, concrete works very well under compression, then using physics and advanced mathematics, it can be shown that the arch must be parabolic.

EXAMPLE 3   **Parabolic Reflector**

A **paraboloid** is formed by revolving a parabola about its axis. A spotlight in the form of a paraboloid 5 inches deep has its focus 2 inches from the vertex. Find, to one decimal place, the radius R of the opening of the spotlight.

Solution   *Step 1.* Locate a parabolic cross section containing the axis in a rectangular coordinate system, and label all known parts and parts to be found. This is a very important step and can be done in infinitely many ways. Since we are in charge, we can make things simpler for ourselves by locating the vertex at the origin and choosing a coordinate axis as the axis. We choose the y axis as the axis of the parabola with the parabola opening up. See the figure.

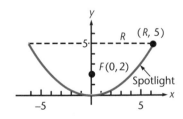

Step 2. Find the equation of the parabola in the figure. Since the parabola has the y axis as its axis and the vertex at the origin, the equation is of the form

$$x^2 = 4ay$$

We are given $F(0, a) = F(0, 2)$; thus, $a = 2$, and the equation of the parabola is

$$x^2 = 8y$$

Step 3. Use the equation found in step 2 to find the radius R of the opening. Since $(R, 5)$ is on the parabola, we have

$$R^2 = 8(5)$$
$$R = \sqrt{40} \approx 6.3 \text{ inches}$$

Problem 3 Repeat Example 3 with a paraboloid 12 inches deep and a focus 9 inches from the vertex.

Answers to Matched Problems

1. Focus: $(-2, 0)$
Directrix: $x = 2$

x	0	-2
y	0	± 4

2. (A) $y^2 = 16x$ (B) Focus: $(4, 0)$; Directrix: $x = -4$
3. $R = 20.8$ inches

EXERCISE 9-1

A _____

In Problems 1–12, graph each equation, and locate the focus and directrix.

1. $y^2 = 4x$ **2.** $y^2 = 8x$ **3.** $x^2 = 8y$ **4.** $x^2 = 4y$

5. $y^2 = -12x$ **6.** $y^2 = -4x$ **7.** $x^2 = -4y$ **8.** $x^2 = -8y$

9. $y^2 = -20x$ **10.** $x^2 = -24y$ **11.** $x^2 = 10y$ **12.** $y^2 = 6x$

Find the coordinates to two decimal places of the focus for each parabola in Problems 13–18.

13. $y^2 = 39x$ **14.** $x^2 = 58y$ **15.** $x^2 = -105y$ **16.** $y^2 = -93x$

17. $y^2 = -77x$ **18.** $x^2 = -205y$

B _____

In Problems 19–26, find the equation of a parabola with vertex at the origin, axis the x or y axis, and:

19. Directrix $y = -3$ **20.** Directrix $y = 4$ **21.** Focus $(0, -7)$ **22.** Focus $(0, 5)$

23. Directrix $x = 6$ **24.** Directrix $x = -9$ **25.** Focus $(2, 0)$ **26.** Focus $(-4, 0)$

In Problems 27–32, find the equation of the parabola having its vertex at the origin, its axis as indicated, and passing through the indicated point.

27. y axis; $(4, 2)$ **28.** x axis; $(4, 8)$ **29.** x axis; $(-3, 6)$ **30.** y axis; $(-5, 10)$

31. y axis; $(-6, -9)$ **32.** x axis; $(-6, -12)$

In Problems 33–36, find all first-quadrant points of intersection for each system of equations to three decimal places:

(A) Algebraically
(B) Using a graphic calculator or computer

33. $x^2 = 4y$
   $y^2 = 4x$

34. $y^2 = 3x$
   $x^2 = 3y$

35. $y^2 = 6x$
   $x^2 = 5y$

36. $x^2 = 7y$
   $y^2 = 2x$

37. The line segment AB through the focus in the figure is called the **focus chord** for the parabola. Find the coordinates of A and B.

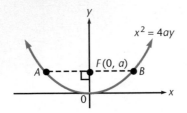

38. The line segment AB through the focus in the figure is called the **focus chord** for the parabola. Find the coordinates of A and B.

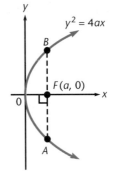

C ────────────────────────────────

In Problems 39–42, use the definition of a parabola and the distance formula to find the equation of a parabola with:

39. Directrix $y = -4$ and focus $(2, 2)$

40. Directrix $y = 2$ and focus $(-3, 6)$

41. Directrix $x = 2$ and focus $(6, -4)$

42. Directrix $x = -3$ and focus $(1, 4)$

───────────────────────────────────

APPLICATIONS

43. Engineering. The parabolic arch in the concrete bridge in the figure must have a clearance of 50 feet above the water and span a distance of 200 feet. Find the equation of the parabola after inserting a coordinate system with the origin at the vertex of the parabola and the vertical y axis (pointing upward) along the axis of the parabola.

Concrete bridge

44. Astronomy. The cross section of a parabolic reflector with 6-inch diameter is ground so that its vertex is 0.15 inch below the rim (see the figure).

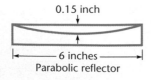
Parabolic reflector

(A) Find the equation of the parabola after inserting an xy coordinate system with the vertex at the origin, the y axis (pointing upward) the axis of the parabola.
(B) How far is the focus from the vertex?

45. Space Science. A designer of a 200-foot-diameter parabolic electromagnetic antenna for tracking space probes wants to place the focus 100 feet above the vertex (see the figure).

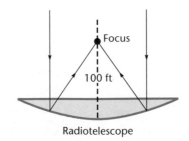
Radiotelescope

(A) Find the equation of the parabola using the axis of the parabola as the y axis (up positive) and vertex at the origin.

(B) Determine the depth of the parabolic reflector.

46. Signal Light. A signal light on a ship is a spotlight with parallel reflected light rays (see the figure to the right). Suppose the parabolic reflector is 12 inches in diameter and the light source is located at the focus, which is 1.5 inches from the vertex.

(A) Find the equation of the parabola using the axis of the parabola as the x axis (right positive) and vertex at the origin.

(B) Determine the depth of the parabolic reflector.

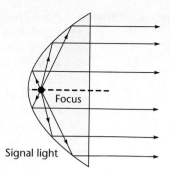

SECTION 9-2 Ellipse

- Definition of an Ellipse
- Drawing an Ellipse
- Standard Equations and Their Graphs
- Applications

We start our discussion of the ellipse with a coordinate-free definition. Using this definition, we show how an ellipse can be drawn and we derive standard equations for ellipses specially located in a rectangular coordinate system.

- **Definition of an Ellipse**

The following is a coordinate-free definition of an ellipse:

DEFINITION 1   **Ellipse**

An **ellipse** is the set of all points P in a plane such that the sum of the distances of P from two fixed points in the plane is constant. Each of the fixed points, F' and F, is called a **focus**, and together they are called **foci**. Referring to the figure, the line segment $V'V$ through the foci is the **major axis**. The perpendicular bisector $B'B$ of the major axis is the **minor axis**. Each end of the major axis, V' and V, is called a **vertex**. The midpoint of the line segment $F'F$ is called the **center** of the ellipse.

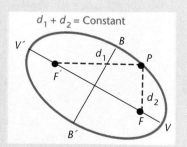

• **Drawing an Ellipse**   An ellipse is easy to draw. All you need is a piece of string, two thumbtacks, and a pencil or pen (see Figure 1). Place the two thumbtacks in a piece of cardboard. These form the foci of the ellipse. Take a piece of string longer than the distance between the two thumbtacks—this represents the constant in the definition—and tie each end to a thumbtack. Finally, catch the tip of a pencil under the string and move it while keeping the string taut. The resulting figure is by definition an ellipse. Ellipses of different shapes result, depending on the placement of thumbtacks and the length of the string joining them.

FIGURE 1   Drawing an ellipse.

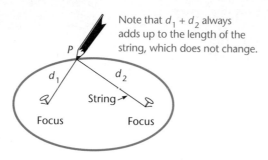

Note that $d_1 + d_2$ always adds up to the length of the string, which does not change.

• **Standard Equations and Their Graphs**   Using the definition of an ellipse and the distance-between-two-points formula, we can derive standard equations for an ellipse located in a rectangular coordinate system. We start by placing an ellipse in the coordinate system with the foci on the x axis equidistant from the origin at $F'(-c, 0)$ and $F(c, 0)$, as in Figure 2.

For reasons that will become clear soon, it is convenient to represent the constant sum $d_1 + d_2$ by $2a$, $a > 0$. Also, the geometric fact that the sum of the lengths of any two sides of a triangle must be greater than the third side can be applied to Figure 2 to derive the following useful result:

$$d(F', P) + d(P, F) > d(F', F)$$

$$d_1 + d_2 > 2c$$

$$2a > 2c$$

$$a > c \qquad (1)$$

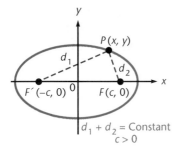

FIGURE 2   Ellipse with foci on x axis.

We will use this result in the derivation of the equation of an ellipse, which we now begin.

Referring to Figure 2, the point $P(x, y)$ is on the ellipse if and only if

$$d_1 + d_2 = 2a$$

$$d(P, F') + d(P, F) = 2a$$

$$\sqrt{(x + c)^2 + (y - 0)^2} + \sqrt{(x - c)^2 + (y - 0)^2} = 2a$$

After eliminating radicals and simplifying, a good exercise for you, we obtain

$$(a^2 - c^2)x^2 + a^2y^2 = a^2(a^2 - c^2) \qquad (2)$$

$$\frac{x^2}{a^2} + \frac{y^2}{a^2 - c^2} = 1 \qquad (3)$$

Dividing both sides of equation (2) by $a^2(a^2 - c^2)$ is permitted, since neither a^2 nor $a^2 - c^2$ is 0. From equation (1), $a > c$, thus $a^2 > c^2$ and $a^2 - c^2 > 0$. The constant a was chosen positive at the beginning.

To simplify equation (3) further, we let

$$b^2 = a^2 - c^2 \qquad b > 0 \tag{4}$$

to obtain

$$\frac{x^2}{a^2} + \frac{y^2}{b^2} = 1 \tag{5}$$

From equation (5) we see that the x intercepts are $x = \pm a$ and the y intercepts are $y = \pm b$. The x intercepts are also the vertices. Thus,

> **Major axis length = 2a**
>
> **Minor axis length = 2b**

To see that the major axis is longer than the minor axis, we show that $2a > 2b$. Returning to equation (4),

$$b^2 = a^2 - c^2 \qquad\qquad a, b, c > 0$$
$$b^2 + c^2 = a^2$$
$$b^2 < a^2 \qquad\qquad \text{Definition of } <$$
$$b^2 - a^2 < 0$$
$$(b - a)(b + a) < 0$$
$$b - a < 0 \qquad\qquad \text{Since } b + a \text{ is positive, } b - a \text{ must be negative.}$$
$$b < a$$
$$2b < 2a$$
$$2a > 2b$$
$$\begin{pmatrix} \text{Length of} \\ \text{major axis} \end{pmatrix} > \begin{pmatrix} \text{Length of} \\ \text{minor axis} \end{pmatrix}$$

If we start with the foci on the y axis at $F(0, c)$ and $F'(0, -c)$ as in Figure 3, instead of on the x axis as in Figure 2, then, following arguments similar to those used for the first derivation, we obtain

$$\frac{x^2}{b^2} + \frac{y^2}{a^2} = 1 \qquad a > b \tag{6}$$

where the relationship among a, b, and c remains the same as before:

$$b^2 = a^2 - c^2 \tag{7}$$

The center is still at the origin, but the major axis is now along the y axis and the minor axis is along the x axis.

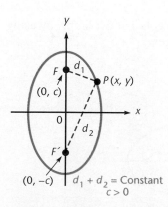

FIGURE 3 Ellipse with foci on y axis.

To sketch graphs of equations of the form (5) or (6) is an easy matter. We find the x and y intercepts and sketch in an appropriate ellipse. Since replacing x with $-x$, or y with $-y$ produces an equivalent equation, we conclude that the graphs are symmetric with respct to the x axis, y axis, and origin. If further accuracy is required, additional points can be found with the aid of a calculator and the use of symmetry properties.

Given an equation of the form (5) or (6), how can we find the coordinates of the foci without memorizing or looking up the relation $b^2 = a^2 - c^2$? There is a simple geometric relationship in an ellipse that enables us to get the same result using the Pythagorean theorem. To see this relationship, refer to Figure 4(a). Then, using the definition of an ellipse and $2a$ for the constant sum, as we did in deriving the standard equations, we see that

$$d + d = 2a$$
$$2d = 2a$$
$$d = a$$

Thus:

> **The length of the line segment from the end of a minor axis to a focus is the same as half the length of a major axis.**

This geometric relationship is illustrated in Figure 4(b). Using the Pythagorean theorem for the triangle in Figure 4(b), we have

$$b^2 + c^2 = a^2$$

or

$$b^2 = a^2 - c^2 \qquad \text{Equations (4) and (7)}$$

or

$$c^2 = a^2 - b^2 \qquad \text{Useful for finding the foci, given } a \text{ and } b$$

Thus, we can find the foci of an ellipse given the intercepts a and b simply by using the triangle in Figure 4(b) and the Pythagorean theorem.

FIGURE 4 Geometric relationships.

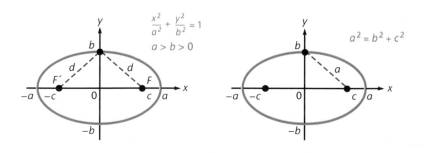

We summarize all of these results for convenient reference in Theorem 1.

Theorem 1 **Standard Equations of an Ellipse with Center at (0, 0)**

1. $\dfrac{x^2}{a^2} + \dfrac{y^2}{b^2} = 1 \qquad a > b > 0$

   x intercepts: $\pm a$ (vertices)
   y intercepts: $\pm b$
   Foci: $F'(-c, 0), F(c, 0)$

   $$c^2 = a^2 - b^2$$

   Major axis length $= 2a$
   Minor axis length $= 2b$

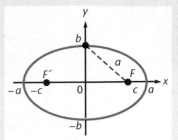

2. $\dfrac{x^2}{b^2} + \dfrac{y^2}{a^2} = 1 \qquad a > b > 0$

   x intercepts: $\pm b$
   y intercepts: $\pm a$ (vertices)
   Foci: $F'(0, -c), F(0, c)$

   $$c^2 = a^2 - b^2$$

   Major axis length $= 2a$
   Minor axis length $= 2b$

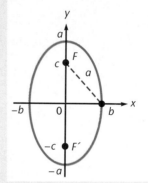

[*Note:* Both graphs are symmetric with respect to the x axis, y axis, and origin. Also, the major axis is always longer than the minor axis.]

EXAMPLE 1 **Graphing Ellipses**

Sketch the graph of each equation, find the coordinates of the foci, and find the lengths of the major and minor axes.

(A) $9x^2 + 16y^2 = 144$ (B) $2x^2 + y^2 = 10$

Solutions (A) First, write the equation in standard form by dividing both sides by 144:

$$9x^2 + 16y^2 = 144$$

$$\boxed{\dfrac{9x^2}{144} + \dfrac{16y^2}{144} = \dfrac{144}{144}}$$

$$\dfrac{x^2}{16} + \dfrac{y^2}{9} = 1 \qquad a^2 = 16 \text{ and } b^2 = 9$$

Locate the intercepts:

$$x \text{ intercepts:} \quad \pm 4$$

$$y \text{ intercepts:} \quad \pm 3$$

and sketch in the ellipse, as shown in the margin.

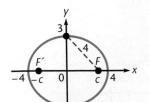

$$\text{Foci:} \quad c^2 = a^2 - b^2$$

$$= 16 - 9$$

$$= 7$$

$$c = \sqrt{7} \quad \text{c is positive}$$

Thus, the foci are $F'(-\sqrt{7}, 0)$ and $F(\sqrt{7}, 0)$.

$$\text{Major axis length} = 2(4) = 8$$

$$\text{Minor axis length} = 2(3) = 6$$

(B) Write the equation in standard form by dividing both sides by 10:

$$2x^2 + y^2 = 10$$

$$\boxed{\frac{2x^2}{10} + \frac{y^2}{10} = \frac{10}{10}}$$

$$\frac{x^2}{5} + \frac{y^2}{10} = 1 \qquad a^2 = 10 \text{ and } b^2 = 5$$

Locate the intercepts:

$$x \text{ intercepts:} \quad \pm\sqrt{5} \approx \pm 2.24$$

$$y \text{ intercepts:} \quad \pm\sqrt{10} \approx \pm 3.16$$

and sketch in the ellipse, as shown in the margin.

$$\text{Foci:} \quad c^2 = a^2 - b^2$$

$$= 10 - 5$$

$$= 5$$

$$c = \sqrt{5}$$

Thus, the foci are $F'(0, -\sqrt{5})$ and $F(0, \sqrt{5})$.

$$\text{Major axis length} = 2\sqrt{10} \approx 6.32$$

$$\text{Minor axis length} = 2\sqrt{5} \approx 4.47$$

Problem 1 Sketch the graph of each equation, find the coordinates of the foci, and find the lengths of the major and minor axes.

(A) $x^2 + 4y^2 = 4$ (B) $3x^2 + y^2 = 18$

EXAMPLE 2   Finding the Equation of an Ellipse

Find an equation of an ellipse in the form

$$\frac{x^2}{M} + \frac{y^2}{N} = 1 \qquad M, N > 0$$

if the center is at the origin, the major axis is along the y axis, and:

(A) Length of major axis = 20   (B) Length of major axis = 10
    Length of minor axis = 12   Distance of foci from center = 4

Solutions   (A) Compute x and y intercepts and make a rough sketch of the ellipse.

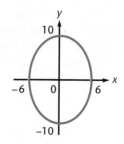

$$\frac{x^2}{b^2} + \frac{y^2}{a^2} = 1$$

$$a = \frac{20}{2} = 10 \qquad b = \frac{12}{2} = 6$$

$$\frac{x^2}{36} + \frac{y^2}{100} = 1$$

(B) Make a rough sketch of the ellipse; locate the foci and y intercepts, then determine the x intercepts using the special triangle relationship discussed earlier.

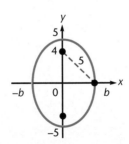

$$\frac{x^2}{b^2} + \frac{y^2}{a^2} = 1$$

$$a = \frac{10}{2} = 5 \qquad b^2 = 5^2 - 4^2 = 25 - 16 = 9$$
$$b = 3$$

$$\frac{x^2}{9} + \frac{y^2}{25} = 1$$

Problem 2   Find an equation of an ellipse in the form

$$\frac{x^2}{M} + \frac{y^2}{N} = 1 \qquad M, N > 0$$

if the center is at the origin, the major axis is along the x axis, and:

(A) Length of major axis = 50   (B) Length of minor axis = 16
    Length of minor axis = 30   Distance of foci from center = 6

● Applications   You are no doubt aware of many occurrences and uses of elliptical forms: orbits of satellites, planets, and comets; shapes of galaxies; gears and cams; some airplane wings, boat keels, and rudders; tabletops; public fountains; and domes in buildings

FIGURE 5 Uses of elliptical forms.

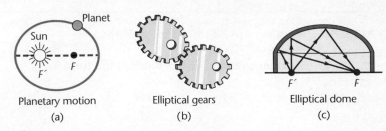

Planetary motion
(a)

Elliptical gears
(b)

Elliptical dome
(c)

are a few examples (see Figure 5). A fairly recent application in medicine is the use of elliptical reflectors and ultrasound to break up kidney stones.

Johannes Kepler (1571–1630), a German astronomer, discovered that planets move in elliptical orbits, with the sun at a focus, and not in circular orbits as had been thought before [Figure 5(a)]. Figure 5(b) shows a pair of elliptical gears with pivot points at foci. Such gears transfer constant rotational speed to variable rotational speed, and vice versa. Figure 5(c) shows an elliptical dome. An interesting property of such a dome is that a sound or light source at one focus will reflect off the dome and pass through the other focus. One of the chambers in the Capitol Building in Washington, D.C., has such a dome, and is referred to as a whispering room because a whispered sound at one focus can be easily heard at the other focus.

Answers to Matched Problems

1. (A)

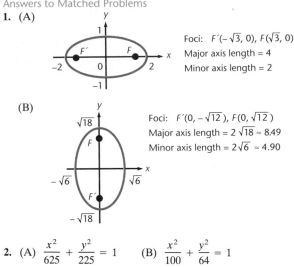

Foci: $F'(-\sqrt{3}, 0), F(\sqrt{3}, 0)$
Major axis length = 4
Minor axis length = 2

(B)

Foci: $F'(0, -\sqrt{12}), F(0, \sqrt{12})$
Major axis length = $2\sqrt{18} \approx 8.49$
Minor axis length = $2\sqrt{6} \approx 4.90$

2. (A) $\dfrac{x^2}{625} + \dfrac{y^2}{225} = 1$ (B) $\dfrac{x^2}{100} + \dfrac{y^2}{64} = 1$

EXERCISE 9-2

A _____

In Problems 1–6, sketch a graph of each equation, find the coordinates of the foci, and find the lengths of the major and minor axes.

1. $\dfrac{x^2}{25} + \dfrac{y^2}{4} = 1$

2. $\dfrac{x^2}{9} + \dfrac{y^2}{4} = 1$

3. $\dfrac{x^2}{4} + \dfrac{y^2}{25} = 1$

4. $\dfrac{x^2}{4} + \dfrac{y^2}{9} = 1$

5. $x^2 + 9y^2 = 9$

6. $4x^2 + y^2 = 4$

B

In Problems 7–12, sketch a graph of each equation, find the coordinates of the foci, and find the lengths of the major and minor axes.

7. $25x^2 + 9y^2 = 225$

8. $16x^2 + 25y^2 = 400$

9. $2x^2 + y^2 = 12$

10. $4x^2 + 3y^2 = 24$

11. $4x^2 + 7y^2 = 28$

12. $3x^2 + 2y^2 = 24$

In Problems 13–20, find an equation of an ellipse in the form

$$\frac{x^2}{M} + \frac{y^2}{N} = 1 \qquad M, N > 0$$

if the center is at the origin, and:

13. Major axis on x axis
Major axis length $= 8$
Minor axis length $= 6$

14. Major axis on x axis
Major axis length $= 14$
Minor axis length $= 10$

15. Major axis on y axis
Major axis length $= 22$
Minor axis length $= 16$

16. Major axis on y axis
Major axis length $= 24$
Minor axis length $= 18$

17. Major axis on x axis
Major axis length $= 16$
Distance of foci from center $= 6$

18. Major axis on x axis
Major axis length $= 24$
Distance of foci from center $= 10$

19. Major axis on y axis
Minor axis length $= 20$
Distance of foci from center $= \sqrt{70}$

20. Major axis on y axis
Minor axis length $= 14$
Distance of foci from center $= \sqrt{200}$

In Problems 21–24, graph each system of equations in the same rectangular coordinate system and find the coordinates of any points of intersection. Find noninteger coordinates to three decimal places.

21. $16x^2 + 25y^2 = 400$
$\quad 2x - 5y = 10$

22. $25x^2 + 16y^2 = 400$
$\quad 5x + 8y = 20$

23. $25x^2 + 16y^2 = 400$
$\quad 25x^2 - 36y = 0$

24. $16x^2 + 25y^2 = 400$
$\quad 3x^2 - 20y = 0$

In Problems 25–28, find all first-quadrant points of intersection for each system of equations to three decimal places:

(A) Algebraically
(B) Using a graphic calculator or a computer

25. $5x^2 + 2y^2 = 63$
$\quad 2x - y = 0$

26. $3x^2 + 4y^2 = 57$
$\quad x - 2y = 0$

27. $2x^2 + 3y^2 = 33$
$\quad x^2 - 8y = 0$

28. $3x^2 + 2y^2 = 43$
$\quad x^2 - 12y = 0$

C

29. Find an equation of the set of points in a plane, each of whose distance from $(2, 0)$ is one-half its distance from the line $x = 8$. Identify the geometric figure.

30. Find an equation of the set of points in a plane, each of whose distance from $(0, 9)$ is three-fourths its distance from the line $y = 16$. Identify the geometric figure.

APPLICATIONS

31. Engineering. The semielliptical arch in the concrete bridge in the figure on the next page must have a clear-

ance of 12 feet above the water and span a distance of 40 feet. Find the equation of the ellipse after inserting a coordinate system with the center of the ellipse at the origin and the major axis on the x axis. The y axis points up, and the x axis points to the right. How much clearance above the water is there 5 feet from the bank?

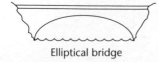

Elliptical bridge

32. Design. A 4 × 8 foot elliptical tabletop is to be cut out of a 4 × 8 foot rectangular sheet of teak plywood (see the figure). To draw the ellipse on the plywood, how far should the foci be located from each edge and how long a piece of string must be fastened to each focus to produce the ellipse (see Figure 1 in the text)? Compute the answer to two decimal places.

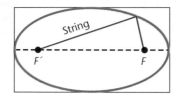

Elliptical table

★ **33. Aeronautical Engineering.** Of all possible wing shapes, it has been determined that the one with the least drag along the trailing edge is an ellipse. The leading edge may be a straight line, as shown in the figure. One of the most famous planes with this design was the World War II British Spitfire. The plane in the figure has a wingspan of 48.0 feet.

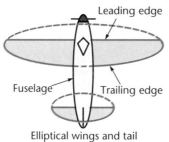

Elliptical wings and tail

(A) If the straight-line leading edge is parallel to the major axis of the ellipse (outlined in color in the figure) and is 1.14 feet in front of it, and if the leading edge is 46.0 feet long (including the width of the fuselage), find the equation of the ellipse. Let the x axis lie along the major axis (positive right), and let the y axis lie along the minor axis (positive forward).

(B) How wide is the wing in the center of the fuselage (assuming the wing passes through the fuselage)?

Compute quantities to 3 significant digits.

★ **34. Naval Architecture.** Currently, many high-performance racing sailboats use elliptical keels, rudders, and main sails for the same reasons stated in Problem 33—less drag along the trailing edge. In the accompanying figure, the ellipse containing the keel has a 12.0 foot major axis. The straight-line leading edge is parallel to the major axis of the ellipse and 1.00 foot in front of it. The chord is 1.00 foot shorter than the major axis.

Rudder   Keel

(A) Find the equation of the ellipse. Let the y axis lie along the minor axis of the ellipse, and let the x axis lie along the major axis, both with positive direction upward.

(B) What is the width of the keel, measured perpendicular to the major axis, 1 foot up the major axis from the bottom end of the keel?

Compute quantities to 3 significant digits.

SECTION $9\text{-}3$ **Hyperbola**

- Definition of a Hyperbola
- Drawing a Hyperbola
- Standard Equations and Their Graphs
- Applications

As before, we start with a coordinate-free definition of a hyperbola. Using this definition, we show how a hyperbola can be drawn and we derive standard equations for hyperbolas specially located in a rectangular coordinate system.

• Definition of a Hyperbola

The following is a coordinate-free definition of a hyperbola:

DEFINITION 1

Hyperbola

A **hyperbola** is the set of all points P in a plane such that the absolute value of the difference of the distances of P to two fixed points in the plane is a positive constant. Each of the fixed points, F' and F, is called a **focus**. The intersection points V' and V of the line through the foci and the two branches of the hyperbola are called **vertices**, and each is called a **vertex**. The line segment $V'V$ is called the **transverse axis**. The midpoint of the transverse axis is the **center** of the hyperbola.

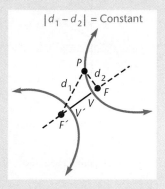

• Drawing a Hyperbola

Thumbtacks, a straightedge, string, and a pencil are all that are needed to draw a hyperbola (see Figure 1). Place two thumbtacks in a piece of cardboard—these form the foci of the hyperbola. Rest one corner of the straightedge at the focus F' so that it is free to rotate about this point. Cut a piece of string shorter than the length of the straightedge, and fasten one end to the straightedge corner A and the other end to the thumbtack at F. Now push the string with a pencil up against the straightedge at B. Keeping the string taut, rotate the straightedge about F', keeping the corner at F'. The resulting curve will be part of a hyperbola. Other parts of the hyperbola can be drawn by changing the position of the straightedge and string. To see that the resulting curve meets the conditions of the definition, note that the difference of the distances BF' and BF is

$$
\begin{aligned}
BF' - BF &= BF' + BA - BF - BA \\
&= AF' - (BF + BA) \\
&= \left(\begin{matrix} \text{Straightedge} \\ \text{length} \end{matrix} \right) - \left(\begin{matrix} \text{String} \\ \text{length} \end{matrix} \right) \\
&= \text{Constant}
\end{aligned}
$$

FIGURE 1   Drawing a hyperbola.

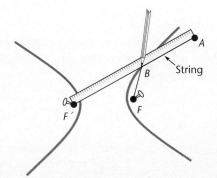

● **Standard Equations and Their Graphs**

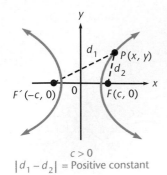

$c > 0$
$|d_1 - d_2|$ = Positive constant

FIGURE 2 Hyperbola with foci on the x axis.

Using the definition of a hyperbola and the distance-between-two-points formula, we can derive the standard equations for a hyperbola located in a rectangular coordinate system. We start by placing a hyperbola in the coordinate system with the foci on the x axis equidistant from the origin at $F'(-c, 0)$ and $F(c, 0)$, $c > 0$, as in Figure 2.

Just as for the ellipse, it is convenient to represent the constant difference by $2a$, $a > 0$. Also, the geometric fact that the difference of two sides of a triangle is always less than the third side can be applied to Figure 2 to derive the following useful result:

$$|d_1 - d_2| < 2c$$

$$2a < 2c$$

$$a < c \qquad (1)$$

We will use this result in the derivation of the equation of a hyperbola, which we now begin.

Referring to Figure 2, the point $P(x, y)$ is on the hyperbola if and only if

$$|d_1 - d_2| = 2a$$

$$|d(P, F') - d(P, F)| = 2a$$

$$|\sqrt{(x + c)^2 + y^2} - \sqrt{(x - c)^2 + y^2}| = 2a$$

After eliminating radicals and absolute value signs by appropriate use of squaring and simplifying, another good exercise for you, we have

$$(c^2 - a^2)x^2 - a^2 y^2 = a^2(c^2 - a^2) \qquad (2)$$

$$\frac{x^2}{a^2} - \frac{y^2}{c^2 - a^2} = 1 \qquad (3)$$

Dividing both sides of equation (2) by $a^2(c^2 - a^2)$ is permitted, since neither a^2 nor $c^2 - a^2$ is 0. From equation (1), $a < c$; thus, $a^2 < c^2$ and $c^2 - a^2 > 0$. The constant a was chosen positive at the beginning.

To simplify equation (3) further, we let

$$b^2 = c^2 - a^2 \qquad b > 0 \qquad (4)$$

to obtain

$$\frac{x^2}{a^2} - \frac{y^2}{b^2} = 1 \qquad (5)$$

From equation (5) we see that the x intercepts, which are also the vertices, are $x = \pm a$ and there are no y intercepts. To see why there are no y intercepts, let $x = 0$ and solve for y:

$$\frac{0^2}{a^2} - \frac{y^2}{b^2} = 1$$

$$y^2 = -b^2$$

$$y = \pm \sqrt{-b^2} \qquad \text{An imaginary number}$$

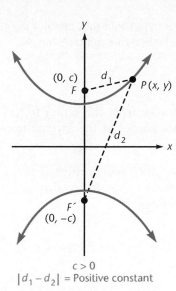

$c > 0$
$|d_1 - d_2|$ = Positive constant

FIGURE 3 Hyperbola with foci on the y axis.

If we start with the foci on the y axis at $F'(0, -c)$ and $F(0, c)$ as in Figure 3, instead of on the x axis as in Figure 2, then, following arguments similar to those used for the first derivation, we obtain

$$\frac{y^2}{a^2} - \frac{x^2}{b^2} = 1 \tag{6}$$

where the relationshiip among a, b, and c remains the same as before:

$$b^2 = c^2 - a^2 \tag{7}$$

The center is still at the origin, but the transverse axis is now on the y axis.

As an aid to graphing equation (5), we solve the equation for y in terms of x, another good exercise for you, to obtain

$$y = \pm\frac{b}{a}x\sqrt{1 - \frac{a^2}{x^2}} \tag{8}$$

As x changes so that $|x|$ becomes larger, the expression $1 - (a^2/x^2)$ within the radical approaches 1. Hence, for large values of $|x|$, equation (5) behaves very much like the lines

$$y = \pm\frac{b}{a}x \tag{9}$$

These lines are **asymptotes** for the graph of equation (5). The hyperbola approaches these lines as a point $P(x, y)$ on the hyperbola moves away from the origin (see Figure 4). An easy way to draw the asymptotes is to first draw the rectangle as in Figure 4, then extend the diagonals. We refer to this rectangle as the **asymptote rectangle**.

FIGURE 4 Asymptotes.

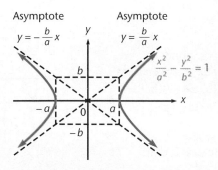

Starting with equation (6) and proceeding as we did for equation (5), we obtain the asymptotes for the graph of equation (6):

$$y = \pm\frac{a}{b}x \tag{10}$$

The perpendicular bisector of the transverse axis, extending from one side of the asymptote rectangle to the other, is called the **conjugate axis** of the hyperbola.

Given an equation of the form (5) or (6), how can we find the coordinates of the foci without memorizing or looking up the relation $b^2 = c^2 - a^2$? Just as with the ellipse, there is a simple geometric relationship in a hyperbola that enables us to get the same result using the Pythagorean theorem. To see this relationship, we rewrite $b^2 = c^2 - a^2$ in the form

$$c^2 = a^2 + b^2 \qquad (11)$$

Note in the figures in Theorem 1 below that the distance from the center to a focus is the same as the distance from the center to a corner of the asymptote rectangle. Stated in another way:

> A circle, with center at the origin, that passes through all four corners of the asymptote rectangle also passes through all foci of hyperbolas with asymptotes determined by the diagonals of the rectangle.

We summarize all the preceding results in Theorem 1 for convenient reference.

Theorem 1   **Standard Equations of a Hyperbola with Center at (0, 0)**

1. $\dfrac{x^2}{a^2} - \dfrac{y^2}{b^2} = 1$

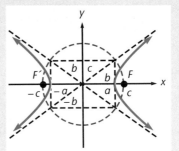

   x intercepts: $\pm a$ (vertices)
   y intercepts: none
   Foci: $F'(-c, 0)$, $F(c, 0)$

   $$c^2 = a^2 + b^2$$

   Transverse axis length $= 2a$
   Conjugate axis length $= 2b$

2. $\dfrac{y^2}{a^2} - \dfrac{x^2}{b^2} = 1$

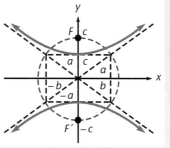

   x intercepts: none
   y intercepts: $\pm a$ (vertices)
   Foci: $F'(0, -c)$, $F(0, c)$

   $$c^2 = a^2 + b^2$$

   Transverse axis length $= 2a$
   Conjugate axis length $= 2b$

[*Note:* Both graphs are symmetric with respect to the x axis, y axis, and origin.]

EXAMPLE 1   Graphing Hyperbolas

Sketch the graph of each equation, find the coordinates of the foci, and find the lengths of the transverse and conjugate axes.

(A) $9x^2 - 16y^2 = 144$ (B) $16y^2 - 9x^2 = 144$ (C) $2x^2 - y^2 = 10$

Solutions (A) First, write the equation in standard form by dividing both sides by 144:

$$9x^2 - 16y^2 = 144$$

$$\frac{x^2}{16} - \frac{y^2}{9} = 1 \qquad a^2 = 16 \text{ and } b^2 = 9$$

Locate x intercepts, $x = \pm 4$; there are no y intercepts. Sketch the asymptotes using the asymptote rectangle, then sketch in the hyperbola.

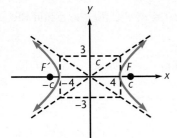

$$\text{Foci:} \quad c^2 = a^2 + b^2$$
$$= 16 + 9$$
$$= 25$$
$$c = 5$$

Thus, the foci are $F'(-5, 0)$ and $F(5, 0)$.

$$\text{Transverse axis length} = 2(4) = 8$$

$$\text{Conjugate axis length} = 2(3) = 6$$

(B) $$16y^2 - 9x^2 = 144$$

$$\frac{y^2}{9} - \frac{x^2}{16} = 1 \qquad a^2 = 9 \text{ and } b^2 = 16$$

Locate y intercepts, $y = \pm 3$; there are no x intercepts. Sketch the asymptotes using the asymptote rectangle, then sketch in the hyperbola. It is important to note that the transverse axis and the foci are on the y axis.

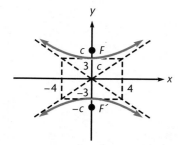

$$\text{Foci:} \quad c^2 = a^2 + b^2$$
$$= 9 + 16$$
$$= 25$$
$$c = 5$$

Thus, the foci are $F'(0, -5)$ and $F(0, 5)$.

$$\text{Transverse axis length} = 2(3) = 6$$

$$\text{Conjugate axis length} = 2(4) = 8$$

(C) $$2x^2 - y^2 = 10$$

$$\frac{x^2}{5} - \frac{y^2}{10} = 1 \qquad a^2 = 5 \text{ and } b^2 = 10$$

Locate x intercepts, $x = \pm\sqrt{5}$; there are no y intercepts. Sketch the asymptotes using the asymptote rectangle, then sketch in the hyperbola.

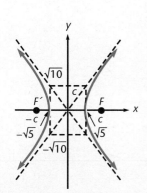

$$\text{Foci:} \quad c^2 = a^2 + b^2$$
$$= 5 + 10$$
$$= 15$$
$$c = \sqrt{15}$$

Thus, the foci are $F'(-\sqrt{15}, 0)$ and $F(\sqrt{15}, 0)$.

$$\text{Transverse axis length} = 2\sqrt{5} \approx 4.47$$

$$\text{Conjugate axis length} = 2\sqrt{10} \approx 6.32$$

Problem 1 Sketch the graph of each equation, find the coordinates of the foci, and find the lengths of the transverse and conjugate axes.

(A) $16x^2 - 25y^2 = 400$ (B) $25y^2 - 16x^2 = 400$ (C) $y^2 - 3x^2 = 12$

Hyperbolas of the form

$$\frac{x^2}{M} - \frac{y^2}{N} = 1 \quad \text{and} \quad \frac{y^2}{N} - \frac{x^2}{M} = 1 \quad M, N > 0$$

are called **conjugate hyperbolas**. In Example 1 and Problem 1, the hyperbolas in parts (A) and (B) are conjugate hyperbolas—they share the same asymptotes.

CAUTION   When making a quick sketch of a hyperbola, it is a common error to have the hyperbola opening up and down when it should open left and right, or vice versa. The mistake can be avoided if you first locate the intercepts accurately.

EXAMPLE 2   Finding the Equation of a Hyperbola

Find an equation of a hyperbola in the form

$$\frac{y^2}{M} - \frac{x^2}{N} = 1 \quad M, N > 0$$

if the center is at the origin, and:

(A) Length of transverse axis is 12   (B) Length of transverse axis is 6
        Length of conjugate axis is 20   Distance of foci from center is 5

Solutions   (A) Start with

$$\frac{y^2}{a^2} - \frac{x^2}{b^2} = 1$$

and find a and b:

$$a = \frac{12}{2} = 6 \quad \text{and} \quad b = \frac{20}{2} = 10$$

Thus, the equation is

$$\frac{y^2}{36} - \frac{x^2}{100} = 1$$

(B) Start with

$$\frac{y^2}{a^2} - \frac{x^2}{b^2} = 1$$

and find a and b:

$$a = \frac{6}{2} = 3$$

To find b, sketch the asymptote rectangle, label known parts, and use the Pythagorean theorem:

$$b^2 = 5^2 - 3^2$$
$$= 16$$
$$b = 4$$

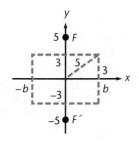

Thus, the equation is

$$\frac{y^2}{9} - \frac{x^2}{16} = 1$$

Problem 2 Find an equation of a hyperbola in the form

$$\frac{x^2}{M} - \frac{y^2}{N} = 1 \qquad M, N > 0$$

if the center is at the origin, and:

(A) Length of transverse axis is 50
Length of conjugate axis is 30

(B) Length of conjugate axis is 12
Distance of foci from center is 9

● **Applications** You may not be aware of the many important uses of hyperbolic forms. They are encountered in the study of comets; the loran system of navigation for pleasure boats, ships, and aircraft; sundials; capillary action; nuclear cooling towers; optical and radiotelescopes; and contemporary architectural structures. The TWA building at Kennedy Airport is a *hyperbolic paraboloid*, and the St. Louis Science Center Planetarium is a *hyperboloid*. (See Figure 5 on the next page).

Some comets from outer space occasionally enter the sun's gravitational field, follow a hyperbolic path around the sun (with the sun at a focus), and then leave, never to be seen again [Figure 5(a)]. In the loran system of navigation, transmitting stations in three locations, S_1, S_2, and S_3 [see Figure 5(b)], send out signals simultaneously. A ship with a receiver records the difference in the arrival times of the signals from S_1 and S_2 and also records the difference in arrival times of the signals from S_2 and S_3. The difference in arrival times can be transformed into differences of the distances that the ship is to S_1 and S_2 and to S_2 and S_3. Plotting all points so that these differences in distances remain constant produces two branches, p_1

FIGURE 5 Uses of hyperbolic forms.

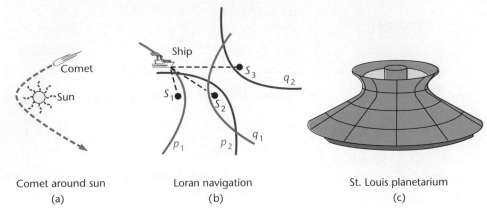

Comet around sun	Loran navigation	St. Louis planetarium
(a)	(b)	(c)

and p_2, of a hyperbola with foci S_1 and S_2 and two branches, q_1 and q_2, of a hyperbola with foci S_2 and S_3. It is easy to tell which branches the ship is on by noting the arrival times of the signals from each station. The intersection of a branch from each hyperbola locates the ship. Most of these calculations are now done by shipboard computers, and positions in longitude and latitude are given. This system of navigation is widely used for coastal navigation. Inexpensive loran units are now found on many small pleasure boats. Figure 5(c) illustrates a hyperboloid used architecturally. With such structures, thin concrete shells can span large spaces.

Answers to Matched Problems   **1.** (A)

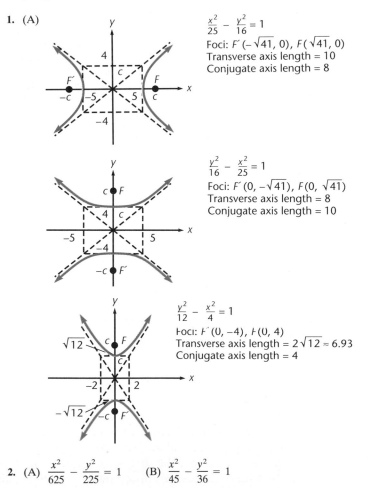

$\dfrac{x^2}{25} - \dfrac{y^2}{16} = 1$
Foci: $F'(-\sqrt{41}, 0)$, $F(\sqrt{41}, 0)$
Transverse axis length $= 10$
Conjugate axis length $= 8$

$\dfrac{y^2}{16} - \dfrac{x^2}{25} = 1$
Foci: $F'(0, -\sqrt{41})$, $F(0, \sqrt{41})$
Transverse axis length $= 8$
Conjugate axis length $= 10$

$\dfrac{y^2}{12} - \dfrac{x^2}{4} = 1$
Foci: $F'(0, -4)$, $F(0, 4)$
Transverse axis length $= 2\sqrt{12} \approx 6.93$
Conjugate axis length $= 4$

2. (A) $\dfrac{x^2}{625} - \dfrac{y^2}{225} = 1$ (B) $\dfrac{x^2}{45} - \dfrac{y^2}{36} = 1$

EXERCISE 9-3

A

Sketch a graph of each equation in Problems 1–8, find the coordinates of the foci, and find the lengths of the transverse and conjugate axes.

1. $\dfrac{x^2}{9} - \dfrac{y^2}{4} = 1$ 　　**2.** $\dfrac{x^2}{9} - \dfrac{y^2}{25} = 1$ 　　**3.** $\dfrac{y^2}{4} - \dfrac{x^2}{9} = 1$ 　　**4.** $\dfrac{y^2}{25} - \dfrac{x^2}{9} = 1$

5. $4x^2 - y^2 = 16$ 　　**6.** $x^2 - 9y^2 = 9$ 　　**7.** $9y^2 - 16x^2 = 144$ 　　**8.** $4y^2 - 25x^2 = 100$

B

Sketch a graph of each equation in Problems 9–12, find the coordinates of the foci, and find the lengths of the transverse and conjugate axes.

9. $3x^2 - 2y^2 = 12$ 　　**10.** $3x^2 - 4y^2 = 24$ 　　**11.** $7y^2 - 4x^2 = 28$ 　　**12.** $3y^2 - 2x^2 = 24$

In Problems 13–20, find an equation of a hyperbola in the form

$$\frac{x^2}{M} - \frac{y^2}{N} = 1 \quad or \quad \frac{y^2}{N} - \frac{x^2}{M} = 1 \quad M, N > 0$$

if the center is at the origin, and:

13. Transverse axis on x axis
Transverse axis length = 14
Conjugate axis length = 10

14. Transverse axis on x axis
Transverse axis length = 8
Conjugate axis length = 6

15. Transverse axis on y axis
Transverse axis length = 24
Conjugate axis length = 18

16. Transverse axis on y axis
Transverse axis length = 16
Conjugate axis length = 22

17. Transverse axis on x axis
Transverse axis length = 18
Distance of foci from center = 11

18. Transverse axis on x axis
Transverse axis length = 16
Distance of foci from center = 10

19. Conjugate axis on x axis
Conjugate axis length = 14
Distance of foci from center = $\sqrt{200}$

20. Conjugate axis on x axis
Conjugate axis length = 10
Distance of foci from center = $\sqrt{70}$

In Problems 21–24, graph each system of equations in the same rectangular coordinate system and find the coordinates of any points of intersection.

21. $3y^2 - 4x^2 = 12$
$y^2 + x^2 = 25$

22. $y^2 - x^2 = 3$
$y^2 + x^2 = 5$

23. $2x^2 + y^2 = 24$
$x^2 - y^2 = -12$

24. $2x^2 + y^2 = 17$
$x^2 - y^2 = -5$

In Problems 25–28, find all points of intersection for each system of equations to three decimal places:

(A) Algebraically
(B) Using a graphic calculator or a computer

25. $y^2 - x^2 = 9$
$2y - x = 8$
$y \geq 0$

26. $y^2 - x^2 = 4$
$y - x = 6$
$y \geq 0$

27. $y^2 - x^2 = 4$
$y^2 + 2x^2 = 36$
$y \geq 0$

28. $y^2 - x^2 = 1$
$2y^2 + x^2 = 16$
$y \geq 0$

C

*Eccentricity. Problems 29 and 30 below and Problems 29 and 30 in Exercise 9-2 are related to a property of conics called **eccentricity**, which is denoted by a positive real number E. Parabolas, ellipses, and hyperbolas all can be defined in terms of E, a fixed point called a focus, and a fixed line not containing the focus called a directrix as follows: The set of points in a plane each of whose distance from a fixed point is E times its distance from a fixed line is an ellipse if 0 < E < 1, a parabola if E = 1, and a hyperbola if E > 1.*

29. Find an equation of the set of points in a plane each of whose distance from (3, 0) is three-halves its distance from the line $x = \frac{4}{3}$. Identify the geometric figure.

30. Find an equation of the set of points in a plane each of whose distance from (0, 4) is four-thirds its distance from the line $y = \frac{9}{4}$. Identify the geometric figure.

APPLICATIONS

31. Architecture. An architect is interested in designing a thin-shelled dome in the shape of a hyperbolic paraboloid, as shown in figure (a). Find the equation of the hyperbola located in a coordinate system [figure (b)] satisfying the indicated conditions. How far is the hyperbola above the vertex 6 feet to the right of the vertex? Compute the answer to two decimal places.

If the tower is 500 feet tall, the top is 150 feet above the center of the hyperbola, and the base is 350 feet below the center, what is the radius of the top and the base? What is the radius of the smallest circular cross section in the tower? Compute answers to 3 significant digits.

33. Space Science. In tracking space probes to the outer planets, NASA uses large parabolic reflectors with di-

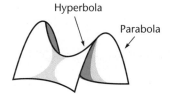

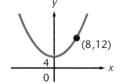

Hyperbolic paraboloid
(a)

Hyperbola part of dome
(b)

32. Nuclear Power. A nuclear cooling tower is a **hyperboloid**, that is, a hyperbola rotated around its conjugate axis, as shown in figure (a). The equation of the hyperbola in figure (b) used to generate the hyperboloid is

$$\frac{x^2}{100^2} - \frac{y^2}{150^2} = 1$$

Nuclear cooling tower
(a)

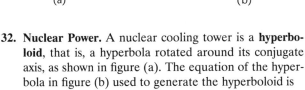

(b)

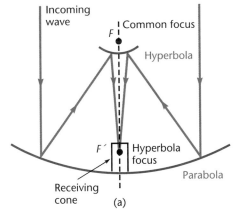

Receiving cone

(a)

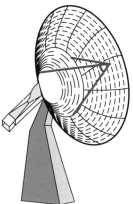

Radiotelescope
(b)

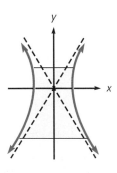

ameters equal to two-thirds the length of a football field. Needless to say, many design problems are created by the weight of these reflectors. One weight problem is solved by using a hyperbolic reflector sharing the parabola's focus to reflect the incoming electromagnetic waves to the other focus of the hyperbola where receiving equipment is installed (see the figure).

For the receiving antenna shown in the figure, the common focus F is located 120 feet above the vertex of the parabola, and focus F' (for the hyperbola) is 20 feet above the vertex. The vertex of the reflecting hyperbola is 110 feet above the vertex for the parabola. Introduce a coordinate system by using the axis of the parabola as the y axis (up positive), and let the x axis pass through the center of the hyperbola (right positive). What is the equation of the reflecting hyperbola? Write y in terms of x.

SECTION 9-4 Translation of Axes

- Translation of Axes
- Standard Equations of Translated Conics
- Graphing Equations of the Form $Ax^2 + Cy^2 + Dx + Ey + F = 0$
- Finding Equations of Conics

In the last three sections we found standard equations for parabolas, ellipses, and hyperbolas located with their axes on the coordinate axes and centered relative to the origin. What happens if we move conics away from the origin while keeping their axes parallel to the coordinate axes? We will show that we can obtain new standard equations that are special cases of the equation $Ax^2 + Cy^2 + Dx + Ey + F = 0$, where A and C are not both zero. The basic mathematical tool used in this endeavor is *translation of axes*. The usefulness of translation of axes is not limited to graphing conics, however. Translation of axes can be put to good use in many other graphing situations.

• Translation of Axes

A **translation of coordinate axes** occurs when the new coordinate axes have the same direction as and are parallel to the original coordinate axes. To see how coordinates in the original system are changed when moving to the translated system, and vice versa, refer to Figure 1.

FIGURE 1 Translation of coordinates.

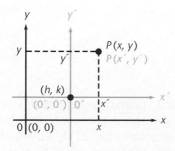

A point P in the plane has two sets of coordinates: (x, y) in the original system and (x', y') in the translated system. If the coordinates of the origin of the translated system are (h, k) relative to the original system, then the old and new coordinates are related as given in Theorem 1.

Theorem 1	**Translation Formulas**
	1. $x = x' + h$ **2.** $x' = x - h$
	$ y = y' + k$ $ y' = y - k$

It can be shown that these formulas hold for (h, k) located anywhere in the original coordinate system.

EXAMPLE 1   **Equation of a Curve in a Translated System**

A curve has the equation

$$(x - 4)^2 + (y + 1)^2 = 36$$

If the origin is translated to $(4, -1)$, find the equation of the curve in the translated system and identify the curve.

Solution   Since $(h, k) = (4, -1)$, use translation formulas

$$x' = x - h = x - 4$$
$$y' = y - k = y + 1$$

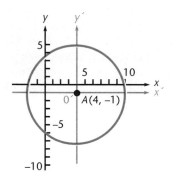

to obtain, after substitution,

$$x'^2 + y'^2 = 36$$

This is the equation of a circle of radius 6 with center at the new origin. The coordinates of the new origin in the original coordinate system are $(4, -1)$. Note that this result agrees with our general treatment of the circle in Section 3-1.

Problem 1   A curve has the equation $(y + 2)^2 = 8(x - 3)$. If the origin is translated to $(3, -2)$, find an equation of the curve in the translated system and identify the curve.

● **Standard Equations of Translated Conics**   We now proceed to find standard equations of conics translated away from the origin. We do this by first writing the standard equations found in earlier sections in the $x'y'$ coordinate system with $0'$ at (h, k). We then use translation equations to find the standard forms relative to the original xy coordinate system. The equations of translation in all cases are

$$x' = x - h$$
$$y' = y - k$$

For parabolas we have

$$x'^2 = 4ay' \qquad (x - h)^2 = 4a(y - k)$$
$$y'^2 = 4ax' \qquad (y - k)^2 = 4a(x - h)$$

For circles we have

$$x'^2 + y'^2 = r^2 \qquad (x - h)^2 + (y - k)^2 = r^2$$

For ellipses we have for $a > b > 0$

$$\frac{x'^2}{a^2} + \frac{y'^2}{b^2} = 1 \qquad \frac{(x - h)^2}{a^2} + \frac{(y - k)^2}{b^2} = 1$$

$$\frac{x'^2}{b^2} + \frac{y'^2}{a^2} = 1 \qquad \frac{(x - h)^2}{b^2} + \frac{(y - k)^2}{a^2} = 1$$

For hyperbolas we have

$$\frac{x'^2}{a^2} - \frac{y'^2}{b^2} = 1 \qquad \frac{(x - h)^2}{a^2} - \frac{(y - k)^2}{b^2} = 1$$

$$\frac{y'^2}{a^2} - \frac{x'^2}{b^2} = 1 \qquad \frac{(y - k)^2}{a^2} - \frac{(x - h)^2}{b^2} = 1$$

Table 1 on the next page summarizes these results with appropriate figures and some properties discussed earlier.

● Graphing Equations of the Form $Ax^2 + Cy^2 + Dx + Ey + F = 0$

It can be shown that the graph of

$$Ax^2 + Cy^2 + Dx + Ey + F = 0 \qquad (1)$$

where A and C are not both zero, is a conic or a degenerate conic or that there is no graph. If we can transform equation (1) into one of the standard forms in Table 1, then we will be able to identify its graph and sketch it rather quickly. The process of completing the square discussed in Section 2-6 will be our primary tool in accomplishing this transformation. A couple of examples should help make the process clear.

EXAMPLE 2 **Graphing a Translated Conic**

Transform

$$y^2 - 6y - 4x + 1 = 0 \qquad (2)$$

into one of the standard forms in Table 1. Identify the conic and graph it.

TABLE 1 Standard Equations for Translated Conics

Parabolas

$$(x - h)^2 = 4a(y - k)$$

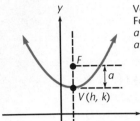

Vertex (h, k)
Focus $(h, k + a)$
$a > 0$ opens up
$a < 0$ opens down

$$(y - k)^2 = 4a(x - h)$$

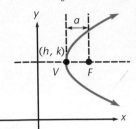

Vertex (h, k)
Focus $(h + a, k)$
$a < 0$ opens left
$a > 0$ opens right

Circles

$$(x - h)^2 + (y - k)^2 = r^2$$

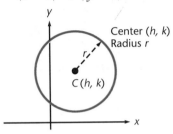

Center (h, k)
Radius r

Ellipses

$$\frac{(x - h)^2}{a^2} + \frac{(y - k)^2}{b^2} = 1 \qquad a > b > 0$$

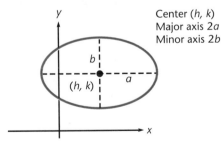

Center (h, k)
Major axis $2a$
Minor axis $2b$

$$\frac{(x - h)^2}{b^2} + \frac{(y - k)^2}{a^2} = 1$$

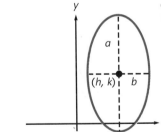

Center (h, k)
Major axis $2a$
Minor axis $2b$

Hyperbolas

$$\frac{(x - h)^2}{a^2} - \frac{(y - k)^2}{b^2} = 1$$

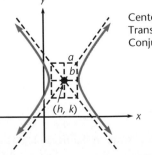

Center (h, k)
Transverse axis $2a$
Conjugate axis $2b$

$$\frac{(y - k)^2}{a^2} - \frac{(x - h)^2}{b^2} = 1$$

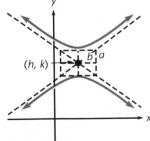

Center (h, k)
Transverse axis $2a$
Conjugate axis $2b$

Solution   *Step 1.* Complete the square in equation (2) relative to each variable that is squared—in this case y:

$$y^2 - 6y - 4x + 1 = 0$$
$$y^2 - 6y \qquad = 4x - 1$$
$$y^2 - 6y \quad + 9 = 4x + 8 \qquad \text{Add 9 to both sides to complete the square on the left side.}$$
$$(y - 3)^2 = 4(x + 2) \tag{3}$$

From Table 1 we recognize equation (3) as an equation of a parabola opening to the right with vertex at $(h, k) = (-2, 3)$.

Step 2. Find the equation of the parabola in the translated system with origin $0'$ at $(h, k) = (-2, 3)$. The equations of translation are read directly from equation (3):

$$x' = x + 2$$
$$y' = y - 3$$

Making these substitutions in equation (3) we obtain

$$y'^2 = 4x' \tag{4}$$

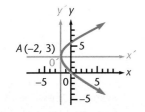

the equation of the parabola in the $x'y'$ system.

Step 3. Graph equation (4) in the $x'y'$ system following the process discussed in Section 9-1. The resulting graph is the graph of the original equation relative to the original xy coordinate system.

Problem 2   Transform

$$x^2 + 4x + 4y - 12 = 0$$

into one of the standard forms in Table 1. Identify the conic and graph it.

EXAMPLE 3   **Graphing a Translated Conic**

Transform

$$9x^2 - 4y^2 - 36x - 24y - 36 = 0$$

into one of the standard forms in Table 1. Identify the conic and graph it. Find the coordinates of any foci relative to the original system.

Solution   *Step 1.* Complete the square relative to both x and y.

$$9x^2 - 4y^2 - 36x - 24y - 36 = 0$$
$$9x^2 - 36x \quad - 4y^2 - 24y \quad = 36$$
$$9(x^2 - 4x \quad) - 4(y^2 + 6y \quad) = 36$$
$$9(x^2 - 4x + 4) - 4(y^2 + 6y + 9) = 36 + 36 - 36$$
$$9(x - 2)^2 - 4(y + 3)^2 = 36$$
$$\frac{(x - 2)^2}{4} - \frac{(y + 3)^2}{9} = 1$$

From Table 1 we recognize the last equation as an equation of a hyperbola opening left and right with center at $(h, k) = (2, -3)$.

Step 2. Find the equation of the hyperbola in the translated system with origin $0'$ at $(h, k) = (2, -3)$. The equations of translation are read directly from the last equation in step 1:

$$x' = x - 2$$
$$y' = y + 3$$

Making these substitutions, we obtain

$$\frac{x'^2}{4} - \frac{y'^2}{9} = 1$$

the equation of the hyperbola in the $x'y'$ system.

Step 3. Graph the equation obtained in step 2 in the $x'y'$ system following the process discussed in Section 9-3. The resulting graph is the graph of the original equation relative to the original xy coordinate system.

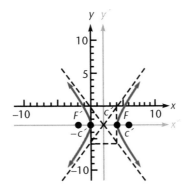

Step 4. Find the coordinates of the foci. To find the coordinates of the foci in the original system, first find the coordinates in the translated system:

$$c'^2 = 2^2 + 3^2 = 13$$
$$c' = \sqrt{13}$$
$$-c' = -\sqrt{13}$$

Thus, the coordinates in the translated system are

$$F'(-\sqrt{13}, 0) \qquad \text{and} \qquad F(\sqrt{13}, 0)$$

Now, use

$$x = x' + h = x' + 2$$
$$y = y' + k = y' - 3$$

to obtain

$$F'(-\sqrt{13} + 2, -3) \qquad \text{and} \qquad F(\sqrt{13} + 2, -3)$$

as the coordinates of the foci in the original system.

Problem 3   Transform

$$9x^2 + 16y^2 + 36x - 32y - 92 = 0$$

into one of the standard forms in Table 1. Identify the conic and graph it. Find the coordinates of any foci relative to the original system.

● **Finding Equations of Conics**   We now reverse the problem: Given certain information about a conic in a rectangular coordinate system, find its equation.

EXAMPLE 4   **Finding the Equation of a Translated Conic**

Find the equation of a hyperbola with vertices on the line $x = -4$, conjugate axis on the line $y = 3$, length of the transverse axis = 4, and length of the conjugate axis = 6.

Solution   Locate the vertices, asymptote rectangle, and asymptotes in the original coordinate system [figure (a)], then sketch the hyperbola and translate the origin to the center of the hyperbola [figure (b)].

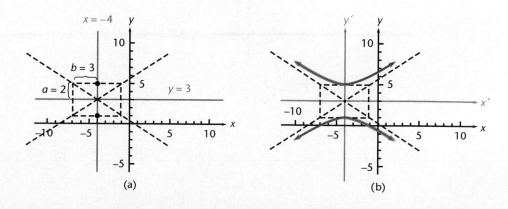

(a)   (b)

Next write the equation of the hyperbola in the translated system:

$$\frac{y'^2}{4} - \frac{x'^2}{9} = 1$$

The origin in the translated system is at $(h, k) = (-4, 3)$, and the translation formulas are

$$x' = x - h = x - (-4) = x + 4$$
$$y' = y - k = y - 3$$

Thus, the equation of the hyperbola in the original system is

$$\frac{(y - 3)^2}{4} - \frac{(x + 4)^2}{9} = 1$$

or, after simplifying and writing in the form of equation (1),

$$4x^2 - 9y^2 + 32x + 54y + 19 = 0$$

Problem 4   Find the equation of an ellipse with foci on the line $x = 4$, minor axis on the line $y = -3$, length of the major axis $= 8$, and length of the minor axis $= 4$.

Answers to Matched Problems
1. $y'^2 = 8x'$; a parabola
2. $(x + 2)^2 = -4(y - 4)$; a parabola

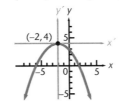

3. $\dfrac{(x + 2)^2}{16} + \dfrac{(y - 1)^2}{9} = 1$; ellipse   Foci: $F'(-\sqrt{7} - 2, 1), F(\sqrt{7} - 2, 1)$

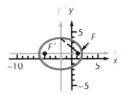

4. $\dfrac{(x - 4)^2}{4} + \dfrac{(y + 3)^2}{16} = 1$, or $4x^2 + y^2 - 32x + 6y + 57 = 0$

EXERCISE $9\text{-}4$

A

In Problems 1–8:
(A) Find translation formulas that translate the origin to the indicated point (h, k).
(B) Write the equation of the curve for the translated system.
(C) Identify the curve.

1. $(x - 3)^2 + (y - 5)^2 = 81; (3, 5)$

2. $(x - 3)^2 = 8(y + 2); (3, -2)$

3. $\dfrac{(x + 7)^2}{9} + \dfrac{(y - 4)^2}{16} = 1; (-7, 4)$

4. $(x + 2)^2 + (y + 6)^2 = 36; (-2, -6)$

5. $(y + 9)^2 = 16(x - 4); (4, -9)$

6. $\dfrac{(y - 9)^2}{10} - \dfrac{(x + 5)^2}{6} = 1; (-5, 9)$

7. $\dfrac{(x + 8)^2}{12} + \dfrac{(y + 3)^2}{8} = 1; (-8, -3)$

8. $\dfrac{(x + 7)^2}{25} - \dfrac{(y - 8)^2}{50} = 1; (-7, 8)$

In Problems 9–14:
(A) Write each equation in one of the standard forms listed in Table 1.
(B) Identify the curve.

9. $16(x - 3)^2 - 9(y + 2)^2 = 144$

10. $(y + 2)^2 - 12(x - 3) = 0$

11. $6(x + 5)^2 + 5(y + 7)^2 = 30$

12. $12(y - 5)^2 - 8(x - 3)^2 = 24$

13. $(x + 6)^2 + 24(y - 4) = 0$

14. $4(x - 7)^2 + 7(y - 3)^2 = 28$

B

In Problems 15–22, transform each equation into one of the standard forms in Table 1. Identify the curve and graph it.

15. $4x^2 + 9y^2 - 16x - 36y + 16 = 0$

16. $16x^2 + 9y^2 + 64x + 54y + 1 = 0$

17. $x^2 + 8x + 8y = 0$

18. $y^2 + 12x + 4y - 32 = 0$

19. $x^2 + y^2 + 12x + 10y + 45 = 0$

20. $x^2 + y^2 - 8x - 6y = 0$

21. $-9x^2 + 16y^2 - 72x - 96y - 144 = 0$

22. $16x^2 - 25y^2 - 160x = 0$

In Problems 23–30, use the given information to find the equation of each conic. Express the answer in the form $Ax^2 + Cy^2 + Dx + Ey + F = 0$ with integer coefficients and $A > 0$.

23. A parabola with vertex at $(2, 5)$, axis the line $x = 2$, and passing through the point $(-2, 1)$.

24. A parabola with vertex at $(4, -1)$, axis the line $y = -1$, and passing through the point $(2, 3)$.

25. An ellipse with major axis on the line $y = -3$, minor axis on the line $x = -2$, length of major axis $= 8$, and length of minor axis $= 4$.

26. An ellipse with major axis on the line $x = -4$, minor axis on the line $y = 1$, length of major axis $= 4$, and length of minor axis $= 2$.

27. An ellipse with vertices $(4, -7)$ and $(4, 3)$ and foci $(4, -6)$ and $(4, 2)$.

28. An ellipse with vertices $(-3, 1)$ and $(7, 1)$ and foci $(-1, 1)$ and $(5, 1)$.

29. A hyperbola with transverse axis on the line $x = 2$, length of transverse axis $= 4$, conjugate axis on the line $y = 3$, and length of conjugate axis $= 2$.

30. A hyperbola with transverse axis on the line $y = -5$, length of transverse axis $= 6$, conjugate axis on the line $x = 2$, and length of conjugate axis $= 6$.

C _____

In Problems 31–36, find the coordinates of any foci relative to the original coordinate system:

31. Problem 15   **32.** Problem 16   **33.** Problem 17

34. Problem 18   **35.** Problem 21   **36.** Problem 22

Chapter 9 Review

9-1   CONIC SECTIONS; PARABOLA

The plane curves obtained by intersecting a right circular cone with a plane are called **conic sections**. If the plane cuts clear through one nappe, then the intersection curve is called a **circle** if the plane is perpendicular to the axis and an **ellipse** if the plane is not perpendicular to the axis. If a plane cuts only one nappe, but does not cut clear through, then the intersection curve is called a **parabola**. If a plane cuts through both nappes, but not through the vertex, the resulting intersection curve is called a **hyperbola**. A plane passing through the vertex of the cone produces a **degenerate conic**—a point, a line, or a pair of lines. The figure illustrates the four nondegenerate conics.

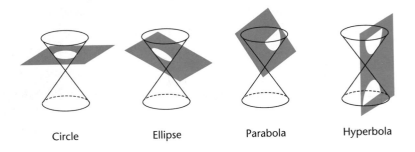

Circle   Ellipse   Parabola   Hyperbola

The graph of

$$Ax^2 + Bxy + Cy^2 + Dx + Ey + F = 0$$

where A, B, and C are not all 0, is a conic.

The following is a coordinate-free definition of a parabola:

Parabola

A **parabola** is the set of all points in a plane equidistant from a fixed point F and a fixed line L in the plane. The fixed point F is called the **focus**, and the fixed line L is called the **directrix**. A line through the focus perpendicular to the directrix is called the **axis**, and the point on the axis halfway between the directrix and focus is called the **vertex**.

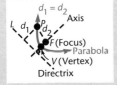

From the definition of a parabola, we can obtain the following standard equations:

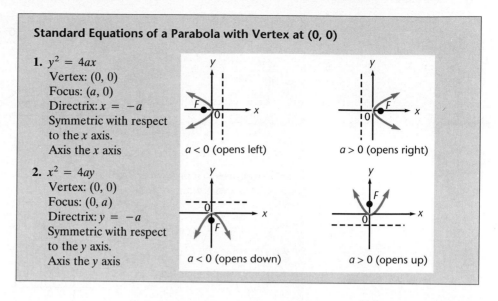

Standard Equations of a Parabola with Vertex at (0, 0)

1. $y^2 = 4ax$
Vertex: $(0, 0)$
Focus: $(a, 0)$
Directrix: $x = -a$
Symmetric with respect to the x axis.
Axis the x axis

$a < 0$ (opens left) $a > 0$ (opens right)

2. $x^2 = 4ay$
Vertex: $(0, 0)$
Focus: $(0, a)$
Directrix: $y = -a$
Symmetric with respect to the y axis.
Axis the y axis

$a < 0$ (opens down) $a > 0$ (opens up)

9-2   ELLIPSE

The following is a coordinate-free definition of an ellipse:

Ellipse

An **ellipse** is the set of all points P in a plane such that the sum of the distances of P from two fixed points in the plane is constant. Each of the fixed points, F' and F, is called a **focus**, and together they are called **foci**. Referring to the figure, the line segment $V'V$ through the foci is the **major axis**. The perpendicular bisector $B'B$ of the major axis is the **minor axis**. Each end of the major axis, V' and V, is called a **vertex**. The midpoint of the line segment $F'F$ is called the **center** of the ellipse.

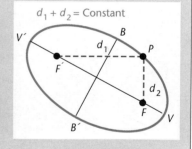

$d_1 + d_2 = \text{Constant}$

From the definition of an ellipse, we can obtain the following standard equations:

Standard Equations of an Ellipse with Center at (0, 0)

1. $\dfrac{x^2}{a^2} + \dfrac{y^2}{b^2} = 1 \qquad a > b > 0$

   x intercepts: $\pm a$ (vertices)
   y intercepts: $\pm b$
   Foci: $F'(-c, 0)$, $F(c, 0)$
   $$c^2 = a^2 - b^2$$
   Major axis length $= 2a$
   Minor axis length $= 2b$

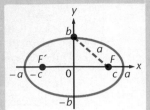

2. $\dfrac{x^2}{b^2} + \dfrac{y^2}{a^2} = 1 \qquad a > b > 0$

   x intercepts: $\pm b$
   y intercepts: $\pm a$ (vertices)
   Foci: $F'(0, -c)$, $F(0, c)$
   $$c^2 = a^2 - b^2$$
   Major axis length $= 2a$
   Minor axis length $= 2b$

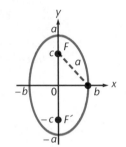

[*Note:* Both graphs are symmetric with respect to the x axis, y axis, and origin. Also, the major axis is always longer than the minor axis.]

9-3   HYPERBOLA

The following is a coordinate-free definition of a hyperbola:

Hyperbola

A **hyperbola** is the set of all points P in a plane such that the absolute value of the difference of the distances of P to two fixed points in the plane is a positive constant. Each of the fixed points, F' and F, is called a **focus**. The intersection points V' and V of the line through the foci and the two branches of the hyperbola are called **vertices**, and each is called a **vertex**. The line segment $V'V$ is called the **transverse axis**. The midpoint of the transverse axis is the **center** of the hyperbola.

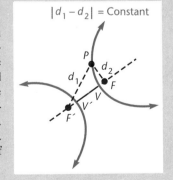

From the definition of a hyperbola, we can obtain the following standard equations:

Standard Equations of a Hyperbola with Center at (0, 0)

1. $\dfrac{x^2}{a^2} - \dfrac{y^2}{b^2} = 1$

 x intercepts: $\pm a$ (vertices)
 y intercepts: none
 Foci: $F'(-c, 0)$, $F(c, 0)$

 $$c^2 = a^2 + b^2$$

 Transverse axis length $= 2a$
 Conjugate axis length $= 2b$

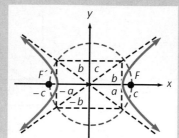

2. $\dfrac{y^2}{a^2} - \dfrac{x^2}{b^2} = 1$

 x intercepts: none
 y intercepts: $\pm a$ (vertices)
 Foci: $F'(0, -c)$, $F(0, c)$

 $$c^2 = a^2 + b^2$$

 Transverse axis length $= 2a$
 Conjugate axis length $= 2b$

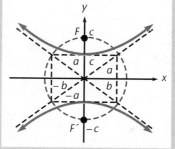

[*Note:* Both graphs are symmetric with respect to the x axis, y axis, and origin.]

9-4   TRANSLATION OF AXES

In the last three sections we found standard equations for parabolas, ellipses, and hyperbolas located with their axes on the coordinate axes and centered relative to the origin. We now move the conics away from the origin while keeping their axes parallel to the coordinate axes. In this process we obtain new standard equations that are special cases of the equation $Ax^2 + Cy^2 + Dx + Ey + F = 0$, where A and C are not both zero. The basic mathematical tool used is *translation of axes*.

A **translation of coordinate axes** occurs when the new coordinate axes have the same direction as and are parallel to the original coordinate axes. **Translation formulas** are as follows:

1. $x = x' + h$ **2.** $x' = x - h$
  $y = y' + k$ $y' = y - k$

where (h, k) are the coordinates of the origin $0'$ relative to the original system.

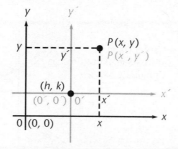

Table 1 lists the standard equations for translated conics.

TABLE 1 Standard Equations for Translated Conics

Parabolas

$$(x - h)^2 = 4a(y - k)$$

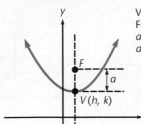

Vertex (h, k)
Focus $(h, k + a)$
$a > 0$ opens up
$a < 0$ opens down

$$(y - k)^2 = 4a(x - h)$$

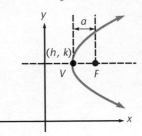

Vertex (h, k)
Focus $(h + a, k)$
$a < 0$ opens left
$a > 0$ opens right

Circles

$$(x - h)^2 + (y - k)^2 = r^2$$

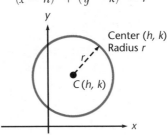

Center (h, k)
Radius r

Ellipses

$$\frac{(x - h)^2}{a^2} + \frac{(y - k)^2}{b^2} = 1 \qquad a > b > 0$$

$$\frac{(x - h)^2}{b^2} + \frac{(y - k)^2}{a^2} = 1$$

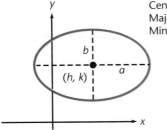

Center (h, k)
Major axis $2a$
Minor axis $2b$

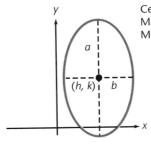

Center (h, k)
Major axis $2a$
Minor axis $2b$

Hyperbolas

$$\frac{(x - h)^2}{a^2} - \frac{(y - k)^2}{b^2} = 1$$

$$\frac{(y - k)^2}{a^2} - \frac{(x - h)^2}{b^2} = 1$$

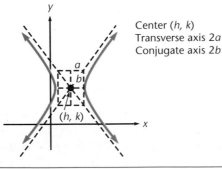

Center (h, k)
Transverse axis $2a$
Conjugate axis $2b$

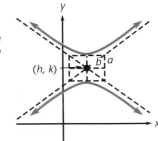

Center (h, k)
Transverse axis $2a$
Conjugate axis $2b$

Chapter 9 Review Exercise

Work through all the problems in this chapter review and check answers in the back of the book. Answers to all review problems are there, and following each answer is a number in italics indicating the section in which that type of problem is discussed. Where weaknesses show up, review appropriate sections in the text.

A

In Problems 1–3, graph each equation and locate foci. Locate the directrix for any parabolas. Find the lengths of major, minor, transverse, and conjugate axes where applicable.

1. $9x^2 + 25y^2 = 225$

2. $x^2 = -12y$

3. $25y^2 - 9x^2 = 225$

In Problems 4–6:
(A) Write each equation in one of the standard forms listed in Table 1 of the review.
(B) Identify the curve.

4. $4(y + 2)^2 - 25(x - 4)^2 = 100$

5. $(x + 5)^2 + 12(y + 4) = 0$

6. $16(x - 6)^2 + 9(y - 4)^2 = 144$

B

7. Find the equation of the parabola having its vertex at the origin, its axis the x axis, and $(-4, -2)$ on its graph.

8. Find an equation of an ellipse in the form

$$\frac{x^2}{M} + \frac{y^2}{N} = 1 \qquad M, N > 0$$

if the center is at the origin, the major axis is on the y axis, the minor axis length is 6, and the distance of the foci from the center is 4.

9. Find an equation of a hyperbola in the form

$$\frac{y^2}{M} - \frac{x^2}{N} = 1 \qquad M, N > 0$$

if the center is at the origin, the conjugate axis length is 8, and the foci are 5 units from the center.

In Problems 10–12, graph each system of equations in the same coordinate system and find the coordinates of any points of intersection.

10. $x^2 + 4y^2 = 32$
$\quad\;\; x + 2y = 0$

11. $16x^2 + 25y^2 = 400$
$\quad\;\; 16x^2 - 45y = 0$

12. $\quad\; x^2 + y^2 = 10$
$\quad 16x^2 + y^2 = 25$

In Problems 13–15, transform each equation into one of the standard forms in Table 1 in the review. Identify the curve and graph it.

13. $16x^2 + 4y^2 + 96x - 16y + 96 = 0$

14. $x^2 - 4x - 8y - 20 = 0$

15. $4x^2 - 9y^2 + 24x - 36y - 36 = 0$

C

16. Use the definition of a parabola and the distance formula to find the equation of a parabola with directrix $x = 6$ and focus at $(2, 4)$.

17. Find an equation of the set of points in a plane each of whose distance from $(4, 0)$ is twice its distance from the line $x = 1$. Identify the geometric figure.

18. Find an equation of the set of points in a plane each of whose distance from $(4, 0)$ is two-thirds its distance from the line $x = 9$. Identify the geometric figure.

In Problems 19–21, find the coordinates of any foci relative to the original coordinate system.

19. Problem 13   20. Problem 14   21. Problem 15

APPLICATIONS

22. **Communications.** A parabolic satellite television antenna has a diameter of 8 feet and is 1 foot deep. How far is the focus from the vertex?

23. **Engineering.** An elliptical gear is to have foci 8 centimeters apart and a major axis 10 centimeters long. Letting the x axis lie along the major axis (right positive) and the y axis lie along the minor axis (up positive), write the equation of the ellipse in the standard form

$$\frac{x^2}{a^2} + \frac{y^2}{b^2} = 1$$

24. **Space Science.** A hyperbolic reflector for a radio-telescope (such as that illustrated in Problem 33, Exercise 9-3) has the equation

$$\frac{y^2}{40^2} - \frac{x^2}{30^2} = 1$$

If the reflector has a diameter of 30 feet, how deep is it? Compute the answer to 3 significant digits.

An Introduction to Probability

Probability theory, like many branches of mathematics, evolved out of practical considerations. Girolamo Cardano (1501–1576), a gambler and physician, produced some of the best mathematics of his time, including a systematic analysis of gambling problems. In 1654, another gambler, the Chevalier de Méré, plagued with bad luck, approached the well-known French philosopher and mathematician Blaise Pascal (1623–1662) regarding certain dice problems. Pascal became interested in these problems, studied them, and discussed them with Pierre de Fermat (1601–1665). Thus, out of the gaming rooms of western Europe the study of probability was born.

In spite of this lowly birth, probability theory has matured into a highly respected and immensely useful branch of mathematics. Its use is found in practically every field. Probability can be thought of as the science of uncertainty. If, for example, a single die is rolled, it is uncertain which number will turn up. But if a die is rolled many times, the percentage of times a particular number, say 2, will appear is approximately predictable. Probability theory is concerned with determining the long-run percentage of occurrences of a given event.

How are probabilities assigned to events? There are two basic approaches to this problem, one theoretical and the other empirical. An example will illustrate the difference between the two approaches.

Returning to our original example, suppose you were asked, "What is the probability of obtaining a 2 on a single throw of a die?" Using a *theoretical approach*, we would reason as follows: Since there are six equally likely ways the die can turn up, assuming it is fair, and there is only one way a 2 can turn up, then the probability of obtaining a 2 is $\frac{1}{6}$. Here, we have arrived at a probability assignment using certain assumptions and a reasoning process. What does the result have to do with reality? We would, of course, expect that in the long run the 2 would appear approximately one-sixth of the time. With the *empirical approach*, we make no assumption about the equally likely ways the die can turn up. We simply set up an experiment and roll the die a large number of times. Then we compute the percentage of times the 2 appears and use this number as an estimate of the probability of obtaining a 2 on a single roll of the die.

We will start our study by considering the theoretical approach and developing procedures that lead to the solution of a variety of interesting problems. The procedures we discuss require counting the number of ways certain events can occur, and this is not always easy. However, there are effective mathematical tools that can assist us in this counting task. The development of these tools is the subject matter of the first section.

SECTION 10-1   **Multiplication Principle, Permutations, and Combinations**

- Multiplication Principle
- Permutations
- Combinations

If there are 10 teams in a league, and each team is to play every other team exactly once, how many games must be scheduled? From a standard 52-card deck, how many 5-card hands will have all hearts? How many ways can 3 letters appear in a row on a license plate if no letter is repeated? These are not particularly easy questions to answer without some thought. In this section we develop counting tools that will enable us to answer these questions, as well as many more.

● **Multiplication Principle**

We start with an example.

EXAMPLE 1   **Combined Outcomes**

Suppose we flip a coin and then throw a single die (see Figure 1). What are the possible combined outcomes?

Solution   To solve this problem, we use a **tree diagram**:

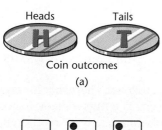

Coin outcomes

(a)

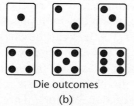

Die outcomes

(b)

FIGURE 1 Coin and die outcomes.

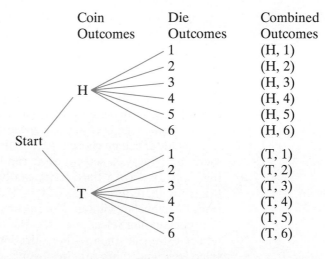

Thus, there are twelve possible combined outcomes—two ways in which the coin can come up followed by six ways in which the die can come up.

Problem 1 Use a tree diagram to determine the number of possible outcomes of throwing a single die followed by flipping a coin.

Now suppose you are asked, "From the 26 letters in the alphabet, how many ways can 3 letters appear in a row on a license plate if no letter is repeated?" To try to count the possibilities using a tree diagram would be extremely tedious, to say the least. The following **multiplication principle**, also called the **fundamental counting principle**, enables us to solve this problem easily. In addition, it forms the basis for several other counting techniques developed later in this section.

Multiplication Principle

1. If two operations O_1 and O_2 are performed in order, with N_1 possible outcomes for the first operation and N_2 possible outcomes for the second operation, then there are

$$N_1 \cdot N_2$$

possible combined outcomes of the first operation followed by the second.

2. In general, if n operations $O_1, O_2, \ldots, O_n$ are performed in order, with possible number of outcomes $N_1, N_2, \ldots, N_n$, respectively, then there are

$$N_1 \cdot N_2 \cdot \cdots \cdot N_n$$

possible combined outcomes of the operations performed in the given order.

In Example 1, we see that there are two possible outcomes from the first operation of flipping a coin and six possible outcomes from the second operation of throwing a die. Hence, by the multiplication principle, there are $2 \cdot 6 = 12$ possible combined outcomes of flipping a coin followed by throwing a die. Use the multiplication principle to solve Problem 1.

To answer the license plate question, we reason as follows: There are 26 ways the first letter can be chosen. After a first letter is chosen, 25 letters remain; hence there are 25 ways a second letter can be chosen. And after 2 letters are chosen, there are 24 ways a third letter can be chosen. Hence, using the multiplication principle, there are $26 \cdot 25 \cdot 24 = 15{,}600$ possible ways 3 letters can be chosen from the alphabet without allowing any letter to repeat. By not allowing any letter to repeat, earlier selections affect the choice of subsequent selections. If we allow letters to repeat, then earlier selections do not affect the choice in subsequent selections, and there are 26 possible choices for each of the 3 letters. Thus, if we allow letters to repeat, there are $26 \cdot 26 \cdot 26 = 26^3 = 17{,}576$ possible ways the 3 letters can be chosen from the alphabet.

EXAMPLE 2   Computer-Generated Tests

Many universities and colleges are now using computer-assisted testing procedures. Suppose a screening test is to consist of 5 questions, and a computer stores 5 equivalent questions for the first test question, 8 equivalent questions for the second, 6 for the third, 5 for fourth, and 10 for the fifth. How many different 5-question tests can the computer select? Two tests are considered different if they differ in one or more questions.

Solution

O_1: Select the first question N_1: 5 ways

O_2: Select the second question N_2: 8 ways

O_3: Select the third question N_3: 6 ways

O_4: Select the fourth question N_4: 5 ways

O_5: Select the fifth question N_5: 10 ways

Thus, the computer can generate

$$5 \cdot 8 \cdot 6 \cdot 5 \cdot 10 = 12{,}000 \text{ different tests}$$

Problem 2   Each question on a multiple-choice test has 5 choices. If there are 5 such questions on a test, how many different response sheets are possible if only 1 choice is marked for each question?

EXAMPLE 3   Counting Code Words

How many 3-letter code words are possible using the first 8 letters of the alphabet if:

(A) No letter can be repeated?   (B) Letters can be repeated?
(C) Adjacent letters cannot be alike?

Solutions   (A) No letter can be repeated.

O_1: Select first letter N_1: 8 ways

O_2: Select second letter N_2: 7 ways   Because 1 letter has been used

O_3: Select third letter N_3: 6 ways   Because 2 letters have been used

Thus, there are

$$8 \cdot 7 \cdot 6 = 336 \text{ possible code words}$$

(B) Letters can be repeated.

O_1: Select first letter N_1: 8 ways

O_2: Select second letter N_2: 8 ways   Repeats are allowed.

O_3: Select third letter N_3: 8 ways   Repeats are allowed.

Thus, there are

$$8 \cdot 8 \cdot 8 = 8^3 = 512 \text{ possible code words}$$

(C) Adjacent letters cannot be alike.

O_1: Select first letter N_1: 8 ways

O_2: Select second letter N_2: 7 ways   Cannot be the same as the first

O_3: Select third letter N_3: 7 ways   Cannot be the same as the second, but can be the same as the first

Thus, there are

$$8 \cdot 7 \cdot 7 = 392 \text{ possible code words}$$

Problem 3 How many 4-letter code words are possible using the first 10 letters of the alphabet under the three conditions stated in Example 3?

The multiplication principle can be used to develop two additional methods for counting that are extremely useful in more complicated counting problems. Both of these methods use the factorial function, which was introduced in Section 8-5.

● **Permutations**

Suppose 4 pictures are to be arranged from left to right on one wall of an art gallery. How many arrangements are possible? Using the multiplication principle, there are 4 ways of selecting the first picture. After the first picture is selected, there are 3 ways of selecting the second picture. After the first 2 pictures are selected, there are 2 ways of selecting the third picture. And after the first 3 pictures are selected, there is only 1 way to select the fourth. Thus, the number of arrangements possible for the 4 pictures is

$$4 \cdot 3 \cdot 2 \cdot 1 = 4! \quad \text{or} \quad 24$$

In general, we refer to a particular arrangement, or **ordering**, of n objects without repetition as a **permutation** of the n objects. How many permutations of *n* objects are there? From the reasoning above, there are *n* ways in which the first object can be chosen, there are $n - 1$ ways in which the second object can be chosen, and so on. Applying the multiplication principle, we have Theorem 1:

Theorem 1

Permutations of *n* Objects

The number of permutations of *n* objects, denoted by $P_{n,n}$, is given by

$$P_{n,n} = n \cdot (n - 1) \cdot \cdots \cdot 1 = n!$$

Now suppose the director of the art gallery decides to use only 2 of the 4 available pictures on the wall, arranged from left to right. How many arrangements of 2 pictures can be formed from the 4? There are 4 ways the first picture can be selected. After selecting the first picture, there are 3 ways the second picture can be selected. Thus, the number of arrangements of 2 pictures from 4 pictures, denoted by $P_{4,2}$, is given by

$$P_{4,2} = 4 \cdot 3 = 12$$

Or, in terms of factorials, multiplying $4 \cdot 3$ by 1 in the form $2!/2!$, we have

$$P_{4,2} = 4 \cdot 3 = \frac{4 \cdot 3 \cdot 2!}{2!} = \frac{4!}{2!}$$

This last form gives $P_{4,2}$ in terms of factorials, which is useful in some cases.

A **permutation of a set of *n* objects taken *r* at a time** is an arrangement of the *r* objects in a specific order. Thus, reasoning in the same way as in the example above, we find that the number of permutations of *n* objects taken *r* at a time, $0 \le r \le n$, denoted by $P_{n,r}$, is given by

$$P_{n,r} = n(n - 1)(n - 2) \cdot \cdots \cdot (n - r + 1)$$

Multiplying the right side of this equation by 1 in the form $(n - r)!/(n - r)!$, we obtain a factorial form for $P_{n,r}$:

$$P_{n,r} = n(n - 1)(n - 2) \cdot \cdots \cdot (n - r + 1) \frac{(n - r)!}{(n - r)!}$$

But

$$n(n - 1)(n - 2) \cdot \cdots \cdot (n - r + 1)(n - r)! = n!$$

Hence, we have Theorem 2:

Theorem 2

Permutation of n Objects Taken r at a Time

The number of permutations of n objects taken r at a time is given by

$$P_{n,r} = \underbrace{n(n-1)(n-2) \cdots \cdots (n-r+1)}_{r \text{ factors}}$$

or

$$P_{n,r} = \frac{n!}{(n-r)!} \qquad 0 \le r \le n$$

Note that if $r = n$, then the number of permutations of n objects taken n at a time is

$$P_{n,n} = \frac{n!}{(n-n)!} = \frac{n!}{0!} = n! \qquad \text{Recall, } 0! = 1.$$

which agrees with Theorem 1, as it should.

The permutation symbol $P_{n,r}$ also can be denoted by P_r^n, $_nP_r$, or $P(n, r)$. Your calculator may have a key labeled $\boxed{P_{n,r}}$ or $\boxed{nP_r}$ that can be used to evaluate the permutation symbol. If not, then it probably has a factorial function key $\boxed{x!}$ that can be used to facilitate the evaluation of $P_{n,r}$.

EXAMPLE 4   Selecting Officers

From a committee of 8 people, in how many ways can we choose a chair and a vice-chair, assuming one person cannot hold more than one position?

Solution   We are actually asking for the number of permutations of 8 objects taken 2 at a time—that is, $P_{8,2}$:

$$P_{8,2} = \frac{8!}{(8-2)!} = \frac{8!}{6!} = \frac{8 \cdot 7 \cdot 6!}{6!} = 56$$

Problem 4   From a committee of 10 people, in how many ways can we choose a chair, vice-chair, and secretary, assuming one person cannot hold more than one position?

CAUTION   Remember to use the definition of factorial when simplifying fractions involving factorials.

$$\frac{6!}{3!} \ne 2! \qquad \frac{6!}{3!} = \frac{6 \cdot 5 \cdot 4 \cdot 3!}{3!} = 120$$

EXAMPLE 5   Evaluating $P_{n,r}$

Find the number of permutations of 25 objects taken 8 at a time. Compute the answer to 4 significant digits using a calculator.

Solution

$$P_{25,8} = \frac{25!}{(25 - 8)!} = \frac{25!}{17!} = 4.361 \times 10^{10} \quad \text{A very large number}$$

Problem 5   Find the number of permutations of 30 objects taken 4 at a time. Compute the answer exactly using a calculator.

• **Combinations**   Now suppose that an art museum owns 8 paintings by a given artist and another art museum wishes to borrow 3 of these paintings for a special show. How many ways can 3 paintings be selected for shipment out of the 8 available? Here, the order of the items selected doesn't matter. What we are actually interested in is how many subsets of 3 objects can be formed from a set of 8 objects. We call such a subset a **combination** of 8 objects taken 3 at a time. The total number of combinations is denoted by the symbol

$$C_{8,3} \qquad \text{or} \qquad \binom{8}{3}$$

To find the number of combinations of 8 objects taken 3 at a time, $C_{8,3}$, we make use of the formula for $P_{n,r}$ and the multiplication principle. We know that the number of permutations of 8 objects taken 3 at a time is given by $P_{8,3}$, and we have a formula for computing this quantity. Now suppose we think of $P_{8,3}$ in terms of two operations:

O_1:   Select a subset of 3 objects (paintings)

N_1: $C_{8,3}$ ways

O_2:   Arrange the subset in a given order

N_2:   3! ways

The combined operation, O_1 followed by O_2, produces a permutation of 8 objects taken 3 at a time. Thus,

$$P_{8,3} = C_{8,3} \cdot 3!$$

To find $C_{8,3}$, we replace $P_{8,3}$ in the above equation with $8!/(8 - 3)!$ and solve for $C_{8,3}$:

$$\frac{8!}{(8 - 3)!} = C_{8,3} \cdot 3!$$

$$C_{8,3} = \frac{8!}{3!(8 - 3)!} = \frac{8 \cdot 7 \cdot 6 \cdot 5!}{3 \cdot 2 \cdot 1 \cdot 5!} = 56$$

Thus, the museum can make 56 different selections of 3 paintings from the 8 available.

A **combination of a set of n objects taken r at a time** is an r-element subset of the n objects. Reasoning in the same way as in the example, the number of combinations of n objects taken r at a time, $0 \leq r \leq n$, denoted by $C_{n,r}$, can be obtained by solving for $C_{n,r}$ in the relationship

$$P_{n,r} = C_{n,r} \cdot r!$$

$$C_{n,r} = \frac{P_{n,r}}{r!}$$

$$= \frac{n!}{r!(n-r)!} \qquad P_{n,r} = \frac{n!}{(n-r)!}$$

Theorem 3

Combination of n Objects Taken r at a Time

The number of combinations of n objects taken r at a time is given by

$$C_{n,r} = \binom{n}{r} = \frac{P_{n,r}}{r!} = \frac{n!}{r!(n-r)!} \qquad 0 \leq r \leq n$$

Note that we used the combination formula in Section 8-5 to represent binomial coefficients.

The combination symbols $C_{n,r}$ and $\binom{n}{r}$ also can be denoted by C_r^n, $_nC_r$, or $C(n, r)$. If n and r are other than small numbers, a calculator with an $\boxed{x!}$ function key will simplify computations involving combinations, and one with a $\boxed{C_{n,r}}$ or $\boxed{_nC_r}$ function key will simplify computations even further.

EXAMPLE 6 Selecting Subcommittees

From a committee of 8 people, in how many ways can we choose a subcommittee of 2 people?

Solution Notice how this example differs from Example 4, where we wanted to know how many ways a chair and a vice-chair can be chosen from a committee of 8 people. In Example 4, ordering matters. In choosing a subcommittee of 2 people, the ordering does not matter. Thus, we are actually asking for the number of combinations of 8 objects taken 2 at a time. The number is given by

$$C_{8,2} = \binom{8}{2} = \frac{8!}{2!(8-2)!} = \frac{8 \cdot 7 \cdot 6!}{2 \cdot 1 \cdot 6!} = 28$$

Problem 6 How many subcommittees of 3 people can be chosen from a committee of 8 people?

EXAMPLE 7   Evaluating $C_{n,r}$

Find the number of combinations of 25 objects taken 8 at a time. Compute the answer to 4 significant digits using a calculator.

Solution
$$C_{25,8} = \binom{25}{8} = \frac{25!}{8!(25-8)!} = \frac{25!}{8!17!} = 1.082 \times 10^6$$

Compare this result with that obtained in Example 5.

Problem 7   Find the number of combinations of 30 objects taken 4 at a time. Compute the answer exactly using a calculator.

Remember: In a permutation, order counts. In a combination, order does not count.

To determine whether a permutation or combination is needed, decide whether rearranging the collection or listing makes a difference. If so, use permutations. If not, use combinations.

A standard deck of 52 cards involves four suits, hearts, spades, diamonds, and clubs, as shown in Figure 2. Example 8, as well as other examples and exercises in this chapter, refer to this standard deck.

FIGURE 2 A standard deck of cards.

EXAMPLE 8   Counting Card Hands

Out of a standard 52-card deck, how many 5-card hands will have 3 aces and 2 kings?

Solution
O_1: Choose 3 aces out of 4 possible   Order is not important.

N_1: $C_{4,3}$

O_2: Choose 2 kings out of 4 possible   Order is not important.

N_2: $C_{4,2}$

Using the multiplication principle, we have

$$\text{Number of hands} = C_{4,3} \cdot C_{4,2} = 4 \cdot 6 = 24$$

Problem 8   From a standard 52-card deck, how many 5-card hands will have 3 hearts and 2 spades?

EXAMPLE 9   Counting Serial Numbers

Serial numbers for a product are to be made using 2 letters followed by 3 numbers. If the letters are to be taken from the first 8 letters of the alphabet with no repeats and the numbers from the 10 digits 0 through 9 with no repeats, how many serial numbers are possible?

Solution O_1:   Choose 2 letters out of 8 available   Order is important.

N_1: $P_{8,2}$

O_2:   Choose 3 numbers out of 10 available   Order is important.

N_2: $P_{10,3}$

Using the multiplication principle, we have

$$\text{Number of serial numbers} = P_{8,2} \cdot P_{10,3} = 40{,}320$$

Problem 9   Repeat Example 9 under the same conditions, except the serial numbers are now to have 3 letters followed by 2 digits with no repeats.

Answers to Matched Problems

1. 12;

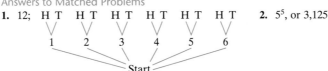

2. 5^5, or 3,125

3. (A) $10 \cdot 9 \cdot 8 \cdot 7 = 5{,}040$ (B) $10 \cdot 10 \cdot 10 \cdot 10 = 10{,}000$ (C) $10 \cdot 9 \cdot 9 \cdot 9 = 7{,}290$

4. $P_{10,3} = \dfrac{10!}{(10-3)!} = 720$ **5.** $P_{30,4} = \dfrac{30!}{(30-4)!} = 657{,}720$ **6.** $C_{8,3} = \dfrac{8!}{3!(8-3)!} = 56$

7. $C_{30,4} = \dfrac{30!}{4!(30-4)!} = 27{,}405$ **8.** $C_{13,3} \cdot C_{13,2} = 22{,}308$ **9.** $P_{8,3} \cdot P_{10,2} = 30{,}240$

EXERCISE 10-1

A

Evaluate Problems 1–16.

1. $\dfrac{11!}{8!}$ **2.** $\dfrac{14!}{12!}$ **3.** $\dfrac{5!}{2!3!}$ **4.** $\dfrac{6!}{4!2!}$

5. $\dfrac{7!}{4!(7-4)!}$ **6.** $\dfrac{8!}{3!(8-3)!}$ **7.** $\dfrac{7!}{7!(7-7)!}$ **8.** $\dfrac{8!}{0!(8-0)!}$

9. $P_{5,3}$ **10.** $P_{4,2}$ **11.** $P_{52,4}$ **12.** $P_{52,2}$

13. $C_{5,3}$ **14.** $C_{4,2}$ **15.** $C_{52,4}$ **16.** $C_{52,2}$

17. A particular new car model is available with 5 choices of color, 3 choices of transmission, 4 types of interior, and 2 types of engine. How many different variations of this model car are possible?

18. A deli serves sandwiches with the following options: 3 kinds of bread, 5 kinds of meat, and lettuce or sprouts. How many different sandwiches are possible, assuming one item is used out of each category?

19. In a horse race, how many different finishes among the first 3 places are possible for a 10-horse race? Exclude ties.

20. In a long-distance foot race, how many different finishes among the first 5 places are possible for a 50-person race? Exclude ties.

21. How many ways can a subcommittee of 3 people be selected from a committee of 7 people? How many ways can a president, vice president, and secretary be chosen from a committee of 7 people?

22. Suppose 9 cards are numbered with the 9 digits from 1 to 9. A 3-card hand is dealt, 1 card at a time. How many hands are possible where:
(A) Order is taken into consideration?
(B) Order is not taken into consideration?

23. There are 10 teams in a league. If each team is to play every other team exactly once, how many games must be scheduled?

24. Given 7 points, no 3 of which are on a straight line, how many lines can be drawn joining 2 points at a time?

B

25. How many 4-letter code words are possible from the first 6 letters of the alphabet, with no letter repeated? Allowing letters to repeat?

26. How many 5-letter code words are possible from the first 7 letters of the alphabet, with no letter repeated? Allowing letters to repeat?

27. A combination lock has 5 wheels, each labeled with the 10 digits from 0 to 9. How many opening combinations of 5 numbers are possible, assuming no digit is repeated? Assuming digits can be repeated?

28. A small combination lock on a suitcase has 3 wheels, each labeled with digits from 0 to 9. How many opening combinations of 3 numbers are possible, assuming no digit is repeated? Assuming digits can be repeated?

29. From a standard 52-card deck, how many 5-card hands will have all hearts?

30. From a standard 52-card deck, how many 5-card hands will have all face cards? All face cards, but no kings? Consider only jacks, queens, and kings to be face cards.

31. How many different license plates are possible if each contains 3 letters followed by 3 digits? How many of these license plates contain no repeated letters and no repeated digits?

32. How many 5-digit zip codes are possible? How many of these codes contain no repeated digits?

33. From a standard 52-card deck, how many 7-card hands have exactly 5 spades and 2 hearts?

34. From a standard 52-card deck, how many 5-card hands will have 2 clubs and 3 hearts?

35. A catering service offers 8 appetizers, 10 main courses, and 7 desserts. A banquet chairperson is to select 3 appetizers, 4 main courses, and 2 desserts for a banquet. How many ways can this be done?

36. Three research departments have 12, 15, and 18 members, respectively. If each department is to select a delegate and an alternate to represent the department at a conference, how many ways can this be done?

C

37. A sporting goods store has 12 pairs of ski gloves of 12 different brands thrown loosely in a bin. The gloves are all the same size. In how many ways can a left-hand glove and a right-hand glove be selected that do not match relative to brand?

38. A sporting goods store has 6 pairs of running shoes of 6 different styles thrown loosely in a basket. The shoes are all the same size. In how many ways can a left shoe and a right shoe be selected that do not match?

39. Eight distinct points are selected on the circumference of a circle.
  (A) How many chords can be drawn by joining the points in all possible ways?
  (B) How many triangles can be drawn using these 8 points as vertices?
  (C) How many quadrilaterals can be drawn using these 8 points as vertices?

41. How many ways can 2 people be seated in a row of 5 chairs? 3 people? 4 people? 5 people?

43. A basketball team has 5 distinct positions. Out of 8 players, how many starting teams are possible if:
  (A) The distinct positions are taken into consideration?
  (B) The distinct positions are not taken into consideration?
  (C) The distinct positions are not taken into consideration, but either Mike or Ken, but not both, must start?

40. Five distinct points are selected on the circumference of a circle.
  (A) How many chords can be drawn by joining the points in all possible ways?
  (B) How many triangles can be drawn using these 5 points as vertices?

42. Each of 2 countries sends 5 delegates to a negotiating conference. A rectangular table is used with 5 chairs on each long side. If each country is assigned a long side of the table, how many seating arrangements are possible? [*Hint:* Operation 1 is assigning a long side of the table to each country.]

44. How many committees of 4 people are possible from a group of 9 people if:
  (A) There are no restrictions?
  (B) Both Juan and Mary must be on the committee?
  (C) Either Juan or Mary, but not both, must be on the committee?

SECTION 10-2 Sample Spaces and Probability

- Experiments
- Sample Spaces and Events
- Probability of an Event
- Equally Likely Assumption

This section provides an introduction to probability, a topic that has whole books and courses devoted to it. Probability studies involve many subtle notions, and care must be taken at the beginning to understand the fundamental concepts on which the studies are based. First, we develop a mathematical model for probability studies. Our development, because of space, must be somewhat informal. More formal and precise treatments can be found in books on probability.

• Experiments

Our first step in constructing a mathematical model for probability studies is to describe the type of experiments on which probability studies are based. Some types of experiments do not yield the same results, no matter how carefully they are repeated under the same conditions. These experiments are called **random experiments**. Familiar examples of random experiments are flipping coins, rolling dice, observing the frequency of defective items from an assembly line, or observing the frequency of deaths in a certain age group.

Probability theory is a branch of mathematics that has been developed to deal with outcomes of random experiments, both real and conceptual. In the work that follows, the word **experiment** will be used to mean a random experiment.

● **Sample Spaces and Events**   Associated with outcomes of experiments are *sample spaces* and *events*. Our second step in constructing a mathematical model for probability studies is to define these two terms. Set concepts will be useful in this regard.

Consider the experiment, "A single six-sided die is rolled." What outcomes might we observe? We might be interested in the number of dots facing up, or whether the number of dots facing up is an even number, or whether the number of dots facing up is divisible by 3, and so on. The list of possible outcomes appears endless. In general, there is no unique method of analyzing all possible outcomes of an experiment. Therefore, before conducting an experiment, it is important to decide just what outcomes are of interest.

In the die experiment, suppose we limit our interest to the number of dots facing up when the die comes to rest. Having decided what to observe, we make a list of outcomes of the experiment, called **simple events**, such that in each trial of the experiment, one and only one of the results on the list will occur. The set of simple events for the experiment is called a **sample space** for the experiment. The sample space S we have chosen for the die-rolling experiment is

$$S = \{1, 2, 3, 4, 5, 6\}$$

Now consider the outcome, "The number of dots facing up is an even number." This outcome is not a simple event, since it will occur whenever 2, 4, or 6 dots appear, that is, whenever an element in the subset

$$E = \{2, 4, 6\}$$

occurs. Subset E is called a *compound event*. In general, we have the following definition:

DEFINITION 1   **Event**

Given a sample space S for an experiment, we define an **event E** to be any subset of S. If an event E has only one element in it, it is called a **simple event**. If event E has more than one element, it is called a **compound event**. We say that **an event E occurs** if any of the simple events in E occurs.

EXAMPLE 1   Choosing a Sample Space

A nickel and a dime are tossed. How will we identify a sample space for this experiment?

Solution   There are a number of possibilities, depending on our interest. We will consider three.

(A) If we are interested in whether each coin falls heads (H) or tails (T), then, using a tree diagram, we can easily determine an appropriate sample space for the experiment:

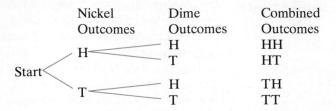

Nickel Outcomes	Dime Outcomes	Combined Outcomes
H	H	HH
	T	HT
T	H	TH
	T	TT

Thus,

$$S_1 = \{HH, HT, TH, TT\}$$

and there are four simple events in the sample space.

(B) If we are interested only in the number of heads that appear on a single toss of the two coins, then we can let

$$S_2 = \{0, 1, 2\}$$

and there are three simple events in the sample space.

(C) If we are interested in whether the coins match (M) or don't match (D), then we can let

$$S_3 = \{M, D\}$$

and there are only two simple events in the sample space.

In Example 1, sample space S_1 contains more information than either S_2 or S_3. If we know which outcome has occurred in S_1, then we know which outcome has occurred in S_2 and S_3. However, the reverse is not true. In this sense, we say that S_1 is a more **fundamental sample space** than either S_2 or S_3.

Important Remark: There is no one correct sample space for a given experiment. When specifying a sample space for an experiment, we include as much detail as necessary to answer *all* questions of interest regarding the outcomes of the experiment. If in doubt, include more elements in the sample space rather than fewer.

Problem 1   An experiment consists of recording the boy–girl composition of families with 2 children.

(A) What is an appropriate sample space if we are interested in the sex of each child in the order of their births? Draw a tree diagram.

(B) What is an appropriate sample space if we are interested only in the number of girls in a family?

(C) What is an appropriate sample space if we are interested only in whether the sexes are alike (A) or different (D)?

(D) What is an appropriate sample space for all three interests expressed above?

Now let's return to the 2-coin problem in Example 1 and the sample space

$$S_1 = \{HH, HT, TH, TT\}$$

Suppose we are interested in the outcome, "Exactly 1 head is up." Looking at S_1, we find that it occurs if either of the two simple events HT or TH occurs.* Thus, to say that the event, "Exactly 1 head is up" occurs is the same as saying the experiment has an outcome in the set

$$E = \{HT, TH\}$$

This is a subset of the sample space S_1. The event E is a compound event.

EXAMPLE 2   **Rolling Two Dice**

Consider an experiment of rolling 2 dice. A convenient sample space that will enable us to answer many questions about interesting events is shown in Figure 1. Let S be the set of all ordered pairs listed in the figure. Note that the simple event (3, 2) is to be distinguished from the simple event (2, 3). The former indicates a 3 turned up on the first die and a 2 on the second, while the latter indicates that a 2 turned up on the first die and a 3 on the second. What is the event that corresponds to each of the following outcomes?

(A) A sum of 7 turns up.   (B) A sum of 11 turns up.
(C) A sum less than 4 turns up.   (D) A sum of 12 turns up.

FIGURE 1 A sample space for rolling two dice.

SECOND DIE

	⚀	⚁	⚂	⚃	⚄	⚅ (a)
⚀	(1, 1)	(1, 2)	(1, 3)	(1, 4)	(1, 5)	(1, 6)
⚁	(2, 1)	(2, 2)	(2, 3)	(2, 4)	(2, 5)	(2, 6)
⚂	(3, 1)	(3, 2)	(3, 3)	(3, 4)	(3, 5)	(3, 6)
⚃	(4, 1)	(4, 2)	(4, 3)	(4, 4)	(4, 5)	(4, 6)
⚄	(5, 1)	(5, 2)	(5, 3)	(5, 4)	(5, 5)	(5, 6)
⚅	(6, 1)	(6, 2)	(6, 3)	(6, 4)	(6, 5)	(6, 6)

FIRST DIE

(b)

Solutions   (A) By "A sum of 7 turns up," we mean that the sum of all dots on both turned-up faces is 7. This outcome corresponds to the event

$$\{(6, 1), (5, 2), (4, 3), (3, 4), (2, 5), (1, 6)\}$$

*Technically, we should write {HT} and {TH}, since there is a logical distinction between an element of a set and a subset consisting of only that element. But we will just keep this in mind and drop the braces for simple events to simplify the notation.

(B) "A sum of 11 turns up" corresponds to the event

$$\{(6, 5), (5, 6)\}$$

(C) "A sum less than 4 turns up" corresponds to the event

$$\{(1, 1), (2, 1), (1, 2)\}$$

(D) "A sum of 12 turns up" corresponds to the event

$$\{(6, 6)\}$$

Problem 2   Refer to the sample space in Example 2 (Figure 1). What is the event that corresponds to each of the following outcomes?

(A) A sum of 5 turns up.
(B) A sum that is a prime number greater than 7 turns up.

Informally, to facilitate discussion, we often use the terms *event* and *outcome of an experiment* interchangeably. Thus, in Example 2 we might say "the event 'A sum of 11 turns up' " in place of "the outcome 'A sum of 11 turns up,' " or even write

$$E = \text{A sum of 11 turns up} = \{(6, 5), (5, 6)\}$$

Technically speaking, an event is the mathematical counterpart of an outcome of an experiment.

● **Probability of an Event**   The next step in developing our mathematical model for probability studies is the introduction of a *probability function*. This is a function that assigns to an arbitrary event associated with a sample space a real number between 0 and 1, inclusive. We start by discussing ways in which probabilities are assigned to simple events in S.

DEFINITION 2   **Probabilities for Simple Events**

Given a sample space

$$S = \{e_1, e_2, \ldots, e_n\}$$

with n simple events, to each simple event e_i we assign a real number, denoted by $P(e_i)$, that is called the **probability of the event e_i**. These numbers may be assigned in an arbitrary manner as long as the following two conditions are satisfied:

1. $0 \leq P(e_i) \leq 1$

2. $P(e_1) + P(e_2) + \cdots + P(e_n) = 1$ The sum of the probabilities of all simple events in the sample space is 1.

Any probability assignment that meets conditions 1 and 2 is said to be an **acceptable probability assignment**.

Our mathematical theory does not explain how acceptable probabilities are assigned to simple events. These assignments are generally based on the expected or actual percentage of times a simple event occurs when an experiment is repeated a large number of times. Assignments based on this principle are called *reasonable*.

Let an experiment be the flipping of a single coin, and let us choose a sample space S to be

$$S = \{H, T\}$$

If a coin appears to be fair, we are inclined to assign probabilities to the simple events in S as follows:

$$P(H) = \tfrac{1}{2} \quad \text{and} \quad P(T) = \tfrac{1}{2}$$

These assignments are based on reasoning that, since there are two ways a coin can land, in the long run, a head will turn up half the time and a tail will turn up half the time. These probability assignments are acceptable, since both of the conditions for acceptable probability assignments in Definition 2 are satisfied:

1. $0 \leq P(H) \leq 1, 0 \leq P(T) \leq 1$

2. $P(H) + P(T) = \tfrac{1}{2} + \tfrac{1}{2} = 1$

But there are other acceptable assignments. Maybe after flipping a coin 1,000 times we find that the head turns up 376 times and the tail turns up 624 times. With this result, we might suspect that the coin is not fair and assign the simple events in the sample space S the probabilities

$$P(H) = .376 \quad \text{and} \quad P(T) = .624$$

This is also an acceptable assignment. But the probability assignment

$$P(H) = 1 \quad \text{and} \quad P(T) = 0$$

though acceptable, is not reasonable, unless the coin has 2 heads. The assignment

$$P(H) = .6 \quad \text{and} \quad P(T) = .8$$

is not acceptable, since $.6 + .8 = 1.4$, which violates condition 2 in Definition 2.

In probability studies, the 0 to the left of the decimal is usually omitted. Thus, we write .8 and not 0.8.

It is important to keep in mind that out of the infinitely many possible acceptable probability assignments to simple events in a sample space, we are generally inclined to choose one assignment over another based on reasoning or experimental results.

Given an acceptable probability assignment for simple events in a sample space S, how do we define the probability of an arbitrary event E associated with S?

DEFINITION 3

Probability of an Event E

Given an acceptable probability assignment for the simple events in a sample space S, we define the **probability of an arbitrary event E**, denoted by $P(E)$, as follows:

1. If E is the empty set, then $P(E) = 0$.

2. If E is a simple event, then $P(E)$ has already been assigned.

3. If E is a compound event, then $P(E)$ is the sum of the probabilities of all the simple events in E.

4. If E is the sample space S, then $P(E) = P(S) = 1$. This is a special case of 3.

EXAMPLE 3   **Finding Probabilities of Events**

Let's return to Example 1, the tossing of a nickel and dime, and the sample space

$$S = \{HH, HT, TH, TT\}$$

Since there are four simple outcomes and the coins are assumed to be fair, it appears that each outcome should occur in the long run 25% of the time. Let's assign the same probability of $\frac{1}{4}$ to each simple event in S:

Simple event, e_i	HH	HT	TH	TT
$P(e_i)$	$\frac{1}{4}$	$\frac{1}{4}$	$\frac{1}{4}$	$\frac{1}{4}$

This is an acceptable assignment according to Definition 2 and a reasonable assignment for ideal coins that are perfectly balanced or coins close to ideal.

(A) What is the probability of getting exactly 1 head?
(B) What is the probability of getting at least 1 head?
(C) What is the probability of getting a head or a tail?
(D) What is the probability of getting 3 heads?

Solutions (A) $$E_1 = \text{Getting 1 head} = \{HT, TH\}$$

Since E_1 is a compound event, we use item 3 in Definition 3 and find $P(E_1)$ by adding the probabilities of the simple events in E_1. Thus,

$$P(E_1) = P(HT) + P(TH) = \tfrac{1}{4} + \tfrac{1}{4} = \tfrac{1}{2}$$

(B) $$E_2 = \text{Getting at least 1 head} = \{HH, HT, TH\}$$

$$P(E_2) = P(HH) + P(HT) + P(TH)$$
$$= \tfrac{1}{4} + \tfrac{1}{4} + \tfrac{1}{4} = \tfrac{3}{4}$$

(C) $$E_3 = \{HH, HT, TH, TT\} = S$$

$$P(E_3) = P(S) = 1 \qquad \qquad \tfrac{1}{4} + \tfrac{1}{4} + \tfrac{1}{4} + \tfrac{1}{4} = 1$$

(D) $$E_3 = \text{Getting 3 heads} = \varnothing \quad \text{Empty set}$$

$$P(\varnothing) = 0$$

Steps for Finding Probabilities of Events

Step 1. Set up an appropriate sample space S for the experiment.

Step 2. Assign acceptable probabilities to the simple events in S.

Step 3. To obtain the probability of an arbitrary event E, add the probabilities of the simple events in E.

The function P defined in steps 2 and 3 is called a **probability function**. The domain of this function is all possible events in the sample space S, and the range is a set of real numbers between 0 and 1, inclusive.

Problem 3   Suppose we toss a nickel and dime 1,000 times and find that HH turns up 273 times, HT 206 times, TH 312 times, and TT 209 times. On the basis of this evidence, we assign the following probabilities to the simple events in the sample space $S = \{HH, HT, TH, TT\}$:

Simple event, e_i	HH	HT	TH	TT
$P(e_i)$	.273	.206	.312	.209

This is an acceptable and reasonable probability assignment for the simple events in S. What are the probabilities of the following events?

(A) $E_1 = $ Getting at least 1 tail   (B) $E_2 = $ Getting 2 tails
(C) $E_3 = $ Getting either a head or a tail

Example 3 and Problem 3 illustrate two important ways in which acceptable and reasonable probability assignments are made for simple events in a sample space S:

1. **Theoretical.** We use assumptions and a deductive reasoning process to assign probabilities to simple events. No experiments are actually conducted. This is what we did in Example 3.

2. **Empirical.** We assign probabilities to simple events based on the results of actual experiments. This is what we did in Problem 3. As an experiment is repeated without end, the percentage of times an event occurs may get closer and closer to a single fixed number. If so, the single fixed number is generally called the **actual probability of the event**.

Each approach has its advantages in certain situations. For the rest of this section, we emphasize the theoretical approach. In the next section, we consider the empirical approach in more detail.

● **Equally Likely Assumption**

In tossing a nickel and dime (Example 3), we assigned the same probability, $\frac{1}{4}$, to each simple event in the sample space $S = \{$HH, HT, TH, TT$\}$. By assigning the same probability to each simple event in S, we are actually making the assumption that each simple event is as likely to occur as any other. We refer to this as an **equally likely assumption**. In general, we have Definition 4.

DEFINITION 4   **Probability of a Simple Event Under an Equally Likely Assumption**

If, in a sample space

$$S = \{e_1, e_2, \ldots, e_n\}$$

with n elements, we assume each simple event e_i is as likely to occur as any other, then we assign the probability $1/n$ to each. That is,

$$P(e_i) = \frac{1}{n}$$

Under an equally likely assumption, we can develop a very useful formula for finding probabilities of arbitrary events associated with a sample space S. Consider the following example.

If a single die is rolled and we assume each face is as likely to come up as any other, then for the sample space

$$S = \{1, 2, 3, 4, 5, 6\}$$

we assign a probability of $\frac{1}{6}$ to each simple event, since there are 6 simple events. Then the probability of

$$E = \text{Rolling a prime number} = \{2, 3, 5\}$$

is

$$P(E) = P(2) + P(3) + P(5) = \tfrac{1}{6} + \tfrac{1}{6} + \tfrac{1}{6} = \tfrac{3}{6} = \tfrac{1}{2}$$

Thus, under the assumption that each simple event is as likely to occur as any other, the computation of the probability of the occurrence of any event E in a sample space S is the number of elements in E divided by the number of elements in S.

Theorem 1   **Probability of an Arbitrary Event Under an Equally Likely Assumption**

If we assume each simple event in sample space S is as likely to occur as any other, then the probability of an arbitrary event E in S is given by

$$P(E) = \frac{\text{Number of elements in } E}{\text{Number of elements in } S} = \frac{n(E)}{n(S)}$$

EXAMPLE 4   **Finding Probabilities of Events**

If in rolling 2 dice we assume each simple event in the sample space shown in Figure 1 (page 637) is as likely as any other, find the probabilities of the following events:

(A) $E_1 = $ A sum of 7 turns up   (B) $E_2 = $ A sum of 11 turns up
(C) $E_3 = $ A sum less than 4 turns up   (D) $E_4 = $ A sum of 12 turns up

Solutions   Referring to Figure 1, we see that:

(A) $P(E_1) = \dfrac{n(E_1)}{n(S)} = \dfrac{6}{36} = \dfrac{1}{6}$ (B) $P(E_2) = \dfrac{n(E_2)}{n(S)} = \dfrac{2}{36} = \dfrac{1}{18}$

(C) $P(E_3) = \dfrac{n(E_3)}{n(S)} = \dfrac{3}{36} = \dfrac{1}{12}$ (D) $P(E_4) = \dfrac{n(E_4)}{n(S)} = \dfrac{1}{36}$

Problem 4   Under the conditions in Example 4, find the probabilities of the following events:

(A) $E_5 = $ A sum of 5 turns up
(B) $E_6 = $ A sum that is a prime number greater than 7 turns up

---

We now turn to some examples that make use of the counting techniques developed in Section 10-1.

EXAMPLE 5   Drawing Cards

In drawing 5 cards from a 52-card deck without replacement, what is the probability of getting 5 spades?

Solution   Let the sample space S be the set of all 5-card hands from a 52-card deck. Since the order in a hand does not matter, $n(S) = C_{52,5}$. The event we seek is

$$E = \text{Set of all 5-card hands from 13 spades}$$

Again, the order does not matter and $n(E) = C_{13,5}$. Thus, assuming each 5-card hand is as likely as any other,

$$P(E) = \frac{n(E)}{n(S)} = \frac{C_{13,5}}{C_{52,5}} = \frac{13!/5!8!}{52!/5!47!} = \frac{13!}{5!8!} \cdot \frac{5!47!}{52!} \approx .0005$$

Problem 5   In drawing 7 cards from a 52-card deck without replacement, what is the probability of getting 7 hearts?

EXAMPLE 6   Selecting Committees

The board of regents of a university is made up of 12 men and 16 women. If a committee of 6 is chosen at random, what is the probability that it will contain 3 men and 3 women?

Solution   Let S = Set of all 6-person committees out of 28 people:

$$n(S) = C_{28,6}$$

Let E = Set of all 6-person committees with 3 men and 3 women. To find $n(E)$, we use the multiplication principle and the following two operations:

O_1: Select 3 men out of the 12 available N_1: $C_{12,3}$

O_2: Select 3 women out of the 16 available N_2: $C_{16,3}$

Thus,

$$n(E) = C_{12,3} \cdot C_{16,3}$$

and

$$P(E) = \frac{n(E)}{N(S)} = \frac{C_{12,3} \cdot C_{16,3}}{C_{28,6}} \approx .327$$

Problem 6 What is the probability that the committee in Example 6 will have 4 men and 2 women?

It needs to be pointed out that there are many counting problems for which it is not possible to produce a simple formula that will yield the number of possible cases. In situations of this type, we often must revert back to drawing tree diagrams and counting branches.

Answers to Matched Problems

1. (A) $S_1 = \{BB, BG, GB, GG\}$;

Sex of First Child	Sex of Second Child	Combined Outcomes
B	B	BB
	G	BG
G	B	GB
	G	GG

    (B) $S_2 = \{0, 1, 2\}$ (C) $S_3 = \{A, D\}$ (D) The sample space in part (A)

2. (A) $\{(4, 1), (3, 2), (2, 3), (1, 4)\}$ (B) $\{(6, 5), (5, 6)\}$ **3.** (A) .727   (B) .209   (C) 1

4. (A) $P(E_5) = \frac{1}{9}$ (B) $P(E_6) = \frac{1}{18}$ **5.** $C_{13,7}/C_{52,7} \approx .000013$

6. $C_{12,4} \cdot C_{16,2}/C_{28,6} \approx .158$

EXERCISE 10-2

A

1. How would you interpret $P(E) = 1$?

2. How would you interpret $P(E) = 0$?

3. In a family with 2 children, excluding multiple births, what is the probability of having 2 children of the opposite sex? Assume a girl is as likely as a boy at each birth.

4. In a family with 2 children, excluding multiple births, what is the probability of having 2 girls? Assume a girl is as likely as a boy at each birth.

5. A spinner can land on 4 different colors: red (R), green (G), yellow (Y), and blue (B). If we do not assume each color is as likely to turn up as any other, which of the following probability assignments have to be rejected, and why?
   (A) $P(R) = .15, P(G) = -.35, P(Y) = .50,$
       $P(B) = .70$
   (B) $P(R) = .32, P(G) = .28, P(Y) = .24,$
       $P(B) = .30$
   (C) $P(R) = .26, P(G) = .14, P(Y) = .30,$
       $P(B) = .30$

6. Under the probability assignments in Problem 5C, what is the probability that the spinner will not land on blue?

7. Under the probability assignments in Problem 5C, what is the probability that the spinner will land on red or yellow?

8. Under the probability assignments on Problem 5C, what is the probability that the spinner will not land on red or yellow?

B

9. In a family with 3 children, excluding multiple births, what is the probability of having 2 boys and 1 girl, in that order? Assume a boy is as likely as a girl at each birth.

10. In a family with 3 children, excluding multiple births, what is the probability of having 2 boys and 1 girl, in any order? Assume a boy is as likely as a girl at each birth.

11. A small combination lock on a suitcase has 3 wheels, each labeled with the 10 digits from 0 to 9. If an opening combination is a particular sequence of 3 digits with no repeats, what is the probability of a person guessing the right combination?

12. A combination lock has 5 wheels, each labeled with the 10 digits from 0 to 9. If an opening combination is a particular sequence of 5 digits with no repeats, what is the probability of a person guessing the right combination?

An experiment consists of dealing 5 cards from a standard 52-card deck. In Problem 13–16, what is the probability of being dealt each of the following hands?

13. 5 black cards

14. 5 hearts

15. 5 face cards if aces are considered to be face cards

16. 5 nonface cards if an ace is considered to be 1 and not a face card

17. If 4-digit numbers less than 5,000 are randomly formed from the digits 1, 3, 5, 7, and 9, what is the probability of forming a number divisible by 5? Digits may be repeated; for example, 1,355 is acceptable.

18. If code words of 4 letters are generated at random using the letters A, B, C, D, E, and F, what is the probability of forming a word without a vowel in it? Letters may be repeated.

19. Suppose 5 thank-you notes are written and 5 envelopes are addressed. Accidentally, the notes are randomly inserted into the envelopes and mailed without checking the addresses. What is the probability that all 5 notes will be inserted into the correct envelopes?

20. Suppose 6 people check their coats in a checkroom. If all claim checks are lost and the 6 coats are randomly returned, what is the probability that all 6 people will get their own coats back?

An experiment consists of rolling 2 fair dice and adding the dots on the 2 sides facing up. Using the sample space shown in Figure 1 (page 637) and assuming each simple event is as likely as any other, find the probabilities of the sums of dots indicated in Problems 21–36.

21. Sum is 2.

22. Sum is 10.

23. Sum is 6.

24. Sum is 8.

25. Sum is less than 5.

26. Sum is greater than 8.

27. Sum is not 7 or 11.

28. Sum is not 2, 4, or 6.

29. Sum is 1.

30. Sum is not 13.

31. Sum is divisible by 3.

32. Sum is divisible by 4.

33. Sum is 7 or 11 (a "natural").

34. Sum is 2, 3, or 12 ("craps").

35. Sum is divisible by 2 or 3.

36. Sum is divisible by 2 and 3.

An experiment consists of tossing 3 fair coins, but one of the 3 coins has a head on both sides. Compute the probabilities of obtaining the indicated results in Problems 37–42.

37. 1 head

38. 2 heads

39. 3 heads

40. 0 heads

41. More than 1 head

42. More than 1 tail

C

An experiment consists of rolling 2 fair dice and adding the dots on the 2 sides facing up. Each die has 1 dot on two opposite faces, 2 dots on two opposite faces, and 3 dots on two opposite faces. Compute the probabilities of obtaining the indicated sums in Problems 43–50.

43. 2

44. 3

45. 4

46. 5

47. 6

48. 7

49. An odd sum

50. An even sum

An experiment consists of dealing 5 cards from a standard 52-card deck. In Problems 51–58, what is the probability of being dealt the following cards?

51. 5 cards, jacks through aces

52. 5 cards, 2 through 10

53. 4 aces

54. 4 of a kind

55. Straight flush, ace high; that is, 10, jack, queen, king, ace in one suit

56. Straight flush, starting with 2; that is, 2, 3, 4, 5, 6 in one suit

57. 2 aces and 3 queens

58. 2 kings and 3 aces

SECTION 10-3   Empirical Probability

- Theoretical versus Empirical Probability
- Statistics versus Probability Theory

We conclude this chapter with a brief comparison of theoretical and empirical probability, which leads to a comparison of statistics and probability theory.

• Theoretical versus Empirical Probability

In Section 10-2 we indicated that acceptable and reasonable probability assignments are made for events in a sample space in two common ways: theoretical and empirical. Let's look at another example and compare the two approaches.

There are 20,000 students registered in a state university. Students are legally considered to be state residents, out-of-state residents, or foreign residents. What is the probability that a student chosen at random is a state resident? An out-of-state resident? A foreign resident? How do we proceed to find these probabilities?

Theoretical Approach

Suppose resident information is available in the registrar's office and can be obtained from a computer printout. Scanning the printout, we find the following:

Event	Description	Number of elements
E_1	State residents	12,000
E_2	Out-of-state residents	5,000
E_3	Foreign residents	3,000
S	All registered students	20,000

Looking at the total structure, we reason as follows: We choose the total list of registered students with resident status indicated as our sample space S. We assume any one student is as likely to be chosen as any other in a random sample of 1. Thus, we assign the probability $\frac{1}{20,000}$ to each simple event in S. This is an acceptable assignment. Under the equally likely assumption,

$$P(E_1) = \frac{n(E_1)}{n(S)} = \frac{12,000}{20,000} = .60$$

$$P(E_2) = \frac{n(E_2)}{n(S)} = \frac{5,000}{20,000} = .25$$

$$P(E_3) = \frac{n(E_3)}{n(S)} = \frac{3,000}{20,000} = .15$$

Our approach here is analogous to that used in assigning a probability of $\frac{1}{4}$ to the drawing of a heart in a single draw of one card from a 52-card deck.

Empirical Approach Suppose residency status was not recorded during registration and the information is not available through the registrar. Not having the time, inclination, or money to interview each student, we choose a random sample of 200 students and find the following:

Description	Number in sample
State residents	128
Out-of-state residents	47
Foreign residents	25
Size of sample	200

Based on this information, it seems reasonable to say that

$$P(E_1) \approx \frac{128}{200} = .640$$

$$P(E_2) \approx \frac{47}{200} = .235$$

$$P(E_3) \approx \frac{25}{200} = .125$$

As we increase the sample size, our confidence in the probability assignments should increase. We refer to these probability assignments as **approximate empirical probabilities** and use them to approximate the actual probabilities for the total population.

In general, if we conduct an experiment n times and an event E occurs with **frequency** $f(E)$, then the ratio $f(E)/n$ is called the **relative frequency** of the occurrence of event E in n trials. We define the **empirical probability** of E, denoted by $P(E)$, by the number, if it exists, that the relative frequency $f(E)/n$ approaches as n gets larger and larger. Of course, for any particular n, the relative frequency $f(E)/n$ is generally only approximately equal to $P(E)$. However, as n increases in size, we expect the approximation to improve.

DEFINITION 1   **Empirical Probability Approximation**

$$P(E) \approx \frac{\text{Frequency of occurrence of } E}{\text{Total number of trials}} = \frac{f(E)}{n}$$

The larger n is, the better the approximation.

If equally likely assumptions used to obtain theoretical probability assignments are actually warranted, then we also expect corresponding approximate empirical probabilities to approach the theoretical ones as the number of trials n of actual experiments becomes very large.

EXAMPLE 1   **Finding Empirical and Theoretical Probabilities**

Two coins are tossed 1,000 times with the following frequencies of outcomes:

2 heads: 200

1 head: 560

0 heads: 240

(A) Compute the approximate empirical probability for each type of outcome.
(B) Compute the theoretical probability for each type of outcome.

Solutions   (A)

$$P(2 \text{ heads}) \approx \frac{200}{1,000} = .20$$

$$P(1 \text{ head}) \approx \frac{560}{1,000} = .56$$

$$P(0 \text{ heads}) \approx \frac{240}{1,000} = .24$$

(B) A sample space of equally likely simple events is $S = \{HH, HT, TH, TT\}$. Let

$$E_1 = 2 \text{ heads} = \{HH\}$$
$$E_2 = 1 \text{ head} = \{HT, TH\}$$
$$E_3 = 0 \text{ heads} = \{TT\}$$

Then

$$P(E_1) = \frac{n(E_1)}{n(S)} = \frac{1}{4} = .25$$

$$P(E_2) = \frac{n(E_2)}{n(S)} = \frac{2}{4} = .50$$

$$P(E_3) = \frac{n(E_3)}{n(S)} = \frac{1}{4} = .25$$

Problem 1   One die is rolled 1,000 times with the following frequencies of outcomes:

$$
\begin{array}{llll}
1: & 180 & 4: & 138 \\
2: & 140 & 5: & 175 \\
3: & 152 & 6: & 215
\end{array}
$$

(A) Calculate the approximate empirical probability for each indicated outcome.
(B) Do the indicated outcomes seem equally likely?
(C) Assuming the indicated outcomes are equally likely, compute their theoretical probabilities.

EXAMPLE 2   **Empirical Probabilities for an Insurance Company**

An insurance company selected 1,000 drivers at random in a particular city to determine a relationship between age and accidents. The data obtained are listed in Table 1. Compute the approximate empirical probabilities of the following events for a driver chosen at random in the city:

(A) E_1: being under 20 years old *and* having exactly 3 accidents in 1 year
(B) E_2: being 30–39 years old *and* having 1 or more accidents in 1 year
(C) E_3: having no accidents in 1 year
(D) E_4: being under 20 years old *or* having exactly 3 accidents in 1 year

TABLE 1

Age	Accidents in 1 year				
---	0	1	2	3	Over 3
Under 20	50	62	53	35	20
20–29	64	93	67	40	36
30–39	82	68	32	14	4
40–49	38	32	20	7	3
Over 49	43	50	35	28	24

Solutions   (A) $P(E_1) \approx \dfrac{35}{1,000} = .035$

(B) $P(E_2) \approx \dfrac{68 + 32 + 14 + 4}{1,000} = .118$

(C) $P(E_3) \approx \dfrac{50 + 64 + 82 + 38 + 43}{1,000} = .277$

(D) $P(E_4) \approx \dfrac{50 + 62 + 53 + 35 + 20 + 40 + 14 + 7 + 28}{1,000} = .309$

Notice that in this type of problem, which is typical of many realistic problems, approximate empirical probabilities are the only type we can compute.

Problem 2   Referring to Table 1 in Example 2, compute the approximate empirical probabilities of the following events for a driver chosen at random in the city:

(A) E_1: being under 20 years old with no accidents in 1 year
(B) E_2: being 20–29 years old and having fewer than 2 accidents in 1 year
(C) E_3: not being over 49 years old

Approximate empirical probabilities are often used to test theoretical probabilities. As we said before, equally likely assumptions may not be justified in reality. In addition to this use, there are many situations in which it is either very difficult or impossible to compute the theoretical probabilities for given events. For example, insurance companies use past experience to establish approximate empirical probabilities to predict future accident rates, baseball teams use batting averages, which are approximate empirical probabilities based on past experience, to predict the future performance of a player, and pollsters use approximate empirical probabilities to predict outcomes of elections.

● Statistics versus Probability Theory

We are now entering the area of *mathematical statistics*, a subject that we will not pursue very far. Mathematical statistics starts with a known sample and proceeds to describe certain characteristics of the total population that are not known. For instance, in Example 2, the insurance company may use the approximate empirical probability .035, computed from the sample, as an approximation for the actual probability of a person drawn at random from the total population being under 20 years old and having exactly 3 accidents in 1 year.

Probability theory, on the other hand, starts with a known composition of a population and from this deduces the probable composition of a sample. For example, knowing the composition of a standard deck of 52 cards and assuming each 5-card hand has the same probability of being dealt as any other, we can deduce that the probability of being dealt 5 cards of the same suit is given by $4C_{13,5}/C_{52,5} = .00198$.

In short, statistics proceeds from a sample to the population, while probability theory proceeds from a population to a sample.

Answers to Matched Problems
1. (A) $P(1) \approx .180, P(2) \approx .140, P(3) \approx .152, P(4) \approx .138, P(5) \approx .175, P(6) \approx .215$
   (B) No   (C) $\frac{1}{6} \approx .167$ for each
2. (A) $P(E_1) \approx .05$ (B) $P(E_2) \approx .157$ (C) $P(E_3) \approx .82$

EXERCISE 10-3

A

1. A ski jumper has jumped over 300 feet in 25 out of 250 jumps. What is the approximate empirical probability of the next jump being over 300 feet?

2. In a certain city there are 4,000 youths between 16 and 20 years old who drive cars. If 560 of them were involved in accidents last year, what is the approximate empirical probability of a youth in this age group being involved in an accident this year?

3. Out of 420 times at bat, a baseball player gets 189 hits. What is the approximate empirical probability that the player will get a hit next time at bat?

4. In a medical experiment, a new drug is found to help 2,400 out of 3,000 people. If a doctor prescribes the drug for a particular patient, what is the approximate empirical probability that the patient will be helped?

5. A thumbtack is tossed 1,000 times with the following outcome frequencies: point down, 389; point up, 611. Compute the approximate empirical probability for each outcome. Does each outcome appear to be equally likely?

6. Toss a thumbtack 100 times, or 10 thumbtacks 10 times, or 25 thumbtacks 4 times, or 100 thumbtacks once. Count the total number of times a tack lands point down. What is the approximate empirical probability of the tack landing point down?

B

7. A random sample of 10,000 families with 2 children, excluding those with twins, produced the following frequencies:

> 2,351 families with 2 girls
>
> 5,435 families with 1 girl
>
> 2,214 families with 0 girls

(A) Compute the approximate empirical probability for each outcome.
(B) Compute the theoretical probability for each outcome, assuming a boy is as likely as a girl at each birth.

8. If we multiply the probability of the occurrence of an event E by the total number of trials n, we obtain the **expected frequency** of the occurrence of E in n trials. Using the theoretical probabilities found in Problem 7B, compute the expected frequency of each outcome in Problem 7 from the sample of 10,000.

9. Three coins are flipped, 1,000 times with the following frequencies of outcomes:

> 3 heads: 132   2 heads: 368
>
> 1 head: 380   0 heads: 120

(A) Compute the approximate empirical probabilities for each outcome.
(B) Compute the theoretical probability for each outcome, assuming fair coins.
(C) Using the theoretical probabilities computed in part (B), compute the expected frequency for each outcome. See Problem 8 above for a definition of expected frequency.

10. Toss 3 coins 50 times and compute the approximate empirical probability for 3 heads, 2 heads, 1 head, and 0 heads, respectively.

C _____

11. If 4 fair coins are tossed 80 times, what is the expected frequency of 4 heads turning up? 3 heads? 2 heads? 1 head? 0 heads? See Problem 8 for a definition of expected frequency.

12. Actually toss 4 coins 80 times, and tabulate the frequencies of the outcomes indicated in Problem 11. What are the approximate empirical probabilities for these outcomes?

_____

13. **Market Analysis.** A company selected 1,000 households at random and surveyed them to determine a relationship between income level and the number of television sets in a home. The information gathered is listed in the table:

Yearly income	Televisions per household				
	0	1	2	3	Above 3
Less than $12,000	0	40	51	11	0
$12,000–19,999	0	70	80	15	1
$20,000–39,999	2	112	130	80	12
$40,000–59,999	10	90	80	60	21
$60,000 or more	30	32	28	25	20

Compute the approximate empirical probabilities:
(A) Of a household earning $12,000–$19,999 per year *and* owning exactly 3 television sets
(B) Of a household earning $20,000–$39,999 per year *and* owning more than 1 television set
(C) Of a household earning $60,000 or more per year *or* owning more than 3 television sets
(D) Of a household not owning 0 television sets

14. **Market Analysis.** Use the sample results in Problem 13 to compute the approximate empirical probabilities:
(A) Of a household earning $40,000–$59,999 per year *and* owning 0 television sets
(B) Of a household earning $12,000–$39,999 per year *and* owning more than 2 television sets
(C) Of a household earning less than $20,000 per year *or* owning exactly 2 television sets
(D) Of a household not owning more than 3 television sets

Chapter 10 Review

10-1 MULTIPLICATION PRINCIPLE, PERMUTATIONS, AND COMBINATIONS

Given a sequence of operations, **tree diagrams** are often used to list all the possible combined outcomes. To count the number of combined outcomes without actually listing them, we use the **multiplication principle**:

1. If operations O_1 and O_2 are performed in order with N_1 possible outcomes for the first operation and N_2 possible outcomes for the second operation, then there are

$$N_1 \cdot N_2$$

possible outcomes of the first operation followed by the second.

2. In general, if n operations $O_1, O_2, \ldots, O_n$ are performed in order, with possible number of outcomes $N_1, N_2, \ldots, N_n$, respectively, then there are

$$N_1 \cdot N_2 \cdot \cdots \cdot N_n$$

possible combined outcomes of the operations performed in the given order.

A particular arrangement or ordering of n objects without repetition is called a **permutation**. The number of permutations of n objects is given by

$$P_{n,n} = n \cdot (n - 1) \cdot \cdots \cdot 1 = n!$$

and the number of permutations of n objects taken r at a time is given by

$$P_{n,r} = \frac{n!}{(n - r)!} \qquad 0 \le r \le n$$

A **combination of a set of n elements taken r at a time** is an r-element subset of the n objects. The number of combinations of n objects taken r at a time is given by

$$C_{n,r} = \binom{n}{r} = \frac{P_{n,r}}{r!} = \frac{n!}{r!(n - r)!} \qquad 0 \le r \le n$$

In a permutation, order is important. In a combination, order is not important.

10-2   SAMPLE SPACES AND PROBABILITY

A list of outcomes of an experiment are called **simple events** if one and only one of these results will occur in each trial of the experiment. The set of all simple events is called the **sample space**. Any subset of the sample space is called an **event**. An event is a **simple event** if it has only one element in it and a **compound event** if it has more than one element in it. We say that **an event E occurs** if any of the simple events in E occurs. A sample space S_1 is **more fundamental** than a second sample space S_2 if knowledge of which event occurs in S_1 tells us which event in S_2 occurs, but not conversely.

Given a sample space $S = \{e_1, e_2, \ldots, e_n\}$ with n simple events, to each simple event e_i we assign a real number, denoted by $P(e_i)$, that is called the **probability of the event e_i** and satisfies:

1. $0 \le P(e_i) \le 1$

2. $P(e_1) + P(e_2) + \cdots + P(e_n) = 1$

Any probability assignment that meets conditions 1 and 2 is said to be an **acceptable probability assignment**.

Given an acceptable probability assignment for the simple events in a sample space S, the **probability of an arbitrary event E** is defined as follows:

1. If E is the empty set, then $P(E) = 0$.

2. If E is a simple event, then $P(E)$ has already been assigned.

3. If E is a compound event, then $P(E)$ is the sum of the probabilities of all the simple events in E.

4. If E is the sample space S, then $P(E) = P(S) = 1$.

If each of the simple events in a sample space $S = \{e_1, e_2, \ldots, e_n\}$ with n simple events is **equally likely** to occur, then we assign the probability $1/n$ to each. If E is an arbitrary event in S, then

$$P(E) = \frac{\text{Number of elements in } E}{\text{Number of elements in } S} = \frac{n(E)}{n(S)}$$

10-3   EMPIRICAL PROBABILITY

If we conduct an experiment n times and event E occurs with **frequency** $f(E)$, then the ratio $f(E)/n$ is called the **relative frequency** of the occurrence of event E in n trials. As n increases, $f(E)/n$ usually approaches a number that is called the **empirical probability** $P(E)$. Thus, $f(E)/n$ is used as an **approximate empirical probability** for $P(E)$.

Chapter 10 Review Exercise

Work through all the problems in this chapter review and check answers in the back of the book. Answers to all review problems are there, and following each answer is a number in italics indicating the section in which that type of problem is discussed. Where weaknesses show up, review appropriate sections in the text.

A

1. A single die is rolled and a coin is flipped. How many combined outcomes are possible? Solve:
   (A) By using a tree diagram
   (B) By using the multiplication principle

2. Evaluate $C_{6,2}$ and $P_{6,2}$.

3. How many seating arrangements are possible with 6 people and 6 chairs in a row? Solve by using the multiplication principle.

4. Solve Problem 3 using permutations or combinations, whichever is applicable.

5. In a single deal of 5 cards from a standard 52-card deck, what is the probability of being dealt 5 clubs?

6. Betty and Bill are members of a 15-person ski club. If the president and treasurer are selected by lottery, what is the probability that Betty will be president and Bill will be treasurer? A person cannot hold more than one office.

7. Each letter of the first 10 letters of the alphabet is printed on 1 of 10 different cards. Three cards are drawn in succession without replacement. What is the probability of drawing the code word *dig* by drawing *d* on the first draw, *i* on the second draw, and *g* on the third draw? What is the probability of being dealt a 3-card hand containing the letters *d*, *i*, and *g* in any order?

8. A drug has side effects for 50 out of 1,000 people in a test. What is the approximate empirical probability that a person using the drug will have side effects?

B

9. Someone tells you that the following approximate empirical probabilities apply to the sample space $\{e_1, e_2, e_3, e_4\}$: $P(e_1) \approx .1$, $P(e_2) \approx -.2$, $P(e_3) \approx .6$, $P(e_4) \approx 2$. There are three reasons why P cannot be a probability function. Name them.

10. Six distinct points are selected on the circumference of a circle. How many triangles can be formed using these points as vertices?

11. How many 3-letter code words are possible using the first 8 letters of the alphabet if no letter can be repeated? If letters can be repeated? If adjacent letters cannot be alike?

12. Solve the following problems using $P_{n,r}$ or $C_{n,r}$, as appropriate:
    (A) How many 3-digit opening combinations are possible on a combination lock with 6 digits if the digits cannot be repeated?
    (B) Suppose 5 tennis players have made the finals. If each of the 5 players is to play every other player exactly once, how many games must be scheduled?

13. Two coins are flipped 1,000 times with the following frequencies:

$$2 \text{ heads:} \quad 210$$
$$1 \text{ head:} \quad 480$$
$$0 \text{ heads:} \quad 310$$

(A) Compute the empirical probability for each outcome.

(B) Compute the theoretical probability for each outcome.

(C) Using the theoretical probabilities computed in part (B), compute the expected frequency of each outcome, assuming fair coins. See Problem 8, Exercise 10-3, for a definition of expected frequency.

15. A group of 10 people includes one married couple. If 4 people are selected at random, what is the probability that the married couple is selected?

14. From a standard deck of 52 cards, what is the probability of obtaining a 5-card hand:
(A) Of all diamonds?
(B) Of 3 diamonds and 2 spades?
Write answers in terms of $C_{n,r}$ or $P_{n,r}$, as appropriate. Do not evaluate.

16. A spinning device has three numbers, 1, 2, 3, each as likely to turn up as the other. If the device is spun twice, what is the probability that:
(A) The same number turns up both times?
(B) The sum of the numbers turning up is 5?

C

17. How many different families with 5 children are possible, excluding multiple births, where the sex of each child in the order of their birth is taken into consideration? How many families are possible if the order pattern is not taken into account?

19. How many ways can 2 people be seated in a row of 4 chairs?

18. If 3 people are selected from a group of 7 men and 3 women, what is the probability that at least 1 woman is selected?

20. Three fair coins are tossed 1,000 times with the following frequencies of outcomes:

Number of heads	0	1	2	3
Frequency	120	360	350	170

(A) What is the approximate empirical probability of obtaining 2 heads?

(B) What is the theoretical probability of obtaining 2 heads?

(C) What is the expected frequency of obtaining 2 heads? See Problem 8, Exercise 10-3, for a definition of expected frequency.

21. Transportation. A distribution center A wishes to distribute its products to 5 different retail stores, $B, C, D, E,$ and F, in a city. How many different route plans can be constructed so that a single truck can start from A,

deliver to each store exactly once, and then return to the center?

22. Market Analysis. A videocassette company selected 1,000 persons at random and surveyed them to determine a relationship between age of purchaser and annual videocassette purchases. The results are given in the table:

Age	Cassettes purchased annually				Totals
	0	1	2	Above 2	
Under 12	60	70	30	10	170
12–18	30	100	100	60	290
19–25	70	110	120	30	330
Over 25	100	50	40	20	210
Totals	260	330	290	120	1,000

Find the empirical probability that a person selected at random:

(A) Is over 25 *and* buys exactly 2 cassettes annually

(B) Is 12–18 years old *and* buys more than 1 cassette annually

(C) Is 12–18 years old *or* buys more than 1 cassette annually

★**23. Quality Control.** Twelve precision parts, including 2 that are substandard, are sent to an assembly plant. The plant manager selects 4 at random and will return the whole shipment if 1 or more of the sample are found to be substandard. What is the probability that the shipment will be returned?

Cumulative Review Exercise
Chapters 8–10

Work through all the problems in this cumulative review and check answers in the back of the book. Answers to all review problems are there, and following each answer is a number in italics indicating the section in which that type of problem is discussed. Where weaknesses show up, review appropriate sections in the text.

A

1. Determine whether each of the following can be the first three terms of an arithmetic sequence, a geometric sequence, or neither.

(A) $20, 15, 10, \ldots$ (B) $5, 25, 125, \ldots$

(C) $5, 25, 50, \ldots$ (D) $27, -9, 3, \ldots$

(E) $-9, -6, -3, \ldots$

In Problems 2–4:
(A) Write the first four terms of each sequence.
(B) Find a_8.   (C) Find S_8.

2. $a_n = 2 \cdot 5^n$

3. $a_n = 3n - 1$

4. $a_1 = 100; a_n = a_{n-1} - 6, n \geq 2$

5. Evaluate each of the following:

(A) $8!$ (B) $\dfrac{32!}{30!}$ (C) $\dfrac{9!}{3!(9-3)!}$

6. Evaluate each of the following:

(A) $\dbinom{7}{2}$ (B) $C_{7,2}$ (C) $P_{7,2}$

In Problems 7–9, graph each equation and locate foci. Locate the directrix for any parabolas. Find the lengths of major, minor, transverse, and conjugate axes where applicable.

7. $25x^2 - 36y^2 = 900$

8. $25x^2 + 36y^2 = 900$

9. $25x^2 - 36y = 0$

10. A coin is flipped three times. How many combined outcomes are possible? Solve:

(A) By using a tree diagram

(B) By using the multiplication principle

11. How many ways can 4 distinct books be arranged on a shelf? Solve:

(A) By using the multiplication principle

(B) By using permutations or combinations, whichever is applicable

12. In a single deal of 3 cards from a standard 52-card deck, what is the probability of being dealt 3 diamonds?

13. Each of the 10 digits 0 through 9 is printed on 1 of 10 different cards. Four of these cards are drawn in succession without replacement. What is the probability of drawing the digits 4, 5, 6, and 7 by drawing 4 on the first draw, 5 on the second draw, 6 on the third draw, and 7 on the fourth draw? What is the probability of drawing the digits 4, 5, 6, and 7 in any order?

14. A thumbtack lands point down in 38 out of 100 tosses. What is the approximate empirical probability of the tack landing point up?

Verify Problems 15 and 16 for n = 1, 2, and 3.

15. P_n: $1 + 5 + 9 + \cdots + (4n - 3) = n(2n - 1)$

16. P_n: $n^2 + n + 2$ is divisible by 2

In Problems 17 and 18, write P_k and P_{k+1}.

17. For P_n in Problem 15

18. For P_n in Problem 16

B

19. Find the equation of the parabola having its vertex at the origin, its axis the y axis, and $(2, -8)$ on its graph.

20. Find an equation of an ellipse in the form

$$\frac{x^2}{M} + \frac{y^2}{N} = 1 \qquad M, N > 0$$

if the center is at the origin, the major axis is the x axis, the major axis length is 10, and the distance of the foci from the center is 3.

21. Find an equation of a hyperbola in the form

$$\frac{x^2}{M} - \frac{y^2}{N} = 1 \qquad M, N > 0$$

if the center is at the origin, the transverse axis length is 16, and the distance of the foci from the center is $\sqrt{89}$.

22. Write $\displaystyle\sum_{k=1}^{5} k^k$ without summation notation and find the sum.

23. Write the series $\dfrac{2}{2!} - \dfrac{2^2}{3!} + \dfrac{2^3}{4!} - \dfrac{2^4}{5!} + \dfrac{2^5}{6!} - \dfrac{2^6}{7!}$ using summation notation with the summation index k starting at $k = 1$.

24. Find S_∞ for the geometric series $108 - 36 + 12 - 4 + \cdots$.

25. How many 4-letter code words are possible using the first 6 letters of the alphabet if no letter can be repeated? If letters can be repeated? If adjacent letters cannot be alike?

26. A basketball team with 12 members has two centers. If 5 players are selected at random, what is the probability that both centers are selected? Express the answer in terms of $C_{n,r}$ or $P_{n,r}$, as appropriate, and evaluate.

27. A single die is rolled 1,000 times with the frequencies of outcomes shown in the table to the right:
(A) What is the approximate empirical probability that the number of dots showing is divisible by 3?
(B) What is the theoretical probability that the number of dots showing is divisible by 3?

Number of dots facing up	1	2	3	4	5	6
Frequency	160	155	195	180	140	170

28. Evaluate each of the following:

(A) $P_{25,5}$ (B) $C(25, 5)$ (C) $\begin{pmatrix} 25 \\ 20 \end{pmatrix}$

29. Expand $(a + \frac{1}{2}b)^6$ using the binomial formula.

30. Find the fifth and the eighth terms in the expansion of $(3x - y)^{10}$.

Establish each statement in Problems 31 and 32 for all positive integers using mathematical induction.

31. P_n in Problem 15

32. P_n in Problem 16

33. Find the sum of all the odd integers between 50 and 500.

34. Use the formula for the sum of an infinite geometric series to write $2.\overline{45} = 2.454\ 545\ \cdots$ as the quotient of two integers.

In Problems 35–37, use a translation of coordinates to transform each equation into a standard equation for a nondegenerate conic. Identify the curve and graph it.

35. $4x + 4y - y^2 + 8 = 0$

36. $x^2 + 2x - 4y^2 - 16y + 1 = 0$

37. $4x^2 - 16x + 9y^2 + 54y + 61 = 0$

38. How many 9-digit zip codes are possible? How many of these have no repeated digits?

39. Three-digit numbers are randomly formed from the digits 1, 2, 3, 4, and 5. What is the probability of forming an even number if digits cannot be repeated? If digits can be repeated?

40. Use mathematical induction to prove that the following statement holds for all positive integers:

$$P_n: \quad \frac{1}{1 \cdot 3} + \frac{1}{3 \cdot 5} + \frac{1}{5 \cdot 7} + \cdots + \frac{1}{(2n - 1)(2n + 1)} = \frac{n}{2n + 1}$$

C

41. Use the binomial formula to expand $(x - 2i)^6$, where i is the imaginary unit.

42. Use the definition of a parabola and the distance formula to find the equation of a parabola with directrix $y = 3$ and focus $(6, 1)$.

43. An ellipse has vertices $(\pm 4, 0)$ and foci $(\pm 2, 0)$. Find the y intercepts.

44. A hyperbola has vertices $(2, \pm 3)$ and foci $(2, \pm 5)$. Find the length of the conjugate axis.

45. Seven distinct points are selected on the circumference of a circle. How many triangles can be formed using these 7 points as vertices?

46. A box of 12 light bulbs contains 4 defective bulbs. If 3 bulbs are selected at random, what is the probability of selecting at least one defective bulb?

47. Use mathematical induction to prove that $2^n < n!$ for all integers $n > 3$.

48. Use mathematical induction to show that $\{a_n\} = \{b_n\}$, where $a_1 = 3$, $a_n = 2a_{n-1} - 1$ for $n > 1$, and $b_n = 2^n + 1$, $n \geq 1$.

APPLICATIONS

49. **Economics.** The government, through a subsidy program, distributes $2,000,000. If we assume that each individual or agency spends 75% of what it receives, and 75% of this is sold, and so on, how much total increase in spending results from this government action?

50. **Engineering.** An automobile headlight contains a parabolic reflector with a diameter of 8 inches. If the light source is located at the focus, which is 1 inch from the vertex, how deep is the reflector?

51. Architecture. A sound whispered at one focus of a whispering chamber can be easily heard at the other focus. Suppose that a cross section of this chamber is a semi-elliptical arch which is 80 feet wide and 24 feet high (see the figure). How far is each focus from the center of the arch? How high is the arch above each focus?

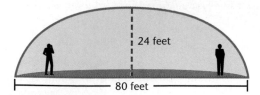

24 feet

80 feet

52. Political Science. A random survey of 1,000 residents in a state produced the following results:

| | **Party Affiliation** | | | |
Age	**Democrat**	**Republican**	**Independent**	**Totals**
Under 30	130	80	40	250
30–39	120	90	20	230
40–49	70	80	20	170
50–59	50	60	10	120
Over 59	90	110	30	230
Totals	460	420	120	1,000

Find the empirical probability that a person selected at random:

(A) Is under 30 *and* a Democrat

(B) Is under 40 *and* a Republican

(C) Is over 59 *or* is an Independent

Significant Digits

Most calculations involving problems of the real world deal with figures that are only approximate. It therefore seems reasonable to assume that a final answer should not be any more accurate than the least accurate figure used in the calculation. This is an important point, since calculators tend to give the impression that greater accuracy is achieved than is warranted.

Suppose we wish to compute the length of the diagonal of a rectangular field from measurements of its sides of 237.8 meters and 61.3 meters. Using the Pythagorean theorem and a calculator, we find

$$d = \sqrt{237.8^2 + 61.3^2}$$
$$= 245.573\ 878 \cdots$$

The calculator answer suggests an accuracy that is not justified. What accuracy is justified? To answer this question, we introduce the idea of *significant digits*.

Whenever we write a measurement such as 61.3 meters, we assume that the measurement is accurate to the last digit written. Thus, the measurement 61.3 meters indicates that the measurement was made to the nearest tenth of a meter. That is, the actual width is between 61.25 meters and 61.35 meters. In general, the digits in a number that indicate the accuracy of the number are called **significant digits**. If all the digits in a number are nonzero, then they are all significant. Thus, the measurement 61.3 meters has 3 significant digits, and the measurement 237.8 meters has 4 significant digits.

What are the significant digits in the number 7,800? The accuracy of this number is not clear. It could represent a measurement with any of the following accuracies:

Between 7,750 and 7,850	Correct to the hundreds place
Between 7,795 and 7,805	Correct to the tens place
Between 7,799.5 and 7,800.5	Correct to the units place

In order to give a precise definition of significant digits that resolves this ambiguity, we use scientific notation.

DEFINITION 1

Significant Digits

If a number x is written in scientific notation as

$$x = a \times 10^n \qquad 1 \le a < 10, n \text{ an integer}$$

then the number of significant digits in x is the number of digits in a.

Thus,

$$7.8 \times 10^3 \qquad \text{has 2 significant digits}$$
$$7.80 \times 10^3 \qquad \text{has 3 significant digits}$$
$$7.800 \times 10^3 \qquad \text{has 4 significant digits}$$

All three of these measurements have the same decimal representation (7,800), but each represents a different accuracy.

Definition 1 tells us how to write a number so that the number of significant digits is clear, but it does not tell us how to interpret the accuracy of a number that is not written in scientific notation. We will use the following convention for numbers that are written as decimal fractions:

Significant Digits in Decimal Fractions

The number of significant digits in a number with no decimal point is found by counting the digits from left to right, starting with the first digit and ending with the last *nonzero* digit.

The number of significant digits in a number containing a decimal point is found by counting the digits from left to right, starting with the first *nonzero* digit and ending with the last digit.

Applying this rule to the number 7,800, we conclude that this number has 2 significant digits. If we want to indicate that it has 3 or 4 significant digits, we must use scientific notation. The significant digits in the following numbers are underlined:

$$\underline{70,00}7 \qquad \underline{82},000 \qquad \underline{5.600} \qquad 0.000\underline{8} \qquad 0.000\ \underline{830}$$

In calculations involving multiplication, division, powers, and roots, we adopt the following convention:

Rounding Calculated Values

The result of a calculation is rounded to the same number of significant digits as the number used in the calculation that has the least number of significant digits.

Thus, in computing the length of the diagonal of the rectangular field shown earlier, we write the answer rounded to 3 significant digits because the width has 3 significant digits and the length has 4 significant digits:

$$d = 246 \text{ meters} \quad \text{3 significant digits}$$

One Final Note: In rounding a number that is exactly halfway between a larger and a smaller number, we use the convention of making the final result even.

EXAMPLE 1   Rounding Numbers

Round each number to 3 significant digits:

(A) 43.0690   (B) 48.05   (C) 48.15   (D) $8.017\ 632 \times 10^{-3}$

Solutions   (A) 43.1

(B) 48.0⎤ Use the convention of making the digit before the
(C) 48.2⎦ 5 even if it is odd, or leaving it alone if it is even.

(D) 8.02×10^{-3}

Problem 1   Round each number to 3 significant digits:

(A) 3.1495   (B) 0.004 135   (C) 32,450   (D) $4.314\ 764\ 09 \times 10^{12}$

---

1. (A) 3.15   (B) 0.004 14   (C) 32,400   (D) 4.31×10^{12}

APPENDIX B
Variation

- Direct Variation
- Inverse Variation
- Joint Variation
- Combined Variation

In reading scientific material, it is common to find statements such as, "The pressure of an enclosed gas varies directly as the absolute temperature," or, "The frequency of vibration of air in an organ pipe varies inversely as the length of the pipe," or even more complicated statements such as, "The force of attraction between two bodies varies jointly as their masses and inversely as the square of the distance between the two bodies." These statements have precise mathematical meaning in that they represent particular types of functions. The purpose of this appendix is to investigate these special functions.

● Direct Variation

The statement *y* **varies directly as** *x* means

$$y = kx \qquad k \neq 0$$

where k is a constant called the **constant of variation**. Similarly, "*y* varies directly as the square of *x*" means

$$y = kx^2 \qquad k \neq 0$$

and so on. The first equation defines a linear function, and the second defines a quadratic function.

Direct variation is illustrated by the familiar formulas

$$C = \pi D \qquad \text{and} \qquad A = \pi r^2$$

where the first formula asserts that the circumference of a circle varies directly as the diameter, and the second that the area of a circle varies directly as the square of the radius. In both cases, π is the constant of variation.

EXAMPLE 1   Finding Equations of Variation

Translate each statement into an appropriate equation, and find the constant of variation if $y = 16$ when $x = 4$.

(A) y varies directly as x.   (B) y varies directly as the cube of x.

Solutions   (A) y varies directly as x.

$$y = kx \quad \text{Don't forget } k.$$

To find the constant of variation k, substitute $x = 4$ and $y = 16$ and solve for k.

$$y = kx$$
$$16 = k \cdot 4$$
$$k = \tfrac{16}{4} = 4$$

Thus, $k = 4$ and the equation of variation is

$$y = 4x$$

(B) y varies directly as the cube of x.

$$y = kx^3 \quad \text{Don't forget } k$$

To find k, substitute $x = 4$ and $y = 16$:

$$y = kx^3$$
$$16 = k \cdot 4^3$$
$$k = \tfrac{16}{64} = \tfrac{1}{4}$$

Thus, the equation of variation is

$$y = \tfrac{1}{4}x^3$$

Problem 1   If $y = 4$ when $x = 8$, find the equation of variation for each statement.

(A) y varies directly as x.   (B) y varies directly as the cube root of x.

● **Inverse Variation**   The statement **y varies as x** means

$$y = \frac{k}{x} \quad k \neq 0$$

where k is the constant of variation. As in the case of direct variation, we also discuss y varying inversely as the square of x, and so on.

An illustration of inverse variation is given in the distance–rate–time formula $d = rt$ in the form $t = d/r$ for a fixed distance d. In driving a fixed distance, say $d = 400$ miles, time varies inversely as the rate; that is,

$$t = \frac{400}{r}$$

where 400 is the constant of variation—as the rate increases, the time decreases, and vice versa.

EXAMPLE 2   Finding Equations of Variation

Translate each statement into an appropriate equation, and find the constant of variation if $y = 16$ when $x = 4$.

(A) y varies inversely as x.   (B) y varies inversely as the square root of x.

Solutions   (A) y varies inversely as x.

$$y = \frac{k}{x}$$

To find k, substitute $x = 4$ and $y = 16$:

$$y = \frac{k}{x}$$

$$16 = \frac{k}{4}$$

$$k = 64$$

Thus, the equation of variation is

$$y = \frac{64}{x}$$

(B) y varies inversely as the square root of x.

$$y = \frac{k}{\sqrt{x}}$$

To find k, substitute $x = 4$ and $y = 16$:

$$y = \frac{k}{\sqrt{x}}$$

$$16 = \frac{k}{\sqrt{4}}$$

$$k = 32$$

Thus, the equation of variation is

$$y = \frac{32}{\sqrt{x}}$$

Problem 2   If $y = 4$ when $x = 8$, find the equation of variation for each statement.

(A) y varies inversely as x.   (B) y varies inversely as the square of x.

● **Joint Variation**   The statement w **varies jointly as x and y** means

$$w = kxy \qquad k \neq 0$$

where k is the constant of variation. Similarly, if

$$w = kxyz^2 \qquad k \neq 0$$

we say that w varies jointly as x, y, and the square of z, and so on. For example, recall that the area A of a rectangle is given by $A = lw$. Thus, the area varies jointly as the length l and the width w, and the constant of variation is 1. Similarly, since the volume V of a right circular cylinder is given by $V = \pi r^2 h$, the volume varies jointly as the height h and the square of the radius r. This time the constant of variation is π.

● **Combined Variation**   The basic types of variation introduced above are often combined. For example, the statement, "w varies jointly as x and y and inversely as the square of z," means

$$w = k\frac{xy}{z^2} \qquad k \neq 0 \quad \text{We do not write: } w = \frac{kxy}{kz^2}$$

Thus, the statement, "The force of attraction F between two bodies varies jointly as their masses m_1 and m_2 and inversely as the square of the distance d between the two bodies," means

$$F = k\frac{m_1 m_2}{d^2} \qquad k \neq 0$$

Assuming k is positive, if either of the two masses is increased, the force of attraction increases. On the other hand, if the distance is increased, the force of attraction decreases.

EXAMPLE 3   **Pressure of an Enclosed Gas**

The pressure P of an enclosed gas varies directly as the absolute temperature T and inversely as the volume V. If 500 cubic feet of gas yields a pressure of 10 pounds per square foot at a temperature of 300 K (absolute temperature*), what will be the pressure of the same gas if the volume is decreased to 300 cubic feet and the temperature is increased to 360 K?

Solution   *Method I.* Write the equation of variation,

$$P = k\left(\frac{T}{V}\right)$$

and find k using the first set of values:

$$10 = k\left(\tfrac{300}{500}\right)$$

$$k = \tfrac{50}{3}$$

Hence, the equation of variation for this particular gas is $P = \tfrac{50}{3}(T/V)$.
  Now find the new pressure P using the second set of values:

$$P = \tfrac{50}{3}\left(\tfrac{360}{300}\right) = 20 \text{ pounds per square foot}$$

Method II. Write the equation of variation $P = k(T/V)$; then convert to the equivalent form

$$\frac{PV}{T} = k$$

If P_1, V_1, and T_1 are the first set of values for the gas and P_2, V_2, and T_2 are the second set, then

$$\frac{P_1V_1}{T_1} = k \qquad \text{and} \qquad \frac{P_2V_2}{T_2} = k$$

Hence,

$$\frac{P_1V_1}{T_1} = \frac{P_2V_2}{T_2}$$

Since all values are known except P_2, substitute and solve. Thus,

$$\frac{(10)(500)}{300} = \frac{P_2(300)}{360}$$

$$P_2 = 20 \text{ pounds per square foot}$$

Problem 3   The length L of skid marks of a car's tires, when brakes are applied, varies directly as the square of the speed v of the car. If skid marks of 20 feet are produced at 30 miles per hour, how fast would the same car be going if it produced skid marks of 80 feet? Solve in two ways (see Example 3).

---

*A Kelvin (absolute) and a Celsius degree are the same size, but 0 on the Kelvin scale is −273° on the Celsius scale. This is the point at which molecular action is supposed to stop and is called *absolute zero*.

EXAMPLE 4   Frequency of Pitch

The frequency of pitch f of a given musical string varies directly as the square root of the tension T and inversely as the length L. What is the effect on the frequency if the tension is increased by a factor of 4 and the length is cut in half?

Solution   Write the equation of variation:

$$f = \frac{k\sqrt{T}}{L} \qquad \text{or equivalently} \qquad \frac{f_2 L_2}{\sqrt{T_2}} = \frac{f_1 L_1}{\sqrt{T_1}}$$

We are given that $T_2 = 4T_1$ and $L_2 = 0.5L_1$. Substituting in the second equation and solving for f_2, we have

$$\frac{f_2 0.5L_1}{\sqrt{4T_1}} = \frac{f_1 L_1}{\sqrt{T_1}}$$

$$\frac{f_2 0.5L_1}{2\sqrt{T_1}} = \frac{f_1 L_1}{\sqrt{T_1}}$$

$$f_2 = \frac{2\sqrt{T_1} f_1 L_1}{0.5L_1 \sqrt{T_1}} = 4f_1$$

Thus, the frequency of pitch is increased by a factor of 4.

Problem 4   The weight w of an object on or above the surface of the Earth varies inversely as the square of the distance d between the object and the center of the Earth. If an object on the surface of the Earth is moved into space so as to double its distance from the Earth's center, what effect will this move have on its weight?

Answers to Matched Problems

1. (A) $y = \frac{1}{2}x$ (B) $y = 2\sqrt[3]{x}$ 2. (A) $y = 32/x$ (B) $y = 256/x^2$
3. $v = 60$ mph   4. It will be one-fourth as heavy.

EXERCISE B

A

In Problems 1–12, translate each statement into an equation using k as the constant of variation.

1. F varies directly as the square of v.

2. u varies directly as v.

3. The pitch, or frequency, f of a guitar string of a given length varies directly as the square root of the tension T of the string.

4. Geologists have found in studies of erosion that the erosive force, or sediment-carrying power, P of a swiftly flowing stream varies directly as the sixth power of the velocity v of the water.

5. y varies inversely as the square root of x.

6. I varies inversely as t.

7. The biologist Reaumur suggested in 1735 that the length of time t that it takes fruit to ripen during the growing season varies inversely as the sum T of the average daily temperatures during the growing season.

8. In a study on urban concentration, F. Auerbach discovered an interesting law. After arranging all the cities of a given country according to their population size, starting with the largest, he found that the population P of a city varied inversely as the number n indicating its position in the ordering.

9. R varies jointly as S, T, and V.

10. g varies jointly as x and the square of y.

11. The volume of a cone V varies jointly as its height h and the square of the radius r of its base.

12. The amount of heat put out by an electrical appliance, in calories, varies jointly as time t, resistance R in the circuit, and the square of the current I.

Solve Problems 13–16 using either of the two methods illustrated in Example 3.

13. u varies directly as the square root of v. If $u = 2$ when $v = 2$, find u when $v = 8$.

14. y varies directly as the square of x. If $y = 20$ when $x = 2$, find y when $x = 5$.

15. L varies inversely as the square root of M. If $L = 9$ when $M = 9$, find L when $M = 3$.

16. I varies inversely as the cube of t. If $I = 4$ when $t = 2$, find I when $t = 4$.

B _____

In Problems 17–20, translate each statement into an equation using k as the constant of variation.

17. U varies jointly as a and b and inversely as the cube of c.

18. w varies directly as the square of x and inversely as the square root of y.

19. The maximum safe load L for a horizontal beam varies jointly as its width w and the square of its height h, and inversely as its length l.

20. Joseph Cavanaugh, a sociologist, found that the number of long-distance phone calls n between two cities in a given time period varied jointly as the populations P_1 and P_2 of the two cities, and inversely as the distance d between the two cities.

Solve Problems 21–26 using either of the two methods illustrated in Example 3.

21. Q varies jointly as m and the square of n, and inversely as P. If $Q = -4$ when $m = 6$, $n = 2$, and $P = 12$, find Q when $m = 4$, $n = 3$, and $P = 6$.

22. w varies jointly as x, y, and z and inversely as the square of t. If $w = 2$ when $x = 2$, $y = 3$, $z = 6$, and $t = 3$, find w when $x = 3$, $y = 4$, $z = 2$, and $t = 2$.

23. The weight w of an object on or above the surface of the Earth varies inversely as the square of the distance d between the object and the center of the Earth. If a girl weighs 100 pounds on the surface of the Earth, how much would she weigh, to the nearest pound, 400 miles above the Earth's surface? Assume the radius of the Earth is 4,000 miles.

24. A child was struck by a car in a crosswalk. The driver of the car had slammed on his brakes and left skid marks 160 feet long. He told the police he had been driving at 30 miles per hour. The police know that the length of skid marks L, when brakes are applied, varies directly as the square of the speed of the car v, and that at 30 miles per hour, under ideal conditions, skid marks would be 40 feet long. How fast was the driver actually going before he applied his brakes?

25. Ohm's law states that the current I in a wire varies directly as the electromotive force E and inversely as the resistance R. If $I = 22$ amperes when $E = 110$ volts and $R = 5$ ohms, find I if $E = 220$ volts and $R = 11$ ohms.

26. Anthropologists, in their study of race and human genetic groupings, often use an index called the *cephalic index*. The cephalic index C varies directly as the width w of the head and inversely as the length l of the head, both when viewed from the top. If an Indian in Baja California, Mexico, has measurements of $C = 75$, $w = 6$ inches, and $l = 8$ inches, what is C for an Indian in northern California with $w = 8.1$ inches and $l = 9$ inches?

C

27. If the horsepower P required to drive a speedboat through water varies directly as the cube of the speed v of the boat, what change in horsepower is required to double the speed of the boat?

28. The intensity of illumination E on a surface varies inversely as the square of its distance d from a light source. What is the effect on the total illumination on a book if the distance between the light source and the book is doubled?

29. The frequency of vibration f of a musical string varies directly as the square root of the tension T and inversely as the length L of the string. If the tension of the string is increased by a factor of 4 and the length of the string is doubled, what is the effect on the frequency?

30. In an automobile accident the destructive force F of a car varies approximately jointly as the weight w of the car and the square of the speed v of the car. This is why accidents at high speed are generally so serious. What would be the effect on the destructive force of a car if its weight were doubled and its speed were doubled?

ADDITIONAL APPLICATIONS

These problems are not grouped from easy (A) to difficult or theoretical (C). Instead, they are grouped according to subject area. As before, the most difficult problems are marked with two stars (★★), the moderately difficult problems are marked with one star (★), and the easier problems are not marked.

Astronomy

31. The square of the time t required for a planet to make one orbit around the sun varies directly as the cube of its mean (average) distance d from the sun. Write the equation of variation, using k as the constant of variation.

★ 32. The centripetal force F of a body moving in a circular path at constant speed varies inversely as the radius r of the path. What happens to F if r is doubled?

33. The length of time t a satellite takes to complete a circular orbit of the Earth varies directly as the radius r of the orbit and inversely as the orbital velocity v of the satellite. If $t = 1.42$ hours when $r = 4,050$ miles, and $v = 18,000$ miles per hour (*Sputnik I*), find t for $r = 4,300$ miles and $v = 18,500$ miles per hour.

Life Science

34. The number N of gene mutations resulting from x-ray exposure varies directly as the size of the x-ray dose r. What is the effect on N if r is quadrupled?

35. In biology there is an approximate rule, called the *bioclimatic rule* for temperate climates, which states that the difference d in time for fruit to ripen (or insects to appear) varies directly as the change in altitude h. If $d = 4$ days when $h = 500$ feet, find d when $h = 2,500$ feet.

Physics—Engineering

36. Over a fixed distance d, speed r varies inversely as time t. Police use this relationship to set up speed traps. If in a given speed trap $r = 30$ miles per hour when $t = 6$ seconds, what would be the speed of a car if $t = 4$ seconds?

★ 37. The length L of skid marks of a car's tires, when the brakes are applied, varies directly as the square of the speed v of the car. How is the length of skid marks affected by doubling the speed?

38. The time t required for an elevator to lift a weight varies jointly as the weight w and the distance d through which it is lifted, and inversely as the power P of the motor. Write the equation of variation, using k as the constant of variation.

39. The total pressure P of the wind on a wall varies jointly as the area of the wall A and the square of the velocity of the wind v. If $P = 120$ pounds when $A = 100$ square feet and $v = 20$ miles per hour, find P if $A = 200$ square feet and $v = 30$ miles per hour.

★★ 40. The thrust T of a given type of propeller varies jointly as the fourth power of its diameter d and the square of the number of revolutions per minute n it is turning. What happens to the thrust if the diameter is doubled and the number of revolutions per minute is cut in half?

Psychology

41. In early psychological studies on sensory perception (hearing, seeing, feeling, and so on), the question was asked: "Given a certain level of stimulation S, what is the minimum amount of added stimulation ΔS that can be detected?" A German physiologist, E. H. Weber (1795–1878) formulated, after many experiments, the

famous law that now bears his name: "The amount of change ΔS that will be just noticed varies directly as the magnitude S of the stimulus."

(A) Write the law as an equation of variation.

(B) If a person lifting weights can just notice a difference of 1 ounce at the 50-ounce level, what will be the least difference she will be able to notice at the 500-ounce level?

(C) Determine the just noticeable difference in illumination a person is able to perceive at 480 candlepower if he is just able to perceive a difference of 1 candlepower at the 60-candlepower level.

42. Psychologists, in their study of intelligence, often use an index called *IQ*. This index varies directly as mental age MA and inversely as chronological age CA (up to the age of 15). If a 12-year-old boy with a mental age of 14.4 has an IQ of 120, what will be the IQ of an 11-year-old girl with a mental age of 15.4?

Music

43. The frequency of vibration of air in an open organ pipe varies inversely as the length of the pipe. If the air column in an open 32-foot pipe vibrates 16 times per second (low C), then how fast would the air vibrate in a 16-foot pipe?

44. The frequency of pitch f of a musical string varies directly as the square root of the tension T and inversely as the length l and the diameter d. Write the equation of vibration using k as the constant of variation. It is interesting to note that if pitch depended on only length, then pianos would have to have strings varying from 3 inches to 38 feet.

Photography

45. The f-stop numbers N on a camera, known as *focal ratios*, vary directly as the focal length F of the lens and inversely as the diameter d of the diaphragm opening. Write the equation of variation using k as the constant of variation.

★46. In taking pictures using flashbulbs, the f-stop number N, which measures the lens opening, varies inversely as the distance d from the object being photographed. What adjustment should you make on the f-stop number if the distance between the camera and the object is doubled?

Chemistry

★47. Atoms and molecules that make up the air constantly fly about like microscopic missiles. The velocity v of a particular particle at a fixed temperature varies inversely as the square root of its molecular weight w. If an oxygen molecule in air at room temperature has an average velocity of 0.3 mile per second, what will be the average velocity of a hydrogen molecule, given that the hydrogen molecule is one-sixteenth as heavy as the oxygen molecule?

48. The Maxwell–Boltzmann equation says that the average velocity v of a molecule varies directly as the square root of the absolute temperature T and inversely as the square root of its molecular weight w. Write the equation of variation using k as the constant of variation.

Business

49. The amount of work A completed varies jointly as the number of workers W used and the time t they spend. If 10 workers can finish a job in 8 days, how long will it take 4 workers to do the same job?

50. The simple interest I earned in a given time varies jointly as the principal P and the interest rate r. If $100 at 4% interest earns $8, how much will $150 at 3% interest earn in the same period?

Geometry

★51. The volume of a sphere varies directly as the cube of its radius r. What happens to the volume if the radius is doubled?

★52. The surface area S of a sphere varies directly as the square of its radius r. What happens to the area if the radius is cut in half?

TABLES

TABLE I   Common Logarithms

x	0	1	2	3	4	5	6	7	8	9
1.0	0.0000	0.004321	0.008600	0.01284	0.01703	0.02119	0.02531	0.02938	0.03342	0.03743
1.1	0.04139	0.04532	0.04922	0.05308	0.05690	0.06070	0.06446	0.06819	0.07188	0.07555
1.2	0.07918	0.08279	0.08636	0.08991	0.09342	0.09691	0.1004	0.1038	0.1072	0.1106
1.3	0.1139	0.1173	0.1206	0.1239	0.1271	0.1303	0.1335	0.1367	0.1399	0.1430
1.4	0.1461	0.1492	0.1523	0.1553	0.1584	0.1614	0.1644	0.1673	0.1703	0.1732
1.5	0.1761	0.1790	0.1818	0.1847	0.1875	0.1903	0.1931	0.1959	0.1987	0.2014
1.6	0.2041	0.2068	0.2095	0.2122	0.2148	0.2175	0.2201	0.2227	0.2253	0.2279
1.7	0.2304	0.2330	0.2355	0.2380	0.2405	0.2430	0.2455	0.2480	0.2504	0.2529
1.8	0.2553	0.2577	0.2601	0.2625	0.2648	0.2673	0.2695	0.2718	0.2742	0.2765
1.9	0.2788	0.2810	0.2833	0.2856	0.2878	0.2900	0.2923	0.2945	0.2967	0.2989
2.0	0.3010	0.3032	0.3054	0.3075	0.3096	0.3118	0.3139	0.3160	0.3181	0.3201
2.1	0.3222	0.3243	0.3263	0.3284	0.3304	0.3324	0.3345	0.3365	0.3385	0.3404
2.2	0.3424	0.3444	0.3464	0.3483	0.3502	0.3522	0.3541	0.3560	0.3579	0.3598
2.3	0.3617	0.3636	0.3655	0.3674	0.3692	0.3711	0.3729	0.3747	0.3766	0.3784
2.4	0.3802	0.3820	0.3838	0.3856	0.3874	0.3892	0.3909	0.3927	0.3945	0.3962
2.5	0.3979	0.3997	0.4014	0.4031	0.4048	0.4065	0.4082	0.4099	0.4116	0.4133
2.6	0.4150	0.4166	0.4183	0.4200	0.4216	0.4232	0.4249	0.4265	0.4281	0.4298
2.7	0.4314	0.4330	0.4346	0.4362	0.4378	0.4393	0.4409	0.4425	0.4440	0.4456
2.8	0.4472	0.4487	0.4502	0.4518	0.4533	0.4548	0.4564	0.4579	0.4594	0.4609
2.9	0.4624	0.4639	0.4654	0.4669	0.4683	0.4698	0.4713	0.4728	0.4742	0.4757
3.0	0.4771	0.4786	0.4800	0.4814	0.4829	0.4843	0.4857	0.4871	0.4886	0.4900
3.1	0.4914	0.4928	0.4942	0.4955	0.4969	0.4983	0.4997	0.5011	0.5024	0.5038
3.2	0.5051	0.5065	0.5079	0.5092	0.5105	0.5119	0.5132	0.5145	0.5159	0.5172
3.3	0.5185	0.5198	0.5211	0.5224	0.5237	0.5250	0.5263	0.5276	0.5289	0.5302
3.4	0.5315	0.5328	0.5340	0.5353	0.5366	0.5378	0.5391	0.5403	0.5416	0.5428
3.5	0.5441	0.5453	0.5465	0.5478	0.5490	0.5502	0.5514	0.5527	0.5539	0.5551
3.6	0.5563	0.5575	0.5587	0.5599	0.5611	0.5623	0.5635	0.5647	0.5658	0.5670
3.7	0.5682	0.5694	0.5705	0.5717	0.5729	0.5740	0.5752	0.5763	0.5775	0.5786
3.8	0.5798	0.5809	0.5821	0.5832	0.5843	0.5855	0.5866	0.5877	0.5888	0.5899
3.9	0.5911	0.5922	0.5933	0.5944	0.5955	0.5966	0.5977	0.5988	0.5999	0.6010
4.0	0.6021	0.6031	0.6042	0.6053	0.6064	0.6075	0.6085	0.6096	0.6107	0.6117
4.1	0.6128	0.6138	0.6149	0.6160	0.6170	0.6180	0.6191	0.6201	0.6212	0.6222
4.2	0.6232	0.6243	0.6253	0.6263	0.6274	0.6284	0.6294	0.6304	0.6314	0.6325
4.3	0.6335	0.6345	0.6355	0.6365	0.6375	0.6385	0.6395	0.6405	0.6415	0.6425
4.4	0.6435	0.6444	0.6454	0.6464	0.6474	0.6484	0.6493	0.6503	0.6513	0.6522
4.5	0.6532	0.6542	0.6551	0.6561	0.6571	0.6580	0.6590	0.6599	0.6609	0.6618
4.6	0.6628	0.6637	0.6646	0.6656	0.6665	0.6675	0.6684	0.6693	0.6702	0.6712
4.7	0.6721	0.6730	0.6739	0.6749	0.6758	0.6767	0.6776	0.6785	0.6794	0.6803
4.8	0.6812	0.6821	0.6830	0.6839	0.6848	0.6857	0.6866	0.6875	0.6884	0.6893
4.9	0.6902	0.6911	0.6920	0.6928	0.6937	0.6946	0.6955	0.6964	0.6972	0.6981
5.0	0.6990	0.6998	0.7007	0.7016	0.7024	0.7033	0.7042	0.7050	0.7059	0.7067
5.1	0.7076	0.7084	0.7093	0.7101	0.7110	0.7118	0.7126	0.7135	0.7143	0.7152
5.2	0.7160	0.7168	0.7177	0.7185	0.7193	0.7202	0.7210	0.7218	0.7226	0.7235
5.3	0.7243	0.7251	0.7259	0.7267	0.7275	0.7284	0.7292	0.7300	0.7308	0.7316
5.4	0.7324	0.7332	0.7340	0.7348	0.7356	0.7364	0.7372	0.7380	0.7388	0.7396

TABLE I (*Continued*)

x	0	1	2	3	4	5	6	7	8	9
5.5	0.7404	0.7412	0.7419	0.7427	0.7435	0.7443	0.7451	0.7459	0.7466	0.7474
5.6	0.7482	0.7490	0.7497	0.7505	0.7513	0.7520	0.7528	0.7536	0.7543	0.7551
5.7	0.7559	0.7566	0.7574	0.7582	0.7589	0.7597	0.7604	0.7612	0.7619	0.7627
5.8	0.7634	0.7642	0.7649	0.7657	0.7664	0.7672	0.7679	0.7686	0.7694	0.7701
5.9	0.7709	0.7716	0.7723	0.7731	0.7738	0.7745	0.7752	0.7760	0.7767	0.7774
6.0	0.7782	0.7789	0.7796	0.7803	0.7810	0.7818	0.7825	0.7832	0.7839	0.7846
6.1	0.7853	0.7860	0.7868	0.7875	0.7882	0.7889	0.7896	0.7903	0.7910	0.7917
6.2	0.7924	0.7931	0.7938	0.7945	0.7952	0.7959	0.7966	0.7973	0.7980	0.7987
6.3	0.7993	0.8000	0.8007	0.8014	0.8021	0.8028	0.8035	0.8041	0.8048	0.8055
6.4	0.8062	0.8069	0.8075	0.8082	0.8089	0.8096	0.8102	0.8109	0.8116	0.8122
6.5	0.8129	0.8136	0.8142	0.8149	0.8156	0.8162	0.8169	0.8176	0.8182	0.8189
6.6	0.8195	0.8202	0.8209	0.8215	0.8222	0.8228	0.8235	0.8241	0.8248	0.8254
6.7	0.8261	0.8267	0.8274	0.8280	0.8287	0.8293	0.8299	0.8306	0.8312	0.8319
6.8	0.8325	0.8331	0.8338	0.8344	0.8351	0.8357	0.8363	0.8370	0.8376	0.8382
6.9	0.8388	0.8395	0.8401	0.8407	0.8414	0.8420	0.8426	0.8432	0.8439	0.8445
7.0	0.8451	0.8457	0.8463	0.8470	0.8476	0.8482	0.8488	0.8494	0.8500	0.8506
7.1	0.8513	0.8519	0.8525	0.8531	0.8537	0.8543	0.8549	0.8555	0.8561	0.8567
7.2	0.8573	0.8579	0.8585	0.8591	0.8597	0.8603	0.8609	0.8615	0.8621	0.8627
7.3	0.8633	0.8639	0.8645	0.8651	0.8657	0.8663	0.8669	0.8675	0.8681	0.8686
7.4	0.8692	0.8698	0.8704	0.8710	0.8716	0.8722	0.8727	0.8733	0.8739	0.8745
7.5	0.8751	0.8756	0.8762	0.8768	0.8774	0.8779	0.8785	0.8791	0.8797	0.8802
7.6	0.8808	0.8814	0.8820	0.8825	0.8831	0.8837	0.8842	0.8848	0.8854	0.8859
7.7	0.8865	0.8871	0.8876	0.8882	0.8887	0.8893	0.8899	0.8904	0.8910	0.8915
7.8	0.8921	0.8927	0.8932	0.8938	0.8943	0.8949	0.8954	0.8960	0.8965	0.8971
7.9	0.8976	0.8982	0.8987	0.8993	0.8998	0.9004	0.9009	0.9015	0.9020	0.9025
8.0	0.9031	0.9036	0.9042	0.9047	0.9053	0.9058	0.9063	0.9069	0.9074	0.9079
8.1	0.9085	0.9090	0.9096	0.9101	0.9106	0.9112	0.9117	0.9122	0.9128	0.9133
8.2	0.9138	0.9143	0.9149	0.9154	0.9159	0.9165	0.9170	0.9175	0.9180	0.9186
8.3	0.9191	0.9196	0.9201	0.9206	0.9212	0.9217	0.9222	0.9227	0.9232	0.9238
8.4	0.9243	0.9248	0.9253	0.9258	0.9263	0.9269	0.9274	0.9279	0.9284	0.9289
8.5	0.9294	0.9299	0.9304	0.9309	0.9315	0.9320	0.9325	0.9330	0.9335	0.9340
8.6	0.9345	0.9350	0.9355	0.9360	0.9365	0.9370	0.9375	0.9380	0.9385	0.9390
8.7	0.9395	0.9400	0.9405	0.9410	0.9415	0.9420	0.9425	0.9430	0.9435	0.9440
8.8	0.9445	0.9450	0.9455	0.9460	0.9465	0.9469	0.9474	0.9479	0.9484	0.9489
8.9	0.9494	0.9499	0.9504	0.9509	0.9513	0.9518	0.9523	0.9528	0.9533	0.9538
9.0	0.9542	0.9547	0.9552	0.9557	0.9562	0.9566	0.9571	0.9576	0.9581	0.9586
9.1	0.9590	0.9595	0.9600	0.9605	0.9609	0.9614	0.9619	0.9624	0.9628	0.9633
9.2	0.9638	0.9643	0.9647	0.9652	0.9657	0.9661	0.9666	0.9671	0.9675	0.9680
9.3	0.9685	0.9689	0.9694	0.9699	0.9703	0.9708	0.9713	0.9717	0.9722	0.9727
9.4	0.9731	0.9736	0.9741	0.9745	0.9750	0.9754	0.9759	0.9763	0.9768	0.9773
9.5	0.9777	0.9782	0.9786	0.9791	0.9795	0.9800	0.9805	0.9809	0.9814	0.9818
9.6	0.9823	0.9827	0.9832	0.9836	0.9841	0.9845	0.9850	0.9854	0.9859	0.9863
9.7	0.9868	0.9872	0.9877	0.9881	0.9886	0.9890	0.9894	0.9899	0.9903	0.9908
9.8	0.9912	0.9917	0.9921	0.9926	0.9930	0.9934	0.9939	0.9943	0.9948	0.9952
9.9	0.9956	0.9961	0.9965	0.9969	0.9974	0.9978	0.9983	0.9987	0.9991	0.9996

TABLE II   Natural Logarithms ($\ln x = \log_e x$)

ln 10 = 2.3026	6 ln 10 = 13.8155
2 ln 10 = 4.6052	7 ln 10 = 16.1181
3 ln 10 = 6.9078	8 ln 10 = 18.4207
4 ln 10 = 9.2103	9 ln 10 = 20.7233
5 ln 10 = 11.5130	10 ln 10 = 23.0259

Note: $\ln 35{,}200 = \ln (3.52 \times 10^4) = \ln 3.52 + 4 \ln 10$
$\ln 0.008\ 64 = \ln (8.64 \times 10^{-3}) = \ln 8.64 - 3 \ln 10$

x	.00	.01	.02	.03	.04	.05	.06	.07	.08	.09
1.0	0.0000	0.0100	0.0198	0.0296	0.0392	0.0488	0.0583	0.0677	0.0770	0.0862
1.1	0.0953	0.1044	0.1133	0.1222	0.1310	0.1398	0.1484	0.1570	0.1655	0.1740
1.2	0.1823	0.1906	0.1989	0.2070	0.2151	0.2231	0.2311	0.2390	0.2469	0.2546
1.3	0.2624	0.2700	0.2776	0.2852	0.2927	0.3001	0.3075	0.3148	0.3221	0.3293
1.4	0.3365	0.3436	0.3507	0.3577	0.3646	0.3716	0.3784	0.3853	0.3920	0.3988
1.5	0.4055	0.4121	0.4187	0.4253	0.4318	0.4383	0.4447	0.4511	0.4574	0.4637
1.6	0.4700	0.4762	0.4824	0.4886	0.4947	0.5000	0.5068	0.5128	0.5188	0.5217
1.7	0.5306	0.5365	0.5423	0.5481	0.5539	0.5596	0.5653	0.5710	0.5766	0.5822
1.8	0.5878	0.5933	0.5988	0.6043	0.6098	0.6152	0.6206	0.6259	0.6313	0.6366
1.9	0.6419	0.6471	0.6523	0.6575	0.6627	0.6678	0.6729	0.6780	0.6831	0.6881
2.0	0.6931	0.6981	0.7031	0.7080	0.7129	0.7178	0.7227	0.7275	0.7324	0.7372
2.1	0.7419	0.7467	0.7514	0.7561	0.7608	0.7655	0.7701	0.7747	0.7793	0.7839
2.2	0.7885	0.7930	0.7975	0.8020	0.8065	0.8109	0.8154	0.8198	0.8242	0.8286
2.3	0.8329	0.8372	0.8416	0.8459	0.8502	0.8544	0.8587	0.8629	0.8671	0.8713
2.4	0.8755	0.8796	0.8838	0.8879	0.8920	0.8961	0.9002	0.9042	0.9083	0.9123
2.5	0.9163	0.9203	0.9243	0.9282	0.9322	0.9361	0.9400	0.9439	0.9478	0.9517
2.6	0.9555	0.9594	0.9632	0.9670	0.9708	0.9746	0.9783	0.9821	0.9858	0.9895
2.7	0.9933	0.9969	1.0006	1.0043	1.0080	1.0116	1.0152	1.0188	1.0225	1.0260
2.8	1.0296	1.0332	1.0367	1.0403	1.0438	1.0473	1.0508	1.0543	1.0578	1.0613
2.9	1.0647	1.0682	1.0716	1.0750	1.0784	1.0818	1.0852	1.0886	1.0919	1.0953
3.0	1.0986	1.1019	1.1053	1.1086	1.1119	1.1151	1.1184	1.1217	1.1249	1.1282
3.1	1.1314	1.1346	1.1378	1.1410	1.1442	1.1474	1.1506	1.1537	1.1569	1.1600
3.2	1.1632	1.1663	1.1694	1.1725	1.1756	1.1787	1.1817	1.1848	1.1878	1.1909
3.3	1.1939	1.1969	1.2000	1.2030	1.2060	1.2090	1.2119	1.2149	1.2179	1.2208
3.4	1.2238	1.2267	1.2296	1.2326	1.2355	1.2384	1.2413	1.2442	1.2470	1.2499
3.5	1.2528	1.2556	1.2585	1.2613	1.2641	1.2669	1.2698	1.2726	1.2754	1.2782
3.6	1.2809	1.2837	1.2865	1.2892	1.2920	1.2947	1.2975	1.3002	1.3029	1.3056
3.7	1.3083	1.3110	1.3137	1.3164	1.3191	1.3218	1.3244	1.3271	1.3297	1.3324
3.8	1.3350	1.3376	1.3403	1.3429	1.3455	1.3481	1.3507	1.3533	1.3558	1.3584
3.9	1.3610	1.3635	1.3661	1.3686	1.3712	1.3737	1.3762	1.3788	1.3813	1.3838
4.0	1.3863	1.3888	1.3913	1.3938	1.3962	1.3987	1.4012	1.4036	1.4061	1.4085
4.1	1.4110	1.4134	1.4159	1.4183	1.4207	1.4231	1.4255	1.4279	1.4303	1.4327
4.2	1.4351	1.4375	1.4398	1.4422	1.4446	1.4469	1.4493	1.4516	1.4540	1.4563
4.3	1.4586	1.4609	1.4633	1.4656	1.4679	1.4702	1.4725	1.4748	1.4770	1.4793
4.4	1.4816	1.4839	1.4861	1.4884	1.4907	1.4929	1.4951	1.4974	1.4996	1.5019
4.5	1.5041	1.5063	1.5085	1.5107	1.5129	1.5151	1.5173	1.5195	1.5217	1.5239
4.6	1.5261	1.5282	1.5304	1.5326	1.5347	1.5369	1.5390	1.5412	1.5433	1.5454
4.7	1.5476	1.5497	1.5518	1.5539	1.5560	1.5581	1.5602	1.5623	1.5644	1.5665
4.8	1.5686	1.5707	1.5728	1.5748	1.5769	1.5790	1.5810	1.5831	1.5851	1.5872
4.9	1.5892	1.5913	1.5933	1.5953	1.5974	1.5994	1.6014	1.6034	1.6054	1.6074
5.0	1.6094	1.6114	1.6134	1.6154	1.6174	1.6194	1.6214	1.6233	1.6253	1.6273
5.1	1.6292	1.6312	1.6332	1.6351	1.6371	1.6390	1.6409	1.6429	1.6448	1.6467
5.2	1.6487	1.6506	1.6525	1.6544	1.6563	1.6582	1.6601	1.6620	1.6639	1.6658
5.3	1.6677	1.6696	1.6715	1.6734	1.6752	1.6771	1.6790	1.6808	1.6827	1.6845
5.4	1.6864	1.6882	1.6901	1.6919	1.6938	1.6956	1.6974	1.6993	1.7011	1.7029

TABLE II (*Continued*)

x	.00	.01	.02	.03	.04	.05	.06	.07	.08	.09
5.5	1.7047	1.7066	1.7084	1.7102	1.7120	1.7138	1.7156	1.7174	1.7192	1.7210
5.6	1.7228	1.7246	1.7263	1.7281	1.7299	1.7317	1.7334	1.7352	1.7370	1.7387
5.7	1.7405	1.7422	1.7440	1.7457	1.7475	1.7492	1.7509	1.7527	1.7544	1.7561
5.8	1.7579	1.7596	1.7613	1.7630	1.7647	1.7664	1.7681	1.7699	1.7716	1.7733
5.9	1.7750	1.7766	1.7783	1.7800	1.7817	1.7834	1.7851	1.7867	1.7884	1.7901
6.0	1.7918	1.7934	1.7951	1.7967	1.7984	1.8001	1.8017	1.8034	1.8050	1.8066
6.1	1.8083	1.8099	1.8116	1.8132	1.8148	1.8165	1.8181	1.8197	1.8213	1.8229
6.2	1.8245	1.8262	1.8278	1.8294	1.8310	1.8326	1.8342	1.8358	1.8374	1.8390
6.3	1.8405	1.8421	1.8437	1.8453	1.8469	1.8485	1.8500	1.8516	1.8532	1.8547
6.4	1.8563	1.8579	1.8594	1.8610	1.8625	1.8641	1.8656	1.8672	1.8687	1.8703
6.5	1.8718	1.8733	1.8749	1.8764	1.8779	1.8795	1.8810	1.8825	1.8840	1.8856
6.6	1.8871	1.8886	1.8901	1.8916	1.8931	1.8946	1.8961	1.8976	1.8991	1.9006
6.7	1.9021	1.9036	1.9051	1.9066	1.9081	1.9095	1.9110	1.9125	1.9140	1.9155
6.8	1.9169	1.9184	1.9199	1.9213	1.9228	1.9242	1.9257	1.9272	1.9286	1.9301
6.9	1.9315	1.9330	1.9344	1.9359	1.9373	1.9387	1.9402	1.9416	1.9430	1.9445
7.0	1.9459	1.9473	1.9488	1.9502	1.9516	1.9530	1.9544	1.9559	1.9573	1.9587
7.1	1.9601	1.9615	1.9629	1.9643	1.9657	1.9671	1.9685	1.9699	1.9713	1.9727
7.2	1.9741	1.9755	1.9769	1.9782	1.9796	1.9810	1.9824	1.9838	1.9851	1.9865
7.3	1.9879	1.9892	1.9906	1.9920	1.9933	1.9947	1.9961	1.9974	1.9988	2.0001
7.4	2.0015	2.0028	2.0042	2.0055	2.0069	2.0082	2.0096	2.0109	2.0122	2.0136
7.5	2.0149	2.0162	2.0176	2.0189	2.0202	2.0215	2.0229	2.0242	2.0255	2.0268
7.6	2.0281	2.0295	2.0308	2.0321	2.0334	2.0347	2.0360	2.0373	2.0386	2.0399
7.7	2.0412	2.0425	2.0438	2.0451	2.0464	2.0477	2.0490	2.0503	2.0516	2.0528
7.8	2.0541	2.0554	2.0567	2.0580	2.0592	2.0605	2.0618	2.0631	2.0643	2.0656
7.9	2.0669	2.0681	2.0694	2.0707	2.0719	2.0732	2.0744	2.0757	2.0769	2.0782
8.0	2.0794	2.0807	2.0819	2.0832	2.0844	2.0857	2.0869	2.0882	2.0894	2.0906
8.1	2.0919	2.0931	2.0943	2.0956	2.0968	2.0980	2.0992	2.1005	2.1017	2.1029
8.2	2.1041	2.1054	2.1066	2.1078	2.1090	2.1102	2.1114	2.1126	2.1138	2.1150
8.3	2.1163	2.1175	2.1187	2.1199	2.1211	2.1223	2.1235	2.1247	2.1258	2.1270
8.4	2.1282	2.1294	2.1306	2.1318	2.1330	2.1342	2.1353	2.1365	2.1377	2.1389
8.5	2.1401	2.1412	2.1424	2.1436	2.1448	2.1459	2.1471	2.1483	2.1494	2.1506
8.6	2.1518	2.1529	2.1541	2.1552	2.1564	2.1576	2.1587	2.1599	2.1610	2.1622
8.7	2.1633	2.1645	2.1656	2.1668	2.1679	2.1691	2.1702	2.1713	2.1725	2.1736
8.8	2.1748	2.1759	2.1770	2.1782	2.1793	2.1804	2.1815	2.1827	2.1838	2.1849
8.9	2.1861	2.1872	2.1883	2.1894	2.1905	2.1917	2.1928	2.1939	2.1950	2.1961
9.0	2.1972	2.1983	2.1994	2.2006	2.2017	2.2028	2.2039	2.2050	2.2061	2.2072
9.1	2.2083	2.2094	2.2105	2.2116	2.2127	2.2138	2.2148	2.2159	2.2170	2.2181
9.2	2.2192	2.2203	2.2214	2.2225	2.2235	2.2246	2.2257	2.2268	2.2279	2.2289
9.3	2.2300	2.2311	2.2322	2.2332	2.2343	2.2354	2.2364	2.2375	2.2386	2.2396
9.4	2.2407	2.2418	2.2428	2.2439	2.2450	2.2460	2.2471	2.2481	2.2492	2.2502
9.5	2.2513	2.2523	2.2534	2.2544	2.2555	2.2565	2.2576	2.2586	2.2597	2.2607
9.6	2.2618	2.2628	2.2638	2.2649	2.2659	2.2670	2.2680	2.2690	2.2701	2.2711
9.7	2.2721	2.2732	2.2742	2.2752	2.2762	2.2773	2.2783	2.2793	2.2803	2.2814
9.8	2.2824	2.2834	2.2844	2.2854	2.2865	2.2875	2.2885	2.2895	2.2905	2.2915
9.9	2.2925	2.2935	2.2946	2.2956	2.2966	2.2976	2.2986	2.2996	2.3006	2.3016

ANSWERS

Exercise 1-1

1. T   **3.** F   **5.** F   **7.** T   **9.** $7 + x$ **11.** $(xy)z$ **13.** $9m$ **15.** Commutative $(\cdot)$ **17.** Distributive
19. Inverse $(\cdot)$ **21.** Inverse $(+)$ **23.** Identity $(+)$ **25.** Negatives   **27.** $\{-2, 0, 2, 4\}$ **29.** $\{a, s, t, u\}$ **31.** $\varnothing$
33. Commutative $(+)$ **35.** Associative $(+)$ **37.** Distributive   **39.** Zero   **41.** Yes   **43.** (A) T   (B) F   (C) T
45. $\frac{3}{5}$ and -1.43 are two examples of infinitely many   **47.** (A) Z, Q, R (B) Q, R (C) R (D) Q, R
49. (A) $0.888\ 888\ 88 \cdots$ (B) $0.272\ 727\ 27 \cdots$ (C) $2.236\ 067\ 97 \cdots$ (D) $1.375\ 000\ 00 \cdots$
51. (B) is false, since, for example, $5 - 3 \neq 3 - 5$ (D) is false, since, for example, $9 \div 3 \neq 3 \div 9$
53. (A) $\{1, 2, 3, 4, 6\}$ (B) $\{2, 4\}$ **55.** $\frac{1}{11}$ **57.**

$$\begin{array}{r} 23 \\ \times 12 \\ \hline 46 \\ 230 \\ \hline 276 \end{array} \qquad \begin{aligned} 23 \cdot 12 &= 23(2 + 10) \\ &= 23 \cdot 2 + 23 \cdot 10 \\ &= 46 + 230 \\ &= 276 \end{aligned}$$

Exercise 1-2

1. 3 **3.** $2x^3 - x^2 + 2x + 4$ **5.** $2x^3 - 5x^2 + 6$ **7.** $6x^4 - 13x^3 + 9x^2 + 13x - 10$ **9.** $4x - 6$ **11.** $6y^2 - 16y$
13. $m^2 - n^2$ **15.** $4t^2 - 11t + 6$ **17.** $3x^2 - 7xy - 6y^2$ **19.** $4m^2 - 49$ **21.** $30x^2 - 2xy - 12y^2$ **23.** $9x^2 - 4y^2$
25. $16x^2 - 8xy + y^2$ **27.** $a^3 + b^3$ **29.** $9x^2 + 12x + 4$ **31.** $-x + 27$ **33.** $32a - 34$ **35.** $2x^4 - 5x^3 + 5x^2 + 11x - 10$
37. $h^4 + h^2k^2 + k^4$ **39.** $-5x^2 - 4x + 5$ **41.** $2m^2 + 15mn$ **43.** $8m^3 - 12m^2n + 6mn^2 - n^3$
45. $8h^3 + 36h^2k + 54hk^2 + 27k^3$ **47.** $5h$ **49.** $6hx + 2h + 3h^2$ **51.** $-4xh - 3h - 2h^2$ **53.** $m^3 - 3m^2 - 5$
55. $2x^3 - 13x^2 + 25x - 18$ **57.** $9x^3 - 9x^2 - 18x$ **59.** m **61.** Perimeter $= 2x + 2(x - 5) = 4x - 10$
63. Value $= 5x + 10(x - 5) + 25(x - 3) = 40x - 125$ **65.** Volume $= \frac{4}{3}\pi(x + 0.3)^3 - \frac{4}{3}\pi x^3 = 1.2\pi x^2 + 0.36\pi x + 0.036\pi$

Exercise 1-3

1. $2x^2(3x^2 - 4x - 1)$ **3.** $5xy(2x^2 + 4xy - 3y^2)$ **5.** $(5x - 3)(x + 1)$ **7.** $(2w - x)(y - 2z)$ **9.** $(x + 3)(x - 2)$
11. $(2m - 1)(3m + 5)$ **13.** $(2x - 3y)(x - 2y)$ **15.** $(4a - 3b)(2c - d)$ **17.** $(2x - 1)(x + 3)$ **19.** $(x - 6y)(x + 2y)$
21. Prime   **23.** $(5m - 4n)(5m + 4n)$ **25.** $(x + 5y)^2$ **27.** Prime   **29.** $6(x + 2)(x + 6)$ **31.** $2y(y - 3)(y - 8)$
33. $y(4x - 1)^2$ **35.** $(3s - t)(2s + 3t)$ **37.** $xy(x - 3y)(x + 3y)$ **39.** $3m(m^2 - 2m + 5)$ **41.** $\overline{(m + n)}(m^2 - mn + n^2)$
43. $(c - 1)(c^2 + c + 1)$ **45.** $2(3x - 5)(2x - 3)(12x - 19)$ **47.** $9x^4(9 - x)^3(5 - x)$ **49.** $2(x + 1)(x^2 - 5)(3x + 5)(x - 1)$
51. $[(a - b) - 2(c - d)][(a - b) + 2(c - d)]$ **53.** $(2m - 3n)(a + b)$ **55.** Prime   **57.** $(x + 3)(x - 3)^2$
59. $(a - 2)(a + 1)(a - 1)$ **61.** $[4(A + B) + 3][(A + B) - 2]$ **63.** $(m - n)(m + n)(m^2 + n^2)$
65. $st(st - 2)(s^2t^2 + 2st + 4)$ **67.** $(m + n)(m + n - 1)$ **69.** $2a[3a - 2(x + 4)][3a + 2(x + 4)]$
71. $(x^2 - x + 1)(x^2 + x + 1)$

Exercise 1-4

1. $\dfrac{a^2}{2}$ **3.** $\dfrac{22y + 9}{252}$ **5.** $\dfrac{x^2 + 8}{8x^3}$ **7.** $\dfrac{1}{2x - 1}$ **9.** $\dfrac{1}{m}$ **11.** $\dfrac{2a}{(a + b)^2(a - b)}$ **13.** $\dfrac{m^2 - 6m + 7}{m - 2}$ **15.** $\dfrac{7}{x - 3}$
17. $\dfrac{3}{y + 3}$ **19.** $\dfrac{x + y}{x}$ **21.** $\dfrac{1}{m}$ **23.** $\dfrac{4(x + 1)(x - 1)(x^2 + 2)^2}{x^3}$ **25.** $\dfrac{x(2 + 3x)}{(1 - 3x)^4}$ **27.** $\dfrac{(x + 1)(x - 9)}{(x + 4)^4}$
29. $\dfrac{y + 3}{(y - 2)(y + 7)}$ **31.** -1 **33.** $\dfrac{7y - 9x}{xy(a - b)}$ **35.** $\dfrac{x^2 - x + 1}{2(x - 9)}$ **37.** $\dfrac{(x - y)^2}{y^2(x + y)}$ **39.** $\dfrac{1}{x - 4}$ **41.** $\dfrac{1}{xy}$ **43.** $\dfrac{x - 3}{x - 1}$
45. $\dfrac{-1}{x(x + h)}$ **47.** $\dfrac{x^2 + xh + 4x + 2h}{(x + h + 2)(x + 2)}$ **49.** $\dfrac{-x(x + y)}{y}$ **51.** $\dfrac{a - 2}{a}$ **53.** $\dfrac{x - 1}{x}$

Exercise 1-5

1. 1 **3.** $6x^9$ **5.** $9x^6/y^4$ **7.** $a^4b^{12}/(c^8d^4)$ **9.** 10^{17} **11.** $2x/y^2$ **13.** n^2 **15.** 4×10^8 **17.** 3.225×10^7
19. 8.5×10^{-2} **21.** 7.29×10^{-8} **23.** 0.005 **25.** $26,900,000$ **27.** $0.000\ 000\ 000\ 59$ **29.** $3y^4/2$ **31.** x^{12}/y^8
33. $4x^8/y^6$ **35.** $w^{12}/(u^{20}v^4)$ **37.** $1/(x + y)^2$ **39.** $x/(x - 1)$ **41.** $-1/(xy)$ **43.** $2x - 6x^{-1}$ **45.** $\frac{5}{3}x - \frac{2}{3}x^{-2}$
47. $x - \frac{3}{2} + \frac{1}{2}x^{-1}$ **49.** 6.65×10^{-17} **51.** 1.54×10^{12} **53.** 1.0295×10^{11} **55.** -4.3647×10^{-18}
57. 9.4697×10^{29} **59.** $2(a + 2b)^5$ **61.** $(x^2 + xy + y^2)/(xy)$ **63.** $(y - x)/y$ **65.** 1.3×10^{25} lb
67. 10^8 or 100 million, 10^{10} or 10 billion; 6×10^9 or 6 billion, 6×10^{11} or 600 billion
69. 1.44×10^3 dollars per person; \$1,440 per person

Exercise 1-6

1. 4 **3.** 64 **5.** -6 **7.** Not a real number **9.** $\frac{8}{125}$ **11.** $\frac{1}{27}$ **13.** $y^{3/5}$ **15.** $d^{1/3}$ **17.** $1/y^{1/2}$ **19.** $2x/y^2$
21. $1/(a^{1/4}b^{1/3})$ **23.** $xy^2/2$ **25.** $2b^2/(3a^2)$ **27.** $2/(3x^{7/12})$ **29.** $a^{1/3}/b^2$ **31.** $6m - 2m^{19/3}$ **33.** $a - a^{1/2}b^{1/2} - 6b$
35. $4x - 9y$ **37.** $x + 4x^{1/2}y^{1/2} + 4y$ **39.** 29.52 **41.** 0.030 93 **43.** 5.421 **45.** 107.6
47. $3 - \frac{3}{4}x^{-1/2}$ **49.** $\frac{3}{5}x^{-1/3} + \frac{1}{5}x^{-1/2}$ **51.** $\frac{1}{2}x^{5/3} - 2x^{1/6}$ **53.** $a^{1/n}b^{1/m}$ **55.** $1/(x^{3m}y^{4n})$
57. (A) $x = -2$, for example (B) $x = 2$, for example **59.** $(x - 3)/(2x - 1)^{3/2}$ **61.** $(4x - 3)/(3x - 1)^{4/3}$
63. 1,920 units **65.** 428 ft

Exercise 1-7

1. $\sqrt[3]{m^2}$ or $(\sqrt[3]{m})^2$ (first preferred) **3.** $6\sqrt[5]{x^3}$ (not $\sqrt[5]{6x^3}$) **5.** $\sqrt[5]{(4xy^3)^2}$ **7.** $\sqrt{x + y}$ **9.** $b^{1/5}$ **11.** $5x^{3/4}$
13. $(2x^2y)^{3/5}$ **15.** $x^{1/3} + y^{1/3}$ **17.** -2 **19.** $3x^4y^2$ **21.** $2mn^2$ **23.** $2ab^2\sqrt{2ab}$ **25.** $2xy^2\sqrt[3]{2xy}$ **27.** $\sqrt{m}$
29. $\sqrt[15]{xy}$ **31.** $3x\sqrt[3]{3}$ **33.** $\sqrt{5}/5$ **35.** $2\sqrt{3x}$ **37.** $2\sqrt{2} + 2$ **39.** $\sqrt{3} - \sqrt{2}$ **41.** $3x^2y^2\sqrt[5]{3x^2y}$ **43.** $n\sqrt[4]{2m^3n}$
45. $\sqrt[3]{a^2(b - a)}$ **47.** $\sqrt[4]{a^3b}$ **49.** $x^2y\sqrt[3]{6xy^2}$ **51.** In simplified form **53.** $\sqrt{2}/2$ or $\frac{1}{2}\sqrt{2}$ **55.** $2a^2b\sqrt[3]{4a^2b}$
57. $\sqrt[4]{12xy^3}/2x$ or $[1/(2x)]\sqrt[4]{12xy^3}$ **59.** $(6y + 9\sqrt{y})/(4y - 9)$ **61.** $(38 + 11\sqrt{10})/117$ **63.** $\sqrt{x^2 + 9} + 3$
65. $1/(\sqrt{t} + \sqrt{x})$ **67.** $1/(\sqrt{x + h} + \sqrt{x})$ **69.** 0.2222 **71.** 1.934 **73.** 0.069 79 **75.** 2.073
77. Both are 1.059. **79.** Both are 0.6300. **81.** $x \leq 0$ **83.** All real numbers **85.** $\dfrac{\sqrt[3]{a^2} + \sqrt[3]{ab} + \sqrt[3]{b^2}}{a - b}$
87. $\dfrac{(\sqrt{x} - \sqrt{y} - \sqrt{z})[(x + y - z) + 2\sqrt{xy}]}{(x + y - z)^2 - 4xy}$ **89.** $\dfrac{1}{\sqrt[3]{(x + h)^2} + \sqrt[3]{x(x + h)} + \sqrt[3]{x^2}}$
91. $\sqrt[kn]{x^{km}} = (x^{km})^{1/kn} = x^{km/kn} = x^{m/n} = \sqrt[n]{x^m}$

Chapter 1 Review Exercise*

1. (A) T (B) T (C) F (D) T (E) F (F) F *(1-1)* **2.** (A) $(y + z)x$ (B) $(2 + x) + y$ (C) $2x + 3x$ *(1-1)*
3. $x^3 + 3x^2 + 5x - 2$ *(1-2)* **4.** $x^3 - 3x^2 - 3x + 22$ *(1-2)* **5.** $3x^5 + x^4 - 8x^3 + 24x^2 + 8x - 64$ *(1-2)* **6.** 3 *(1-2)*
7. 1 *(1-2)* **8.** $14x^2 - 30x$ *(1-2)* **9.** $9m^2 - 25n^2$ *(1-2)* **10.** $6x^2 - 5xy - 4y^2$ *(1-2)* **11.** $4a^2 - 12ab + 9b^2$ *(1-2)*
12. $(3x - 2)^2$ *(1-3)* **13.** Prime *(1-3)* **14.** $3n(2n - 5)(n + 1)$ *(1-3)* **15.** $(12a^3b - 40b^2 - 5a)/(30a^3b^2)$ *(1-4)*
16. $(7x - 4)/[6x(x - 4)]$ *(1-4)* **17.** $(y + 2)/[y(y - 2)]$ *(1-4)* **18.** u *(1-4)* **19.** $6x^5y^{15}$ *(1-5)* **20.** $3u^4/v^2$ *(1-5)*
21. 6×10^2 *(1-5)* **22.** x^6/y^4 *(1-5)* **23.** $u^{7/3}$ *(1-6)* **24.** $3a^2/b$ *(1-6)* **25.** $3\sqrt[5]{x^2}$ *(1-7)* **26.** $-3(xy)^{2/3}$ *(1-7)*
27. $3x^2y\sqrt[3]{x^2y}$ *(1-7)* **28.** $6x^2y^3\sqrt{xy}$ *(1-7)* **29.** $2b\sqrt{3a}$ *(1-7)* **30.** $(3\sqrt{5} + 5)/4$ *(1-7)* **31.** $\sqrt[4]{y^3}$ *(1-7)*
32. $\{-3, -1, 1\}$ *(1-1)* **33.** Subtraction *(1-1)* **34.** Commutative $(+)$ *(1-1)* **35.** Distributive *(1-1)*
36. Associative $(\cdot)$ *(1-1)* **37.** Negatives *(1-1)* **38.** Identity $(+)$ *(1-1)* **39.** (A) T (B) F *(1-1)*
40. 0 and -3 are two examples of infinitely many. *(1-1)* **41.** (A) (a) and (d) (B) None *(1-2)* **42.** $4xy - 2y^2$ *(1-2)*
43. $m^4 - 6m^2n^2 + n^4$ *(1-2)* **44.** $10xh + 5h^2 - 7h$ *(1-2)* **45.** $2x^3 - 4x^2 + 12x$ *(1-2)*
46. $x^3 - 6x^2y + 12xy^2 - 8y^3$ *(1-2)* **47.** $(x - y)(7x - y)$ *(1-3)* **48.** Prime *(1-3)* **49.** $3xy(2x^2 + 4xy - 5y^2)$ *(1-3)*
50. $(y - b)(y - b - 1)$ *(1-3)* **51.** $3(x + 2y)(x^2 - 2xy + 4y^2)$ *(1-3)* **52.** $(y - 2)(y + 2)^2$ *(1-3)*
53. $x(x - 4)^2(5x - 8)$ *(1-3)* **54.** $\dfrac{(x - 4)(x + 2)^2}{x^3}$ *(1-4)* **55.** $2m/[(m + 2)(m - 2)^2]$ *(1-4)* **56.** y^2/x *(1-4)*
57. $(x - y)/(x + y)$ *(1-4)* **58.** $-ab/(a^2 + ab + b^2)$ *(1-4)* **59.** $\frac{1}{4}$ *(1-5)* **60.** $\frac{5}{9}$ *(1-5)* **61.** $3x^2/(2y^2)$ *(1-6)*
62. $27a^{1/6}/b^{1/2}$ *(1-6)* **63.** $x + 2x^{1/2}y^{1/2} + y$ *(1-6)* **64.** $6x + 7x^{1/2}y^{1/2} - 3y$ *(1-6)* **65.** 2×10^{-7} *(1-5)*
66. 3.213×10^6 *(1-5)* **67.** 4.434×10^{-5} *(1-5)* **68.** -4.541×10^{-6} *(1-5)* **69.** 128,800 *(1-6)* **70.** 0.01507 *(1-6)*
71. 0.3664 *(1-7)* **72.** 1.640 *(1-7)* **73.** 0.08726 *(1-6)* **74.** $-6x^2y^2\sqrt[3]{3x^2y}$ *(1-7)* **75.** $x\sqrt[3]{2x^2}$ *(1-7)*
76. $\sqrt[5]{12x^3y^2}/(2x)$ *(1-7)* **77.** $y\sqrt[3]{2x^2y}$ *(1-7)* **78.** $\sqrt[3]{2x^2}$ *(1-7)* **79.** $2x - 3\sqrt{xy} - 5y$ *(1-7)*
80. $(6x + 3\sqrt{xy})/(4x - y)$ *(1-7)* **81.** $(4u - 12\sqrt{uv} + 9v)/(4u - 9v)$ *(1-7)* **82.** $\sqrt{y^2 + 4} + 2$ *(1-7)*
83. $1/(\sqrt{t} + \sqrt{5})$ *(1-7)* **84.** $2 - \frac{3}{2}x^{-1/2}$ *(1-7)* **85.** $\frac{6}{11}$; rational *(1-1)* **86.** (A) $\{-4, -3, 0, 2\}$ (B) $\{-3, 2\}$ *(1-1)*
87. 0 *(1-7)* **88.** $x^3 + 8x^2 - 6x + 1$ *(1-2)* **89.** $x(2a + 3x - 4)(2a - 3x - 4)$ *(1-3)* **90.** $(x - 2)(x + 1)/(2x)$ *(1-4)*
91. $\frac{2}{3}(x - 2)(x + 3)^4$ *(1-5)* **92.** $a^2b^2/(a^3 + b^3)$ *(1-5)* **93.** $x - y$ *(1-6)* **94.** x^{m-1} *(1-6)*
95. $(1 + \sqrt[3]{x} + \sqrt[3]{x^2})/(1 - x)$ *(1-7)* **96.** $1/(\sqrt[3]{t^2} + \sqrt[3]{5t} + \sqrt[3]{25})$ *(1-7)* **97.** x^{n+1} *(1-7)*
98. Volume $= 3\pi(x + 2)^2 - 3\pi x^2 = 12\pi x + 12\pi$ ft^3 *(1-2)* **99.** $9.60 \times 10^3 = 9,600$ kg/person *(1-5)*

CHAPTER 2 ## Exercise 2-1

1. $x = 18$ **3.** $t = 9$ **5.** $x = 6$ **7.** $x = 9$ **9.** $x = \frac{11}{2}$, or 5.5 **11.** $x = 10$ **13.** $x = 8$ **15.** $x = \frac{5}{3}$ **17.** $m = 3$
19. No solution **21.** $x = \frac{8}{5}$ **23.** $s = 2$ **25.** No solution **27.** $t = -4$ **29.** $x = \frac{2}{3}$ **31.** $x = 1.83$
33. $x = -8.55$ **35.** $d = (a_n - a_1)/(n - 1)$ **37.** $f = d_1d_2/(d_1 + d_2)$ **39.** $a = (A - 2bc)/(2b + 2c)$

---

*The number in parentheses after each answer to a chapter review problem refers to the section in which that type of problem is discussed.

41. $x = (5y + 3)/(2 - 3y)$ **43.** 4   **45.** All real numbers except 0 and 1   **47.** No solution
49. $x = (by + cy - ac)/(a - y)$

Exercise 2-2

1. 24   **3.** 8, 10, 12, 14   **5.** 17 by 10 m   **7.** 42 ft   **9.** $90   **11.** 5,200   **13.** $7,200 at 10% and $4,800 at 15%
15. (A) $T = 30 + 25(x - 3)$ (B) 330°C   (C) 13 km   **17.** 90 mi   **19.** 5,000 trout   **21.** 10 gal
23. 30 liters of 20% solution and 70 liters of 80% solution   **25.** 1.5 hr   **27.** (A) 216 mi   (B) 225 mi   **29.** 330 Hz; 396 Hz
31. 150 cm   **33.** 150 ft   **35.** $5\frac{5}{11}$ min after 1 P.M.

Exercise 2-3

1. $-8 \le x \le 7$ **3.** $-6 \le x < 6$ **5.** $x \ge -6$
7. $(-2, 6]$ **9.** $(-7, 8)$ **11.** $(-\infty, -2]$
13. $[-7, 2); -7 \le x < 2$ **15.** $(-\infty, 0]; x \le 0$ **17.** $x < 5$ or $(-\infty, 5)$ **19.** $x \ge 3$ or $[3, \infty)$
21. $N < -8$ or $(-\infty, -8)$ **23.** $t > 2$ or $(2, \infty)$ **25.** $m > 3$ or $(3, \infty)$
27. $B \ge -4$ or $[-4, \infty)$ **29.** $-2 < t \le 3$ or $(-2, 3]$ **31.** $-5 < x \le 7$
33. $2 < x < 4$ **35.** $-\infty < x < \infty$ **37.** $x < -1$ or $3 \le x < 7$
39. $1 < x < 5$ **41.** $x \le 6$ **43.** $q < -14$ or $(-\infty, -14)$
45. $x \ge 4.5$ or $[4.5, \infty)$ **47.** $-20 \le x \le 20$ or $[-20, 20]$
49. $-30 \le x < 18$ or $[-30, 18)$ **51.** $-8 \le x < -3$ or $[-8, -3)$
53. $-14 < x \le 11$ or $(-14, 11]$ **55.** $x \ge -0.60$ **57.** $-0.259 < x < 0.357$ **59.** $x \le 1$ **61.** $x \ge -\frac{5}{3}$
63. $x > -\frac{3}{2}$ **65.** (A) and (C) $a > 0$ and $b > 0$, or $a < 0$ and $b < 0$ (B) and (D) $a > 0$ and $b < 0$, or $a < 0$ and $b > 0$
67. (A) >   (B) <   **69.** Positive   **71.** (A) F   (B) T   (C) T   **77.** $9.8 \le x \le 13.8$ (from 9.8 to 13.8 km)
79. $x > 600$ **81.** $2 \le I \le 25$ or $[2, 25]$ **83.** $\$2,060 \le$ Benefit reduction $\le \$3,560$

Exercise 2-4

1. $\sqrt{5}$ **3.** 4   **5.** $5 - \sqrt{5}$ **7.** $5 - \sqrt{5}$ **9.** 12   **11.** 12   **13.** 9   **15.** 4   **17.** 4   **19.** 9
21. $x = \pm 7$ **23.** $-7 \le x \le 7$ or $[-7, 7]$ **25.** $x \le -7$ or $x \ge 7$ or $(-\infty, -7] \cup [7, \infty)$
27. $y = 2, 8$ **29.** $2 < y < 8$ or $(2, 8)$ **31.** $y < 2$ or $y > 8$ or $(-\infty, 2) \cup (8, \infty)$
33. $u = -11, -5$ **35.** $-11 \le u \le -5$ or $[-11, -5]$
37. $u \le -11$ or $u \ge -5$ or $(-\infty, -11] \cup [-5, \infty)$ **39.** $x = -4, \frac{4}{3}$ **41.** $-\frac{9}{5} \le x \le 3$ or $[-\frac{9}{5}, 3]$
43. $y < 3$ or $y > 5$ or $(-\infty, 3) \cup (5, \infty)$ **45.** $t = -\frac{4}{5}, \frac{18}{5}$ **47.** $-\frac{5}{7} < u < \frac{23}{7}$ or $(-\frac{5}{7}, \frac{23}{7})$
49. $x \le -6$ or $x \ge 9$ or $(-\infty, -6] \cup [9, \infty)$ **51.** $-35 < C < -\frac{5}{9}$ or $(-35, -\frac{5}{9})$ **53.** $-2 < x < 2$ or $(-2, 2)$
55. $t \le -4$ or $t \ge 4$ or $(-\infty, -4] \cup [4, \infty)$ **57.** $-\frac{1}{3} \le t \le 1$ or $[-\frac{1}{3}, 1]$ **59.** $t < 0$ or $t > 3$ or $(-\infty, 0) \cup (3, \infty)$
61. m is 4 units from 2.   **63.** x is 7 units from -1.   **65.** u is less than 5 units from 4.   **67.** z is no more than 3 units from 2.
69. $|x - 3| = 4$ **71.** $|m + 2| = 5$ **73.** $|x - 3| < 5$ **75.** $|p + 2| > 6$ **77.** $|q - 1| \ge 2$ **79.** $(2.9, 3) \cup (3, 3.1)$
81. $(c - d, c) \cup (c, c + d)$ **83.** $x \ge 5$ **85.** $x \le -8$ **87.** $x \ge -\frac{3}{4}$ **89.** $x \le \frac{2}{5}$ **91.** $x = -2, 2$
105. $42.2 < x < 48.6$ **107.** $|P - 500| \le 20$ **109.** $|N - 2.37| \le 0.005$

Exercise 2-5

1. $7 + 5i$ **3.** $5 + 3i$ **5.** $2 + 4i$ **7.** $5 + 9i$ **9.** $4 - 3i$ **11.** -24 or $-24 + 0i$ **13.** $-12 - 6i$ **15.** $15 - 3i$
17. $-4 - 33i$ **19.** 65 or $65 + 0i$ **21.** $\frac{2}{5} - \frac{1}{5}i$ **23.** $\frac{3}{13} + \frac{11}{13}i$ **25.** $5 + 3i$ **27.** $7 - 5i$ **29.** $-3 + 2i$
31. $8 + 25i$ **33.** $\frac{5}{7} - \frac{2}{7}i$ **35.** $\frac{2}{13} + \frac{3}{13}i$ **37.** $-\frac{2}{5}i$ or $0 - \frac{2}{5}i$ **39.** $\frac{3}{2} - \frac{1}{2}i$ **41.** $-6i$ or $0 - 6i$ **43.** 0 or $0 + 0i$
45. $i^{18} = -1, i^{32} = 1, i^{67} = -i$ **47.** $x = 3, y = -2$ **49.** $x > 3$ **51.** $x > \frac{2}{3}$ **53.** $33.89 - 20.38i$ **55.** $0.85 - 0.89i$
57. $(a + c) + (b + d)i$ **59.** $a^2 + b^2$ or $(a^2 + b^2) + 0i$ **61.** $(ac - bd) + (ad + bc)i$
63. $i^{4k} = (i^4)^k = (i^2 \cdot i^2)^k = [(-1)(-1)]^k = 1^k = 1$
65. (1) Definition of addition; (2) Commutative $(+)$ property for R; (3) Definition of addition

Exercise 2-6

1. $u = 0, 2$ **3.** $y = \frac{2}{3}$ (double root)   **5.** $x = \frac{3}{2}, 4$ **7.** $m = \pm 2\sqrt{3}$ **9.** $x = \pm 5i$ **11.** $y = \pm \frac{4}{3}$ **13.** $x = \pm \frac{5}{2}i$
15. $n = -2, -8$ **17.** $d = 3 \pm 2i$ **19.** $x = 5 \pm 2\sqrt{7}$ **21.** $x = 2 \pm 2i$ **23.** $x = (2 \pm \sqrt{2})/2$ **25.** $x = \frac{1}{5} \pm \frac{3}{5}i$

27. $x = 3 \pm 2\sqrt{3}$ **29.** $y = (3 \pm \sqrt{3})/2$ **31.** $x = (1 \pm \sqrt{7})/3$ **33.** $x = (-m \pm \sqrt{m^2 - 4n})/2$ **35.** $x = -\frac{5}{4}, \frac{2}{3}$

37. $y = (3 \pm \sqrt{5})/2$ **39.** $x = (3 \pm \sqrt{13})/2$ **41.** $n = -\frac{4}{7}, 0$ **43.** $x = 2 \pm 2i$ **45.** $m = -50, 2$

47. $x = (-5 \pm \sqrt{57})/2$ **49.** $x = (-3 \pm \sqrt{57})/4$ **51.** $u = 1, 2, (-3 \pm \sqrt{17})/2$ **53.** $t = \sqrt{2s/g}$

55. $I = (E + \sqrt{E^2 - 4RP})/2R$ **57.** $x = 1.35, 0.48$ **59.** $x = -1.05, 0.63$

61. Has real solutions, since discriminant is positive   **63.** Has no real solutions, since discriminant is negative

65. $x = \frac{4}{3}\sqrt{6} \pm \frac{2}{3}\sqrt{15}$ or $(4\sqrt{6} \pm 2\sqrt{15})/3$ **67.** $x = \sqrt{2} - i, -\sqrt{2} - i$ **69.** $x = 1, -\frac{1}{2} \pm \frac{1}{2}i\sqrt{3}$ **71.** F

73. $[(-b + \sqrt{b^2 - 4ac})/2a] \times [(-b - \sqrt{b^2 - 4ac})/2a] = [b^2 - (b^2 - 4ac)]/4a^2 = c/a$

75. The $\pm$ in front still yields the same two numbers even if a is negative.   **77.** 8, 13   **79.** 12, 14   **81.** 5.12 by 3.12 in

83. 20%   **85.** 100 mph, 240 mph   **87.** 13.09 h and 8.09 h   **89.** 50 mph

91. 50 ft wide and 300 ft long or 150 ft wide and 100 ft long   **93.** 52 mi

Exercise 2-7

1. $x = 22$ **3.** $n = 8$ **5.** No solution   **7.** $x = 0, 4$ **9.** $y = \pm 2, \pm i\sqrt{2}$ **11.** $x = -\sqrt[5]{5}, \sqrt[5]{2}$ **13.** $x = \frac{1}{8}, -8$

15. $m = -2, 3, \frac{1}{2} \pm \frac{1}{2}i\sqrt{7}$ **17.** No solution   **19.** $y = 1$ **21.** $x = 2$ **23.** $n = -\frac{3}{4}, \frac{1}{5}$ **25.** $y = \pm 1, \pm 3$

27. $y = 1, 16$ **29.** $m = 2, 3, 7, 8$ **31.** $x = -2$ **33.** $y = \pm\sqrt{\dfrac{3 \pm \sqrt{3}}{2}}$ (four roots)   **35.** $m = 9, 16$ **37.** $t = 4, 81$

39. 13.1 in by 9.1 in   **41.** 1.65 ft or 3.65 ft

Exercise 2-8

1. $-5 < x < 2$ **3.** $x < 3$ or $x > 7$ **5.** $0 \le x \le 8$ **7.** $-5 \le x \le 0$ **9.** $x < -2$ or $x > 2$
$(-5, 2)$ $(-\infty, 3) \cup (7, \infty)$ $[0, 8]$ $[-5, 0]$ $(-\infty, -2) \cup (2, \infty)$

11. $-4 < x \le 2$ **13.** $x \le -4$ or $x > 1$ **15.** $-5 \le x \le 0$ or $x > 3$ **17.** $-3 < x < 1$ **19.** $x < 0$ or $x > \frac{1}{4}$
$(-4, 2]$ $(-\infty, -4] \cup (1, \infty)$ $[-5, 0] \cup (3, \infty)$ $(-3, 1)$ $(-\infty, 0) \cup (\frac{1}{4}, \infty)$

21. $-4 < x \le \frac{3}{2}$ **23.** $-1 < x < 2$ or $x \ge 5$ **25.** $x \le -4$ or $0 \le x \le 2$ **27.** $x \le -3$ or $x \ge 3$ **29.** $x \le -2$ or $x \ge \frac{3}{2}$
$(-4, \frac{3}{2}]$ $(-1, 2) \cup [5, \infty)$ $(-\infty, -4] \cup [0, 2]$

31. $-7 \le x < 3$ **33.** No solution; $\varnothing$ **35.** No solution; $\varnothing$ **37.** $x \le 2 - \sqrt{5}$ or $x \ge 2 + \sqrt{5}$
$(-\infty, 2 - \sqrt{5}] \cup [2 + \sqrt{5}, \infty)$

39. $1 - \sqrt{2} < x < 0$ or $x > 1 + \sqrt{2}$ **41.** $-2 \le x \le -\frac{1}{2}$ or $\frac{1}{2} \le x \le 2$ **43.** $-2 \le x \le 2$
$(1 - \sqrt{2}, 0) \cup (1 + \sqrt{2}, \infty)$ $[-2, -\frac{1}{2}] \cup [\frac{1}{2}, 2]$ $[-2, 2]$

45. (A) Profit: $4 < p < \$7$ or $(\$4, \$7)$ (B) Loss: $\$0 \le p < \$4$ or $p > \$7$ or $[\$0, \$4) \cup (\$7, \infty)$ **47.** $2 \le t \le 5$

49. $v > 75$ mph   **51.** $5 \le t \le 20$

Chapter 2 Review Exercise

1. $x = 21$ *(2-1)*   **2.** $x = \frac{30}{11}$ *(2-1)*   **3.** $x \ge 1$ *(2-3)*   **4.** $-14 < y < -4$ *(2-4)*   **5.** $-1 \le x \le 4$ *(2-4)*
$[1, \infty)$ $(-14, -4)$ $[-1, 4]$

6. $-5 < x < 4$ *(2-8)*   **7.** $x \le -3$ or $x \ge 7$ *(2-8)*   **8.** (A) $3 - 6i$ (B) $15 + 3i$ (C) $2 + i$ *(2-5)*
$(-5, 4)$ $(-\infty, -3] \cup [7, \infty)$

9. $x = \pm\sqrt{\frac{7}{2}}$ or $\pm\frac{1}{2}\sqrt{14}$ *(2-6)*   **10.** $x = 0, 2$ *(2-6)*   **11.** $x = \frac{1}{2}, 3$ *(2-6)*   **12.** $m = -\frac{1}{2} \pm (\sqrt{3}/2)i$ *(2-6)*

13. $y = (3 \pm \sqrt{33})/4$ *(2-6)*   **14.** $x = 2, 3$ *(2-7)*   **15.** $x \le \frac{3}{5}$ *(2-3)*   **16.** $x = -15$ *(2-1)*   **17.** No solution   *(2-1)*

18. $x \geq -19$ *(2-3)*   **19.** $x < 2$ or $x > \frac{10}{3}$ *(2-4)*   **20.** $x < 0$ or $x > \frac{1}{2}$ *(2-8)*   **21.** $x \leq 1$ or $3 < x < 4$ *(2-8)*
$[-19, \infty)$ $(-\infty, 2) \cup (\frac{10}{3}, \infty)$ $(-\infty, 0) \cup (\frac{1}{2}, \infty)$ $(-\infty, 1] \cup (3, 4)$

22. $-1 \leq m \leq 2$ *(2-4)*   **23.** $-4 \leq x < 2$ or $[-4, 2)$ *(2-8)*   **24.** (A) 6   (B) 6 *(2-4)*
$[-1, 2]$

25. (A) $5 + 4i$ (B) $-i$ *(2-5)*   **26.** (A) $-1 + i$ (B) $\frac{4}{13} - \frac{7}{13}i$ (C) $\frac{5}{2} - 2i$ *(2-5)*   **27.** $u = (-5 \pm \sqrt{5})/2$ *(2-6)*
28. $u = 1 \pm i\sqrt{2}$ *(2-6)*   **29.** $x = (1 \pm \sqrt{43})/3$ *(2-6)*   **30.** $x = -\frac{27}{8}, 64$ *(2-7)*   **31.** $m = \pm 2, \pm 3i$ *(2-7)*
32. $y = \frac{9}{4}, 3$ *(2-7)*   **33.** $x = 0.45$ *(2-1)*   **34.** $-2.24 \leq x \leq 1.12$ or $[-2.24, 1.12]$ *(2-3)*   **35.** $0.89 - 0.32i$ *(2-5)*
36. $x = -1.64, 0.89$ *(2-6)*   **37.** $M = P/(1 - dt)$ *(2-1)*   **38.** $I = (E \pm \sqrt{E^2 - 4PR})/2R$ *(2-6)*
39. $y = (5 - x)/(2x - 4)$ *(2-1)*   **40.** All real b and all negative a *(2-3)*   **41.** Less than 1 *(2-3)*   **42.** $x = 1/(1 - y)$ *(2-1)*
43. $6 - d < x < 6 + d, x \neq 6$ *(2-4)*   **44.** $x = \frac{1}{4}\sqrt{3} \pm \frac{1}{4}i$ *(2-6)*   **45.** $x = \pm\sqrt{(2 + \sqrt{5})/2}$ (two real roots) *(2-7)*
$(6 - d, 6) \cup (6, 6 + d)$

46. 1 *(2-5)*   **47.** No solution *(2-8)*   **48.** Set of all real numbers *(2-8)*
49. $x \leq -4$ or $-2 \leq x < 0$ or $0 < x \leq 2$ or $x \geq 4$; $(-\infty, -4] \cup [-2, 0) \cup (0, 2] \cup [4, \infty)$ *(2-8)*   **50.** $\frac{5}{3}$ or $-\frac{3}{5}$ *(2-6)*
51. (A) $H = 0.7(220 - A)$ (B) 140 beats/min   (C) 40 years old *(2-2)*
52. 20 mL of 30% solution, 30 mL of 80% solution *(2-2)*   **53.** 3 mph *(2-6)*   **54.** (A) 2,000 and 8,000   (B) 5,000 *(2-6)*
55. $x = (13 \pm \sqrt{45})/2$ thousand, or approx. 3,146 and 9,854 *(2-6)*
56. $(13 - \sqrt{45})/2 < x < (13 + \sqrt{45})/2$ or approx. $3.146 < x < 9.854$, x in thousands *(2-8)*   **57.** $|T - 110| \leq 5$ *(2-4)*
58. 20 cm by 24 cm *(2-6)*   **59.** 6.58 ft or 14.58 ft *(2-7)*

Cumulative Review Exercise: Chapters 1 and 2

1. $x = \frac{5}{2}$ *(2-1)*   **2.** (A) $ca + cb$ (B) $a + (b + c)$ (C) $c(a + b)$ *(1-1)*   **3.** $4x^2 - 6x + 10$ *(1-2)*
4. $3x^2 - 26x + 9$ *(1-2)*   **5.** $6x^2 - 7xy - 20y^2$ *(1-2)*   **6.** $4a^2 - 9b^2$ *(1-2)*   **7.** $25m^2 + 20mn + 4n^2$ *(1-2)*
8. $y \geq 5$ *(2-3)*   **9.** $-5 < x < 9$ *(2-4)*   **10.** $x \leq -5$ or $x \geq 2$ *(2-8)*   **11.** Prime *(1-3)*   **12.** $(3t + 5)(2t - 1)$ *(1-3)*
$[5, \infty)$ $(-5, 9)$ $(-\infty, -5] \cup [2, \infty)$

13. $\dfrac{2}{(x - 2)(x - 3)}$ *(1-4)*   **14.** $2y$ *(1-4)*   **15.** (A) $7 - 10i$ (B) $23 + 7i$ (C) $1 - i$ *(2-5)*   **16.** $3x^7y^{10}$ *(1-5)*
17. $\dfrac{2y^2}{x}$ *(1-5)*   **18.** $\dfrac{a^6}{b^9}$ *(1-6)*   **19.** $x = -4, 0$ *(2-6)*   **20.** $x = -\sqrt{5}, \sqrt{5}$ *(2-6)*   **21.** $x = 3 \pm \sqrt{7}$ *(2-6)*
22. $x = 3$ *(2-7)*   **23.** $5\sqrt[4]{a^3}$ *(1-7)*   **24.** $2x^{2/5}y^{3/5}$ *(1-7)*   **25.** $x^2y^4\sqrt[3]{xy^2}$ *(1-7)*   **26.** $2xy^2\sqrt{3x}$ *(1-7)*   **27.** $\frac{1}{2}\sqrt{2}$ *(1-7)*
28. $x \geq -\frac{2}{3}$ or $[-\frac{2}{3}, \infty)$ *(2-3)*   **29.** $\{2, 3, 5\}$ *(1-1, 1-2)*   **30.** (A) F   (B) T   (C) F *(1-1)*   **31.** No solution *(2-1)*
32. $x = \frac{1}{2}, 3$ *(2-6)*   **33.** $x = 1, \frac{5}{2}$ *(2-7)*   **34.** (a) is a second-degree polynomial   (c) is a fourth-degree polynomial *(1-2)*
35. $10ab$ *(1-2)*   **36.** $6hx + 3h^2 - 4h$ *(1-2)*   **37.** $64m^3 + 144m^2n + 108mn^2 + 27n^3$ *(1-2)*   **38.** $(y + 4)(3y + 4)$ *(1-3)*
39. $(a + 3)(a + 2)(a - 2)$ *(1-3)*   **40.** $x^3(x + 1)^2(7x + 4)$ *(1-3)*   **41.** $\dfrac{x^3(x + 4)}{(x + 1)^4}$ *(1-4)*   **42.** $\dfrac{5}{a - 2b}$ *(1-4)*
43. $\dfrac{x + 2y}{x - 2y}$ *(1-4)*   **44.** $x < \frac{3}{2}$ or $x > 3$ *(2-4)*   **45.** $\frac{2}{3} \leq m \leq 2$ *(2-4)*   **46.** $-1 < x < 2$ or $x \geq 5$ *(2-8)*   **47.** $x \geq 2, x \neq 4$ *(2-3)*
$(-\infty, \frac{3}{2}) \cup (3, \infty)$ $[\frac{2}{3}, 2]$ $(-1, 2) \cup [5, \infty)$ $[2, 4) \cup (4, \infty)$

48. (A) 0   (B) $\frac{6}{5}$ (C) $-i$ *(2-5)*   **49.** (A) $3 + 18i$ (B) $-2.9 + 10.7i$ (C) $-4 - 6i$ *(2-5)*
50. (A) 0.1767   (B) 1.434   (C) 1.435   (D) 5.724×10^{14} *(1-5, 1-7)*   **51.** $6x^2y^4\sqrt[3]{2x^2y}$ *(1-7)*   **52.** $\dfrac{1}{2b}\sqrt[5]{14a^3b}$ *(1-7)*
53. $y\sqrt{2y}$ *(1-7)*   **54.** $\sqrt{t + 9} + 3$ *(1-7)*   **55.** (A) $\frac{2}{5}x + 2x^{-2}$ (B) $\frac{1}{2} - \frac{5}{4}x^{-1/2}$ *(1-5, 1-7)*   **56.** $y = 3 \pm i\sqrt{5}$ *(2-6)*
57. $x = -\frac{1}{8}, \frac{27}{8}$ *(2-7)*   **58.** $u = \pm\sqrt{3}, \pm 2i$ *(2-7)*   **59.** $t = \frac{9}{4}$ *(2-7)*   **60.** $-18.36 \leq x < 16.09$; $[-18.36, 16.09)$ *(2-3)*
61. $x = -5.68, 1.23$ *(2-6)*   **62.** $y = \dfrac{3 - 3x}{x + 4}$ *(2-1)*   **63.** 0 *(1-7)*   **64.** 0 *(2-5)*   **65.** $6x^2 + 2$ *(1-2)*
66. All a and b such that $a < b$ *(2-3)*   **67.** $b^2(3b + 4a - 4)(3b - 4a + 4)$ *(1-3)*   **68.** $\dfrac{2 - m}{4m}$ *(1-4)*

69. $y = \dfrac{x + x^2}{x - 1}$ *(2-1)*   **70.** $x = (\sqrt{2} \pm i)/3$ *(2-6)*   **71.** $x = \pm\sqrt{3\sqrt{2} + 3}$ (two real roots)   *(2-7)*

72. (A) $a - b$ (B) $\dfrac{x^2 - y^2}{x^2 + y^2}$ *(1-5, 1-7)*   **73.** $\dfrac{1}{(8 + h)^{2/3} + 2(8 + h)^{1/3} + 4}$ *(1-7)*   **74.** $\dfrac{a^2 - b^2}{a^2 + b^2} + \dfrac{2ab}{a^2 + b^2}\,i$ *(2-5)*

75. $x < -1$ or $x > 2$; $(-\infty, -1) \cup (2, \infty)$ *(2-8)*   **76.** x^2 *(1-7)*   **77.** 2 or $-\frac{1}{2}$ *(2-6)*   **78.** 10.5 min   *(2-2)*

79. 2.5 mph   *(2-6)*   **80.** 12 gal   *(2-2)*   **81.** 8,800 books   *(2-2)*   **82.** $|p - 200| \le 10$ *(2-4)*

83. $x^2 - (x - 1)^2 = 2x - 1$ ft^3 *(1-2)*

84. (A)   Profit: $\$5.5 < p < \$8$ or $(\$5.5, \$8)$ (B)   Loss: $\$0 \le p < 5.5$ or $p > \$8$ or $[\$0, \$5.5) \cup (\$8, \infty)$ *(2-8)*

85. 40 mi from A to B and 75 mi from B to C or 75 mi from A to B and 40 mi from B to C *(2-6)*

CHAPTER 3   Exercise 3-1

1.   **3.** Symmetric with respect to the origin   **5.** Symmetric with respect to the x axis   **7.** Symmetric with respect to the x axis, y axis, and origin   **9.** $\sqrt{145}$

  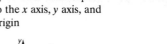

11. $\sqrt{68}$ **13.** $x^2 + y^2 = 49$ **15.** $(x - 2)^2 + (y - 3)^2 = 36$ **17.** $(x + 4)^2 + (y - 1)^2 = 7$ **19.** $(x + 3)^2 + (y + 4)^2 = 2$

21. Symmetric with respect to the x axis   **23.** Symmetric with respect to the y axis   **25.** Symmetric with respect to the x axis, y axis, and origin   **27.** Symmetric with respect to the x axis, y axis, and origin

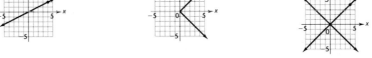

29. Symmetric with respect to the x axis, y axis, and origin   **31.** Symmetric with respect to the x axis, y axis, and origin   **33.** Symmetric with respect to the origin   **35.** Symmetric with respect to the y axis

37. Symmetric with respect to the y axis   **39.** Symmetric with respect to the y axis   **41.** A right triangle   **43.** 18.11   **45.** $x = -3, 7$

 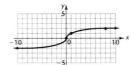

47. Center: $(3, 5)$; radius: 7   **49.** Center: $(-4, 2)$; radius: $\sqrt{7}$ **51.** Center: $(3, 2)$; radius: 7   **53.** Center: $(-4, 3)$; radius: $\sqrt{17}$

 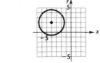

55. Symmetric with respect to the y axis   **57.** Symmetric with respect to the origin   **59.** 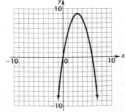   **61.** $5x + 3y = -2$

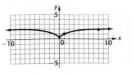

63. $(x - 4)^2 + (y - 2)^2 = 34$ **65.** $y = \pm\sqrt{3 - x^2}$ **67.** $y = -1 \pm \sqrt{2 - (x + 3)^2}$ **69.** Center: $(1, 0)$; radius: 1; $(x - 1)^2 + y^2 = 1$

        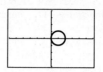

71. Center: $(2, 1)$; radius: 3; $(x - 2)^2 + (y - 1)^2 = 9$ **73.**   **75.** 2.5 ft   **77.** (A) $(x + 12)^2 + (y + 5)^2 = 26^2$; center: $(-12, -5)$; radius: 26

    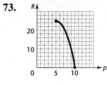

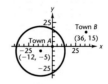

(B) 13.5 mi

Exercise 3-2

1. x intercept: -2; y intercept: 2; slope: 1   **3.** x intercept: -2; y intercept: -4; slope: -2 **5.** x intercept: 3; y intercept: -1; slope: $\frac{1}{3}$
7. Slope $= -\frac{3}{5}$ **9.** Slope $= -\frac{3}{4}$ **11.** Slope $= \frac{2}{3}$ **13.** Slope $= \frac{4}{5}$ **15.** Slope $= 2$

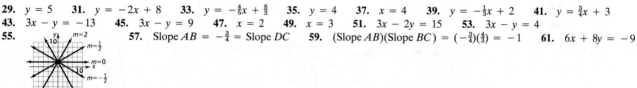

17. Slope not defined   **19.** Slope $= 0$ **21.** $y = x$ **23.** $y = -\frac{2}{3}x - 4$ **25.** $y = -3x + 4$ **27.** $y = -\frac{2}{5}x + 2$

29. $y = 5$ **31.** $y = -2x + 8$ **33.** $y = -\frac{4}{5}x + \frac{8}{3}$ **35.** $y = 4$ **37.** $x = 4$ **39.** $y = -\frac{1}{3}x + 2$ **41.** $y = \frac{3}{4}x + 3$
43. $3x - y = -13$ **45.** $3x - y = 9$ **47.** $x = 2$ **49.** $x = 3$ **51.** $3x - 2y = 15$ **53.** $3x - y = 4$
55.   **57.** Slope $AB = -\frac{3}{4} =$ Slope DC **59.** (Slope AB)(Slope BC) $= \left(-\frac{3}{4}\right)\left(\frac{4}{3}\right) = -1$ **61.** $6x + 8y = -9$

63. $3x + 4y = 25$ **65.** $x - y = 10$ **67.** $5x - 12y = 232$ **69.**   **71.**

73.   **79.** (A) $F = \frac{9}{5}C + 32$ (B) 68°F; 30°C   (C) $\frac{9}{5}$

81. (A) $V = -1,600t + 8,000, 0 \le t \le 5$ (B) \$3,200   (C) $-1,600$
83. (A) $T = -5A + 70, A \ge 0$ (B) 14,000 ft   (C) -5; the temperature changes -5°F for each 1,000-ft rise in altitude
85. (A) $h = 1.13t + 12.8$ (B) 32.9 h
87. (A) $R = 0.001\ 52C - 0.159, C \ge 210$ (B) 0.236
(C) $0.001\ 52$; coronary risk increases $0.001\ 52$ per unit increase in cholesterol above the 210 cholesterol level

Exercise 3-3

1. Function   **3.** Not a function   **5.** Function   **7.** Function; domain = {2, 3, 4, 5}; range = {4, 6, 8, 10}
9. Not a function   **11.** Function; domain = {0, 1, 2, 3, 4, 5}; range = {1, 2}   **13.** Function   **15.** Not a function
17. Not a function   **19.** -8 **21.** -6 **23.** 1   **25.** 10   **27.** $-\frac{30}{17}$ **29.** 3   **31.** Function   **33.** Function
35. Not a function   **37.** Function   **39.** Not a function   **41.** Function   **43.** Function   **45.** Function   **47.** Function
49. Domain: all real numbers   **51.** Domain: $x \geq -2$ **53.** Domain: all real numbers except 4
55. Domain: all real numbers except -2 and 4   **57.** Domain: $-2 \leq x \leq 2$ **59.** Domain: $x \leq -1$ or $x \geq 4$
61. Domain: $2 < x \leq 5$ **63.** Domain: all real numbers except -1 and 1   **65.** 3   **67.** $-6 - h$ **69.** h **71.** $-2h + 11$
73. (A) 3   (B) 3   **75.** (A) $2x + h$ (B) $x + a$ **77.** (A) $-6x - 3h + 9$ (B) $-3x - 3a + 9$
79. (A) $3x^2 + 3xh + h^2$ (B) $x^2 + ax + a^2$ **81.** $P(w) = 2w + (128/w)$ **83.** $h(b) = \sqrt{b^2 + 25}$ **85.** $C(x) = 300 + 1.75x$
87. (A) $s(0) = 0, s(1) = 16, s(2) = 64, s(3) = 144$ (B) $64 + 16h$
(C) Value of expression tends to 64; this number appears to be the speed of the object at the end of 2 s.

89. $V(x) = x(8 - 2x)(12 - 2x)$; domain: $0 < x < 4$ **91.** $F(x) = 8x + (250/x) - 12$;

x	4	5	6	7
$F(x)$	82.5	78	77.7	79.7

93. $C(x) = 10{,}000(20 - x) + 15{,}000\sqrt{x^2 + 64}$; domain: $0 \leq x \leq 20$ **95.** $C(v) = 100v + (200{,}000/v)$

Exercise 3-4

1. (A) $(-\infty, \infty)$ (B) $[-4, \infty)$ (C) $-3, 1$ (D) -3 (E) $[-1, \infty)$ (F) $(-\infty, -1]$ (G) None   (H) None
3. (A) $(-\infty, 2) \cup (2, \infty)$ (B) $(-\infty, -1) \cup [1, \infty)$ (C) None   (D) 1   (E) None   (F) $(-\infty, -2], (2, \infty)$
(G) $[-2, 2)$ (H) $x = 2$
5. Slope $= 2$, x intercept $= -2$,   **7.** Slope $= -\frac{1}{2}$, x intercept $= -\frac{10}{3}$,   **9.** $f(x) = -3x + 5$ **11.** $f(x) = -\frac{1}{2}x + 4$
y intercept $= 4$ y intercept $= -\frac{5}{3}$

13. $f(x) = -\frac{3}{2}x + 4$ **15.** Min $f(x) = f(3) = 2$ **17.** Max $f(x) = f(-3) = -2$ **19.** x intercepts: $-1, 5$
Range $= [2, \infty)$ Range $= (-\infty, -2]$ y intercept $= -5$

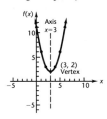

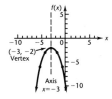

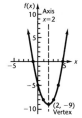

21. x intercepts: 0, 6   **23.** Increasing on $[-3, \infty)$ **25.** Increasing on $(-\infty, 3]$ **27.** Domain: $[-1, 1]$
y intercept $= 0$ Decreasing on $(-\infty, -3]$ Decreasing on $[3, \infty)$ Range: $[0, 1]$

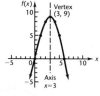

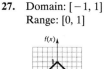

29. Domain: $[-3, -1) \cup (-1, 2]$ **31.** Domain: all real numbers   **33.** Domain: $x \neq 0$, or $(-\infty, 0) \cup (0, \infty)$
Range: $\{-2, 4\}$ (a set, not an interval)   Range: all real numbers   Range: $(-\infty, -1) \cup (1, \infty)$
Discontinuous at $x = -1$ Discontinuous at $x = -1$ Discontinuous at $x = 0$

35. Domain: all real numbers
Range: [1, 3)
Discontinuous at $x = 2$

37. Domain: all real numbers
Range: [0, 4]
Discontinuous at $x = 0$

39. Min $f(x) = f(-2) = 1$
Range: [1, ∞)
No x intercepts
y intercept $= f(0) = 3$
Increasing on $[-2, ∞)$
Decreasing on $(-∞, -2]$

41. Min $f(x) = f(\frac{3}{2}) = 0$
Range: [0, ∞)
x intercept $= \frac{3}{2}$
y intercept $= f(0) = 9$
Increasing on $[\frac{3}{2}, ∞)$
Decreasing on $(-∞, \frac{3}{2}]$

43. Max $f(x) = f(-2) = 6$
Range: $(-∞, 6]$
x intercepts: $-2 ± \sqrt{3}$
y intercept $= f(0) = -2$
Increasing on $(-∞, -2]$
Decreasing on $[-2, ∞)$

45. $f(x) = \begin{cases} -1 & \text{if } x < 0 \\ 1 & \text{if } x > 0 \end{cases}$

Domain: $x ≠ 0$, or $(-∞, 0) \cup (0, ∞)$
Range: $\{-1, 1\}$ (a set, not an interval)
Discontinuous at $x = 0$

47. $f(x) = \begin{cases} x - 1 & \text{if } x < 1 \\ x + 1 & \text{if } x > 1 \end{cases}$

Domain: $x ≠ 1$, or $(-∞, 1) \cup (1, ∞)$
Range: $(-∞, 0) \cup (2, ∞)$
Discontinuous at $x = 1$

49. $f(x) = \begin{cases} 2 - 2x & \text{if } x < 0 \\ 2 & \text{if } 0 ≤ x < 2 \\ -2 + 2x & \text{if } x ≥ 2 \end{cases}$

Domain: all real numbers
Range: [2, ∞)
No discontinuities

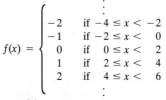

51. Domain: all real numbers
Range: all integers
Discontinuous at the even integers

$f(x) = \begin{cases} \vdots \\ -2 & \text{if } -4 ≤ x < -2 \\ -1 & \text{if } -2 ≤ x < 0 \\ 0 & \text{if } 0 ≤ x < 2 \\ 1 & \text{if } 2 ≤ x < 4 \\ 2 & \text{if } 4 ≤ x < 6 \\ \vdots \end{cases}$

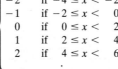

53. Domain: all real numbers
Range: all integers
Discontinuous at rational numbers of the form $k/3$, where k is an integer

$f(x) = \begin{cases} \vdots \\ -2 & \text{if } -\frac{2}{3} ≤ x < -\frac{1}{3} \\ -1 & \text{if } -\frac{1}{3} ≤ x < 0 \\ 0 & \text{if } 0 ≤ x < \frac{1}{3} \\ 1 & \text{if } \frac{1}{3} ≤ x < \frac{2}{3} \\ 2 & \text{if } \frac{2}{3} ≤ x < 1 \\ \vdots \end{cases}$

55. Domain: all real numbers
Range: [0, 1)
Discontinuous at all integers

$f(x) = \begin{cases} \vdots \\ x + 2 & \text{if } -2 ≤ x < -1 \\ x + 1 & \text{if } -1 ≤ x < 0 \\ x & \text{if } 0 ≤ x < 1 \\ x - 1 & \text{if } 1 ≤ x < 2 \\ x - 2 & \text{if } 2 ≤ x < 3 \\ \vdots \end{cases}$

57. Axis: $x = 2$; vertex: (2, 4); range: [4, ∞); no x intercepts

59. $y = 2x - 1$

61. (A) $h + 1$ (B)

h	1	0.1	0.01	0.001
Slope	2	1.1	1.01	1.001

; approaching 1

63. Graphs of f and g **65.** Graphs of f and g **67.** Graphs of f and g **69.** $m(x) = \max[f(x), g(x)]$

Graph of m Graph of m Graph of m

Graph of n Graph of n Graph of n

71. (A) $s = f(w) = w/10$
(B) $f(15) = 1.5$ in, $f(30) = 3$ in
(C) Slope $= \frac{1}{10}$
(D)

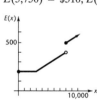

73. $E(x) = \begin{cases} 200 & \text{if } 0 \le x \le 3{,}000 \\ 80 + 0.04x & \text{if } 3{,}000 < x < 8{,}000 \\ 180 + 0.04x & \text{if } x \ge 8{,}000 \end{cases}$

Discontinuous at $x = 8{,}000$
$E(5{,}750) = \$310$, $E(9{,}200) = \$548$

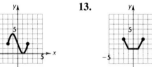

75. (A) $A(x) = 50x - x^2$
(B) Domain: $0 < x < 50$
(C)

(D) 25 by 25 ft

77.

x	4	-4	6	-6	24	25	247	-243	-245	-246
$f(x)$	0	0	10	-10	20	30	250	-240	-240	-250

; f rounds numbers to the tens place.

79. $f(x) = [\![0.5 + 100x]\!]/100$ **81.** \$50 per day; maximum income $= \$12{,}500$ **83.** (A) $32\sqrt{5} \approx 71.55$ ft/s   (B) 15 ft

Exercise 3-5

1. Odd   **3.** Even   **5.** Neither   **7.** Even   **9.** Neither   **11.**   **13.**

15.   **17.**   **19.**   **21.**

            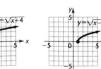

23.   **25.**

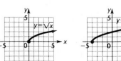

27.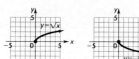

29. It is the same as the graph of $y = x^2$ reflected with respect to the x axis and shifted to the left 2 units.

31. It is the same as the graph of $y = |x|$ reflected with respect to the x axis and shifted to the left 2 units.

33. It is the same as the graph of $y = \sqrt{x}$ reflected with respect to the x axis and shifted to the right 1 unit.

35.

37.

39. Even

41.

43.

45. Odd

47.

49. It is the same as the graph of $p(x) = x^2$ shifted to the right 2 units, expanded by a factor of 2, and shifted down 4 units.

51. It is the same as the graph of $p(x) = x^2$ shifted to the right 2 units, contracted by a factor of $\frac{1}{2}$, reflected with respect to the x axis, and shifted up 3 units.

53. Discontinuous at the integers

55. Discontinuous at the integers

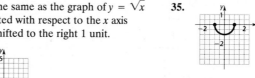

57. Discontinuous at the integers

61. Graph of $f(x)$

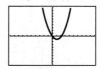

Graph of $f(|x|)$

Graph of $f(-|x|)$

63. Graph of $f(x)$

Graph of $f(|x|)$

Graph of $f(-|x|)$

65. The right side of the graph of $y = f(|x|)$ is the same as the right side of the graph of $y = f(x)$, and the left side is the reflection of the right side of the graph of $y = f(x)$ with respect to the y axis.

67. Graph of $f(x)$

Graph of $|f(x)|$

Graph of $-|f(x)|$

69. Graph of $f(x)$

Graph of $|f(x)|$

Graph of $-|f(x)|$

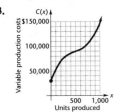

71. The graph of $y = |f(x)|$ is the same as the graph of $y = f(x)$ whenever $f(x) \geq 0$ and is the reflection of the graph of $y = f(x)$ with respect to the x axis whenever $f(x) < 0$.

73.

75.

77.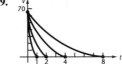

79.

Exercise 3-6

1. Domain: $(-\infty, -1) \cup (-1, \infty)$; x intercept: 2   **3.** Domain: $(-\infty, -4) \cup (-4, 4) \cup (4, \infty)$; x intercepts: $-1, 1$

5. Domain: $(-\infty, -3) \cup (-3, 4) \cup (4, \infty)$; x intercepts: $-2, 3$ **7.** Domain: all real numbers; x intercept: 0

9. Vertical asymptote: $x = 4$; horizontal asymptote: $y = 2$ **11.** Vertical asymptotes: $x = -4, x = 4$; horizontal asymptote: $y = \frac{2}{3}$

13. No vertical asymptotes; horizontal asymptote: $y = 0$ **15.** Vertical asymptotes: $x = -1, x = \frac{5}{3}$; no horizontal asymptote

17. **19.** **21.** **23.** **25.**

27. **29.** **31.** **33.** **35.**

37. **39.** **41.** **43.** **45.**

47. Vertical asymptote: $x = 1$; oblique asymptote: $y = 2x + 2$ **49.** Vertical asymptote: $x = -1$; horizontal asymptote: $y = 0$

51. Oblique asymptote: $y = x$ **53.** Vertical asymptote: $x = 0$; oblique asymptote: $y = 2x - 3$ **55.**

57.

59.

61. Domain: $x \neq 2$, or $(-\infty, 2) \cup (2, \infty)$; $f(x) = x + 2$

63. Domain: $x \neq 2, -2$, or $(-\infty, -2) \cup (-2, 2) \cup (2, \infty)$; $f(x) = \dfrac{1}{x - 2}$

65. $N \to 50$ as $t \to \infty$

67. $N \to 5$ as $t \to \infty$

69. (A) $\overline{C}(n) = 25n + 175 + \dfrac{2{,}500}{n}$

(B) 10 yr

(C)

71. (A) $L(x) = 2x + \dfrac{450}{x}$

(B) $(0, \infty)$

(C) 15 ft by 15 ft

(D)

Exercise 3-7

1. $(f + g)(x) = 5x + 1$, $(f - g)(x) = 3x - 1$, $(fg)(x) = 4x^2 + 4x$, $(f/g)(x) = 4x/(x + 1)$; Domain of $f + g$ = Domain of $f - g$ = Domain of $fg = (-\infty, \infty)$, Domain of $f/g = (-\infty, -1) \cup (-1, \infty)$

3. $(f + g)(x) = 3x^2 + 1$, $(f - g)(x) = x^2 - 1$, $(fg)(x) = 2x^4 + 2x^2$, $(f/g)(x) = 2x^2/(x^2 + 1)$; Domain of each function $= (-\infty, \infty)$

5. $(f + g)(x) = x^2 + 3x + 4$, $(f - g)(x) = -x^2 + 3x + 6$, $(fg)(x) = 3x^3 + 5x^2 - 3x - 5$, $(f/g)(x) = (3x + 5)/(x^2 - 1)$; Domain of $f + g$ = Domain of $f - g$ = Domain of $fg = (-\infty, \infty)$, Domain of $f/g = (-\infty, -1) \cup (-1, 1) \cup (1, \infty)$

7. $(f \circ g)(x) = (x^2 - x + 1)^3$, $(g \circ f)(x) = x^6 - x^3 + 1$; Domain of $f \circ g$ = Domain of $g \circ f = (-\infty, \infty)$

9. $(f \circ g)(x) = |2x + 4|$, $(g \circ f)(x) = 2|x + 1| + 3$; Domain of $f \circ g$ = Domain of $g \circ f = (-\infty, \infty)$

11. $(f \circ g)(x) = (2x^3 + 4)^{1/3}$, $(g \circ f)(x) = 2x + 4$; Domain of $f \circ g$ = Domain of $g \circ f = (-\infty, \infty)$

13. $(f + g)(x) = \sqrt{4 - x^2} + \sqrt{4 + x^2}$, $(f - g)(x) = \sqrt{4 - x^2} - \sqrt{4 + x^2}$, $(fg)(x) = \sqrt{16 - x^4}$, $(f/g)(x) = \sqrt{(4 - x^2)/(4 + x^2)}$; Domain of each function $= [-2, 2]$

15. $(f + g)(x) = \sqrt{2 - x} + \sqrt{x + 3}$, $(f - g)(x) = \sqrt{2 - x} - \sqrt{x + 3}$, $(fg)(x) = \sqrt{6 - x - x^2}$, $(f/g)(x) = \sqrt{(2 - x)/(x + 3)}$; Domain of $f + g$ = Domain of $f - g$ = Domain of $fg = [-3, 2]$, Domain of $f/g = (-3, 2]$

17. $(f + g)(x) = 2\sqrt{x} - 2$, $(f - g)(x) = 6$, $(fg)(x) = x - 2\sqrt{x} - 8$, $(f/g)(x) = (\sqrt{x} + 2)/(\sqrt{x} - 4)$; Domain of $f + g$ = Domain of $f - g$ = Domain of $fg = [0, \infty)$, Domain of $f/g = [0, 16) \cup (16, \infty)$

19. $(f + g)(x) = \sqrt{x^2 + x - 6} + \sqrt{7 + 6x - x^2}$, $(f - g)(x) = \sqrt{x^2 + x - 6} - \sqrt{7 + 6x - x^2}$, $(fg)(x) = \sqrt{-x^4 + 5x^3 + 19x^2 - 29x - 42}$, $(f/g)(x) = \sqrt{(x^2 + x - 6)/(7 + 6x - x^2)}$; Domain of $f + g$ = Domain of $f - g$ = Domain $fg = [2, 7]$, Domain of $f/g = [2, 7)$

21. $(f \circ g)(x) = x$, $(g \circ f)(x) = x$; Domain of $f \circ g$ = Domain of $g \circ f = (-\infty, \infty)$

23. $(f \circ g)(x) = \sqrt{x} - 4$, $(g \circ f)(x) = \sqrt{x - 4}$; Domain of $f \circ g = [4, \infty)$, Domain of $g \circ f = [0, \infty)$

25. $(f \circ g)(x) = (1/x) + 2$, $(g \circ f)(x) = 1/(x + 2)$; Domain of $f \circ g = (-\infty, 0) \cup (0, \infty)$, Domain of $g \circ f = (-\infty, -2) \cup (-2, \infty)$

27. $(f \circ g)(x) = 1/|x - 1|$, $(g \circ f)(x) = 1/(|x| - 1)$; Domain of $f \circ g = (-\infty, 1) \cup (1, \infty)$, Domain of $g \circ f = (-\infty, -1) \cup (-1, 1) \cup (1, \infty)$

29. $h(x) = (f \circ g)(x)$; $f(x) = x^4$, $g(x) = 2x - 7$

31. $h(x) = (f \circ g)(x)$; $f(x) = x^{1/2}$, $g(x) = 4 + 2x$

33. $h(x) = (g \circ f)(x)$; $g(x) = 3x - 5$, $f(x) = x^7$

35. $h(x) = (g \circ f)(x)$; $g(x) = 4x + 3$, $f(x) = x^{-1/2}$

37. $(f + g)(x) = 2x$, $(f - g)(x) = 2/x$, $(fg)(x) = x^2 - (1/x^2)$, $(f/g) = (x^2 + 1)/(x^2 - 1)$; Domain of $f + g$ = Domain of $f - g$ = Domain of $fg = (-\infty, 0) \cup (0, \infty)$, Domain of $f/g = (-\infty, -1) \cup (-1, 0) \cup (0, 1) \cup (1, \infty)$

39. $(f + g)(x) = 2x$, $(f - g)(x) = 16/\sqrt{x}$, $(fg)(x) = x^2 - (64/x)$, $(f/g) = (x^{3/2} + 8)/(x^{3/2} - 8)$; Domain of $f + g$ = Domain of $f - g$ = Domain of $fg = (0, \infty)$, Domain of $f/g = (0, 4) \cup (4, \infty)$

41. $(f + g)(x) = 2$, $(f - g)(x) = -2x/|x|$, $(fg)(x) = 0$, $(f/g) = 0$; Domain of $f + g$ = Domain of $f - g$ = Domain of $fg = (-\infty, 0) \cup (0, \infty)$, Domain of $f/g = (0, \infty)$

43. $(f \circ g)(x) = \sqrt{4 - x^2}$, $(g \circ f)(x) = 4 - x$; Domain of $f \circ g = [-2, 2]$, Domain of $g \circ f = (-\infty, 4]$

45. $(f \circ g)(x) = |x|$, $(g \circ f)(x) = x$; Domain of $f \circ g = (-\infty, \infty)$, Domain of $g \circ f = [0, \infty)$

47. $(f \circ g)(x) = (6x - 10)/x$, $(g \circ f)(x) = (x + 5)/(5 - x)$; Domain of $f \circ g = (-\infty, 0) \cup (0, 2) \cup (2, \infty)$, Domain of $g \circ f = (-\infty, 0) \cup (0, 5) \cup (5, \infty)$

49. $(f \circ g)(x) = \sqrt{4 - x}$, $(g \circ f)(x) = \sqrt[4]{4 - x^2}$; Domain of $f \circ g = [0, 4]$, Domain of $g \circ f = [-2, 2]$

51. $(f \circ g)(x) = \sqrt{16 - x^2}$, $(g \circ f)(x) = \sqrt{34 - x^2}$; Domain of $f \circ g = [-4, 4]$, Domain of $g \circ f = [-5, 5]$

53. $(f \circ g)(x) = \sqrt{2 + x}$; Domain of $f \circ g$ $= [-2, 3]$; the first graph is correct.

55. $(f \circ g)(x) = \sqrt{x^2 + 1}$; Domain of $f \circ g$ $= (-\infty, -2] \cup [2, \infty)$; the first graph is correct.

57. $(f \circ g)(x) = \sqrt{16 - x^2}$; Domain of $f \circ g = [-3, 3]$; the first graph is correct.

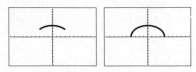

59. $P(p) = -70{,}000 + 6{,}000p - 200p^2$ **61.** $V(t) = 0.016\pi t^{2/3}$

63. (A) $r(h) = \frac{1}{2}h$ (B) $V(h) = \frac{1}{12}\pi h^3$ (C) $V(t) = \frac{0.125}{12}\pi t^{3/2}$

Exercise 3-8

1. One-to-one   **3.** Not one-to-one   **5.** One-to-one   **7.** Not one-to-one   **9.** One-to-one   **11.** Not one-to-one
13. One-to-one   **15.** One-to-one   **17.** One-to-one   **19.** Not one-to-one   **21.** One-to-one
23. Domain of $f^{-1} = [1, 5]$ **25.** Domain of $f^{-1} = [-3, 5]$ **27.**   **29.**   **31.**
  Range of $f^{-1} = [-4, 4]$ Range of $f^{-1} = [-5, 3]$

33. $f^{-1}(x) = \frac{1}{3}x$ **35.** $f^{-1}(x) = \frac{1}{4}x + \frac{3}{4}$ **37.** $f^{-1}(x) = 10x - 6$ **39.** $f^{-1}(x) = (2 + x)/x$ **41.** $f^{-1}(x) = 2x/(1 - x)$
43. $f^{-1}(x) = (4x + 5)/(3x - 2)$ **45.** $f^{-1}(x) = \sqrt[3]{x - 1}$ **47.** $f^{-1}(x) = (4 - x)^5 - 2$ **49.** $f^{-1}(x) = 16 - 4x^2, x \geq 0$
51. $f^{-1}(x) = (3 - x)^2 + 2, x \leq 3$ **53.** $f^{-1}(x) = 1 + \sqrt{x - 2}$ **55.** $f^{-1}(x) = -1 - \sqrt{x + 3}$
57. $f^{-1}(x) = \sqrt{9 - x^2}$ **59.** $f^{-1}(x) = -\sqrt{9 - x^2}$ **61.** $f^{-1}(x) = \sqrt{2x - x^2}$ **63.** $f^{-1}(x) = -\sqrt{2x - x^2}$
  Domain of $f^{-1} = [-3, 0]$ Domain of $f^{-1} = [0, 3]$ Domain of $f^{-1} = [1, 2]$ Domain of $f^{-1} = [0, 1]$
  Range of $f^{-1} = [0, 3]$ Range of $f^{-1} = [-3, 0]$ Range of $f^{-1} = [0, 1]$ Range of $f^{-1} = [-1, 0]$

65. $f^{-1}(x) = (x - b)/a$ **69.** (A) $f^{-1}(x) = 2 - \sqrt{x}$ (B) $f^{-1}(x) = 2 + \sqrt{x}$
71. (A) $f^{-1}(x) = 2 - \sqrt{4 - x^2}, 0 \leq x \leq 2$ (B) $f^{-1}(x) = 2 + \sqrt{4 - x^2}, 0 \leq x \leq 2$
73. One-to-one   **75.** Not one-to-one   **77.** Not one-to-one   **79.** One-to-one

Chapter 3 Review Exercise

1. (A) $\sqrt{45}$ (B) $-\frac{1}{2}$ (C) 2   *(3-1, 3-2)*   **2.** (A) $x^2 + y^2 = 7$ (B) $(x - 3)^2 + (y + 2)^2 = 7$ *(3-1)*
3. Center: $(-3, 2)$; radius $= \sqrt{5}$ *(3-1)*   **4.** Slope $= -\frac{3}{2}$ *(3-2)*   **5.** $2x + 3y = 12$ *(3-2)*   **6.** $y = -\frac{2}{3}x + 2$ *(3-2)*

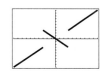

7. Vertical: $x = -3$, slope not defined; horizontal: $y = 4$, slope $= 0$ *(3-2)*   **8.** (A) and (C) specify functions.   *(3-3)*
9. 16   *(3-3)*   **10.** 1   *(3-3)*   **11.** 3   *(3-3)*   **12.** $-2a - h$ *(3-3)*   **13.** $9 + 3x - x^2$ *(3-7)*   **14.** $1 + 3x + x^2$ *(3-7)*
15. $20 + 12x - 5x^2 - 3x^3$ *(3-7)*   **16.** $(3x + 5)/(4 - x^2), x \neq \pm 2$ *(3-7)*   **17.** $17 - 3x^2$ *(3-7)*
18. $-21 - 30x - 9x^2$ *(3-7)*   **19.** Min $f(x) = f(3) = 2$; vertex: $(3, 2)$ *(3-4)*

20. (A) Reflected across x axis   (B) Shifted down 3 units   (C) Shifted left 3 units *(3-5)*

21. (A) Domain: $(-\infty, -4) \cup (-4, \infty)$; x intercept: $\frac{3}{2}$ (B) Domain: $(-\infty, -2) \cup (-2, 3) \cup (3, \infty)$; x intercept: 0 *(3-6)*

22. (A) Horizontal asymptote: $y = 2$; vertical asymptote: $x = -4$
(B) Horizontal asymptote: $y = 0$; vertical asymptotes: $x = -2, x = 3$ *(3-6)*

23. Domain $= (-\infty, \infty)$, range $= (-3, \infty)$ *(3-4)*   **24.** $[-2, -1], [1, \infty)$ *(3-4)*   **25.** $[-1, 1)$ *(3-4)*   **26.** $(-\infty, -2)$ *(3-4)*

27. $x = -2, x = 1$ *(3-4)*   **28.** (A) $3x + 2y = -6$ (B) $\sqrt{52}$ *(3-1, 3-2)*

29. (A) $y = -2x - 3$ (B) $y = \frac{1}{2}x + 2$ *(3-2)*   **30.** It is symmetric with respect to all three. *(3-1)*   **31.** $(-\infty, 3)$ *(3-3)*

32. Range $= [-4, \infty)$ *(3-4)*   **33.** *(3-6)*
Intercepts: $x = 1$ and $x = 5, y = 5$
Min $f(x) = f(3) = -4$

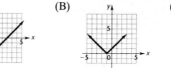

34. (A) $(f \circ g)(x) = \sqrt{|x|} - 8, (g \circ f)(x) = |\sqrt{x} - 8|$ (B) Domain of $f \circ g = (-\infty, \infty)$, domain of $g \circ f = [0, \infty)$ *(3-7)*

35. Functions (A), (C), and (D) are one-to-one *(3-8)*   **36.** (A) $(x + 7)/3$ (B) 4   (C) x (D) Increasing *(3-8)*

37. Domain $= [-1, 1]$ *(3-4)*   **38.** (A)   (B)   (C) *(3-5)*
Range $= [0, 1] \cup (2, 3]$
Discontinuous at $x = 0$

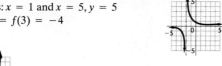

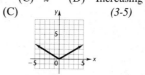

39. (A) Domain: $(-\infty, -1) \cup (-1, \infty)$ **40.** (A) $f^{-1}(x) = x^2 + 1$
x intercept: 1; y intercept: $-\frac{1}{2}$ (B) Domain of $f = [1, \infty) =$ Range of f^{-1}
(B) Vertical asymptote: $x = -1$; horizontal asymptote: $y = \frac{1}{2}$ Range of $f = [0, \infty) =$ Domain of f^{-1}
(C) *(3-6)*   (C) *(3-8)*

41. $(x - 3)^2 + y^2 = 32$ *(3-1)*   **42.** Center: $(-2, 3)$, radius $= 4$ *(3-1)*   **43.** Symmetric with respect to the origin *(3-1)*

44. Decreasing *(3-2, 3-3)*

45. (A) Domain of $f = [0, \infty) =$ Range of f^{-1}; Range of $f = [-1, \infty) =$ Domain of f^{-1} (B) $\sqrt{x + 1}$
(C) 2   (D) 4   (E) x *(3-8)*

46. It is the same as the graph of g shifted to the right 2 units and down 1 unit; then turned upside down. *(3-5)*   **47.** *(3-5)*

48. $[-5, 5]$ *(2-5, 3-3)*

49. (A) $x^2\sqrt{1 - x}$; domain $= (-\infty, 1]$ (B) $x^2/\sqrt{1 - x}$, domain $= (-\infty, 1)$
(C) $1 - x$, domain $= (-\infty, 1]$ (D) $\sqrt{1 - x^2}$, domain $= [-1, 1]$ *(3-7)*

50. (A) $(3x + 2)/(x - 1)$ (B) $\frac{11}{2}$ (C) x *(3-8)*

51. $f(x) = \begin{cases} -2 & \text{if } x < -1 \\ 2x & \text{if } -1 \le x < 1 \\ 2 & \text{if } x \ge 1 \end{cases}$; domain $= (-\infty, \infty)$, range $= [-2, 2]$ *(3-4)*

52. $x - y = 3$; a line *(3-1, 3-2)*   **54.**   *(3-6)*   **55.** (A)   (B) *(3-5)*

56. (A) $V = -1,250t + 12,000$ (B) $5,750 *(3-2)*   **57.** (A) $R = 1.6C$ (B) $168 *(3-2)*

58. $E(x) = \begin{cases} 200 & \text{if } 0 \le x \le 3,000 \\ 0.1x - 100 & \text{if } x > 3,000 \end{cases}$; $E(2,000) = \$200, E(5,000) = \$400$ *(3-4)*

59. $P(p) = -14,000 + 700p - 10p^2$ *(3-7)*   **60.** 5 ft *(3-1)*

61. $A(x) = 60x - \frac{3}{2}x^2$ (B) $0 < x < 40$ (C) $x = 20, y = 15$ *(3-4)*

62. (A) 0   (B) 1   (C) 2   (D) 0   (E) 1   (F) 0 *(3-5)*

CHAPTER 4   Exercise 4-1

1. $2m + 1$ **3.** $4x - 5, R = 11$ **5.** $x^2 + x + 1$ **7.** $2y^2 - 5y + 13, R = -27$ **9.** $\dfrac{x^2 + 3x - 7}{x - 2} = x + 5 + \dfrac{3}{x - 2}$

11. $\dfrac{4x^2 + 10x - 9}{x + 3} = 4x - 2 - \dfrac{3}{x + 3}$ **13.** $\dfrac{2x^3 - 3x + 1}{x - 2} = 2x^2 + 4x + 5 + \dfrac{11}{x - 2}$ **15.** $3x^3 - 3x^2 + 3x - 4$

17. $x^4 - x^3 + x^2 - x + 1$ **19.** $3x^3 - 7x^2 + 21x - 67, R = 200$ **21.** $2x^5 - 3x^4 - 15x^3 + 2x + 10, R = 0$

23. $4x^3 - 6x - 2, R = 2$ **25.** $4x^2 - 2x - 4, R = 0$ **27.** $3x^3 - 0.8x^2 + 1.68x - 2.328, R = 0.0688$

29. $3x^4 - 0.4x^3 + 5.32x^2 - 4.256x - 3.5952, R = -0.123\ 84$ **31.** $2.14x^2 + 1.98x - 2.03, R = 0.00$

33. $0.96x^3 - 1.32x^2 + 5.89x + 1.37, R = -3.74$ **35.** $2x^2 - 3x + 2, R = 0$ **37.** $x^2 + (-3 + i)x - 3i$

39. (A) In both cases, the coefficient of x is a_2, the constant term is $a_2r + a_1$, and the remainder is $(a_2r + a_1)r + a_0$.
     (B) The remainder expanded is $a_2r^2 + a_1r + a_0 = P(r)$.

Exercise 4-2

1. 4   **3.** 3   **5.** -6 **7.** $3, -5$ **9.** $-\frac{1}{2}, 8, -2$ **11.** Yes   **13.** Yes   **15.** -3 **17.** 0.427

19.   **21.** 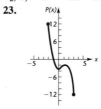   **23.**   **25.**   **27.** $-4, -8, 1$

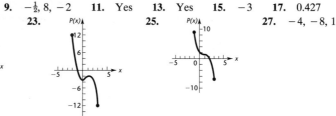

29. $\frac{1}{8}, -\frac{3}{5}, -4$ **31.** $P(x) = \left(x - \dfrac{3 + \sqrt{5}}{2}\right)\left(x - \dfrac{3 - \sqrt{5}}{2}\right)$ **33.** $P(x) = [x - (3 + i)][x - (3 - i)]$ **35.** Yes   **37.** No

39. 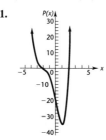   **41.** 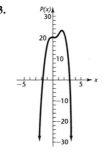   **43.**   **45.**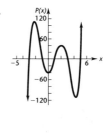

47. $P(-2) = 81; P(1.7) = 6.2$ **49.** (A) $P(r) = (a_2r + a_1)r + a_0$ (B) $P(r) = R = (a_2r + a_1)r + a_0$; same as in part (A)

51. 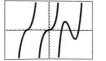   **53.**

Exercise 4-3

1. -8 (multiplicity 3), 6 (multiplicity 2); degree of $P(x)$ is 5   3. -4 (multiplicity 3), 3 (multiplicity 2), -1; degree of $P(x)$ is 6
5. $P(x) = (x - 3)^2(x + 4)$; degree 3   7. $P(x) = (x + 7)^3[x - (-3 + \sqrt{2})][x - (-3 - \sqrt{2})]$; degree 5
9. $P(x) = [x - (2 - 3i)][x - (2 + 3i)](x + 4)^2$; degree 4   11. One real and two imaginary, or three real
13. Six imaginary, two real and four imaginary, four real and two imaginary, or six real
15. One real and four imaginary, three real and two imaginary, or five real
17. Four imaginary, two real and two imaginary, or four real   19. $P(x) = (x + 4)^2(x + 1)$
21. $P(x) = (x - 1)(x + 1)(x - i)(x + i)$ 23. $P(x) = (2x - 1)[x - (4 + 5i)][x - (4 - 5i)]$
25. Four real, two real and two imaginary, or four imaginary   27. Three real, or one real and two imaginary
29. Six real, four real and two imaginary, two real and four imaginary, or six imaginary
31. Five real, three real and two imaginary, or one real and four imaginary   33. $x^2 - 8x + 41$ 35. $x^2 - 6x + 25$
37. $x^2 - 2ax + a^2 + b^2$ 39. $3 + i, -1$ 41. $5i, 3$ 43. $2 - i, \pm\sqrt{2}$
45. (A) 3   (B) $-\frac{1}{2} - (\sqrt{3}/2)i, -\frac{1}{2} + (\sqrt{3}/2)i$ 47. Max. $= n$, min. $= 1$
49. No, since $P(x)$ is not a polynomial with real coefficients (the coefficient of x is the imaginary number $2i$)

Exercise 4-4

1.

+	−	I
1	1	0

3.

+	−	I
0	2	0
0	0	2

5.

+	−	I
3	0	0
1	0	2

7. UB $= 2$, LB $= -1$ 9. UB $= 2$, LB $= -3$

11. UB $= 1$, LB $= -2$ 13. $P(3) = -2, P(4) = 2$ 15. $P(-3) = -13, P(-2) = 3$ 17. $Q(1) = 4, Q(2) = -1$
19. (A)

+	−	I
2	1	0
0	1	2

(B) UB $= 3$, LB $= -3$ (C) 1 is a zero; a real zero in each interval $(-3, -2)$ and $(2, 3)$

21. (A)

+	−	I
1	2	0
1	0	2

(B) UB $= 3$, LB $= -2$ (C) One real zero in the interval $(2, 3)$

23. (A)

+	−	I
1	3	0
1	1	2

(C) Each interval contains exactly one real zero: $(-4, -3), (-2, -1), (-1, 0)$, and $(1, 2)$

(B) UB $= 2$, LB $= -5$

25. (A)

+	−	I
3	2	0
3	0	2
1	2	2
1	0	4

(B) UB $= 2$, LB $= -2$ (C) At least one real zero in the interval $(1, 2)$

27. (A)

+	−	I
1	2	2
1	0	4

(B) UB $= 2$, LB $= -3$ (C) Each interval contains exactly one real zero: $(-3, -2), (-2, -1)$, and $(1, 2)$

29. (A)

+	−	I
3	1	2
1	1	4

(B) UB $= 2$, LB $= -2$ (C) Exactly one real zero in $(-2, -1)$ and at least one real zero in $(1, 2)$

Exercise 4-5

1. (A) $\pm 1, \pm 2, \pm 3, \pm 6$ (B) $-2, 1, 3$ **3.** (A) $\pm 1, \pm 2, \pm 4, \pm\frac{1}{3}, \pm\frac{2}{3}, \pm\frac{4}{3}$ (B) 2 (double zero), $-\frac{1}{3}$
5. (A) $\pm 1, \pm 3, \pm\frac{1}{2}, \pm\frac{3}{2}, \pm\frac{1}{3}, \pm\frac{1}{4}, \pm\frac{3}{4}, \pm\frac{1}{6}, \pm\frac{1}{12}$ (B) $-\frac{1}{2}, \frac{1}{3}, \frac{3}{2}$ **7.** (A) $\pm 1, \pm 2, \pm 4, \pm\frac{1}{3}, \pm\frac{2}{3}, \pm\frac{4}{3}$ (B) $-\frac{1}{3}$
9. (A) $\pm 1, \pm 2, \pm 3, \pm 6$ (B) No rational zeros   **11.** (A) $\pm 1, \pm 2, \pm 4, \pm 8$ (B) ± 2
13. (A) $\pm 1, \pm 2, \pm 3, \pm 6, \pm\frac{1}{3}, \pm\frac{2}{3}$ (B) $-\frac{1}{3}, 2$ **15.** $\frac{1}{2}, 1 \pm \sqrt{2}$ **17.** -2 (double root), $\pm\sqrt{5}$ **19.** $\pm 2, 1 \pm \sqrt{2}$
21. $\pm 1, \frac{3}{2}, \pm i$ **23.** $-5, 2, 3$ **25.** $-\frac{2}{5}, 0, \frac{1}{2}, 2$ **27.** $\frac{1}{2} \pm \frac{1}{2}\sqrt{3}, 2$ (double zero)   **29.** -1 (double zero), $-\frac{1}{3}, 2 \pm i$
31. $P(x) = (x + 2)(2x - 1)(3x + 2)$ **33.** $P(x) = (x + 4)[x - (1 + \sqrt{2})][x - (1 - \sqrt{2})]$
35. $P(x) = (2x - 1)(2x + 1)(x - 2)(x + 1)$ **37.** Inequality notation: $2 - \sqrt{3} \le x \le 2 + \sqrt{3}$; interval notation: $[2 - \sqrt{3}, 2 + \sqrt{3}]$
39. Inequality notation: $x \le -1$ or $1 \le x \le 3$; interval notation: $(-\infty, -1] \cup [1, 3]$
41. Inequality notation: $-3 \le x \le \frac{1}{2}$ or $x \ge 2$; interval notation: $[-3, \frac{1}{2}] \cup [2, \infty)$
43. Inequality notation: $-\frac{5}{2} < x < -1$ or $x > 1$; interval notation: $(-\frac{5}{2}, -1) \cup (1, \infty)$
45. Inequality notation: $x \le -2$ or $-1 < x < 2$ or $3 < x \le 5$; interval notation: $(-\infty, -2] \cup (-1, 2) \cup (3, 5]$
47. $\sqrt{6}$ is a zero of $P(x) = x^2 - 6$, but $P(x)$ has no rational zeros.   **49.** $\sqrt[3]{5}$ is a zero of $P(x) = x^3 - 5$, but $P(x)$ has no rational zeros.
51. $\frac{1}{3}, 6 \pm 2\sqrt{3}$ **53.** $-\frac{5}{2}, \frac{3}{2}, \pm 4i$ **55.** $\frac{3}{2}$ (double zero), $4 \pm \sqrt{6}$ **57.** 2 ft   **59.** 0.5 by 0.5 in or 1.59 by 1.59 in

Exercise 4-6

1. 1.4   **3.** 0.8   **5.** 2.3   **7.** -0.7 **9.** 2, 1.2   **11.** $-\frac{1}{2}, 3, 1.9$ **13.** -1.13 **15.** 0.71, 1.49   **17.** 1.14, 1.71
19. 1.1206, 3.3473, 4.5321   **21.** 0.7336, 1.9305   **23.** $-1.3820, 0.6180, 1.6180, 3.6180$ **25.** 0.3438, 0.6820, 1.5991
27. (1, 1) and (1.66, 2.76)   **29.** 1.49 ft

Exercise 4-7

1. $A = 2, B = 5$ **3.** $A = 7, B = -2$ **5.** $A = 1, B = 2, C = 3$ **7.** $A = 2, B = 1, C = 3$
9. $A = 0, B = 2, C = 2, D = -3$ **11.** $\dfrac{3}{x - 4} - \dfrac{4}{x + 2}$ **13.** $\dfrac{3}{3x + 4} - \dfrac{1}{2x - 3}$ **15.** $\dfrac{2}{x} - \dfrac{1}{x - 3} - \dfrac{3}{(x - 3)^2}$
17. $\dfrac{2}{x} + \dfrac{3x - 1}{x^2 + 2x + 3}$ **19.** $\dfrac{2x}{x^2 + 2} + \dfrac{3x + 5}{(x^2 + 2)^2}$ **21.** $x - 2 + \dfrac{3}{x - 2} - \dfrac{2}{x - 3}$ **23.** $\dfrac{2}{x - 3} + \dfrac{2x + 5}{x^2 + 3x + 3}$
25. $\dfrac{2}{x - 4} - \dfrac{1}{x + 3} + \dfrac{3}{(x + 3)^2}$ **27.** $\dfrac{2}{x - 2} - \dfrac{3}{(x - 2)^2} - \dfrac{2x}{x^2 - x + 1}$ **29.** $x + 2 + \dfrac{1}{2x - 1} - \dfrac{2}{x + 2} + \dfrac{x - 1}{2x^2 - x + 1}$

Chapter 4 Review Exercise

1. $2x^3 + 3x^2 - 1 = (x + 2)(2x^2 - x + 2) - 5$ *(4-1)*   **2.** $P(3) = -8$ *(4-1, 4-2)*   **3.** $2, -4, -1$ *(4-2)*   **4.** $1 - i$ *(4-3)*
5. (A)

+	−	I
2	1	0
0	1	2

(B)

+	−	I
0	1	4

(4-4)   **6.** LB $= -2, -1$, UB $= 4$ *(4-4)*

7. $P(1) = -5$ and $P(2) = 1$ are of opposite sign. *(4-4)*   **8.** $\pm 1, \pm 2, \pm 3, \pm 6$ *(4-5)*   **9.** $-1, 2, 3$ *(4-4, 4-5)*
10. 1.4 *(4-6)*   **11.** $\dfrac{2}{x - 3} + \dfrac{5}{x + 2}$ *(4-7)*   **12.** $Q(x) = 8x^3 - 12x^2 - 16x - 8, R = 5; P(\frac{1}{4}) = R = 5$ *(4-1, 4-2)*
13. -4 *(4-1, 4-2)*   **14.** $P(x) = [x - (1 + \sqrt{2})][x - (1 - \sqrt{2})]$ *(4-2)*
15. Yes, since $P(-1) = 0, x - (-1) = x + 1$ must be a factor. *(4-2)*
16. (A)

+	−	I
2	2	0
2	0	2
0	2	2
0	0	4

(B) UB $= 4$, LB $= -3$ (C) -2 is a zero; $(-1, 0), (0, 1)$, and $(3, 4)$ each contains a zero. *(4-4)*

17. $-2, -\frac{1}{2}, 4$ *(4-3, 4-4)*   **18.** $P(x) = (x + 2)(2x + 1)(x - 4)$ *(4-2)*   **19.** No rational zeros *(4-4, 4-5)*
20. $-1, \frac{1}{2}, \dfrac{1 \pm i\sqrt{3}}{2}$ *(4-4, 4-5)*   **21.** $(x + 1)(2x - 1)\left(x - \dfrac{1 + i\sqrt{3}}{2}\right)\left(x - \dfrac{1 - i\sqrt{3}}{2}\right)$ *(4-2)*
22. Inequality notation: $x \le -3$ or $-\frac{1}{2} \le x \le 2$; interval notation $(-\infty, -3] \cup [-\frac{1}{2}, 2]$ *(4-5, 2-8)*   **23.** 1.3 *(4-6)*
24. $\dfrac{1}{x} - \dfrac{2}{x - 2} + \dfrac{3}{(x - 2)^2}$ *(4-7)*   **25.** $\dfrac{3}{x} + \dfrac{2x - 1}{2x^2 - 3x + 3}$ *(4-7)*
26. $P(x) = [x - (1 + i)][x^2 + (1 + i)x + (3 + 2i)] + (3 + 5i)$ *(4-1)*   **27.** $P(x) = (x + \frac{1}{2})^2(x + 3)(x - 1)^3$, degree 6 *(4-2)*
28. $P(x) = (x + 5)[x - (2 - 3i)][x - (2 + 3i)]$, degree 3 *(4-2)*   **29.** $\frac{1}{2}, \pm 2, 1 \pm \sqrt{2}$ *(4-4, 4-5)*
30. $(x - 2)(x + 2)(2x - 1)[x - (1 - \sqrt{2})][x - (1 + \sqrt{2})]$ *(4-2)*

31. Inequality notation: $-3 < x \le -\frac{3}{2}$ or $-\frac{1}{2} < x \le \frac{1}{2}$ or $x > 2$; interval notation: $(-3, -\frac{3}{2}] \cup (-\frac{1}{2}, \frac{1}{2}] \cup (2, \infty)$ *(4-5, 2-8)*

32. $x = -3, 1.34, 2.82$ *(4-6)*   **33.** $\dfrac{2}{x-3} - \dfrac{3}{x} + \dfrac{x-1}{x^2+1}$ *(4-7)*   **34.** 4×12 ft or 5.2×9.2 ft *(4-5)*   **35.** 4.8 ft *(4-6)*

CHAPTER 5   Exercise 5-1

1.

x	y
-3	0.04
-2	0.11
-1	0.33
0	1
1	3
2	9
3	27

3.

x	y
-3	27
-2	9
-1	3
0	1
1	0.33
2	0.11
3	0.04

5.

x	$g(x)$
-3	-27
-2	-9
-1	-3
0	-1
1	-0.33
2	-0.11
3	-0.04

7.

x	$h(x)$
-3	0.19
-2	0.56
-1	1.67
0	5
1	15
2	45
3	135

9.

x	y
-6	-4.96
-5	-4.89
-4	-4.67
-3	-4
-2	-2
-1	4
0	22

11. 10^{2x+3} **13.** 3^{2x-1} **15.** $\dfrac{4^{3xz}}{5^{3yz}}$ **17.** $x = 2$

19. $x = -1, 3$ **21.** $x = \frac{2}{3}$ **23.** $x = -2$ **25.** $\frac{1}{2}$ **27.** $\frac{1}{2}, 1$ **29.** 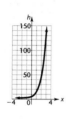   **31.**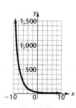

33. 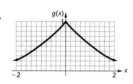   **35.** 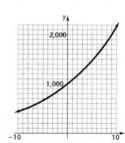   **37.** 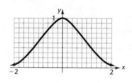   **39.** 16.24   **41.** 5.047

43. 4.469   **45.** $6^{2x} - 6^{-2x}$ **47.** 4   **49.**   **51.**   **53.**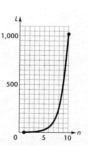

55. (A) 76 flies   (B) 570 flies   **57.** (A) 19 lb   (B) 7.9 lb   **59.** (A) $4,225.92   (B) $12,002.71   **61.** $9,841

Exercise 5-2

1.

x	y
-3	-0.05
-2	-0.14
-1	-0.37
0	-1
1	-2.72
2	-7.39
3	-20.09

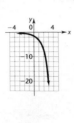

3.

x	y
-5	3.68
-4	4.49
-3	5.49
-2	6.7
-1	8.19
0	10
1	12.21
2	14.92
3	18.22
4	22.26
5	27.18

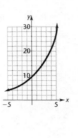

5.

t	$f(t)$
-5	164.87
-4	149.18
-3	134.99
-2	122.14
-1	110.52
0	100
1	90.48
2	81.87
3	74.08
4	67.03
5	60.65

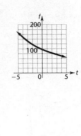

7. e^{-x} **9.** e^{3x} **11.** e^{3x-1} **13.** **15.** 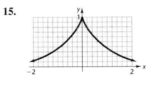 **17.**

19. 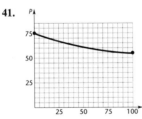 **21.** $\dfrac{e^{-2x}(-2x-3)}{x^4}$ **23.** $2e^{2x}+2e^{-2x}$ **25.** $2e^{2x}$ **27.** $x=0$ **29.** $x=0,5$ **31.**

33. (A)

s	$f(s)$	s	$f(s)$
-0.5	4.0000	0.5	2.2500
-0.2	3.0518	0.2	2.4883
-0.1	2.8680	0.1	2.5937
-0.01	2.7320	0.01	2.7048
-0.001	2.7196	0.001	2.7169
-0.0001	2.7184	0.0001	2.7181

(B) $2.718\cdots \approx e$ **35.** **37.**

39. 7.1 billion **41.**  **43.** (A) 62% (B) 39% **45.** (A) $10,691.81 (B) $36,336.69

47. Gill Savings: $1,230.60; Richardson S&L: $1,231.00; USA Savings: $1,229.03 **49.** $12,197.09 **51.** (A) 190,000 (B) 270,000
53. 40 boards **55.** 50°F **57.** 0.0009 coulomb **59.** 100 deer **61.**

Exercise 5-3

1. $81=3^4$ **3.** $0.001=10^{-3}$ **5.** $3=81^{1/4}$ **7.** $16=\left(\frac{1}{2}\right)^{-4}$ **9.** $\log_{10}0.0001=-4$ **11.** $\log_4 8=\frac{3}{2}$
13. $\log_{32}\left(\frac{1}{2}\right)=-\frac{1}{5}$ **15.** $\log_{49}7=\frac{1}{2}$ **17.** 0 **19.** 1 **21.** 4 **23.** -2 **25.** $\frac{1}{3}$ **27.** $\sqrt{x}$ **29.** x^2 **31.** $x=4$
33. $y=2$ **35.** $b=4$ **37.** $b=$ any positive real number except 1. **39.** $x=2$ **41.** $y=-2$ **43.** $b=100$
45. $2\log_b u+7\log_b v$ **47.** $\frac{2}{3}\log_b m-\frac{1}{2}\log_b n$ **49.** $\log_b u-\log_b v-\log_b w$ **51.** $-2\log_b a$ **53.** $\frac{1}{3}\log_b(x^2-y^2)$

55. $\frac{1}{3}\log_b N - 2\log_b p - 3\log_b q$ **57.** $\frac{1}{4}(2\log_b x + 3\log_b y - \frac{1}{2}\log_b z)$ **59.** $\log_b (x^2/y)$ **61.** $\log_b (w/xy)$
63. $\log_b (x^3 y^2/\sqrt[4]{z})$ **65.** $\log_b (\sqrt{u}/v^2)^5$ **67.** $\log_b \sqrt[5]{x^2 y^3}$ **69.** $5\log_b (x + 3) + 2\log_b (2x - 7)$
71. $7\log_b (x + 10) - 2\log_b (1 + 10x)$ **73.** $2\log_b x - \frac{1}{2}\log_b (x + 1)$ **75.** $2\log_b x + \log_b (x + 5) + \log_b (x - 4)$
77. $x = 4$ **79.** $x = \frac{1}{3}$ **81.** $x = \frac{8}{7}$ **83.** $x = 2$ **85.** $x = 2$ **87.** 3.40   **89.** -0.92 **91.** 3.30   **93.** 0.23   **95.** -0.05
97. 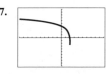   **99.**   **101.** (A)   (B) Domain $f = (-\infty, \infty)$ = Range f^{-1}
Range $f = (0, \infty)$ = Domain f^{-1}
(C) $f^{-1}(x) = \log_{1/2} x = -\log_2 x$

103. $f^{-1}(x) = \frac{1}{3}[1 + \log_5 (x - 4)]$ **105.** $g^{-1}(x) = \frac{1}{5}(e^{x/3} + 2)$ **107.** $x = 100e^{-0.08t}$

Exercise 5-4

1. 4.9177   **3.** -2.8419 **5.** 3.7623   **7.** -2.5128 **9.** 200,800   **11.** 0.000 664 8   **13.** 47.73   **15.** 0.6760
17. 4.959   **19.** 7.861   **21.** 3.301   **23.** 4.561   **25.** $x = 12.725$ **27.** $x = -25.715$ **29.** $x = 1.1709 \times 10^{32}$
31. $x = 4.2672 \times 10^{-7}$ **33.**   **35.**   **37.**   **39.**

41. The inequality sign in the last step reverses because $\log \frac{1}{3}$ is negative.   **43.** 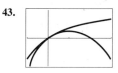   **45.**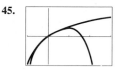

47. (A) 0 decibel   (B) 120 decibels   **49.** 30 decibels more   **51.** 8.6   **53.** 1,000 times more powerful   **55.** 7.67 km/s
57. (A) 8.3, basic   (B) 3.0, acid   **59.** 6.3×10^{-6} mole/liter

Exercise 5-5

1. $x = 1.46$ **3.** $x = 0.321$ **5.** $x = 1.29$ **7.** $x = 3.50$ **9.** $x = 1.80$ **11.** $x = 2.07$ **13.** $x = 20$ **15.** $x = 5$
17. $x = \frac{11}{9}$ **19.** $x = 14.2$ **21.** $x = -1.83$ **23.** $x = 11.7$ **25.** $x = \pm 1.21$ **27.** $x = 5$ **29.** $x = 2 + \sqrt{3}$
31. $x = \frac{1}{4}(1 + \sqrt{89})$ **33.** $x = 1, e^2, e^{-2}$ **35.** $x = e^e$ **37.** $x = 0.1, 100$ **39.** 3.6776   **41.** -1.6094 **43.** -1.7372
45. $r = \dfrac{\ln (A/P)}{t}$ **47.** $I = I_0(10^{D/10})$ **49.** $I = I_0[10^{(6-M)/2.5}]$ **51.** $t = \dfrac{-L}{R}\ln\left(1 - \dfrac{RI}{E}\right)$ **53.** $x = \ln (y \pm \sqrt{y^2 - 1})$
55. $x = \dfrac{1}{2}\ln\dfrac{1 + y}{1 - y}$ **57.**   **59.**   **61.** 0.38   **63.** 0.55   **65.** 0.57   **67.** 0.85

69. 0.43   **71.** 0.27   **73.** Approx. 5 years   **75.** 9.16%   **77.** (A) 6   (B) 100 times brighter   **79.** Approx. 35 years
81. 18,600 years old   **83.** 7.52 s   **85.** $k = 0.40; t = 2.9$ h   **87.** 10 years

Chapter 5 Review Exercise

1. $\log m = n$ (5-3)   **2.** $\ln x = y$ (5-3)   **3.** $x = 10^y$ (5-3)   **4.** $y = e^x$ (5-3)   **5.** 7^{2x} (5-1)   **6.** e^{2x^2} (5-1)
7. $x = 8$ (5-3)   **8.** $x = 5$ (5-3)   **9.** $x = 3$ (5-3)   **10.** $x = 1.24$ (5-3)   **11.** $x = 11.9$ (5-3)   **12.** $x = 0.984$ (5-3)
13. $x = 103$ (5-3)   **14.** $x = 4$ (5-3)   **15.** $x = 2$ (5-3)   **16.** $x = -1, 3$ (5-2)   **17.** $x = 1$ (5-1)
18. $x = \pm 3$ (5-2)   **19.** $x = -2$ (5-3)   **20.** $x = \frac{1}{3}$ (5-3)   **21.** $x = 64$ (5-3)   **22.** $x = e$ (5-3)   **23.** $x = 33$ (5-3)
24. $x = 1$ (5-3)   **25.** 1.145 (5-3)   **26.** Not defined (5-3)   **27.** 2.211 (5-3)   **28.** 11.59 (5-3)   **29.** $x = 41.8$ (5-1)
30. $x = 1.95$ (5-3)   **31.** $x = 0.0400$ (5-3)   **32.** $x = -6.67$ (5-3)   **33.** $x = 1.66$ (5-3)   **34.** $x = 2.32$ (5-5)
35. $x = 3.92$ (5-5)   **36.** $x = 92.1$ (5-5)   **37.** $x = 2.11$ (5-5)   **38.** $x = 0.881$ (5-5)   **39.** $x = 300$ (5-5)
40. $x = 2$ (5-5)   **41.** $x = 1$ (5-5)   **42.** $x = \frac{1}{2}(3 + \sqrt{13})$ (5-5)   **43.** $x = 1, 10^3, 10^{-3}$ (5-5)   **44.** $x = 10^e$ (5-5)

45. $e^{-x} - 1$ *(5-2)* **46.** $2 - 2e^{-2x}$ *(5-2)* **47.** *(5-1)* **48.** *(5-2)* 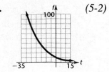 **49.** *(5-3)*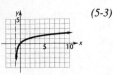

50. *(5-2)* **51.** $I = I_0(10^{D/10})$ *(5-5)* **52.** $x = \pm\sqrt{-2\ln(\sqrt{2\pi}\,y)}$ *(5-5)* **53.** $I = I_0(e^{-kx})$ *(5-5)*

54. $n = -\dfrac{\ln[1 - (Pi/r)]}{\ln(1 + i)}$ *(5-5)* **55.** $f^{-1}(x) = e^{x/2} + 1$ *(5-5, 3-7)* **56.** $f^{-1}(x) = \ln(x + \sqrt{x^2 + 1})$ *(5-5, 3-7)*

57. $y = ce^{-5t}$ *(5-3, 5-5)* **58.** Domain $f = (0, \infty) = $ Range f^{-1} *(5-3)*
Range $f = (-\infty, \infty) = $ Domain f^{-1}

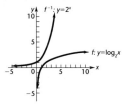

59. If $\log_1 x = y$, then we have $1^y = x$; that is, $1 = x$ for arbitrary positive x, which is impossible. *(5-3)* **61.** 23.4 years *(5-5)*
62. 23.1 years *(5-5)* **63.** 37,100 years *(5-5)* **64.** (A) $N = 2^{2t}$ or $N = 4^t$ (B) 15 days *(5-5)* **65.** $\$1.1 \times 10^{26}$ *(5-2)*
66. (A) (B) 0 *(5-2)* **67.** 6.6 *(5-4)* **68.** 7.08×10^{16} joules *(5-4)* **69.** 50 decibels more *(5-4)*

70. $k = 0.009\ 42$; 489 ft *(5-2)* **71.** 3 years *(5-5)*

Cumulative Review Exercise: Chapters 3–5

1. (A) $2\sqrt{5}$ (B) 2 (C) $-\tfrac{1}{2}$ *(3-1, 3-2)* **2.** (A) $x^2 + y^2 = 2$ (B) $(x + 3)^2 + (y - 1)^2 = 2$ *(3-1)*
3. Slope: $\tfrac{2}{3}$; y intercept: -2; **4.** (A) 20 (B) $x^2 + x + 3$ (C) $9x^2 - 18x + 13$ (D) $2a + h - 2$ *(3-2, 3-7)*
x intercept: 3 *(3-2)*

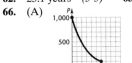

5. (A) Expanded by a factor of 2 (B) Shifted right 2 units (C) Shifted down 2 units *(3-5)*
6. $3x^3 + 5x^2 - 18x - 3 = (x + 3)(3x^2 - 4x - 6) + 15$ *(4-1)* **7.** $-2, 3, 5$ *(4-2)*

8. $P(1) = -5$ and $P(2) = 5$ are of opposite sign *(4-4)* **9.** $1, 2, -4$ *(4-5)* **10.** $\dfrac{3}{x + 1} + \dfrac{2}{x - 2}$ *(4-7)*

11. (A) $x = \log y$ (B) $x = e^y$ *(5-4)* **12.** (A) $8e^{3x}$ (B) e^{5x} *(5-2)* **13.** (A) 9 (B) 4 (C) $\tfrac{1}{2}$ *(5-3)*
14. (A) 0.371 (B) 11.4 (C) 0.0562 (D) 15.6 *(5-4)*
15. (A) All real numbers (B) $\{-2\} \cup [1, \infty)$ (C) 1 (D) $[-3, -2]$ and $[2, \infty)$ (E) $-2, 2$ *(3-3, 3-4)*
16. (A) $y = -\tfrac{3}{2}x - 8$ (B) $y = \tfrac{2}{3}x + 5$ *(3-2)* **17.** $[-4, \infty)$ *(3-3)* **18.** Range: $[-9, \infty)$ *(3-4)*
Intercepts: $x = -2$ and $x = 4, y = -8$
Min $f(x) = f(1) = -9$

19. $(f \circ g)(x) = \dfrac{x}{3 - x}$;

Domain: $x \neq 0, 3$ *(3-7)*

20. $f^{-1}(x) = \frac{1}{2}x - \frac{5}{2}$ *(3-8)*

21. Domain: all real numbers *(3-4)*
Range: $(-\infty, -1) \cup [1, \infty)$
Discontinuous at $x = 0$

22. (A) (B) *(3-5)*

23. (A) Domain: $x \neq -2$; x intercept: -4;
y intercept: 4
(B) Vertical asymptote: $x = -2$
Horizontal asymptote: $y = 2$
(C) *(3-6)*

24. (A) $f^{-1}(x) = x^2 - 4, x \geq 0$
(B) Domain $f = [-4, \infty) = $ Range f^{-1};
Range $f = [0, \infty) = $ Domain f^{-1}
(C) *(3-8)*

25. Center: $(3, -1)$; radius: $\sqrt{10}$ *(3-1)*

26. Symmetric with respect to the origin *(3-1)*

27. $P\left(\frac{1}{2}\right) = \frac{5}{2}$ *(4-2)*

28. (B) *(4-2)*

29. (A)

+	−	I
1	1	2
3	1	0

(B) LB = -1, UB = 5

(C) 1 is a zero; $(-1, 0)$, $(1, 2)$, and $(2, 3)$ each contains a zero *(4-4)*

30. $3, 1 \pm \frac{1}{2}i$ *(4-5)*

31. $-4, -1, \pm\sqrt{3}; P(x) = (x + 4)(x + 1)(x - \sqrt{3})(x + \sqrt{3})$ *(4-5)*

32. $x \leq -2$ or $3 \leq x \leq 6$; $(-\infty, -2] \cup [3, 6]$ *(4-5)*

33. 2.2 *(4-6)*

34. $\dfrac{1}{x} + \dfrac{2}{x + 1} - \dfrac{5}{(x + 1)^2}$ *(4-7)*

35. $-\dfrac{2}{x} + \dfrac{3x - 1}{x^2 - x + 1}$ *(4-7)*

36. $x = -2, 4$ *(5-1)*

37. $x = -1, \frac{1}{2}$ *(5-2)*

38. $x = 2.5$ *(5-3)*

39. $x = 10$ *(5-3)*

40. $x = \frac{1}{27}$ *(5-3)*

41. $x = 5$ *(5-5)*

42. $x = 7$ *(5-5)*

43. $x = 5$ *(5-5)*

44. $x = e^{0.1}$ *(5-4)*

45. $x = 1, e^{0.5}$ *(5-5)*

46. $x = 3.38$ *(5-5)*

47. $x = 4.26$ *(5-4)*

48. $x = 2.32$ *(5-4)*

49. $x = 3.67$ *(5-5)*

50. $x = 0.549$ *(5-5)*

51. *(5-1)*

52. *(5-4)*

53. *(5-2)*

54. *(3-5)*

55. *(5-2)*

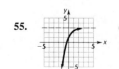

56. $f(x) = \begin{cases} -2x & \text{if } x < -2 \\ 4 & \text{if } -2 \leq x \leq 2 \\ 2x & \text{if } x > 2 \end{cases}$

Domain: all real numbers; Range: $[4, \infty)$ *(3-4)*

57. (A) Domain g: $[-2, 2]$
(B) $\left(\dfrac{f}{g}\right)(x) = \dfrac{x^2}{\sqrt{4 - x^2}}$; Domain $\left(\dfrac{f}{g}\right)$: $(-2, 2)$
(C) $(f \circ g)(x) = 4 - x^2$; Domain $(f \circ g)$: $[-2, 2]$ *(3-7)*

58. (A) $f^{-1}(x) = 1 + \sqrt{x + 4}$
   (B) Domain f^{-1}: $[-4, \infty)$;
       Range f^{-1}: $[1, \infty)$
   (C)

(3-8)

59. Vertical asymptote: $x = -2$
   Oblique asymptote: $y = x + 2$

(3-6)

60. $f(x) = \begin{cases} 2x + 2 & \text{if } -1 \le x < -\frac{1}{2} \\ 2x + 1 & \text{if } -\frac{1}{2} \le x < 0 \\ 2x & \text{if } 0 \le x < \frac{1}{2} \\ 2x - 1 & \text{if } \frac{1}{2} \le x < 1 \\ 2x - 2 & \text{if } 1 \le x < \frac{3}{2} \\ 2x - 3 & \text{if } \frac{3}{2} \le x < 2 \end{cases}$

Domain: all real numbers; Range: $[0, 1)$;
discontinuous at $x = k/2$, k an integer

(3-4)

61. $P(x) = (x + 1)^2 x^3 (x - 3 - 5i)(x - 3 + 5i)$; degree 7   *(4-3)*

62. -1 (double root), 2, $2 \pm i\sqrt{2}$, $P(x) = (x + 1)^2 (x - 2)(x - 2 - i\sqrt{2})(x - 2 + i\sqrt{2})$ *(4-5)*   **63.** -2 (double root), 0.45   *(4-6)*

64. $\dfrac{-2}{(x - 1)} + \dfrac{2}{(x - 1)^2} + \dfrac{2x + 3}{x^2 + x + 2}$ *(4-7)*   **65.** (A) $f^{-1}(x) = e^{x/3} + 2$
   (B) Domain $f = (2, \infty) = $ Range f^{-1}
       Range $f = $ Domain $f^{-1} = (-\infty, \infty)$
   (C)

(3-8, 5-5)

66. $n = \dfrac{\ln(1 + Ai/P)}{\ln(1 + i)}$ *(5-5)*

67. $y = Ae^{5x}$ *(5-5)*   **68.** $x = \ln(y + \sqrt{y^2 + 2})$ *(5-5)*   **69.** $x = -900p + 4{,}571$; 1,610 bottles   *(3-2)*

70. $C(x) = \begin{cases} 0.06x & \text{if } 0 \le x \le 60 \\ 0.05x + 0.6 & \text{if } 60 < x \le 150 \\ 0.04x + 2.1 & \text{if } 150 < x \le 300 \\ 0.03x + 5.1 & \text{if } 300 < x \end{cases}$

(3-4)

71. (A) $A(x) = 80x - 2x^2$ (B) $0 < x < 40$
   (C) 20×40 ft

(3-4)

72. (A) $f(1) = f(3) = 1$, $f(2) = f(4) = 0$
   (B) $f(n) = \begin{cases} 1 & \text{if } n \text{ is an odd integer} \\ 0 & \text{if } n \text{ is an even integer} \end{cases}$ *(3-4)*

73. $x = 2$ ft and $y = 2$ ft or $x = 1.3$ ft and $y = 4.8$ ft   *(4-5)*

74. 1.8 by 3.3 ft   *(4-6)*   **75.** (A) 46.8 million   (B) 103 million   *(5-1)*   **76.** 10.2 years   *(5-5)*   **77.** 9.90 years   *(5-5)*
78. 63.1 times more powerful   *(5-4)*   **79.** 6.31×10^{-4} W/m^2 *(5-4)*

CHAPTER 6   Exercise 6-1

1. $(5, 2)$ **3.** $(2, -3)$ **5.** No solution (parallel lines)   **7.** $(6, 2)$ **9.** $(2, -1)$ **11.** $(2{,}500, 200)$ **13.** $(2, -1)$

15. $(5, 2)$ **17.** $(1, -\frac{2}{3})$ **19.** $(3, -1)$ **21.** $\begin{bmatrix} 4 & -6 & | & -8 \\ 1 & -3 & | & 2 \end{bmatrix}$ **23.** $\begin{bmatrix} -4 & 12 & | & -8 \\ 4 & -6 & | & -8 \end{bmatrix}$ **25.** $\begin{bmatrix} 1 & -3 & | & 2 \\ 8 & -12 & | & -16 \end{bmatrix}$

27. $\begin{bmatrix} 1 & -3 & | & 2 \\ 0 & 6 & | & -16 \end{bmatrix}$ **29.** $\begin{bmatrix} 1 & -3 & | & 2 \\ 2 & 0 & | & -12 \end{bmatrix}$ **31.** $\begin{bmatrix} 1 & -3 & | & 2 \\ 3 & -3 & | & -10 \end{bmatrix}$ **33.** $x_1 = 3, x_2 = 2$ **35.** $x_1 = 3, x_2 = 1$

37. $x_1 = 2, x_2 = 1$ **39.** $x_1 = 2, x_2 = 4$ **41.** No solution   **43.** $x_1 = 1, x_2 = 4$
45. Infinitely many solutions: $x_2 = s, x_1 = 2s - 3$ for any real number s
47. Infinitely many solutions: $x_2 = s, x_1 = \frac{1}{2}s + \frac{1}{2}$ for any real number s **49.** $x_1 = 2, x_2 = -1$ **51.** $x_1 = 2, x_2 = -1$
53. $x_1 = 1.1, x_2 = 0.3$ **55.** Twenty-five 29¢ stamps, fifty 23¢ stamps   **57.** Airspeed $= 330$ mph, wind rate $= 90$ mph
59. 40 mL 50% solution, 60 mL 80% solution   **61.** $3\frac{1}{3}$ g 18 carat, $6\frac{2}{3}$ g 12 carat
63. (A) $a = 196, b = -16$ (B) 196 ft   (C) $3.5\, s$ **65.** 40 s, 24 s, 120 mi

Exercise 6-2

1. Yes   3. No   5. No   7. Yes   9. $x_1 = -2, x_2 = 3, x_3 = 0$
11. $x_1 = 2t + 3, x_2 = -t - 5, x_3 = t, t$ any real number   13. No solution
15. $x_1 = 2s + 3t - 5, x_2 = s, x_3 = -3t + 2, x_4 = t, s$ and t any real numbers   17. $\begin{bmatrix} 1 & 0 & | & -7 \\ 0 & 1 & | & 3 \end{bmatrix}$

19. $\begin{bmatrix} 1 & 0 & 0 & | & -5 \\ 0 & 1 & 0 & | & 4 \\ 0 & 0 & 1 & | & -2 \end{bmatrix}$ 21. $\begin{bmatrix} 1 & 0 & 2 & | & -\frac{5}{3} \\ 0 & 1 & -2 & | & \frac{1}{3} \\ 0 & 0 & 0 & | & 0 \end{bmatrix}$ 23. $x_1 = -2, x_2 = 3, x_3 = 1$ 25. $x_1 = 0, x_2 = -2, x_3 = 2$

27. $x_1 = 2t + 3, x_2 = t - 2, x_3 = t, t$ any real number   29. $x_1 = (-4t - 4)/7, x_2 = (5t + 5)/7, x_3 = t, t$ any real number
31. $x_1 = -1, x_2 = 2$ 33. No solution   35. No solution   37. $x_1 = 2t + 4, x_2 = t + 1, x_3 = t, t$ any real number
39. $x_1 = s + 2t - 1, x_2 = s, x_3 = t, s$ and t any real numbers   41. $x_1 = 0, x_2 = 2, x_3 = -3$ 43. $x_1 = 1, x_2 = -2, x_3 = 1$
45. $x_1 = 2s - 3t + 3, x_2 = s + 2t + 2, x_3 = s, x_4 = t, s$ and t any real numbers
47. 15¢ stamps: $3t - 100$, 20¢ stamps: $145 - 4t$, 35¢ stamps: t, where $t = 34, 35,$ or 36
49. 10% containers: $6t - 24$, 20% containers: $48 - 8t$, 50% containers: t, where $t = 4, 5,$ or 6   51. $a = 3, b = 2, c = 1$
53. $a = -2, b = -4, c = -20$ 55. 20 one-person boats, 220 two-person boats, 100 four-person boats
57. One-person boats: $t - 80$, two-person boats: $-2t + 420$, four-person boats: $t, 80 \le t \le 210, t$ an integer
59. No solution; no production schedule will use all the labor-hours in all departments.   61. 8 oz food A, 2 oz food B, 4 oz food C
63. No solution   65. 8 oz food A, $-2t + 10$ oz food B, t oz food C, $0 \le t \le 5$ 67. Company A: 10 h, company B: 15 h

Exercise 6-3

1. $(-12, 5), (-12, -5)$ 3. $(2, 4), (-2, -4)$ 5. $(5, -5), (-5, 5)$ 7. $(4 + 2\sqrt{3}, 1 + \sqrt{3}), (4 - 2\sqrt{3}, 1 - \sqrt{3})$
9. $(2, 4), (2, -4), (-2, 4), (-2, -4)$ 11. $(1, 3), (1, -3), (-1, 3), (-1, -3)$ 13. $(1 + \sqrt{5}, -1 + \sqrt{5}), (1 - \sqrt{5}, -1 - \sqrt{5})$
15. $(\sqrt{2}, \sqrt{2}), (-\sqrt{2}, -\sqrt{2}), (2, 1), (-2, -1)$ 17. $(2, 2i), (2, -2i), (-2, 2i), (-2, -2i)$
19. $(2, \sqrt{2}), (2, -\sqrt{2}), (-1, i), (-1, -i)$ 21. $(3, 0), (-3, 0), (\sqrt{5}, 2), (-\sqrt{5}, 2)$ 23. $(2, 1), (-2, -1), (i, -2i), (-i, 2i)$
25. $(-1, 4), (3, -4)$ 27. $(0, 0), (3, 6)$ 29. $(1, 4), (4, 1)$ 31. $(-1, 3), (4, 8)$ 33. $(-5, -\frac{3}{5}), (-\frac{3}{2}, -2)$
35. $(0, -1), (-4, -3)$ 37. $(2, 2), (-2, -2), (\sqrt{2}, -\sqrt{2}), (-\sqrt{2}, \sqrt{2})$ 39. $(-3, 1), (3, -1), (-i, i), (i, -i)$
41. $(1.50, 1.75), (4.00, 3.00)$ 43. $(1.56, 1.94), (5.44, -1.94)$ 45. $(-1.84, 0.16), (1.15, 3.15)$ 47. $\frac{1}{2}(3 - \sqrt{5}), \frac{1}{2}(3 + \sqrt{5})$
49. 5 in and 12 in   51. 6 by 4.5 in   53. 22 by 26 ft   55. Boat A: 30 mph; boat B: 25 mph

Exercise 6-4

1. 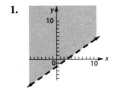   3. 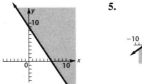   5. 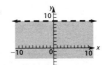   7.    9.

11. Region IV   13. Region I   15. 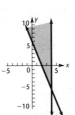   17. 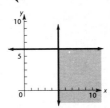   19.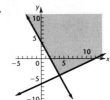

21. Region IV; corner points $(6, 4), (8, 0), (18, 0)$ 23. Region I, corner points: $(0, 16), (6, 4), (18, 0)$
25. Corner points: $(0, 0), (3, 0), (0, 2)$ 27. Corner points: $(5, 0), (0, 4)$ 29. Corner points: $(0, 0), (0, 4), (\frac{12}{5}, \frac{16}{5}), (4, 0)$
Bounded   Unbounded   Bounded

   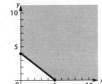

31. Corner points: $(0, 8)$, $(3, 4)$, $(9, 0)$
Unbounded

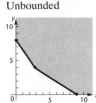

33. Corner points: $(0, 0)$, $(0, 5)$,
$(4, 3)$, $(5, 2)$, $(6, 0)$
Bounded

35. Corner points: $(0, 14)$, $(2, 10)$,
$(8, 4)$, $(16, 0)$
Unbounded

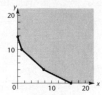

37. Corner points: $(2, 5)$, $(10, 1)$, $(1, 10)$
Bounded

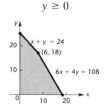

39. The feasible region is empty.

41. Corner points: $(0, 3)$, $(5, 0)$, $(7, 3)$, $(2, 8)$
Bounded

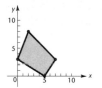

43. $6x + 4y \leq 108$
$x + y \leq 24$
$x \geq 0$
$y \geq 0$

45. $10x + 30y \geq 280$
$30x + 10y \geq 360$
$x \geq 0$
$y \geq 0$

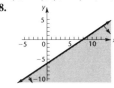

Exercise 6-5

1. Max $z = 16$ at $(7, 9)$ **3.** Max $z = 84$ at $(0, 12)$ and $(7, 9)$ (multiple optimal solutions)   **5.** Min $z = 32$ at $(0, 8)$
7. Min $z = 36$ at $(12, 0)$ and $(4, 3)$ (multiple optimal solutions)   **9.** Max $z = 18$ at $(4, 3)$ **11.** Min $z = 12$ at $(4, 0)$
13. Max $z = 52$ at $(4, 10)$ **15.** Min $z = 44$ at $(4, 4)$
17. Min $z = 1,500$ at $(60, 0)$; max $z = 3,000$ at $(60, 30)$ and $(120, 0)$ (multiple optimal solutions)
19. Min $z = 300$ at $(0, 20)$; max $z = 1,725$ at $(60, 15)$
21. (A) $a > 2b$ (B) $\frac{1}{3}b < a < 2b$ (C) $b > 3a$ (D) $a = 2b$ (E) $b = 3a$ **23.** 6 trick skis, 18 slalom skis; $780
25. 9 model A trucks, 6 model B trucks, $279,000   **27.** (A) 40 tables, 40 chairs; $4,600   (B) 20 tables, 80 chairs; $3,800
29. (A) 40-megabyte disk drives: 18, 80-megabyte disk drives: 6; $19,800
   (B) 40-megabyte disk drives: 6, 80-megabyte disk drives: 12; $21,600
31. (A) 140 bags of brand A, 50 bags of brand B; 1,190 pounds of nitrogen
   (B) 40 bags of brand A, 100 bags of brand B, 940 pounds of nitrogen

Chapter 6 Review Exercise

1. $(2, 3)$ *(6-1)*   **2.** No solution (inconsistent) *(6-1)*   **3.** Infinitely many solutions $(t, (4t + 8)/3)$, for any real number t *(6-1)*
4. $(-1, 3)$, $(5, -3)$ *(6-3)*   **5.** $(1, -1)$, $(\frac{7}{5}, -\frac{1}{5})$ *(6-3)*   **6.** $(1, 3)$, $(1, -3)$, $(-1, 3)$, $(-1, -3)$ *(6-3)*
7. *(6-1)*

8. *(6-4)*

9. *(6-4)*

10. $\begin{bmatrix} 3 & -6 & | & 12 \\ 1 & -4 & | & 5 \end{bmatrix}$ *(6-1)*   **11.** $\begin{bmatrix} 1 & -4 & | & 5 \\ 1 & -2 & | & 4 \end{bmatrix}$ *(6-1)*   **12.** $\begin{bmatrix} 1 & -4 & | & 5 \\ 0 & 6 & | & -3 \end{bmatrix}$ *(6-1)*   **13.** $x_1 = 4$ *(6-2)*
$x_2 = -7$
$(4, -7)$

14. $x_1 - x_2 = 4$ *(6-2)*   **15.** $x_1 - x_2 = 4$ *(6-2)*   **16.** Min $z = 18$ at $(0, 6)$; max $z = 42$ at $(6, 4)$ *(6-5)*
$\qquad 0 = 1$ $\qquad\qquad\qquad\qquad\quad x_1 = t + 4, x_2 = t, t$ any real number
No solution

17. $x_1 = -1, x_2 = 3$ *(6-2)*   **18.** $x_1 = -1, x_2 = 2, x_3 = 1$ *(6-2)*   **19.** $x_1 = 2, x_2 = 1, x_3 = -1$ *(6-2)*

20. Infinitely many solutions: $x_1 = -5t - 12, x_2 = 3t + 7, x_3 = t, t$ any real number *(6-2)*   **21.** No solution *(6-2)*

22. Infinitely many solutions: $x_1 = -\frac{3}{7}t - \frac{4}{7}, x_2 = \frac{5}{7}t + \frac{9}{7}, x_3 = t, t$ any real number *(6-2)*

23. $(2, \sqrt{2}), (2, -\sqrt{2}), (-1, i), (-1, -i)$ *(6-3)*   **24.** $(1, -2), (-1, 2), (2, -1), (-2, 1)$ *(6-3)*

25. $(2, -2), (-2, 2), (\sqrt{2}, \sqrt{2}), (-\sqrt{2}, -\sqrt{2})$ *(6-3)*

26. Corner points: $(0, 0), (0, 4),$ **27.** Corner points: $(0, 8), (\frac{12}{5}, \frac{16}{5}), (12, 0)$ *(6-4)*   **28.** Corner points: $(4, 4), (10, 10),$
$(3, 2), (4, 0)$ *(6-4)*   Unbounded $(20, 0)$ *(6-4)*
Bounded   Bounded

      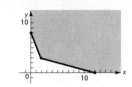   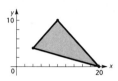

29. Max $z = 46$ at $(4, 2)$ *(6-5)*   **30.** Min $z = 75$ at $(3, 6)$ and $(15, 0)$ (multiple optimal solutions) *(6-5)*

31. Min $z = 44$ at $(4, 3)$; max $z = 82$ at $(2, 9)$ *(6-5)*   **32.** $x_1 = 1,000, x_2 = 4,000, x_3 = 2,000$ *(6-2)*

33. $(2, 2), (-2, -2), (\frac{4}{7}\sqrt{7}, -\frac{2}{7}\sqrt{7}), (-\frac{4}{7}\sqrt{7}, \frac{2}{7}\sqrt{7})$ *(6-3)*   **34.** Max $z = 26,000$ at $(600, 400)$ *(6-5)*

35. $\frac{1}{2}$-lb packages: 48, $\frac{1}{3}$-lb packages: 72 *(6-1, 6-2)*   **36.** 6 by 8 m *(6-3)*   **37.** 40 g mix A, 60 g mix B, 30 g mix C *(6-2)*

38. (A) 22 nickels, 8 dimes   (B) $22 + 3t$ nickels, $8 - 4t$ dimes, and t quarters, $t = 0, 1, 2$ *(6-1, 6-2)*

39. 80 regular sails, 30 competition sails; \$7,800 *(6-5)*

CHAPTER 7   Exercise 7-1

1. B is 2×2; E is 1×4 **3.** 2   **5.** $\begin{bmatrix} 0 & 0 \\ 0 & 0 \end{bmatrix}$ **7.** C, D **9.** A, B **11.** $\begin{bmatrix} -1 & 0 \\ 5 & -3 \end{bmatrix}$ **13.** Not defined   **15.** $\begin{bmatrix} -2 \\ 3 \\ 0 \end{bmatrix}$

17. $\begin{bmatrix} -1 \\ 6 \\ 5 \end{bmatrix}$ **19.** $\begin{bmatrix} -15 & 5 \\ 10 & -15 \end{bmatrix}$ **21.** $\begin{bmatrix} 1 & -1 & 4 & 1 \\ -2 & -6 & 4 & 2 \\ 2 & 7 & 0 & -2 \end{bmatrix}$ **23.** $\begin{bmatrix} 2 & -5 & 0 \\ -5 & 3 & 1 \end{bmatrix}$ **25.** $\begin{bmatrix} 5 & -2 \\ -1 & 0 \\ 0 & 1 \end{bmatrix}$

27. $\begin{bmatrix} 250 & 360 \\ 40 & 350 \end{bmatrix}$ **29.** Not defined   **31.** $\begin{bmatrix} 9 & -5 \\ -6 & 5 \end{bmatrix}$ **33.** $\begin{bmatrix} 34 & 38 \\ 28 & 67 \end{bmatrix}$ **35.** $a = -1, b = 1, c = 3, d = -5$

37. $x = 1, y = 2$ **39.** No   **41.** No   **43.** $\begin{matrix} \text{Guitar} \ \ \text{Banjo} \\ \begin{bmatrix} \$33 & \$26 \\ \$57 & \$77 \end{bmatrix} \end{matrix}$ $\begin{matrix} \text{Materials} \\ \text{Labor} \end{matrix}$

45. $A + B = \begin{bmatrix} 135 & 282 & 50 \\ 55 & 258 & 155 \end{bmatrix}$; $\frac{1}{935}(A + B) = \begin{bmatrix} 0.14 & 0.30 & 0.05 \\ 0.06 & 0.28 & 0.17 \end{bmatrix}$

Exercise 7-2

1. 10   **3.** -1 **5.** $[12 \ \ 13]$ **7.** $\begin{bmatrix} 5 \\ -3 \end{bmatrix}$ **9.** $\begin{bmatrix} 2 & 4 \\ 1 & -5 \end{bmatrix}$ **11.** $\begin{bmatrix} 1 & -5 \\ -2 & -4 \end{bmatrix}$ **13.** $[-14]$ **15.** $\begin{bmatrix} -20 & 10 \\ -12 & 6 \end{bmatrix}$

17. 6   **19.** 15   **21.** $\begin{bmatrix} 0 & 9 \\ 5 & -4 \end{bmatrix}$ **23.** $\begin{bmatrix} 5 & 8 & -5 \\ -1 & -3 & 2 \\ -2 & 8 & -6 \end{bmatrix}$ **25.** $[11]$ **27.** $\begin{bmatrix} 3 & -2 & -4 \\ 6 & -4 & -8 \\ -9 & 6 & 12 \end{bmatrix}$ **29.** Not defined

31. $\begin{bmatrix} 7 \\ 35 \\ 23 \end{bmatrix}$ **33.** $\begin{bmatrix} -4 & 2 & 1 \\ -4 & -13 & 27 \\ 8 & -10 & 11 \end{bmatrix}$ **35.** 0   **37.** $\begin{bmatrix} \frac{1}{3} & \frac{1}{3} \\ \frac{2}{3} & \frac{2}{3} \end{bmatrix}$ **39.** $AB = \begin{bmatrix} 5 & 7 \\ 2 & 3 \end{bmatrix}, BA = \begin{bmatrix} 1 & 3 \\ 2 & 7 \end{bmatrix}$

41. Both sides equal $\begin{bmatrix} 0 & 12 \\ 1 & 5 \end{bmatrix}$ **43.** $A^2 - B^2 = \begin{bmatrix} -2 & 0 \\ -8 & -10 \end{bmatrix}, (A - B)(A + B) = \begin{bmatrix} 2 & 4 \\ -8 & -14 \end{bmatrix}$

45. (A) \$11.80 per boat   (B) $[1.5 \ \ 1.2 \ \ 0.4] \cdot \begin{bmatrix} 9 \\ 12 \\ 6 \end{bmatrix} = \$30.30$ per boat   (C) 3×2 (D) $\begin{matrix} \text{Plant I} \ \ \text{Plant II} \\ \begin{bmatrix} \$11.80 & \$13.80 \\ \$18.50 & \$21.60 \\ \$26.00 & \$30.30 \end{bmatrix} \end{matrix} \begin{matrix} \text{One-person boat} \\ \text{Two-person boat} \\ \text{Four-person boat} \end{matrix}$

Labor costs per boat at each plant

47. (A) $2,025 (B) $[2{,}000 \quad 800 \quad 8{,}000] \cdot \begin{bmatrix} \$0.40 \\ \$0.75 \\ \$0.25 \end{bmatrix} = \$3{,}400$ (C) $\begin{bmatrix} \$2{,}025 \\ \$3{,}400 \end{bmatrix}$ Berkeley Oakland Cost per town

(D) $\begin{matrix} \text{Telephone} & \text{House call} & \text{Letter} \\ [3{,}000 & 1{,}300 & 13{,}000] \end{matrix}$
Number of each type of contact made

Exercise 7-3

1. $\begin{bmatrix} 2 & -3 \\ 4 & 5 \end{bmatrix}$ **3.** $\begin{bmatrix} 2 & -3 \\ 4 & 5 \end{bmatrix}$ **5.** $\begin{bmatrix} -2 & 1 & 3 \\ 2 & 4 & -2 \\ 5 & 1 & 0 \end{bmatrix}$ **7.** $\begin{bmatrix} -2 & 1 & 3 \\ 2 & 4 & -2 \\ 5 & 1 & 0 \end{bmatrix}$ **15.** $\begin{bmatrix} 4 & 1 \\ -1 & 0 \end{bmatrix}$ **17.** $\begin{bmatrix} 3 & -2 \\ -1 & 1 \end{bmatrix}$

19. $\begin{bmatrix} 7 & -3 \\ -2 & 1 \end{bmatrix}$ **21.** $\begin{bmatrix} 7 & 6 & -3 \\ 2 & 2 & -1 \\ -6 & -5 & 3 \end{bmatrix}$ **23.** $\begin{bmatrix} \frac{3}{2} & -\frac{1}{2} & -\frac{1}{2} \\ -\frac{1}{2} & \frac{1}{2} & \frac{1}{2} \\ -\frac{3}{2} & \frac{1}{2} & \frac{3}{2} \end{bmatrix}$ or $\frac{1}{2}\begin{bmatrix} 3 & -1 & -1 \\ -1 & 1 & 1 \\ -3 & 1 & 3 \end{bmatrix}$ **25.** Does not exist

27. $\begin{bmatrix} 2 & -\frac{5}{2} \\ -1 & \frac{3}{2} \end{bmatrix}$ **29.** $\begin{bmatrix} 1 & 0 & 1 \\ 1 & 1 & 2 \\ 3 & 2 & 6 \end{bmatrix}$ **31.** Does not exist **33.** $\begin{bmatrix} -9 & -15 & 10 \\ 4 & 5 & -4 \\ -1 & -1 & 1 \end{bmatrix}$

Exercise 7-4

1. $\begin{aligned} 2x_1 - x_2 &= 3 \\ x_1 + 3x_2 &= -2 \end{aligned}$ **3.** $\begin{aligned} -2x_1 \quad + x_3 &= 3 \\ x_1 + 2x_2 + x_3 &= -4 \\ x_2 - x_3 &= 2 \end{aligned}$ **5.** $\begin{bmatrix} 4 & -3 \\ 1 & 2 \end{bmatrix}\begin{bmatrix} x_1 \\ x_2 \end{bmatrix} = \begin{bmatrix} 2 \\ 1 \end{bmatrix}$ **7.** $\begin{bmatrix} 1 & -2 & 1 \\ -1 & 1 & 0 \\ 2 & 3 & 1 \end{bmatrix}\begin{bmatrix} x_1 \\ x_2 \\ x_3 \end{bmatrix} = \begin{bmatrix} -1 \\ 2 \\ -3 \end{bmatrix}$

9. $x_1 = -8, x_2 = 2$ **11.** $x_1 = 0, x_2 = 4$ **13.** (A) $x_1 = -3, x_2 = 2$
(B) $x_1 = -1, x_2 = 2$
(C) $x_1 = -8, x_2 = 3$

15. (A) $x_1 = 17, x_2 = -5$ **17.** (A) $x_1 = 1, x_2 = 0, x_3 = 0$ **19.** (A) $x_1 = 1, x_2 = 1, x_3 = 3$
(B) $x_1 = 7, x_2 = -2$ (B) $x_1 = -1, x_2 = 0, x_3 = 1$ (B) $x_1 = -1, x_2 = 1, x_3 = -1$
(C) $x_1 = 24, x_2 = -7$ (C) $x_1 = -1, x_2 = -1, x_3 = 1$ (C) $x_1 = 5, x_2 = -1, x_3 = -5$
21. $X = (A - B)^{-1}C$ $[X \neq C(A - B)^{-1}]$ **23.** $X = (A + I)^{-1}C$ **25.** $X = (A + B)^{-1}(C + D)$
27. Concert 1: 6,000 $4-tickets and 4,000 $8-tickets; Concert 2: 5,000 $4-tickets and 5,000 $8-tickets;
Concert 3: 3,000 $4-tickets and 7,000 $8-tickets
29. (A) $I_1 = 4, I_2 = 6, I_3 = 2$ (B) $I_1 = 3, I_2 = 7, I_3 = 4$ (C) $I_1 = 7, I_2 = 8, I_3 = 1$
31. (A) $a = 1, b = 0, c = -3$ (B) $a = -2, b = 5, c = 1$ (C) $a = 11, b = -46, c = 43$
33. Diet 1: 60 oz mix A and 80 oz mix B; Diet 2: 20 oz mix A and 60 oz mix B; Diet 3: 0 oz mix A and 100 oz mix B

Exercise 7-5

1. 8 **3.** -20 **5.** -0.88 **7.** $\begin{vmatrix} a_{22} & a_{23} \\ a_{32} & a_{33} \end{vmatrix}$ **9.** $\begin{vmatrix} a_{11} & a_{12} \\ a_{31} & a_{32} \end{vmatrix}$ **11.** $(-1)^{1+1}\begin{vmatrix} a_{22} & a_{23} \\ a_{32} & a_{33} \end{vmatrix}$ **13.** $(-1)^{2+3}\begin{vmatrix} a_{11} & a_{12} \\ a_{31} & a_{32} \end{vmatrix}$

15. $\begin{vmatrix} 1 & -2 \\ -4 & 8 \end{vmatrix}$ **17.** $\begin{vmatrix} -2 & 0 \\ 5 & -2 \end{vmatrix}$ **19.** $(-1)^{1+1}\begin{vmatrix} 1 & -2 \\ -4 & 8 \end{vmatrix} = 0$ **21.** $(-1)^{3+2}\begin{vmatrix} -2 & 0 \\ 5 & -2 \end{vmatrix} = -4$ **23.** 10 **25.** -21

27. -40 **29.** $(-1)^{1+1}\begin{vmatrix} a_{22} & a_{23} & a_{24} \\ a_{32} & a_{33} & a_{34} \\ a_{42} & a_{43} & a_{44} \end{vmatrix}$ **31.** $(-1)^{4+3}\begin{vmatrix} a_{11} & a_{12} & a_{14} \\ a_{21} & a_{22} & a_{24} \\ a_{31} & a_{32} & a_{34} \end{vmatrix}$ **33.** 22 **35.** -12 **37.** 0 **39.** 6 **41.** 60

43. 114 **45.** $\begin{vmatrix} a & b \\ ka & kb \end{vmatrix} = akb - kab = 0$ **47.** $\begin{vmatrix} a & b \\ c & d \end{vmatrix} = ad - cb = ad - bc = \begin{vmatrix} a & c \\ b & d \end{vmatrix}$ **51.** $49 = (-7)(-7)$

53. $f(x) = x^2 - 4x + 3; 1, 3$ **55.** $f(x) = x^3 + 2x^2 - 8x; -4, 0, 2$

Exercise 7-6

1. Theorem 1 **3.** Theorem 1 **5.** Theorem 2 **7.** Theorem 3 **9.** Theorem 5 **11.** $x = 0$ **13.** $x = 5$ **15.** 25
17. -12 **19.** Theorem 1 **21.** Theorem 2 **23.** Theorem 5 **25.** $x = 5, y = 0$ **27.** $x = -3, y = 10$ **29.** -28
31. 106 **33.** 0 **35.** 6 **37.** 14 **39.** Expand the left side of the equation using minors.
41. Expand both sides of the equation and compare. **43.** This follows from Theorem 4.
45. Expand the determinant about the first row to obtain $(y_1 - y_2)x - (x_1 - x_2)y + (x_1y_2 - x_2y_1) = 0$.
Then show that the two points satisfy this linear equation.
47. If the determinant is 0, then the area of the triangle formed by the three points is 0. The only way this can happen is if the three points are on the same line—that is, the points are collinear.

Exercise 7-7

1. $x = 5, y = -2$ **3.** $x = 1, y = -1$ **5.** $x = -\frac{6}{5}, y = \frac{3}{5}$ **7.** $x = \frac{2}{17}, y = -\frac{20}{17}$ **9.** $x = 6,400, y = 6,600$
11. $x = 760, y = 760$ **13.** $x = 2, y = -2, z = -1$ **15.** $x = \frac{4}{3}, y = -\frac{1}{3}, z = \frac{2}{3}$ **17.** $x = -9, y = -\frac{7}{3}, z = 6$
19. $x = \frac{3}{2}, y = -\frac{7}{6}, z = \frac{2}{3}$ **21.** $x = 4$ **23.** $y = 2$ **25.** $z = \frac{5}{2}$
27. Since $D = 0$, the system has either no solution or infinitely many. Since $x = 0, y = 0, z = 0$ is a solution, the second case must hold.

Chapter 7 Review Exercise

1. $\begin{bmatrix} 3 & 3 \\ 4 & 2 \end{bmatrix}$ *(7-1)*   **2.** Not defined *(7-1)*   **3.** $\begin{bmatrix} -3 & 0 \\ 1 & -1 \end{bmatrix}$ *(7-1)*   **4.** $\begin{bmatrix} 4 & 3 \\ 7 & 4 \end{bmatrix}$ *(7-2)*   **5.** Not defined *(7-2)*

6. $\begin{bmatrix} 5 \\ 5 \end{bmatrix}$ *(7-2)*   **7.** $\begin{bmatrix} 2 & 3 \\ 4 & 6 \end{bmatrix}$ *(7-2)*   **8.** 8 (a real number) *(7-2)*   **9.** Not defined *(7-1)*   **10.** $\begin{bmatrix} 3 & -2 \\ -4 & 3 \end{bmatrix}$ *(7-3)*

11. (A) $x_1 = -1, x_2 = 3$ (B) $x_1 = 1, x_2 = 2$ (C) $x_1 = 8, x_2 = -10$ *(7-4)*   **12.** -17 *(7-5)*   **13.** 0 *(7-5, 7-6)*

14. $x = 2, y = -1$ *(7-7)*   **15.** Not defined *(7-1)*   **16.** $\begin{bmatrix} 10 & -8 \\ 4 & 6 \end{bmatrix}$ *(7-1, 7-2)*   **17.** $\begin{bmatrix} -2 & 8 \\ 8 & 6 \end{bmatrix}$ *(7-1, 7-2)*

18. 9 (a real number) *(7-2)*   **19.** [9] (a matrix) *(7-2)*   **20.** $\begin{bmatrix} 10 & -5 & 1 \\ -1 & -4 & -5 \\ 1 & -7 & -2 \end{bmatrix}$ *(7-1, 7-2)*

21. $\begin{bmatrix} -\frac{5}{2} & 2 & -\frac{1}{2} \\ 1 & -1 & 1 \\ \frac{1}{2} & 0 & -\frac{1}{2} \end{bmatrix}$ or $\frac{1}{2}\begin{bmatrix} -5 & 4 & -1 \\ 2 & -2 & 2 \\ 1 & 0 & -1 \end{bmatrix}$ *(7-3)*

22. (A) $x_1 = 2, x_2 = 1, x_3 = -1$ (B) $x_1 = 1, x_2 = -2, x_3 = 1$ (C) $x_1 = -1, x_2 = 2, x_3 = -2$ *(7-4)*   **23.** $-\frac{11}{12}$ *(7-5)*
24. 35 *(7-5, 7-6)*   **25.** $y = \frac{10}{5} = 2$ *(7-7)*   **26.** $X = (A - C)^{-1}B$ *(7-4)*

27. $\begin{bmatrix} -\frac{11}{12} & -\frac{1}{12} & 5 \\ \frac{10}{12} & \frac{2}{12} & -4 \\ \frac{1}{12} & -\frac{1}{12} & 0 \end{bmatrix}$ or $\frac{1}{12}\begin{bmatrix} -11 & -1 & 60 \\ 10 & 2 & -48 \\ 1 & -1 & 0 \end{bmatrix}$ *(7-3)*   **28.** $x_1 = 1,000, x_2 = 4,000, x_3 = 2,000$ *(7-4)*   **29.** 42 *(7-6)*

30. $\begin{vmatrix} u + kv & v \\ w + kx & x \end{vmatrix} = (u + kv)x - (w + kx)v = ux + kvx - wv - kvx = ux - wv = \begin{vmatrix} u & v \\ w & x \end{vmatrix}$ *(7-6)*

31. (A) 60 tons at Big Bend, 20 tons at Saw Pit
(B) 30 tons at Big Bend, 50 tons at Saw Pit
(C) 40 tons at Big Bend, 40 tons at Saw Pit *(7-4)*

32. (A) $[0.9 \quad 1.8 \quad 0.6] \cdot \begin{bmatrix} 10 \\ 8.50 \\ 4.50 \end{bmatrix} = \$27$ (B)

	North Carolina	South Carolina	
	$46.35	$41.00	Desks
	$30.45	$27.00	Stands

Total labor costs for each item at each plant *(7-1, 7-2)*

33. (A) $\begin{bmatrix} 1,600 & 1,730 \\ 890 & 720 \end{bmatrix}$ (B) $\begin{bmatrix} 200 & 160 \\ 80 & 40 \end{bmatrix}$ (C)

3,150	Desks	
1,550	Stands	

Total production of each item in January *(7-1, 7-2)*

Cumulative Review Exercise: Chapters 6 and 7

1. $x = 2, y = -1$ *(6-1)*   **2.** $(-1, 2)$ *(6-1)*   **3.** $(-\frac{1}{5}, -\frac{7}{5}), (1, 1)$ *(6-3)*   **4.** *(6-4)*

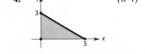

5. Maximum: 33; Minimum: 10 *(6-5)*

6. (A) $\begin{bmatrix} 0 & -3 \\ 3 & -9 \end{bmatrix}$ (B) Not defined   (C) 3   (D) $\begin{bmatrix} 1 & 7 \\ 4 & -7 \end{bmatrix}$ (E) $[-1, 8]$ (F) Not defined *(7-1, 7-2)*

7. -10 *(7-5)*   **8.** (A) $x_1 = 3, x_2 = -4$ (B) $x_1 = 2t + 3, x_2 = t, t$ any real number.   (C) No solution *(6-1)*

9. (A) $\begin{bmatrix} 1 & 1 & | & 3 \\ -1 & 1 & | & 5 \end{bmatrix}$ (B) $\begin{bmatrix} 1 & 0 & | & -1 \\ 0 & 1 & | & 4 \end{bmatrix}$ (C) $x_1 = -1, x_2 = 4$ *(6-1, 6-2)*

10. (A) $\begin{bmatrix} 1 & -3 \\ 2 & -5 \end{bmatrix}\begin{bmatrix} x_1 \\ x_2 \end{bmatrix} = \begin{bmatrix} k_1 \\ k_2 \end{bmatrix}$ (B) $A^{-1} = \begin{bmatrix} -5 & 3 \\ -2 & 1 \end{bmatrix}$ (C) $x_1 = 13, x_2 = 5$ (D) $x_1 = -11, x_2 = -4$ *(7-4)*

11. (A) 2   (B) $x = \frac{1}{2}, y = 0$ *(7-7)*   **12.** $(1, 0, -2)$ *(6-2)*   **13.** No solution *(6-2)*
14. $(t - 3, t - 2, t)$ t any real number *(6-2)*   **15.** $(1, 1), (-1, -1), (\sqrt{3}, \sqrt{3}/3), (-\sqrt{3}, -\sqrt{3}/3)$ *(6-3)*

16. $(0, i), (0, -i), (1, 1), (-1, -1)$ *(6-3)*   **17.** (A) -3 (B) $[-3]$ (C) $\begin{bmatrix} 1 & 2 & -1 \\ -1 & -2 & 1 \\ 2 & 4 & -2 \end{bmatrix}$ *(7-2)*

18. (A) $\begin{bmatrix} -1 & 2 \\ 2 & 3 \end{bmatrix}$ (B) Not defined *(7-2)*   **19.** *(6-4)*   **20.** 63 *(6-5)*

21. (A) $\begin{bmatrix} 1 & 4 & 2 \\ 2 & 6 & 3 \\ 2 & 5 & 2 \end{bmatrix}\begin{bmatrix} x_1 \\ x_2 \\ x_3 \end{bmatrix} = \begin{bmatrix} k_1 \\ k_2 \\ k_3 \end{bmatrix}$ (B) $A^{-1} = \begin{bmatrix} -3 & 2 & 0 \\ 2 & -2 & 1 \\ -2 & 3 & -2 \end{bmatrix}$ (C) $(7, -5, 6)$ (D) $(-6, 3, -2)$ *(7-4)*

22. (A) $D = 1$ (B) $x_1 = 83$ *(7-6, 7-7)*   **23.** $L, M,$ and P *(6-2)*   **27.** \$8,000 at 8% and \$4,000 at 14% *(6-1, 6-2)*
28. 60 g mix A, 50 g mix B, 40 g mix C *(6-2)*   **29.** 8 by 4 m *(6-3)*
30. 400 standard packs and 200 deluxe packs for a maximum profit of \$5,600 *(6-5)*

CHAPTER 8   Exercise 8-1

1. $-1, 0, 1, 2$ **3.** $0, \frac{1}{3}, \frac{1}{2}, \frac{3}{5}$ **5.** $4, -8, 16, -32$ **7.** 6   **9.** $\frac{99}{101}$ **11.** $1 + 2 + 3 + 4 + 5$ **13.** $\frac{1}{10} + \frac{1}{100} + \frac{1}{1,000}$
15. $-1 + 1 - 1 + 1$ **17.** $1, -4, 9, -16, 25$ **19.** $0.3, 0.33, 0.333, 0.3333, 0.333\ 33$ **21.** $1, -\frac{1}{2}, \frac{1}{4}, -\frac{1}{8}, \frac{1}{16}$
23. $7, 3, -1, -5, -9$ **25.** $4, 1, \frac{1}{4}, \frac{1}{16}, \frac{1}{64}$ **27.** $a_n = n + 3$ **29.** $a_n = 3n$ **31.** $a_n = n/(n + 1)$ **33.** $a_n = (-1)^{n+1}$
35. $a_n = (-2)^n$ **37.** $a_n = x^n/n$ **39.** $\frac{4}{1} - \frac{8}{2} + \frac{16}{3} - \frac{32}{4}$ **41.** $x^2 + \frac{x^3}{2} + \frac{x^4}{3}$ **43.** $x - \frac{x^2}{2} + \frac{x^3}{3} - \frac{x^4}{4} + \frac{x^5}{5}$ **45.** $\displaystyle\sum_{k=1}^{4} k^2$
47. $\displaystyle\sum_{k=1}^{5} \frac{1}{2^k}$ **49.** $\displaystyle\sum_{k=1}^{n} \frac{1}{k^2}$ **51.** $\displaystyle\sum_{k=1}^{n} (-1)^{k+1}k^2$
53. (A) $3, 1.83, 1.46, 1.415$ (B) $\sqrt{2} \approx 1.4142$ (C) For $a_1 = 1$: $1, 1.5, 1.417, 1.414$
55. Series approx. of $e^{0.2} = 1.221\ 400\ 0$; calculator value of $e^{0.2} = 1.221\ 402\ 8$

Exercise 8-2

1. Fails at $n = 2$ **3.** Fails at $n = 3$ **5.** $P_1: 2 = 2 \cdot 1^2; P_2: 2 + 6 = 2 \cdot 2^2; P_3: 2 + 6 + 10 = 2 \cdot 3^2$
7. $P_1: a^5 a = a^{5+1}; P_2: a^5 a^2 = a^5(a^1 a) = (a^5 a)a = a^6 a = a^7 = a^{5+2}; P_3: a^5 a^3 = a^5(a^2 a) = a^5(a^1 a)a = [(a^5 a)a]a = a^8 = a^{5+3}$
9. $P_1: 9^1 - 1 = 8$ is divisible by 4; $P_2: 9^2 - 1 = 80$ is divisible by 4; $P_3: 9^3 - 1 = 728$ is divisible by 4
11. $P_k: 2 + 6 + 10 + \cdots + (4k - 2) = 2k^2; P_{k+1}: 2 + 6 + 10 + \cdots + (4k - 2) + (4k + 2) = 2(k + 1)^2$
13. $P_k: a^5 a^k = a^{5+k}; P_{k+1}: a^5 a^{k+1} = a^{5+k+1}$ **15.** $P_k: 9^k - 1 = 4r; P_{k+1}: 9^{k+1} - 1 = 4s; r, s \in N$
39. $2 + 4 + 6 + \cdots + 2n = n(n + 1)$ **41.** $1 + 2 + 3 + \cdots + (n - 1) = n(n - 1)/2, n \geq 2$ **47.** $3^4 + 4^4 + 5^4 + 6^4 \neq 7^4$

Exercise 8-3

1. (B) $d = -0.5; 5.5, 5$ (C) $d = -5; -26, -31$ **3.** $a_2 = -1, a_3 = 3, a_4 = 7$ **5.** $a_{15} = 67, S_{11} = 242$ **7.** $S_{21} = 861$
9. $a_{15} = -21$ **11.** $d = 6, a_{101} = 603$ **13.** $S_{40} = 200$ **15.** $a_{11} = 2, S_{11} = \frac{77}{6}$ **17.** $a_1 = 1$ **19.** $S_{51} = 4,131$
21. $-1,071$ **23.** $4,446$ **27.** $a_n = -3 + (n - 1)3$ **31.** *Hint:* $y = x + d, z = x + 2d$ **33.** $x = -1, y = 2$
35. Firm A: \$501,000; firm B: \$504,000   **37.** \$700 per year; \$115,500   **39.** (A) 336 ft   (B) 1,936 ft   (C) $16t^2$
41. 3,420°

Exercise 8-4

1. (A) $r = -2; -16, 32$ (D) $r = \frac{1}{3}; \frac{1}{54}, \frac{1}{162}$ **3.** $a_2 = 3, a_3 = -\frac{3}{2}, a_4 = \frac{3}{4}$ **5.** $a_{10} = \frac{1}{243}$ **7.** $S_7 = 3,279$ **9.** $r = 0.398$
11. $S_{10} = -1,705$ **13.** $a_2 = 6, a_3 = 4$ **15.** $S_7 = 547$ **17.** $\frac{1,023}{1,024}$ **19.** $x = 2\sqrt{3}$ **21.** $S_\infty = \frac{3}{2}$ **23.** No sum
25. $S_\infty = \frac{8}{5}$ **27.** $\frac{7}{9}$ **29.** $\frac{6}{11}$ **31.** $3\frac{8}{37}$ or $\frac{119}{37}$ **33.** $a_n = (-2)(-3)^{n-1}$ **37.** \$4,000,000
39. $A = P(1 + r)^n$; approx. 12 yr   **41.** 900   **43.** 1,250,000   **45.** $A = A_0 2^{2t}$ **47.** $r = 10^{-0.4} = 0.398$ **49.** 2
51. $\$9.223 \times 10^{16}$; $\$1.845 \times 10^{17}$ **53.** 0.0015 psi

Exercise 8-5

1. 720   **3.** 20   **5.** 720   **7.** 15   **9.** 1   **11.** 28   **13.** 9!/8!   **15.** 8!/5!   **17.** 126   **19.** 6   **21.** 1
23. 2,380   **25.** $m^3 + 3m^2 n + 3mn^2 + n^3$ **27.** $8x^3 - 36x^2 y + 54xy^2 - 27y^3$ **29.** $x^4 - 8x^3 + 24x^2 - 32x + 16$
31. $m^4 + 12m^3 n + 54m^2 n^2 + 108mn^3 + 81n^4$ **33.** $32x^5 - 80x^4 y + 80x^3 y^2 - 40x^2 y^3 + 10xy^4 - y^5$

35. $m^6 + 12m^5n + 60m^4n^2 + 160m^3n^3 + 240m^2n^4 + 192mn^5 + 64n^6$ **37.** $5{,}005u^9v^6$ **39.** $264m^2n^{10}$ **41.** $924w^6$
43. $-48{,}384x^3y^5$ **45.** 1.1046

Chapter 8 Review Exercise

1. (A) Geometric (B) Arithmetic (C) Arithmetic (D) Neither (E) Geometric *(8-1, 8-3, 8-4)*
2. (A) $5, 7, 9, 11$ (B) $a_{10} = 23$ (C) $S_{10} = 140$ *(8-1, 8-3)*
3. (A) $16, 8, 4, 2$ (B) $a_{10} = \frac{1}{32}$ (C) $S_{10} = 31\frac{31}{32}$ *(8-1, 8-4)*
4. (A) $-8, -5, -2, 1$ (B) $a_{10} = 19$ (C) $S_{10} = 55$ *(8-1, 8-3)*
5. (A) $-1, 2, -4, 8$ (B) $a_{10} = 512$ (C) $S_{10} = 341$ *(8-1, 8-4)* **6.** $S_\infty = 32$ *(8-4)* **7.** 720 *(8-5)*
8. $20 \cdot 21 \cdot 22 = 9{,}240$ *(8-5)* **9.** 21 *(8-5)* **10.** $P_1: 5 = 1^2 + 4 \cdot 1; P_2: 5 + 7 = 2^2 + 4 \cdot 2; P_3: 5 + 7 + 9 = 3^2 + 4 \cdot 3$ *(8-2)*
11. $P_1: 2 = 2^{1+1} - 2; P_2: 2 + 4 = 2^{2+1} - 2; P_3: 2 + 4 + 8 = 2^{3+1} - 2$ *(8-2)*
12. $P_1: 49^1 - 1 = 48$ is divisible by 6; $P_2: 49^2 - 1 = 2{,}400$ is divisible by 6; $P_3: 49^3 - 1 = 117{,}648$ is divisible by 6 *(8-2)*
13. $P_k: 5 + 7 + 9 + \cdots + (2k + 3) = k^2 + 4k; P_{k+1}: 5 + 7 + 9 + \cdots + (2k + 3) + (2k + 5) = (k + 1)^2 + 4(k + 1)$ *(8-2)*
14. $P_k: 2 + 4 + 8 + \cdots + 2^k = 2^{k+1} - 2; P_{k+1}: 2 + 4 + 8 + \cdots + 2^k + 2^{k+1} = 2^{k+2} - 2$ *(8-2)*
15. $P_k: 49^k - 1 = 6r$ for some integer r; $P_{k+1}: 49^{k+1} - 1 = 6s$ for some integer s *(8-2)*
16. $S_{10} = -6 - 4 - 2 + 0 + 2 + 4 + 6 + 8 + 10 + 12 = 30$ *(8-3)* **17.** $S_7 = 8 + 4 + 2 + 1 + \frac{1}{2} + \frac{1}{4} + \frac{1}{8} = 15\frac{7}{8}$ *(8-4)*
18. $S_\infty = \frac{81}{5}$ *(8-4)* **19.** $S_n = \sum_{k=1}^{n} \frac{(-1)^{k+1}}{3^k}; S_\infty = \frac{1}{4}$ *(8-4)* **20.** $d = 3, a_5 = 25$ *(8-3)* **21.** $\frac{8}{11}$ *(8-4)* **22.** 190 *(8-5)*
23. $1{,}820$ *(8-5)* **24.** 1 *(8-5)* **25.** $x^5 - 5x^4y + 10x^3y^2 - 10x^2y^3 + 5xy^4 - y^5$ *(8-5)* **26.** $-1{,}760x^3y^9$ *(8-5)*
30. $49\,g/2$ ft; $625\,g/2$ ft *(8-3)* **31.** $x^6 + 6ix^5 - 15x^4 - 20ix^3 + 15x^2 + 6ix - 1$ *(8-5)*

CHAPTER 9

Exercise 9-1

1. 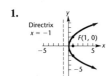 **3.** **5.** **7.** **9.**

11. 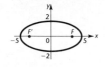 **13.** $(9.75, 0)$ **15.** $(0, -26.25)$ **17.** $(-19.25, 0)$ **19.** $x^2 = 12y$ **21.** $x^2 = -28y$

23. $y^2 = -24x$ **25.** $y^2 = 8x$ **27.** $x^2 = 8y$ **29.** $y^2 = -12x$ **31.** $x^2 = -4y$

33. **35.** 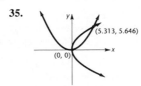 **37.** $A(-2a, a), B(2a, a)$ **39.** $x^2 - 4x - 12y - 8 = 0$

41. $y^2 + 8y - 8x + 48 = 0$ **43.** $x^2 = -200y$ **45.** (A) $y = 0.0025x^2, -100 \le x \le 100$ (B) 25 ft

Exercise 9-2

1. Foci: $F'(-\sqrt{21}, 0), F(\sqrt{21}, 0)$
Major axis length $= 10$
Minor axis length $= 4$

3. Foci: $F'(0, -\sqrt{21}), F(0, \sqrt{21})$
Major axis length $= 10$
Minor axis length $= 4$

5. Foci: $F'(-\sqrt{8}, 0), F(\sqrt{8}, 0)$
Major axis length $= 6$
Minor axis length $= 2$

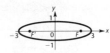

7. Foci: $F'(0, -4), F(0, 4)$
Major axis length = 10
Minor axis length = 6

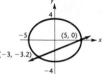

9. Foci: $F'(0, -\sqrt{6}), F(0, \sqrt{6})$
Major axis length = $2\sqrt{12} \approx 6.93$
Minor axis length = $2\sqrt{6} \approx 4.90$

11. Foci: $F'(-\sqrt{3}, 0), F(\sqrt{3}, 0)$
Major axis length = $2\sqrt{7} \approx 5.29$
Minor axis length = 4

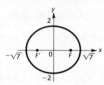

13. $\dfrac{x^2}{16} + \dfrac{y^2}{9} = 1$ **15.** $\dfrac{x^2}{64} + \dfrac{y^2}{121} = 1$ **17.** $\dfrac{x^2}{64} + \dfrac{y^2}{28} = 1$ **19.** $\dfrac{x^2}{100} + \dfrac{y^2}{170} = 1$

21.

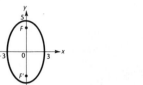

23.

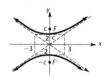

25. (A) $(2.201, 4.402)$ (B) $(2.201, 4.402)$

27. (A) $(3.565, 1.589)$ (B) $(3.565, 1.589)$ **29.** $\dfrac{x^2}{16} + \dfrac{y^2}{12} = 1$; ellipse   **31.** $\dfrac{x^2}{400} + \dfrac{y^2}{144} = 1$; 7.94 ft

33. (A) $\dfrac{x^2}{576} + \dfrac{y^2}{15.9} = 1$ (B) 5.13 ft

Exercise 9-3

1. Foci: $F'(-\sqrt{13}, 0), F(\sqrt{13}, 0)$
Transverse axis length = 6
Conjugate axis length = 4

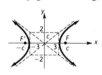

3. Foci: $F'(0, -\sqrt{13}), F(0, \sqrt{13})$
Transverse axis length = 4
Conjugate axis length = 6

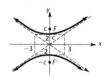

5. Foci: $F'(-\sqrt{20}, 0), F(\sqrt{20}, 0)$
Transverse axis length = 4
Conjugate axis length = 8

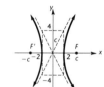

7. Foci: $F'(0, -5), F(0, 5)$
Transverse axis length = 8
Conjugate axis length = 6

9. Foci: $F'(-\sqrt{10}, 0), F(\sqrt{10}, 0)$
Transverse axis length = 4
Conjugate axis length = $2\sqrt{6} \approx 4.90$

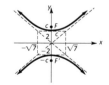

11. Foci: $F'(0, -\sqrt{11}), F(0, \sqrt{11})$
Transverse axis length = 4
Conjugate axis length = $2\sqrt{7} \approx 5.29$

13. $\dfrac{x^2}{49} - \dfrac{y^2}{25} = 1$ **15.** $\dfrac{y^2}{144} - \dfrac{x^2}{81} = 1$ **17.** $\dfrac{x^2}{81} - \dfrac{y^2}{40} = 1$ **19.** $\dfrac{y^2}{151} - \dfrac{x^2}{49} = 1$

21.

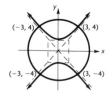

23.

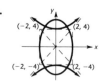

25. (A) $(-1.389, 3.306), (6.722, 7.361)$
(B) $(-1.389, 3.306), (6.722, 7.361)$

27. (A) $(-3.266, 3.830), (3.266, 3.830)$ (B) $(-3.266, 3.830), (3.266, 3.830)$ **29.** $\dfrac{x^2}{4} - \dfrac{y^2}{5} = 1$; hyperbola

31. $\dfrac{y^2}{16} - \dfrac{x^2}{8} = 1$; 5.38 ft above vertex   **33.** $y = \tfrac{4}{3}\sqrt{x^2 + 30^2}$

Exercise 9-4

1. (A) $x' = x - 3, y' = y - 5$ (B) $x'^2 + y'^2 = 81$ (C) Circle

3. (A) $x' = x + 7, y' = y - 4$ (B) $\dfrac{x'^2}{9} + \dfrac{y'^2}{16} = 1$ (C) Ellipse

5. (A) $x' = x - 4, y' = y + 9$ (B) $y'^2 = 16x'$ (C) Parabola

7. (A) $x' = x + 8, y' = y + 3$ (B) $\dfrac{x'^2}{12} + \dfrac{y'^2}{8} = 1$ (C) Ellipse   **9.** (A) $\dfrac{(x - 3)^2}{9} - \dfrac{(y + 2)^2}{16} = 1$ (B) Hyperbola

11. (A) $\dfrac{(x + 5)^2}{5} + \dfrac{(y + 7)^2}{6} = 1$ (B) Ellipse   **13.** (A) $(x + 6)^2 = -24(y - 4)$ (B) Parabola

15. $\dfrac{(x - 2)^2}{9} + \dfrac{(y - 2)^2}{4} = 1$; ellipse   **17.** $(x + 4)^2 = -8(y - 2)$; parabola

   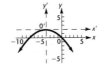

19. $(x + 6)^2 + (y + 5)^2 = 16$; circle   **21.** $\dfrac{(y - 3)^2}{9} - \dfrac{(x + 4)^2}{16} = 1$; hyperbola

23. $x^2 - 4x + 4y - 16 = 0$ **25.** $x^2 + 4y^2 + 4x + 24y + 24 = 0$ **27.** $25x^2 + 9y^2 - 200x + 36y + 211 = 0$
29. $4x^2 - y^2 - 16x + 6y + 11 = 0$ **31.** $F'(-\sqrt5 + 2, 2), F(\sqrt5 + 2, 2)$ **33.** $F(-4, 0)$ **35.** $F'(-4, -2), F(-4, 8)$

Chapter 9 Review Exercise

1. Foci: $F'(-4, 0), F(4, 0)$ *(9-2)*
Major axis length = 10
Minor axis length = 6

2. *(9-1)*

3. Foci: $F'(0, -\sqrt{34}), F(0, \sqrt{34})$ *(9-3)*
Transverse axis length = 6
Conjugate axis length = 10

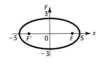

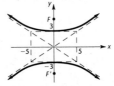

4. (A) $\dfrac{(y + 2)^2}{25} - \dfrac{(x - 4)^2}{4} = 1$ (B) Hyperbola *(9-4)*   **5.** (A) $(x + 5)^2 = -12(y + 4)$ (B) Parabola *(9-4)*

6. (A) $\dfrac{(x - 6)^2}{9} + \dfrac{(y - 4)^2}{16} = 1$ (B) Ellipse *(9-4)*   **7.** $y^2 = -x$ *(9-1)*

8. $\dfrac{x^2}{9} + \dfrac{y^2}{25} = 1$ *(9-2)*   **9.** $\dfrac{y^2}{9} - \dfrac{x^2}{16} = 1$ *(9-3)*

10. *(9-2, 6-3)*   **11.** *(9-1, 9-2, 6-3)*   **12.** *(9-2, 6-3)*

13. $\dfrac{(x + 3)^2}{4} + \dfrac{(y - 2)^2}{16} = 1$;
ellipse *(9-4)*

14. $(x - 2)^2 = 8(y + 3)$;
parabola *(9-4)*

15. $\dfrac{(x + 3)^2}{9} - \dfrac{(y + 2)^2}{4} = 1$;
hyperbola *(9-4)*

16. $(y - 4)^2 = -8(x - 4)$, or $y^2 - 8y + 8x - 16 = 0$ *(9-1)*   **17.** $\dfrac{x^2}{4} - \dfrac{y^2}{12} = 1$; hyperbola *(9-3)*

18. $\dfrac{x^2}{36} + \dfrac{y^2}{20} = 1$; ellipse *(9-2)*   **19.** $F'(-3, -\sqrt{12} + 2), F(-3, \sqrt{12} + 2)$ *(9-4)*   **20.** $F(2, -1)$ *(9-4)*

21. $F'(-\sqrt{13} - 3, -2), F(\sqrt{13} - 3, -2)$ *(9-4)*   **22.** 4 ft *(9-1)*   **23.** $\dfrac{x^2}{5^2} + \dfrac{y^2}{3^2} = 1$ *(9-2)*   **24.** 4.72 ft deep *(9-3)*

CHAPTER 10   Exercise 10-1

1. 990   **3.** 10   **5.** 35   **7.** 1   **9.** 60   **11.** 6,497,400   **13.** 10   **15.** 270,725   **17.** $5 \cdot 3 \cdot 4 \cdot 2 = 120$
19. $P_{10,3} = 10 \cdot 9 \cdot 8 = 720$ **21.** $C_{7,3} = 35$ subcommittees; $P_{7,3} = 210$ **23.** $C_{10,2} = 45$
25. No repeats: $6 \cdot 5 \cdot 4 \cdot 3 = 360$; with repeats: $6 \cdot 6 \cdot 6 \cdot 6 = 1{,}296$ **27.** No repeats: $P_{10,5} = 30{,}240$; with repeats: $10^5 = 100{,}000$
29. $C_{13,5} = 1{,}287$ **31.** $26 \cdot 26 \cdot 26 \cdot 10 \cdot 10 \cdot 10 = 17{,}576{,}000$ possible license plates; no repeats: $26 \cdot 25 \cdot 24 \cdot 10 \cdot 9 \cdot 8 = 11{,}232{,}000$
33. $C_{13,5}C_{13,2} = 100{,}386$ **35.** $C_{8,3}C_{10,4}C_{7,2} = 246{,}960$ **37.** $12 \cdot 11 = 132$
39. (A) $C_{8,2} = 28$ (B) $C_{8,3} = 56$ (C) $C_{8,4} = 70$
41. 2 people: $P_{5,2} = 20$; 3 people: $P_{5,3} = 60$; 4 people: $P_{5,4} = 120$; 5 people: $P_{5,5} = 120$
43. (A) $P_{8,5} = 6{,}720$ (B) $C_{8,5} = 56$ (C) $2 \cdot C_{6,4} = 30$

Exercise 10-2

1. Occurrence of E is certain.   **3.** $\frac{1}{2}$ **5.** (A) No probability can be negative.   (B) $P(R) + P(G) + P(Y) + P(B) \neq 1$
7. $P(R) + P(Y) = .56$ **9.** $\frac{1}{8}$ **11.** $1/P_{10,3} \approx .0014$ **13.** $C_{26,5}/C_{52,5} \approx .025$ **15.** $C_{16,5}/C_{52,5} \approx .0017$
17. $(2 \cdot 5 \cdot 5 \cdot 1)/(2 \cdot 5 \cdot 5 \cdot 5) = .2$ **19.** $1/P_{5,5} = 1/5! = .008\ 33$ **21.** $\frac{1}{36}$ **23.** $\frac{5}{36}$ **25.** $\frac{1}{6}$ **27.** $\frac{7}{9}$ **29.** 0   **31.** $\frac{1}{3}$
33. $\frac{2}{9}$ **35.** $\frac{2}{3}$ **37.** $\frac{1}{4}$ **39.** $\frac{1}{4}$ **41.** $\frac{3}{4}$ **43.** $\frac{1}{9}$ **45.** $\frac{1}{3}$ **47.** $\frac{1}{9}$ **49.** $\frac{4}{9}$ **51.** $C_{16,5}/C_{52,5} \approx .001\ 68$
53. $48/C_{52,5} \approx .000\ 018\ 5$ **55.** $4/C_{52,5} \approx .000\ 001\ 5$ **57.** $C_{4,2}C_{4,3}/C_{52,5} \approx .000\ 009$

Exercise 10-3

1. .1   **3.** .45   **5.** $P(\text{Point down}) = .389$, $P(\text{Point up}) = .611$; no
7. (A) $P(2\ \text{girls}) \approx .2351$, $P(1\ \text{girl}) \approx .5435$, $P(0\ \text{girls}) \approx .2214$
    (B) $P(2\ \text{girls}) = .25$, $P(1\ \text{girl}) = .50$, $P(0\ \text{girls}) = .25$
9. (A) $P(3\ \text{heads}) \approx .132$, $P(2\ \text{heads}) \approx .368$, $P(1\ \text{head}) \approx .380$, $P(0\ \text{heads}) \approx .120$
    (B) $P(3\ \text{heads}) \approx .125$, $P(2\ \text{heads}) \approx .375$, $P(1\ \text{head}) \approx .375$, $P(0\ \text{heads}) \approx .125$
    (C) 3 heads, 125; 2 heads, 375; 1 head, 375; 0 heads, 125
11. 4 heads, 5; 3 heads, 20; 2 heads, 30; 1 head, 20; 0 heads, 5
13. (A) .015   (B) .222   (C) .169   (D) .958

Chapter 10 Review Exercise

1. (A) 12 combined outcomes: (B) $6 \cdot 2 = 12$ *(10-1)*

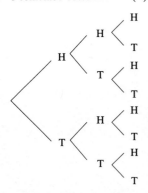

2. $C_{6,2} = 15, P_{6,2} = 30$ *(10-1)*
3. $6 \cdot 5 \cdot 4 \cdot 3 \cdot 2 \cdot 1 = 720$ *(10-1)*   4. $P_{6,6} = 6! = 720$ *(10-1)*
5. $C_{13,5}/C_{52,5} \approx .0005$ *(10-2)*   6. $1/P_{15,2} \approx .0048$ *(10-2)*
7. $1/P_{10,3} \approx .0014; 1/C_{10,3} \approx .0083$ *(10-2)*   8. $.05$ *(10-3)*
9. (*1*) Probability of an event cannot be negative; (*2*) sum of probabilities of simple events must be 1; (*3*) probability of an event cannot be greater than 1 *(10-2)*
10. $C_{6,3} = 20$ *(10-1)*
11. 336; 512; 392 *(10-1)*
12. (A) $P_{6,3} = 120$ (B) $C_{5,2} = 10$ *(10-1)*
13. (A) $P(2 \text{ heads}) = .21, P(1 \text{ head}) = .48, P(0 \text{ heads}) = .31$
    (B) $P(2 \text{ heads}) = .25, P(1 \text{ head}) = .50, P(0 \text{ heads}) = .25$
    (C) 2 heads, 250; 1 head, 500; 0 heads, 250 *(10-3)*
14. (A) $C_{13,5}/C_{52,5}$ (B) $C_{13,3} \cdot C_{13,2}/C_{52,5}$ *(10-2)*
15. $C_{8,2}/C_{10,4} = \frac{2}{15}$ *(10-2)*   16. (A) $\frac{1}{3}$ (B) $\frac{2}{9}$ *(10-2)*
17. $2^5 = 32; 6$ *(10-1)*   18. $1 - C_{7,3}/C_{10,3} = \frac{17}{24}$ *(10-2)*
19. $P_{4,4}/P_{2,2} = 12$ *(10-1)*
20. (A) $.350$ (B) $\frac{3}{8} = .375$ (C) 375 *(10-1)*
21. $P_{5,5} = 120$ *(10-1)*
22. (A) $.04$ (B) $.16$ (C) $.54$ *(10-3)*
23. $1 - C_{10,4}/C_{12,4} \approx .576$ *(10-2)*

Cumulative Review Exercise: Chapters 8–10

1. (A) Arithmetic   (B) Geometric   (C) Neither   (D) Geometric   (E) Arithmetic *(8-3, 8-4)*
2. (A) 10, 50, 250, 1,250   (B) $a_8 = 781,250$ (C) $S_8 = 976,560$ *(8-4)*
3. (A) 2, 5, 8, 11   (B) $a_8 = 23$ (C) $S_8 = 100$ *(8-3)*
4. (A) 100, 94, 88, 82   (B) $a_8 = 58$ (C) $S_8 = 632$ *(8-3)*   5. (A) 40,320   (B) 992   (C) 84 *(8-5)*
6. (A) 21   (B) 21   (C) 42 *(8-5, 10-1)*
7. Foci: $F'(-\sqrt{61}, 0), F(\sqrt{61}, 0)$ *(9-3)*
   Transverse axes length = 12
   Conjugate axes length = 10
8. Foci: $F'(-\sqrt{11}, 0), F(\sqrt{11}, 0)$ *(9-2)*
   Major axis length = 12
   Minor axis length = 10
9. *(9-1)*

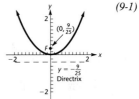

10. 8 combined outcomes: (B) $2 \cdot 2 \cdot 2 = 8$ *(10-1)*   11. (A) $4 \cdot 3 \cdot 2 \cdot 1 = 24$ (B) $P_{4,4} = 4! = 24$ *(10-1)*

12. $C_{13,3}/C_{52,3} \approx .0129$ *(10-2)*   13. $1/P_{10,4} \approx .0002; 1/C_{10,4} \approx .0048$ *(10-2)*   14. $\frac{62}{100} = .62$ *(10-3)*
15. P_1: $1 = 1(1); P_2$: $1 + 5 = 2(3)$
    P_3: $1 + 5 + 9 = 3(5)$ *(8-2)*

16. P_1: $1^2 + 1 + 2 = 4$ is divisible by 2
P_2: $2^2 + 2 + 2 = 8$ is divisible by 2
P_3: $3^2 + 3 + 2 = 14$ is divisible by 2 *(8-2)*

17. P_k: $1 + 5 + 9 + \cdots + (4k - 3) = k(2k - 1)$
P_{k+1}: $1 + 5 + 9 + \cdots + (4k - 3) + (4k + 1) = (k + 1)(2k + 1)$ *(8-2)*

18. P_k: $k^2 + k + 2 = 2r$ for some integer r
P_{k+1}: $(k + 1)^2 + (k + 1) + 2 = 2s$ for some integer s *(8-2)*

19. $y = -2x^2$ *(9-1)* **20.** $\dfrac{x^2}{25} + \dfrac{y^2}{16} = 1$ *(9-2)* **21.** $\dfrac{x^2}{64} - \dfrac{y^2}{25} = 1$ *(9-3)*

22. $1 + 4 + 27 + 256 + 3,125 = 3,413$ *(8-1)* **23.** $\displaystyle\sum_{k=1}^{6} \dfrac{(-1)^{k+1}2^k}{(k + 1)!}$ *(8-1)* **24.** 81 *(8-4)* **25.** 360; 1,296; 750 *(10-1)*

26. $C_{10,3}/C_{12,5} = \frac{5}{33} = .\overline{15}$ *(10-2)* **27.** (A) .365 (B) $\frac{1}{3}$ *(10-3)*
28. (A) 6,375,600 (B) 53,130 (C) 53,130 *(8-5, 10-2)* **29.** $a^6 + 3a^5b + \frac{15}{4}a^4b^2 + \frac{5}{2}a^3b^3 + \frac{15}{16}a^2b^4 + \frac{3}{16}ab^5 + \frac{1}{64}b^6$ *(8-5)*
30. $153,090x^6y^4$; $-3,240x^3y^7$ *(8-5)* **33.** 61,875 *(8-3)* **34.** $\frac{27}{11}$ *(8-4)*

35. $(y - 2)^2 = 4(x + 3)$;
parabola *(9-1)*

36. $\dfrac{(y + 2)^2}{4} - \dfrac{(x + 1)^2}{16} = 1$;
hyperbola *(9-3)*

37. $\dfrac{(x - 2)^2}{9} + \dfrac{(y + 3)^2}{4} = 1$;
ellipse *(9-2)*

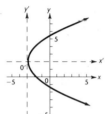

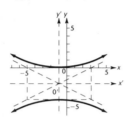

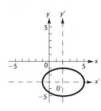

38. $10^9 = 1,000,000,000$; 3,628,800 *(10-1)* **39.** $\frac{2}{5}$; $\frac{2}{5}$ *(10-2)* **41.** $x^6 - 12ix^5 - 60x^4 + 160ix^3 + 240x^2 - 192ix - 64$ *(8-5)*
42. $(x - 6)^2 = -4(y - 2)$ or $x^2 - 12x + 4y + 28 = 0$ *(9-1)* **43.** $\pm 2\sqrt{3}$ *(9-2)* **44.** 8 *(9-2)* **45.** $C_{7,3} = 35$ *(10-1)*
46. $1 - C_{8,3}/C_{12,3} = \frac{41}{55} = .7\overline{45}$ *(10-2)* **49.** \$6,000,000 *(8-4)* **50.** 4 in *(9-1)* **51.** 32 ft, 14.4 ft *(9-2)*
52. (A) .13 (B) .17 (C) .32 *(10-3)*

APPENDIX B

1. $F = kv^2$ **3.** $f = k\sqrt{T}$ **5.** $y = k/\sqrt{x}$ **7.** $t = k/T$ **9.** $R = kSTV$ **11.** $V = khr^2$ **13.** 4 **15.** $9\sqrt{3}$
17. $U = k(ab/c^3)$ **19.** $L = k(wh^2/l)$ **21.** -12 **23.** 83 lb **25.** 20 amperes
27. The new horsepower must be 8 times the original. **29.** No effect **31.** $t^2 = kd^3$ **33.** 1.47 h (approx.) **35.** 20 days
37. Quadrupled **39.** 540 lb **41.** (A) $\Delta S = kS$ (B) 10 oz (C) 8 candlepower **43.** 32 times/s **45.** $N = k(F/d)$
47. 1.2 mi/s **49.** 20 days **51.** The volume is increased by a factor of 8.

INDEX